油气钻采输送装备全国重点实验室科研成果汇编

(2023 年)

中国石油集团工程材料研究院有限公司
油气钻采输送装备全国重点实验室 编
中国石油集团石油管工程重点实验室
陕西省石油管材及装备材料服役行为与结构安全重点实验室

石油工业出版社

内 容 提 要

本书汇编了中国石油集团工程材料研究院有限公司、油气钻采输送装备全国重点实验室、中国石油集团石油管工程重点实验室和陕西省石油管材及装备材料服役行为与结构安全重点实验室在2023年正式发表在国际国内刊物上的论文，反映了近几年石油管工程的科研成果及进展。内容涉及高性能油气钻采装备与材料设计制造、先进油气钻采输送管材制造、装备质量基础与服役安全等方面。

本书可供从事石油管工程相关工作的技术人员和石油高等院校相关专业师生参考。

图书在版编目（CIP）数据

油气钻采输送装备全国重点实验室科研成果汇编. 2023年 / 中国石油集团石油管工程技术研究院等编. 北京：石油工业出版社，2024. 11. -- ISBN 978-7 -5183-7168-6

Ⅰ. TE973-53

中国国家版本馆CIP数据核字第2024CM8086号

出版发行：石油工业出版社
（北京安定门外安华里2区1号楼　100011）
网　　址：www.petropub.com
编辑部：（010）64523687　图书营销中心：（010）64523633
经　　销：全国新华书店
印　　刷：北京中石油彩色印刷有限责任公司

2024年11月第1版　2024年11月第1次印刷
787×1092毫米　开本：1/16　印张：38.5
字数：980千字

定价：200.00元
（如出现印装质量问题，我社图书营销中心负责调换）
版权所有，翻印必究

《油气钻采输送装备全国重点实验室科研成果汇编（2023年）》编辑委员会

顾　　问：黄维和　李鹤林　高德利　陈学东　赵怀斌　张永庶

主　　任：刘亚旭

副 主 任：霍春勇

委　　员：（按姓氏笔画排序）

马秋荣　王华明　王国栋　毛新平　田　鹏　乐　宏
刘洪涛　孙　军　李贺军　张　杲　张对红　张矿生
张忠铧　林永茂　周建良　忽宝民　郑新权　赵新伟
胥志雄　郭小龙　涂善东　黄海霞　曹晓宇　韩恩厚

主　　编：马秋荣

副 主 编：罗金恒　林　凯　尹成先　韩礼红　池　强　王维旭
高　展　宫少涛

编辑组：陈宏远　付安庆　李厚补　马卫锋　雷广进　张春霞
林元华　王宴滨　李天雷　黄桂柏

前 言

2022年11月，国家科学技术部批准重组油气钻采输送装备全国重点实验室（前身为"石油管材及装备材料服役行为与结构安全国家重点实验室"，以下简称"实验室"），与中国石油天然气集团有限公司石油管工程重点实验室和陕西省石油管材及装备材料服役行为与结构安全重点实验室两个省部级重点实验室并行运行，是我国在油气钻采输送装备研究领域的科技创新基地、人才培养基地和学术交流基地。

实验室依托单位为中国石油集团工程材料研究院有限公司、宝鸡石油机械有限责任公司、宝山钢铁股份有限公司，设置高性能油气钻采装备与材料设计制造技术、先进油气钻采输送管材制造技术、装备质量基础与服役安全技术三个研究方向，密切围绕国家能源安全、制造强国重大战略需求，瞄准我国油气钻采输送装备供应链中急需解决的重大科技问题，坚持需求导向和问题导向，重点突破油气钻采输送装备技术的原创性新理论、新方法，建立自主标准和认证体系，提升国内油气装备制造水平和油气开发保障能力，引领油气装备制造产业发展，实现油气钻采输送重大装备核心技术自主可控，打造国际领先、保障有力、特色鲜明的油气钻采输送装备原创技术策源地和国家战略科技力量。

2023年，实验室获省部级和社会团体二等奖以上科技奖励10项，其中中国钢铁工业协会、中国金属学会特等奖1项，省部级一等奖1项，中国石油和化学工业联合会科技进步一等奖1项，中国专利优秀奖1项，省部级二等奖3项，中国石油、宝武集团科技奖励二等奖3项；发布国际标准1项，国家标准4项，行业标准7项；申请发明专利330件，授权发明专利57件；发表论文300余篇，其中SCI收录34篇；研发新材料5种、新产品9种，建立新方法8项，形成新技术6项。取得的一系列创新成果支撑了我国深地塔科1井、深地川科1井万米深井钻探，以及页岩油气勘探开发和重大管道工程建设。

在高性能油气钻采装备与材料设计制造技术研究领域，研究形成特深井自动化钻机集成配套技术，研制了12000m特深井自动化钻机，成功应用在国家深地塔科1井和深地川科1井特深井工程，关键部件全面实现国产化替代，有效解决了核心部件"卡脖子"，提升了我国油气勘探能力，创造我国钻井井深最深新纪录；依托国家能源局"小体积、大功率、智能化压裂装备"科研项目，研究形成了7000型电驱压裂装备、压裂机组运行状态在线监测预警系统、压裂设备全生命周期管理系统和消防安全保障关键装备等多项核心技术，应用在川渝、新疆非常规油气开发工程上，支撑了大功率电驱压裂装备技术进步；依托中国石油天然气集团有限公司"大功率智能电驱压裂装备关键技术研究"科研项目，形成了关键部件国产化、机组全流程协同作业、全生命周期管理和健康评价等技术，应用在川渝深层页岩气开发工程，支撑了压裂装备数智化进步。

在先进油气钻采输送管材制造技术研究领域，依托国家重点研发计划项目"苛刻环境能源井钻采用高性能钛合金管材研发及应用"，发布GB/T 41343—2022《石油天然气工业 钛合金钻杆》国家标准，在国内率先设计开发出105ksi、120ksi两种强度级别的钛合金钻杆，填补了国内120ksi级钛合金钻杆研发、应用空白；设计研发的高强高韧钛合金钻杆在塔里木油田8296m和8579m两口超深井中顺利完成首次下井试验，为超深层复杂、苛刻工况钻采作业提供了重要技术储备；针对塔里木油田深池塔科1井超深、超高温、超高压、超高载荷和超高应力的工况环境，开发出155ksi超高强高韧材料和超高内压气密封扣及其关键生产制造技术，为万米深地工程科学探索研究及未来超深层油气资源的开发提供了核心装备和坚实的技术保障；依托中国石油天然气集团有限公司关键核心技术攻关项目"提高页岩气井抗套损能力的新型套管研发"，成功研制出125klb/in^{2}级大壁厚高频焊热轧套管，填补了国内空白，在威204H页岩气钻井平台等顺利完成6口井（平均井深6000m）的下井应用，为深层页岩气钻采工程等提供有力技术支撑；针对我国重点管道工程西气东输四线建设需求，完成国家管网φ1219mm×33mm厚规格X80M HD2大应变管线钢及UOE焊管开发，完成1100t焊管生产及供货，成功应用于西气东输四线工程，保障了我国能源战略通道建设；依托国家重点研发计划"高应变海洋管线管研制"项目，建立基于应变海洋管道设计方法，形成高应变海洋管线管全产业链"设计—制造—连接—检测—评价"成套生产应用技术，为实现国产化提供支撑。

在装备质量基础与服役安全技术研究领域，依托国家重点研发计划"高压力高钢级管道失效机理及全生命周期可靠性评价技术研究"课题，开展了长自然时效和加速时效下X80管线钢力学性能试验研究，为中俄管道安全保障和运行智能决策提供科学依据；依托中国石油天然气集团有限公司技术攻关项目"提高页岩气井抗套损能力的新型套管研发"，首次提出水泥环空心化改性增塑新思路，研发了高强空心玻璃微珠新材料，形成了拥有自主知识产权且操作性强的专有技术，实现了页岩油气井复杂压裂套管变形的有效控制，该技术在西南油气田工程应用23井次，套变预防效果良好；依托中国石油天然气集团有限公司重大科技专项"二氧化碳规模化捕集、驱油与埋存全产业链关键技术研究及示范",发布 GB/T 42797—2023《二氧化碳捕集、输送和地质封存 管道输送系统》国家标准,在国内率先开展了超临界二氧化碳管道全尺寸爆破试验,开发以含氮类和有机硫类为主剂的超临界 CO_2 缓释阻垢一体剂,研发 DLC 涂层、Ni-W 镀层及常温无溶剂固化涂层等系列抗 CO_2 腐蚀涂层,形成 CO_2 驱环境管材的腐蚀防控技术及选材规范体系,为长庆油田、吐哈油田、塔里木油田等重点油气田 CCUS 工程提供了技术支撑;开展了纯氢/掺氢管道输送适用性评价技术、纯氢输送管道关键技术指标、输送工艺、管道失效后果及防护技术、完整性评价技术研究,形成了输氢管道标准和规范,建立了输氢管道设计、管材与装备、失效防护与运行等成套技术,为中长距离输氢管道工程设计、产品开发、工程建设和安全运行提供技术支撑;依托国家自然科学基金项目"基于气体渗透的热塑性塑料内衬管径向屈曲失效机制及定量风险评价研究",形成聚合物管材的气体渗透与阻隔改性理论,开发多层共挤高阻隔管材制造和铝箔阻隔层缠绕成型技术,以及全尺寸管材综合性能评价技术,完成了国内首次非金属管材的高压纯氢输送试验和纯氢爆破试验,为输氢非金属管材的制造和应用提供理论支撑和技术储备;牵头制定发布了 ISO 24139-2《石油天然气工业 管道输送系统用耐蚀合金内覆复合弯管和管件 第 2 部分:复合管件》,该国际标准的发布是中国在管道输送关键产品领域的科技创新及国际标准化工作取得的新突破,对推动我国高性能绿色石油管材产业健康发展意义重大。

  本书汇编收集了实验室于 2023 年在国内外重要刊物和学术会议上发表的 47 篇论文,并介绍了 2023 年授权专利及省部级以上获奖成果,这些内容从一个侧面反映了实验室近期所取得的研究成果,可以为从事油气钻采装备、油气输送装备、油气管道工程、油气井工程、石油工程材料、安全工程等方面的工程技术人员、研究人员和管理人员提供参考。

  由于编者水平有限,经验不足,加之时间仓促,不足之处在所难免,敬请广大读者批评指正。

目 录

第一篇 论文篇

一、高性能油气钻采装备与材料设计制造

钻机电控系统再制造升级标准研制与应用 …………… 董兴华 孙 娟 孟庆滨 等（4）
压裂泵凡尔阀动态排液性能仿真研究 ………………… 雷广进 刘树前 刘宏亮 等（10）
无线 Profinet 通信技术在石油钻机中的应用 ………… 李博洋 张志伟 李亚辉 等（17）
高精度深孔钻井补偿系统集成设计研究 ……………… 李 鹏 樊春明 袁 亮 等（21）
机械式旋冲工具对水平井送钻摩阻的影响规律研究 … 闫 炎 韩礼红 刘永红 等（29）
压裂泵曲轴的疲劳裂纹扩展研究 ……………………… 余朋伟 雷广进 刘树前 等（37）
一种基于振动分析的钻井泵故障诊断方法 …………… 袁 方 夏 辉 邹 涛 等（43）
Arrhenius Constitutive Equation and Artificial Neural Network Model of Flow Stress
　　in Hot Deformation of Offshore Steel with High Strength and Toughness
　　………………………………… Li Fangpo　Li Ning　Ren Xiaojian et al（50）
Failure Analysis of 2205 Duplex Stainless Steel Connecting Pipe for Composite Plate
　　Pressure Vessel in a High Pressure Gas Field
　　………………………………… Li Lei　Chen Qingguo　Fang Yan et al（65）
Failure Analysis of Casing in Shale Oil Wells under Multistage Fracturing Conditions
　　………………………………… Mou Yisheng　Zhao Han　Cui Jian et al（81）
Development of Design Method for Casing and Tubing Strings under Complex Alternating Loads
　　………………………………… Wang Jianjun　Li Dening　Du Xu et al（103）
Research on Casing Deformation Prevention Technology Based on Cementing Slurry
　　System Optimization ………………… Yan Yan　Cai Meng　Ma Wenhai et al（111）

二、先进油气钻采输送管材制造

不同 Nb 含量 X80 钢管环焊热影响区的微观组织与性能
　　………………………………………………………… 何小东 杨耀彬 陈越峰 等（128）
油气管道工程用大口径 TE555 增材制造三通的开发 … 吉玲康 胡美娟 田 野 等（137）
L485 高应变海洋管道的焊接技术研究 ………………… 李为卫 栾陈杰 邱新杰 等（146）
无缝钢管在线测厚系统的测量原理与应用实践 ……………… 王久刚 岳世斌 孙建安（155）

燕尾形超高抗扭特殊螺纹接头研究 …………………………… 詹先觉　高　展（161）
X80 管线钢与氢环境相容性试验研究 ……………… 张伟卫　封　辉　陈越峰　等（165）
热处理工艺对膨胀管组织和力学性能的影响 ……………… 朱兴华　董晓明　高　展（173）
Effect of Nb Content and Second Heat Cycle Peak Temperatures on Toughness of
　　X80 Pipeline Steel ……………… Chen Yuefeng　Yang Yaobin　He Xiaodong et al（179）
Safety Analysis of Casing String after Sidetracking in Existing Production-resumed Wells
　　……………………………………… Zhou Sheng　Wang Jianjun　Hao Xuelei et al（198）
Creep Crack Opening Tip Displacement（CCTOD）of X80 Pipeline Steel
　　at Room Temperature ………………… Wang Peng　Yang Sen　Chen Fuxing　et al（209）
Prediction of Collapsing Strength of High Collapse-resistant Casing Based
　　on Machine Learning ……………… Wang Peng　Zhong Chengxu　Fan Shuai et al（228）
Effect of Deviation of Welding Parameters on Mechanical Properties of
　　X80 Steel Girth Weld ……………… Zhu Lixia　Luo Jinheng　Jia Haidong et al（250）

三、装备质量基础与服役安全

从技术视角看北溪管道泄漏事件 ……………………… 刘亚旭　李为卫　霍春勇　等（262）
基于 FCE-SEW 模型的某工程项目风险评估研究 …… 梁晓斌　马卫锋　谭朝成　等（268）
Corrosion Behavior and Failure Mechanism of V150 Drill Pipe in HPHT and
　　Ultra-deep Drilling Process ………… Chen Zihan　Sun Shixuan　Huang Hao et al（280）
Failure Analysis of Corrosion and Fracture of P110 Tubing in a Development Well
　　……………………………………………… Han Yan　Ren Hong　Zhang Wenbin et al（293）
Collapse Failure Analysis of S13Cr-110 Tubing in a High-pressure and
　　High-temperature Gas Well ………… Ji Nan　Zhao Mifeng　Wu Zhenjiang et al（304）
Investigation of Corrosion Behavior of 2205 Duplex Stainless Steel Coiled Tubing
　　in Complex Operation Environments of Oil and Gas Wells
　　………………………………………… Luo Jinheng　Yan Pai　Fan Yujie et al（325）
The Application of Terahertz Nondestructive Testing Technology in the Detection of
　　Polyethylene Pipe Defects ………… Nie Hailiang　Hao Fengdan　Wang Litao et al（343）
Failure Analysis of the Crack and Leakage of a Crude Oil Pipeline
　　under CO_2-Steam Flooding ………… Song Chengli　Li Yuanpeng　Wu Fan et al（359）
Failure Risk Prediction Model for Girth Welds in High-strength Steel Pipeline Based
　　on Historical Data and Artificial Neural Network
　　……………………………………………… Wang Ke　Zhang Min　Guo Qiang et al（373）
Internal Localized Corrosion of X65-grade Crude Oil Pipeline Caused by the Synergy
　　of Deposits and Microorganisms ……… Yuan Juntao　Tian Lu　Zhu Wenxu et al（393）

Failure Analysis of the Leakage in Girth Welds of Bimetal Composite Pipe
………………………………… Zhang Shuxin　Xie Faqin　Li Xianming et al（409）
Corrosion Behavior of Tubing in High-salinity Formation Water Environment Containing
　H_2S/CO_2 in Yingzhong Block ……… Zhao Xuehui　Liu Junlin　Yao Baisheng et al（425）
Failure Analysis of S13Cr-110 Telescopic Tube Used in an Ultra-deep Gas Well ………………
　……………………………………… Zhu Lixia　Luo Jinheng　Long Yan et al（440）

四、其他

A Novel Inner Wall Coating-insulated Oil Pipeline for Scale and Wax Prevention
　…………………………………… Cao Jing　Ma Wenhai　Huang Weiming et al（452）
Effect of Carbon-foam Composite-coated Electrode on the Power Storage Performance
　of Soluble Lead Flow Batteries ……… Ji Dongdong　Liu Zheng　Li Liwei et al（466）
Cracking Failure Analysis of a Steel Wire Reinforced Thermoplastic Composite Pipe
　Used in an Oily Sewage Conveying System …… Kong Lushi　Li Houbu　Wei Bin et al（486）
Oxidation Behavior of Boron-containing （Zr, Ti）C_xB_y Solid Solution Ceramic
　at 1600℃ in Air ……………………… Lun Huilin　Zeng Yi　Xiong Xiang et al（498）
Research on Property and Burning Behavior of Flammable Casing for
　Underground Coal Gasification ……… Ren Xiangyi　Wu Jianjun　Wang Cankun et al（518）
Study on Spray Cooling of Ultra-high Temperature Production Wellbore
　in Underground Coal Gasification … Wang Jianjun　Zhao Huanzhen　Zhang Chao et al（534）
Permeability Model of Liquid Microcapsule Based on Multiple Linear Regression Method
　………………………………………… Xu Xiuqing　Li Fagen　Zhao Xuehui et al（544）
Experimental Study on Erosion Behavior of Fracturing Pipelines Involving Fluctuating Stress
　…………………………………………… Yang Siqi　Fan Jianchun　Zhao Sheng et al（553）
Quantum Dots Bridge Enabling Highly Efficient Carbon-based HTM-free Perovskite Solar Cells
　……………………………………… Yang Yuanbo　Wang Shuo　Li Simiao et al（570）
Understanding Hydrogen Diffusion Mechanisms in Doped α-Fe through DFT Calculation
　……………………………………… Zhu Lixia　Luo Jinheng　Zheng Shunli et al（584）

第二篇　成果篇

一、2023年获得省部级（含社会力量）科技奖励二等奖以上成果

1. 第三代超大输量低温高压管线用钢关键技术开发及产业化
　（中国钢铁工业协会、中国金属学会冶金科学技术特等奖）……………………（597）
2. 大口径高压力非金属复合管及其制备方法（中国专利奖优秀奖）………………（598）

3. 能源开发用钛合金石油管材料、配套技术研发及推广应用

（辽宁省科技进步一等奖） ……………………………………………（598）

4. 油气田集输管网服役安全关键技术与工业化应用

（中国石油和化学工业联合会科技进步一等奖） ……………………（599）

二、授权发明专利目录

第一篇 论文篇

一、高性能油气钻采装备与材料设计制造

钻机电控系统再制造升级标准研制与应用

董兴华[1,2]　孙娟[1,2]　孟庆滨[3]　罗震[1,2]　李洪波[1,2]　刘晶晶[1]

(1. 宝鸡石油机械有限责任公司；2. 中油国家油气钻井装备工程技术研究中心有限公司；
3. 山东祺龙海洋石油钢管股份有限公司)

摘　要：为了规范再制造市场，提高再制造产品质量，引领新技术在再制造领域的发展和应用。因此有必要针对电控系统再制造的特殊性，对其再制造升级各个环节进行细化，对标准构建的基准、结构框架、再制造升级与传统维修的区别等领域开展研究，并在国外某项目上实践，应用效果获得用户认可和国际报刊报道。该标准的实施为助推制造业绿色发展给出新的思路。

关键词：电控系统；标准；再制造升级；油气勘探

标准是推动创新成果产业化、加速成果市场化应用的桥梁[1]；同时标准的推行有利于产业的良性发展。许多调查研究表明细化专业领域的标准可以有效避免技术层面的安全风险，提高技术应用过程的操作效率，降低产品的生产成本，进而获得产品相关产业链发展的最佳秩序，保障相关产业的可持续发展。

根据国务院印发的《中国制造2025》，中国提出"大力发展再制造产业，实施高端再制造、在役再制造"[2]，标志着中国对再制造产业的重视上升到战略高度。石油钻机是野外油气勘探开发的必备装备，电气控制系统(以下简称电控系统)是钻机的重要组成部分[3]。电控系统自从20世纪末逐步在中国推广[4]，现在临近产品退役周期[5-6]，钻机配套的电控系统如果随着钻机一同退役，会因为这些系统中的资源没有得到最大限度地开发和利用而造成浪费，考虑到电控系统具有较大投资和较高技术含量，新电控系统的采购成本都是百万元以上，如果对现有废旧电控系统进行有效的延寿改造(以下简称再制造[7])会造就一个几十亿元的市场，因此需要构建相关标准为这个市场提供支撑，引导其良性发展。

1 电控系统再制造升级标准制定的意义

1.1 再制造升级标准制定的必要性

钻机电控系统现行的再制造相关技术标准有国际标准、国家标准、行业标准三大类(详见表1)，这些标准的实施对电控系统新品设计、制造等相关工作的规范化起到了非常重要的作用。但电控系统的再制造有其特殊性，随着高新技术的快速发展，电控系统升级换代的速度与资源浪费之间的矛盾正在日益凸显，值得关注的是，产品更新换代的速度越快，再制造恢复原型新品的效果就越差，再制造升级[8]新性能产品的呼声就越高，若一味推行恢复原有性能为目的的再制造行为，将造成再制造产品满足不了当前市场的需求，甚至造成产品

作者简介：董兴华，高级工程师，博士研究生，主要研究方向为再制造、AI视觉识别和海洋钻井平台自动化控制技术。

堆积，无法销售，给企业造成损失(十几年前的电控系统无论是自动化水平还是安全性，在当今市场上与新品都没有可比性)。同时由于现行技术标准难以满足实际电控系统再制造工作需要，开展相关新技术(如信息化、智能化[9])及现行标准的适应性研究势在必行。

表1 "电控系统再制造升级"相关标准统计表(部分节选)

行业标准		国家标准		国际标准	
标准编号	名称	标准编号	名称	标准编号	名称
JB/T 14263	电子电气产品可再制造性评价通则	GB/T 35980	机械产品再制造工程设计导则	ISO 13534	石油和天然气工业 钻井和生产设备 提升设备的检查，维护，修理和再制造
SN/T 3696	进口再制造用途机电产品检验风险评估方法指南	GB/T 32810	再制造 机械产品拆解技术规范	ISO 10987	土方机械 持久性 第2部分：再制造
JB/T 12993	三相异步电动机再制造技术规范	GB/T 41352	再制造 机械产品质量评价通则		

1.2 现有再制造标准存在的问题

研究表明，相关标准化工作主要存在以下问题：(1)目前再制造标准主要针对机械结构件的"恢复性再制造"(也就是以恢复设备原出厂性能为目的)。例如：GB/T 35980—2018《机械产品再制造工程设计 导则》和GB/T 32810—2016《再制造 机械产品拆解技术规范》等，对电控系统再制造升级(采用类似再制造的工艺步骤，同时通过增加新模块和利用新技术来提升产品的性能和功能)的内容欠缺。(2)再制造升级各个环节缺乏量化衡量标准(例如：JB/T 14263—2022《电子电气产品 可再制造性评价通则》和ISO 13534《石油和天然气工业 钻井和生产设备 提升设备的检查，维护，修理和再制造》给出了宏观的要求，在具体细节上对电气设备再制造升级研究存在缺失，无法满足市场竞争和相关法规要求)，由此造成各方对标准的理解存在差异，导致最终产品质量无法得到有效控制。(3)钻机电控系统再制造升级方向缺乏引领，影响新技术的推广应用。因此，编制再制造升级标准势在必行，须尽快建立一套科学完整、应用合理的标准来规范钻机再制造市场，引领技术快速发展，满足钻探开发的绿色、安全、高质量需求。

2 钻机电控系统再制造升级的标准研制

2.1 再制造升级标准的定义

为了提升再制造产品质量、规范、再制造升级过程和引导再制造升级相关新技术的推广应用，而起草该标准。该标准主要关注废旧产品，并通过模块化嵌入、结构优化和各类新技术应用来确保再制造升级各个环节(如状态评估、拆解、清洗、制造、试验等)的规范化和标准化，使其最终性能超过原型新品的制造过程，并做到相关要求有据可查、资料准备有章可循，从而规范再制造市场，满足油田用户对产品更高性能的需要。

2.2 再制造升级与传统维修的区别

针对市场上"再制造升级"定义模糊，用户无法辨别"再制造升级"和"维修"区别，从而无法保证再制造产品质量，针对这种情况，本标准对两者的区别进行对比说明(详见表2)，从标准的高度规范电控系统再制造升级市场。

表 2　再制造升级和传统电控系统维修对比表

项目	电控系统再制造升级	传统电控系统维修
使役寿命	改造后产品整体使役寿命等于或超过原新产品的使役寿命，属于新产品	一般增加功能后并不延长原产品的剩余使役寿命，属于旧产品
应用对象	报废或淘汰的产品	有故障的产品
拆解要求	所有零部件的全面拆解，一般宜拆解至最小不拆解状态	部分零部件的浅层拆解
清洗要求	(1) 采用专业的清洗工具处理零部件表面的污垢、锈蚀、变色、变形等问题； (2) 进行专业的探伤和漆膜厚度测量等质量检测	去除简单污垢，如表面灰尘、泥点等
检测	检测所有零部件	仅检测有故障的零件
零部件分类	零部件按照维修的难易程度和附加值高低分为三类，并分别判定采用哪种模式：如弃用、再制造加工、更新件、可用件	凭经验分类成可用件和废弃件
产品升级	原系统作为一个整体考虑，升级过程中通过增加新模块和利用新技术来提升产品的整体性能和功能	无
加工	(1) 对原结构件进行加工，满足再制造升级安装要求； (2) 对相关设备表面应用再制造技术进行恢复，确保产品质量不低于原型新品，如断路器的动触头导杆、电动机的转子和定子、集成控制单元等	根据需要对原系统的结构件进行加工，满足维修安装要求
装配	产品重新装配	零部件重新安装
试验	严格按照同类新品的使用标准进行试验，确保产品性能和质量不低于同类新品	进行简单的性能测试，确定能达到用户使用要求即可
价值	赋予旧电控系统不低于同类新品的价值，如原旧系统估值 10 万元，购买升级后同类新系统费用 400 万元，则通过再制造升级为用户创造 40 倍的价值	只是按照原产品设计寿命，继续维持产品性能
技术定位	再制造是一种产业	维修是常规性工作

2.3　标准构建的基准

需要明确钻机电控系统是为油田现场用户服务，所以要满足用户需求和相关的法律法规要求，为此应遵循以下原则：

(1) 质量等同新型新品：标准的构建必须严把质量关，以钻探行业市场需求为导向，严控准入门槛，要求电控系统再制造升级的质量特性和安全性能应不低于现阶段新型新品。

(2) 内容量化可实施：针对电控系统再制造升级的每个环节都提供量化且可操作的具体内容，确保具备可实施性。

(3) 贯穿再制造过程始终：该标准作为检查、评审产品设计再制造性的依据，应贯穿到设计、生产、试验和后期使用的相关环节，确保引导行业良性发展。

(4) 可扩展性：编制该标准时，不仅要考虑到目前的技术需求和当前的技术发展水平，也要对未来的技术发展方向进行科学预判，便于后期技术升级扩展。

(5) 持续改进：加入再制造升级后的定期检查环节，以确保使用过程中发现的问题能及时反馈，帮助持续改进再制造设计方案。

2.4　具体标准内容

2.4.1　标准的结构框架

通过对用户的需求分析，该标准(详见图 1)将内容分为三大块：基本要求、工艺流程和其他。并对相关内容都进行了规范和标准量化;对相关附属要求进行了细化。

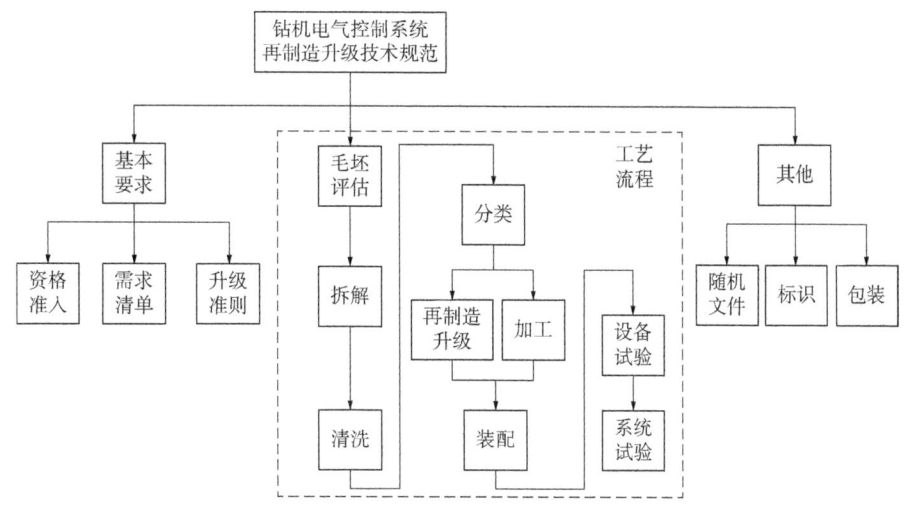

图 1 钻机电控系统再制造升级标准结构框架

2.4.2 基本要求

为了规范市场,本标准从"资格准入""需求清单"和"升级准则"三个方面提出约束,设置准入门槛。

2.4.3 工艺流程

(1)为了确保要求具备可实施性,本标准依据每个环节的特点,逐级进行约束。例如,"零部件的拆解"环节就提出"总体原则"和"通用要求",并针对电气设备特点,提出"绝缘电阻专项要求""工频耐受电压专项要求"和"电气柜(箱)、操作台壳体专项要求"等。

(2)有关再制造升级新技术引领:对目前成熟的技术(如保护功能升级和控制系统升级等)和前沿技术(如信息化和智能化、节能和环保要求等)都做了列示,确保相关技术的推广和良性发展。

(3)试验验证:按照不同设备特点,严格按照新品要求提出试验内容。并依据"先设备再系统"的原则,分别细化了试验要求。

2.4.4 其他要求

有关完工文件等内容,也分为"标识、包装、随机文件等"逐项提出了可操作的约束要求。

3 钻机电控系统再制造升级标准的应用

宝鸡石油机械有限责任公司自提出加快推进服务型制造转型的理念后,经过多年的积累和多个项目的实际运用,几经修改完善,并汇总各方专家的意见,形成团体标准《钻机电气控制系统再制造升级技术规范》,这也是石油钻探行业内第一部关于"电控系统再制造升级"的标准,并在实际项目中得到应用。

截至 2023 年中,该团体标准已在国外、国内等多个改造项目上进行实践,相关系统目前均运行良好,稳定可靠。尤其是国外项目,宝鸡石油机械有限责任公司根据用户需求,按照该标准要求对某旧电控系统(图 2)进行改造,该旧电控系统如果不进行改造会因性能不满足最新钻井要求而功能性退役,相关电气设备因没有达到使用寿命而造成资源浪费。为此宝鸡石油机械有限责任公司引入嵌入式模块化、无损清洗、寿命评估和预测等技术，对该电控系统进行再制造升级改造。再制造升级前后对比详见表3。

图 2　旧电控系统

表 3　再制造升级前后对比表

电控系统	功能对比	备注
20年前第一代直流电控系统(升级前)	人员必须手动操作，工作强度大。 仅有主要设备的安全互锁(3个设备之间)。 无法实现对发电机组燃油消耗控制	系统折旧后，估值仅值10万元。第三方机构评估因功能不满足要求，建议半年内退役
第三代电控系统（升级后）	实现自动化控制，人员工作强度较低。 多设备安全互锁(扩展到12个设备之间)。 实现对发电机组燃油消耗控制，提高燃油能耗率，降低排放	如购买同类新系统费用400万元。再制造升级费用(国外往返运费+再制造升级花费)共计98万元。为用户节约292万元

此次改造特点：

（1）严控产品质量，改造后质量与第三代新品一致。

（2）本标准提出的新技术获得国外用户认可，促进了国内技术和标准在国际市场的推广应用。

（3）改造周期相对于新品制造缩短60%的时间。

（4）改造后设备在国外沙漠高温恶劣环境下工作性能优良，获得用户和第三方监造的好评，被国外报刊报道（图3[10]）。

图 3　国外报刊报道

4 结语

通过对电控系统相关技术和再制造技术应用的介绍，探讨"再制造升级"技术如何在电控系统中应用，对电控系统最新技术、标准等内容进行调查分析研究，构建了电控系统再制造升级标准，通过区分"再制造升级"和"维修"，有助于规范再制造市场，让那些有实力的再制造企业做大做强。电控系统再制造升级标准的建立，使今后电控系统再制造升级工作和标准化建设"有章可循，有标可依"，有效指导相关装备制造企业应用再制造技术，确保再制造产品质量，同时为未来新技术的推广和良性发展创造条件。可以预见该标准的实施必将对石油钻机电控系统的再制造各环节规范化和再制造产品质量提升起到重要的指导作用，为相关行业的发展提供技术和标准支持。

参 考 文 献

[1] 徐风. 架起标准促进科技成果转化应用的桥梁[N]. 中国质量报, 2016-10-14(001).
[2] 国家制造强国建设战略咨询委员会. 中国制造2025蓝皮书[M]. 北京：电子工业出版社, 2016.
[3] 张奇志, 李琳, 等. 电动钻机自动化技术[M]. 北京：石油工业出版社, 2006.
[4] 赵宇. 电动钻机动力控制系统的设计[D]. 西安：西安石油大学, 2013.
[5] 丁卫峰. 石油钻井装备的运行管理[J]. 化学工程与装备, 2018, 256(5)：220-221.
[6] 赵备. 石油钻井机械设备故障预防与维护保养[J]. 中国机械, 2014(5)：119-120.
[7] DONG X H, ZHANG Z W, SUN J, et al. Remanufacturing an evaluation system for electrical control systems of drilling rig based on the improved FCE and ANN. PLoSONE17(5)：e0268788. https://doi.org/10.1371/journal.pone.0268788.
[8] 姚巨坤, 朱胜. 再制造升级[M]. 北京：机械工业出版社, 2017.
[9] 董兴华, 张文英, 杨琨, 等. 海工平台电缆路径智能设计方法和应用研究[J]. 中国造船, 2021, 62(2)：267-274.
[10] 董兴华, 孙娟, 李洪波, 等. "分级服务+再制造"标准助推钻探绿色安全发展[J]. 中国标准化, 2022(2)：72-76.

本论文原发表于《标准应用研究》2023年。

压裂泵凡尔阀动态排液性能仿真研究

雷广进[1,2]　刘树前[1]　刘宏亮[1,2]　高　龙[1,2]　郝建旭[1,2]　李　晨[1,2]

（1．宝鸡石油机械有限责任公司；
2．中油国家油气钻井装备工程技术研究中心有限公司）

摘　要：针对凡尔阀关闭滞后引起泵排量减少及凡尔阀开启阻力过大使排液性能变差等问题，在虚拟试验中，以某型压裂泵的凡尔阀为研究对象，利用系统性能仿真方法建立压裂泵液力端仿真模型，对压裂泵排液动态特性进行仿真。研究了弹簧刚度、阀盘质量对泵排出流量的影响及弹簧刚度、阀盘质量、阀座孔径对泵排出压力的影响。结果表明：泵排出流量、排出压力受阀座孔径影响较大而受弹簧刚度和阀盘质量影响较小；在各弹簧刚度和阀盘质量取值下泵的平均排出流量均在 4000L/min 左右，平均排出压力均在 140MPa 左右；泵的瞬时排出流量和瞬时排出压力随弹簧刚度、阀盘质量的增大而增大，随阀座孔径的增大而减小。研究结果对于优化凡尔阀设计，提高压裂泵排液性能具有重要的参考价值。

关键词：凡尔阀；压裂泵；仿真；滞后角；动态特性

压裂泵用于油气井压裂和酸化施工中，其性能好坏直接决定了压裂施工的成败[1]。据统计，压裂泵运行时出现的泄漏及工作寿命短等问题 80% 源于凡尔阀[2]。作为压裂泵液力端中控制压裂液进出的关键部件，凡尔阀的结构参数对其工作特性（包括启闭滞后角和开启阻力），以及压裂泵的排液性能（包括排出压力和排出流量）影响很大。凡尔阀的启闭相对于曲轴回转的滞后（凡尔阀启闭滞后）会造成泵的容积损失，导致泵排出流量减少[2]。此外，凡尔阀的结构参数（如弹簧刚度）过大，会导致阀盘开启阻力过大进而影响液体的吸入和排出[3]。

目前，对压裂泵凡尔阀的研究主要集中在阀运动规律、液动力模拟等，对凡尔阀影响压裂泵排液性能的研究较少[4-6]。本文运用 AMESim 系统性能仿真方法建立压裂泵仿真模型，对压裂泵动态特性进行仿真，以期获得凡尔阀参数对泵排液性能的影响规律，为优化凡尔阀设计提供参考依据。

1　凡尔阀参数对泵性能的影响

压裂泵正常工作时，凡尔阀的结构参数必须设置合理，以保证凡尔阀运动的灵敏度及吸排性能，否则无法减弱由压裂液可压缩性带来的凡尔阀运动滞后[7]。滞后会导致凡尔阀开启和关闭的速度过小进而造成压裂泵的流量损失，降低泵的容积效率。此外，凡尔阀的结构参数除了会影响阀盘的运动外，还会影响流体流动的能量损失情况及凡尔阀开启（关闭）阻力的大小，进而影响泵排出压力大小[8]。

基金项目：中国石油科技基础条件平台建设项目"大功率超高压压裂泵试验台建设"（2019D-5006-55）。
作者简介：雷广进（1968—）男，本科，高级工程师，从事石油钻采机械技术研究工作。

1.1 凡尔阀吸排性能影响因素分析

1.1.1 影响开启(关闭)滞后角的因素

凡尔阀的滞后程度用开启(关闭)滞后角来衡量，滞后角越大，凡尔阀开启和关闭滞后程度越大，根据韦斯特法尔公式[9]（阀在非稳定状态下的运动规律）得到的滞后角公式为：

$$\varphi_0 = \arctan\left(\frac{\omega A_{阀}}{\mu d_K \pi}\sqrt{\frac{\rho A_{阀}}{2(mg+Ch+F_0)}}\right) \quad (1)$$

式中：$A_{阀}$ 为凡尔阀当量面积，m^2；ω 为曲柄旋转角速度，rad/s；μ 为阀的流量系数；d_K 为阀座孔径，mm；ρ 为介质密度，kg/m^3；m 为阀盘在介质中的质量，kg；F_0 为阀弹簧的预紧力，N；C 为阀弹簧刚度，N/mm；h 为阀升程，mm。

由式(1)可看出凡尔阀的阀盘质量、弹簧刚度和阀座孔径会影响其启闭滞后角的大小。

1.1.2 影响开启(关闭)阻力的因素

压裂泵工作时压裂液顶开凡尔阀需要克服凡尔阀的开启(关闭)阻力，包括弹簧力和重力，开启(关闭)阻力的大小是反映凡尔阀吸排性能的重要参数之一。凡尔阀在运动稳定状态下的开启阻力公式[10]为：

$$\Delta P = \frac{4G\omega^2}{\pi g d_K}\left\{\frac{g}{d_K\omega^2}\left[\left(1-\frac{\gamma_j}{\gamma_f}\right)+\frac{R}{G}\right]+\frac{SD^2}{2d_k d_{阀}^2}(1+\lambda)\right\} \quad (2)$$

式中：λ 为杆径比；d_K 为阀孔直径，mm；D 为柱塞直径，mm；γ_j 为介质重度，Pa；γ_f 为凡尔阀重度，Pa；S 为柱塞冲程，mm。

由式(2)看出影响开启阻力的主要因素为阀盘质量、阀座孔径和弹簧刚度。

1.2 凡尔阀关闭滞后对泵排出流量的影响

在不考虑柱塞和缸体密封泄漏及死区流量损失的前提下，泵的流量损失主要是由于凡尔阀的关闭滞后产生压裂液回流所导致，根据流量计算公式[11]，吸入过程的损失流量为：

$$\Delta Q_a = rAt\left(1-\cos\varphi_a-\frac{\lambda}{2}\sin^2\varphi_a\right) \quad (3)$$

排出过程的流量损失为：

$$\Delta Q_b = rAt\left(1-\cos\varphi_b-\frac{\lambda}{2}\sin2\varphi_b\right) \quad (4)$$

式中：r 为曲柄半径，mm；A 为柱塞横截面积，mm^2；φ_a 和 φ_b 分别为吸入过程凡尔阀关闭滞后角和排出过程凡尔阀关闭滞后角。

考虑流量损失后的泵的实际排出流量为：

$$Q_p = Q_{rp} - \Delta Q_a - \Delta Q_b \quad (5)$$

式中：Q_{rp} 为泵理论排出流量，L/min。

由式(3)至式(5)可知，凡尔阀关闭滞后角决定了泵的损失流量大小，进而决定泵的实际排出流量大小。

1.3 凡尔阀开启(关闭)阻力对泵压力的影响

对泵一个工作周期内吸排压力及液缸内压力的变化进行分析。阀盘工作时的受力如图1所示。

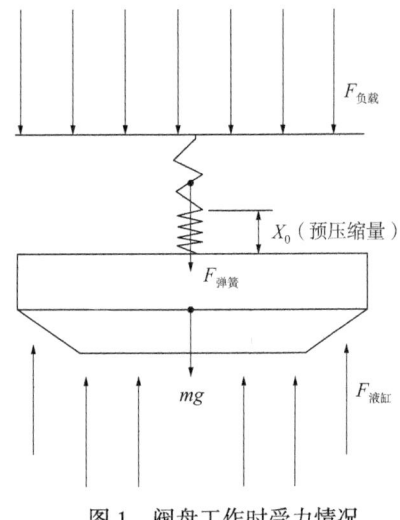

图 1 阀盘工作时受力情况

对于排液过程，当吸入阀关闭后，柱塞压缩液体的运动使得液缸内压力继续增大，当液缸内压力增大到与排出口负载压力之差刚好平衡排出阀的开启阻力时，排出阀开启。

随着液缸内液体的排出，液缸内压力逐渐减小[12]。当柱塞运动到排出冲程的极限位置时，排出阀由于运动滞后并不能及时关闭，柱塞返回一段距离后，液缸内压力才能减小到与负载压力之差平衡排出阀的关闭阻力，此时排出阀关闭，有：

$$(P_{负载} - P_{缸}) A_{阀} = \Delta P \tag{6}$$

由以上分析可知，凡尔阀开启（关闭）阻力是压裂液吸入和排出需要克服的阻力之一，因此有必要分析其对泵工作压力的影响。

2 仿真模型建立与分析

针对某型五缸压裂泵，为分析凡尔阀参数对其排液性能的影响（包括排出压力、排出流量），利用液压系统仿真软件 AMESim 建模分析[13-14]。

2.1 液力端液压模型的建立

利用 AMESim 建立的液力端液压模型如图 2 所示，为了便于建模和分析，假设系统不存在泄漏并且忽略机械摩擦的影响。

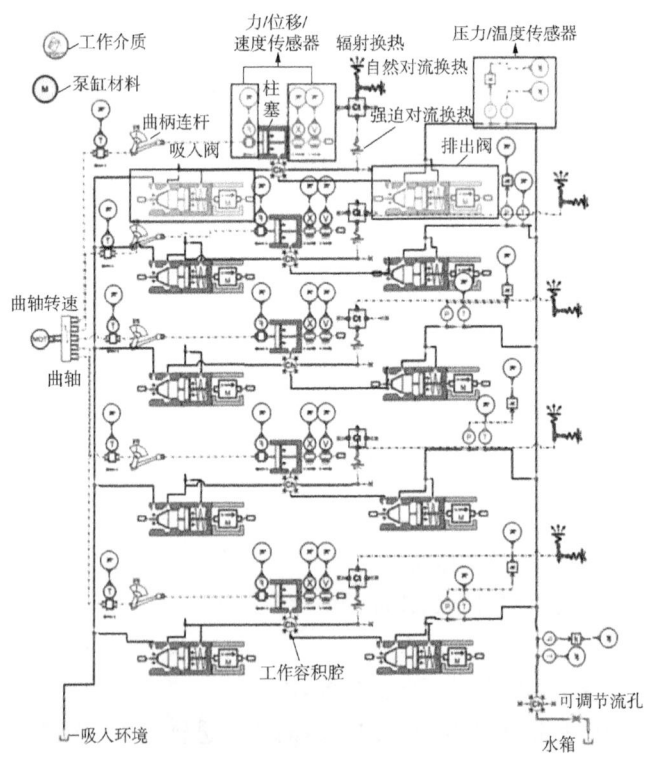

图 2 某一型五缸压裂泵液压模型

在参数模式下,其模型主要工况参数设置见表1。在仿真模式中,设定仿真时间为1s,运行时间间隔为0.001s,运行仿真。

表1 某一型压裂泵主要工况参数

参数	参数值	参数	参数值
额定工作压力/MPa	140	吸入压力/MPa	0.3
曲轴转速/(r/min)	240	工作介质	清水

2.2 仿真结果分析

2.2.1 凡尔阀参数对泵排量影响分析

(1)弹簧刚度对泵排量的影响分析。阀弹簧刚度一般在8.75~26.27N/mm的范围内[15],取4个弹簧刚度值,以2N/mm作为一个步长进行仿真得到不同弹簧刚度下泵排出流量随曲柄转角的变化曲线,如图3所示。

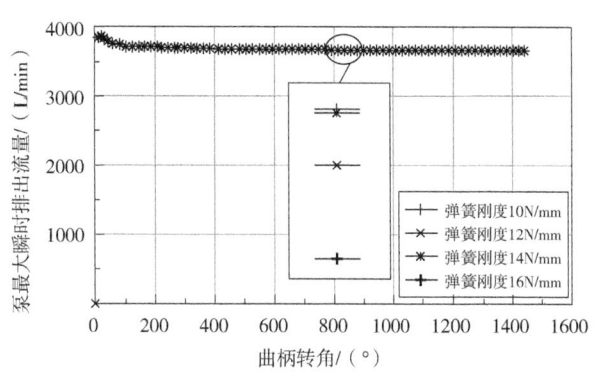

图3 不同弹簧刚度下泵的排量曲线

据图3结果可知:不同弹簧刚度下,泵排量的变化趋势基本相同,最大瞬时排量在4000L/min上下波动且最大瞬时排量随弹簧刚度的增加而增加。

凡尔阀开启(关闭)滞后角随着弹簧刚度的增大而趋于减小[16],随着阀关闭滞后角的减小,因关闭滞后引起的回流量减少,泵的排出压力增加。

(2)阀盘质量对泵排量的影响分析。

取4个阀盘质量值,以0.2kg作为一个步长进行仿真,得到不同阀盘质量下泵排出流量随曲柄转角的变化曲线,如图4所示。

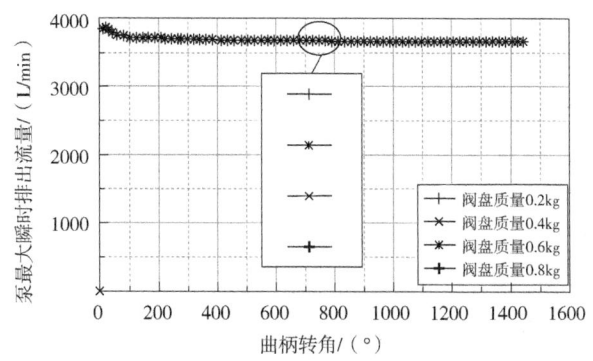

图4 不同阀盘质量下泵的排量曲线

据图4结果可知：不同阀盘质量下泵排量都在4000L/min上下波动且波动幅度较大；泵的最大瞬时排量随阀盘质量的增大而增大。

根据1.1.1节阀关闭滞后角公式可知，阀盘质量增加会使阀的关闭滞后角减小，从而使泵排量增加。

2.2.2 凡尔阀参数对泵排出压力影响分析

（1）阀座孔径对泵排出压力的影响分析。

阀座内孔是泵吸入液体和排出液体的通道，其直径与泵的理论平均流量和通过阀座孔的最大瞬时流速有关[15]。通常最大瞬时流速范围为1~3m/s，把泵参数代入式(7)算得阀座孔径范围为95.1~164.7mm：

$$d_K = 2\sqrt{\frac{\pi D^2 Sn}{240 v_{Kmax}}} \tag{7}$$

式中：d_K为阀座孔径，mm；v_{Kmax}为最大瞬时流速，m/s。

取14个阀座孔径值，以5mm作为一个步长进行仿真得到不同阀座孔径下泵的平均排出压力随时间的变化曲线，如图5所示。

泵最大瞬时排出压力与阀座孔径的关系如图6所示。

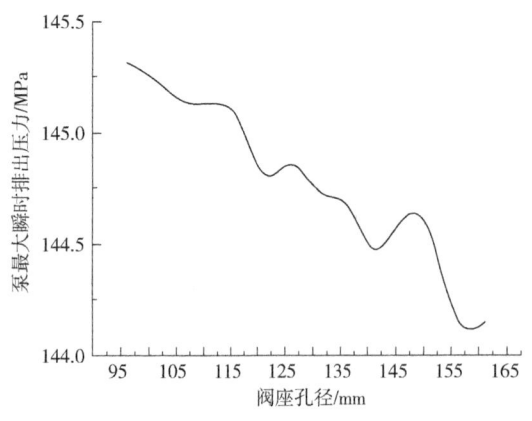

图5 不同阀座孔径下泵排出压力曲线

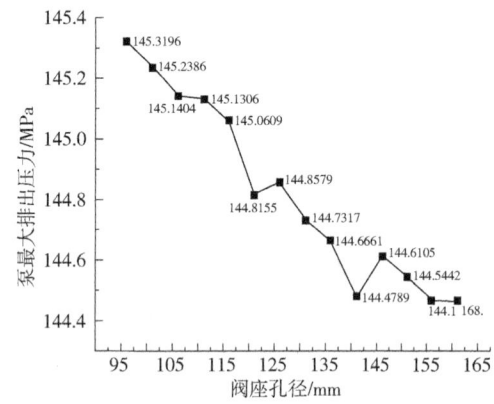

图6 泵最大排出压力与阀座孔径的关系

据图5和图6结果可知：随着仿真进行，泵排出压力从0MPa增加至约140MPa，增至140MPa后发生周期性波动，波动幅度小于5MPa；泵的最大排出压力随阀座孔径的增大呈减小趋势，但在阀座孔径为121~126mm和141~146mm两区间内又有一定幅度增大。

分析1.1.2节中阀开启阻力公式可知，开启阻力随阀座孔径的增大而减小，但是存在两个"极值"点，即存在两个阀座孔径取值区间，开启阻力增加，与仿真结果一致。

（2）阀盘质量对泵排出压力的影响分析。

取2.2.1节中10组阀盘质量值进行仿真，得到不同阀盘质量下泵平均排出压力随时间的变化曲线，如图7所示。

泵最大瞬时排出压力与阀盘质量的关系如图8所示。

据图7和图8结果可知：泵最大排出压力随阀盘质量的增加而增大，阀盘质量的增加使阀的开启阻力增加，泵提供压力增大。

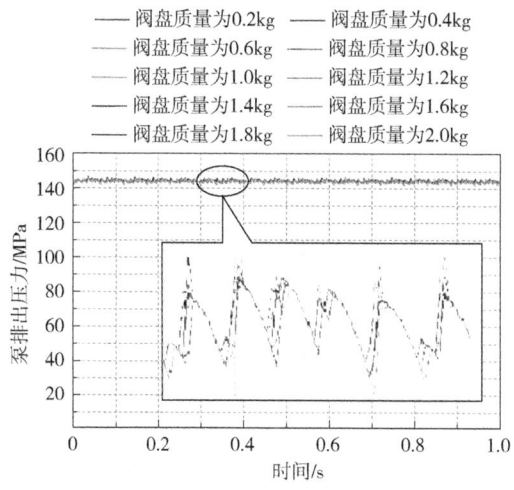

图7 不同阀盘质量下泵排出压力曲线

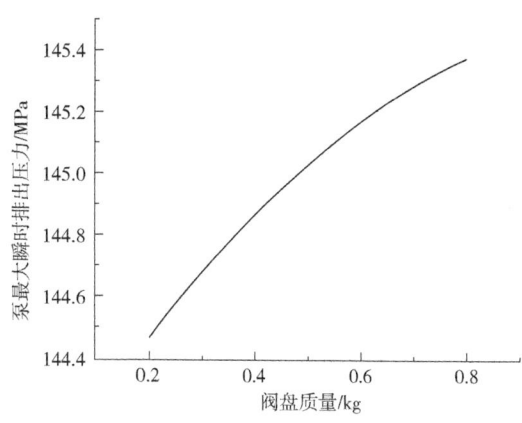

图8 泵最大瞬时排出压力与阀盘质量的关系

(3) 弹簧刚度对泵排出压力的影响分析。

取2.2.1节中4组弹簧刚度值进行仿真,得到不同弹簧刚度下泵平均排出压力随时间的变化曲线,如图9所示。

泵最大瞬时排出压力与弹簧刚度的关系如图10所示。

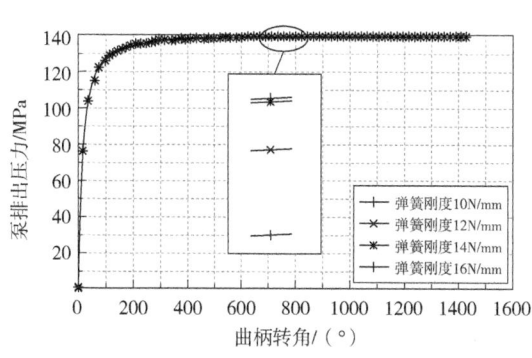

图9 不同弹簧刚度下泵排出压力曲线

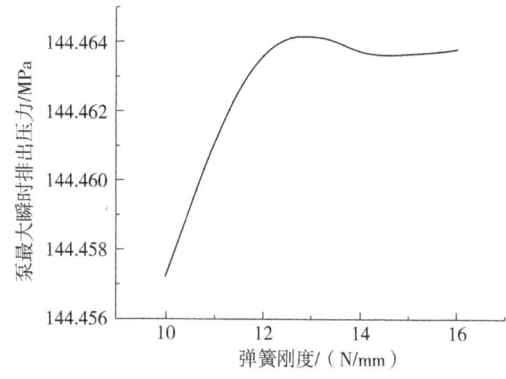

图10 泵最大瞬时排出压力与弹簧刚度的关系

据图9和图10结果可知:弹簧刚度为10～14N/mm时泵最大排出压力随弹簧刚度增加而增大,当弹簧刚度超过14N/mm后,泵最大排出压力稳定为144.4638MPa且不再增大。

3 结论

通过理论分析凡尔阀参数对凡尔阀吸排性能及泵排出压力、排出流量的影响,利用AMESim软件对某型压裂泵的排液动态特性进行仿真,得到以下结论:(1)不同弹簧刚度下泵排量的变化趋势基本相同,平均排量在4000L/min上下波动,泵的最大瞬时排量随弹簧刚度的增大而增大。(2)不同阀盘质量下泵排量都在4000L/min上下波动;泵的最大瞬时排量随阀盘质量的增大而增大。(3)泵的最大瞬时排出压力随阀座孔径的增大呈减小趋势,但存在两个"极值点",在阀座孔径为121～126mm和141～146mm两区间内泵的最大排出压力为增大趋势。(4)泵最大瞬时排出压力随阀盘质量或弹簧刚度的增大而增大,但弹簧刚度超过一定值后,泵最大排出压力不再增大。

参 考 文 献

[1] LIU Z, XIAO W, CUI J, et al. Study on the Laminated Frame Strength of Heavy-duty fracturing Pump Truck Considering Slip Effect[J]. International Journal of Vehicle Systems Modelling and Testing, 2020, 14(1): 40-50.

[2] 许旭. 基于同步挤压S变换和深度残差网络的压裂车压裂泵故障诊断[D]. 北京: 北京建筑大学, 2020.

[3] 曹春玲, 潘源源, 李海宁. 吸液管路对乳化液泵性能影响及改善研究[J]. 机床与液压, 2020, 48(4): 154-158.

[4] VOORDE J V, VIERENDEELS J, DICK E. Flow Simulations in Rotary Volumetric Pumps and Compressors with the Fictitious Domain Method[J]. Journal of Computational & Applied Math-ematics, 2004, 168(1-2): 491-499.

[5] YANG R. CFD Simulations of Oil Flow and Flow Induced Forces Inside Hydraulic Valves[C]. International Off-highway & Powerplant Congress. 2002.

[6] BARMAN P. Computational Fluid Dynamics(CFD) Analysis to Predict and Control the Cavitation Erosion in a Hydraulic Control Valve[C]. SAE World Congress & Exhibition. 2002.

[7] 徐静. 3500五缸柱塞泵液力端的设计与研究[D]. 青岛: 中国石油大学(华东), 2013.

[8] 郭志奇, 肖晓华, 李蓉. 基于AMESim的某型压裂泵建模与仿真研究[J]. 重庆科技学院学报(自然科学版), 2014, 16(5): 150-152.

[9] QIAN J Y, GAO Z X, WANG J K, et al. Experimental and Numeri-cal Analysis of Spring Stiffness on Flow and Valve Core Move-ment in Pilot Control Globe Valve[J]. International Journal of Hydrogen Energy, 2017, 42(27): 17192-17201.

[10] 闫国军, 赵军明, 董泳. 往复凡尔阀运动规律的研究[J]. 中国机械工程, 2004(18): 21-23.

[11] 贾铭新. 液压传动与控制[M]. 北京: 电子工业出版社, 2017.

[12] PEI J, HE C, LV M, et al. The Valve Motion Characteristics of a Reciprocating Pump[J]. Mechanical Systems & Signal Processing, 2016: 657-664.

[13] IMAGINE. AMESim 4.2 User Manual[Z]. IMAGINESA., 2004.

[14] 窦美玲. 基于虚拟样机技术的乳化液泵特性仿真[D]. 西安: 西安科技大学, 2009.

[15] 杜晓旭. 高压压裂泵液力端设计与研究[D]. 大连: 大连理工大学, 2014.

[16] 雷中清. 2000型车载压裂泵液力端工作机理研究[D]. 成都: 西南石油大学, 2012.

本论文原发表于《机械工程师》2023年第2期。

无线 Profinet 通信技术在石油钻机中的应用

李博洋[1,2] 张志伟[1,2] 李亚辉[1,2] 李西方[1,2]
孔永超[1,2] 王永鹏[1,2] 范 磊[1,2] 曹 童[1,2] 张 猛[1,2]

(1. 宝鸡石油机械有限责任公司；2. 国家油气钻井装备工程技术研究中心)

摘 要：为了解决石油钻机控制系统中，有线工业以太网可扩展性差、需要额外空间等缺点，研发无线传输工业以太网，应用无线 Profinet 技术，以钻井管柱堆场吊运装置子模块作为实验载体，依托无线客户端、无线接收器、S7-1200 PLC、Profinet 通信协议，实现指令在司钻房和各子模块之间的无线传输。在石油钻机系统中，该方法简化了线路，提高石油钻机智能化，提高了设备的可扩展性、提高了生产效率，降低了维护成本。使得信息传输速率高、传输可靠性高、故障率低。

关键词：石油钻机；Profinet；无线传输；工业以太网

近年来，我国自主研发的石油钻采设备以其可靠的质量、良好的信誉和相对低廉的价格，日渐受到国内外市场青睐。石油钻井的网络控制系统作为石油钻机的核心控制部件，扮演着重要角色，被业界誉为石油钻机的"心脏"。

随着网络通信技术和计算机技术的发展，无线网络化控制系统在钻机中日趋普及，为了满足石油钻机特殊化的控制需求和考虑到石油钻机自身的工作环境，有线连接的工业以太网存在安装复杂、灵活性较差、扩展性较差等缺点，对钻机的通信系统产生了一定阻碍。

工业无线局域网采用工业级无线传输设备搭建。工业级无线传输设备针对工业环境采取了特别措施，针对工业通信增加了额外的功能。组网虽然需要更大的初始投资，但能增加正常运行时间、提高网络性能并降低维护成本。宝鸡石油机械有限公司(以下简称宝石机械)的管柱堆场中，管柱堆场吊运装置负责整个堆场钻杆、钻铤的移动吊运工作，工作量大，移动频繁，且无固定停靠位置，采用有线工业以太网作为通信介质，会存在故障率高、布线难度高、故障点难找、妨碍其他设备排布等缺点。

Profinet 技术是新一代基于工业以太网技术的自动化总线标准。Profinet 具有多制造商产品之间的通信能力、自动化和工程模式，并针对分布式智能自动化系统进行了优化。

采用无线 Profinet 技术可以实现石油钻机控制系统工业以太网的无线传输。在石油钻机系统中，该方法简化了线路，提高石油钻机智能化，提高了生产效率，降低了维护成本，增加了可扩展性，使得信息传输速率高、传输可靠性高、故障率低。

1 石油钻机的 Profinet 无线通信技术总体方案设计

以石油钻机中管柱堆场吊装移运装置为例，司钻房作为上位机通过以太网交换机 X150 与总集成控制单元连接，通过无线网桥终端以无线 AP 的方式发送数据包。管柱堆场吊装移

作者简介：李博洋，宝鸡石油机械有限责任公司。

运装置通过无线网桥终端接收发送过来的数据进行解码、读取,再传输给吊装移运装置的控制器,通过有线 Profinet 通信控制变频器和电磁铁,从而进一步控制行车机构、绞车电机的相应动作。图 1 为无线 Profinet 的管子堆场吊装移运装置总体结构图。

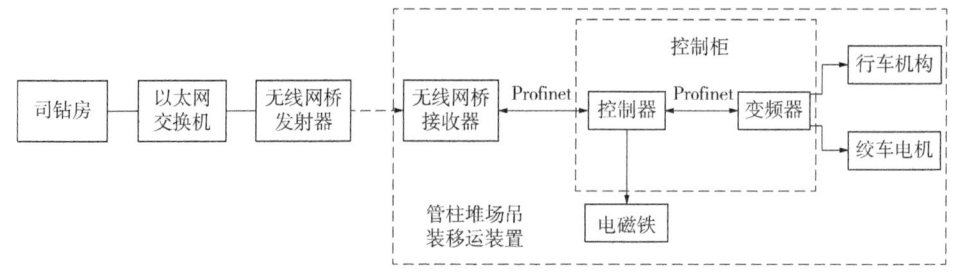

图 1　无线 Profinet 的管柱堆场吊装移运装置总体结构图

2　硬件设计与器件选型

司钻房作为上位机通过工业交换机将指令信息传递给无线客户端 AP01,通过天线 2 将信号发出,终端设备通过天线 1 捕捉到信号,无线 AP 解码后采用 Profinet 协议将信号送至控制箱,进而完成相应的操作。无线客户端 AP01 输入电压 24V,输入电流 0.6A,功耗 7.2W,如图 2 所示。

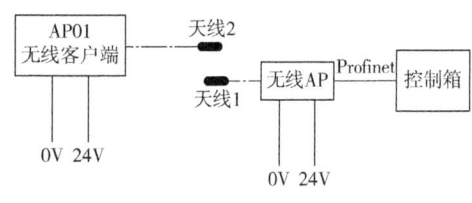

图 2　系统的通信电路图

该产品的接入点适合建立 2.4GHz 或 5GHz 的工业无线局域网,可以用于需要较高可靠性的场所。802.11a 协议工作在 5GHz 频段,802.11g 协议工作在 2.4GHz 频段,在 802.11a 和 802.11g 模式下,总传输率可达到 54Mbps;802.11b 协议工作在 2.4GHz 频带,其总传输率最高可达 11Mbps;802.11n 标准不仅适用于 2.4GHz 频段(802.11g/b),而且也适用于 5GHz 频段(802.11a),其采用多输入多输出的多重传播技术(MIMO)、20MHz 和 40MHz 信道及双频(2GHz 和 5GHz),以提供高速无线通信,同时仍然能够与传统 802.11a/b/g 设备通信。该产品具有 DFS 信道支持,在 AP 模式下,当检测到雷达信号时,设备将自动切换到另一个信道。根据规定,在切换频道后,需要经过 60s 的可用性检查后才能启动服务。

3　软件设计与配置

3.1　无线网络平台配置

将无线网桥发射器通过交叉网线与控制器 PLC1 相连,配置控制器 PLC1 与无线网桥发射器在同一子网 IP 地址。

打开计算机浏览器,在地址字段中输入无线网桥发射器地址,访问基于 Web 的网络管理器主页。选择工作模式默认为"AP"模式。

将无线网桥接收器通过交叉网线与控制器 PLC2 相连,配置控制器 PLC2 与无线网桥接收器在同一子网 IP 地址。

打开连接 PLC2 计算机的浏览器,在地址字段中输入无线网桥接收器的 IP 地址,访问基于 Web 的网络管理器主页。选择工作模式默认为"客户端"模式。

PLC1 与无线网桥发射器作为主机,PLC2 与无线网桥接收器作为从机。图 3 为无线网桥客户端设备配置界面。

3.2 无线网络通信测试

PLC2 打开 DOS 窗口,连接 PLC1 的 IP 地址,测试验证是否通信成功,访问是否正常,确保无线网络通信成功建立。图 4 为无线通信示意图。

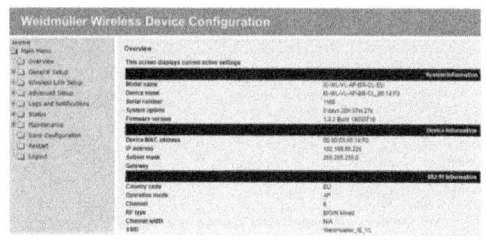

图 3　IE-WL-VL-AP-BR-CL-EU 配置界面

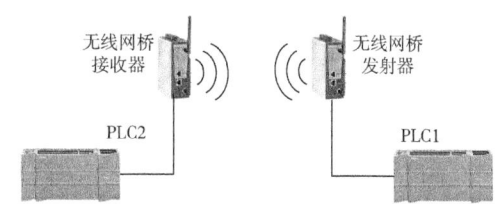

图 4　无线通信示意图

4　工艺对象模型搭建

如图 5 所示,为管柱堆场吊装移运装置控制箱内部控制结构,左上侧为中央控制单元 S7-1200 PLC,右上侧为变频器,右下角左侧为工业交换机 X150,右下角右侧的无线网桥终端作为发射端,发射端的天线接在距离 50m 的远端距离控制箱,如图 6 所示。

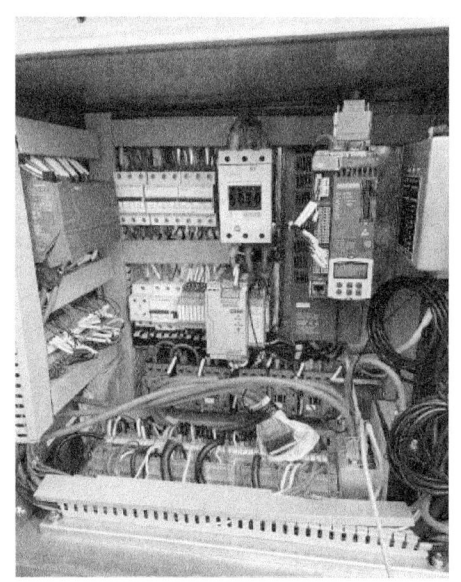

图 5　管柱堆场吊装移运装置控制柜内部结构

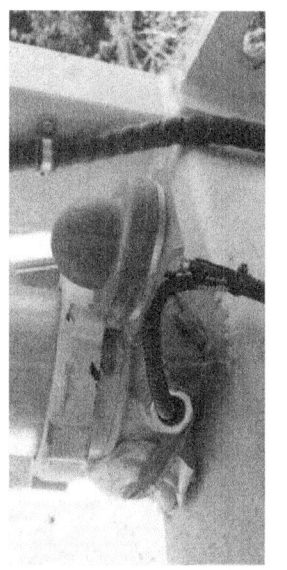

图 6　发射端天线安装位置

5　系统测试与分析

为验证无线 Profinet 通信方式在石油钻机系统中的有效性,将无线接收器分别安装在距离发射端 50m、100m、150m 的位置,测试在不同距离下通信速率的变化和稳定性的变化。

如表 1 所示,在距离 50m 的实验条件下,无线 Profinet 平均传输速率可达 30.51Mbps,且相对稳定。通信距离 100m 和 150m,传输速率会随着传输距离的增加而下降,稳定性相对变差。无线 Profinet 技术传输速率和稳定性满足石油钻机系统要求,同时具备扩展性强、灵活性强、无需额外空间的优点。

表 1　不同距离的通信传输速率

时间/s	通信速率/Mbps		
	距离 50m	距离 100m	距离 150m
0	30.2	21.5	11.5
40	31.8	23.1	12.1
80	29.5	22.7	15.7
120	31.6	21.6	7.6
160	30.8	19.3	9.3
200	29.2	19.6	10.6

6　结论

将无线 Profinet 技术用于石油钻机数据传输与信号控制中去，可以简化线路，提高了传输速率、传输可靠性、可扩展性，降低故障率。提高石油钻机智能化，提高了生产效率，降低了维护成本。Profinet 无线通信方式已在石油钻机控制系统中得到了广泛的应用，适应效果稳定，随着钻机智能化越来越高，无线 Profinet 技术将在钻机数据通信方面发挥更大的作用。

参 考 文 献

[1] 刘辉荣，付如愿，韩志成，等. 石油钻机变频器控制系统网络冗余研究[J]. 大功率变流技术，2013(2)：27-31，52.
[2] 罗磊，田德宝，周海磊，等. 网络环境下钻机集成控制系统研究[J]. 化工自动化及仪表，2020，47(5)：425-428，434.
[3] 杨建军，崔天鑫. 基于 Profinet 的无线通信技术在生活垃圾电厂的应用[J]. 化工自动化及仪表，2020，47(6)：546-548.
[4] 王静. PROFINET 用于大众高货架仓库的无线存储管理[J]. 中国仪器仪表，2019(9)：37-38.
[5] 杨珍珍，侯波，王戈. 基于环形工业以太网基础矿井泵房自动化控制系统设计与研究[J]. 机械制造与自动化，2022，51(4)：182-183，210.
[6] 汪君明，袁鸿轶，韦丁午. 基于 LoRa 射频的石油钻机智能监控系统设计[J]. 机械工程师，2022(10)：173-175.
[7] 王施权，钟志伟，刘冰. 现代先进技术在石油钻机中的应用及展望[J]. 中国设备工程，2022(18)：193-195.
[8] 樊勇利，朱星元，张洪，等. TCP 通信技术在钻机控制系统中的开发应用[J]. 设备管理与维修，2022(15)：116-118.
[9] 吴建华. 基于西门子工业无线通信在电解行车上的应用[J]. 设备管理与维修，2020(13)：108-110.
[10] 杨双业，于兴军，张鹏飞，等. 钻机司钻集成控制系统技术现状及发展建议[J]. 石油机械，2017，45(9)：1-7.
[11] 方原柏. 采用工业无线通信系统之前要考虑的 10 件事[J]. 冶金自动化，2017，41(1)：7-12.

本论文原发表于《设备管理与维修》2023 年第 5 期(下)。

高精度深孔钻井补偿系统集成设计研究

李 鹏[1,2] 樊春明[1,2] 袁 亮[3] 白兰昌[1,2] 李 欢[1,2] 郑万里[1,2]

(1. 中油国家油气钻井装备工程技术研究中心有限公司；
2. 宝鸡石油机械有限责任公司；3. 中国石油集团川庆钻探工程有限公司新疆分公司)

摘 要：深孔钻井补偿系统是海洋深水资源开发的关键装备。本文以最大补偿载荷4500kN、补偿精度90%，适应全球海域全天候作业要求，开展补偿系统集成设计研究。通过分析五种补偿类型和三种补偿机理对补偿系统综合性能的影响，以及与钻探系统的适应性，选择适应钻采船作业要求的补偿形式。通过补偿系统整体配置计算，确定系统主要参数。通过对比分布式和集中式控制两种方案，确定控制形式。通过布置方案和结构优化，以提高作业效率、降低作业成本、提高操作维护性为目标，形成高效作业的深孔钻进补偿系统集成设计解决方案。最后通过试验验证了设计方案的合理性。

关键词：深孔钻井；补偿形式；补偿机理；半主动补偿系统；死绳端液缸补偿

1 前言

浮式钻井补偿系统是海洋深水资源开发的关键装备，主要用于克服波浪升沉运动对钻采作业的影响。我国深海资源开发起步较晚，国内公司于2009年开始浮式钻井补偿系统相关产品研究。近年来，国产浮式钻井补偿系统有了突飞猛进的发展，从研究阶段进入样机研制阶段，于2014年成功研制国内首套游车补偿系统。目前，浅海平台(船)配套的被动式游车补偿已经实现工程化应用。但上述应用的产品，补偿载荷较小，补偿精度较低，无法满足深孔钻井高精度补偿的要求。目前2000m水深以上的深海钻探配套的补偿系统，尚未实现工程应用。

本文以适应全球海域全天候作业要求的钻井平台/船配套，满足3000m水深，12000m钻深的深孔钻井补偿系统为目标，设计与举升液缸方案相结合，实现最大载荷9000kN、补偿载荷4500kN、补偿精度90%的补偿装置。本补偿系统配套的钻探系统，除去常规井架、天车、游车等装置，采用举升液缸进行起下钻，井架不承担大钩载荷，只承担摇摆造成的偏载，同时起上下运动导向作用，大幅降低了钻探系统的整体重量和重心。

基金项目：国家工信部高技术船舶项目"天然气水合物钻采船工程开发"（工信部装函〔2018〕473号）；中石油集团科研项目"适应南海天然气水合物的丛式多分支水平井组钻采技术和装置研究及试验"（2021DJ1903）。

作者简介：李鹏(1981—)，男，陕西咸阳人，高级工程师，现从事海洋油气钻采装备研发工作，E-mail：lp2192@163.com。

2 深孔钻井补偿系统形式分析

补偿类型和机理对补偿系统综合性能影响大,与钻机举升方案匹配性亦不同,对钻探设备总体配套、承载、空间占用、总体重量及重心等方面均有影响。其内部配置、布置及结构等因素对系统配套、运行成本、安全性、可靠性、噪声控制、系统热损耗、维护难度等各方面各有优劣。根据深孔钻探工况对补偿系统的综合需求,综合评估不同形式和机理的补偿系统优缺点,选择适应钻井平台/船作业要求的补偿形式。

2.1 补偿形式分析

浮式钻井补偿系统具有多种形式,从结构形式和安装位置来看,可分为天车式补偿、游车式补偿、补偿钻井绞车、液缸式补偿和伸缩钻杆等[1],各有其优缺点。国内外钻井平台/船配套产品主要以 Aker MH、NOV、Cameron 为主。

游车式钻柱升沉补偿系统成功应用较早,产品主要以 NOV 和 Aker MH 为代表,用于浅水钻井或用于深水勘察船勘探取样。天车式钻柱升沉补偿系统是继游车式钻柱升沉补偿系统之后发展起来的,主要以 Aker MH 为代表,天车补偿系统的补偿载荷一般在 300t 以上,主要用于深海大吨位的钻井平台/船。钻井升沉补偿绞车主要以 NOV 为代表,其分别开发了电动主动式补偿绞车和液压驱动补偿绞车系统,但由于消耗功率过大,应用受到一定限制。

钢丝绳式举升液缸补偿系统是近年来开发的产品,主要在深水浮式钻井系统中应用。由 Aker MH 于 1987 年提出 Ramrig 钻机设计理念,用举升液缸代替钻井绞车系统,取消了钻井绞车、天车及游吊系统、VFD 房,井架不需承担钻井主载荷,大大简化了钻井系统,扩大了钻台面空间,降低了钻探设备的重量和重心,对重量、重心要求敏感的浮式钻井平台/船具有明显的优势,成为浮式深水钻探系统发展趋势。与举升液缸方案配套的补偿系统,包括集成式(举升和补偿)液缸补偿和死绳端式液缸补偿两种方案。其中,集成式(举升和补偿)液缸补偿如图1所示,以 Aker MH 的 Ramrig 钻机为代表;死绳端式液缸补偿如图2所示,以 NOV 的 Cylinder Rig 钻机为代表。

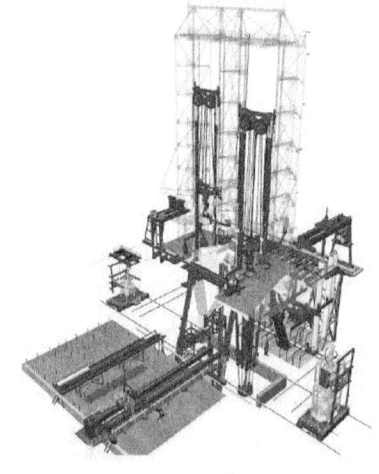

图1 集成式液缸补偿

图2 死绳端式液缸补偿

表1对天车式、游车式、绞车式、集成式(举升和补偿)及死绳端式,进行参数性能对比。从对比结果可看出:

对于4500kN补偿载荷，游车式补偿由于尺寸过大导致占用井架空间，降低起升载荷而无法应用。天车式补偿和绞车式补偿均满足最大补偿载荷4500kN的要求。但是由于绞车式补偿为主动补偿，虽然能小幅增加补偿精度，但其系统能耗为半主动补偿系统的11.4倍，大幅增加系统配套和运行成本，因此天车式补偿为传统绞车提升钻机补偿系统的配套首选。

集成式(举升和补偿)和死绳端式补偿两种方案，具有常规钻机所不具备的优点，除补偿液缸与举升液缸是否独立外，其系统组成与配套基本一致。但是，由于集成式(举升和补偿)补偿方案中的补偿液缸与举升液缸高度集成，增加了设备的制造、控制及维护难度，设备的故障点增多，降低了可靠性。

与天车式补偿相比，死绳端式补偿方案是专为举升液缸式钻探设备而设计，且补偿液缸低位安装，能够大幅降低钻探设备的重量和重心及安装维护难度，因此，以NOV Cylinder Rig钻机配套的死绳端式补偿方案为大吨位补偿系统最优方案。

表1 不同补偿类型性能对比表

补偿形式	游车式	天车式	绞车式	集成式	死绳端式
质量(本体)/t	62	140	120	78	80
质量(整体)/t	134	212	192	150	280(+举升)
重心高/m	30	70	2	8	7
补偿精度/%	85~90	85~90	90~93	85~90	85~90
补偿载荷/kN	4500				
补偿行程/m	7.62	7.62	—	7.62	7.62
补偿速度/(m/s)	1.31				
空间占用	高	中	中	低	低
噪声	高	高	高	低	低
经济性	好	好	差	好	好
控制系统	简单	简单	复杂	复杂	简单
可靠性	高	高	低	低	高
维护保养	方便	困难	方便	困难	方便

2.2 补偿机理分析

从补偿原理来看，可分为被动式、主动式和半主动式补偿[2]。下文针对三种补偿形式的工作机理研究分析。

图3(a)为被动升沉补偿系统工作原理图。被动补偿系统工作原理简单，为早期海洋油气钻井作业的升沉补偿装置所采用。其相当于一个大型液压空气弹簧，依靠海浪的举升力和船自身的重力来压缩和释放蓄能器中的压缩空气，减小升沉幅度以实现补偿[3]。当钻井平台上升时，蓄能器内的气体进一步被压缩以补偿上升位移并储存能量；当钻井平台下沉时，蓄能器内的气体膨胀以补偿下沉位移，蓄能器储存的能量被释放。被动缸承担整个钻柱载荷。

图3(b)为主动升沉补偿系统工作原理图。主动补偿系统通过液压泵连续地给主动补偿缸的两端输入液压油，通过检测液缸活塞及船体的位移，通过PLC控制使液缸活塞的位移与浮式平台的升沉尽可能协调一致。主动补偿系统主要克服摩擦力和惯性载荷等附加力，主要用于需精确定位的工况[4-5]。

由于被动升沉补偿系统和主动升沉补偿系统均存在一定的局限性，半主动升沉补偿系统综合了上述两个系统的特点。图3(c)为半主动升沉补偿系统原理图。半主动升沉补偿系统即在被动升沉补偿系统的基础上叠加主动升沉补偿系统协同工作实现对大钩的补偿过程[6]。半主动升沉补偿系统中气液蓄能器发挥气液弹簧作用承受大部分载荷，消减大部分大钩升沉位移；双向变量泵向主动缸上腔或下腔供油，推动主动缸活塞运动克服其余诸如机械摩擦、气体压力波动产生的附加载荷，进一步提高整个升沉补偿系统的补偿效果。

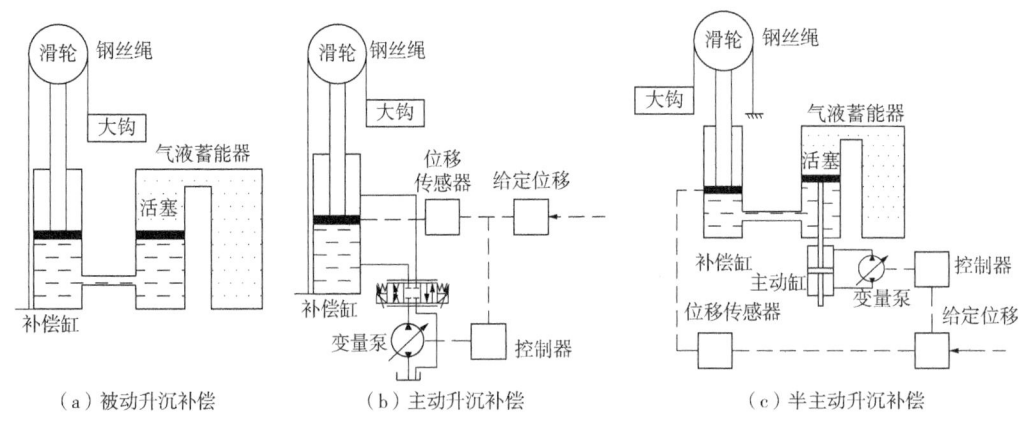

图3 升沉补偿系统原理图

根据对比分析，为满足全球水域大洋钻探及海上油气勘探作业需求，需达到4500kN补偿载荷、90%补偿精度的要求，实现升沉补偿、调节钻压功能，同时系统能耗低，响应速度快。对于4500kN补偿载荷的要求，主动式补偿系统能耗过高，经济性太差；对于90%补偿精度、调节钻压功能及响应速度快的要求，被动式补偿系统性能无法满足要求。因此，选择半主动式补偿方式，无论从经济性和可行性来看都是最优方案。

3 深孔钻井补偿系统集成设计方案

3.1 主要参数的确定

补偿载荷：被动补偿系统承受最大钻深时大部分补偿载荷，因此，被动补偿载荷设为4500kN。主动补偿装置作为辅助补偿配套，主要用于需精确定位的工况，用于克服补偿装置机械摩擦、气瓶组压力波动、补偿装置机械结构惯性力等附加作用力，提高装置补偿精度。

最大补偿行程：根据南海浪高的主要分布规律，钻采平台/船应满足长期作业要求，参考长期在南海的蓝鲸一号和海洋石油981平台补偿器行程，确定钻柱补偿系统补偿行程为7.62m。

最大补偿速度：南海海域在24.3m/s风速下，波浪周期$T=12.1s$。平台升沉运动的周期与波浪周期相同，升沉振幅为最大补偿行程的一半，即3.81m，实际平台的升沉运动规律为：

位移：$S=3.81\times\sin\dfrac{2\pi}{T}t=3.81\times\sin 0.519t$。

速度：$v=1.977\times\cos 0.519t$（对位移求一阶导数）。

补偿缸的运动规律与平台一致，主动缸与补偿缸运动规律相同。由于补偿采取增距式结构，补偿缸速度实际为平台速度的二分之一，因此补偿缸最大补偿速度为0.9885m/s。参考长期在南海的蓝鲸一号和海洋石油981平台补偿器能力,确定钻柱补偿系统最大补偿速度为1.31m/s。

由表2可以看出,本文设计的死绳端液缸式补偿系统,基本性能参数达到了国际主流产品同等水平。

表2 国内外主要同级别补偿产品对比表

项目	地球号(日本)	NOV(液缸式)	MH(Ramrig)	本方案
类型	天车补偿	死绳端补偿	集成液缸补偿	死绳端补偿
形式	半主动	半主动	半主动	半主动
被动补偿载荷/kN	5180	4540	4540	4500
主动补偿载荷/kN	350	350	350	350
补偿行程/m	7.62	7.62	7.62	7.62
补偿速度/(m/s)	1.31	1.31	1.57	1.31
额定静载荷/kN	1250	9000	6750	9000
气瓶体积/m³	30	24	24	24

3.2 补偿控制系统方案选择

控制系统原理:运动参考单元(MRU)采集船体升沉运动的参数,并输出给PLC控制系统,PLC控制系统根据MRU提供的运动参数,实时提取船体的升沉位移并与液缸位移传感器输出的液缸位移进行对比,根据船体的升沉运动位移和液缸位移的差值,控制伺服阀的开启度和液压油的流向,给主动缸供油,控制主动补偿液缸的移动速度和方向,使液缸活塞的运动与船体的升沉运动曲线尽可能吻合,从而减小升沉运动对钻头钻压的影响,提升整体系统的补偿效果[7]。

图4为控制系统分布式采集控制结构图。分布式站点通过Profibus-DP现场总线与PLC进行通信。司钻房触摸屏通过Ethernet的方式与PLC实时地进行数据交换和指令传递。PLC采用冗余设计,控制器热备份,降低停机风险。

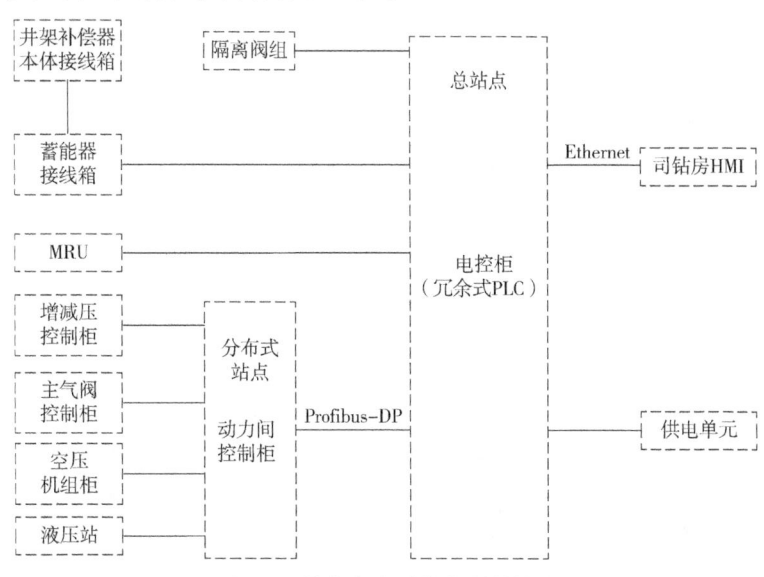

图4 系统分布式采集控制结构图

集中采集控制的方案特点是所有的信号经过电缆直接接入控制柜，由于控制点数太多，一个机架不能安装足够数量的模块，需要进行机架扩展。

两种方案的对比见表3。由于该系统分别布置于平台不同位置，距离较远，采集信号点多，通过对比，电控系统采用分布式采集控制方案。

表3 控制系统方案对比表

方案	优点	缺点
集中采集控制	（1）全部信号低位安装，便于检查； （2）各信号独立走线，不会交叉影响	（1）需要扩展机架； （2）不利于功能扩展； （3）电缆多，布线复杂； （4）故障检测、维护麻烦
分布采集控制	（1）电缆规格少、电缆用量少、布线方便； （2）总线通信干扰少； （3）故障检测、维护方便； （4）功能扩展容易实现	（1）井架站点的维护不方便； （2）风险集中（总线故障会导致该总线所有信号故障）

3.3 系统方案设计

在确定主参数的基础上，结合被动补偿系统和主动补偿系统配置计算及布置方案设计，进行目标钻井平台/船高精度深孔钻井补偿系统集成方案设计。高精度深孔钻井补偿系统集成设计方案如图5所示。

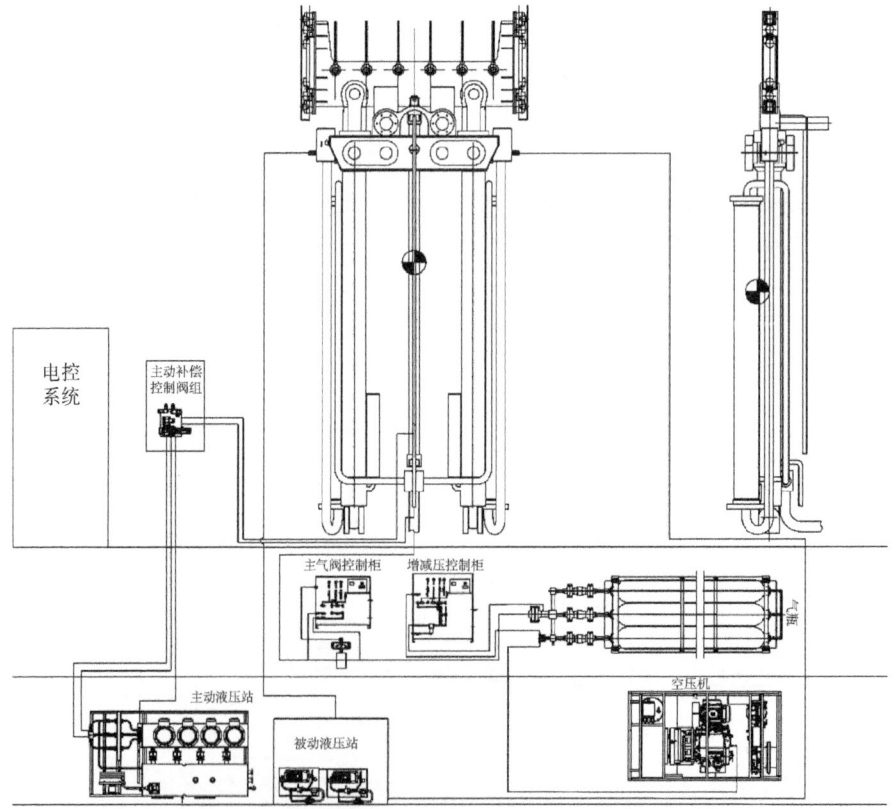

图5 死绳端液缸补偿系统总体方案图

（1）采用半主动死绳端式补偿方案，被动补偿系统承担大部分载荷，单独开启补偿精度为80%左右；主动补偿系统克服其余诸如机械摩擦、气体压力波动产生的附加载荷，同时开启可达90%左右补偿精度。

（2）作业时，浮动架体和固定架体锁销打开，补偿系统可承担4500kN的负荷；移运或处理事故时，浮动架体和固定架体锁销锁紧，可承担最大9000kN的负荷。

（3）补偿缸采用并列竖直布置方式，低位安装，减小重量，降低安装和维护难度，节约空间，利于补偿本体装置。

（4）补偿缸活塞杆向上伸出，采用拉缸形式，降低压杆屈曲，提高设备可靠性。

（5）被动补偿采用蓄能器加高压气瓶形成空气弹簧，补偿效果可靠，同时达到节能的效果。主动补偿形式采用并行等速缸调节升沉量和钻压，提高补偿精度。

（6）控制系统采用冗余式PLC分布式站点控制，保证系统控制的可靠、灵活。主动补偿采用泵控和阀控并联方式控制，提高效率和动态响应性能。

4 深孔钻井补偿系统试验验证

钻井补偿装置作为一种特殊的运动补偿装备，其结构形式、功能要求为海洋浮式钻井所特有。其验证试验项目包括：最大补偿载荷试验、额定静载荷试验、补偿功能试验。

最大补偿载荷试验、额定静载荷试验分别检验补偿工况和非补偿工况最大承载强度。通过载荷测试应力最大值、残余变形并与试验大纲要求数值比较，各组数据应力最大值均小于265MPa，残余变形小于0.2%，满足设计验证要求。试验完成后将产品解体，对主要承载件进行无损检测，各承载件强度均符合设计要求。

功能试验的目的是通过海况模拟试验，检验被动补偿的补偿载荷及主动补偿的补偿位移精度。模拟平台升沉运动的正向试验方案较为直观，但运动模拟平台造价高昂，波浪模拟难度高，因此采用反向试验方案，即保证补偿装置不动，给定一个输入的升沉位移信号，控制活塞杆实时跟随升沉信号运动，在预定正弦波激励和额定补偿载荷作用下，检测活塞杆的运动和输入位移信号的偏差，检测补偿精度，进行控制策略评价，验证控制系统设计合理性及安全可靠性。试验结果表明，在外部输入三组不同特性（幅值和周期）人工模拟海浪的激励下，配挂负载4500kN时最大位移偏差±544mm（国外同类产品为±533.4～±1143mm），补偿精度为92.86%，符合设计要求。

5 总结

（1）根据分析对比游车式、天车式、绞车式、集成式（举升和补偿）和死绳端式补偿方案，针对液缸举升钻机，选择死绳端式液缸补偿方案。

（2）对被动式、主动式和半主动式补偿三种补偿形式的工作机理进行分析对比，确定了半主动式补偿方案。

（3）根据全球海域全天候作业要求，确定了钻柱补偿行程、最大运动速度、最大补偿载荷、最大大钩载荷、补偿精度等主参数。

（4）进行死绳端式液缸补偿的钻柱补偿系统整体设计，举升液缸和补偿液缸分离，主动补偿和被动补偿分离，举升和补偿分别控制。采用分布式采集控制策略，降低制造难度，减少系统干扰，提高控制精度。

（5）通过负荷和海况模拟试验，验证了补偿系统总体设计和控制方案的合理性和可靠性。

参 考 文 献

[1] 任克忍，沈大春，王定亚，等.海洋钻井升沉补偿系统技术分析[J].石油机械，2009，37(9)：125-128.

[2] 方华灿.海洋石油钻采装备与结构[M].北京：石油工业出版社，1990.

[3] 姜浩，刘衍聪，张彦廷，等.浮式钻井平台被动升沉补偿装置设计[J].液压与气动，2011(10)：50-54.

[4] 张彦廷，刘振东，姜浩，等.浮式钻井平台升沉补偿系统主动力研究[J].石油矿场机械，2010，39(4)：1-4.

[5] KORDE U A. Active Heave Compensation on Drill Ships in Irregular Waves[J]. Ocean Engineering，1998，25(7)：541-561.

[6] ROBICHAUX L R，HATLESKOG J T. Semi-active Heave Compensation System for Marine Vessels：United States，005209302A[P]. 1993-05-11.

[7] 方华灿.海洋钻井船升沉补偿装置工作理论的初步研究[J].华东石油学院学报，1978(3)：56-67.

本论文原发表于《石油矿物机械》2023年第52卷第6期。

机械式旋冲工具对水平井送钻摩阻的影响规律研究

闫 炎[1]　韩礼红[1]　刘永红[2]　杨尚谕[1]　王建军[1]

(1. 中国石油集团工程材料研究院有限公司；2. 中国石油大学(华东)机电工程学院)

摘 要： 托压问题是制约水平井滑动钻进机械钻速的主要因素之一，使用振动减阻工具是水平井减阻最直接有效的方式。为分析振动减阻工具在水平井中的减阻效果，本文针对一种螺杆驱动的机械式旋转冲击钻井工具，对工具在不同钻进工况下的钻压波动、位移变化进行了模拟计算，解释了低频冲击工具可以实现振动减阻的根本原因。计算结果表明：旋冲工具降低摩阻需要借助钻压；激励频率越大，钻头处的钻压波动越剧烈；存在最佳的激励频率使得钻头处钻压整体处于最高水平。旋冲工具工作时，钻头处的钻压波动不仅包含激励本身的频率，还会包含激励频率的二倍甚至三倍等更高倍数的频率。研究结果可以为旋冲工具工作参数的优化，以及工具在水平井中实现最佳提速效果提供理论指导。

关键词： 水平井；摩阻；旋转冲击；冲击频率；钻压波动

1 引言

托压问题是制约定向井、水平井滑动钻进机械钻速的关键因素之一[1-4]，钻柱与井壁间的摩擦力导致钻压无法完全施加到钻头上，使得钻头的破岩效率降低。工程上通常通过降低钻柱与井壁之间的摩擦系数[5-9]、优化井眼轨迹[10-11]、使用降摩减阻工具[12-15]等方式来缓解托压问题。

在上述方法中，使用振动减阻工具是水平井减阻最直接有效的治理方式，不同石油公司与科研院所根据不同的工作原理研制了不同振动方式的振动减阻工具[16-19]。在这些振动减阻工具中，以螺杆驱动的减阻工具居多。为此，本文以一种螺杆驱动的机械式旋转冲击钻井工具[20]为例，进行水平井钻进过程送钻摩阻分析。工具通过螺杆带动下凸轮旋转，下凸轮带动啮合的上凸轮及整个钻柱做轴向运动，完成对下凸轮及钻头的冲击作用。当在直井中使用时，该工具可以为钻头提供冲击力，提高钻头破岩效率。当在定向井、水平井中使用时，该工具可以有效解决托压问题，并在现场应用中取得了很好的提速效果[21]，但工具在水平井中的减阻效果及影响因素还需要进一步厘清。

基金项目：中国石油天然气集团有限公司科学研究与技术开发项目"油气井套损评价与修复关键技术研究"(2021DJ2705)；"井筒完整性关键理论与控制技术"(2021DJ4403)。

作者简介：闫炎(1993—)，男，博士后，毕业于中国石油大学(华东)油气井工程专业，主要从事油气井岩石力学研究工作。地址：(030600)陕西省西安市雁塔区锦业二路89号，E-mail：yanyan3@cnpc.com.cn。

间隙元理论[22]、分段计算方法[23]、非线性有限单元法等方法都是研究水平井振动减阻技术的常用方法,这些方法中大多忽略井底钻头,钻压的施加、井壁对钻柱的摩擦力、钻柱底部的激励等均简化假设为一定的外部边界条件;简化后虽然可获得钻压、摩擦系数、激励频率等参数对减阻的影响效果,但由于模型过于简化,导致某些重要参数无法通过计算结果直观地反映出来,如钻进过程中使用振动减阻工具前后钻头处的钻压增幅,钻压波动的幅值和频率是否变化等。基于此,本文针对螺杆驱动的机械式旋转冲击钻井工具工作原理,基于有限元软件平台 ABAQUS 建立了水平井轴向振动减阻模型,模型考虑了钻柱与井壁的接触、实际钻井过程中上部钻柱的钻压作用,并且把钻头上方旋冲工具对钻柱的激励直接设置为一个凸轮结构,以模拟出更加真实的激励方式,对该类型旋冲工具在水平井中的减阻作用提供理论参考。

2 数值模型

2.1 几何模型

根据机械式旋转冲击钻井工具举升钻柱的工作原理(图1)设计了仿真的几何模型,如图2所示。模型包含井筒、钻柱、钻头、凸轮、凸轮轴,钻柱的轴线与井筒中心线重合,由于钻柱在井筒内没有旋转运动,且在仿真过程中只受重力、左端施加的钻压,以及右端施加的位移激励,所以在钻柱与凸轮之间加了一个限位块,用来约束钻柱与钻头的横向相对运动,且限位块可将凸轮的旋转运动转化为对钻柱的位移激励。工具本体使用的是上下啮合的凸轮,工具对钻柱的激励频率由凸轮齿数与钻头转速共同决定,为了研究方便,有限元模型使用图2中所示的盘形凸轮,该凸轮绕凸轮轴旋转一周即对钻柱形成一次激励,改变激励频率只需改变凸轮绕凸轮轴的旋转速度即可。模型各部件的尺寸如下:钻头为 8½in 六刀翼 PDC 钻头,外径 215.9mm;5½in 钻杆长 1000m;井筒长 1001m,内径 216mm;凸轮基圆半径 12mm,升程为 8mm;凸轮轴直径 10mm。

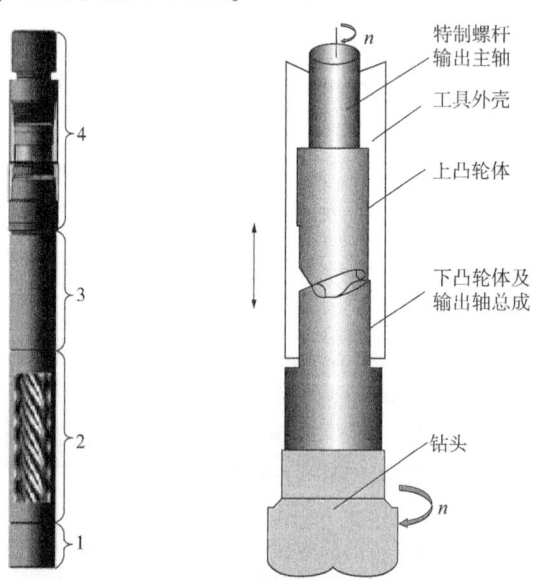

图1 机械式旋转冲击钻井工具及工作原理示意图

1—旁通阀总成;2—马达动力总成;3—马达驱动轴总成;4—传动轴总成和冲击发生机构总成

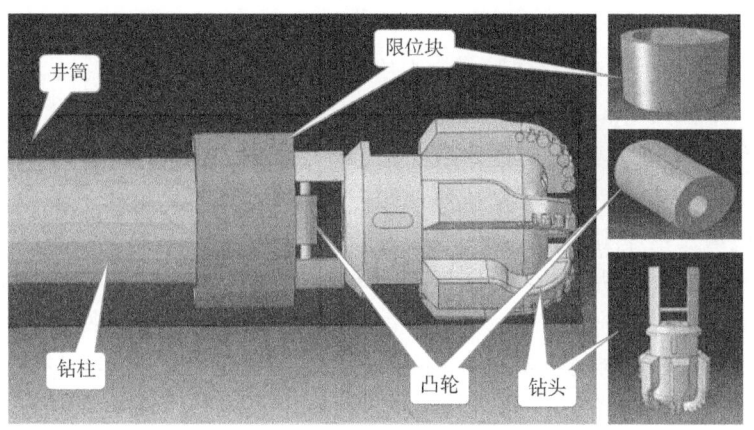

图 2 几何模型

2.2 控制方程

钻柱在水平井中的运动属于动力学问题,模型中的每个变形体的运动都满足拉格朗日方程:

$$\frac{\mathrm{d}}{\mathrm{d}t}\left[\frac{\partial L}{\partial \dot{u}}\right]-\frac{\partial L}{\partial u}+\frac{\partial F_c}{\partial \dot{u}}=0 \tag{1}$$

式中:$L=T-U+W$ 为拉格朗日函数;T 为物体的动能;U 为物体的应变能;W 为外力做的功;F_c 为耗散函数。

在有限单元法中,变形体的每个单元都满足拉格朗日方程(1),变形体的单元节点位移向量代入拉格朗日方程之后可以得到单元的运动方程,组合各个单元的方程,可得到变形体的整体运动方程为:

$$M\ddot{u}+C\dot{u}+Ku=P \tag{2}$$

式中:M 为质量矩阵;C 为阻尼矩阵;K 为刚度矩阵;P 为外力;u,$\dot{u}$,$\ddot{u}$ 分别为位移向量,速度向量,加速度向量。

该方程组为常微分方程组,本文模拟过程中使用中心差分法求解该方程组,该算法为显式算法,以应力波传播的方式求解运动方程。在中心差分法中,节点速度向量、加速度向量分别表示为:

$$\dot{u}_t=\frac{1}{2\Delta t}(-u_{t-\Delta t}+u_{t+\Delta t}) \tag{3}$$

$$\ddot{u}_t=\frac{1}{\Delta t^2}(u_{t-\Delta t}-2u_t+u_{t+\Delta t}) \tag{4}$$

本文模拟过程中,钻柱与井壁、钻柱右端与凸轮均存在接触碰撞,所以在模拟过程中需要进行接触设置。为了模拟钻柱右端与凸轮的接触碰撞行为,在钻柱的右端与凸轮之间加了一个实体限位块,且凸轮和限位块均设置成刚体。因为模型中存在刚体与刚体的接触行为,所以本文模拟过程中,接触算法使用罚函数法。

在罚接触算法中,弹簧单元作用在从节点上的法向接触力为:

$$f = -lk\boldsymbol{n} \tag{5}$$

式中：f 为接触点上的法向接触力；l 为从节点与主面之间的距离；k 为弹簧刚度；$\boldsymbol{n}$ 为接触点处的法向单位矢量。

罚函数法允许物体间发生一定的穿透。在求解过程中，在每一个时间步中，先检查各从节点是否穿透主面，没有穿透则不做任何处理，否则在该从节点与被穿透的主表面之间引入一个大小与穿透量及主片刚度成正比的接触力。

采用显式算法求解运动方程时，在每个时间步中先不考虑物体间的接触和碰撞，完全独立地对各物体的运动方程进行求解，得到各物体运动状态的试探解，如果试探解满足物体之间的无接触条件，则在本时间步内物体之间没有发生接触，因此物体之间的运动完全独立，试探解即为真实解。如果试探解不满足物体之间的无接触条件，则需要修正试探解，使其满足接触界面条件。

2.3 材料参数与网格划分

由于本文只关注钻柱在受到重力、钻压，以及位移激励情况下的变形和运动，除钻柱外其他部件均设置为刚体。钻柱的材料为钢，密度 7850kg/m³，弹性模量为 210GPa，泊松比为 0.3。由于钻柱长 1000m，且要与井筒发生碰撞，因此使用三维一阶铁木辛柯梁单元 B31。仿真过程中，假设钻头、凸轮、凸轮轴、限位块均不会发生变形，因此 PDC 钻头、凸轮和凸轮轴均使用 4 节点减缩积分壳单元 S4R，并进行刚体约束，限位块使用三维 8 节点减缩积分实体单元 C3D8R，并进行刚体约束。模型各部件的网格划分如图 3 所示。

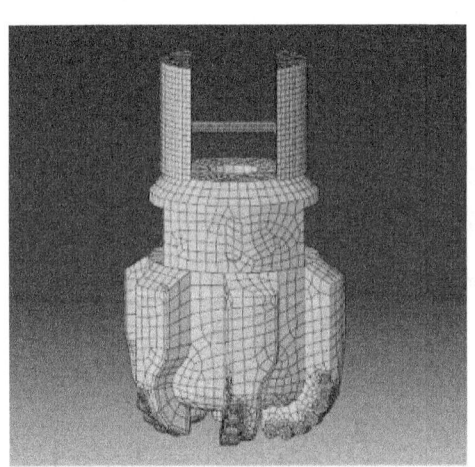
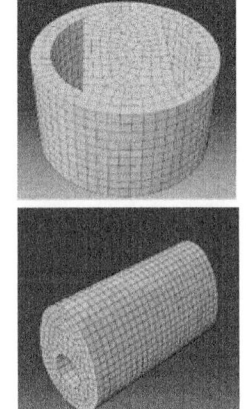

图 3　模型各部件的网格划分

2.4 分析步、边界条件及接触设置

实际钻井过程中，钻柱一开始就因受重力和上部钻柱所施加的钻压影响而贴于下井壁且整个钻柱中都存在轴向力。因此在仿真过程中，为了真实反映实际钻井工况，仿真过程包含三个显式动力学分析步：（1）对整个钻柱施加重力，使其贴于下井壁，重力施加时间为 65s；（2）在钻柱左端缓慢施加钻压，为了不使钻柱产生过大振动，钻压施加时间设定为 65s；（3）使凸轮绕凸轮轴旋转，凸轮带动限位块做周期运动，对钻柱施加位移激励，激励时间设置为 20s。

模型中井筒的自由度全部约束；对限位块施加刚体约束；钻柱右端、钻头与凸轮轴的自由度全部约束；约束凸轮与凸轮轴的相对自由度，使凸轮只能绕凸轮轴转动，研究在没有进尺的情况下，旋冲钻井工具能产生的最大冲击力与冲击频率和所施加钻压之间的关系。限位块与凸轮外表面、限位块与钻头外表面之间设置面面接触，且无摩擦系数；在模拟过程中钻柱与井壁发生接触碰撞，因此设置通用接触，摩擦系数设为0.3。

3 旋冲工具对水平井送钻摩阻的影响

图4为钻柱左端施加钻压100kN、激励频率8Hz的工况下，钻压施加过程中[分析步(2)]钻柱左端的位移与钻头处钻压随时间的变化规律曲线。由图4可以看出，施加钻压后钻柱左端向右移动了0.0866m，钻头处的最终钻压只有23.3kN，说明此时井筒内出现托压。

图5为钻柱左端施加钻压100kN、激励频率8Hz工况下，旋冲工具工作过程中[分析步(3)]钻柱左端位移与钻头处钻压随时间的变化规律曲线。当旋冲工具工作后，即凸轮给钻柱右端一个幅值8mm、频率8Hz的激励位移，钻柱左

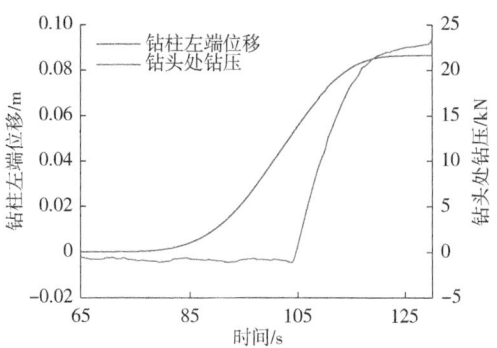

图4 钻压施加过程中钻柱左端位移与钻头处钻压变化规律

端的位移在5s内从0.0866m逐渐增大到0.1361m，向右移动了0.0495m，之后保持稳定并伴随有微小的波动；而钻头处的钻压也从23.3kN逐渐增大，直到保持一个稳定的波动状态，钻压平均值为94.3kN，谷值为45.5kN，峰值为148.5kN；说明旋冲工具可以降低水平井中的送钻摩阻，有效解决托压问题。

图5中钻压在135s之后保持一个稳定的波动状态，因此取135~150s这15s的钻压信号进行分析，对其进行快速傅里叶变换，提取钻压的频谱，如图6所示。可以看出，在旋冲工具稳定工作的时候，钻压波动不仅包含凸轮本身对钻柱的激励频率8Hz，而且还包括16Hz、24Hz等二倍、三倍的倍频。说明旋冲工具单一频率的激励可以使钻柱对钻头产生多种倍频的冲击。

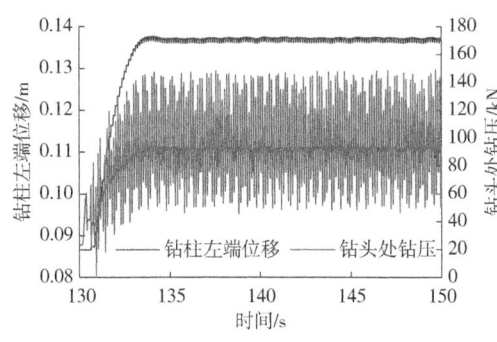

图5 旋冲工具工作过程中的钻柱左端位移与钻头处钻压

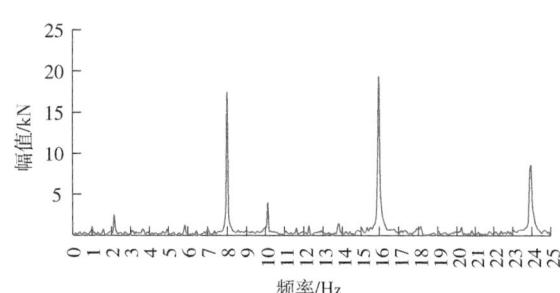

图6 钻头处钻压频谱

4 水平井送钻摩阻影响因素分析

4.1 左端钻压不同时激励频率对钻头处钻压幅值的影响

图7为钻头处钻压峰值、谷值和平均值随激励频率的变化规律。由图7(a)和图7(b)可知，相同钻压下，随着旋冲工具激励频率的增大，钻头处钻压峰值逐渐增大，钻压谷值逐渐减小，说明激励频率的升高加剧了钻柱底端的振动；而钻压平均值随着激励频率的升高先增

后减,但均在激励频率为8Hz时最大,分别为94.3kN和72.8kN,说明此时钻头整体处于高压状态。

由图7(c)可看出,在钻压60kN工况下,无论激励频率为多少,钻压的谷值和平均值均接近于0,而钻压峰值约为30kN;说明在钻压60kN工况下,由于送钻摩阻的存在,钻柱左端的钻压无法传递到钻头处,此时旋冲工具无法解决托压问题,但由于凸轮对钻柱的激励作用,钻头处依然会产生30kN左右的冲击力。

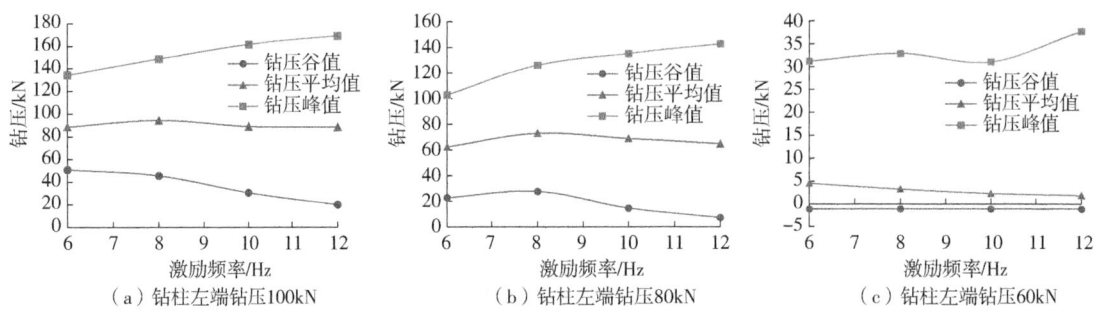

图7 不同激励频率时钻头处钻压的峰值、谷值和平均值

4.2 左端钻压不同时激励频率对钻柱左端位移的影响

图8为不同激励频率时钻柱左端位移随时间的变化规律。图9为不同钻压下旋冲工具激励引起的钻柱左端最大位移与激励频率的关系。由图8(a)、图8(b)和图9(a)、图9(b)可知,相同钻压下,钻柱左端位移随激励频率的升高先增后减,且均在8Hz时最大,分别为0.0495m和0.0493m;说明旋冲工具在水平井中的振动减阻作业中存在最佳工作频率。

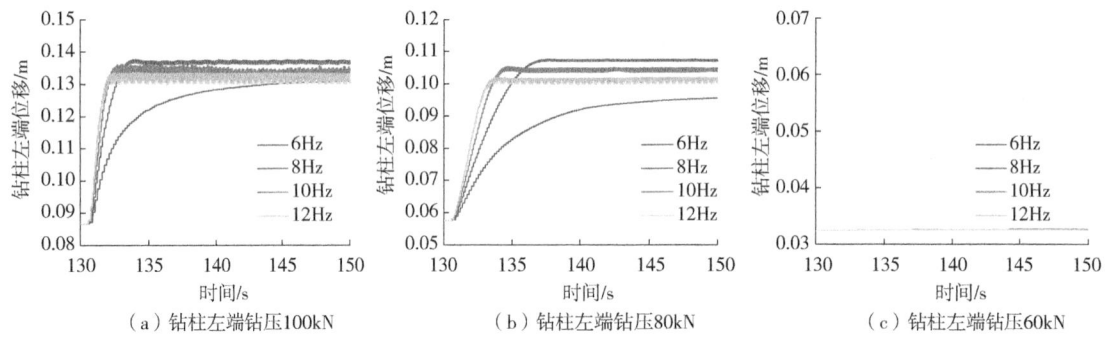

图8 不同激励频率时钻柱左端的位移随时间的变化规律

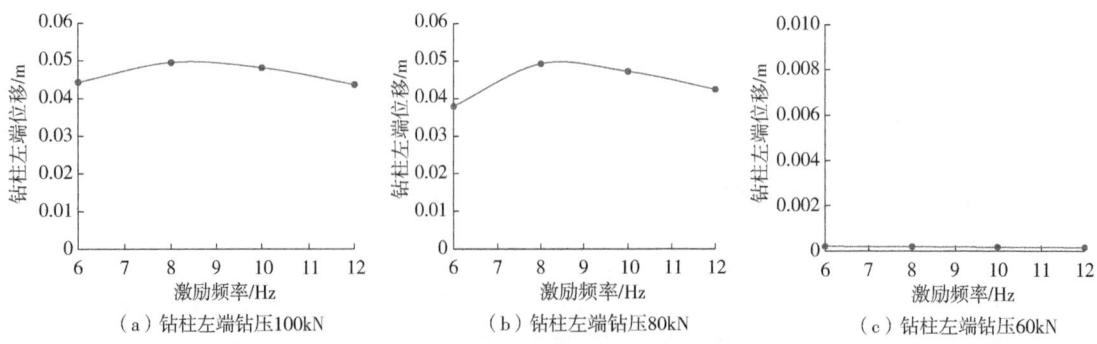

图9 不同钻压下激励引起的钻柱左端位移与激励频率的关系

由图8(c)和图9(c)可知,在钻压60kN工况下,无论激励频率为多少,钻柱左端位移均无变化,说明此时旋冲工具无法解决托压问题;这也是图7(c)中钻头处钻压平均值并没有随着激励而增大的原因。

4.3 左端钻压不同时激励频率对钻头处钻压波动频率的影响

图10为不同钻压条件下不同激励频率时钻头处钻压的频谱图。由图10(a)和图10(b)可知,在100kN和80kN钻压工况下,钻头处的钻压波动不仅包含旋冲工具本身的激励频率,而且还包含激励频率的2倍、3倍等高倍数频率,说明低频的旋冲工具也会带来高频冲击,其幅值甚至不低于低频冲击的幅值。由图10(c)可知,在60kN钻压工况下,钻头处钻压波动各频率的幅值均接近于0,结合图7(c)可知,此时钻头处只存在幅值不大的冲击力作用。

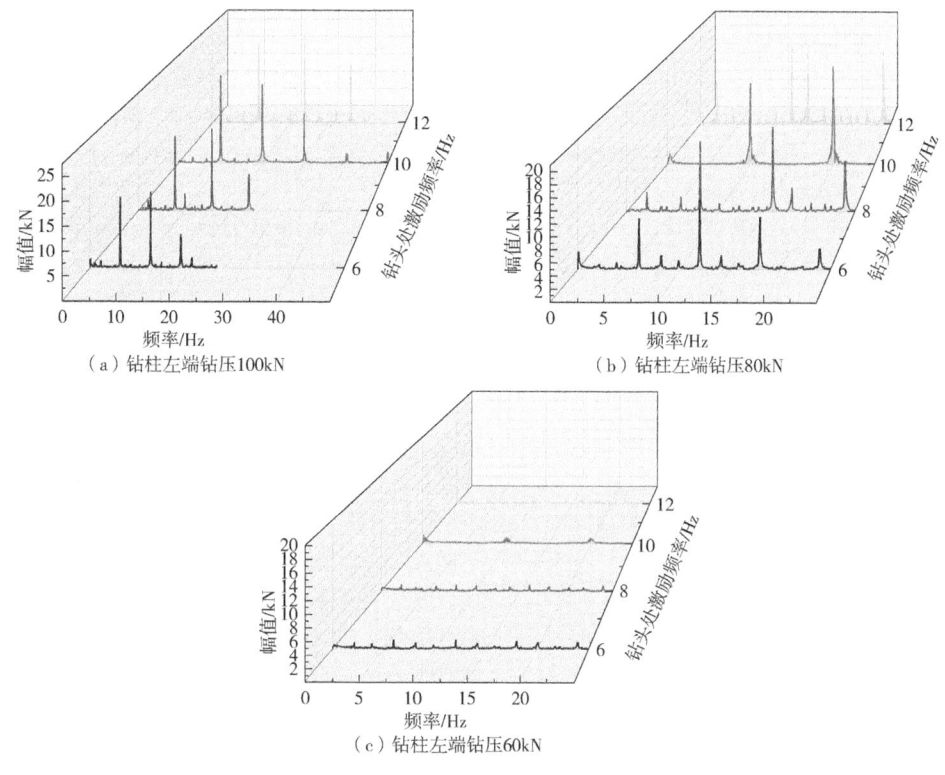

图10 不同激励频率时钻头处钻压频谱

5 结论

(1)旋冲工具在水平井中的减摩阻作用受钻压影响,当钻压可以传递到钻头处时,旋冲工具可以有效解决托压问题;当钻压无法传递到钻头处时,旋冲工具无法解决托压问题,但是激励会使钻头处产生一定的冲击力。

(2)钻压一定时随着激励频率的增大,钻头处钻压峰值逐渐增大,钻压谷值逐渐减小,钻压波动会更加剧烈;钻压平均值随着激励频率的增大先升高后降低,存在最佳的激励频率使得钻头处钻压整体处于最高水平。

(3)在有旋冲的工况下,钻头处的钻压波动不仅包含旋冲工具本身的激励频率,还包含二倍甚至数倍的高频,说明低频的旋冲工具也会带来高频冲击,其幅值甚至不低于低频冲击幅值。

参 考 文 献

[1] NWEMAM K R, BURNETT T G, PURSELL J C, et al. Modeling the affect of a downhole vibrator[C]. The SPE/ICoTA Coiled Tubing Well Intervention Conference and Exhibition. Texas, USA, April 1-3, 2009.

[2] GOICOECHEA H E, BUEZAS F S, ROSALES M B. A novel tool to improve the rate of penetration by transferring drilling string vibration energy to hydraulic energy[J]. International Journal of Mechanical Sciences, 2019, 157: 98-110.

[3] CHEN P J, GAO D L, WANG Z H, et al. Study on aggressively working casing string in extended-reach well [J]. Journal of Petroleum Science and Engineering, 2017, 157: 604-616.

[4] 李勇政, 陈涛, 江川, 等. 四川盆地磨溪—高石梯区块定向钻井关键技术[J]. 石油钻探技术, 2021, 49(2): 26-31.

[5] 陈乐亮. 水平井钻井液降摩阻问题[J]. 钻井液与完井液, 1992, 9(4): 1-6.

[6] 赵虎. 钻井液润滑性评价方法对比研究[J]. 石油工业技术监督, 2021, 37(4): 24-27.

[7] LIU W L, NI H J, WANG P, et al. Investigation on the tribological performance of mico-dimples textured surface combined with longitudinal or transverse vibration under hydrodynamic lubrication[J]. International Journal of Mechanical Sciences, 2020, 174: 105474.

[8] LIU W L, NI H J, WANG P, et al. Analytical investigation of the friction reduction performance of longitudinal vibration based on the modeified elastoplastic contact model[J]. Tribology International, 2020, 146: 106237.

[9] 白冬青. 最小曲率法计算中的几个问题[J]. 断块油气田, 2007, 14(5): 67-69.

[10] 韦龙贵, 王赟, 陈立强. 基于熵权法的大位移井井眼轨迹设计方案优选[J]. 石油钻采工艺, 2021, 43(3): 295-301.

[11] 赵静. YT25-1S油田大位移井岩屑床清除器优化研究[D]. 北京: 中国石油大学(北京), 2016.

[12] 王建龙, 郑峰, 刘学松, 等. 井眼清洁工具研究进展及展望[J]. 石油机械, 2018, 46(9): 18-23.

[13] 孔令镕, 王瑜, 邹俊, 等. 水力振荡减阻钻进技术发展现状与展望[J]. 石油钻采工艺, 2019, 41(1): 23-30.

[14] 余志清. 降摩阻短节在定向钻井及水平钻井中的应用[J]. 钻采工艺, 1999, 22(1): 66-67.

[15] CHEN J K, LIAO H L, ZHANG Y T, et al. A torsional-axial vibration analysis of drill string endowed with kinematic coupling and stochastic approach[J]. Journal of Petroleum Science and Engineering, 2021, 198: 108157.

[16] 张辉, 吴仲华, 蔡文军. 水力振荡器的研制及现场试验[J]. 石油机械, 2014, 42(6): 12-15.

[17] 张会增, 管志川, 刘永旺, 等. 基于旋转激励的钻柱激振减阻工具的研制[J]. 石油机械, 2015, 43(5): 9-12.

[18] 王甲昌, 滕春明, 张海平, 等. 机械式旋转冲击钻井工具研制及试验[J]. 钻采工艺, 2020, 43(6): 68-71.

[19] 席传明, 穆总结, 罗翼, 等. GCY-I型冲击螺杆钻井提速技术研究与试验[J]. 石油机械, 2020, 48(10): 39-43.

[20] 张学鸿, 董振刚, 张雄, 等. 探井水平井整体钻柱摩阻力分析的间隙元法[J]. 石油学报, 2002, 23(5): 105-109.

[21] 秦永和, 付胜利, 高德利. 大位移井摩阻扭矩力学分析新模型[J]. 天然气工业, 2006, 26(11): 77-79.

[22] 闫铁, 李庆明, 王岩, 等. 水平井钻柱摩阻扭矩分段计算模型[J]. 大庆石油学院学报, 2011, 35(5): 69-72.

[23] 李文哲, 代锋, 曾光, 等. 长宁超长水平井钻井摩阻数值模拟分析[J]. 钻采工艺, 2021, 44(5): 39-44.

本论文原发表于《石油机械》2023年第51卷第4期。

压裂泵曲轴的疲劳裂纹扩展研究

余朋伟[1,2]　雷广进[1,2]　刘树前[1]　侯晓东[1,2]　李　晨[1,2]　王亚丽[1,2]

(1. 宝鸡石油机械有限责任公司；
2. 中油国家油气钻井装备工程技术研究中心有限公司)

摘　要：为研究曲轴疲劳裂纹及其寿命，建立曲轴刚柔耦合动力学模型，使用 ABAQUS 和 FRANC3D 对曲轴裂纹进行引入及扩展分析，基于裂纹剩余寿命计算和初始裂纹形状比对，对应力强度因子进行研究。结果表明，当裂纹深度扩展至 18.2mm 时，等效应力强度因子达到曲轴材料断裂韧性，裂纹达到临界深度，剩余寿命为 8.6×10^5 次；在裂纹深度一定时，初始裂纹的表面长度越长，裂纹扩展寿命越低，初始裂纹表面长度越短，裂纹扩展寿命越高；裂纹初始角度越大，裂纹前缘最深处的应力强度因子越低，裂纹初始角度越大，裂纹扩展后得到的裂纹长度越短，裂纹扩展寿命越高。

关键词：曲轴；疲劳裂纹；裂纹扩展；剩余寿命

1　引言

压裂泵工作时，连续长时间高负荷工作时会发生共振和疲劳，曲轴作为最关键的零部件之一，其最容易产生裂纹，裂纹的存在和发展直接关乎曲轴的使用寿命的长短，很多产品安全事故都是由裂纹存在和长期发展导致的，所以对曲轴裂纹的研究变得非常有必要[1-2]。

目前，多位专家学者基于各种数值模拟软件已经对曲轴疲劳裂纹进行了大量的研究分析，王勇等通过刚柔耦合动力学仿真分析，利用全寿命方法(S—N法)得出疲劳寿命。何畏等运用应变能密度准则对含裂纹的曲轴进行了应力强度因子计算，分析了裂纹长短轴之比对等效应力强度因子的影响。董超群对曲轴表面裂纹进行了研究，分析计算了拉弯复合加载下裂纹尖端应力强度因子及剩余寿命[3-11]。

本文通过对曲轴进行刚柔耦合动力学仿真分析，采用软件对曲轴中关键部位所产生的裂纹在生命周期内变化过程进行模拟和数值分析，从而进一步推算出曲轴的剩余寿命，文中随后也研究了裂纹初始条件与裂纹变化拓展的关系。

2　曲轴模态分析

将在 UG 软件中建好的压裂泵曲轴模型导入有限元分析软件中，再通过优化精简模型，选取材料为 42CrMoA，设定密度、杨氏模量等相关参数。

基金项目：中国石油科技基础条件平台建设项目"大功率超高压压裂泵试验台建设"(2019D-5006-55)。
作者简介：余朋伟(1988—)，男，工程师，硕士研究生，现从事石油钻采机械技术研究工作，E-mail：yupengwei1998@126.com。

按照工况，对曲轴主轴颈施加径向约束，在曲轴连接电机端面处施加位移约束。对曲轴进行网格划分，如图1所示。通过计算得出固有频率及振型，见表1和图2。

表1 曲轴前6阶固有频率及振型

阶次	固有频率	振型	阶次	固有频率	振型
1阶	118.7	一阶弯曲振动	4阶	352.2	二阶弯曲振动
2阶	138.4	一阶弯曲振动	5阶	441.1	二阶弯曲振动
3阶	312.6	一阶扭转振动	6阶	532.2	二阶扭转振动

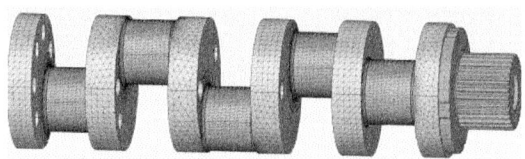

图1 曲轴有限元模型

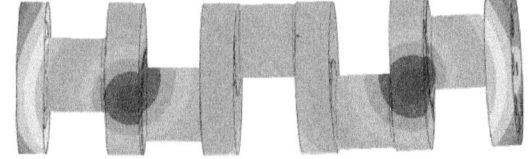

图2 曲轴1阶振型

3 曲轴裂纹扩展仿真分析

为方便计算，对曲轴模型进行简化，仅取一个曲拐部分进行仿真分析。曲轴轴颈上所受的载荷，基于传统的油膜压力应力分布规律，通过使用面载荷的形式施加在曲轴轴颈上，采用公式加载的方式进行加载，轴颈两侧采用固定约束，如图3所示。

裂纹的萌生和扩展均是在应力集中的区域发生，由应力云图可知，曲轴第5轴颈受力最大，轴颈圆角处出现应力集中，通过疲劳分析计算，其轴颈圆角部位安全系数最小，容易产生裂纹。如图4所示，可在第5曲轴轴颈圆角处引入一条半椭圆裂纹，沿着与轴肩呈45°夹角的方向向内部拓展，可对初始裂纹进行分析。

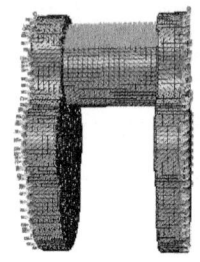

图3 曲轴轴颈约束

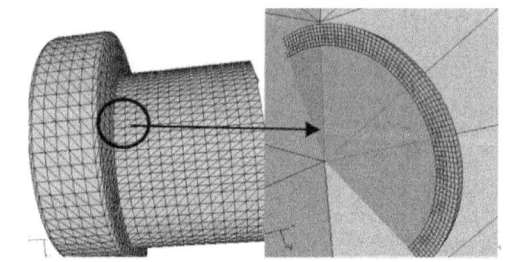

图4 初始裂纹引入

4 初始裂纹对裂纹扩展行为和裂纹扩展寿命的影响

通过联合使用ABAQUS和FRANC3D软件来模拟曲轴裂纹的扩展过程时，最重要的步骤是嵌入初始裂纹，初始裂纹的尺寸和裂纹初始角度对裂纹扩展行为有较大影响，从而影响裂纹扩展寿命。

本文以裂纹的初始尺寸和裂纹初始角度为变量，研究初始裂纹对裂纹扩展行为和裂纹扩展寿命的影响。

4.1 不同几何尺寸的初始裂纹

建立三种不同尺寸的初始裂纹模型，初始裂纹的形状都是半椭圆形，参数见表2。

表2 初始裂纹尺寸参数

初始裂纹编号	初始裂纹长度 a/mm	初始裂纹深度 b/mm
初始裂纹1	0.50	0.2
初始裂纹2	0.35	0.2
初始裂纹3	0.20	0.2

嵌入初始裂纹后，三种初始裂纹下的裂纹前缘应力强度因子(K_I)计算结果如图5和表3所示。在嵌入初始裂纹1下，裂纹前缘应力强度因子随着裂纹前缘的归一化值的增大，先增大到极大值，然后减小到极小值；在嵌入初始裂纹2下，裂纹前缘应力强度因子随着归一化值的增大，先减小到极小值，然后增大到极大值，再减小到极小值，最后有一定的增加。在嵌入初始裂纹3下，裂纹前缘应力强度因子随着归一化值的增大，先减小到极小值，然后增大到极大值。

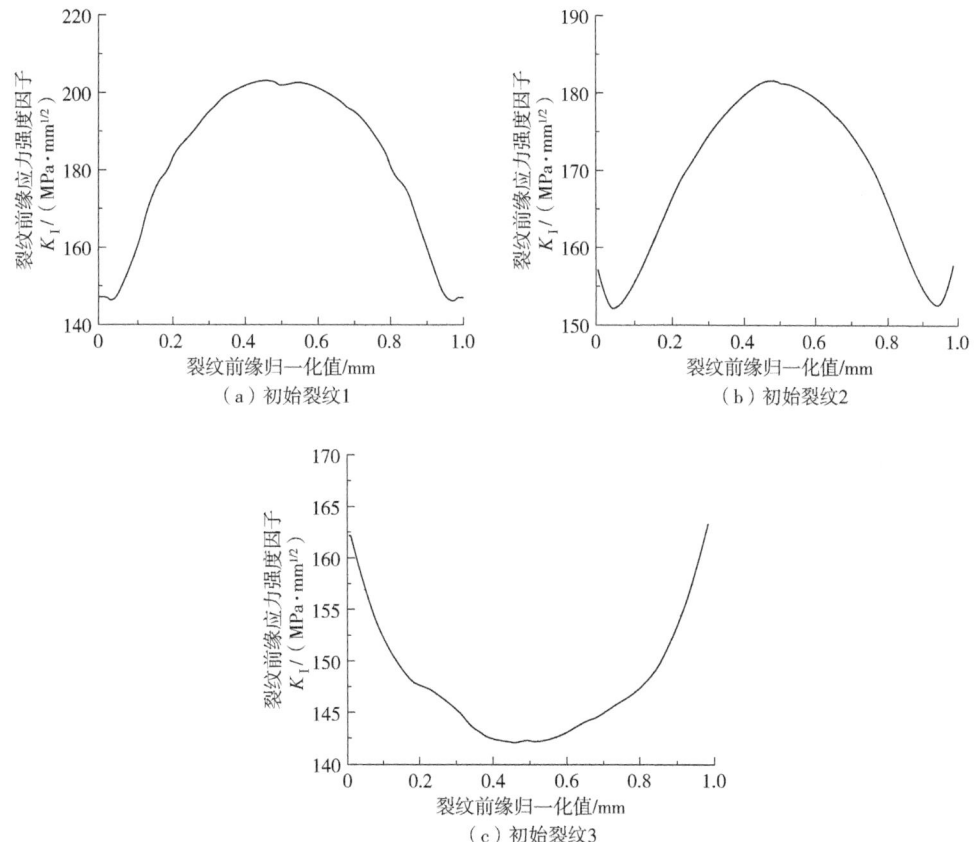

图5 嵌入初始裂纹后 K_I 的变化曲线

表3 三种初始裂纹下的裂纹前缘应力强度因子极值

初始裂纹编号	K_{Imax}/(MPa·mm$^{1/2}$)	K_{Imin}/(MPa·mm$^{1/2}$)
初始裂纹1	202.3	146.1
初始裂纹2	181.5	152.2
初始裂纹3	162.1	142.1

通过对比嵌入初始裂纹 1 和 3，可以得出：裂纹深度相同时，随着裂纹长度增加，最深处的应力强度因子越小，两边点的应力强度因子越大。

由表 4 可以得知，嵌入初始裂纹 1 时，最后在曲轴模型上留下的裂纹长度最长，裂纹扩展寿命最短；嵌入初始裂纹 3 时，最后在曲轴模型上留下的裂纹长度最短，裂纹扩展寿命最长。

表 4 三种不同尺寸初始裂纹下裂纹扩展寿命

初始裂纹编号	裂纹长度 a/mm	裂纹扩展寿命 $N/10^4$ 次
初始裂纹 1	2.10	3.60
初始裂纹 2	1.85	4.13
初始裂纹 3	1.73	4.95

综上所述，在裂纹深度一定时，随着初始裂纹的表面长度增加，裂纹前缘的应力强度因子也随之变大，初始裂纹尺寸发生改变，裂纹前缘的应力强度因子的变化趋势会发生很大的变化。在裂纹深度一定时，初始裂纹的表面长度越长，裂纹扩展寿命越低，初始裂纹表面长度越短，裂纹扩展寿命越高。

4.2 不同的裂纹初始角度

以 XZ 平面作为基准平面，建立了裂纹面与 XZ 平面呈现 0°、10°、20°、30°、40° 的五种不同裂纹初始角度的曲轴裂纹扩展模型，五种裂纹的尺寸参数均是 $a = 0.5$mm，$b = 0.2$mm。五种初始裂纹尺寸参数和裂纹嵌入位置见表 5 和图 6，为了在图 6 中突出裂纹的初始角度不同，图 6 中的裂纹尺寸比实际计算时放大了 10 倍。

表 5 初始裂纹参数

初始裂纹编号	尺寸参数		裂纹初始角度/(°)
	初始裂纹长度 a/mm	初始裂纹深度 b/mm	
初始裂纹 1	0.5	0.2	0
初始裂纹 2	0.5	0.2	10
初始裂纹 3	0.5	0.2	20
初始裂纹 4	0.5	0.2	30
初始裂纹 5	0.5	0.2	40

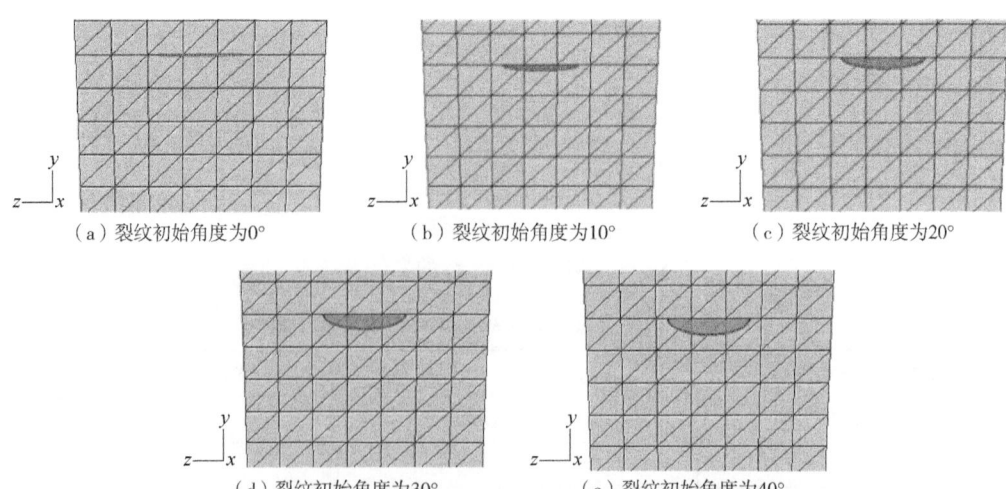

图 6 五种初始裂纹位置图

嵌入初始裂纹后，通过 FRANC3D 静态计算出五种不同裂纹初始角下的裂纹前缘应力强度因子，计算结果如图 7 所示。五种不同裂纹初始角下的裂纹前缘应力强度因子都是呈现先增大后减小的趋势，且都是裂纹前缘最深处的应力强度因子最大，裂纹前缘两边最小；通过对各种不同裂纹初始角下的应力强度因子比对，裂纹初始角度越大，裂纹前缘最深处的应力强度因子越小。

表 6 是五种不同裂纹初始角下的裂纹前缘应力强度因子极值，K_{Imax} 为 202.3MPa·$mm^{1/2}$，K_{Imin} 为 78.2MPa·$mm^{1/2}$。

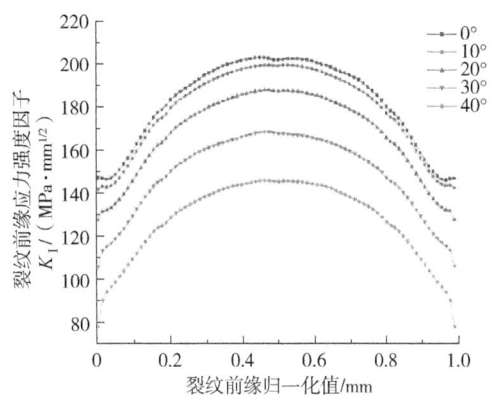

图 7 五种不同裂纹初始角下 K_I 的变化曲线

表 6 五种不同裂纹初始角下的裂纹前缘应力强度因子极值

裂纹初始角度/(°)	K_{Imax}/(MPa·$mm^{1/2}$)	K_{Imin}/(MPa·$mm^{1/2}$)
0	202.3	146.1
10	199.3	142.0
20	187.8	127.7
30	168.5	106.1
40	146.0	78.2

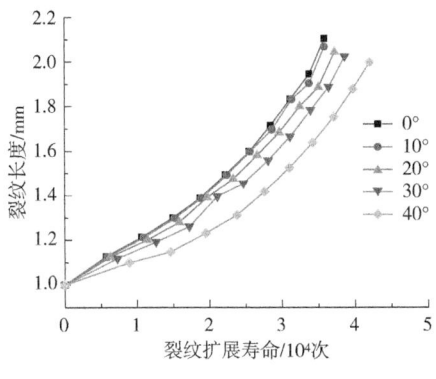

图 8 五种不同裂纹初始角下裂纹扩展寿命 a—N 曲线

完成五种不同裂纹角下的曲轴裂纹扩展模拟计算，设定裂纹扩展 10 步后停止裂纹自动扩展，每次裂纹扩展量取 0.06mm，得到裂纹扩展寿命 a—N 曲线如图 8 和表 7 所示。曲轴裂纹自动扩展 10 步以后，五种不同裂纹角下得到的裂纹长度不一样，这是由于应力强度因子在裂纹扩展前后存在差异和裂纹前缘各个节点的扩展量也存在差异；初始裂纹角度由 0°增加到 40°，裂纹长度减小了 4.8%，裂纹扩展寿命增加了 16%；裂纹初始角度越大，裂纹扩展结束时在曲轴杆身留下的裂纹长度越短，裂纹扩展寿命也越高。

表 7 五种不同裂纹初始角下裂纹扩展寿命

裂纹初始角度/(°)	裂纹长度 a/mm	裂纹扩展寿命 N/10^4 次
0	2.10	3.60
10	2.07	3.61
20	2.04	3.72
30	2.02	3.86
40	2.00	4.20

5 结论

本文通过动力学分析，得到曲轴连杆轴颈载荷谱，确定曲轴裂纹易萌生位置，进行了裂纹扩展分析及剩余寿命计算，分析裂纹不同长度和角度对裂纹扩展寿命的影响，结论如下：

（1）轴颈圆角处裂纹扩展深度达到18.2mm时，等效应力强度因子接近该材料破坏韧性极限，应力循环次数为$8.6×10^5$次。

（2）初始裂纹尺寸不同，裂纹前缘的应力强度因子的变化趋势会发生很大的不同，在裂纹深度一定时，初始裂纹的表面长度越长，裂纹扩展寿命越低，初始裂纹表面长度越短，裂纹扩展寿命越高；裂纹初始角度在0°~40°的变化范围内，裂纹初始角度越大，裂纹前缘最深处的应力强度因子越低；裂纹初始角度越大，裂纹扩展后得到的裂纹长度越短，裂纹扩展寿命越高。

参 考 文 献

[1] 贺运初，胡祖汉，戴新西，等. 大型往复压缩机曲轴开裂的原因及改造[J]. 流体机械，2005(4)：39-41.

[2] 周世秀. 压缩机曲轴运动中断裂的失效分析[J]. 科技传播，2014，6(9)：197，208.

[3] 游孟平，李斌，谢春晖. 压缩机曲轴疲劳裂纹扩展及寿命研究[J]. 塑性工程学报，2021，28(11)：196-203.

[4] LD2—CS铝材中心穿透裂纹平板裂纹扩展速率试验报告[J]. 西北工业大学学报，1978(1)：23-40.

[5] 何畏，肖祥，陈波，等. 基于EQ4H型内燃机含裂纹曲轴的应力强度因子计算[J]. 机械强度，2016，38(2)：369-373.

[6] 董超群. 整体往复式压缩机曲轴力学性能研究与安全评价[D]. 成都：西南石油大学，2014.

[7] 何芝仙，干洪，韩后祥. 含裂纹的曲轴-轴承系统动力学、摩擦学和断裂力学耦合分析[J]. 内燃机工程，2013，34(2)：35-41.

[8] 程仁庆. 基于多体动力学的曲轴疲劳强度分析及含裂纹曲轴动态特性研究[D]. 南宁：广西大学，2014.

[9] VILLANUEVA J A B, ESPADAFOR F J, CRUZ-PERAGÓN F, et al. A methodology for cracks identification in large crankshafts[J]. Mechanical Systems and Signal Processing, 2011, 25(8).

[10] SOLA J F, ALINEJAD F, RAHIMIDEHGOLAN F, et al. Fatigue life assessment of crankshaft with increased horsepower[J]. International Journal of Structural Integrity, 2019, 10(1).

[11] 赵国文，赵启元，蔡振雄，等. 船舶柴油机曲轴裂纹的有限元模态分析[J]. 舰船科学技术，2008(3)：148-151.

本论文原发表于《机械工程师》2023年第3期。

一种基于振动分析的钻井泵故障诊断方法

袁 方[1,2]　夏 辉[1,2]　邹 涛[3]　周小明[1,2]　秦羿涵[1,2]　王 强[3]

(1.宝鸡石油机械有限责任公司；
2.中油国家油气钻井装备工程技术研究中心有限公司；3.西部钻探工程有限公司)

摘　要：本文介绍了一种振动分析方法在钻井泵上的实际应用方案。采用压电式振动加速度传感器采集钻井泵振动数据，经采集板卡频谱变换后输出时域波形、频域波形、包络谱波形和一系列特征值，边缘计算网关进行数据初步处理提取各故障敏感特征，特征量上传平台后触发告警规则，实现实时数据的智能诊断。积累足够数据后根据大量案例数据统计分析建模，实现钻井泵的准确监测和状态预测。

关键字：钻井泵；智能诊断；振动监测

随着时代的发展，工业企业对设备的要求越来越多，机械设备呈多样化发展，诸如自动化、智能化、大型化、复杂化等是机械设备的几个重要发展方向。在石油钻井行业中，机械设备的重要性不容忽视，尤其在自动化、智能化的新一代钻机平台上，机械设备的故障可能导致整个钻机系统的瘫痪，可谓牵一发而动全身，机械设备中钻井泵作为至关重要的一环，更不容小视。钻井泵故障轻则导致设备损坏、生产停机，重则可能造成人员伤亡，因此钻井泵需要准确、可靠、智能的方法实时监测运行状态、及时预警故障信息，减少设备故障导致的意外损失，避免不必要的损失。

1　研究背景

在常用的正循环钻探中，钻井泵是将地表冲洗介质——清水、钻井液或聚合物冲洗液在一定的压力下，经过高压软管、水龙头及钻杆柱中心孔直送钻头的底端，以达到冷却钻头、将切削下来的岩屑清除并输送到地表的目的。常用的钻井泵是活塞式或柱塞式的，由动力机带动泵的输入轴和曲轴回转，曲轴通过连杆十字头将旋转运动转化为活塞或柱塞在泵缸中的往复运动。在吸入阀和排出阀的交替作用下，实现压送与循环冲洗液的目的。

F-1600钻井泵是众多石油钻井作业用泵的主流泵型，其主要作用是在钻井过程中以高压向井底输送高黏度、大相对密度和含砂量较高的钻井液，用于冷却钻头、冲刷井底、破碎岩石，从井底返回时携带出岩屑。其主要外形及结构如图1所示。

钻井泵目前较多的是三缸泵和五缸泵，主要由动力端和液力端两部分构成。动力端包括电机、机座、联轴器、曲轴总成、齿轮组总成、连杆、十字头组件等，液力端包括液缸、吸入阀、排出阀、缸套、活塞、活塞杆、空气包、安全阀等。三缸泵技术指标见表1。

基金项目：陕西省科协企业创新争先青年人才托举计划项目"Enterprise top innovative young talents support plan"(20090101)。

作者简介：袁方，宝鸡石油机械有限责任公司。

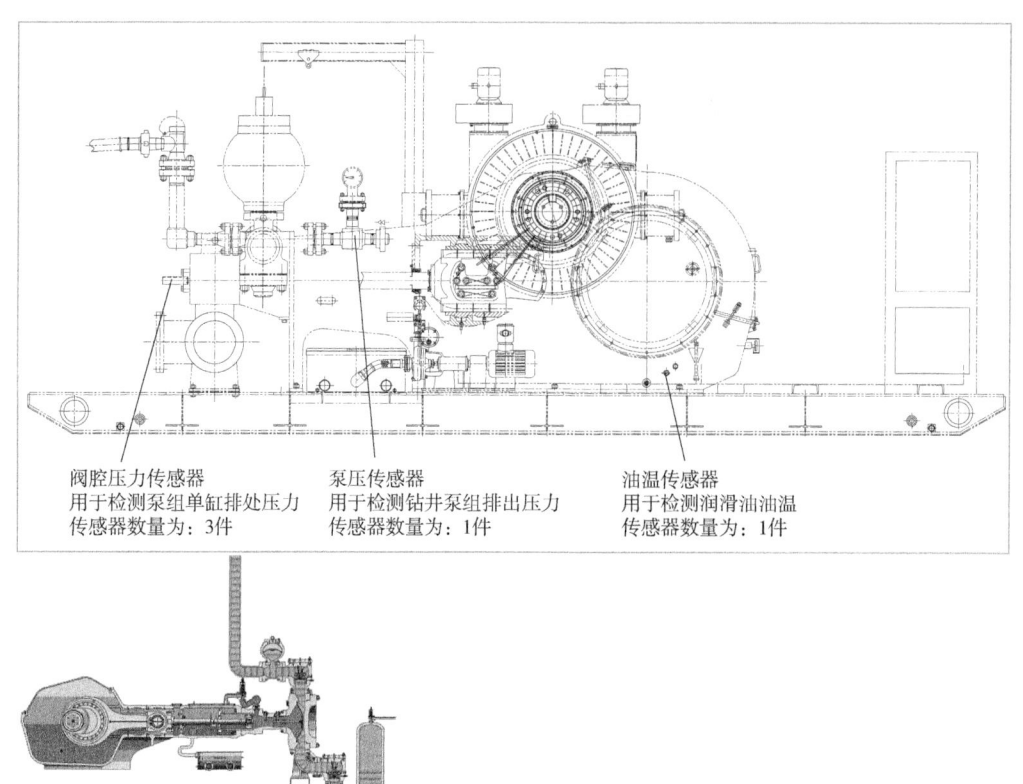

图 1 钻井泵结构和外观图

表 1 钻井泵技术指标

型 号	F-1300	F-1600	型 号	F-1300	F-1600
形式	三缸单作用活塞泵		冲程长度/mm	305	305
最大缸径/mm	180	180	齿轮速比	4.206	4.206
额定功率/kW	969	1193	阀腔	API 7#	API 7#
额定冲数/(次/min)	120	120	质量/kg	26570	27020

钻井泵常见故障及原因为：

(1) 泵不吸水，原因：灌注引水不够、泵内空气无法排出、吸水管漏气。

(2) 泵磨损快，原因：施工环境差(颗粒大)、输送距离远、进水管路长。

(3) 轴承发热，原因：油管或油孔堵死、润滑油太脏或变质、滚动轴承磨损或损坏。

(4) 动力端有敲击声，原因：十字头导板已严重磨损、轴承磨损、导板松动。

2 钻井泵振动采集方案

振动传感器的种类很多，且有不同的分类方法。按工作原理的不同，可分为电涡流式、磁电式(电动式)、压电式；按参考坐标的不同，可分为相对式与绝对式(惯性式)；按是否与被测物体接触，可分为接触式与非接触式；按测量的振动参数的不同，可分为位移传感器、速度传感器、加速度传感器；以及由电涡流式传感器和惯性式传感器组合而成的复合式传感器等。

在现场实际振动检测中，常用的传感器有磁电式速度传感器（其中又以绝对式应用较多）、压电式加速度传感器和电涡流式位移传感器。

本方案采用压电式加速度传感器，振动采集板卡包含 A/D 转换、信号放大、高频采集、基础处理等功能，汇总到边缘计算网关后进行关键特征提取，经 4G/5G 无线信号传输到云平台建模诊断（图 2）。

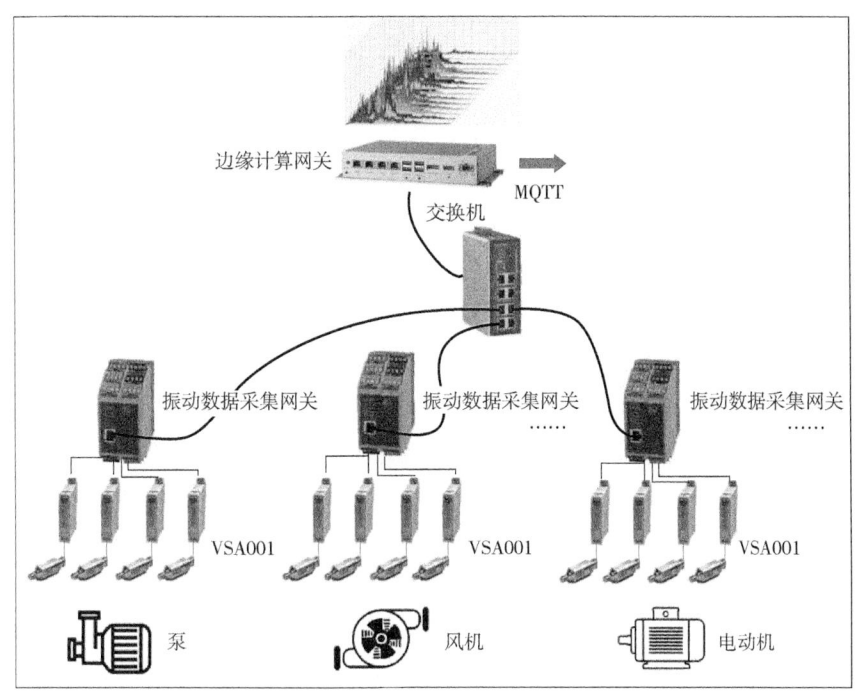

图 2　振动数据采集流程

钻井泵设置 10 个振动测点，分别是输入轴两侧轴承、曲轴两侧轴承、3 处下导板、吸入管路、排出管路、底座，可有效监控钻井泵机械结构的整体状态，覆盖电机、轴承、主轴、齿轮组、十字头组件、管道等重点部件，可监控加工类问题如轴不平衡、转子偏心，装配类问题如不对中、地脚松动，发展类故障如轴承磨损、齿轮组断齿等（图 3）。

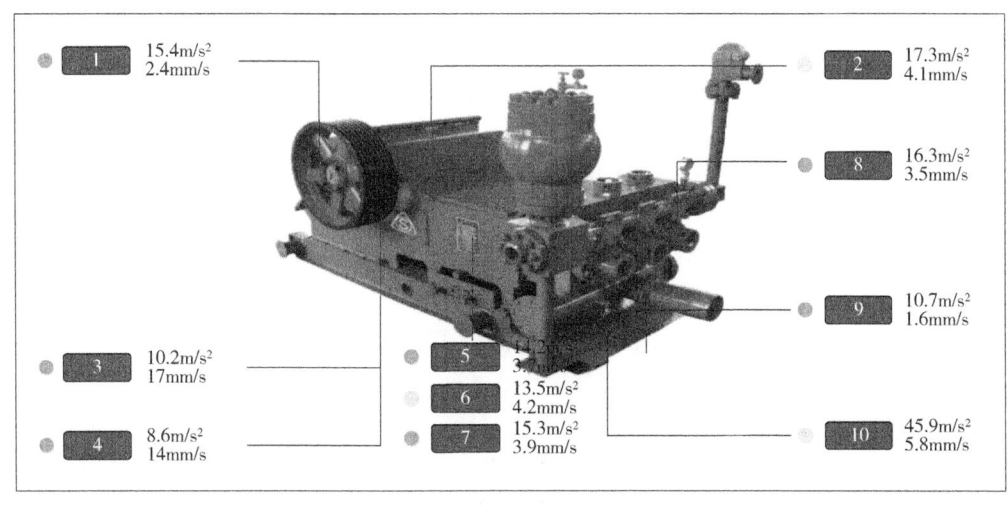

图 3　钻井泵振动测点分布

3 钻井泵故障分析方法

传统振动分析是对设备所产生的机械振动进行信号采集、数据处理后,根据振幅、频率、相位及相关图谱进行故障分析。分析以人为主,凭借大量现场经验进行分析判断。为了实现钻井泵智能故障诊断,本方案采用传统振动监测技术和大数据分析相融合的方法,提炼传统振动分析关注的敏感特征,借鉴振动分析人工经验,对接实际数据进行大数据训练,实现钻井泵各部件的准确、可靠监控(图4)。

图4 振动分析流程图

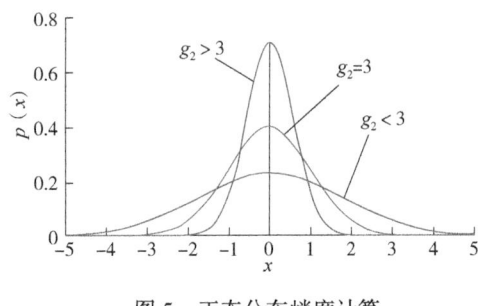

图5 正态分布峭度计算

3.1 特征提取

振动特征提炼分时域、频域、包络、趋势等几大维度。

时域特征提取,以峭度特征提取为例,峭度因子是表示波形平缓程度的参数,用于描述变量的分布。正态分布的峭度等于3,峭度小于3时分布的曲线会较"平",大于3时分布的曲线较"陡"(图5)。峭度计算公式如下:

$$K_4 = \frac{E[(X-\mu)^4]}{\delta^4} \tag{1}$$

式中:μ 为波形均值;δ 为波形标准差,计算公式为:

$$\delta_x = \sqrt{\lim_{N\to\infty} \frac{1}{N} \sum_{i=1}^{N} [X_i(t) - \mu_i(t)]^2} \tag{2}$$

频域特征提取,以轴承特征频率提取为例,轴承主要故障类型分外环故障、内环故障、滚动体故障、保持架故障4大类,不同结构的轴承按照尺寸参数可计算对应的频率特征,具体方法和波形如图6所示。

包络特征提取,低转速部件故障引起的冲击信号往往会激起高频固有频率,在频谱上表现为出现共振带,即低频故障信号作为某高频载波的边频出现。因此,对于这种出现调制现象的故障信号,往往需要通过包络进行分析诊断。包络机理如图7所示(具体特征提取方式同频谱提取)。

趋势分析常用图谱为瀑布图(图8)和趋势图,可以根据振动总值和主要峰值的变化量,进行故障跟踪训练。

3.2 数据分析方法

本方案采用的DAStudio是面向工程数据分析人员的全流程数据建模分析平台,聚焦工业领域,对接丰富的海量工业数据源,快速构建数据模型和数据对象,支持托拉拽、零代码、敏捷式的算法模型开发，提供海量数据预处理、机器学习和人工智能建模分析，以及模型在线部署能力，帮助工程技术人员快速从繁杂的数据中，通过智能的分析建模，挖掘数据的价值。具体流程如图9所示。

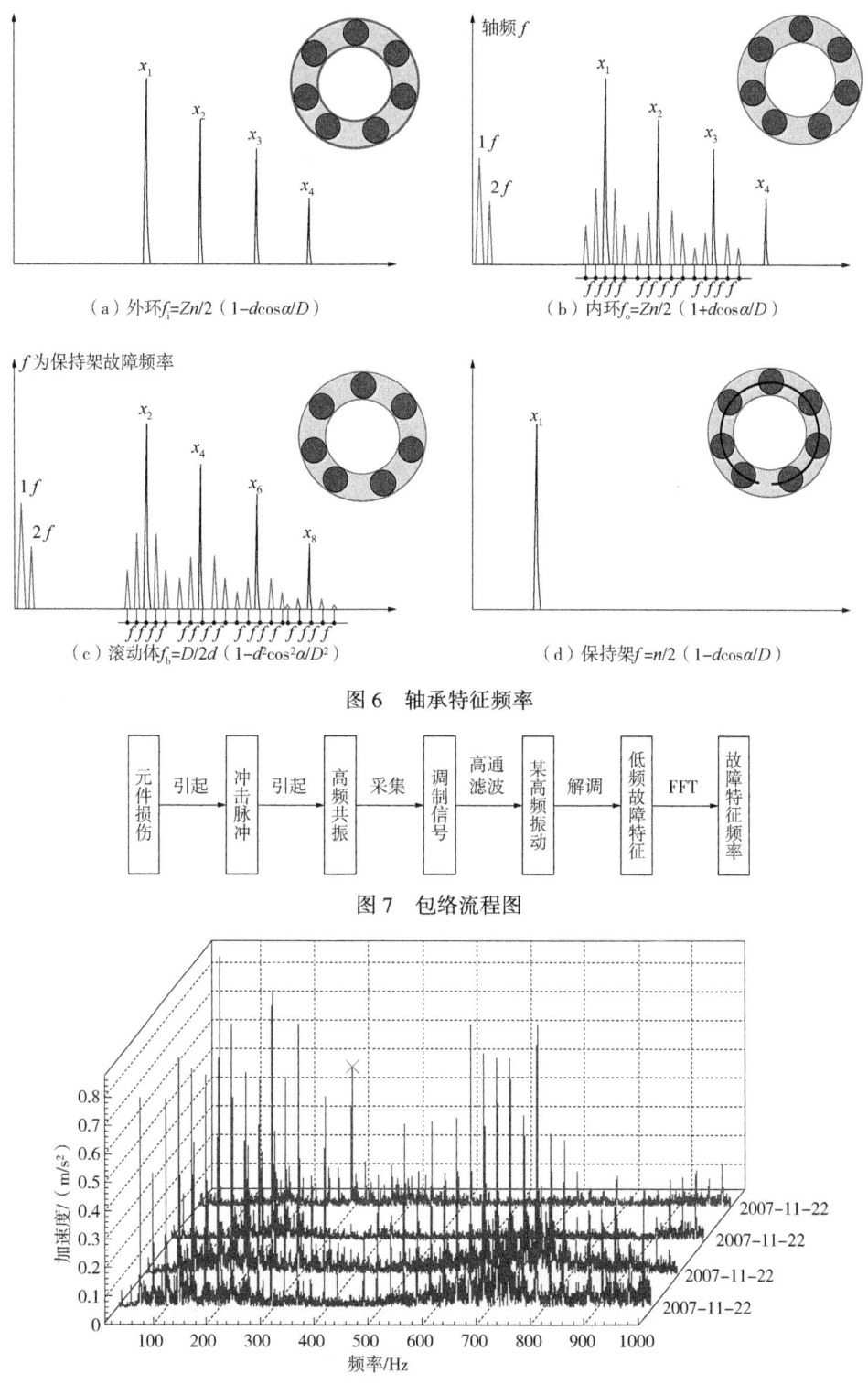

（a）外环 $f_i=Zn/2(1-d\cos\alpha/D)$

（b）内环 $f_o=Zn/2(1+d\cos\alpha/D)$

（c）滚动体 $f_b=D/2d(1-d^2\cos^2\alpha/D^2)$

（d）保持架 $f=n/2(1-d\cos\alpha/D)$

图6　轴承特征频率

图7　包络流程图

图8　瀑布图

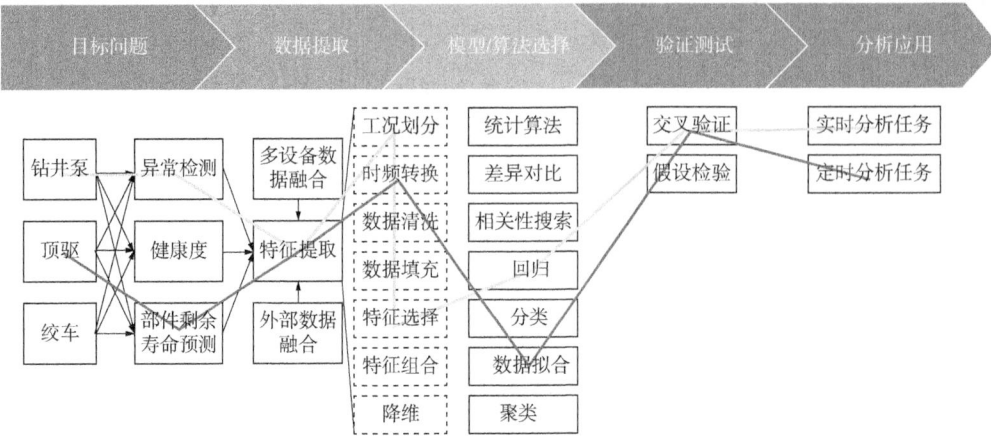

图 9　大数据分析平台处理流程

（1）数据准备：对接包括文件、结构化数据库、实时数据库在内的多种数据源，并构建模板+实例化的数据对象。

（2）数据预处理：提供丰富的预处理算法组件，实现快速的数据清洗和特征工程。

（3）数据分析和 AI 建模：提供包括统计、聚类、分类、机器学习在内的丰富的算法组件，提供快速的托拉拽画布式模型开发，并支持对模型结果进行实时评估。

（4）部署应用：模型可以直接同实时数据进行对接部署，实现实时的判决和预测。

4　诊断结论

根据钻井泵各部件结构，振动分析方法提炼了电机、传动机构、装配结构三大部分的故障特征，具体如图 10 所示。

部位	故障类型	故障特征
	轴承故障	轴承故障特征频率
	转子动平衡异常	径向一倍频上升
	电气故障	50Hz、100Hz

部位	故障类型	故障特征
 输入轴轴承	轴承润滑不良	加速度上升，波形无冲击特征或少量特征，2000~4000Hz 频段出现茅草堆，轴承温度上升
输入轴轴承	轴承损伤	轴承特征频率信号，早期出现在包络谱，中期出现在加速谱，末期出现在速度谱
输入轴轴承	轴与轴承间隙大	水平方向振动大，出现3~8倍频
 输入主轴	旋转不平衡	一倍频大，振幅随转速上升
输入主轴	轴不对中	一、二倍频偏大，二倍频甚至超过一倍频，轴向振动大
输入主轴	配合不良	水平方向振动大，出现3~8倍频
 齿轮	齿轮疲劳损伤	啮合频率被转速调制后特征信号，包络谱出现啮合频率特征
齿轮	断齿	一倍频上升、啮合频率特征，振动幅值突变

部位	故障类型	故障特征
	地脚螺栓松动	竖直方向振动上升，出现3~8倍频
	十字头滑道磨损	幅值上升，冲击特征信号明显
	紧固件松动	出现松动特征信号，3~8倍频幅值上升
	撞缸	敲击信号出现，幅值剧烈上升

图 10　常见故障特征

将时域、频域、包络、趋势中提炼的 20 余种特征量结合故障知识库和现场实际案例，合计开发 8 大类故障诊断模型，分别为不平衡故障、不对中故障、松动类故障、电机电气故障、轴承故障、齿轮箱故障、十字销组件异常、入口振动异常。目前故障监控诊断率达到 90%以上。

本论文原发表于《设备管理与维修》2023 年第 4 期(上)。

Arrhenius Constitutive Equation and Artificial Neural Network Model of Flow Stress in Hot Deformation of Offshore Steel with High Strength and Toughness

Li Fangpo[1]　Li Ning[2]　Ren Xiaojian[3]　Qiao Song[4]
Lu Caihong[1]　Wang Jianjun[1]　Xu Yang[3]　Wang bin[3]

(1. CNPC Tubular Goods Research Institute; 2. Institute for Hygiene of Ordance Industry;
3. State Key Laboratory of Rolling and Automation, Northeastern University;
4. Shandong Iron and Steel Group Rizhao Co., Ltd)

Abstract: In this study, the thermal deformation behavior of a high strength offshore steel at different temperatures and rates was investigated through thermal compression experiments. An Arrhenius constitutive model and a back propagation artificial neural network (BP-ANN) were established to address more complex deformation characteristics. The performances of both models were was evaluated using stardard ststistical parameters such as the correlation coefficient (R) and average absolute relative error (AARE). The results showed that both models can accurately predict the rheological stresses generated during deformation. The BP-ANN outperforms the Arrhenius equation model with correlation coefficients of fit greater than 99.9% and less than 0.8% relative error. At a strain rate of $0.01s^{-1}$ and $10s^{-1}$, the accuracy of the ANN decreases slightly due to the fact that it exceeds the strain rate range of the training set, as compared to the Arrhenius constitutive equations as these are more accurately predicted.

Keywords: High strength offshore steel; Compression experiment; Arrhenius constitutive equation; Back-propagation artificial neural network; Flow stress

1 Introduction

The thermal deformation behavior of metals is significantly influenced by the deformation parameters (temperature, deformation rate, etc.). During plastic deformation under different conditions, the phenomena of work hardening (WH), dynamic recovery (DRV), and dynamic recrystallization (DRX) occur[1-2]. In WH the plastic properties of the material are reduced, and the rheological stress is increased. However, softening phenomena such as DRV and DRX reduce the rheological

Corresponding author: Li Fangpo, lifangpo@163.com.

stress and restore plasticity[3-4]. Therefore, the interaction under thermal deformation conditions is an important reason for complex deformation behavior[5-6].

In recent years, researchers have proposed various intrinsic constitutive equations to describe the high-temperature deformation behavior of metallic materials. One of the most typical constitutive equations is the Arrhenius intrinsic constitutive equation, which considers the deformation temperature and strain rate[7]. However, the traditional Arrhenius equation does not consider the effect of degree of deformation. Shi et al.[8] combined the strain degree and temperature in the equation to obtain the flow stress constitutive equation of material strain parameters, which is widely used to predict the flow stress of various metals. Examples include P91 steel[9], Ti-deformed austenitic stainless steel[10], 800H high-temperature alloy[11], TC4-DT titanium alloy[12], AZ81 magnesium alloy[13], and 2124 aluminum alloy[14]. In these studies, the average relative error (AARE) between the calculated and experimental values was mostly concentrated within the range of 5%~9%, and artificial neural network (ANN) models that did not require repeated regression calculations were introduced to address the problem of low prediction accuracy[15-19].

ANNs, a newer artificial intelligence technique, are more effective in solving highly complex problems compared to regression methods. The typical structure of a BP-ANN consists of an input layer, output layer, and one or more hidden layers with artificial neurons that select and weigh the inputs. The input layer receives the input data, and after processing, sends the data to the hidden layer. The hidden layer acts as a complex network structure to model the nonlinear relationship between the input and output layers. It processes the data computations, and sends the responses to the output layer, which produces outputs[20-21].

In recent years, an increasing number of researchers have used ANN models to dissect the relationship between material properties. Lin et al.[22] established optimal hot-forming process parameters for 42CrMo steel based on ANN models. Zhao et al.[23] characterized the thermal deformation of Ti600 titanium alloy using an instantaneous equation and ANN. Quan et al.[24] developed an ANN model for cast Ti-6Al-2Zr-1Mo-1V alloy over a wide temperature range involving phase transformation, and predicted the high-temperature flow behavior of 20MnNiMo alloys using an ANN[25].

Intrinsic constitutive equations and ANNs have been applied to characterize the rheological behavior of materials. Examples include titanium alloys[26-27], stainless steels[28-29], high-temperature alloys[30], aluminum[31-32] and magnesium alloys[33-34]. The relationship between the process, microstructure, and properties can be accurately characterized after considering the effects of the parameters together. For the offshore steels, the thermal deformation behavior also has a strong influence on the combination of strength and toughness[35-43], which is of great significance for applications in the offshore engineering environments. In this study, the hot deformation stresses of high-strength offshore steels under different models were predicted separately using an Arrhenius equation-based model and a BP-ANN. Error analysis was performed to evaluate the performance of these models in predicting rheological stresses.

2 Material and experiments

Experiments were conducted using forged ingots with a mass of approximately 65kg, and the smelting process was conducted under complete vacuum (Table 1). From the ingots, φ8mm× 12mm sized specimens were machined and introduced in a Gleeble-3800 testing machine for single pass hot compression tests. The specimens were heated at a rate of 5℃/s to 1200℃ for 5min, cooled at a rate of 5℃/s to the corresponding deformation temperatures (800℃, 900℃, 1000℃, 1100℃), held for 30s, compressed at the corresponding strain rates (0.1s^{-1}, 0.5s^{-1}, 1s^{-1}, 5s^{-1}) with a 55% depression, and then water quenched after compression to retain the deformation morphology. Hot-compression tests were performed under Ar gas to avoid metal oxidation during heating and deformation.

Table1 Chemical composition of high-strength offshore steel

Element	C	Cr	Ni	Mo	Si	Mn	Al	Fe
Mass fraction/%	0.29	0.70	0.60	0.45	0.30	1.30	0.02	Balance

The true stress-true strain curves of the experimental specimens deformed at different rates and temperatures for a single pass of compression, as shown in Figure 1. As the deformation temperature decreases and the strain rate increases, the rheological stress of the material shows an increasing trend. The slow decrease in the rheological stress after the strain-hardening stage is due to dynamic softening caused by the occurrence of DRV and DRX. The rheological stress gradually tends to a steady state at 1100℃ and 0.1s^{-1}, indicating a more balanced state between DRV, DRX and WH. These findings show that the stresses in high-strength offshore steel exhibit highly nonlinear behavior in response to the three deformation parameters (temperature, strain, and strain rate).

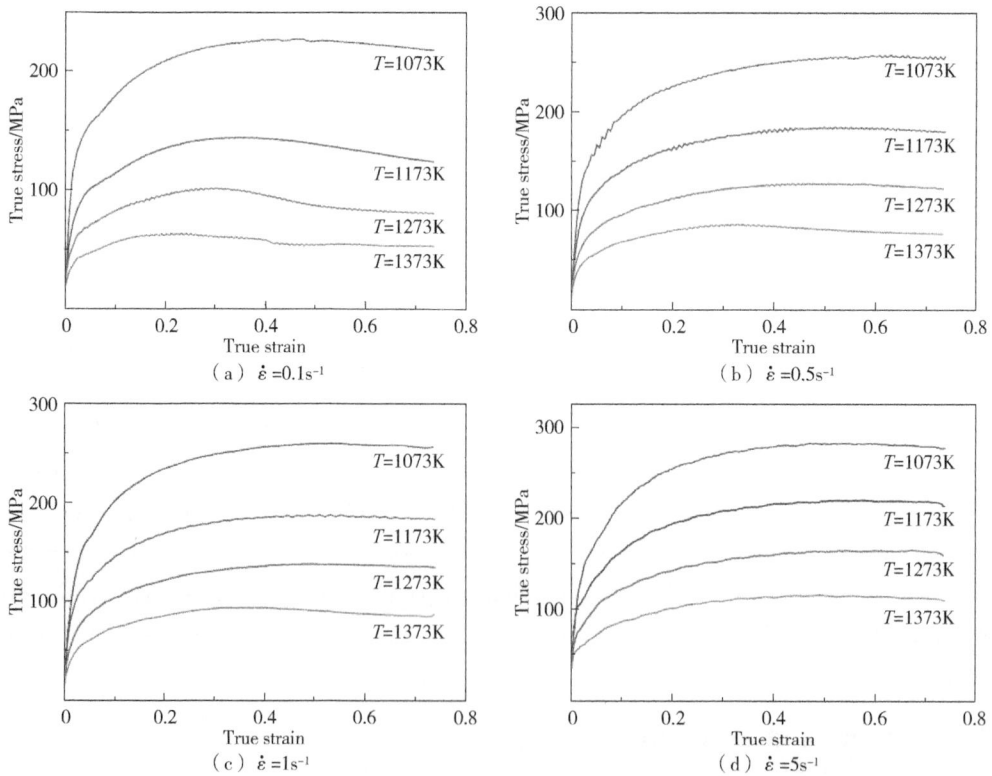

Figure 1 True stress-strain curves at different temperatures and rates

3 Comparison between improved Arrhenius-type constitutive equation and BP-ANN model

3.1 Improved Arrheniue constitutive equation

The Arrhenius-type constitutive equation is widely used as an image-only model to describe the intrinsic constitutive relationship of materials. However, strain, which has an important influence on the flow behavior, has been neglected in equations proposed in previous studies. Thus, the previous Arrhenius-type constitutive equations could not predict flow behavior accurately. Lin et al[44]. first proposed a modified model considering the effect of strain, and used it to accurately describe the deformation behavior of 42CrMo steel at high temperatures. Owing to the superiority of predictions of the revised model, the modified Arrhenius-type intrinsic constitutive equation has been successfully applied to describe the intrinsic constitutive relationships of various materials[45-47].

$$\dot{\varepsilon} = A_1 \sigma^{n_1} \cdot \exp[-Q/(RT)], \quad \alpha\sigma < 0.8 \quad (1)$$

$$\dot{\varepsilon} = A_2 \exp(\beta\sigma) \cdot \exp[-Q/(RT)], \quad \alpha\sigma > 1.2 \quad (2)$$

$$\dot{\varepsilon} = A[\sinh(\alpha\sigma)]^n \cdot \exp[-Q/(RT)], \quad \text{for all } \sigma \quad (3)$$

In this study, the coefficients of the intrinsic constitutive equations are solved at a strain of 0.5. The natural logarithms of Equations (1) and (2) can be expressed as:

$$\ln \dot{\varepsilon} = \ln A_1 + n_1 \ln \sigma - Q/(RT) \quad (4)$$

$$\ln \dot{\varepsilon} = \ln A_2 + \beta\sigma - Q(RT) \quad (5)$$

As shown in Figure 2, the linear regression of the data under different deformation temperatures using the least-squares method can yield the value of n_1 at different temperatures, and the average value is 17.299. Similarly, β is 0.131 MPa^{-1}, and α is obtained as 7.55×10^{-5} MPa^{-1}.

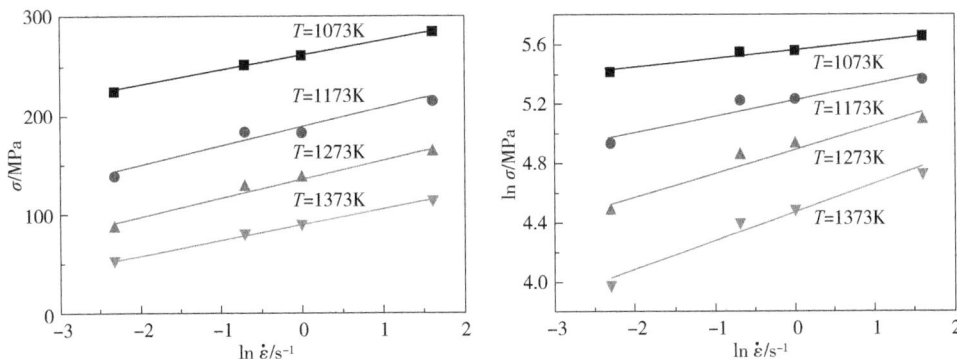

Figure 2 Experimental rheological stress versus strain rate

Equation (6) is obtained by taking the natural logarithm of both sides of Equation (3) and applying a partial differential.

$$Q = R \cdot \{\partial \ln/\partial \ln[\sinh(\alpha\sigma)]\}|_T \cdot \{\partial \ln[\sinh(\alpha\sigma)]/\partial(1/T)\}|_{\dot{\varepsilon}} \quad (6)$$

As shown in Figure 3, the slope and reciprocal are evaluated as 6.001 and 7.385, respectively.

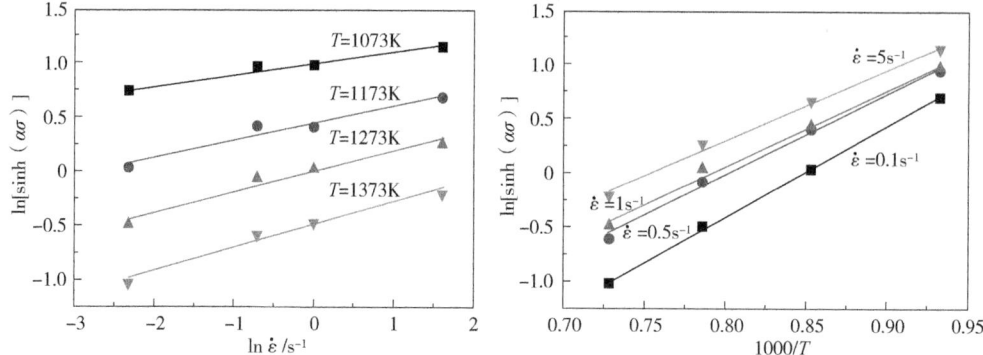

Figure 3 Experimental rheological stress as a function of strain rate and deformation temperature

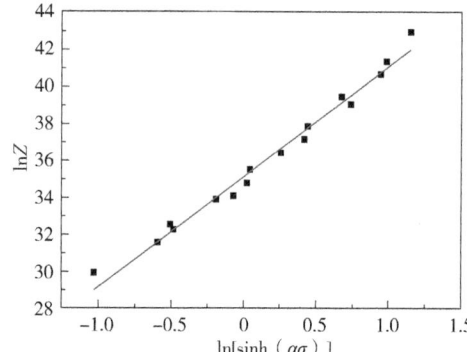

Figure 4 Z-parameter versus rheological stress of the experimental steel

The regression coefficient obtained from Figure 3 is introduced into Equation (6), and the heat deformation activation energy Q is calculated to be 368.469kJ/mol for the experimental steel at a strain level of 0.5. By inserting Q into the above Equation (3) and (8), the following relationship is obtained. In Equation (9), the constants are the intercepts and the slope of the regression line, which are 35.1464 and 5.9051, respectively, as shown in Figure 4.

$$Z = \dot{\varepsilon} \cdot \exp[368469.736/(RT)] \quad (7)$$

$$Z = \dot{\varepsilon} \cdot \exp[Q/(RT)] = A[\sinh(\alpha\sigma)]^n \quad (8)$$

$$\ln Z = \ln A + n\ln[\sinh(\alpha\sigma)] \quad (9)$$

Repeating the method of solving the coefficients of each principal equation for a strain of 0.5, the coefficients of each stress-strain principal equation (α, β, n, Q and $\ln A$) can be obtained for different strains (0.1, 0.2, 0.3, 0.4, 0.45, 0.55, 0.6, 0.65, 0.7), and the results are shown in Table 2.

Table 2 Coefficients of the equation at different degrees of deformation

True strain	α/MPa^{-1}	β/MPa^{-1}	n	Q/(J/mol)	$\ln A$
0.10	0.00868	0.0888	8.34682	492.169	46.492
0.20	0.00759	0.0893	8.29395	487.920	45.789
0.30	0.00740	0.0995	7.40676	450.600	42.024
0.40	0.00740	0.1140	6.42610	402.502	37.292
0.45	0.00756	0.1256	5.97244	380.486	35.136
0.50	0.00757	0.1305	5.90510	368.469	35.146
0.55	0.00753	0.1347	5.32672	363.609	33.339
0.60	0.00756	0.1375	5.27868	355.703	33.032
0.65	0.00763	0.1407	5.07219	347.493	32.422
0.70	0.00766	0.1411	5.17656	348.285	32.315

Therefore, in order to describe the intrinsic structure of the experimental steel under the aforementioned deformation conditions more precisely, the relationship between the equation coefficients (α, β, n, Q and $\ln A$) and strain is established by polynomial equations (of order 2 to 9) considering the degree of deformation. The fitting curve of polynomial coefficients is shown in Figure 5. The constitutive equation (11) is obtained from Table 3 and Equation (10). Figure 6 and 7 show the high correlation between the predicted and actual stress values.

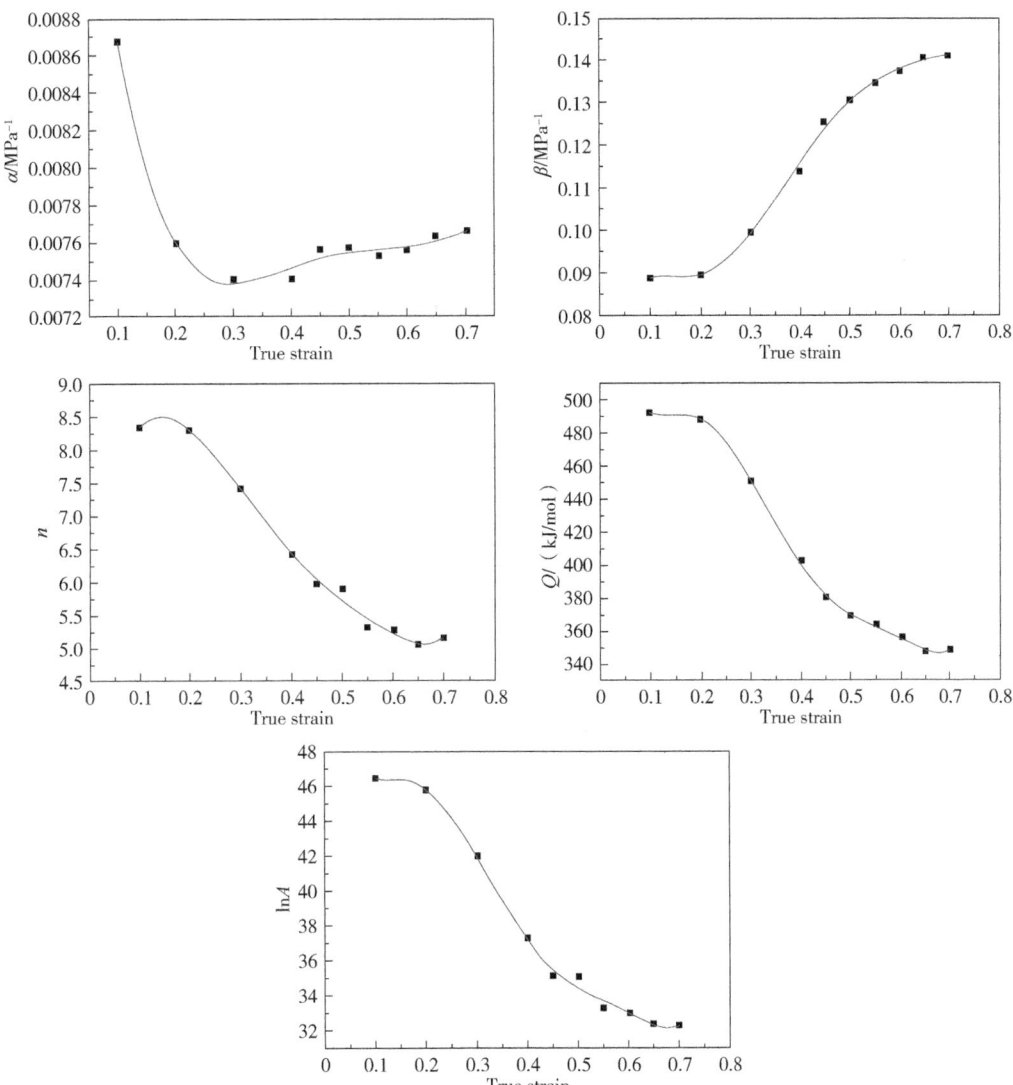

Figure 5 Variation of regression coefficients with strain

$$\begin{cases} \alpha = B_0 + B_1\varepsilon^1 + B_2\varepsilon^2 + B_3\varepsilon^3 + B_4\varepsilon^4 + B_5\varepsilon^5 \\ \beta = C_0 + C_1\varepsilon^1 + C_2\varepsilon^2 + C_3\varepsilon^3 + C_4\varepsilon^4 + C_5\varepsilon^5 \\ n = D_0 + D_1\varepsilon^1 + D_2\varepsilon^2 + D_3\varepsilon^3 + D_4\varepsilon^4 + D_5\varepsilon^5 \\ Q = E_0 + E_1\varepsilon^1 + E_2\varepsilon^2 + E_3\varepsilon^3 + E_4\varepsilon^4 + E_5\varepsilon^5 \\ \ln A = F_0 + F_1\varepsilon^1 + F_2\varepsilon^2 + F_3\varepsilon^3 + F_4\varepsilon^4 + F_5\varepsilon^5 \end{cases} \quad (10)$$

Table 3 Polynomial coefficients of α, β, n, Q and $\ln A$

α	β	n	Q	$\ln A$
$B_0 = 0.08783$	$C_0 = -0.71381$	$D_0 = 16.52829$	$E_0 = -3346.82229$	$F_0 = -237.13491$
$B_1 = -0.39230$	$C_1 = -6.59852$	$D_1 = -6.34715$	$E_1 = 33006.85263$	$F_1 = 2411.83505$
$B_2 = 0.57528$	$C_2 = 27.88814$	$D_2 = -500.30071$	$E_2 = -148222.42632$	$F_2 = -11449.40994$
$B_3 = -0.38483$	$C_3 = -55.25173$	$D_3 = 1765.65152$	$E_3 = 322246.31304$	$F_3 = 25777.18778$
$B_4 = -1.66470$	$C_4 = 52.03427$	$D_4 = -2348.24096$	$E_4 = -336998.50511$	$F_4 = -27617.26188$
$B_5 = 1.08389$	$C_5 = -18.96104$	$D_5 = 1121.59614$	$E_5 = 136590.81012$	$F_5 = 11404.34674$

$$\begin{cases} \sigma = \dfrac{1}{\alpha}\left\{\left(\dfrac{Z}{A}\right)^{\frac{1}{n}} + \left[\left(\dfrac{Z}{A}\right)^{\frac{2}{n}} + 1\right]^{\frac{1}{2}}\right\} \\ Z = \dot{\varepsilon}\exp\left(\dfrac{Q}{RT}\right) \\ \alpha = 0.01120 - 0.03357\varepsilon + 0.08547\varepsilon^2 + 0.01818\varepsilon^3 - 0.38097\varepsilon^4 + 0.54783\varepsilon^5 - 0.24440\varepsilon^6 \\ \beta = 0.06052 + 0.71381\varepsilon - 6.59852\varepsilon^2 + 27.88814\varepsilon^3 - 55.25173\varepsilon^4 + 52.03427\varepsilon^5 - 18.96014\varepsilon^6 \\ n = 7.10303 + 16.52829\varepsilon - 6.34175\varepsilon^2 - 500.30071\varepsilon^3 + 1765.65152\varepsilon^4 - 2348.20496\varepsilon^5 \\ \qquad + 1121.59614\varepsilon^6 \\ Q = 626.04636 - 3346.82229\varepsilon + 33006.85263\varepsilon^2 - 148222.46232\varepsilon^3 + 322246.31304\varepsilon^4 \\ \qquad - 336998.50511\varepsilon^5 + 1365900.81012\varepsilon^6 \\ \ln A = 55.2217 + 237.13491\varepsilon + 2411.83505\varepsilon^2 - 11449.40994\varepsilon^3 + 25777.18778\varepsilon^4 \\ \qquad - 27617.26188\varepsilon^5 + 11404.34674\varepsilon^6 \end{cases} \quad (11)$$

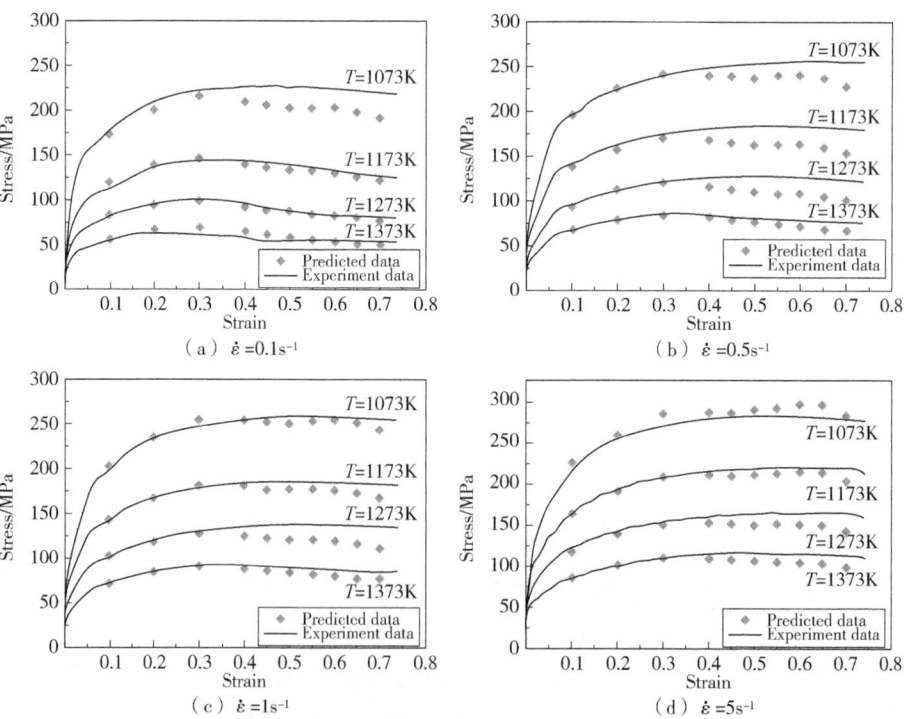

Figure 6 Calculated and measured rheological stresses from the Arrhenius constitutive model at different deformation temperatures and strain rates

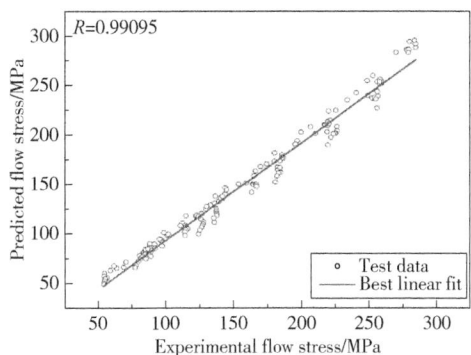

Figure 7 Correlation between the rheological stress calculated
by the Arrhenius constitutive model and the experimentally measured rheological stress

3.2 Development of BP-ANN model

In this study, a BP – ANN model for the correlation of hot compression rheological stress with strain, strain rate and temperature for high-strength offshore steel is developed. The deformation temperature (T), strain rate ($\dot{\varepsilon}$), and strain (ε) are the input variables and the actual stress (σ) is the only output. A schematic of the structure is shown in Figure 8. To develop an accurate BP-ANN model, it is crucial to determine the appropriate number of hidden layers and the number of

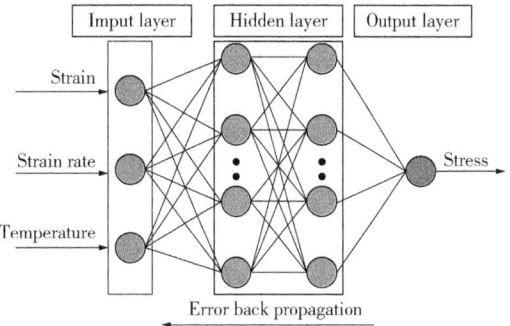

Figure 8 Schematic diagram of the BP-ANN structure

neurons in each hidden layer. As experimental data, such as temperature, strain, strain rate, and stress, are measured in different units, these data must be normalized to dimensionless units before training, which reduces the convergence speed and accuracy of the model. Assuming that one and two hidden layers are used for testing, the appropriate number of hidden layers is determined by evaluating the predicted data from the training and testing using the experimental data with appropriate tolerances. In addition, an empirical equation is proposed to determine the range of E-values, i.e., the range of number of neurons per hidden layer, and the range is calculated accordingly as 3~12. Finally, the accuracy of the neural network model is guaranteed using a two-layer hidden layer structure.

$$e = \sqrt{n+m} + a \tag{12}$$

where e is the number of neurons in each hidden layer; n and m are the number of neurons in the input and output layers of the network, respectively: $n=3$, $m=1$. and a is a ranging from 1 to 10.

The values of the input and output variables are distributed over different ranges and uniform dimensions, which leads to poor convergence speed and prediction accuracy of ANN models. Therefore, normalization of the initial experimental stress – strain data is essential to ensure dimensionless and approximately similar quantitative values of the input and output variables. In this study, the magnitude of the normalized data is narrowed to adjust the parameters to within 0~0.3, with coefficients of 0.05 and 0.25 in the regression equation. The pilot algorithm demonstrates that such magnitudes can improve convergence speed and forecast accuracy. Excessive errors caused by a wide range are avoided. In addition, it should be noted that the initial value of the true strain rate has a large magnitude. Thus, we take the logarithm to convert the true strain rate data prior to normalization:

$$y_n = 0.05 + 0.25 \times \frac{y - 0.95 y_{min}}{1.05 y_{max} - 0.95 y_{min}} \quad (13)$$

$$y_n = 0.05 + 0.25 \times \frac{(3+y) - 0.95(3 + y_{min})}{1.05(3 + y_{max}) - 0.95(3 + y_{min})} \quad (14)$$

where y_n is the normalized value of y, y is the experimental data, y_{max} and and y_{min} are the maximum and minimum values of, respectively.

In this study, R and AARE were used as statistical indicators to comprehensively evaluate the predictive ability of the ANN model, and are expressed as Equations (15) and (16), respectively. A high R value close to 1 indicates that the predicted value agrees well with the experimental value, while a low AARE value close to 0 indicates that the sum of the errors between the predicted and experimental values tends to zero.

$$R = \frac{\sum_{i=1}^{N}(E_i - \overline{E})(P_i - \overline{P})}{\sqrt{\sum_{i=1}^{N}(E_i - \overline{E})^2 \sum_{i=1}^{N}(P_i - \overline{P})^2}} \quad (15)$$

$$AARE = \frac{1}{N} \sum_{i=1}^{N} \left| \frac{P_i - E_i}{E_i} \right| \times 100\% \quad (16)$$

where E and P are the experimental and predicted values of the true stress, respectively, $\overline{E}$ and $\overline{P}$ are the average values of E and P, respectively, P_i is the predicted stress value, E_i is the corresponding experimental stress value, and N is the number of predicted points.

The relative error (δ) in the equation represents the percentage error of each stress-strain prediction relative to the corresponding experimental value, and is introduced for a more in-depth and detailed evaluation of the ANN model. The parameters μ and w are the mean and standard deviation, respectively. For a smaller δ value, the precision is higher. Information on the values is provided by calculating and counting the δ values of all prediction points, including the training and testing points. Figure 9 shows the correlation between predicted and true stress values for the training and testing data. Figure 10 shows a histogram of the relative errors in the training and testing parts. The relative errors are within a narrow range of 0~6% for both the training and testing parts. More notably, most of the δ values are concentrated around the ideal value of 0. In the training part, 79.03% of the points with approximately 6% error are within the [-1%, 1%] interval. In the test part, 78.47% of the points are concentrated within the [-1%, 1%] interval. These results, generated using statistical data, provide direct evidence that the ANN model achieves high accuracy in both the training and testing phases.

$$\delta = \frac{P_i - E_i}{E_i} \times 100\% \tag{17}$$

$$\mu = \frac{1}{N} \sum_{i=1}^{N} \delta_i \tag{18}$$

$$w = \sqrt{\frac{1}{N-1} \sum_{i=1}^{N} (\delta_i - \mu)^2} \tag{19}$$

where P_i is the predicted stress value and E_i is the corresponding experimental stress value. δ_i is the relative error; μ and w are the mean and standard deviation of δ, respectively; and N is the number of relative errors.

As shown in Figure 9, the data points with the true stress values as the horizontal axis and the predicted stress as the vertical axis are consistent with the best linear fit line, indicating excellent prediction performance. The correlation coefficients, R, for the training and test points are 0.99974 and 0.9998, respectively. In addition, the $AARE$ values calculated from the training and test parts are 0.74575% and 0.70758%, respectively. The small errors in both cases fully illustrate the high-accuracy prediction performance of the BP-ANN in training and testing.

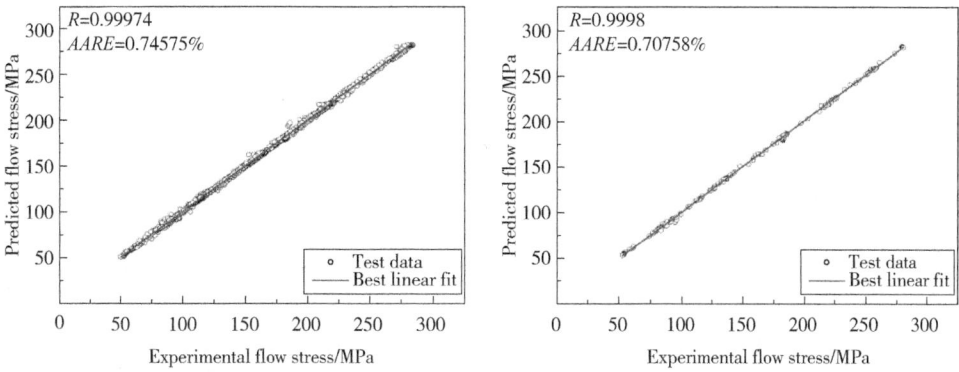

Figure 9 Correlation between predicted and true stress values for the training and test data

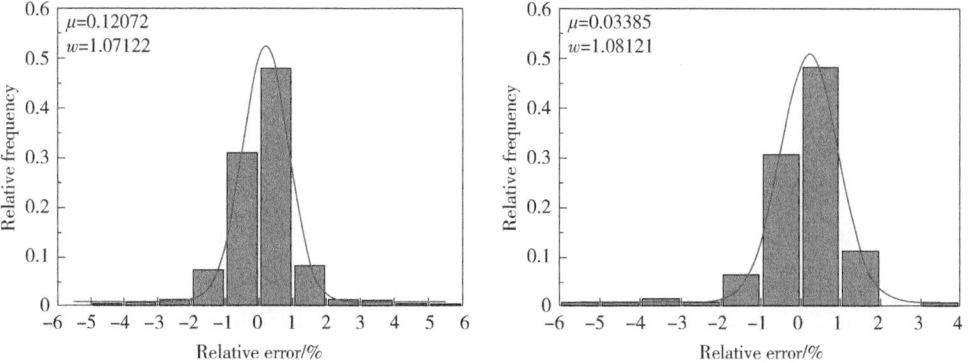

Figure 10 Relative error distribution of predicted truestress values for the training and test points

Based on the well-trained ANN model, the actual stress values under experimental conditions are included to predict the deformation conditions corresponding to the previously trained and tested points. Figure 11 compares the stress values predicted by the BP-ANN model with the experimental stress values. The predicted values follow the same trend as the true values, and increase with decreasing temperature or increasing strain rate. This indicates that the BP-ANN model is more accurate compared to the Arrhenius-type constitutive equation in determining the variation law of the rheological stress.

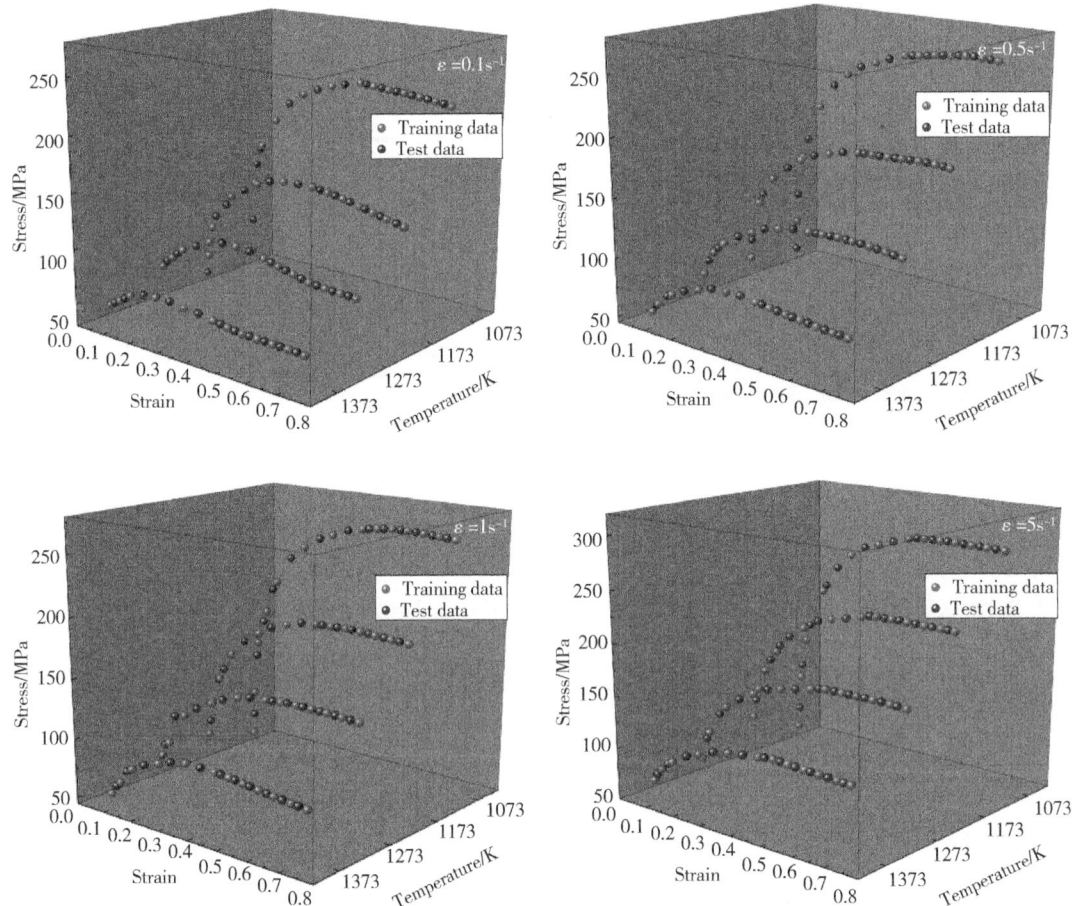

Figure 11 Comparison of predicted stress values from the BP-ANN model
with experimental values at differenttemperatures and strain rates

The experimental stress-strain curves, predictions from the BP-ANN model and Arrhenius constitutive equation at 1123K and strain rates $0.01s^{-1}$, $0.03s^{-1}$ and $10s^{-1}$, are shown in Figure 12. The new deformation temperature and deformation rates are introduced. Among the three curves predicted by BP-ANN, when the strain rate is $3s^{-1}$, it shows good predictive tracking ability in the range of $0.1s^{-1}$ to $5s^{-1}$ of the training set. However, as the rates of $0.01s^{-1}$ and $10s^{-1}$ exceeded the training range of the training set, although the trend was consistent, the accuracy of the ANN model is lower than that of the Arrhenius constitutive mode.

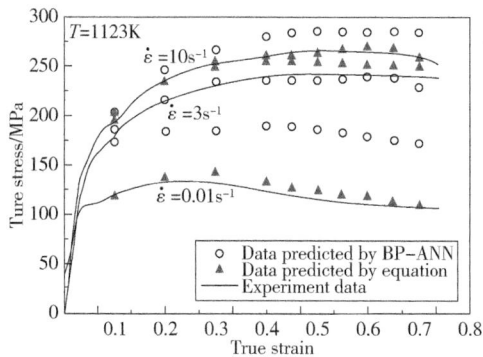

Figure 12 Comparison of predicted stress values of the Arrhenius constitutive model and BP-ANN model with experimental values at 1123K and different strain rates

4 Conclusions

(1) A BP-ANN model of high-strength offshore steel is developed based on isothermal compression test data from a Gleeble 3800 thermal simulator. The established neural network model can effectively simulate complex thermal deformation behavior and has good generalization ability over a wide range of temperatures and strain rates.

(2) In the BP-ANN model, the deformation temperature (T), strain rate ($\dot{\varepsilon}$), and strain (ε) are the input variables, and true stress (σ) is the output variable. The higher R values, lower $AARE$ values, and stable percentage error distribution results indicate that the BP-ANN model has a better predictive performance than the Arrhenius-type intrinsic constitutive equation under limited experimental conditions.

(3) The ANN model has excellent potential for application in investigating the thermal deformation processes. The accuracy of the predicted rheological stress provides strong theoretical support for the determination of parameters for processes such as rolling and for the heat treatment of high-strength offshore steel.

Acknowledgments

This research was financially supported by Science and Technology Research Projects of the CNPC(No. 2021ZG14).

References

[1] Chen MS, Lin YC, Ma XS. The kinetics of dynamic recrystallization of 42CrMo steel. Mater Sci Eng A. 2012; 556: 260-266. doi:10.1016/j.msea.2012.06.084.

[2] Zhu H, Chen F, Zhang H, et al. Review on modeling and simulation of microstructure evolution during dynamic recrystallization using cellular automaton method. Sci China Technol Sci. 2020; 63(3): 357-396. doi:10.1007/s11431-019-9548-x.

[3] Liu G, Du D, Wang K, et al. Epitaxial growth behavior and stray grains formation mechanism during laser surface re-melting of directionally solidified nickel-based superalloys. J Alloys Compd. 2021; 853: 157325. doi:10.1016/j.jallcom.2020.157325.

[4] Guo C, Liu SH, Hu RZ, et al. Adv Mater Sci Eng. 2020; 2020: 8470738.

[5] Lin YC, Chen XM. A critical review of experimental results and constitutive descriptions for metals and alloys in hot working. Mater Des. 2011; 32(4): 1733-1759. doi:10.1016/j.matdes.2010.11.048.

[6] Wu B, Li MQ, Ma DW. The flow behavior and constitutive equations in isothermal compression of 7050 aluminum alloy. Mater Sci Eng A. 2012; 542: 79-87. doi:10.1016/j.msea.2012.02.035.

[7] Sellars CM, McTegart WJ. On the mechanism of hot deformation. Acta Metall. 1966; 14(9): 1136-1138. doi:10.1016/0001-6160(66)90207-0.

[8] Shi Z, Yan X, Duan C. Characterization of hot deformation behavior of GH925 superalloy using constitutive equation, processing map and microstructure observation. J Alloys Compd. 2015; 652: 30-38. doi:10.1016/j.jallcom.2015.08.118.

[9] Samantaray D, Mandal S, Bhaduri AK. A comparative study on Johnson Cook, modified Zerilli-Armstrong and Arrhenius-type constitutive models to predict elevated temperature flow behaviour in modified 9Cr-1Mo steel. Comput Mater Sci. 2009; 47(2): 568-576. doi:10.1016/j.commatsci.2009.09.025.

[10] Mandal S, Rakesh V, Sivaprasad PV, et al. Constitutive equations to predict high temperature flow stress in a Ti-modified austenitic stainless steel. Mater Sci Eng A. 2009; 500(1-2): 114-121. doi:10.1016/j.msea.2008.09.019.

[11] Cao Y, Di HS, Misra RDK, et al. Hot deformation behavior of alloy 800H at intermediate temperatures: constitutive models and microstructure analysis. J Mater Eng Perform. 2014; 23(12): 4298-4308. doi:10.1007/s11665-014-1220-4.

[12] Peng X, Guo H, Shi Z, et al. Constitutive equations for high temperature flow stress of TC4-DT alloy incorporating strain, strain rate and temperature. Mater Des. 2013; 50: 198-206. doi:10.1016/j.matdes.2013.03.009.

[13] Changizian P, Zarei-Hanzaki A, Roostaei AA. The high temperature flow behavior modeling of AZ81 magnesium alloy considering strain effects. Mater Des. 2012; 39: 384-389. doi:10.1016/j.matdes.2012.02.049.

[14] Lin YC, Xia YC, Chen XM, et al. Constitutive descriptions for hot compressed 2124-T851 aluminum alloy over a wide range of temperature and strain rate. Comput Mater Sci. 2010; 50(1): 227-233. doi:10.1016/j.commatsci.2010.08.003.

[15] Zhao J, Ding H, Zhao W, et al. Modelling of the hot deformation behaviour of a titanium alloy using constitutive equations and artificial neural network. Comput Mater Sci. 2014; 92: 47-56. doi:10.1016/j.commatsci.2014.05.040.

[16] Ashtiani HRR, Shahsavari P. A comparative study on the phenomenological and artificial neural network models to predict hot deformation behavior of AlCuMgPb alloy. J Alloys Compd. 2016; 687: 263-273. doi:10.1016/j.jallcom.2016.04.300.

[17] Peng WW, Zeng WD, Wang QJ, et al. Comparative study on constitutive relationship of as-cast Ti60 titanium alloy during hot deformation based on Arrhenius-type and artificial neural network models. Mater Des. 2013; 51: 95-104. doi:10.1016/j.matdes.2013.04.009.

[18] Sabokpa O, Zarei-Hanzaki A, Abedi HR, et al. Artificial neural network modeling to predict the high temperature flow behavior of an AZ81 magnesium alloy. Mater Des. 2012; 39: 390-396. doi:10.1016/j.matdes.2012.03.002.

[19] Liu Y, Li HY, Jiang HF, et al. Artificial neural network modelling to predict hot deformation behaviour of zinc-aluminium alloy. Mater Sci Technol. 2013; 29(2): 184-189. doi:10.1179/1743284712Y.0000000127.

[20] Zhu YC, Zeng WD, Sun Y, et al. Artificial neural network approach to predict the flow stress in the isothermal compression of as-cast TC21 titanium alloy. Computational Materials Science. 2011; 50(5): 1785-1790. doi:10.1016/j.commatsci.2011.01.015.

[21] Sheikh H, Serajzadeh S, Mater J. Estimation of flow stress behavior of AA5083 using artificial neural networks with regard to dynamic strain ageing effect. Process Technol. 2008; 196(1-3): 115-119. doi:10.1016/j.jmatprotec.2007.05.027.

[22] Lin YC, Zhang J, Zhong J. Application of neural networks to predict the elevated temperature flow behavior of a low alloy steel. Comput Mater Sci. 2008; 43(4): 752-758. doi:10.1016/j.commatsci.2008.01.039.

[23] Zhao JW, Ding H, Zhao WJ, et al. Modelling of the hot deformation behaviour of a titanium alloy using constitutive equations and artificial neural network. Comput Mater Sci. 2014; 92: 47-56. doi:10.1016/j.com matsci.2014.05.040.

[24] Quan GZ, Lv WQ, Mao YP, et al. Prediction of flow stress in a wide temperature range involving phase transformation for as-cast Ti-6Al-2Zr-1Mo-1V alloy by artificial neural network. Mater Des. 2013; 50: 51-61. doi:10.1016/j.matdes.2013.02.033.

[25] Quan GZ, Yu CT, Liu YY, et al. A comparative study on improved Arrhenius-type and artificial neural network models to predict high-temperature flow behaviors in 20MnNiMo alloy. Sci World J. 2014; 2014: 1-12. doi:10.1155/2014/108492.

[26] Zhao JW, Ding H, Zhao WJ, et al. Modelling of the hot deformation behaviour of a titanium alloy using constitutive equations and artificial neural network. Comput Mater Sci. 2014; 92: 47-56. doi:10.1016/j.com matsci.2014.05.040.

[27] Sun Y, Zeng WD, Zhao YQ, et al. Modeling constitutive relationship of Ti40 alloy using artificial neural network. Mater Des. 2011; 32(3): 1537-1541. doi:10.1016/j.matdes.2010.10.004.

[28] Gupta AK, Singh SK, Reddy S, et al. Prediction of flow stress in dynamic strain aging regime of austenitic stainless steel 316 using artificial neural network. Mater Des. 2012; 35: 589-595. doi:10.1016/j.matdes.2011.09.060.

[29] Xiao X, Liu GQ, Hu BF, et al. A comparative study on Arrhenius-type constitutive equations and artificial neural network model to predict high-temperature deformation behaviour in 12Cr3WV steel. A Comput Mater Sci. 2012; 62: 227-234. doi:10.1016/ j.commatsci.2012.05.053.

[30] Bariani PF, Bruschi S, Dal Negro T, et al. Prediction of nickel-base superalloys' rheological behaviour under hot forging conditions using artificial neural networks. Process Technol. 2004; 152(3): 395-400. doi:10.1016/j.jmatprotec.2004.04.416.

[31] Toros S, Ozturk F. Flow curve prediction of Al-Mg alloys under warm forming conditions at various strain rates by ANN. Appl Soft Comput. 2011; 11(2): 1891-1898. doi:10.1016/j.asoc.2010.06.004.

[32] Quan GZ, Wang T, Li YL, et al. Artificial neural network modeling to evaluate the dynamic flow stress of 7050 aluminum alloy. J Mater Eng Perform. 2016; 25(2): 553-564. doi:10.1007/s11665-016-1884-z.

[33] Yan LM, An D, Shi G, et al. A comparative study of constitutive and neural network models for flow behavior of mg-5.9Zn-1.6Zr-1.6Nd-0.9Y alloy and processing maps. J Mater Eng Perform. 2017; 26(5): 2368-2376. doi:10.1007/s11665-017-2643-5.

[34] Kappatos V, Chamos AN, Pantelakis SG. Assessment of the effect of existing corrosion on the tensile behaviour of magnesium alloy AZ31 using neural networks. Mater Des. 2010; 31(1): 336-342. doi:10.1016/j.matdes.2009.06.009.

[35] Ning H, Li X, Meng L, et al. Effect of Ni and Mo on microstructure and mechanical properties of grey cast iron. Mater Technol Adv Perform Mater. 2023; 38(1): 2172991. doi:10.1080/10667857.2023.2172991.

[36] Misra RDK, Challa VSA, Injeti YS. Phase reversion-induced nanostructured austenitic alloys: an overview. Mater Technol Adv Perform Mater. 2022; 37(7): 437-449. doi:10.1080/10667857.2022.2065621.

[37] Guo L, Su X, Dai L, et al. Strain ageing embrittlement behaviour of X80 self-shielded flux-cored girth weld metal. Mater Technol Adv Perform Mater. 2023; 38(1): 2164978. doi:10.1080/10667857.2023.2164978.

[38] Li SF, Misra RDK, Liu ZQ. Towards strength-ductility synergy in nanosheets strengthened titanium matrix composites through laser power bed fusion of MXene/Ti composite powder. Mater Technol Adv Perform Mater. 2023; 38(1): 2181680. doi:10.1080/ 10667857.2023.2181680.

[39] Niu G, Zurpb H, Guyuen M, et al. Superior fracture toughness in a high-strength austenitic steel with heterogeneous lamellar microstructure. Acta Mater. 2022; 226: 117462. doi:10.1016/j.actamat.2022.117642.

[40] Yang C, Xu H, Wang Y, et al. Hot tearing analysis and process optimisation of the fire face of al-cu alloy cylinder head based on MAGMA numerical simulation. Mater Technol: Adv Perform Mater. 2023; 38(1): 2165245. doi:10.1080/10667857.2023.2165245.

[41] Li Q, Zuo H, Feng J, et al. Strain rate and temperature sensitivity on the flow behaviour of a duplex stainless steel during hot deformation. Mater Technol: Adv Perform Mater. 2023; 38(1): 2166216. doi:10.1080/ 10667857.2023.2166216.

[42] Misra RDK. Strong and ductile texture-free ultrafine-grained magnesium alloy via three-axial forging. Mate Lett. 2023; 31: 133443. doi:10.1016/j.matlet.2022.133443.

[43] Misra RDK. Enabling manufacturing of multi-axial forging-induced ultrafine-grained strong and ductile magnesium alloys: a perspective of process-structure-property paradigm. Mater Technol Adv Perform Mater. 2023; 38(1): 2189769. doi:10.1080/10667857.2023.2189769.

[44] Lin YC, Chen M-S, Zhong J. Constitutive modeling for elevated temperature flow behavior of 42CrMo steel. Comp Mater Sci. 2008; 42(3): 470–477. doi:10.1016/j.commatsci.2007.08.011.

[45] Pu ZJ, Wu KH, Shi J, et al. Development of constitutive relationships for the hot deformation of boron microalloying TiAl? Cr? V alloys. Mater Sci Eng A. 1995; 192: 780–787. doi:10.1016/0921-5093(94) 03314-5.

[46] Slooff FA, Zhou J, Duszczyk J, et al. Constitutive analysis of wrought magnesium alloy Mg-Al4-Zn1. Scripta Mater. 2007; 57(8): 759–762. doi:10.1016/j.scriptamat.2007.06.023.

[47] Haghdadi N, Zarei-Hanzaki A, Abedi HR. The flow behavior modeling of cast A356 aluminum alloy at elevated temperatures considering the effect of strain. Mater Sci Eng A. 2012; 535: 252–257. doi:10.1016/j.msea.2011.12.076.

本论文原发表于《Materials Technology》2023 年第 38 卷第 1 期。

Failure Analysis of 2205 Duplex Stainless Steel Connecting Pipe for Composite Plate Pressure Vessel in a High Pressure Gas Field

Li Lei[1]　Chen Qingguo[2]　Fang Yan[2]　Li Xuanpeng[1]
Luo Jinheng[1]　Song Chengli[1]　Wang Shuai[1]

(1. State Key Laboratory for Performance and Structure Safety of
Petroleum Tubular Goods and Equipment Materials,
CNPC Tubular Goods Research Institute;
2. Oil and Gas Engineering Research Institute of Petro China Tarim Oilfield Company)

Abstract: Pitting and cracking failure of 2205 duplex stainless steel connecting pipe in a high-pressure gas field occurred. Macro inspection, non-destructive testing, physical and chemical tests, corrosion tests and heat treatment were carried out for the pipe in order to identify the failure cause and mechanism. The results showed that large-size non-metallic inclusions composed of MnS and (Si, Mn, Ca, Al, Cr, Mg) oxides and existence of tiny precipitates didn't meet the requirements of Order Technical Agreement, which resulted in a serious reduction in the impact toughness and pitting resistance. The main reason for the failure is the existence of the harmful precipitates produced by stress relief annealing and large-size non-metallic inclusions. The rapid deterioration of the corrosion environment in the past three years has played a role in inducing and promoting the failure. Fine precipitates mainly reduce the impact toughness, while large-size non-metallic inclusions mainly reduce the pitting resistance. The formation of pits and the large-size non-metallic inclusions are the key factors for the generation of micro cracks, and the fine precipitates accelerate the crack growth.

Keywords: Composite plate pressure vessel; Connecting pipe; 2205 duplex stainless steel; Pitting; Crack

A growing number of gas fields have been found around the world with the deepening of exploration and development of oil and gas resources, such as North America, the Gulf of Mexico, Africa, the Middle East and western China[1]. These gas fields are generally characterized by high pressure, high temperature, high production and complex produced media, which lead to higher corrosion environment. 2205 duplex stainless steel has advantages of both ferritic (α) stainless steel

---

Corresponding author: Li Lei, Senior engineer, 623549473@ qq. com.

and austenitic (γ) stainless steel, exhibiting high strength, good toughness, excellent welding performance and corrosion resistance. It is usually applied in high-pressure gas fields, as well as in petrochemical, electric power and others[2-6]. However, its larger scale application is restricted by its high one-time use cost[7]. 2205 duplex stainless steel composite plate not only retains the excellent corrosion resistance of stainless steel, but also reduces the primary use cost, which is the best choice for selecting materials of highly corrosive equipment[8]. In particular, the manufacturing technology of 2205-16MnR explosive composite plate is mature, which has been widely used for manufacturing pressure vessels, chemical vessels and heat exchangers[9-10].

2205-16MnR explosive composite plate pressure vessels have been abundantly used in a high-pressure gas field in China with the safe operation time of nearly 20 years. However, 2205 duplex stainless steel connecting pipe of the pressure vessel suffered pitting corrosion and cracking for the first time, recently. Studies[2,4,7] have confirmed that 2205 duplex stainless steel has excellent resistance to pitting corrosion and stress corrosion cracking in the high chloride and CO_2 corrosion environments of these high-pressure gas fields. Hence, the failure reasons may be more complex. The composite plate pressure vessels normally need stress relief annealing heat treatment to improve the interface bonding property in order to reduce the residual stress caused by the instant explosion[2,11]. It should be noted that 2205 duplex stainless steel connecting pipes also undergo the annealing heat treatment, which probably results in the presence of harmful phase[12-14]. Literatures have mainly reported[10,15] the effect of the harmful phase on the structure and properties of 2205-16MnR explosive composite plate, but no researches on the 2205 duplex stainless steel connecting pipe. Besides, stress corrosion cracking refers to the brittle fracture of metal materials under tensile stress in some specific media. 2205 duplex stainless steel has the risk of chloride stress corrosion cracking under high pressure and high chloride corrosion environment. Thus, determining the cause of cracking is critical to ensure safety of the equipment. What's more, considering that this kind of failure is the first time in the gas field, it is extremely necessary to find out the cause of the failure. It is hoped that the work in this paper will clarify the cause and mechanism of failure, so as to provide a basis for the failure control and prevention of similar connecting pipe.

1 Samples and experimental methods

1.1 Samples

Fig. 1 shows the site photographic view of the failed connecting pipe. The external structure is complete, but there are pits observed at two locations (Fig. 1). The larger one has a circumferential length of 55mm, a radial length of 35mm, and a depth of 8mm (the lower pitting pit in Fig. 1b). The smaller one has a circumferential length of 10mm, a radial length of 8mm, and a depth of 5mm (the upper pitting pit in Fig. 1b). Through penetrant testing after polishing, there are several discontinuous cracks at the bottom of two pits (the enlarged image of the lower left corner and upper right corner in Fig. 1b), and the longitudinal depth of the cracks has exceeded 40mm. No other defects are found.

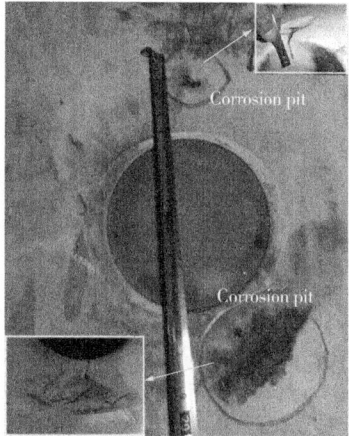

(a) external　　　　　　　　(b) internal

Fig. 1　The site photographic view of the failed connecting pipe

The failed connecting pipe is made of 2205 Ⅲ forging and its inner diameter and wall thickness are 98.3mm and 45mm respectively, which is connected to the gas gathering manifold by welding. The longitudinal section diagram of the connection is shown in Fig. 2. The cylinder of the gas gathering manifold is made of 2205-16MnR explosive composite plate. The thickness of the base layer and the cladding layer is 60mm and 4mm separately. Before leaving the factory, the connecting pipe has been welded. Then, they are subject to annealing heat treatment in the same furnace. The specific process is to

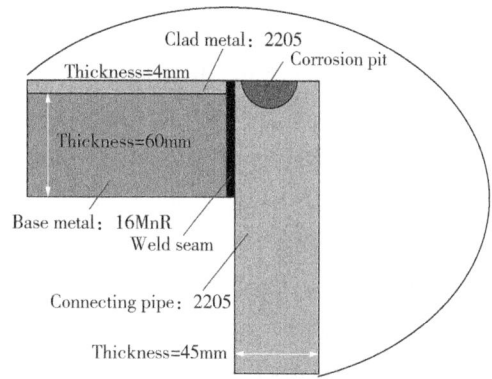

Fig. 2　Longitudinal section of the connection between the failed connecting pipe and the gas gathering manifold

keep the temperature at 580℃ for 2.5h. After the furnace is cooled to 400℃, it is discharged and air cooled to room temperature.

Mark the failed connecting pipe as failed sample. The newly manufactured connecting pipe named as comparison sample was fabricated by the same forging process with failed sample, which was in the state of solid solution.

1.2　Experimental methods

1.2.1　Physical and chemical tests

The chemical composition analysis of the sample was carried out by ARL4460 direct reading spectrometer and TC600 oxygen and nitrogen analyzer according to ASTM A751-20. The non-metallic inclusions, content of α phase and precipitation were detected by OLS4100 laser confocal microscope in accordance with GB/T 13298—2015, GB/T 10561—2005 and GB/T 13305—2008. The hardness was tested by RB2002T Rockwell hardness tester on the basis of GB/T 230.1—2018. Charpy V-type impact test was conducted by ZBC2752-B impact tester in accordance with GB/T 229—2007, which the specification of samples was the thickness of 10mm, the width of 10mm and the length of 55mm.

1.2.2 Corrosion tests

Intergranular corrosion test was conducted by method E (16% sulfuric acid copper sulfate corrosion test) of GB/T 4334—2020. The bending method was used to determine the intergranular corrosion tendency. Two samples in each group were bent 180° after 24h boiling test to observe whether there were cracks. 25% $MgCl_2$ boiling stress corrosion cracking test was carried out by the method of four point bending according to ASTM G36-94 (R 2018). The tensile stress was 50% of the minimum yield strength specified in the standard, i.e. 224MPa, and the test period was 96h. 6% $FeCl_3$ corrosion test was conducted in accordance with ASTM A923—2014, which the test temperature was 25℃ and the test period is 24h.

The potentiodynamic polarization curve test was carried out by CS370 electrochemical workstation, which the solution composition was shown in Tab. 1. The test temperature and pressure were 60℃ and 0.1MPa respectively. A three electrode system was used, with Pt sheet as the counter electrode, saturated calomel electrode (SCE) as the reference electrode, and the sample as the working electrode. The test process was divided into four steps. The first step was to test the open circuit potential and the acquisition frequency was 2s/point. The second step was potentiostatic polarization test to remove the original surface passive film formed by the sample in the air. The polarization potential was relative open circuit potential -1V, and the polarization time was 100s. Third, measure the open circuit potential again until the open circuit potential was stable, that is, the potential change per minute didn't exceed 2mV, and the acquisition frequency was 2s/point. The fourth step was the dynamic potential polarization test. The starting scanning potential was the relative open circuit potential of -250mV, and the ending scanning potential was the potential when the current was greater than 2mA. The scanning rate and the acquisition frequency were 0.33mV/s and 2.44mV separately. The breakdown potential and dimensional passive current density could be obtained from the polarization curve of the driven potential.

Tab. 1 Chemical composition of gas field water

SO_4^{2-}/(mg/L)	$Na^+ + K^+$/(mg/L)	Ca^{2+}/(mg/L)	Cl^-/(mg/L)	Mg^{2+}/(mg/L)	HCO_3^-/(mg/L)	Total mineralization/(mg/L)	pH value
439.1	48450	7092	88300	476.5	93.2	144900	6.54

1.2.3 Micro analysis tests

The micro morphology and composition of the cracks and fractures of the failure sample were analyzed by TESCAN VEGA scanning electron microscope (SEM) and INCA-350 X-ray Energy Dispersive Spectrometer (EDS) respectively.

1.2.4 Heat treatments

Solution treatment on failed sample was carried out by KLS 0513 muffle furnace, which was marked as failed sample after solution treatment. The solution treatment process was to keep the temperature at 1080℃ for 30min, and to cool the sample rapidly with water. Subsequently, the physical and chemical tests and corrosion tests were conducted according to Section 1.2.1 and 1.2.2 respectively.

2 Results and analysis

2.1 Chemical composition and metallographic structure

As shown in Tab. 2, chemical composition and pitting resistance equivalent (PREN) of failed sample and comparison sample comply with the provisions of Order Technical Agreement, where the calculation formula of PREN is shown in Formula (1). There is no significant difference in PREN and the content of Cr, Ni and Mo. Optical micrographs of metallographic structure are shown in Fig. 3. There are few precipitates which are mainly distributed in α phase and phase boundary of α/γ and large-size non-metallic inclusions in the failed sample. After solution treatment, the precipitates is eliminated, but the non-metallic inclusions have not changed. There are no precipitates and small-size non-metallic inclusions in the comparison sample. As shown in Tab. 3, non-metallic inclusions of failed sample are more than grade 3 for category D (spherical oxide), 0.5 for category DS (single particle spherical oxide), and 4.5 for the total. The metallographic structure is α+γ and the content of α phase is 58%~60%. Grain boundaries in α phase are obviously present. The grade of non-metallic inclusions and the precipitates of failed sample don't conform to the provisions of Order Technical Agreement. The biggest change of failed sample after solution treatment is the elimination of precipitates. In addition, non-metallic inclusions of comparison sample is 3.0 for the total. The metallographic structure is α+γ and the content of α phase is 58%~60%. There are no obvious precipitates. It comply with the provisions of Order Technical Agreement.

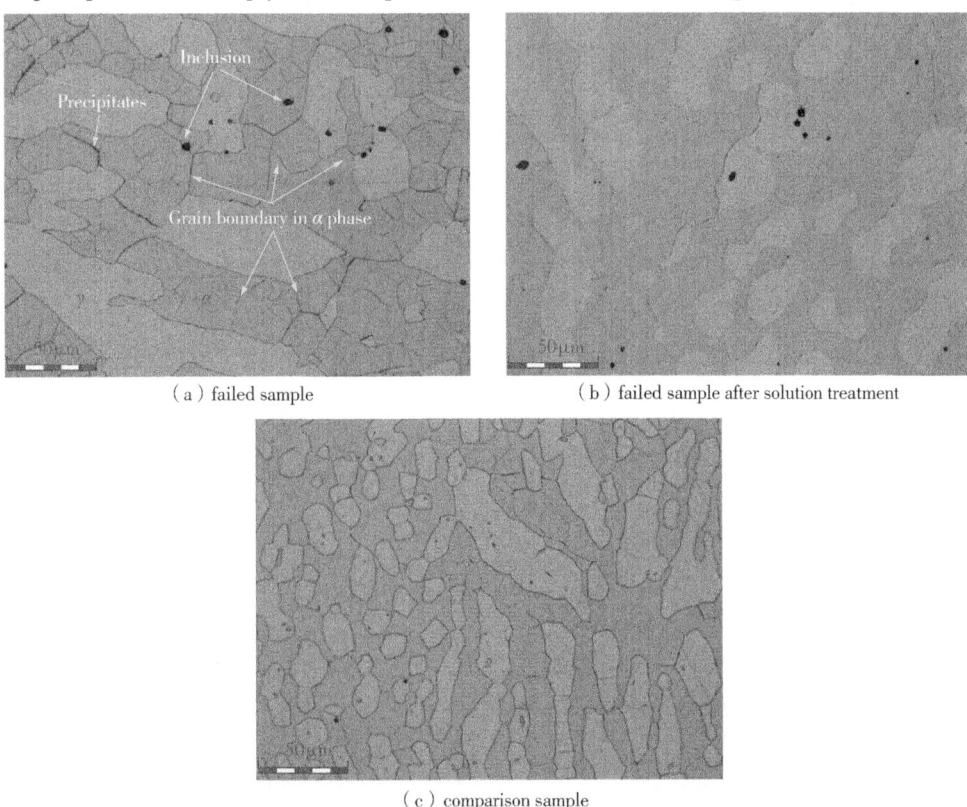

(a) failed sample (b) failed sample after solution treatment

(c) comparison sample

Fig. 3 Optical micrographs of metallographic structure

$$PREN = \%Cr + 3.3\%Mo + 16\%N \tag{1}$$

Tab. 2 Analysis result of chemical composition(%, mass fraction)

Sample description	C	Si	Mn	P	S	Cr	Mo	Ni	N	PREN
Failed sample	0.020	0.53	0.64	0.023	0.0190	23.00	3.1	4.50	0.20	36.43
Comparison sample	0.015	0.40	1.26	0.023	0.0008	22.28	3.1	5.03	0.19	35.55
Order Technical Agreement	≤0.020	≤1.00	≤2.00	≤0.025	≤0.0200	22.00~23.00	3.0~3.5	4.50~6.50	0.15~0.20	≥35.00

Tab. 3 Results of metallographic analysis

Sample description	Non-metallic inclusions	Metallographic structure	Content of α phase/%	Precipitation
Failed sample	A0.5 B0.5 D3.0, DS0.5	α+γ(Fig. 3a)	58~60	Have
Failed sample after solution treatment	A0.5 B0.5 D3.0, DS0.5	α+γ(Fig. 3b)	55~57	Nothing
Comparison sample	A0.5 B0.5 D2.0	α+γ(Fig. 3c)	55~57	Nothing
Order Technical Agreement	Total≤3.5	α+γ	40~60	Nothing

2.2 mechanical properties

Mechanical properties test results are shown in Fig. 4. The hardness of failed sample is 21±0.6HRC (Fig. 4a), which is in line with the stipulation that the hardness is not higher than 26 HRC in Order Technical Agreement. After solution treatment, the hardness is reduced to 18±0.8HRC, which is basically consistent with the hardness of comparison sample of 19±0.5HRC. Impact energy of failed sample at -20℃ is only 11±2J (Fig. 4b), which is far lower than the stipulation that the impact energy is not less than 56J in Order Technical Agreement. Impact energy reaches 90±18J after solution treatment, which indicates the recovery of impact toughness. From the above test results, the existence of fine precipitates lead to a slight increase in hardness, but a serious decrease in impact energy. Meanwhile, present literatures[13,16-17] show that large-scale preci-pitates of 2205 duplex stainless steel can cause its hardness to increase and impact energy to decrease significantly. This further shows that the fine precipitates have a huge impact on the impact energy, but a small impact on the hardness.

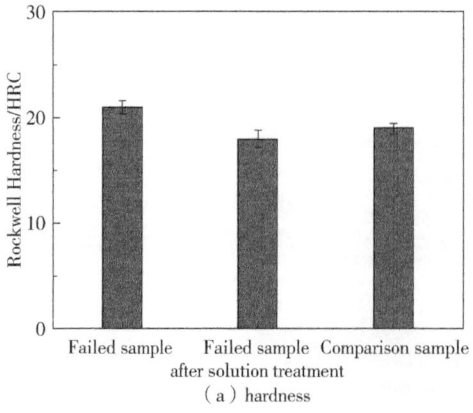
(a) hardness

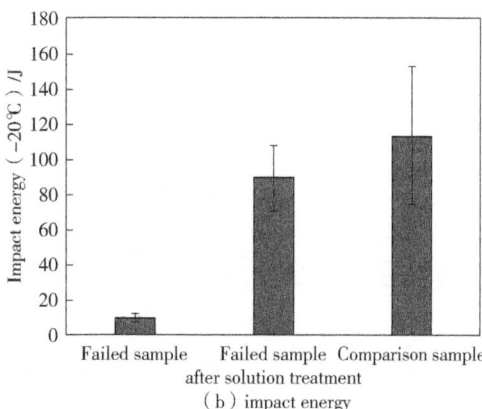
(b) impact energy

Fig. 4 Results of mechanical properties

In addition, compared with impact energy 114±39J of comparison sample at −20℃, impact energy of failed sample after solution treatment is still low. At the same time, impact energy of failed sample after solution treatment at −40℃ is 46±14J, which is also lower than the requirement of ASTM A923 that impact energy of 2205 duplex stainless steel should not be lower than 56J. From these two aspects, it can be determined that the reason for the extremely low impact energy of failed sample is not only affected by the precipitates, but also the large-size non-metallic inclusions[18].

2.3 Corrosion resistance

After the intergranular corrosion test, failed sampleis bent 180°, as shown in Fig. 5a. Failed sample is observed with a 10x magnifying glass, which shows no crack (Fig. 5a). It follows that although there are fine precipitates in grain boundaries, it doesn't lead to obvious corrosion weak areas[17]. What's more, the stress corrosion cracking test result of 25% boiling magnesium chloride (Fig. 5b) shows that failed sample is not cracked. It further indicates that the chloride stress corrosion cracking resistance has not been significantly reduced.

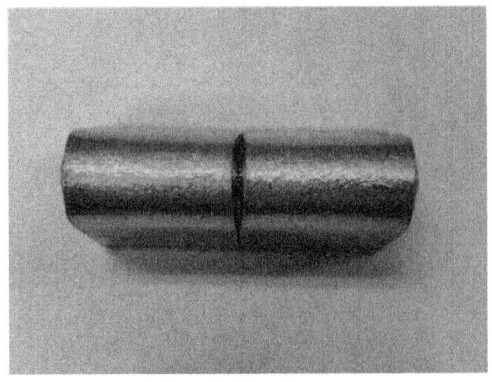
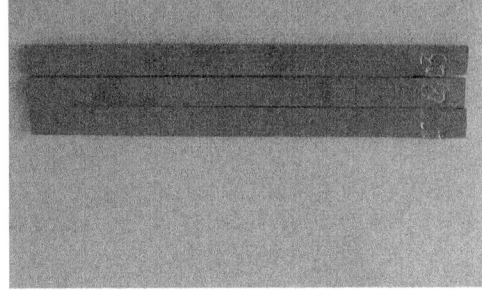

(a) intergranular corrosion　　　　　　　　　(b) stress corrosion cracking

Fig. 5　Macro photos of intergranular corrosion and stress corrosion cracking samples of failed sample

The result of 6% $FeCl_3$ corrosion test is shown in Fig. 6. The corrosion rate of failed sample is 26±5mg/(dm² · d)(mdd), which is reduced to 18±4mdd after solution treatment, with an average reduction of 31%, but still higher than the requirement of ASTM A923-2014 which is not higher than 10mdd. In addition, the corrosion rate of comparison sample is only 1±0.4mdd. It can be seen that the existence of fine precipitates and large-size non-metallic inclusions can significantly improve the corrosion rate, but the influence of the large-size non-metallic inclusions is greater. The macroscopic observation of the sample shows that that there are many pits on the surface of failed sample and failed sample after solution treatment, but no pits are found on the surface of comparison sample. Pits are mostly circular, nearly circular or strip shaped, with uneven distribution, similar to the distribution and morphology characteristics of non-metallic inclusions. This indicates that non-metallic inclusions are more susceptible to corrosion in $FeCl_3$ solution. As curves are presented in Fig. 7. The pitting breakdown potential and passivation current density of failed sample are −0.15V and 13.02μA/cm² respectively, while its values are 0.11V and 4.58μA/cm² individually for comparison sample. The pitting breakdown potential and passivation current density of failed sample after solution treatment are in the middle of them. These show that the corrosion resistance of failed sample is the worst, while that of comparison sample is the best, and the failed sample after solution treatment is in the middle[19]. It can be determined from the difference between the pitting breakdown potential and the passivation current density that large-size non-metallic inclusions have a greater impact on the corrosion resistance.

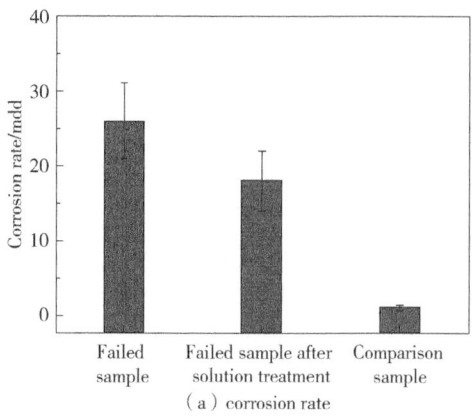

(b) failed sample and   (c) failed sample after solution treatment

Fig. 6 Result of 6%FeCl$_3$ corrosion test

Fig. 7 Potentiodynamic polarization curves for different samples

2.4 Micro analysis

2.4.1 Non-metallic inclusions and precipitates

SEM images of large-size non-metallic inclusions are shown in Fig. 8. The non-metallic inclusions are black, gray black and multi granular, and also have obvious interfaces with the matrix structure. The results of EDS analysis (Tab. 4) show that the main chemical components of the near circular multi granular inclusions (Zone A in Fig. 8) are Mn, S, O, Cr, Si, Ca, Al, which can be inferred as MnS and (Cr, Si, Ca, Al) oxide. The main chemical components of the strip inclusions (Zone B in Fig. 8) are O, Si, Mn, Ca, Al, Cr, Mg, which are inferred as (Si, Mn, Ca, Al, Cr, Mg) oxide. Hence, the large-size non-metallic inclusions are mainly composed of MnS and (Si, Mn, Ca, Al, Cr, Mg) oxide. SEM image of grain boundaries in α phase is shown in Fig. 9. It should be noted that this sample is the aforementioned metallographic analysis sample. As can be seen from Fig. 9, grain boundaries in α phase are clearly visible, which should be caused by corrosion of precipitates gathered at the grain boundary.

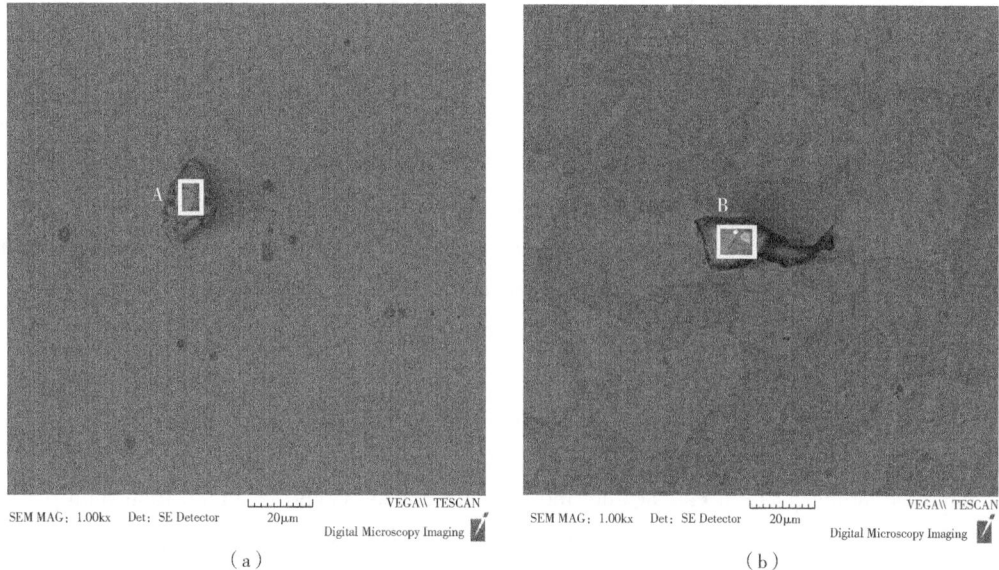

Fig. 8　SEM images of non-metallic inclusions

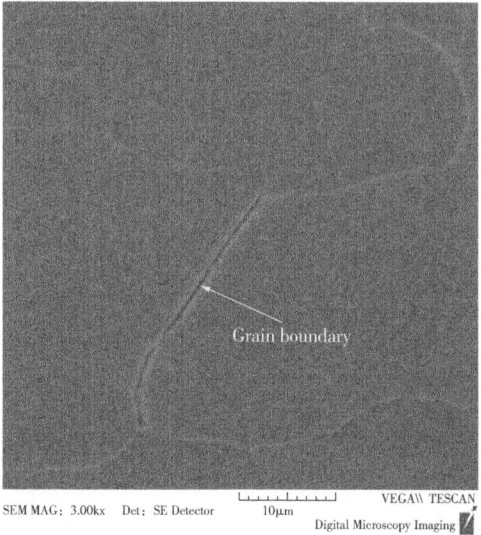

Fig. 9　SEM image of grain boundaries in α phase

Tab. 4　EDS analysis results of non-metallic inclusions(%, mass fraction)

Element	C	O	Mg	Al	Si	S	Ca	Cr	Mn
Zone A	0	18.43	0	2.42	11.15	20.98	3.42	5.58	38.03
Zone B	7.05	43.41	0.76	4.77	21.22	0	5.86	2.99	13.95

2.4.2　Crack and fracture

Metallographic analysis of cracks is shown in Fig. 10. The sample at low magnification has multiple discontinuous cracks with large gaps and more broken particles (Fig. 10a). There are small cracks on both sides of the main crack at high magnification. Some cracks originate from large-size non-metallic inclusions nearby the main crack, but no micro cracks originating from non-metallic inclusions in the area far from the main crack (Fig. 10b and Fig. 10c). SEM images of the metallographic analysis sample of the crack is shown in Fig. 11a. Consistent with the above result of metallographic analysis, there are dendritic micro cracks near the main crack and a large number of broken particles in the main crack (Fig. 11a). Result of EDS analysis (Tab. 5) shows that the main components include C, O and Ca in addition to the base metal components, where C and O should be the components of corrosion products, while Ca is the component of residual non-metallic inclusions. SEM image of the fracture surface is shown in Fig. 11b. It is tongue shaped, mainly cleavage fracture, and there are intergranular cracks in some parts. Result of EDS analysis shows that the chemical composition of the cleavage surface and the intergranular section are similar, but the content of Ni is quite different. It can be judged that the intergranular section with high content of Ni (as shown in zone D of Fig. 11b) is fracture surface of γ phase, while the cleavage surface with low content of Ni (as shown in zone E of Fig. 11b) is fracture surface of α phase. From the above microscopic analysis, it can be inferred that the cracks should originate from multiple non-metallic inclusions and expand to failure along grain boundaries of α phase, α/γ phase boundaries and non-metallic inclusions interface.

Tab. 5　Analysis results of EDS(%, mass fraction)

Element	C	O	Ca	Cr	Ni	Mo	Fe
Zone C	20.05	19.08	1.17	11.92	1.87	2.82	43.09
Zone D	5.12	11.57	0	19.92	5.31	2.30	55.06
Zone E	7.28	10.78	0	19.26	3.19	2.81	56.68

3　Discussion

3.1　Failure cause analysis

Results of macro inspection and penetrant testing show that there are pitting on the base metal surface at two locations, and several discontinuous cracks at the bottom of the pits. The maximum longitudinal depth is more than 40mm. The result of microscopic analysis confirm that cracks originate from multiple non-metallic inclusions and present brittle cracking along grain boundaries of α phase, α/γ phase boundaries and non-metallic inclusions interface. It can be determined that the failure characteristics of the connecting pipe are pitting on the base metal and multisource brittle cracking.

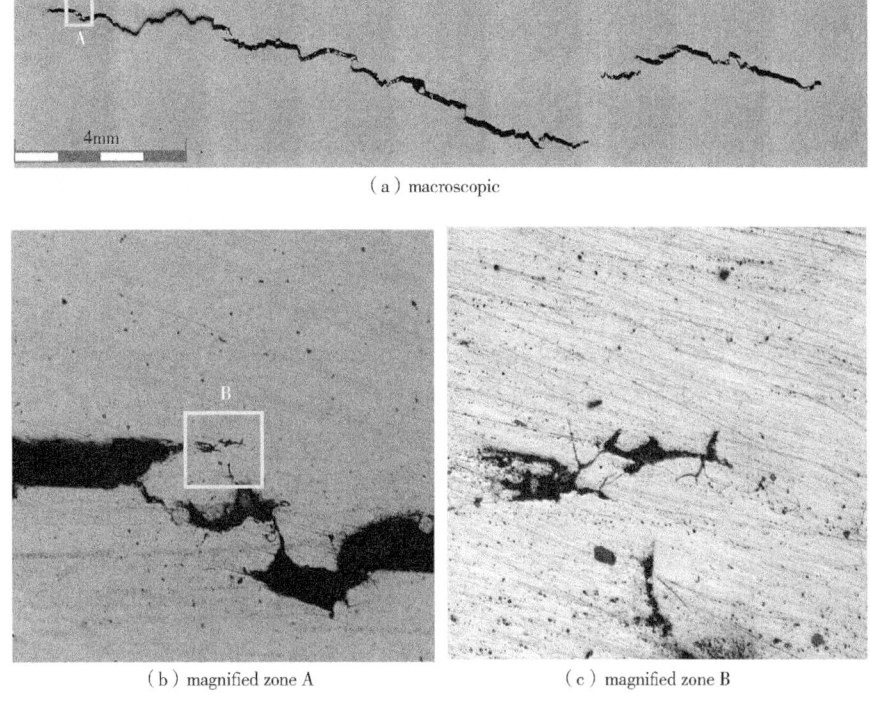

(a) macroscopic

(b) magnified zone A   (c) magnified zone B

Fig. 10 Metallographic analysis photos of crack cross section

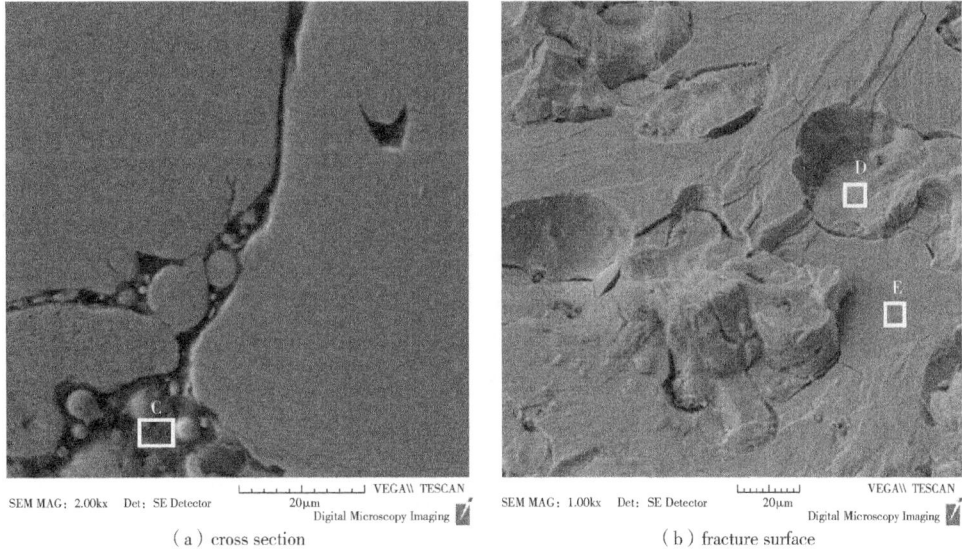

(a) cross section   (b) fracture surface

Fig. 11 Sem images of cracks

Chemical composition of failed sample complies with the provisions of Order Technical Agreement. PREN and the content of Cr, Ni, and Mo have no significant difference compared with comparison sample, indicating that its chemical composition is normal. Results of metallographic structure and microscopic analysis show that non-metallic inclusions of failed sample seriously exceed the standard, mainly distributed in α phase and at α/γ phase boundaries, which are composed of MnS and (Si, Mn, Ca, Al, Cr, Mg) oxides. Grain boundaries in α phase are obviously present. There are black fine precipitates along grain boundaries in α phase and the phase boundaries of α/γ. Due to the existence of large-size non-metallic inclusions and precipitates, the pitting resistance and impact toughness have seriously decreased[20-22], which also can be determined from the results with the impact test, 6% $FeCl_3$ corrosion test and electrochemical corrosion test of different samples. The severe degradation of corrosion resistance of failed sample results in a significant reduction of its resistance to harsh corrosion environments. In the first 13 years of application, the transmission medium does not contain H_2S and the partial pressure of CO_2 is about 0.08MPa. There is little water and no sand. The chloride concentration is 31000mg/L, the operating temperature is 30~40℃, and the operating pressure is 12MPa, which exhibit weak corrosive environment. However, due to the change of gas source in the past three years, the water content has increased from 540t/d to 896t/d, the chloride concentration has increased from 31000mg/L to 61525mg/L, and a large amount of sand has been carried, resulting in rapid deterioration of the corrosion environment. What's more, the connecting pipe is a dead liquid section, which further makes the corrosion environment worse. Therefore, pitting occurs during this period, which can also be determined from the overhaul of the device in recent three years. With the development of pitting, the stress concentration of large-size non-metallic inclusions at the bottom of pitting is more significant, and micro cracks are easily generated. Multiple discontinuous cracks at the bottom of pit and non-metallic inclusion particles in the crack gap are proof. Once micro cracks are initiated, due to the poor impact toughness and the continuous impact of high-pressure gas, they can continue to expand until failure.

From the above analysis, large-size non-metallic inclusions and fine precipitates lead to a serious reduction in the impact toughness and corrosion resistance. Previous studies[23-24] have shown that non-metallic inclusions come from the smelting and casting process, and are the result of a series of physical and chemical reactions occurring during steel melting and solidification. Subsequent molding and heat treatment processes have no significant impact on them. Large-size non-metallic inclusions appear in materials of the failed connecting pipe due to impure molten steel, uneven composition and other reasons during smelting. Results of metallographic and EDS analysis confirm that its main components are MnS and (Si, Mn, Ca, Al, Cr, Mg) oxides, which are the reaction products in the process of desulfurization and deoxidation of molten steel. They are not removed in time in the later smelting and pouring process, and remain in the ingot as blocks or strips[25]. The normal structure of 2205 duplex stainless steel after solution treatment shall be α+γ. The two-phase interface is clear and smooth without precipitates, but it is found in the failed connecting pipe that there are obvious grain boundaries in α phase and precipitates. For 2205 duplex stainless steel, isothermal aging at 300~1000℃ often produces some precipitates, such as $M_{23}C_6$, σ, Cr_2N, R, etc., which the types of precipitates are different at different temperatures[26]. According to the manufacturing process, 2205 forgings can be delivered only after solution treatment, and then they are welded and assembled with the gas gathering manifold. Finally, they are subject to overall stress relief annealing treatment. Under this process condition, the precipitates are completely eliminated after solution treatment at 1080℃, but new precipitates are inevitably produced after stress relieving annealing at 580℃. The metallographic analysis result also shows that the precipitates of failed connecting pipe are mainly distributed in grain boundaries of α phase with very little content. Based on the previous literature[16], it can be inferred that the precipitates mainly come from 580℃ stress relief annealing treatment. After 1080℃ solution treatment is carried out for failed connecting pipe to eliminate the precipitates, average impact energy is increased from 11J to 90J, and impact toughness is greatly improved, which has reached the provisions of Order Technical Agreement. At the same time, the average corrosion rate decreases from 26mdd to 18mdd, the pitting breakdown potential increases from −0.15V to −0.02V, which indicates that the corrosion resistance is significantly improved. However, compared with comparison sample, there is still a significant gap in impact toughness and corrosion resistance, especially corrosion rate and pitting breakdown potential. In conclusion, although large-size non-metallic inclusions and fine precipitations have a significant impact on the impact toughness and pitting resistance, the former have a greater impact on the reduction of corrosion resistance and the latter has a greater impact on the reduction of impact toughness.

To sum up, the large-size non-metallic inclusions generated during smelting and casting and the fine precipitates generated by stress relieving annealing are the root failure causes of the connecting pipe. The rapid deterioration of the corrosion environment in the past three years has played a role in inducing and promoting the failure.

3.2 Failure mechanism

Pitting corrosion of stainless steel is a kind of corrosion form that the corrosion is concentrated in a small range of metal surface and goes deep into the metal interior or even perforation. The initiation of pitting corrosion refers to the corrosion event caused by the rupture of passive film on the stainless steel surface. The occurrence mechanism is mainly divided into two categories[27]. One is the induction mechanism of non-metallic inclusions, and the other is the rupture mechanism of passive film. The common point of both is that the passive film is discontinuous. The pitting corrosion initiation of failed connecting pipe should belong to the first mechanism. Because the large-size non-metallic inclusions existing in the failed connecting pipe are MnS and (Si, Mn, Ca, Al, Cr, Mg) oxides, which can significantly deteriorate the pitting corrosion resistance of 2205 duplex stainless steel and become the starting point of pitting corrosion initiation[18]. From results of 6% $FeCl_3$ corrosion test and potentiodynamic polarization curve test, it is also found that large-size non-metallic inclusions are more harmful to precipitate, which has a greater impact on the reduction of corrosion resistance and are more likely to be corroded. The mechanism of pitting and cracking is shown in Fig. 12. The non-metallic inclusions cause the passivation film to become incomplete (Fig. 12a). Compared with the stainless steel matrix, the large-size non-metallic inclusions have greater activity and are more vulnerable to corrosion[22,28]. With the continuous development of corrosion, large-size non-metallic inclusions such as MnS are preferentially corroded off (Fig. 12b) until they are completely dissolved and leak out of the fresh metal surface (Fig. 12c). At this time, the solution in the pitting hole is constantly acidified due to the formation of the occluded cell, which makes it difficult to passivate the fresh metal surface again and promotes the pitting hole to develop into a "small mouth and large cavity" pitting morphology (Fig. 12d). With the continuous increase of pitting, the stress concentration at the bottom of the pit increases, especially at the place with large-size non-metallic inclusions. Under the action of high-pressure gas, micro cracks are initiated at the interface of large-size non-metallic inclusions (as shown by the red short solid line in Fig. 12d). Results of intergranular corrosion test and 25% boiling magnesium chloride stress corrosion cracking test show that the stress corrosion cracking resistance is not significantly reduced. In addition, the fracture surface of failed sample has no stress corrosion cracking characteristics. So it is ruled out that the cracking is caused by chloride stress corrosion cracking. Due to the fine precipitates, the bonding force of grain boundaries in α phase and phase boundaries of α/γ is weakened. Therefore, micro cracks preferentially propagate along them, especially when the grain boundaries of α phase with more precipitates has worse adhesion, becoming the main expansion path (Fig. 12e and Fig. 12f). Besides, the interface bonding force between non-metallic inclusions and base metal is also weak, and the crack will continue to expand along the interface after extending to non-metallic inclusions (Fig. 12g). Finally, with the continuous development of pitting holes and the continuous impact of high-pressure gas, the cracks continue to develop in depth along grain boundaries of α phase, α/γ phase boundaries and non-metallic inclusions interface.

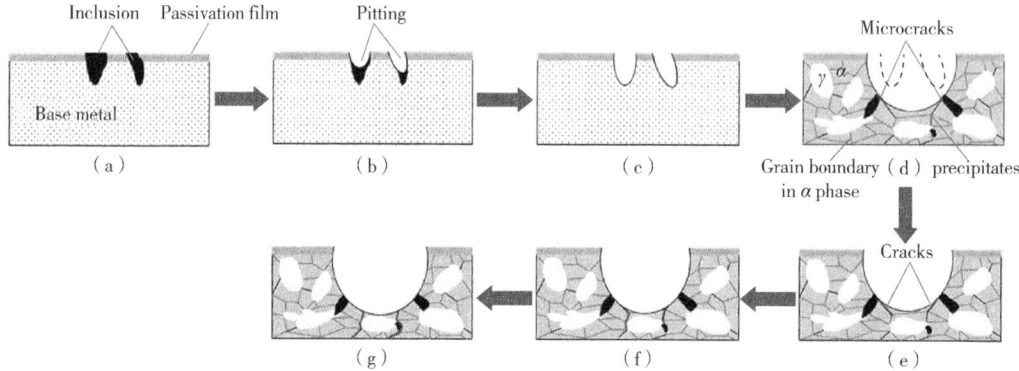

Fig. 12 Schematic diagram of pitting and cracking mechanism of the failed connecting pipe

4 Conclusions

The conclusions from the present work are given below.

(1) The large-size non-metallic inclusions composed of MnS and (Si, Mn, Ca, Al, Cr, Mg) oxides and fine precipitates didn't meet the requirements of Order Technical Agreement, which resulted in a serious reduction in the impact toughness and corrosion resistance of the failed connecting pipe.

(2) Pitting of the base metal and multi-source brittle cracking are the failure behaviors of the failed connecting pipe. The main reason for failure is the existence of large-size non-metallic inclusions produced in the smelting and casting process and the harmful precipitates produced by stress relief annealing. The rapid deterioration of the corrosion environment in the past three years has played a role in inducing and promoting the failure.

(3) Large-size non-metallic inclusions mainly reduce the pitting resistance, while fine precipitates mainly reduce the impact toughness. The initiation of pitting belongs to the non-metallic inclusion inducing mechanism. The formation of pits and the large non-metallic inclusions are the key factors for the generation of micro cracks, and the fine precipitates accelerates the crack growth.

Acknowledgements

This work was supported by theNational Natural Science Foundations of China (52071338).

References

[1] A. Shadravan, M. Amani, What petroleum engineers and geoscientists should know about high pressure high temperature wells environment, Energy Sci. Technol. 4 (2012) 36 - 60. https://doi.org/10.3968/j.est.1923847920120402.635.

[2] J. Wu, Duplex stainless steel, Metallurgical Industry Press, Beijing, 1999.

[3] E. Silva, V. Albuquerque, J. P. Leite, A. Varela, E. Moura, J. Tavares, Phase transformations evaluation on a UNS S31803 duplex stainless steel based on nondestructive testing, Mat. Sci. Eng. A 516(2009)126-130. https://doi.org/10.1016/j.msea.2009.03.004.

[4] L. Peguet, A. Gaugain, C. Dussart, B. Malki, B. Baroux, Statistical study of the critical pitting temperature of 2205 duplex stainless steel, Corros. Sci. 60(2012)280-283. https://doi.org/10.1016/j.corsci.2012.03.025.

[5] L. Y. Li, Z. X. Du, X. Z. Sheng, M. Zhao, L. X. Song, B. Han, X. D. Li, Comparative analysis of GTAW+SMAW and GTAW welded joints of duplex stainless steel 2205 pipe, Int. J. Pres. Ves. Pip. 199 (2022)104748. https://doi.org/10.1016/j.ijpvp.2022.104748.

[6] V. Muthupandi, P. Bala Srinivasan, S. K. Seshadri, S. Sundaresan, Effect of weld metal chemistry and heat input on the structure and properties of duplex stainless steel welds, Mater. Sci. Eng. A 358(2003) 9-16. https://doi.org/10.1016/S0921-5093(03)00077-7.

[7] Platt, J A; Guzman, A; Zuccari, A; Thornburg, D W; Rhodes, B F; Oshida, Y; Moore, B K, Corrosion behavior of 2205 duplex stainless steel, Am. J. Orthod. Dentofac. 112(1997)69-79. https://doi.org/10.1016/S0889-5406(97)70276-2.

[8] R. P. Zhang, Design and manufacture of pressure vessel about S32205+Q345R duplex stainless steel cladding plat, Pet. Chem. Equip. 45(2016)85-89. https://doi.org/10.3969/j.issn.1000-7466.2016.04.017.

[9] R. Kacar, M. Acarer, Microstructure property relationship in explosively weld duplex stainless steel, Mat. Sci. Eng. 36(2003)290-296. https://doi.org/10.1016/S0921-5093(03)00643-9.

[10] L. Y. Yu, H. Luo, X. Y. Li, Effect of annealing process on properties of explosive clad plates of 2205 duplex stainless steel-16MnR steel, T. Metal Heat Treat. 35(2010)50-53. https://doi.org/CNKI:SUN:JSRC.0.2010-03-013.

[11] S. O. Akinwamide, A. Venter, O. J. Akinribide, B. J. Babalola, A. Andrews, P. A. Olubambi, Residual stress impact on corrosion behaviour of hot and cold worked 2205 duplex stainless steel: A study by X-ray diffraction analysis, Eng. Fail. Anal. 131(2022)105913. https://doi.org/10.1016/j.engfailanal.2021.105913.

[12] D. Y. Chao, Microstructure and properties of nuclear power duplex stainless steel(Master's thesis), Yantai University, 2013. https://doi.org/DOI:10.7666/d.D432703.

[13] M. Pohl, O. Storz, T. Glogowski, Effect of intermetallic precipitations on the properties of duplex stainless steel, Mater. Charact. 58(2007)65-71. https://doi.org/10.1016/j.matchar.2006.03.015.

[14] J. K. Sahu, U. Krupp, R. N. Ghosh, H. J. Christ, Effect of 475℃ embrittlement on the mechanical properties of duplex stainless steel, Mat. Sci. Eng. A 508 (2009) 1-14. https://doi.org/10.1016/j.msea.2009.01.039.

[15] L. Pezzato, M. Lago, K. Brunelli, M. Breda, I. Calliari, Effect of the heat treatment on the corrosion resistance of duplex stainless steels, J. of Mat. Eng. Perform. 27 (2018) 3859-3868. https://doi.org/10.1007/s11665-018-3408-5.

[16] J. Singh, A. S. Shahi, Impact toughness, fatigue crack growth and corrosion behavior of thermally aged UNS S32205 duplex stainless steel, Trans. Indian Inst. Met. 72(2019)1497–1502. https://doi.org/10.1007/s12666-019-01574-7.

[17] X. F. Luo, W. L. Zhang, Z. L. Zhai, H. K. Zhang, X. Q. Zha, X. Y. Liu, Test method of intergranular corrosion for austenite-ferrite duplex stainless steel, Phys. Testing Chem. Anal. Part A (Phys. Anal.) 57 (2021)1–10. https://doi.org/10.11973/lhjy-wl202108001.

[18] D. Angeles-Herrera, A. Albiter-Hernandez, R. Cuamatzi-Melendez, A. de J. Morales-Ramirez, Influence of non-metallic inclusions on the fracture-toughness properties on the longitudinal welding of an API 5L steel pipeline, J. Test. Eval. 45(2017)687–694. https://doi.org/10.1520/JTE20150061.

[19] M. Djama, D. Saidi, A. Kadri, N. Kherrouba, B. Mehdi, S. Mathieu, T. Schweitzer, I. Brahim, Correlation between the pitting potential evolution and σ phase precipitation kinetics in the 2205 duplex stainless steel, J. Mat. Eng. Perform. 27(2018)3911–3919. https://doi.org/10.1007/s11665-018-3482-8.

[20] A. Fedorov, A. Zhitenev, V. Karasev, A. Alkhimenko, P. Kovalev, Development of a methodology for the quality management of duplex stainless steels, Mater. 15(2022)6008. https://doi.org/10.3390/ma15176008.

[21] S. M. Zhang, X. F. Zhang, X. P. Wang, Effects of aging temperature on microstructure and impact properties of 2205 duplex stainless steel welded joint, Mater. Mech. Eng. 39(2015)5. https://doi.org/JournalArticle/5b3b8415c095d70f007a45af.

[22] C. C. Liu, L. F. Zhang, Y. Ren, J. Zhang, Review on effect of non-metallic inclusions on pitting corrosion resistance of stainless steel, J. Iron Steel Res. 33(2021)1040–1051. https://doi.org/10.13228/j.boyuan.issn1001-0963.20210078.

[23] D. Bruch, D. Henes, P. Leibenguth, C. Holzapfel, Mechanical properties and corrosion resistance of duplex stainless steel forgings with large wall thicknesses, Metall. Ital. 6(2008)7–13. https://orcid.org/0000-0002-8811-5114.

[24] G. Camicia, M. Longin, P. Ferro, F. Bonollo, Chromium nitrides effects on low temperature impact toughness and durability of duplex stainless steels forgings, Metall. Ital. 9(2017)49–55. https://orcid.org/0000-0001-8008-1557.

[25] J. H. Park, Y. Kang, Inclusions in stainless steels-a review. Steel Res. Int. 88(2017)1700130. https://doi.org/10.1002/srin.201700130.

[26] Y. Liu, R. Q. Gao, X. R. Chen, Z. X. Qian, Analysis of influencing factors of σ phase precipitation in 2205 duplex stainless steel, T. Metal Heat Treat. 46(2021)4. https://doi.org/10.13251/j.issn.0254-6051.2021.03.036.

[27] A. Pardo, M. C. Merino, A. E. Coy, F. Viejo, R. Arrabal, E. Matykina, Pitting corrosion behaviour of austenitic stainless steels-combining effects of Mn and Mo additions. Corros. Sci. 50(2008)1796–1806. https://doi.org/10.1016/j.corsci.2008.04.005.

[28] J. C. Zheng, X. J. Hu, C. Pan, S. P. Fu, P. Lin, K. C. Chou, Effects of inclusions on the resistance to pitting corrosion of S32205 duplex stainless steel. Mat. Corros. 69(2018)572–579. https://doi.org/10.1002/maco.201709723.

本论文原发表于《International Journal of Pressure Vesse's and Piping》2023 年第 202 卷。

Failure Analysis of Casing in Shale Oil Wells under Multistage Fracturing Conditions

Mou Yisheng[1]   Zhao Han[2]   Cui Jian[3]   Wang Zhe[4]
Wei Fengqi[5]   Han Lihong[1]

(1. State key laboratory for performance and structure safety of petroleum tubular goods and equipment materials, CNPC Tubular Goods Research Institute;
2. CNPC Chuanqing Drilling Engineering Co. LTD;
3. Changqing Branch, CNPC China National Logging Corporation;
4. CNPC Tarim Oilfield Company;
5. CNPC Exploration and Production Company)

Abstract: During the multistage fracturing in shale oil and gas wells with tieback and liner, one of the major challenges is the wellbore temperature variation due to the high-rate fracturing. In such case, axial shrinkage trend of the casing string could be caused due to the sudden drop in temperature, but the actual axial length of casing string would not change due to the cement constraints. Therefore, this could lead to cementation damage between casing and cement due to excessive load from casing string. This wellbore seal out of control often leads to irreversible consequences, even well abandonment. In order to study the mechanism of casing deformation in shale oil and gas wells with tieback and liner quantitatively, in this paper, take LS1 well (a typical shale oil and gas well with tieback and liner, and casing deformation is caused) for example, the transient changes of temperature and pressure in the whole wellbore during multistage fracturing are studied. Moreover, the cementing strength test of interface between casing and cement are also tested. Then, the testing results are extended to modelling the finite element (FE) model of whole vertical section casing string with tieback and liner is established, and simulating its internal force changes under fracturing conditions with different stage fracturing. Meanwhile, the cashing deformation mechanism in LS1 well is analyzed and studied in detail. Our simulation results indicated that failure process and mechanism of cementation between casing and cement in shale oil and gas wells with tieback and liner. Our work can provide detailed theoretical reference and basis for field application.

Keywords: Casing deformation; Shale oil and gas wells; Tieback; Liner; Fracturing

---

Corresponding author: Mou Yisheng, mouys@cnpc.com.cn; Han Lihong, hanlihong@cnpc.com.cn.

1 Introduction

In light of the depletion of traditional oil and gas reserves, as well as the sharp rise in global energy demand, unconventional reservoirs characterized by their extremely low permeability have garnered increasing interest within the industry for exploration and production endeavors[1-2]. The utilization of the extensive multi-stage fracturing technique has proven to be an efficacious approach in stimulating these reservoirs. Nonetheless, casing integrity issues induced by the fracturing process, have been progressively exacerbating[3].

Recently, there has been significant research on casing failure during fracturing, with a particular emphasis on utilizing the finite element method to study shear deformation. However, a unified theoretical understanding of the underlying mechanism of this issue has not yet been achieved. In 2016, Liu et al.[4] established a calculation model for casing stress under local load to study casing damage in Changning-Weiyuan shale gas wells in Sichuan Province. They concluded that temperature stress during fracturing and local load caused by periodic changes in casing pressure led to casing deformation in shale gas wells. Yan et al.[5] also investigated the effect of temperature during the fracturing process and found that the temperature inside the casing drops significantly during high-flow hydraulic fracturing, causing the shrinkage of the retained liquid in the cement ring. This leads to a decrease in pressure inside the cement sheath, making it unable to supplement the pore pressure in a short time, resulting in a non-uniform external support state of the casing. Simultaneously, casing experiences high internal pressure during hydraulic fracturing. The combination of uneven external support and high internal pressure causes casing deformation. Yu et al.[2,6-8] proposed a microseismic data inversion method, suggesting that asymmetric fracturing creates a stress gap effect around the wellbore trajectory, leading to lateral extrusion of borehole rock and casing deformation. After repeated fracturing, various effects, such as a decrease in rock properties and an increase in geostress non-uniformity, result in radial elliptic deformation and axial S-shaped deformation of the casing. This represents a new theoretical understanding. In 2017, Liu et al.[9] developed a mechanical model to study the behavior of casing under local loads, and they verified its accuracy using numerical simulation, Nester method, and field data. They discovered that during fracturing, the fracturing fluid flows into the wellbore annulus, becomes sealed and heated, and then enters the original reservoir, causing rock sliding. The local load applied to the casing is a crucial factor that leads to casing damage. With the extensive research conducted by scholars, the study of casing damage in fracturing operations has become more diverse, and many researchers have achieved unique insights. In 2019, Li et al.[10] disputed that the uneven load on the casing in the fracturing process is not the primary cause of casing deformation, as the casing deformation in the fracturing process is not due to casing yield. Instead, the deformation is mainly due to a substantial reduction in casing diameter, which is caused by the activation of existing fractures or faults in the fracturing process. The critical value of casing diameter reduction is a more precise criterion for casing failure than the critical value of casing strength yield. Restrepo et al.[11] investigated wellbore integrity during hydraulic fracturing and concluded that poorly concentrated cement creates drilling fluid voids during fracturing. They proposed a unique method for capturing the additional stresses generated by these voids, which can lead to casing deformation and failure. Additionally, the low-temperature fracturing fluid used during fracturing can cause cavity shrinkage, reduce pressure, and generate local non-uniform loads, all of which can lead to casing deformation. These findings are consistent with Yan et al.'s views in 2016. Li et al.[12] analyzed the Roewei shale gas well and found that the interaction between fracturing fluid and clay during the fracturing process can cause shale expansion. The difference between injection pressure and formation pressure can also lead to stress concentration at the lithologic interface. Furthermore, the accumulation of stress induced by hydraulic fractures increases the maximum principal stress near the wellbore. If this stress exceeds the compressive strength of the casing, it can lead to casing deformation. In 2021, Yang et al.[13] conducted a classification study on casing deformation or damage during fracturing in the Sichuan-Chongqing region of China. They found that collapse failure was more significant than shear failure, and that formation displacement caused by fracturing resulted in casing shear failure. The interaction between the injection rate of fracturing fluid and the wellbore environment causes external load imbalance, leading to casing collapse. They recommended a strain-based casing design instead of the current maximum stress-based casing design for hydraulic fracturing wells in Sichuan and Chongqing areas of China. In 2022, Zorica et al.[14] argued that the increase of axial force caused by active fractures during the hydraulic fracturing stage is the main cause of casing deformation. They found that the shear force generated by migration fractures on the casing reduces the critical force of casing deformation, thereby amplifying the effect of axial force on casing deformation. Improving cementing quality is an effective measure to reduce excessive casing deformation. In February 2023, Zhang et al.[15] used MIT-24 technology to analyze the planning surface of deformed casing in the Weiyuan area. They combined this with the relationship between casing deformation and engineering parameters and found that hydraulic fracturing caused an increase in fluid pressure in intersecting faults and large fractures, resulting in shear slip and asymmetric compression of casing on both sides of faults and large fractures, leading to casing deformation. Therefore, they proposed the integrated technology of temporary plugging fault fissure, multicluster perforation, and fracturing fluid flow-back. After a field test in the Weiyuan area, the casing deformation rate decreased from 54% to 9.1%.

Previous work thus clearly demonstrated the characteristic and mechanism of casing deformation, however the whole well section string model with tieback and liner has not excited in previous research. Furthermore, such technical guidance and supporting are urgently needed in the oil field. Therefore, the FE (finite element) model of tieback-liner-cement ring-rock formation in the well is established in this work for calculation and analysis, so as to guiding the field that using tieback and liner to reduce the occurrence of casing deformation and damage during fracturing.

2 Damage and deformation

2.1 Wellbore configuration

A typical well integrity failure occurred in LS1 well with tieback and liner during fracturing operations. Fig. 1 presents the wellbore configuration of LS1 well and its introduction of cement bond logging results. LS1 well is a sidetrack horizontal well, its measure depth is 4744m and vertical depth is 3274m, its configuration is three hole-ins well. Table 1 statistics the parameters of casing string in LS1 well, it's worth noting that the liner (ϕ139.7mm×10.54mm-P110, depth: 2464~4744m) and tieback (ϕ139.7mm×10.54mm-P110, depth: 0~2465m) are used as production casing. The window depth is 2653m and the window is drilled in intermediate casing (ϕ244.5mm×11.99mm-P110, depth: 0~2748m), the original vertical well section below the window point is filled with cement, geothermal gradient is 3.0℃/100m.

On the other hand, it can be found that the cementing quality is significantly different along tieback according to the cement bond logging results: (1) well depth 0~1626m continuous segment: high cementing quality along whole segment is continuous; (2) well depth 1626~2464m intermittent segment: high cementing quality along whole segment is intermittent, and the longest length of cement segment with high cementing quality is 12.5m and the shortest length is 1.5m, meanwhile, the solid filling rate in the empty space (cement segment with low cementing quality) is less than 25%.

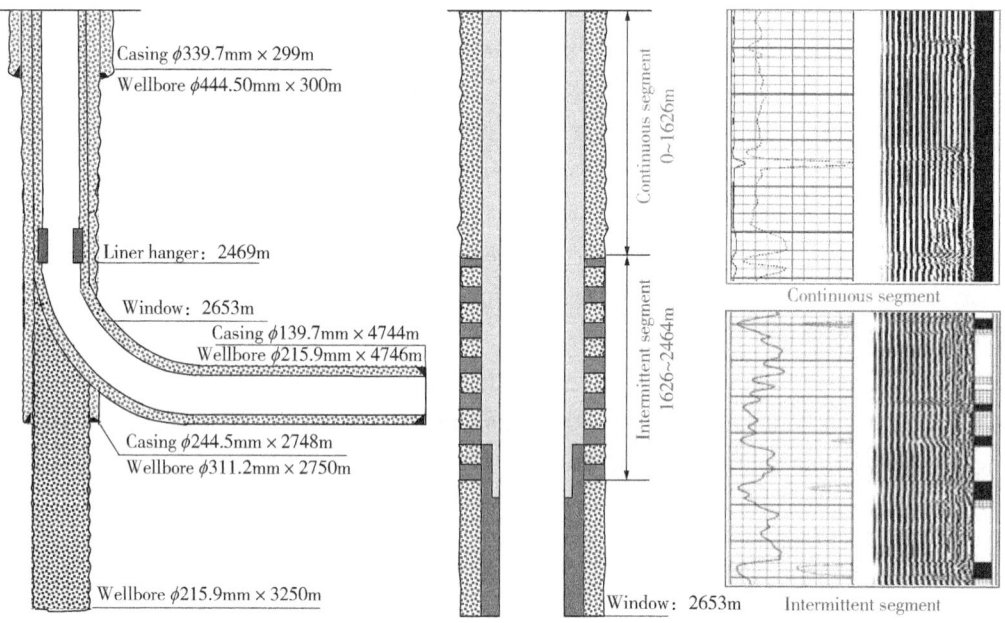

Fig. 1 Wellbore configuration of LS1 well and its introduction of cement bond logging results

Table 1 Parameters of casing string in LS1 well

Type		Depth/m	OD/mm	Thickness/mm	Steel grade	Thread type
Surface casing		0~299	339.7	9.65	J55	BC
Intermediate casing		0~2748	244.5	11.99	P110	BC
Production casing	Tieback	0~2469	139.7	10.54	P110	TS1
	Liner	2469~4744	139.7	10.54	P110	TS1

2.2 Failure introduction

The vertical depth of the reservoir is 3271~3276m and the lithology is shale (the resistance is 232.5~3279.9Ω·m, the porosity is 2.4%~2.9%) in LS1 well. Meanwhile, the number of fracturing stages is 23 (each stage has four fracturing clusters), ranging from 3200m to 4744m in the original fracturing design. Each stage is designed with a flow rate of 14 ~ 20m³/min and the duration is 2h, the sand by volume ratio is 20% and its density is 1.4g/cm³.

After the cementing is complete, the first 12 stages of the operation are successfully fractured, however, there is a sudden drop in wellhead fracturing pressure during No. 13 fracturing operation (well depth: 3998.5 ~ 4018.5m). After a systematic well control operation, a pressure test is conducted, then, the production casing is found to be unable to hold pressure and the wellbore flow path has failed. And then, wellbore inspection work is carried out. Fig. 2 illustrates the 24 arm logging results of producing casing inner diameter between measure depth 2463 ~ 2464.8m. It can be found that the body of the production casing has broken and separated by about 1.2m at the mentioned location. In this 1.2m section, the measured diameter reaches 170mm, much larger than the production casing inner diameter (118.62mm). As a result of this incident, the well was shut in for a long period of time (185 days so far), resulting in a significant financial losses.

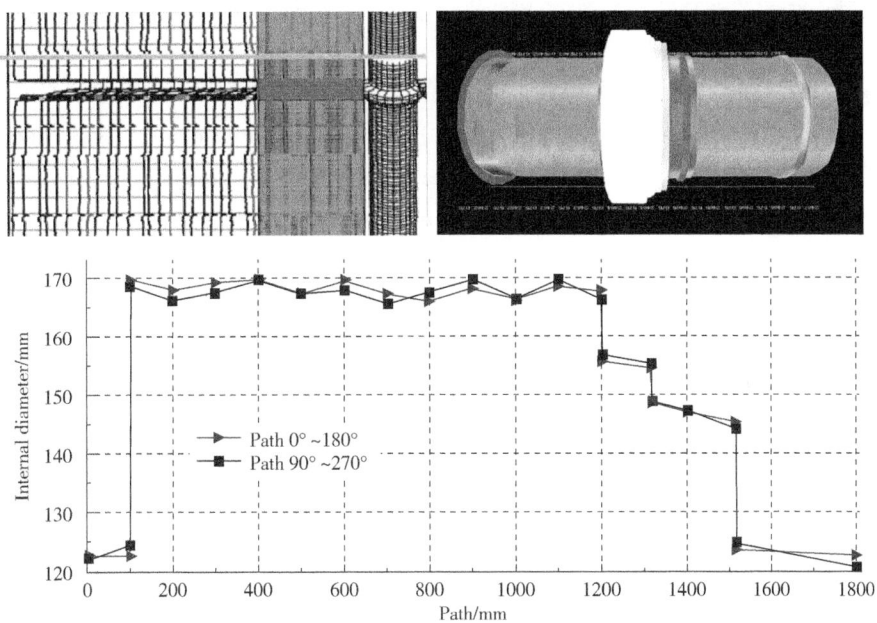

Fig. 2 The 24 arm logging results of producing casing inner diameter between measure depth 2463 ~ 2464.8m

3 Failure factor analysis

3.1 Material

To qualitatively and quantitatively characterize the properties of alloy steel casing pipe in LS1 well, several casing samples from the same manufacturer's production line is tested to verify the quality of the alloy steel material in this work. The studied alloy steel casing pipe is manufactured according to 110ksi (758MPa) nominal yield strength. Hence, a series of experimental measurements of the P110 alloy steel are performed to characterize metallographic structure, mechanical properties, etc. The casing material mentioned in this work is P110. A HCS 140 high frequency infrared ray carbon sulfur analyzer (Shanghai Dekai Instruments Co., Shanghai, China) is used to measure its chemical composition: (%, mass fraction) C-0.28, Si-0.23, Mn-1.32, P-0.01, S-0.011, Cr-0.0015, Mo-0.003, W-0.028, Ni-0.01, V-0.002, Fe-Balance. Fig. 3 presents the optical micrographs revealing the typical microstructures of the base material, the tempered sorbite is a remarkable microstructure feature. The chemical element composition of the casing material is in accordance with ASTM A732-1998, and failure caused by inconsistent chemical composition is ruled out.

Fig. 3　Optical micrographs revealing the typical microstructures of the base material

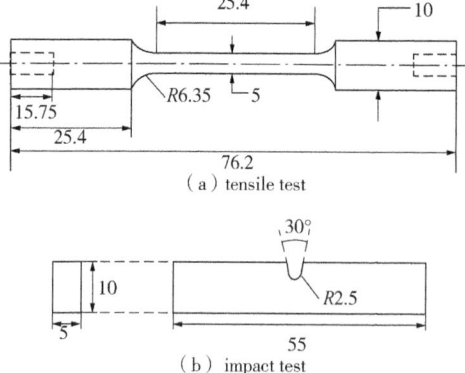

Fig. 4　Schematic of the specimens used for different tests in this study

In order to identify other possible factors, three types of specimens are prepared and characterized via a tensile test and impact test to the mechanical properties of P110 alloy steel. The dimensions of all the types of specimens are shown in Fig. 4. The specimens are ground using an 800-grit SiC paper, followed by mechanical polishing with a suspension of SiO_2 particles of 50nm diameter. Finally, anhydrous ethanol was used to clean the specimen. The test results are also extended to the subsequent FE model.

Fig. 5 illustrates the stress-strain curves of the P110 alloy steel under different temperatures (20℃ and 80℃). According to the key parameters in Table 2, The yield strength, tensile strength, and modulus of elasticity of the P110 alloy steel at 20℃ are greater than those at 80℃ (temperature at well depth 2464m, according to 3.0℃/100m geothermal gradient). Obviously, the percentage elongation at fracture of the P110 alloy steel at 20℃ is less than that at 80℃. It can be observed that there is a slight change in the mechanical properties of the P110 alloy steel under different temperatures (20℃ and 80℃). From the perspective of material mechanics, the temperature increase could result in intense atomic movement, and the corresponding macroscopic manifestation is that the material becomes more prone to plastic deformation, and resulting in a reduction in the strength of the material. However, both the tensile test and service temperature of the tieback casing materials are within 80℃. Temperature thus had little influence on the strength and was not the main factor that resulted in the wellbore failure.

Fig. 6 presents the load/energy versus displacement curves of P110 alloy steel. The related parameters are presented in Table 3. It can be seen that the P110 alloy steel remains reliably tough over a series of temperature changes in the field according to the results of the impact tests. Meanwhile, the mechanical strength and toughness of the casing alloy steel are not obviously weakened in the service environment based on tensile and impact tests. In general, the microstructures and macroscopic mechanical properties of the casing material used in the field could meet the requirements of service design.

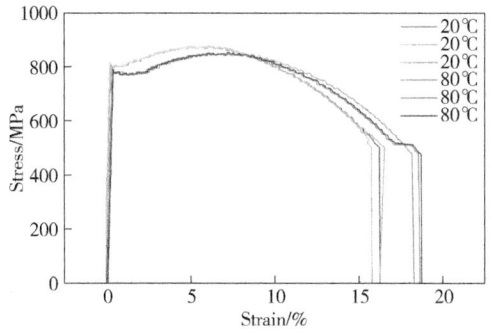

Fig. 5 Stress-strain curves of the P110 alloy steel under different temperatures

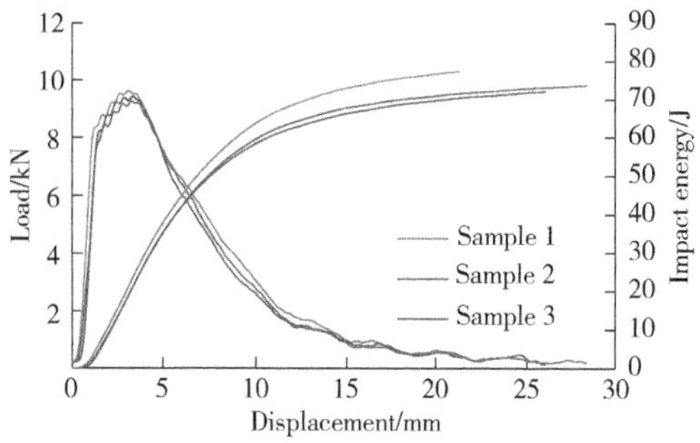

Fig. 6 Load/energy versus displacement curves of P110 alloy steel

Table 2 Static thermal parameters of various materials in wellbore

Property	At 20℃ Average±Standard deviation ($n=3$)	At 80℃ Average±Standard deviation ($n=3$)
Yield strength $Rp_{0.2}$/MPa	786.54±2.05	746.49±2.12
Tensile strength/MPa	901.25±2.15	867.46±3.16
Young's modulus/GPa	208±0.0001	206±0.0001
Elongation δ/%	15.67±1.42	18.48±1.09

Table 3 Dynamic mechanical properties obtained from Charpy impact tests

Property	Average±Standard deviation ($n=3$)	Property	Average±Standard deviation ($n=3$)
Impact energy/J	75.57±3.51	Total displacement/mm	25.95±3.47
Crack initiation energy/J	21.32±2.15	Maximum load/kN	9.54±2.84
Crack propagation energy/J	54.26±3.85		

3.2 Alternation of temperature and pressure

It is well known that casing fracturing is commonly used in shale oil and gas wells, therefore, it has a higher injection rate. The casing string temperature could drop significantly in a short period during fracturing, and this change is the main reason for the accumulation of axial deform and internal force in casing string. Fig. 7 shows the physical model of fracturing fluid and heat transfer in the shale borehole. It can be found that fracturing fluid enters the casing string at the wellhead with a temperature (T_{FRin}) and flows down into the reservoir with a temperature (T_{FRout}). The rate of heat convection between the fracturing fluid and the inner wall of the casing could affect fluid temperature (T_{FR}) significantly. Heat generated is continuously carried out of the inside the casing, and a temperature decrease surrounding the borehole is caused. Finally, the wellbore temperature slowly rises after the fracturing operation is complete, because heat generated in the far distance of the borehole diffuses to the wellbore by heat conduction due to the effect of temperature difference.

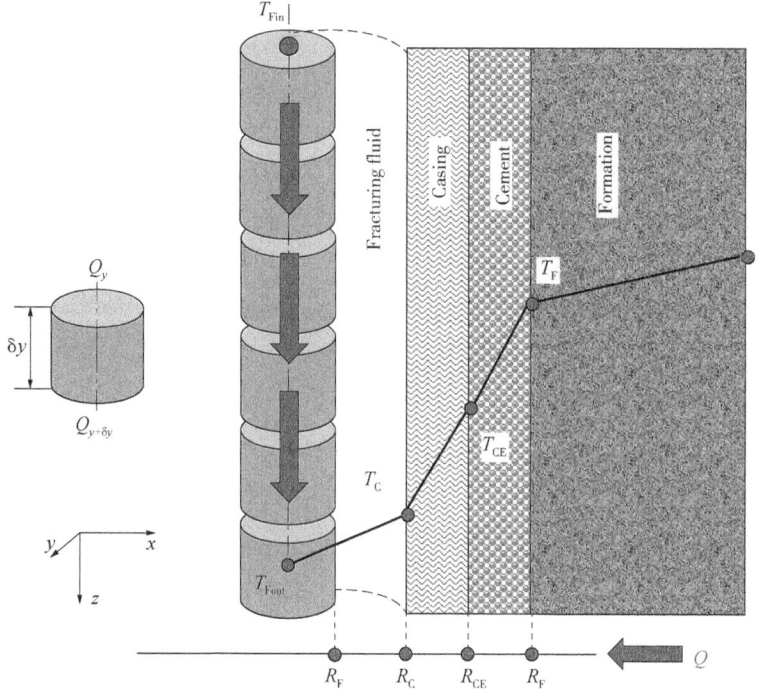

Fig. 7　The physical model of fracturing fluid and heat transfer in the shale borehole

Fracturing fluid flowing in casing could be divided into several units (the Z axis is the axial direction of wellbore), and the heat of a unit consists of four parts: (1) the change of internal energy of fracturing fluid, (2) heat generated by convection heat transfer between fracturing fluid and casing, (3) the heat carried by the down-flowing fracturing fluid, (4) thermal friction between fracturing fluid and casing[16-17].

The internal energy E_{FRI} of fracturing fluid in per unit time dt can be presented as:

$$dE_{FRI} = \pi R_C \delta z \rho_{FR} c_{FR} \frac{\partial T_{FR}}{\partial t} dt \tag{1}$$

The heat generated by convection heat transfer Q_{HT} between fracturing fluid and casing per unit time dt can be obtained as follows:

$$dQ_{HT} = 2\pi R_C h_C (T_C - T_{FR}) \delta z dt \tag{2}$$

The heat Q_{HC} carried by the down-flowing fracturing fluid per unit time dt can be expressed as:

$$dQ_{HC} = Q_z - Q_{z+\delta z} = N_q \rho_{FR} c_{FR} | T_{FRy} - T_{FR(z+\delta z)} | dt \tag{3}$$

The work from thermal friction W_F between fracturing fluid and casing per unit time dt can be illustrated as:

$$dW_F = Q_{FR} \delta z dt \tag{4}$$

The equilibrium equations of fracturing fluid can be expressed as:

$$N_q \rho_{FR} c_{FR} |T_{FRy} - T_{FR(z+\delta z)}| dt + 2\pi R_C h_C (T_C - T_{FR}) \delta z dt + Q_{FR} \delta z dt = \pi R_C^2 \delta z \rho_{FR} c_{FR} \frac{\partial T_{FR}}{\partial t} dt \quad (5)$$

Casing string can be divided into several units, the heat of a unit consists of four parts: (1) axial heat conduction of casing string, (2) convection heat transfer between the casing inner wall and fracturing fluid, (3) heat conduction between the casing outer wall and cement ring, (4) the change of casing unit internal energy.

The axial heat conduction Q_{aC} of a casing unit per unit time dt can be presented as:

$$dQ_{aC} = \mu_C \left(\frac{\partial T_{C(z+\delta z)}}{\partial z} - \frac{\partial T_{Cz}}{\partial z} \right) \pi (R_{CE}^2 - R_C^2) \delta z dt = \mu_C \frac{\partial^2 T_C}{\partial z^2} \pi (R_{CE}^2 - R_C^2) \delta z dt \quad (6)$$

The equilibrium equations including the heat conduction Q_{rC} of the casing outer wall−cement ring and the convection heat transfer of the casing inner wall−fracturing fluid per unit time dt can be expressed as:

$$dQ_{rC} = 2\pi \mu_C (T_{CE} - T_C) \delta z dt - 2\pi R_{CE} h_C (T_C - T_{FR}) \delta z dt \quad (7)$$

The change in internal energy E_{CI} of a casing unit per unit time dt can be illustrated as:

$$dE_{CI} = \rho_C c_C \frac{\partial T_C}{\partial t} \cdot \pi (R_{CE}^2 - R_C^2) \delta z dt \quad (8)$$

The equilibrium equations of casing string can be calculated as follows:

$$\mu_C \frac{\partial^2 T_C}{\partial z^2} + \frac{2R_C h_C (T_C - T_{FR})}{R_{CE}^2 - R_C^2} - \frac{2\mu_C (T_{CE} - T_C)}{R_{CE}^2 - R_C^2} = \rho_C c_C \frac{\partial T_C}{\partial t} \quad (9)$$

The heat transfer relation of casing wall−cement ring (inside radius of casing R_C, temperature T_C) − cement (R_{CE}, T_{CE}) − formation (R_F, T_F) all could be regarded as the heat conduction between multi-layer cylinder wall.

3.3 Wellbore temperature and pressure during fracturing

The casing temperature and internal pressure in casing could change urgently under different stage fracturing conditions. The whole section casing temperature could decrease and the internal pressure in casing could increase under each stage. Meanwhile, casing temperatures could rise toward formation temperature (the reservoir temperature is 165℃) and internal pressure drops toward formation pressure (the wellhead pressure is 0MPa, the reservoir pressure is 58MPa) as the bridge plug is run.

According to the heat transfer theory in section 3, and static thermal parameters of various materials in wellbore are presents in Table 4. Fig. 8a presents the temperature of production casing under different stage fracturing conditions, it can be seen that casing temperature could drop from reservoir temperature significantly during No. 1 fracturing (maximum temperature difference between casing temperature and reservoir temperature reaches 95℃), then, casing temperature could increase during No. 1 bridge plug. Subsequent fracturing operations are consistent with the above rules of temperature change. Simultaneous, according to the dynamic friction resistance theory[16], Fig. 8b presents the internal pressure of production casing under different stage fracturing conditions, it can be seen that internal pressure could increase from reservoir pressure (the wellhead pressure is 0MPa, the reservoir pressure is 58MPa) significantly during No. 1 fracturing (the wellhead pressure is 64MPa, the reservoir pressure is 66MPa). Subsequent fracturing operations are consistent with the above rules of internal pressure change. The calculation of fracturing pressure takes into account wellhead construction pressure, fracturing fluid column pressure and frictional pressure drop (ground test: friction pressure drop is 760kPa in casing with 76mm inner diameter at 4.5m³/min displacement).

Table 4 Static thermal parameters of various materials in wellbore

Material	Density/(kg/m³)	Specific heat/ [J/(kg·℃)]	Thermal conductivity/ [J/(m·℃)]	Coefficient of expansion
Fracturing fluid	1200	4178	0.84	—
Casing/tieback/liner	7849	460	51.9	1.25×10^{-5}
Cement	3000	840	2.1	3.68×10^{-4}
Rock	2650	765	2.5	6.85×10^{-4}

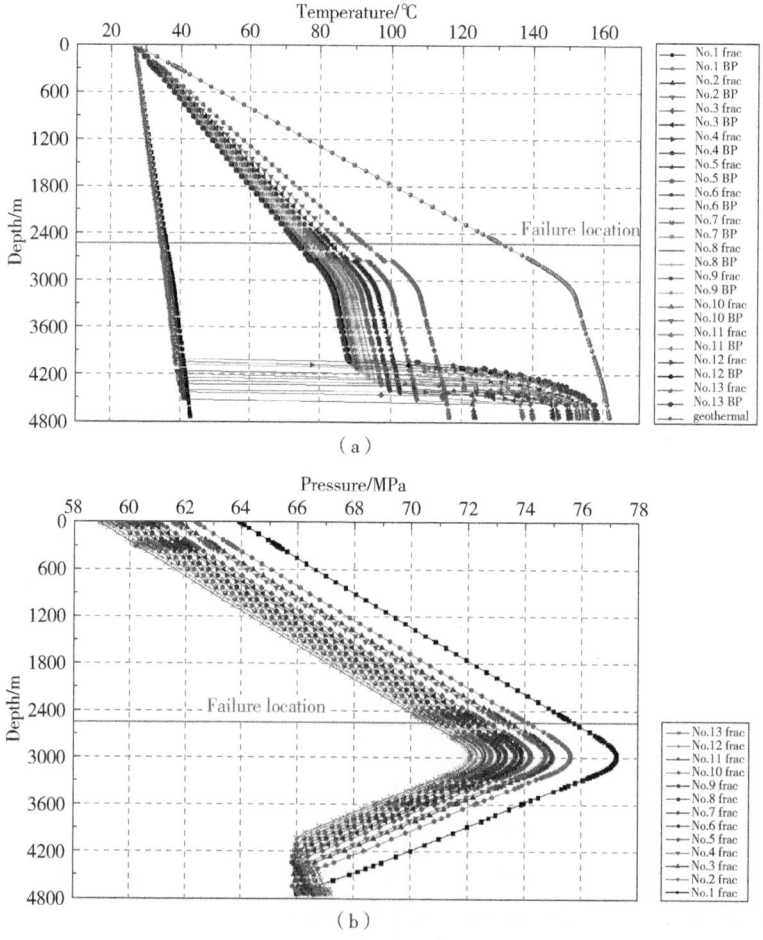

Fig. 8 The temperature and internal pressure of production casing under different stage fracturing conditions

The casing temperature and internal pressure in casing could change urgently under different stage fracturing conditions. The whole section casing temperature could decrease and the internal pressure in casing could increase under each stage. Meanwhile, casing temperatures could rise toward formation temperature (the reservoir temperature is 165℃) and internal pressure drops toward formation pressure (the wellhead pressure is 0MPa, the reservoir pressure is 58MPa) as the bridge plug is run. Fig. 9 describes the service load change at the casing deformation position (measure depth 2464m): (a) casing temperature, (b) internal pressure. It can be seen that the temperature of the casing at the failure point (2464m) fluctuates significantly.

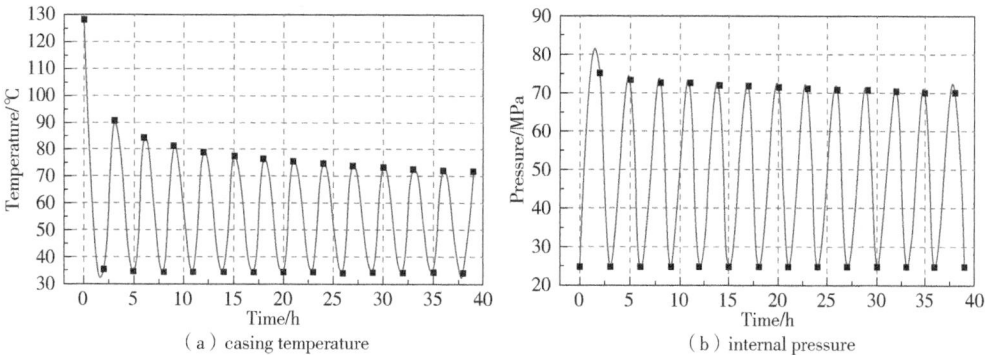

Fig. 9   The service load change at the casing deformation position (measure depth 2464m)

Based on the geothermal gradient, the initial casing temperature is 130℃. During the No. 1 fracturing stage, the temperature drops to 35℃ and ΔT (the difference between maximum temperature and minimum temperature) reaches 95℃. During the No. 1 bridge plug, the casing temperature increases to 91℃. During subsequent multistage fracturing operations, ΔT of casing temperatures between fracturing stage and bridge plug process varies from 40℃ to 45℃.

According to the formation pressure gradient, the initial casing inner pressure is 24MPa. During the No. 1 fracturing stage, the inner pressure rises to 76MPa and Δp (the difference between maximum inner pressure and minimum inner pressure) reaches to 52MPa. During the No. 1 bridge plug, the casing temperature decreases to 24MPa. During subsequent multistage fracturing operations, Δp of casing inner pressure between fracturing stage and bridge plug process varies from 45MPa to 50MPa.

4   Cementing strength test of interface between casing and cement

4.1   Preparation and procedure

In order to quantify the strength and the failure load of cementation, the test of cementing strength of casing-cement ring is carried out. The main equipment used is the wellbore integrity combined load system [capacity: (1) specimen size: ≤9⅝in; (2) non-uniform capacity: 1200t; (3) shear capacity: 600t; (4) axial tensile/compress load: 1000t; (5) internal pressure: 200MPa; (6) bending: <20°/30m; (7) temperature: 200~1000℃; (8) torque≥40000N·m; (9) temperature and pressure alternate load], as shown in Fig. 10. The test equipment can realize the service performance simulation of casing-cement ring in non-uniform collapse, shear, axial tension and compression, bending and internal pressure, high temperature and other combined load conditions. Meanwhile, the specimen involved in this paper is a casing–cement system, which consists of production casing [OD (outer diameter)139.7mm+wall thickness 10.54mm+110ksi], cement ring (OD 215.9mm+ID 139.7mm, G-grade cement), protective casing (OD 244.5mm+ID 215.9mm+110ksi), with length of 1.3m. The cement length of the specimen is 0.3m, 0.5m, 0.8m and 1m respectively.

(a) wellbore integrity combined load system　　　　　(b) casing–cement specimen

Fig. 10　Physical photograph

The production process and loading process of the casing–cement ring specimen are presented in Fig. 11. The production process of the specimen: (1) the G-grade cement with non-Newtonian fluid state is placed between the production casing and the protective casing, and set for 48h, and the upper end of the production casing is exposed 20cm, the lower end of the production casing is exposed 10cm, (2) the rigid dam-board is placed under the cement ring–protective casing, (3) the axial downward push force from the wellbore integrity combined load system is applied to the upper end of the production casing until the cementation of production casing–cement is broken. When the cementation is broken, the production casing moves downward axially, meanwhile the push force value and displacement could be recorded by the wellbore integrity combined load system.

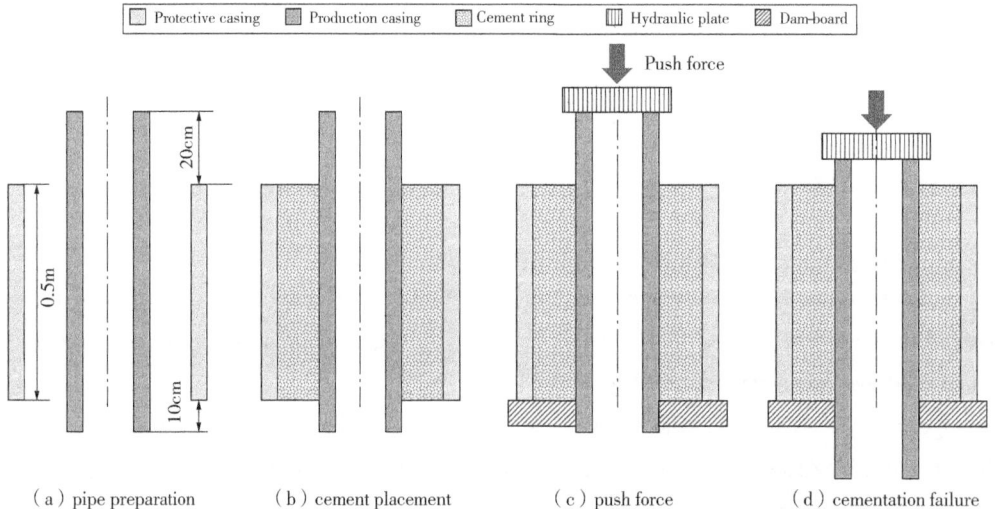

(a) pipe preparation　　(b) cement placement　　(c) push force　　(d) cementation failure

Fig. 11　The production process and loading process of the casing–cement ring specimen

4.2 Experimental result

Fig. 12a illustrates the load from push force vs. dam-board displacement (0.5m cement length specimen), it can be seen that the law of cementation damage is divided into three stages: (1) the elastic stage: taking the specimen with 0.5m cement length as an example, when the dam-board displacement is 0~1.25mm, the push force increases rapidly and linearly to 101kN, the casing-cement cementation is intact at this stage, (2) the plastic stage: when the displacement of the damboard exceeds 1.25mm, the push force is almost constant with the increase of displacement, the casing-cement cementation begins to break down at this stage, meanwhile, when the displacement of the dam-board is increased to 2.5mm, the push force reaches its maximum (110kN) and the cementation begins to tear, and this force is defined as the cementation strength between casing and cement, (3) the failure stage: when the displacement of the dam-board exceeds 3.8mm, the push force decreases rapidly with the increase of displacement, the casing-cement cementation is completely broken at this stage. Meanwhile, Fig. 12b illustrates the cementation strength of casing-cement vs. cement length. It can be seen that the cementation strength increases linearly as the length increases. Additionally, the average cementing strength of specimens with different cement lengths (0.3m, 0.5m, 0.8m, 1m, and there are three duplicate specimens of each length) reaches 72kN, 105kN, 189kN, 215kN, respectively. In addition, the mathematical relationship between cementation strength (S_c) and cement length (L_c) can be fitted as: $S_c = 222.87 L_c + 0.998$. Obviously, the longer the casing-cement cementing length, the stronger the cementing force.

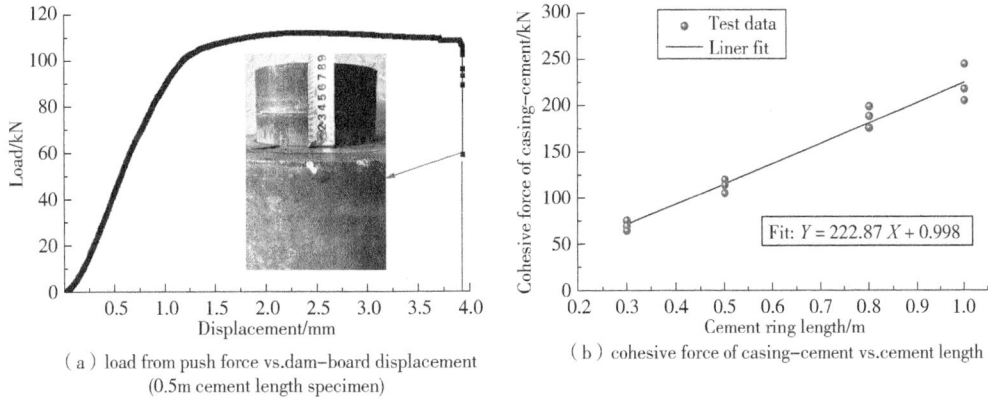

(a) load from push force vs. dam-board displacement
(0.5m cement length specimen)

(b) cohesive force of casing-cement vs. cement length

Fig. 12 Test results of casing-cement ring cementation

To further quantify casing-cement cementation strength, FE model with cohesive element is established to obtain the constitutive relation of the element characterizing the strength. FE model with cohesive element (The size of the model is consistent with that of the test specimen in Fig. 11) consists of production casing (OD 139.7mm+wall thickness 10.54mm+110ksi), cement ring (OD 215.9mm+ID 139.7mm, G-grade cement), protective casing (OD 244.5mm+ID 215.9mm+110ksi). Meanwhile, the cohesive elements are arranged at the interface between the production casing and the cement ring.

Fig. 13 compares the experimental test results of cement cementation strength with the FE results. The constitutive relation of cementation presents four stages: (1) when casing displacement varies from 0mm to 1.5mm, the axial load applied to the casing rapidly increases from 0kN to 105kN, at this time, casing and cement sheath are still in the state of cementation, (2) then, when the casing displacement is between 1.5mm and 2.5mm, the axial load increases slowly, the casing and cement sheath gradually peel off, (3) when the casing displacement exceeds 2.5mm, the axial load decreases slowly, (4) and the axial load decreases abruptly when the casing displacement reaches 4.0mm, the casing is completely separated from the cement. It is well known that the cementation strength of the cohesive element is characterized by the constitutive relationship between the cohesive force and the normal displacement of two opposites. Therefore, by adjusting the constitutive model of the cohesive element between the casing and the cement ring, the results of the FE model are basically consistent with the experimental results. And, the quantified strength is used to compensate for the establishment of FE models of wellbore.

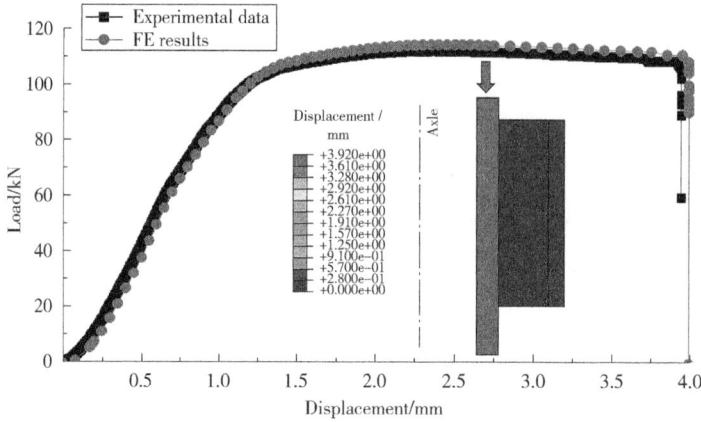

Fig. 13 The comparison between experimental data and FE results

5 Numerical simulation study

5.1 Modeling

To further analyze the quantitative deformation of casing quantitatively, the coupling FE model of tieback-liner-cement ring-rock formation in vertical section LS1 well is established, and Fig. 14 illustrates the coupling FE model. The model consists of tieback (ϕ139.7mm×10.54mm-P110, depth: 0~2465m), liner (ϕ139.7mm×10.54mm-P110, depth: 2464~2653m), cement ring (the cementing quality: (1) well depth 0~1626m continuous segment: high cementing quality along whole segment is continuous; (2) well depth 1626~2464m intermittent segment: high cementing quality along whole segment is intermittent, and the longest length of cement segment with high cementing quality is 12.5m and the shortest length is 1.5m, meanwhile, the solid filling rate in the empty space (cement segment with low cementing quality) is less than 25% and rock formation (Inner diameter 215.9mm, outer diameter 2m).

The material properties of the different components involved in the model are shown in Table 5. The cohesive element is used at the interface between the tieback and the cement ring. The cohesion force of the cohesive element is given according to the material constitutive relation in Fig. 13. The hexahedral scanning element is used to divide the 3D model, and the elements in tieback pad and rubber ring are secondary encrypted.

Load application process: (1) the initial temperature of the whole model is assigned according to the geothermal gradient (the wellhead temperature is 25℃, the reservoir temperature is 165℃), the internal pressure is assigned according to the completion fluid column pressure (the wellhead pressure is 0MPa, the reservoir pressure is 58MPa) and the external pressure is assigned according to the formation brine column pressure (the wellhead pressure is 0MPa, the reservoir pressure is 34MPa), (2) based on the transient heat transfer theory, the temperature (Fig. 8a) and internal pressure (Fig. 8b) of tieback are assigned according to the actual operation period, meanwhile, the external pressure is assigned according to the formation brine column pressure (the wellhead pressure is 0MPa, the reservoir pressure is 34MPa). Significantly, it is expected that the tieback axial force variation and cementation damage between tieback and cement could be obtained by numerical simulation.

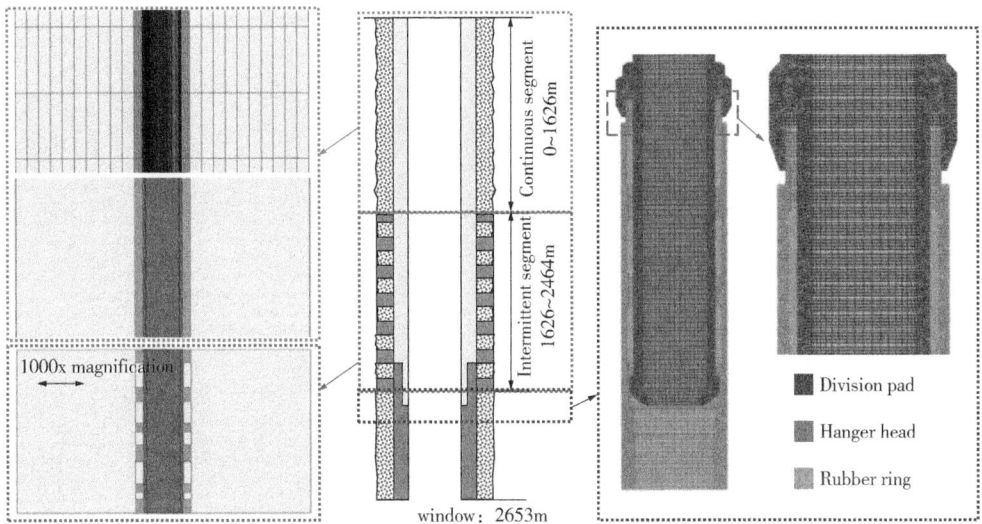

Fig. 14 The coupling FE model of tieback-liner-cement ring-rock formation in in vertical section LS1 well

Table 5 Static mechanical properties of various materials in wellbore

Material	Modulus of elasticity/GPa	Poisson's ratio	Yield strength σ_y, $Rp_{0.2}$/MPa	Ultimate strength σ_u/MPa	Elongation δ/%
Casing/tieback/liner	In Table 2	0.30	In Table 2	In Table 2	In Table 2
Cement	20.0	0.25	—	—	—
Rock	43.5	0.25	—	—	—

5.2 Simulation results

Through numerical simulation, the axial force variation of casing string is quantified during fracturing. Fig. 15a illustrates the change of axial force of tieback vs. measure depth. It can be seen that the axial force of the tieback changes dramatically due to the temperature and pressure changes during multistage fracturing. For example, the casing temperature drops and internal pressure increases during the No. 1 fracturing (flow rate of $18m^3/min$ and the duration is 2h, the sand by volume ratio is 20% and its density is $1.4g/cm^3$) according to the results in Fig. 8, therefore, the axial force of casing string changes from the conventional state (wellhead is 400kN, bottom is −

86kN with compression state, the neutralization point is 2075m) to the whole section tension state (wellhead is 840kN, bottom is 950kN, the neutralization point disappearance). When No. 1 bridge plug is run into the wellbore, the wellbore temperature rise, the pressure drops, and the axial force gradually returned to conventional state. The reason for this change in axial force is: (1) when the casing temperature drops, the casing could shrink if the casing string is unrestrained, but the cementation from the interface between casing and cement ring constrains this axial displacement, resulting in an increase in the tensile axial force of the string, (2) as the internal pressure increases, the casing could swell if the casing string is unrestrained, but the cementation from the interface between casing and cement ring constrains this axial displacement, also resulting in an increase in the tensile axial force of the string. In the subsequent 2~13 stage fracturing process, the change of axial force of the tieback follows the above rules.

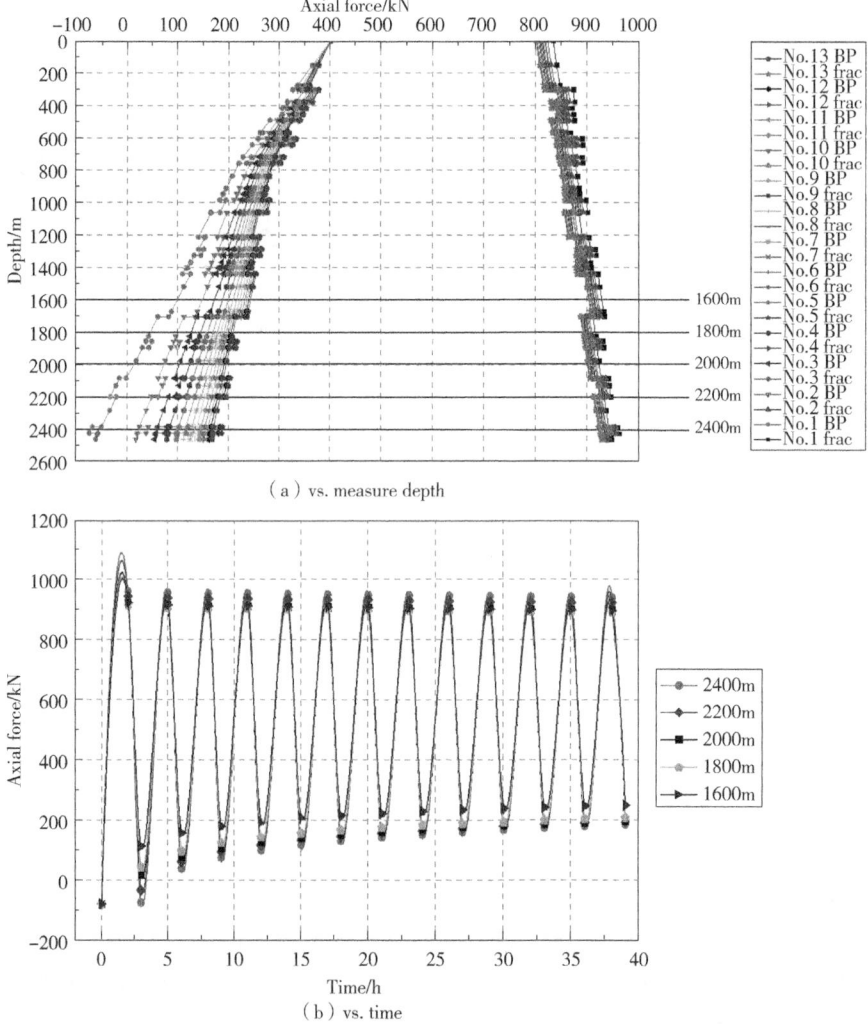

Fig. 15 The change of axial force of tieback during multistage fracturing

Fig. 15b illustrates the change of axial force of tieback vs. time at different well depth positions. It can be seen that the axial force at any position presents the law of alternating change: (1) the amplitude of axial force (ΔF_a) change increases with the increase of well depth (maximum ΔF_a at

1600m is 799kN, maximum ΔF_a at 1800m is 911kN, maximum ΔF_a at 2000m is 1023kN, maximum ΔF_a at 2200m is 1135kN, maximum ΔF_a at 2400m is 1250kN), (2) the amplitude of axial force change decreases with the increase of time (ΔF_a from 799kN decreases to 601kN at 1600m, ΔF_a from 911kN decreases to 653kN at 1800m, ΔF_a from 1023kN decreases to 705kN at 2000m, ΔF_a from 1135kN decreases to 756 kN at 2200m, ΔF_a from 1250kN decreases to 805kN at 2400m).

It can be seen from Fig. 15, the temperature of the casing string decreases during the fracturing process, but the cementation of the cement restricts the casing axial shrinkage because of the temperature drop. Therefore, the axial shrinkage force and the casing-cement cementation force become a pair of resistance. When the axial shrinkage force is greater than the cementation force, the cementation could be destroyed. On the other hand, according to the casing-cement cementation test results, the longer the casing-cement cementing length, the stronger the cementing force. Unfortunately, in well depth 1626 ~ 2464m intermittent segment, high cementing quality along whole segment is intermittent, and the longest length of cement segment with high cementing quality is 12.5m and the shortest length is 1.5m. Therefore, the cementation between casing and cement could be damaged under the alternating axial force. Fig. 16 illustrates the casing axial force (well depth 2464m) and cementation damage of casing-cement interface (well depth 2455.8 ~ 2464m) vs. time. The simulation shows that the cement in the first 8.2m high quality cementing section (from bottom to top in 1626~2464m intermittent segment) is broken during the first fracture. During the first stage (120min), the axial force of casing at 2464m increases linearly with time, from -83kN to 875kN, and then stabilized after 70min. Meanwhile, in the process of change the casing axial force, casing-cement cementation is destroyed synchronously, and the failure sequence of cementation is from bottom to up. And, the length of cementation failure at 10min, 30min, 50min, 70min and 90min is 0.5m, 1.1m, 3.4m, 5.8m and 8.2m, respectively (In the FE model, when the damage factor reaches 1, the cementation is completely destroyed).

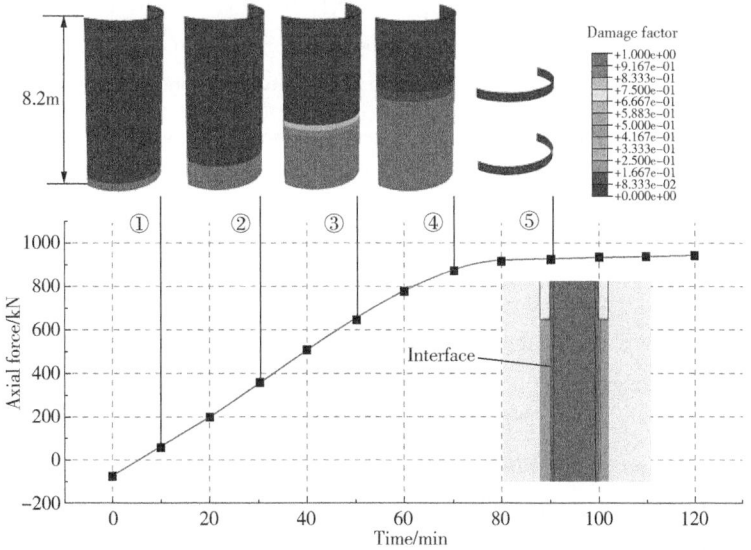

Fig. 16 Casing axial force (well depth 2464m) and cementation damage of casing-cement interface (well depth 2455.8~2464m) vs. time

According to the above studies, the axial force of casing string could frequent change due to the alternating condition of temperature and pressure during the multistage fracturing process. Therefore, the cementation of casing-cement interface is progressively broken. Fig. 17 presents the free deformation section of casing vs. fracturing stage under alternating condition of temperature and pressure. It can be seen that the tieback is normally inserted into the liner before fracturing (initial).

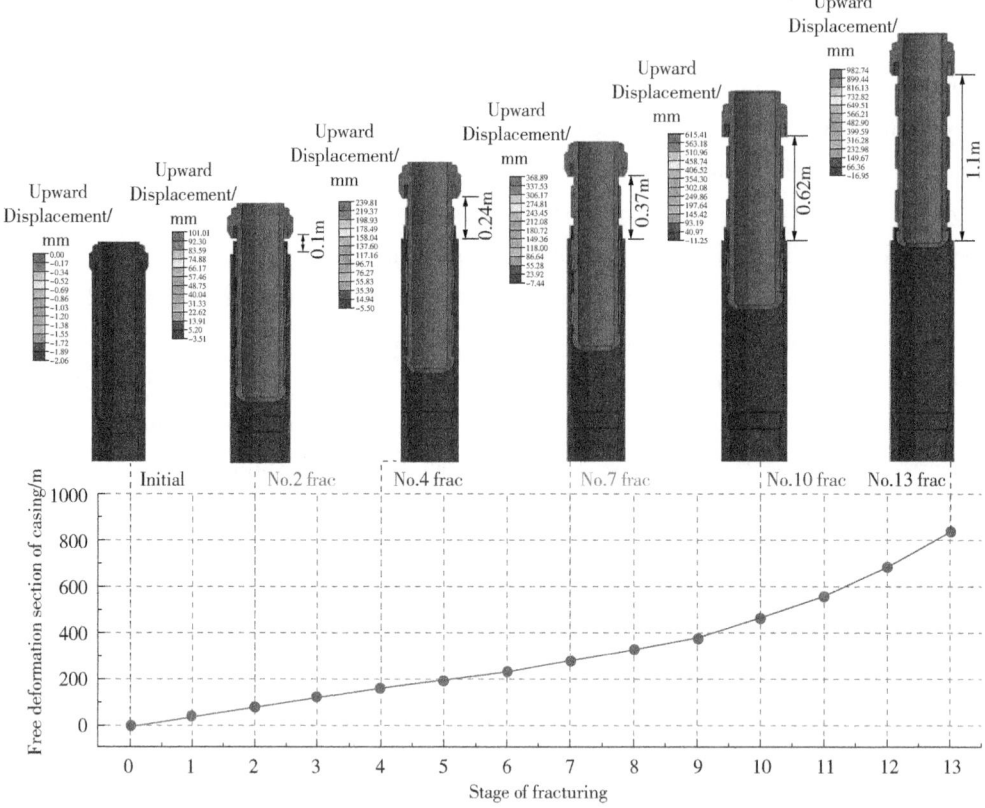

Fig. 17   Free deformation section of casing vs. fracturing stage under alternating condition of temperature and pressure

The positioning ring on the tieback is pressed down by 10t (downward displacement 2.06mm) after contacting the connector, and the length of the broken cementation of casing-cement interface is 0m. During No. 2 fracturing, the positioning ring on the tieback is separated from liner, tieback plug migrates 101.01mm upward, length of broken cementation of casing-cement interface reaches to 78m (above positioning ring on the tieback); During No. 4 fracturing, the tieback plug migrates 239.81mm upward, length of broken cementation of casing-cement interface reaches to 161m; No. 7 fracturing, the tieback plug migrates 368.89mm upward, length of broken cementation of casing-cement interface reaches to 275m; No. 10 fracturing, the tieback plug migrates 615.41mm upward, length of broken cementation of casing-cement interface reaches to 462m; No. 13 fracturing, the tieback plug migrates 982.74mm upward, length of broken cementation of casing-cement interface reaches to 838m.

It is analyzed that under alternating temperature and pressure conditions, if there is no constraint, the casing string could shorten and extend in the axial direction. However, the cement ring could prevent the mentioned axial casing string deformation under high cementing quality. Unfortunately, high cementing quality at well depth 1626~2464m is intermittent, and the longest length of cement segment with high cementing quality is 12.5m and the shortest length is 1.5m, meanwhile, the solid filling rate in the empty space (cement segment with low cementing quality) is less than 25%. So, the axial deformation of the casing string progressively damages the cementation of casing-cement interface from the tieback bottom (2464m) upward.

6 Discussion

Fig. 18 presents the comparative analysis of logging results of casing inner diameter (0°~180° high position and 90°~270° right position) and FE results. It can be seen that according to the wellbore orientation in the horizontal section, logging results of casing inner diameter along high position and right position is obtained (high position is distinguished from right position, 0°~180° is the high position of the wellbore, which indicates the casing position relative to the high or low side of the bore, meanwhile, 90°~270° is the right position of the wellbore, which indicates the casing position relative to the left or right side of the bore). Then, the logging data are compared with the FE results, and the results showed a high degree of consistency, casing inner diameter suddenly increase from 121.4mm to 166mm, and the length of the expanding section reaches 1.1m. Then, a stepped neck appears, which is a change in the inner diameter of the upper liner.

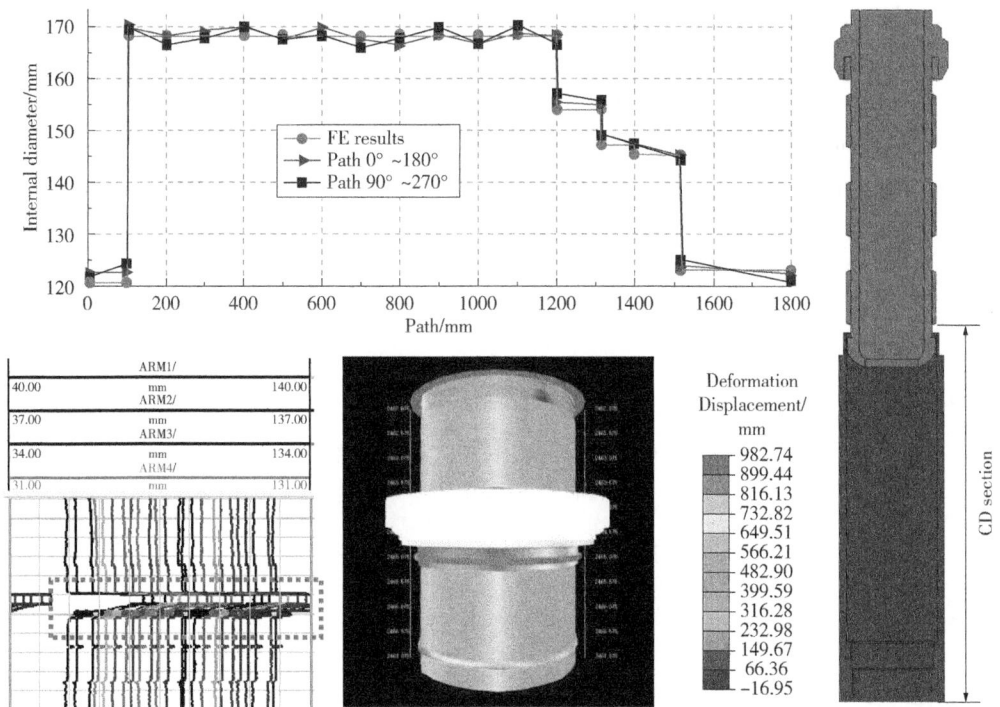

Fig. 18 Comparative analysis of logging results of casing inner diameter
(0°~180° high position and 90°~270° right position) and FE results

According to the comparison results above, it is analyzed that the casing string could shorten and extend in the axial direction under alternating temperature and pressure conditions if there is no constraint. However, the cement ring could prevent the mentioned axial casing string deformation under high cementing quality. Unfortunately, high cementing quality at well depth 1626~2464m is intermittent, and the longest length of cement segment with high cementing quality is 12.5m and the shortest length is 1.5m, meanwhile, the solid filling rate in the empty space (cement segment with low cementing quality) is less than 25%. So, the axial deformation of the casing string progressively damages the cementation of casing-cement interface from the tieback bottom (2464m) upward. During No.13 fracturing, the tieback plug migrates 982.74mm upward, length of broken cementation of casing-cement interface reaches to 838m. Without the constraint of cementation from the cement ring, the 838m casing contracted axially when the temperature drops, resulting in the 1.1m expansion section.

7 Conclusion

(1) During the multistage fracturing process, casing temperature and internal pressure change significantly repeatedly. These alternating loads result in significant changes in the axial force of the casing. As a result, the cementation between casing and cement could be damaged under the alternating axial force.

(2) High cementing quality at well depth 1626~2464m is intermittent in LS1 well, and the longest length of cement segment with high cementing quality is 12.5m and the shortest length is 1.5m. So, the axial deformation of the casing string progressively damages the cementation of casing-cement interface from the tieback bottom (2464m) upward.

(3) During No.13 fracturing, the tieback plug migrates 982.74mm upward, length of broken cementation of casing-cement interface reaches to 838m. Without the constraint of cementation from the cement ring, the 838m casing contracted axially when the temperature drops, resulting in the 1.1m expansion section.

CRediT authorship contribution statement

Yisheng Mou: Methodology, Investigation, Writing-original draft. Han Zhao: Conceptualization, Resources, Writing-review & editing. Jian Cui: Formal analysis, Validation, Investigation, Funding acquisition. Zhe Wang: Writing-review & editing, Investigation. Fengqi Wei: Validation, Investigation. Lihong Han: Writing-review & editing.

Declaration of Competing Interest

The authors declare that there are no conflicts of interest.

Acknowledgments

This research was funded by the Innovative Talents Promotion Program—Young Science and Technology Nova Project (2021KJXX-63), the Research on key technology of casing damage evaluation and repair in oil and gas wells (2021DJ2705) and the Study on key technology of stimulation and modification for Gulong shale oil (2021ZZ10-04).

Abbreviations

R_C	inside radius of the casing	m
h_C	convection coefficient of the inner wall of the casing	W/(m² · ℃)
T_C	casing temperature	℃
T_{FR}	temperature of fracture fluid in the casing	℃
W_F	work from thermal friction between fracturing fluid and casing per unit time	W/m
T_{Ca}	casing temperature	℃
T_{CE}	cement temperature	℃
T_F	formation temperature	℃
ρ_{FR}	density of fracture fluid	kg/m³
c_{FR}	specific heat capacity of fracture fluid	J/(kg · ℃)
N_q	injection rate	m³/s
Q_{HT}	heat generated by convection heat transfer	W/m
Q_{HC}	heat carried by the down-flowing fracturing fluid per unit	W/m
Q_{aC}	axial heat conduction of a casing unit per unit time	W/m
Q_{rC}	radial heat conduction of a casing unit per unit time	W/m
R_{CE}	inside radius of the cement	m
R_F	inside radius of the formation	m
ρ_C	density of casing material	kg/m³
c_C	specific heat capacity of casing material	J/(kg · ℃)
μ_C	thermal conductivity of casing material	W/(m · ℃)
E_{FRI}	internal energy of fracturing fluid in per unit	W/m
E_{CI}	internal energy of casing in per unit	N

References

[1] Hao Yu, Yisheng Mou, Zhanghua Lian, Qiang Zhang. Is Titanium Drillpipe Applicable to Offshore Drilling? A Question from a Corrosion Fatigue Perspective. SPE Journal. 2022, 27(1): 116-132.

[2] Hao Yu, Arash Dahi Taleghani, Zhanghua Lian. On how pumping hesitations may improve complexity of hydraulic fractures, a simulation study. FUEL. 2019, 249(2019): 294-308.

[3] Zhanghua Lian, Hao Yu, Tiejun Lin, Jianhua Guo. A study on casing deformation failure during multi-stage hydraulic fracturing for the stimulated reservoir volume of horizontal shale wells. Journal of Natural Gas Science and Engineering, 2015, 23: 538-546.

[4] Kui Liu, Deli Gao, Yanbin Wang. Effects of local load on shale gas well casing deformation. Natural Gas Industry, 2016, 36(11): 76-82.

[5] Wei Yan, Lingzhan Zou, Hong Li. Investigation of casing deformation during hydraulic fracturing in high geo-stress shale gas play. Journal of Petroleum Science and Engineering, 2016, 150.

[6] Hao Yu, Zhanghua Lian, Tiejun Lin. Study on failure mechanism of casing in stimulated reservoir volume fracturing of shale gas. Journal of Safety Science and Technology, China, 2016, 12(10): 37-43. 5.

[7] Hao Yu, Arash Dahi Taleghani, Zhanghua Lian, Tiejun Lin. On how asymmetric stimulated rock volume in shales may impact casing integrity. Energy Science & Engineering. 2020, 8(5): 1524-1540.

[8] Hao Yu, Arash Dahi Taleghani, Zhanghua Lian, Yisheng Mou. Severe Casing Failure in Multistage Hydraulic Fracturing Using Dual-Scale Modeling Approach. SPE Drilling and Completion, 2022, 37(3): 252-266.

[9] Kui Liu, Deli Gao, Yanbin Wang. Effect of local loads on shale gas well integrity during hydraulic fracturing process. Journal of Natural Gas Science and Engineering, 2017, 37.

[10] Yang Li, Wei Liu, Wei Yan. Mechanism of casing failure during hydraulic fracturing: Lessons learned from a tight-oil reservoir in China[J]. Engineering Failure Analysis, 2019, 98.

[11] Michael Mendez Restrepo, Catalin Teodoriu, Saeed Salehi. A novel way to look at the cement sheath integrity by introducing the existence of empty spaces inside of the cement(voids). Journal of Natural Gas Science and Engineering, 2020, 77(C).

[12] Hongtao Li, Ze Li, Gao Li. Casing deformation mechanisms of horizontal wells in Weirong shale gas field during multistage hydraulic fracturing. Journal of Natural Gas Science and Engineering, 2020, 84.

[13] Shangyu Yang, Lihong Han, Jianjun Wang. Laboratory study on casing deformation during multistage horizontal well fracturing in shale gas development and strain based casing design. Journal of Natural Gas Science and Engineering, 2021, 89.

[14] Radakovic-Guzina Zorica, Damjanac Branko, Savitski Alexei A. Suarez. "Analysis of the Production Casing Deformation Due to Shearing of Offset Hydraulic Fractures." Paper presented at the SPE/AAPG/SEG Unconventional Resources Technology Conference, Houston, Texas, USA, June 2022.

[15] Hongxiang Zhang, Hengmao Tong, Ping Zhang. How can casing deformation be prevented during hydraulic fracturing of shale gas? —A case study of theWeiyuan area in Sichuan, China. Journal of Petroleum Science and Engineering, 2023, 221.

[16] Yisheng Mou, Shangyu Yang, Lihong Han, Jianjun Wang, Zhanghua Lian. Mechanical Behavior and Optimization of Tubing String with Expansion Joint during Fracturing in HTUHP Wells[J]. Processes, 2022, 10(6): 1063.

[17] Yisheng Mou, Jian Cui, Jianjun Wu, Fengqi Wei, Ming Tian, Lihong Han. The mechanism of casing deformation before hydraulic fracturing and mitigation measures in shale gas horizontal wells, processes, 2022, 10(12), 2612.

本论文原发表于《Processes》2023年第2250卷第11期。

Development of Design Method for Casing and Tubing Strings under Complex Alternating Loads

Wang Jianjun[1] Li Dening[2] Du Xu[3] Li Hongjie[2] Yang Shangyu[1]

(1. CNPC Tubular Goods Research Institute, State Key Laboratory of Performance and Structural Safety for Petroleum Tubular Goods and Equipment Materials;
2. CNPC Tianjin Undergroud Gas Storage Company, Dagang Oilfield Co.;
3. CNPC Logging Corporation Changqing)

Abstract: In the increasingly complex environment of downhole working conditions such as frequent opening and closing of oil and gas wells, acidification/multi-stage fracturing, steam huff and puff, and strong injection and production, the traditional casing string design method standards based on strength design have gradually highlighted their limitations. It is necessary to formulate string design method standards that consider the full life cycle of wells such as drilling and completion, fracturing, and production operations. Starting from the analysis of the advantages and disadvantages of traditional casing string design methods, this paper presents the characteristics of the strain design methods and seal design methods that have been formulated for the life cycle of the string. The strain design method breaks through the traditional design concept and allows the design concept of controllable deformation of the pipe string. The sealing design method is currently the only standard method for the design of tubing strings. At the same time, it further proposes to consider the trend of pipe strength deterioration and establish a time dimension-based life cycle pipe string design method standard, which effectively solves the safety problem of pipe string design in production and operation.

Keywords: Oil and gas well; Tubing; Casing; Pipe string design; Method

Due to the international low oil prices and the contradiction between supply and demand of oil and gas resources, The Ministry of Land and Resources of China has put forward the strategic policy of reducing cost and increasing efficiency for oil and gas field[1]. In view of the problems of harsh storage conditions, difficult exploitation, complex and variable load environment, and high casing loss rate in oil and gas wells such as low permeability well, deep well or ultra-deep well, offshore well and unconventional well[2], higher requirements are put forward for the design and evaluation methods of casing string, such as the cyclic dynamic load environment caused by frequent switching,

---

Corresponding author: Wang Jianjun, wangjianjun005@cnpc.com.cn

acidizing, multistage fracturing, steam half and puff, strong injection and production[3-4]. At the same time, because the casing selection dose not highlight the difference of working conditions, the series of failures and hidden risks will further affect the safe production and operation of oil and gas, especially in high speed ($100\times10^6 m^3/d$) gas storage injection-production wells[5].

Before the 21st century, the design standard of tubing string in oil and gas wells mostly focused on the strength design method of casing string, and mostly focused on the calculation and verification of static strength during drilling[6-8], and the design of tubing string was mainly based on numerical calculation without forming a standard design method[9]. At the beginning of the 21st century, scholars began to pay attention to the research on the design method of the whole life cycle of the pipe string, which covers the calculation of casing load and safety check in the process of drilling and production operation[10-11], and formulated the design standard of the string based on the strain design and sealing design[12-13], especially in the aspect of the tubing string for the first time.

This paper focuses on the characteristics and applicability of various casing string design methods, Discuss the gradual establishment of single strength design method of pipe string to strength+strain or strength+seal life cycle design method, In view of the complex and changeable working conditions in the well, the casing strength, sealing, deformation and other factors are comprehensively considered, and the design method of the full life cycle string considering the time dimension is further proposed.

1 Traditional casing string design method

The traditional casing string design method is mainly based on the strength design check, that is, the load on the casing string cannot exceed the casing strength, and the casing load is controlled within the elastic range. Traditional casing string design methods mainly include safety factor method, boundary load method, maximum load method, American AMOCO casing design method, German BEB casing design method, former Soviet casing design method and domestic casing string strength design method[14]. On this basis, the probabilistic reliability design method and the strength margin design method are further developed.

1.1 Safety factor method

For a safe and economical design, the ratio of the strength of the casing against various external loads to the external loads on the casing must be equal to the specified safety factor. Because the axial load increases from bottom to top, and the extrusion pressure increases from top to bottom, the whole casing string should be composed of multiple sections of casing with different strength (determined by different steel grades and wall thickness) in order to achieve both safety and economy. The minimum safety factor of each section shall be equal to the specified safety factor.

1.2 Boundary load method

The internal pressure and collapse resistance design method of this method is the same as that of the equal safety factor method. For the tensile design, the allowable strength calculated by the boundary load determined by the tensile strength and safety factor of the casing in the first section shall be used to select the following sections of casing. In this way, the boundary load between each casing section designed is the same, but not the same safety factor, which avoids excessive residual strength of the selected casing, reduces the total weight of the casing string, and makes the design result more reasonable and economic.

1.3 Maximum load method

The casing string is designed according to the effective load of the casing string under the actual conditions and a certain safety factor. The biggest feature is that the external load calculation has been carefully considered. According to the classification of technical casing, reservoir casing and surface casing, the external load calculation methods of various types of casing are also different, so as to fully show the actual external load in the design.

1.4 Former AMOCO casing design method in the United States

The former AMOCO company of the United States proposed two casing design methods, graphical method and analytical method, which have unique features in both load analysis and design methods. The impact of tensile stress is considered for both anti extrusion and internal pressure, and the shoulder force is considered when calculating external load.

1.5 German BEB casing design method

German BEB Company has proposed a relatively complete set of casing design methods, whose main feature is to provide different calculation methods of external load in detail according to different casing types. For example, when calculating the collapse strength of a technical casing due to leakage, it is necessary to consider buoyancy, calculate the neutral point, and perform a double extraction stress calculation. The formation pore pressure value for collapse calculation should be based on the specific gravity of the mud at the time of casing; When calculating the internal pressure resistance, 40% of the kick volume is specified as the internal pressure basis, and 0.115bar/m is selected as the formation pore pressure gradient; After the composite verification of anti-collapse and internal pressure resistance, the casing string structure is initially selected, and the tension check is conducted. Finally, the casing string structure is selected; Calculate the maximum test pressure at the wellhead while waiting for solidification.

1.6 Casing design method in the former Soviet Union

The former Soviet casing design method provides formulas for calculating the internal and external pressures at various parts of the casing string during different periods and under different downhole operating conditions. When calculating the external pressure of the cementing section, the unloading effect of the cement sheath should be considered. The reduction of tensile strength against collapse under biaxial stress is not considered, but it is specified that when the tensile stress of the pipe body reaches 50% of the yield strength, the safety factor against collapse should be increased by 10%. The buoyancy of the drilling fluid is not considered when calculating the axial tension. When designing the technical casing, there are two methods: considering wear and not considering wear.

1.7 Domestic casing string strength design method

In the 1980s, a group of old petroleum experts began to devote themselves to the study of domestic casing string strength design standards, absorbing advanced foreign practices, and in 1988, they formulated the SY/T 5322—1988《Recommended Method for Casing String Strength Design》. In 2002, they further improved and revised the SY/T 5322—2002《Casing String Strength Design Method》. After revising the casing string load calculation and strength calculation methods in

2008, SY/T 5724—2008 《Casing String Structure and Strength Design》[8] has been formulated in combination with other standards and has been used until now. This standard provides a unified set of casing design procedures, providing a selection method of basic parameters and relevant design conditions for the design, and the design method is gradually improved. However, there are still many disputes among domestic experts regarding the two-way stress calculation, strength calculation, and design methods in the new standard[15].

1.8 Probabilistic reliability design method

Using the stress strength interference theory, it is assumed that the expected load and rated value of the design are functions of two random variables C and L (where C is the load capacity of the casing, and L is the estimated value of the maximum load during drilling). The design goal is to ensure that C is greater than $L(C>L)$. The size of the probability $P(C>L)$ is the probability of design success PS.

1.9 Strength margin design method

The casing strength margin design must be studied using a combination of safety factor design methods for casing loads (such as the maximum load method) and probabilistic reliability design methods. That is, in the design of the casing string, the safety factor method is still used to determine the casing load, and the probability statistical method is used to study the distribution law of the casing strength performance to clarify the existence of strength margin in the casing. Then, the strength margin is used to reduce the safety factor in the design of the casing string. Then, reliability analysis is conducted on the casing string to ensure that the casing string remains reliable after reducing the safety factor, meet the use requirements, and ensure that the drilling cost is reduced.

2 Full life cycle string design method

The above traditional casing string design methods have their respective advantages and disadvantages, and should be comprehensively considered and optimized based on the actual operating conditions of the oilfield. However, these methods consider more drilling conditions and less production conditions. On this basis, considering changes in temperature and pressure during production, research and development have been conducted on full life cycle string design techniques such as strain based casing string design methods for thermal recovery wells, and sealing design methods for gas storage injection production strings.

2.1 Design method of casing string for thermal recovery wells based on strain

The casing string design method for heavy oil thermal recovery wells includes casing string strength design and strain design[16]. That is, when performing casing string strain design for thermal recovery wells, the casing string strength design should be performed first, and then the casing string strain design should be performed. The strength design is to make the casing string meet the requirements of the drilling and completion process of heavy oil thermal recovery wells, and the strain design is to make the casing string meet the requirements of the production process of heavy oil thermal recovery. Therefore, the strain design of the casing string in heavy oil thermal recovery wells follows two design criteria: strength design criteria and strain design criteria.

The strength design criteria for the casing string are shown in Formula (1):

$$\sigma = \begin{Bmatrix} p_{be} \\ p_{ce} \\ T_e \end{Bmatrix} \leqslant [\sigma] = \begin{Bmatrix} \dfrac{p_{bo}}{S_i} \\ \dfrac{p_{co}}{S_c} \\ \dfrac{T_o}{S_t} \end{Bmatrix} \qquad (1)$$

In the formula, σ is the working stress of the casing string, which is obtained by calculating the effective internal pressure p_{be}, effective external pressure p_{ce}, and effective axial force T_e (including bending stress) borne by the casing string according to SY/T 5724 standard; The allowable stress $[\sigma]$ of the casing string is calculated according to ISO 10400 standard, and the internal pressure resistance p_{bo}, collapse resistance p_{co}, and tensile strength of the casing string T_o are divided by the internal pressure safety coefficient S_i, collapse resistance safety coefficient S_c, and tensile safety coefficient S_t specified in SY/T 5724 standard.

The design criteria for casing string strain are shown in Equation (2):

$$\varepsilon_\Sigma \leqslant [\varepsilon] = \frac{\delta}{S_s} \text{ or } S_{sc} = \frac{\delta}{\varepsilon_\Sigma} \geqslant S_s \qquad (2)$$

In the formula, ε_Σ is the working strain of the casing string, %; $[\varepsilon]$ is the allowable strain of the casing string, %; δ is the uniform elongation of the casing material, %; S_s is the strain safety factor.

This design method has formed the Chinese petroleum and natural gas industry standard SY/T 6952.1—2014《Thermal Recovery Well Casing String Based on Strain Design: Part 1 Design Method》[10].

2.2 Sealing design method for injection production string of gas storage

In the design of gas storage injection and production pipe strings, comprehensive consideration should be given to the alternating load caused by changes in parameters such as temperature, pressure, and flow. Especially in the design of pipe strings under alternating tensile/compressive loads, not only the structural strength design of the pipe string should be carried out, but also the sealing design of the pipe string should be carried out, that is, the design of the injection and production pipe string should be carried out from the perspective of structural strength and sealing integrity. In addition to the strength design criteria, the design of the injection and production string for gas storage should also follow the sealing design criteria. The strength design criteria cover the entire process of load and strength analysis such as tubing running, packer setting, annular pressure testing, and injection production operation. The sealing design criteria focus more on the bearing capacity of the string during the injection production alternating process. The sealing design criteria for injection production strings are shown in Equations (3) and (4):

$$\frac{F_{etmax}}{T_{to}} \times 100\% \leqslant \frac{\delta_t}{S_{tt}} \qquad (3)$$

$$\frac{F_{ecmax}}{T_{to}} \times 100\% \leqslant \frac{\delta_c}{S_{tc}} \tag{4}$$

In the formula, F_{etmax} is the maximum axial tensile load of the injection production string, N; F_{ecmax} is the maximum axial compressive load of the injection production string, N; T_{to} is the rated tensile strength of the injection production string, N; δ_t is the tensile efficiency of the injection production string joint under airtight sealing,%; δ_c is the compression efficiency under airtight sealing of the injection production string joint,%; S_{tt} is the tensile safety coefficient of the joint under the gas seal of the injection production string; S_{tc} is the safety factor against compression of the joint under the gas seal of the injection production string.

This design method has formed the Chinese petroleum and natural gas industry standard SY/T 7370—2017《Recommended Practice for Selection and Design of Injection Production String for Underground Gas Storage》[11].

3 Design method of pipe string based on time dimension

Under complex operating conditions such as multi-stage fracturing of unconventionaloil and gas wells, multi cycle injection and production operations of gas reservoirs, deep natural oil and gas acid fracturing, and frequent well opening and closing, there are significant periodic pressure/temperature changes in the wellbore, which degrade the strength of oil and casing materials, further leading to a reduction in the service strength of the string. For example, after 30 weeks of injection and production operation in a gas storage, the test confirms that the yield strength of the injection and production string decreases by 9.2% after 30 weeks(Figure 1), resulting in a decrease of 9.20% in tensile strength, 9.30% in internal pressure strength, and 5.46% in collapse strength of the injection and production string, as shown in Figure 2; The casing material strength of a tight oil well also weakened during multistage fracturing. After 30 fracturing operations, the tensile strength, internal pressure resistance, and crush resistance of the 139.7mm×7.72mm P110 fracturing casing string decreased by 5.0%, 5.0% and 2.43%, as shown in Figure 3.

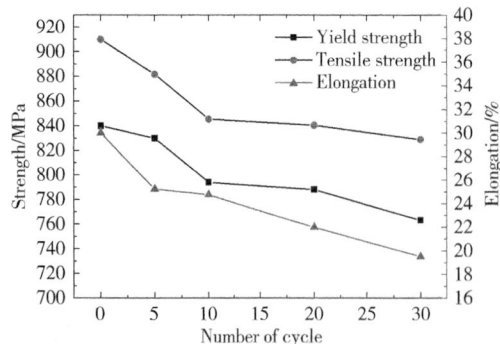
Figure 1　Material properties of tubing after alternating tension and compression cycles

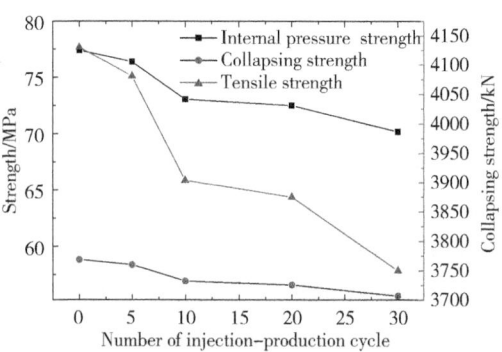
Figure 2　Service performance of tubing in different production periods

The above full life cycle string design method has considered the load factors throughout the drilling and completion, pressure testing, and production pro-cesses, but does not consider the impact of strength changes. At the same time, it lacks a casing strength evaluation method under load/ temperature coupling fluctuations, resulting in a reduction in casing strength margin during service, increasing production safety risks. Therefore, there is an urgent need to further study on this basis and establish a time dimension based design method for pipe strings.

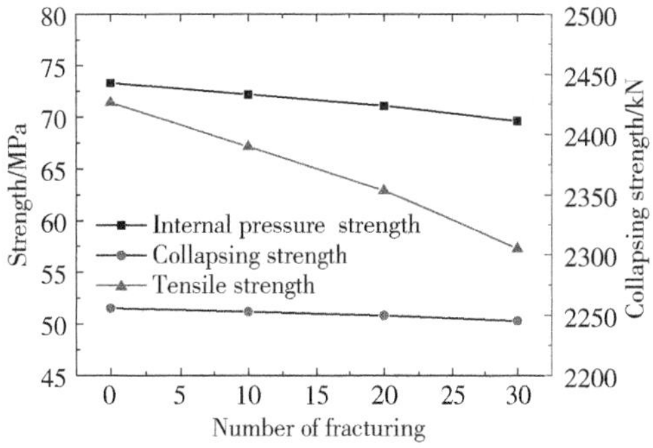

4 Conclusion

(1) The traditional casing string design method focuses on strength design, which should be comprehensively considered and optimized based on the actual operating conditions of the oilfield.

Figure 3 Service performance of ϕ139.7mm×7.72mm P110 casing in different fracturing periods

(2) The full life cycle string design method comprehensively considers drilling, completion, and production operating conditions, improving the accuracy and safety of string design, and forming corresponding industry design standards.

(3) The increasing complexity of downhole conditions, especially the increase in load cycles, has led to deterioration in the strength of oil and casing materials. It is necessary to further study and establish a full life cycle string design method and standard based on the time dimension.

Acknowledgements

This research is supported by CNPC Basic Research Project [2021DQ03(2022Z-03)] and CNPC Key Technology Research Project (2022ZG06).

References

[1] National Development and Reform Commission. Notice on the issue of "Suggestions on accelerating the use of natural gas": NDRC energy[2017] 1217[A/OL]. [2017-06-23]. http://www.ndrc.gov.cn/zcfb/zcfbtz/201707/t20170704_853931.html.

[2] JIANG Tongwen, SUN Xiongwei. Development status and technology development trend of deep natural gas in China[J]. Oil Drilling & Production Technology, 2020, 42(5): 610-621.

[3] HAN Lihong, YANG Shangyu, WEI Fengqi, YE Xinqun, WANG Jianjun, WANG Hang, PAN Zhiyong, ZHANG Huali, YUE Wenhan, XIE Bin, SHU Zhenhui, ZHANG Ping, LU Caihong, YIN Fei. Casing Deformation Mechanism and Controlling Method for Shale Gas Well Under Complex Fracture Environment[J]. Petroleum Tubular Goods & Instruments, 2020, 6(4): 16-23.

[4] Wang Jianjun, Sun Jianhua, Xue Chengwen, Han Jun, Zhang Guohong, Wang Rui. Optimization of gas-tight thread connectors on injection-production strings in underground gas storage wells[J]. Natural Gas Industry, 2017, 37(5): 76-80.

[5] Wang Jianjun, Lu Caihong, He Haijun, Sun Jianhua, Wu Xuehu, Li Fangpo. Selection and Evaluation of piping string in Underground Gas Storage with gas reservoirs[J]. Petroleum Tubular Goods & Instruments, 2019, 5(2): 26-29.

[6] Han Zhiyong. Tubing string mechanics in hydraulic environment[M]. Beijing: Petroleum Industry Press, 2011.

[7] Gao Baokui & Gao Deli. A new method for testing tubing axial load in high-pressure and high-temperature wells [J]. Journal of the University of Petroleum, China, 2002, 26(2): 39-41.

[8] National Development and Reform Commission. SY/T 5724—2008 Design for casing string structure and strength[S]. Beijing: Petroleum Industry Press, 2008.

[9] Editorial Committee of offshore oil and gas field completion Handbook. Offshore well completion[M]. Beijing: Petroleum Industry Press, 1998.

[10] Yu Hao, Zhao Zhaoyang, Lian Zhanghua etc. Analysis on Wellhead Uplift by Tieback Casing String under Temperature Difference Effect[J]. China Petroleum Machinery, 2022, 50(1): 100-107.

[11] Chen Sheng, Fan Mingtao, Li Jun etc. Study on the Mechanical Behavior of Casing during Stimulated Reservoir Volume Fracturing [J]. China Petroleum Machinery, 2019, 47(1): 1-7.

[12] National Energy Administration. SY/T 695.1—2014 Thermal casing string for strain based design Part1: Design method[S]. Beijing: Petroleum Industry Press, 2015.

[13] National Energy Administration. SY/T 7370—2017 Recommended practice for selection and design of the injection-production tubing string in underground gas storage well[S]. Beijing: Petroleum Industry Press, 2017.

[14] "Drilling Handbook" preparation group. Drilling Handbook [M]. Beijing: Petroleum Industry Press, 2013.

[15] Han Zhiyong. Discussing on formula of tri-axial tension strength of casing string[J]. Journal of China University of Petroleum, 2011, 45(4): 77-80.

[16] WANC J J. Steam stimulation wells casing string strain design and security technology research of shallow heavy oil[D]. Dongying: China University of Petroleum(Huadong), 2015.

本论文原发表于《Processes》2023年第2582卷第11期。

Research on Casing Deformation Prevention Technology Based on Cementing Slurry System Optimization

Yan Yan[1,2] Cai Meng[3] Ma Wenhai[3]
Zhang Xiaochuan[3] Han Lihong[1] Liu Yonghong[2]

(1. State Key Laboratory for Performance and Structure Safety of Petroleum Tubular Goods and Equipment Materials, CNPC Tubular Goods Research Institute;
2. College of Mechanical and Electrical Engineering,
China University of Petroleum(East China);
3. CNPC Daqing Oilfield Company)

Abstract: The casing deformation prevention technology based on the optimization of cement slurry is proposed to reduce the casing deformation of shale oil and gas wells during hydraulic fracturing. In this paper, the fracture mechanism of hollow particles in cement sheath was firstly analyzed by discrete element method, and the effect of hollow particles in cement on casing deformation was investigated by laboratory experiment method. Finally, field test was carried out to verify the improvement effect of the casing deformation based on cement slurry modification. The results show that the formation displacement can be absorbed effectively by hollow particles inside the cement transferring the excessive deformation away from casing. The particles in the uncemented state provide deformation space during formation slipping. The casing with diameter of 139.7mm could be passed through by bridge plug with the diameter of 99mm when the mass ratio of particle/cement reaches 1 : 4. According to the field test feedback, the method based on optimization of cement slurry can effectively reduce the risk of casing deformation, and the recommended range of hollow microbeads content in the cement slurry is between 15%~25%.

Keywords: Cement slurry; Hollow ceramsite; Casing deformation; Formation slip; Field test

1 Introduction

Hydraulic fracturing has been widely used in shale oil/gas around the world. During fracturing, casing diameter need be maintained large enough to ensure tools passing through the wellbore (Lian et al., 2015; Yu et al., 2016; Yu et al., 2022). Casing deformation in the horizontal section

Corresponding author: Han Lihong, hanlihong@cnpc.com.cn

severely restricts fracturing operations, resulting in fracturing failure and lower oil/gas recovery. Scholars have carried out a lot of studies on the casing deformation analysis (Yan et al., 2017; Yin et al., 2018; Li et al., 2020; Zhang et al., 2020; Li et al., 2021). It is generally approved that the formation slip and local high pressure are main reasons for casing deformation. Especially, natural fractures or artificial fracture network will lead to uneven stress distribution near the wellbore, and the casing-cement will be sheared or extruded by formation slip, which cause casing deformation and even damage (Xi et al., 2018; Xi et al., 2020; Liu et al., 2019; Meng et al., 2020), resulting in economic losses (Furui et al., 2012; Peng et al., 2007; Guo et al., 2020). Therefore, it is significant to investigate the influence of formation slip on wellbore deformation and recommend the appropriate prevention measures for casing deformation.

A lot of research has been carried out about the optimization of cement slurry, but most of them focus on the mechanical performance to improve the cement integrity such as strength, elastoplasticity and so on (Wei et al., 2022; Cheng et al., 2023). Guo (2020) and Teodoriu (2019) respectively developed composite cement suitable for high temperature environment in geothermal wells, and verified its mechanical properties through a series of experiments. Hudson (2017), Santos (2021), Jafariesfad (2017a, 2017b) et al. developed ductile cement slurry with different formulations to improve the cementing strength of interface and enhance their deformability. Adjei (2021) proposed to use bentonite instead of fly ash to develop a new weighted cement and tested its performance. The results showed that this slurry shortened the thickening time and improved cement compressive strength. In addition, Jafariesfad (2020) summarized the challenges encountered in the cementing process after the addition of nano-scale additives in cement, including expansion nano-scale additives for shrinkage reduction, nano-rubber/flexible particles for modification, and charged nano-particles for cement hardening. The role of nano-material additives is to enhance the sealing and durability of cement (Salehi et al., 2018; Quercia et al., 2018). To sum up, there are many researches on cement additives, but their aim is to improve the cementing interface bonding quality and lower its own damage. The addition of the above additives can only improve the mechanical properties of cement itself, but they will not provide additional protection for casing under the condition of formation slip (Fig. 1).

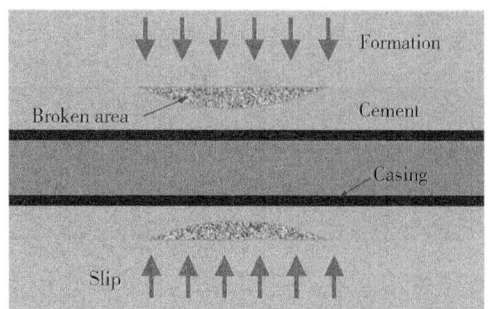

Fig. 1 Casing deformation mitigation by cement sheath

Geology slip has been treated as the main cause of casing deformation (Xi et al., 2019; Yin et al., 2018; Dong et al., 2019; Li et al., 2021). Xi (2021) indicated the casing deformation prevention could not only rely on casing steel grade and thickness, but also optimizing the stage spacing, perforation density or temporarily block. Zhao (2021) believed that casing deformation is caused by the geological migration during fracturing, so the propagation of geological fractures should be traced and measures should be taken such as enhancing cement strength and eliminating micro-annulus in wellbore. Liu (2019) investigated the influence of thickness and ovality of cement sheath on casing deformation. Guo (2019) and Wang (2021) considered that the casing deformation was comprehensively related to the non-uniform mechanics around the wellbore, poor cementing quality, and geological fractures. To sum up, the scholars have done a lot of theoretical analysis, however, the research on methods for controlling casing deformation is relatively few. Therefore, on the basis of hollowing cement, the effective casing deformation prevention method should be formed, which has reference value for preventing casing deformation under the condition of formation slip.

2　Discrete element model

2.1　Geometric model

According to the conventional well structure of shale gas in southwest China, the outer diameter of production casing is 139.7mm, the thickness of casing and cement sheath is 10.5mm and 38.1mm. The commercial software PFC^{2D} was used to establish the numerical model under extrusion condition. In order to improve the calculation speed, the cement length in the model is 20mm. The size of the model is shown in Fig. 2. According to the logging data and previous research (Yang et al., 2023), the order of magnitudes is centimeter for shale slippage, thus the displacement boundary was set as 20mm in this model. The end face of cement is the boundary of constant pressure and the subface of cement is fixed. Mechanical parameters should be defined in PFC^{2D} according to the actual physical parameters of cement and ceramite. The micro-mechanical parameters of these two materials are shown in Table 1.

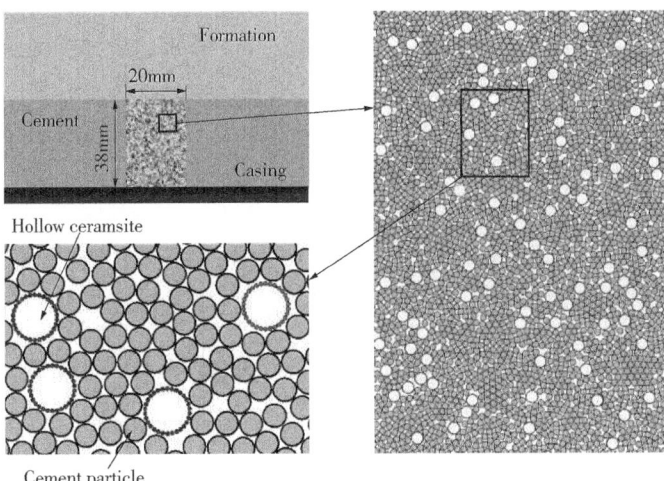

Fig. 2　Discrete element model of cement sheath with hollow particles

Table 1　Microscopic parameters of cement and particle

Material parameter	Cement	Hollow particle
Cement particle density $\rho_s/(\text{kg/m}^3)$	3150	1430
Radius of particle $R_s/\mu m$	50	100
Effective modulus with linear contact E_{mod}/GPa	0.50	0.81
Effective modulus of parallel bonding P_{b_emod}/GPa	0.50	0.81
Stiffness ratio k^*	1.0	1.0
Friction coefficient u	0.200	0.557

continued

Material parameter	Cement	Hollow particle
Tensile strength between particles p_{b_ten}/MPa	6.0	8.2
Cohesion between particles p_{b_coh}/MPa	26.70	6.56
Internal friction angle θ_{b_fa}/(°)	17.797	30.000

2.2 Governing equation

Based on particle flow theory in PFC^{2D}, the parallel bonding model was used for contact between particles in order to get closer mechanical properties of cement particles and hollow particles.

The parallel bonding model provides the mechanical behavior of cement-based material between two contact slices (Li et al., 2020). The parallel bonded component interacts elastically between the slices. The parallel bonds does not exclude the possibility of slipping. They can transfer the forces and torques between the blocks and should be treated as a set of uniform elastic springs with constant normal and shear stiffness.

The contact force F_c and the contact torque M_c in parallel bonding model are:

$$F_c = F_l + F_d + \overline{F} \tag{1}$$

$$M_c = \overline{M} \tag{2}$$

where F_l is the linear force; F_d is the damping force; $\overline{F}$ is the parallel bonding force; $\overline{M}$ is the parallel bonding torque.

The parallel bonding force can be decomposed into normal force and shear force, and the parallel bonding torque can be decomposed into torsional torque and bending torque:

$$\overline{F} = -\overline{F}_n \hat{n}_c + \overline{F}_s \tag{3}$$

$$\overline{M} = \overline{M}_b \tag{4}$$

where $\overline{F}_n$ is the normal parallel bonding force; $\overline{F}$ is the parallel bonding force; $\hat{n}_c$ is unit normal vector in the contact surface; $\overline{F}_s$ is shear force; $\overline{M}_b$ is the bending torque.

The parallel bonding shear force $\overline{F}_s$ and bonding torque $\overline{M}_b$ are acted on the contact surface when the normal parallel bonding force $\overline{F}_n > 0$. They can be represented by the contact surface coordinate system:

$$\overline{F}_s = \overline{F}_{st} \hat{t}_c \tag{5}$$

$$\overline{M}_b = \overline{M}_{bs} \hat{s}_c \tag{6}$$

where $\overline{F}_{st}$ is the parallel bonded shear stiffness; $\overline{M}_{bs}$ is the parallel bonded bending stiffness; $\hat{t}_c$ is the unit direction vector in the tension state; $\hat{s}_c$ is the unit tangent vector of the contact surface.

An interface is set up between these two conceptual surfaces when a parallel bond is established and the parallel bond forces and torques are zeroed out. The parallel bond provides the elastic interaction between two conceptual surfaces, this interaction is invalid when the bonding is broken (Li et al., 2020). Each conceptual surface is rigidly attached to the component. The gap g_s of the parallel bonding surface is defined as the cumulative relative normal displacement of the particle surface:

$$\bar{g}_s = \sum_1^n \Delta \&_n \tag{7}$$

where $\bar{g}_s$ is the relative normal displacement increment of the contact between particles; It is indicated that the parallel bonding contact is not destroyed when $\bar{g}_s > 0$. Contact is divided into ball-to-ball contact and ball-to-surface contact, which is shown in Fig. 3.

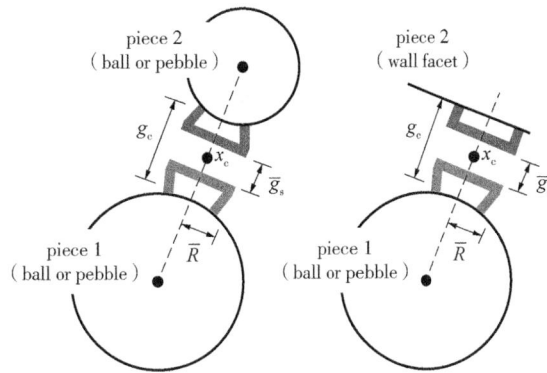

Fig. 3 Schematic diagram of parallel contact model

2.3 Calculation results

Fig. 4 shows the particle geometry under the confining pressure of 30MPa. As can be seen from Fig. 4, 8% high-strength hollow microbeads in cement has been broken before formation slippage due to the downhole hydrostatic pressure. The compressed hollow microbeads are squeezed circumferentially by the cement particles and the adjacent microbeads. The broken form of hollow microbeads is multi-segment strip. It is worth noting that tensile stress is distributed near the broken hollow beads caused by the compressed space. The high strength hollow microbeads randomly distributed in the solidified cement in the form of single broken particles and large pores composed of adjacent microbeads, as shown in Fig. 4. These pores will be collapsed firstly in the process of compression due to the lower strength.

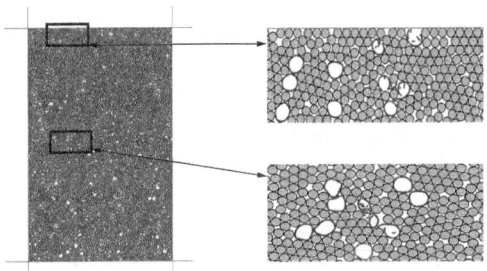

Fig. 4 Geometry of hollow microbeads in cement sheath under confining pressure

A displacement loading of 20mm was applied on the cement to simulate shale extrusion. The broken morphology under the slippage of 5mm, 10mm, 15mm and 20mm was shown in Fig. 5. All the hollow beads have been deformed when the formation slides down 5mm, including the elastoplastic unbroken deformation, the broken uncompaction and the broken compaction. The unbroken hollow beads with elastoplastic deformation are usually elliptical or irregular. The broken hollow microbeads have lost their original supporting function, providing space for the formation slip. When the formation slides down from 5mm to 10mm, more hollow microbeads will be broken and further compactified. Because there are still a large number of unbroken microbeads in cement under the slippage of 5mm. The characteristics of broken particles are shown in Fig. 5(c) when the formation slides down from 10mm to 15mm. The hollow beads are basically broken and compacted in this stage. It can be inferred that there is only compaction effect in cement sheath if the formation continues to slip downward. The cement with hollow beads is completely compacted when the formation slip is 20mm, and the cement shows the characteristics of axial compression and transverse extension due to the limit of constant confining pressure boundary.

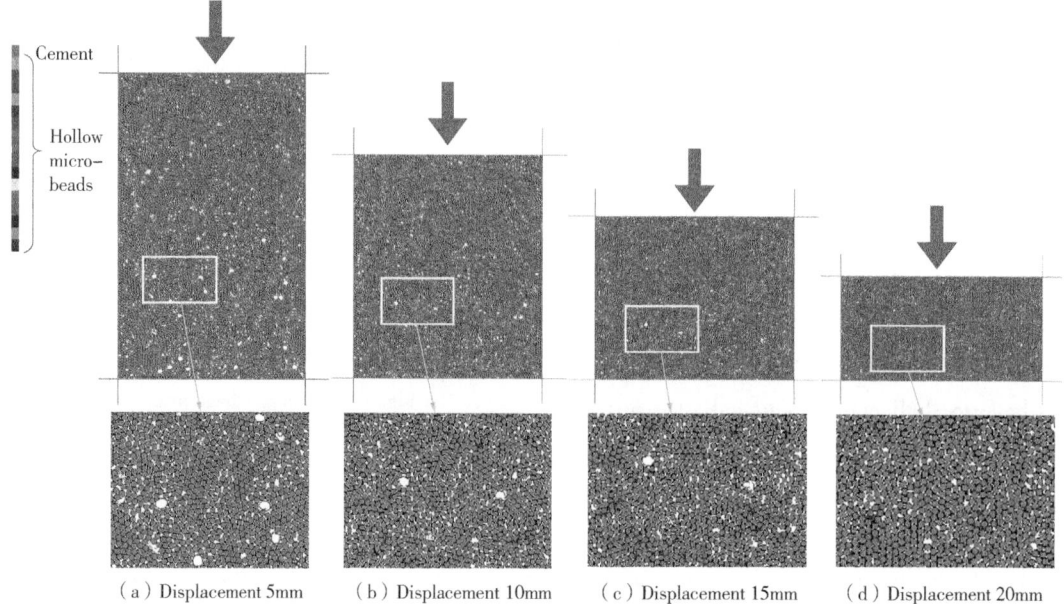

Fig. 5 Compression process of high-strength hollow microbeads in cement sheath

Fig. 6 shows the displacement distribution in cement with different hollow beads contents after formation slip. As can be seen from Fig. 6, the radial displacement becomes more uniform with higher particle content. This means the displacement transposition effect is better. Fig. 7 shows the adhesion distribution of compressed particles with the slippage of 20mm. There are mainly two forms in cement: block and dispersion. As can be seen from Fig. 7, The gray zone is in the uncemented state between particles, which accounts for about 60%. The cemented particles (blue zone) are evenly dispersed in the cement, the supporting and protecting performance of cement is weakened.

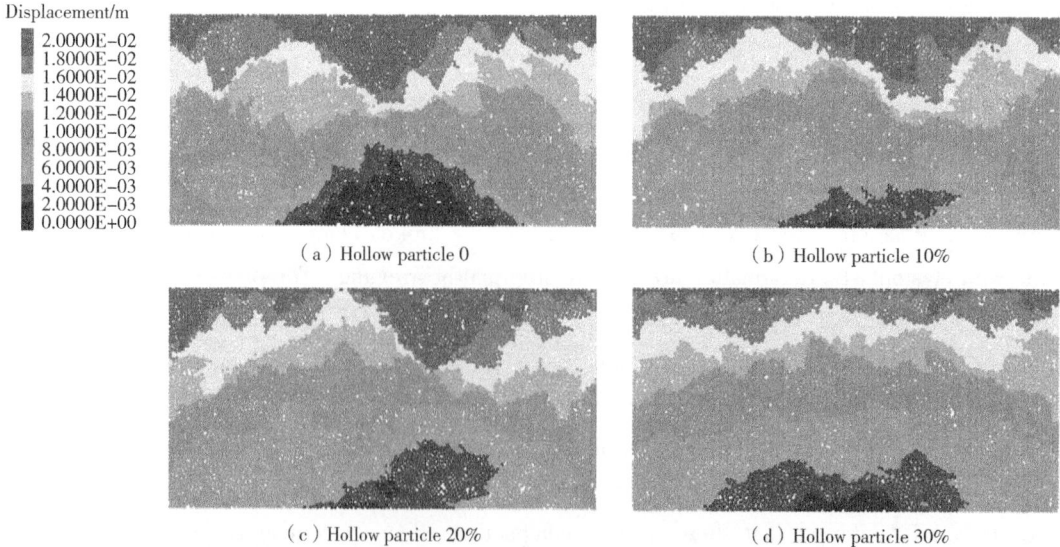

Fig. 6 Displacement distribution of cement with different contents of hollow microbeads after formation slip

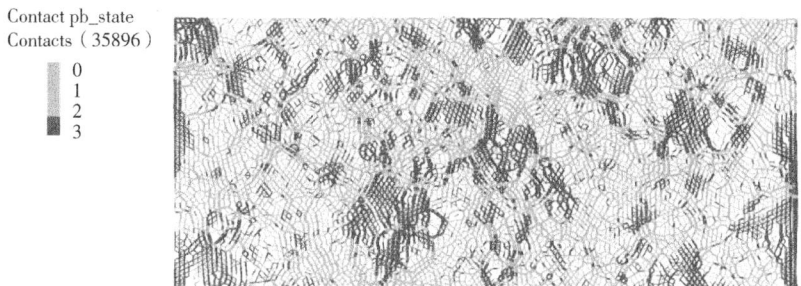

Fig. 7 Adhesion distribution of compressed particles

3 Extrusion test of casing-cement system

3.1 Experimental device

The device for casing deformation simulation in Fig. 8 is mainly composed of host frame, mechanical loading system, temperature control system, strain measurement system, data acquisition system. It can realize the axial tension/compression, shear, external extrusion under different temperature (Yang et al., 2023). A simulated wellbore was prepared before extrusion. The steel grade of casing is 110ICY, 125SG and 140V, the outer diameter is 139.7mm and 1281mm, the thickness is 10.54mm and 9.17mm. The cement thickness is 38.1mm. The water/cement ratio is 0.44. The mixed cement slurry of class G was poured into the annular space between the casing and the iron ring. The curing time was 15 days under the temperature of 80℃. The wellbore was placed on the fixture after preparation, and the extrusion force was applied to the wellbore by the loading system. The outer ring of simulated wellbore was removed to observe deformed casing and broken cement.

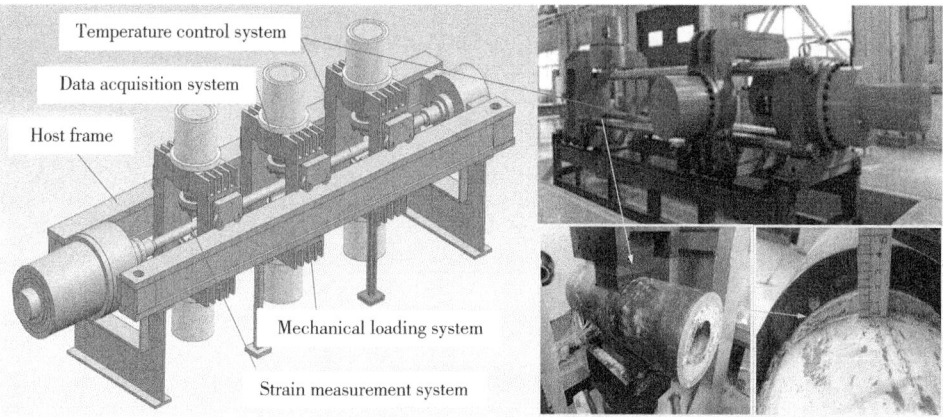

Fig. 8 The experimental device

3.2 Experimental scheme

According to the statistical data in southwestern China, about 77% of casing was damaged by extrusion and 23% was damaged by shear. The casing deformation caused by external extrusion is the main failure mode during hydraulic fracturing. Therefore, the wellbore was extruded under non-uniform loading. The casing length was 0.8m, and the clamps were squeezed 20mm respectively in experiment. The appearance, size and deformation of the casing were measured and analyzed after extrusion. The specific experimental scheme is shown in Table 2. Among them, elastic cement slurry was used in No. 1#, 6# and 11# experiment, cement slurry with 10% high strength hollow microbeads was used in No. 2#, 7# and 12# experiment, and cement slurry with 20% high strength hollow microbeads was used in the other experiments. The properties of cement slurry are shown in Table 3. According to the downhole performance requirements of hollow particles, the compressive strength of hollow particle is selected as 52.7MPa, and the density range is between $0.6 \sim 0.66 \text{g/cm}^3$. It can be seen that the deformation degree is mainly distributed in the range of 10% ~ 20%. Deformation degree increases with the increase of casing diameter and thickness. Deformation degree is decreased with a higher proportion of hollow microbeads. The reason is that the broken hollow microbeads absorb part of formation displacement, which greatly protects the casing string.

Table 2 Thickness and outer diameter of casing-cement sheath with different specifications in experiment

No	Steel grade/ksi	Thickness/mm	Outer diameter/mm	Inner diameter/mm	Hollow particle content/%	Minimum inner diameter after extrusion of 40mm/mm	Value of deformation/mm	Degree of deformation/%
1	110	10.54	139.7	118.62	0	96.55668	22.06332	18.6
2	110	10.54	139.7	118.62	10	98.33598	20.28402	17.1
3	110	10.54	139.7	118.62	20	100.47114	18.14886	15.3
4	110	9.17	139.7	121.36	20	100.97152	20.38848	16.8
5	110	10.59	128.1	106.92	20	95.05188	11.86812	11.1
6	125	10.54	139.7	118.62	0	97.98012	20.63988	17.4
7	125	10.54	139.7	118.62	10	100.58976	18.03024	15.2
8	125	10.54	139.7	118.62	20	103.08078	15.53922	13.1
9	125	9.17	139.7	121.36	20	103.64144	17.71856	14.6
10	125	10.59	128.1	106.92	20	97.19028	9.72972	9.1
11	140	10.54	139.7	118.62	0	100.47114	18.14886	15.3
12	140	10.54	139.7	118.62	10	103.79250	14.82750	12.5
13	140	10.54	139.7	118.62	20	106.04628	12.57372	10.6
14	140	9.17	139.7	121.36	20	106.91816	14.44184	11.9
15	140	10.59	128.1	106.92	20	98.79408	8.12592	7.6

Table 3 Parameters of cement slurry

Slurry type	Temperature/°C	Density/(g/cm³)	Initial consistency/Bc	Thickening time/min	Compressive strength in 48h/MPa
Ordinary cement	100	1.83	23	150	23.4
Cement with 10% particle	100	1.81	24	162	20.8
Cement with 20% particle	100	1.78	28	168	19.6

3.3 Numerical model

According to the non-uniform extrusion test, the casing-cement model in Fig. 9 was set up. Based on experiment 3#, the Poisson's ratio of casing is 0.24, its elastic modulus is 213GPa, and its yield strength is 836MPa. The two clamps were squeezed 20mm respectively in the numerical model. The shape, deformation and stress variation of the casing during extrusion were observed.

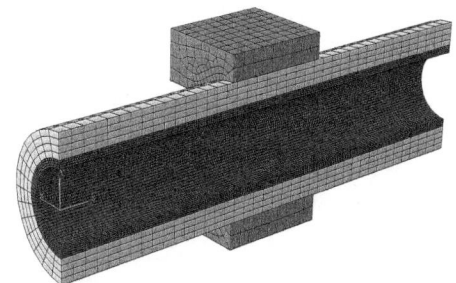

Fig. 9 Numerical model

Fig. 10 shows the casing deformation degree and stress distribution after extrusion. It can be seen from Fig. 10 that the casing is reduced due to non-uniform extrusion. The calculation show that the minimum diameter after extrusion is 117.8mm, and the minimum diameter measured after experiment is 118.3mm, The error is only 2.3%. Compared with the morphology of the extruded cement in Fig. 11, the wellbore is obviously compressed at the extruded position, especially the cement sheath. The thickness of the extruded area is significantly reduced with comminution failure of cement sheath. The shape of cement sheath after extrusion is shown in Fig. 11. There are local cracks in cement under the extrusion force. According to the above numerical and experimental results, the numerical model can effectively trace the casing-cement extrusion process. It shows that the modified cement sheath has a certain protective effect on the casing by comparing casing deformation degree with different hollow particle content.

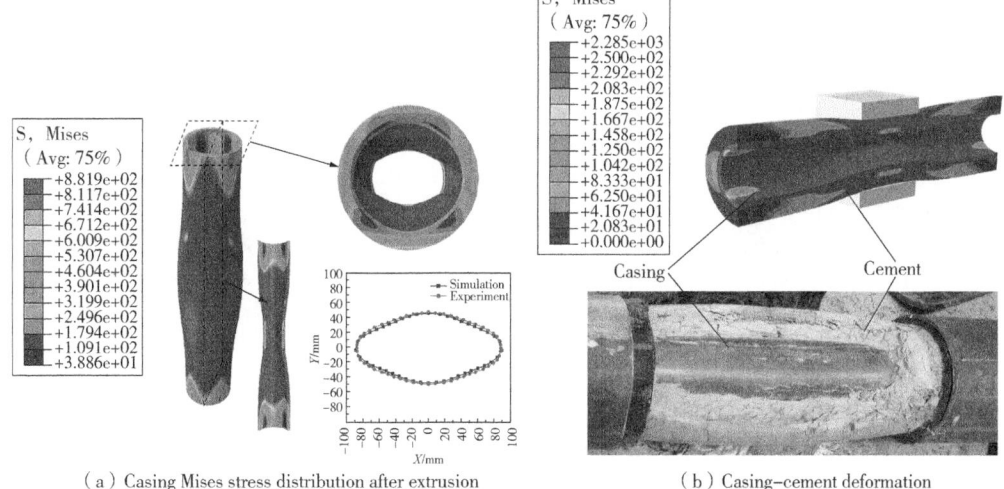

(a) Casing Mises stress distribution after extrusion   (b) Casing-cement deformation

Fig. 10 Comparison between numerical and experimental results

3.4 Analysis of influencing factors

The steel grade, casing thickness, particle size and particle content were varied in experiment to investigate the influence on casing deformation degree after slippage. Fig. 12(a) shows the relationship between casing deformation degree and formation slippage under different casing steel grade. As can be seen from Fig. 12(a), the deformation degree gradually decreases with the increase of casing steel grade. The larger formation slip, the more obvious the reduction degree. Fig. 12(b) shows the relationship between casing deformation degree and formation slippage under different casing thickness. As can be seen from Fig. 12 (b), the deformation degree gradually decreases with the increase of casing thickness. The larger formation slip, the more obvious the reduction degree. However, improving the steel grade and casing thickness will make the well construction cost rise sharply on the field.

Fig. 11 Failure pattern of modified cement sheath after extrusion

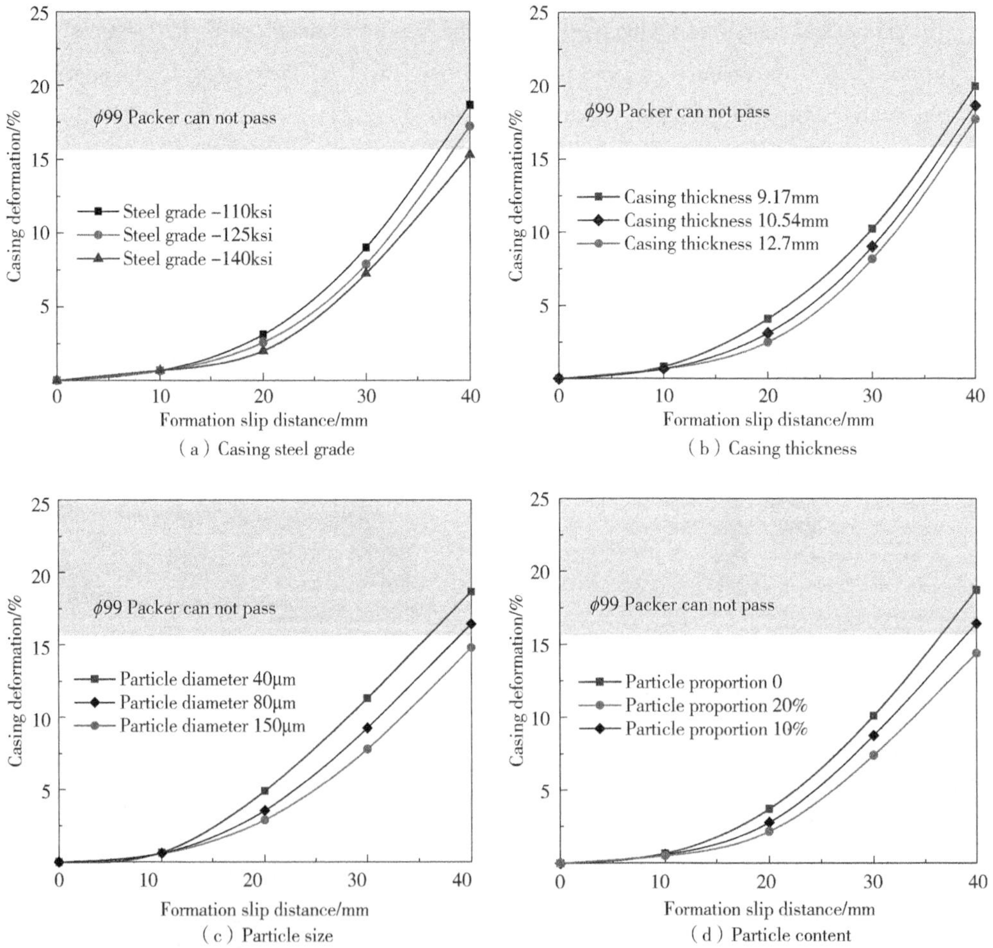

Fig. 12 Influence of different factors on casing inner diameter after formation slip

Fig. 12(c) shows the relationship between casing deformation degree and formation slippage under different hollow particle sizes. As can be seen from Fig. 12(c), the deformation degree gradually decreases with the increase of particle size. The larger formation slip, the more obvious the reduction degree. Fig. 12(d) shows the relationship between casing deformation degree and formation slippage under different hollow particle contents. As can be seen from Fig. 12(d), the deformation degree gradually decreases with the increase of particle content. The larger formation slip, the more obvious the reduction degree. The influence degree of the particle size and content on the casing deformation degree is larger than that of the casing steel grade and thickness. The addition of particles will not affect the drilling and fracturing process on the field, and the prevention effect of casing deformation is more obvious.

4 Field test

The geostress is heterogeneous in Weiyuan gas field of Sichuan Basin in China, and the casing string has been subjected to non-uniform extrusions chronically. In addition, the propagation of artificial fractures and the local pressure in the fracture will change the stress distribution near the wellbore, thus aggravating the extrusion load on the casing (Zhang et al., 2020). The statistical analysis of casing deformation locations for 20 platforms in Block 204 of Weiyuan area shows that the casing deformation is concentrated near the inclined section at point A in Fig. 13, and the proportion of deformation decreases from point A to point B. This rule is also consistent with the engineering practice. Fracturing procedure moves gradually from point B to point A in engineering, and the deformation caused by each stage will gradually accumulate. The closer position is to point A, the larger deformation possibility will be.

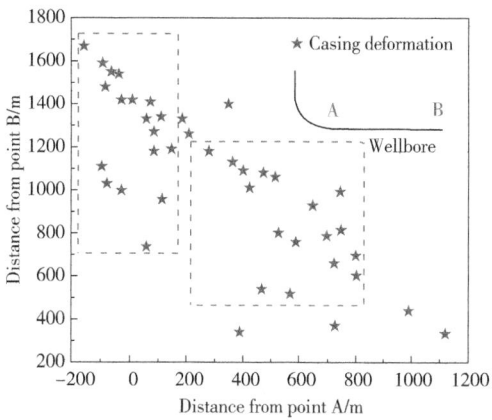

Fig. 13 Location statistics of casing deformation in Block 204 of Weiyuan Zone

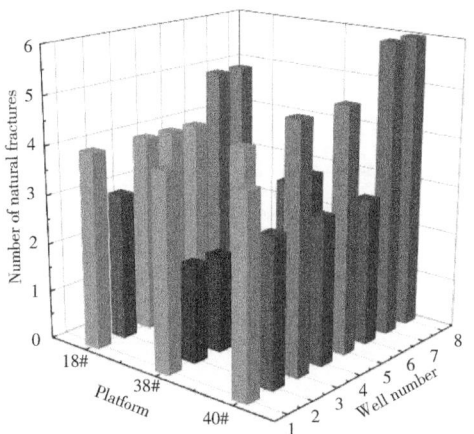

Fig. 14 Fault number encountered in each well

In order to prove the effect of hollow particle on preventing casing deformation, well 18-5, well 38-4 and well 40-3 in Block 204 of Weiyuan area were selected to use cement slurry with 15% hollow particle. Class G oil well cement was used in the other wells of the three platforms. The cement slurry of the offset wells is the ordinary cement system without hollow particles. The water-cement ratio is 0.44, and the its density ranges from 1.80~1.82g/cm^3. The hard trip of downhole tools in the test wells was compared with the surrounding offset wells. Fig. 14 shows the fault numbers encountered by 22 wells in the three platforms. The number of faults

· 121 ·

mainly distributed in 3~6. Table 4 shows the statistical results of the casing deformation in these wells in Weiyuan block. There is no downhole hard trip in W204H38-4 and W204H40-3, while the surrounding wells all have. Besides that, the W204H18-5 encountered downhole sticking three times, W204H18-4 and W204H18-6 encountered sticking five times and four times respectively. The prevention effect of casing deformation in W204H18-5 cannot be judged only by the sticking times. The reason is the fault zone at Platform 18# is widely distributed and the fault is relatively active. Logging data of MIT24 arm in well W204H18-5 and W204H18-6 show that the maximum casing deformation degree in well 5 is 8.7mm. The maximum casing deformation degree of well 6 is 11.2mm. Therefore, the modified cement slurry system can effectively reduce the possibility of downhole hard trip in the fracturing process, and effectively control the downhole casing deformation degree.

Table 4 Casing deformation statistics of test wells in Weiyuan block

No	Well number	Number of designed segments	Whether to use hollow cement	Number of stages which the fracturing tool is blocked
1	W204H18-1	17	No	2
2	W204H18-2	18	No	4
3	W204H18-3	18	No	5
4	W204H18-4	18	No	5
5	W204H18-5	18	Yes	3
6	W204H18-6	18	No	4
7	W204H18-7	18	No	4
8	W204H38-1	17	No	0
9	W204H38-2	18	No	4
10	W204H38-3	17	No	15
11	W204H38-4	17	Yes	0
12	W204H38-5	17	No	6
13	W204H38-6	18	No	4
14	W204H38-7	17	No	0
15	W204H40-1	18	No	2
16	W204H40-2	18	No	0
17	W204H40-3	19	Yes	0
18	W204H40-4	18	No	2
19	W204H40-5	18	No	0
20	W204H40-6	18	No	2
21	W204H40-7	18	No	2
22	W204H40-8	18	No	2

5 Discussion

According to Section 3 and 4, the formation slip can be effectively absorbed by adding hollow particles into the cement slurry and the degree of casing deformation is reduced. The cement breaking process and the stress–strain differences of casing was briefly summarized before and after adding hollow particles in Fig. 15. As the blue line in Fig. 15, the cement steps into the elastic state firstly when the formation begins to extrude the wellbore. The cement transits into the plastic state as the displacement loading continues to increase. Because the particle is more brittle than cement, the particles start breaking once the compressive strength is reached. When the particle and cement damage to a certain extent, the

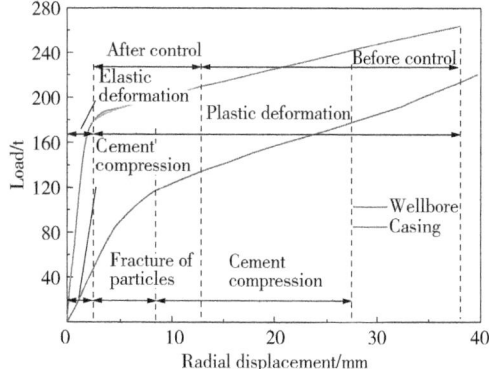

Fig. 15 Illustration of deformation prevention technology based on cement slurry modification

combination steps into a stable plastic stage until the end of slip. Before adding the hollow particles, the radial displacement will increase sharply when the load rises to a certain value due to the poor cement compressibility, as the red line of Fig. 15. However, the load will be shared with the addition of hollow particles due to the strong compressibility and the casing radial displacement will be controlled.

6 Conclusion

The fracture mechanism of hollow particles in cement sheath was analyzed by discrete element method in this paper, and the influence of hollow cement on casing deformation degree was investigated by laboratory experiment. According to the experimental results and field application, the casing deformation prevention technology based on cement slurry modification was summarized, and the conclusions were as follow:

(1) The cement sheath has good compressibility by adding hollow particles into the cement slurry when the formation slips. The particles can effectively absorb the formation displacement and transfer the excessive casing deformation.

(2) The experimental results show that the larger the diameter and content of hollow particles, the lower the casing deformation degree after formation slip. Compared with improving the steel grade and thickness of casing, the parameters of hollow particles in cement have a larger influence on the degree of casing deformation mitigation.

(3) Based on the serious casing deformation of shale gas well in southwest China, a casing deformation control method based on modified cement slurry was set up by means of numerical simulation and laboratory test. According to the feedback of field tests, the modified cement slurry system can decrease the casing deformation degree by 11.9% when the proportion of hollow particles is 15%, ensuring the smooth trip of fracturing tools.

Acknowledgements

The authors would like to gratefully acknowledge the supports of Scientific Research and Technology Development of CNPC (2021DJ2705, 2021DJ4403, 2020B-4020), Study on Key

Technologies of Production Increase and Transformation of Gulong Shale Oil (2021ZZ10-04), Innovative Talents Promotion Plan-Young Science and Technology Star Project (2021KJXX-63).

References

[1] Adjei S, Elkatatny S, Sarmah P, et al. Investigation of dehydroxylated sodium bentonite as a pozzolanic extender in oil-well cement. SPE Drilling & Completion, 2021, 36(3): 730-737.

[2] Cheng Y Q, Liu S Q, Shen J Y, et al. Matching analysis and experimental study of mechanical properties of cement sheath interface. Journal of Petroleum Science and Engineering, 2023, 220: 111138.

[3] Dong K, Liu N Z, Chen Z W, et al. Geomechanical analysis on casing deformation in Longmaxi shale formation. Journal of Petroleum Science and Engineering, 2019, 177: 724-733.

[4] Furui K., Fuh G., Morita N., et al. Casing-and screen-failure analysis in highly compacting sandstone fields. SPE Drilling & Completion, 2012, 27(2): 241-252.

[5] Guo S L, Bu Y H, Zhou A N, et al. A three components thixotropic agent to enhance the thixotropic property of natural gas well cement at high temperatures[J]. Journal of Natural GasScience and Engineering, 2020, 84: 103699.

[6] Guo X L, Li J., Liu G H, et al., Numerical simulation of casing deformation during volume fracturing of horizontal shale gas wells. Journal of Petroleum Science and Engineering. 2019, 172: 731-742.

[7] Guo Y, Blanford M, Candella J. D.. Evaluating the risk of casing failure caused by high-density perforation: A 3D finite-element-method study of compaction-induced casing deformation in a deepwater reservoir, Gulf of Mexico. SPE Drilling & Completion, 30(2): 141-151.

[8] Hudson M D, Sheperd P, Ricci J. Flexible cement slurry survives multistage hydraulic fracturing treatment. SPE Unconventional Resources Conference, February 15-16, 2017. SPE-185068-MS.

[9] Jafariesfad N, Geiker M R, Skalle P. Nanosized Magnesium Oxide with Engineered Expansive Property for Enhanced Cement-System Performance. SPE Journal. 2017, 22(5): 1681-1689.

[10] Jafariesfad N, Sangesland S, Gawel K, et al. New materials and technologies for life-lasting cement sheath: A review of recent advances. SPE Drilling & Completion, 2020, 35(2): 262-278.

[11] Jafariesfad N., Geiker M. R., Gong Y. et al. Cement Sheath Modification Using Nanomaterials for Long-Term Zonal Isolation of Oil Wells: Review. Journal of Petroleum Science and Engineering, 2017, 156: 662-672.

[12] Li H T, Li Z, Li G, et al. Casing deformation mechanisms of horizontal wells in Weirong shale gas field during multistage hydraulic fracturing. Journal of Natural Gas Science and Engineering, 2020, 84, 103646.

[13] Li H, Yang C H, Ma H L, et al. A 3D grain-based creep model(3D-GBCM) for simulating long-term mechanical characteristic of rock salt. Journal of Petroleum Science and Engineering, 2020, 185: 106672.

[14] Li Z, Li G, Li H, et al. A novel investigation on casing deformation during hydraulic fracturing in the Weirong shalegas field, Sichuan basin, China. Arabian Journal of Geosciences, 2021, 14(24): 1-15.

[15] Li Z, Li H, Li G, et al. The influence of shale swelling on casing deformation during hydraulic fracturing. Journal of Petroleum Science and Engineering, 2021, 205(9): 108844.

[16] Lian Z H, Yu H, Lin T J, et al. A study on casing deformation failure during multi-stage hydraulic fracturing for the stimulated reservoir volume of horizontal shale wells. Journal of Natural Gas Science & Engineering, 2015, 23: 538-546.

[17] Liu K, Gao D L, Yang J, et al. Effect of expandable cement on increasing sealing ability of cement sheath in shale gas wells. Journal of Petroleum Science and Engineering, 2019, 176: 850-861.

[18] Liu K, Taleghani A D, Gao D L. Calculation of hydraulic fracture induced stress and corresponding fault slippage in shale formation. 2019, Fuel, 254: 115525

[19] Meng H, Ge H K, Fu D W, et al. Numerical investigation of casing shear deformation due to fracture/fault slip during hydraulic fracturing. Energy Science and Engineering, 2020, 8(10): 1-14.

[20] Peng S P, Fu, J T., Zhang, J C. Borehole casing failure analysis in unconsolidated formations: a case study. Journal of Petroleum Science and Engineering, 2007, 59(4): 226-238.

[21] Quercia G, Brouwers H J H, Garnier A, et al. Influence of Olivine Nano–Silica on Hydration and Performance of Oil-Well Cement Slurries. Materials and Design, 2016, 96: 162-170.

[22] Salehi S, Khattak M J, Ali N, et al. Study and Use of Geopolymer Mixtures for Oil and Gas Well Cementing Applications. Journal of Energy Resource Technology, 2018, 140(1): 012908.

[23] Santos L, Taleghani A D, Li G. Nanosilica–treated shape memory polymer fibers to strengthen wellbore cement. Journal of Petroleum Science and Engineering, 2021, 196: 107646.

[24] Teodoriu C, Yi M C, Salehi S. A Novel Experimental Investigation of Cement Mechanical Properties with Application to Geothermal Wells. Energies, 2019, 12(18): 3426.

[25] Wang Y B, Liu K, Gao D L. Investigation of the interface cracks on the cement sheath stress in shale gas wells during hydraulic fracturing. Journal of Petroleum Science and Engineering, 2021, 205: 108981.

[26] Wei S M, Kuru E, Jin Y, et al. Numerical investigation of the factors affecting the cement sheath integrity in hydraulically fractured wells. Journal of Petroleum Science and Engineering, 2022, 215: 110582.

[27] Xi Y, Jiang J W, Li J, et al. Research on the influence of strike-slip fault slippage on production casing and control methods and engineering application during multistage fracturing in deep shale gas wells. Energy Reports, 2021, 7: 2989-2998.

[28] Xi Y, Li J, Fan L, et al. Mechanism and numerical simulation of a new device of bypass cementing device for controlling casing shear deformation induced by fault slipping. Journal of Petroleum Science and Engineering, 2020, 196(5): 107820.

[29] Xi Y, Li J, Liu G H, et al. A new numerical method for evaluating the variation of casing inner diameter after strike-slip fault sliding during multistage fracturing in shale gas wells. Energy Science and Engineering, 2019, 7: 2046-2058.

[30] Xi Y, Li J, Liu G H, et al. Numerical investigation for different casing deformation reasons in Weiyuan–Changning shale gas field during multistage hydraulic fracturing. Journal of Petroleum Science & Engineering, 2018, 163: 691-702.

[31] Yan W, Zou L Z, Li H, et al. Investigation of casing deformation during hydraulic fracturing in high geo-stress shale gas play. Journal of Petroleum Science & Engineering, 2017, 150: 22-29.

[32] Yang S Y, Zeng B, Yan Y, et al. Research on casing deformation mechanism and prevention technology in salt rock creep formation. Journal of Petroleum Science and Engineering. 2023, 220: 111176.

[33] Yin F, Han L H, Yang S Y, et al. Casing deformation from fracture slip in hydraulic fracturing. Journal of Petroleum Science and Engineering, 2018, 166: 235-241.

[34] Yin F., XiaoY, Han L H, et al. Quantifying the induced fracture slip and casing deformation in hydraulically fracturing shale gas wells. Journal of Natural Gas Science and Engineering, 2018, 60: 103-111.

[35] Yu H, Taleghani A. D., Lian Z H, et al. Severe Casing Failure in Multistage Hydraulic Fracturing Using Dual-Scale Modeling Approach. SPE Drilling & Completion, 2022, 37(3): 252-266.

[36] Yu, H., Lian, Z H, Lin, T J, et al. Study on failure mechanism of casing in stimulated reservoir volume fracturing of shale gas. Journal of Safety Science and Technology, 2016, 12(10), 37-43.

[37] Zhang F S, Yin Z R, Chen Z W, et al. Fault reactivation and induced seismicity during multistage hydraulic

fracturing: Microseismic analysis and geomechanical modeling. SPE Journal, 2020, 25(2): 692-711.

[38] Zhang F, Jiang Z, Chen Z, et al. Hydraulic fracturing induced fault slip and casing shear in Sichuan Basin: A multi-scale numerical investigation. Journal of Petroleum Science and Engineering, 2020, 195: 107797.

[39] Zhao C J, Li J, Zaman F, et al. Investigation of casing deformation characteristics under cycling loads and the effect on casing strength based on full-scale equipment. Journal of Petroleum Science and Engineering, 2021, 205: 108973.

本论文原发表于《Petroleum Science》2023 年。

二、先进油气钻采输送管材制造

不同 Nb 含量 X80 钢管环焊热影响区的微观组织与性能

何小东[1,2]　杨耀彬[1,2]　陈越峰[1,2]　David Han[2]　张永青[2,3]

(1. 中国石油集团工程材料研究院有限公司，石油管材及装备材料服役行为与结构安全国家重点实验室；2. 国际焊接研究中心；3. 中信金属股份有限公司)

摘　要：为研究 Nb 含量对焊接热影响区微观组织和性能的影响，采用熔化极气体保护焊(gas metal arc welding, GMAW)和手工焊条电弧焊(shielded metal arc welding, SMAW)对 0.055%Nb 和 0.075%Nb 含量的 X80 钢管进行环焊。采用夏比冲击试验和金相分析方法，研究热影响区的微观组织差异和夏比冲击韧性。并借助扫描电镜和超高温激光共聚焦显微镜分析不同 Nb 含量 X80 管体的微观组织形貌对热影响区性能的影响。结果表明，在 0℃和-20℃时，0.075%Nb 和 0.055%Nb 的 X80 钢管 GMAW 环焊接头热影响区均具有较高的冲击韧性，其平均冲击吸收能量均高于 150J。但是，0.055%Nb 略高于 0.075%Nb 的 GMAW 环焊接头热影响区夏比冲击吸收能量；焊接热输入较低时，0.055%Nb 低于 0.075%Nb 的 X80 环焊接头粗晶区的韧脆转变温度，具有更好的低温韧性。焊接热输入较高时，与 0.05%Nb X80 相比，0.075%Nb 的 X80 环焊接头粗晶区具有更高的上平台冲击吸收能量，且上平台温度和韧脆转变温度也更低，其低温韧性也更优异；还探讨了 X80 环焊接头热影响区的冲击韧性不仅与热输入大小和热影响区马氏体—奥氏体组织(M-A)的形状、大小、分布有关，而且还受管体中 Nb 含量、原始的强度与韧性、微观组织状态的遗传影响。

关键词：含铌 X80 钢管；环焊接头；热影响区；冲击韧性；M-A 组织

铌微合金化高强度钢应用历史已达 70 多年[1]。特别是在 20 世纪 80 年代早期，人们采用添加铌来设计新一代高强度低合金钢[2]，充分利用铌的固溶和析出行为，结合热机械轧制(thermo-mechanical controlled process, TMCP)工艺，达到细化晶粒、控制相变和析出强化的效果，从而获得具有高强度、高韧性的钢材，如 X80 管线钢。在随后的焊接过程中，根据焊接工艺的不同，含铌管线钢热影响区(heat-affected zone, HAZ)经历了一系列的奥氏体形成和分解循环，尤其是在临界再热影响区(inter-critically reheated HAZ, ICRHAZ)奥氏体的部分形成和分解导致了微观组织的复杂性。因此，X80 管线钢热影响区的微观组织演化受基体中 Nb 含量的强烈影响[3]。

焊接热影响区的微观组织和韧性是影响高钢级油气管道完整性的重要因素。虽然 X80

基金项目：国家重点研发计划项目(2023YFB4707205)。

作者简介：何小东，硕士，正高级工程师/国际焊接工程师；主要从事管线钢焊接工艺、材料性能测试及表征研究；E-mail: xiaodonghe@126.com。

高强度管线钢具有良好的抗延性断裂能力,但焊接过程中形成的热影响区,尤其是粗晶区(coarse-grained heat-affected zone, CGHAZ),其微观组织分布不均匀,且具有高的局部化特征,使得该区域力学性能变差,容易形成诱发裂纹的局部脆性区,是整个焊接结构的薄弱地带。文献[4]研究认为,对于含Nb量为0.1%的X80管线钢,虽然热影响区原始奥氏体晶粒的平均尺寸不会随热输入增加而过于粗大,但当热输入高于40kJ/cm时,会使得原始奥氏体晶粒内粒状贝氏体的晶体取向选择过于单一,大角晶界(大于15°)密度会明显降低,有效晶粒尺寸较大,马氏体—奥氏体组元(M-A)也由于热输入量过大而明显粗化,从而导致粗晶区的韧性明显降低。Teixeira1等[5]研究了高强钢热影响区的组织梯度对焊接接头不稳定断裂行为的影响,认为粗晶区组织基本由粗贝氏体组成,并在大的原始奥氏体晶粒的晶界处有少量马氏体和先共析铁素体。随预制疲劳裂纹前缘侵入CGHAZ,其韧性明显下降。文献[6]通过试验和3D有限元模拟,研究了焊接热模拟X80管线钢在不同温度下的断裂韧度,认为随着温度降低断裂韧性减小,并使钢材由韧性断裂向脆性断裂转变,不同温度下测得的裂纹尖端张开位移值均具有一定的分散性,且分散程度随温度升高而增大。袁军军等[7]认为冲击试样的取样位置、缺口尖端组织状态和缺陷等因素对X70管线钢药芯焊丝多层多道焊接头的冲击性能稳定有一定影响,试样缺口处柱状晶所占比例和粗大晶粒是导致冲击韧性出现波动和低值的主要原因。管线钢热影响区微观组织及其性能还受焊接热输入量(heat input, HI)、其他合金元素和碳当量的影响。Mohsen Mohammadijoo等[8]研究发现,X70管线钢热影响区软化随着Mo,Mn,Ti,N和碳当量的增加,热影响区软化程度逐渐减小,但合金的添加对HAZ韧性产生了不利影响,尤其是对填充焊和盖面焊热影响区的影响。文献[9]采用热模拟试样研究了铌微合金钢焊接热影响粗晶区的微观组织的主要相为含有大量M-A的粒状贝氏体,并利用原子探针断层扫描(atom probe tomography, APT)技术,研究了铌在原始奥氏体晶界(prior austenite grain boundary, PAGB)、铁素体/M-A界面和铁素体/铁素体界面的分布,结果表明,Nb在铁素体/M-A界面处富集最明显,Nb偏析降低了PAGB的吉布斯能。原始奥氏体晶界处Nb的强偏析可以有效地防止高温时奥氏体晶粒的生长,而铁素体/M-A界面处Nb偏析可以抑制冷却时贝氏体、铁素体的生长,进一步解释了焊接后含Nb微合金钢中贝氏体组织较细的原因,Nb原子与空位的结合能预测结果也表明焊接热循环对Nb的偏析是非平衡机制所致。

采用熔化极气体保护焊与手工焊条电弧焊的实焊方法,对接头取样进行夏比冲击试验,且利用高温激光共聚焦显微镜观察组织结构,研究了不同Nb含量X80钢管在低热输入和较高热输入下环焊缝HAZ的微观组织与性能,为高强度管线钢的成分优化设计与焊接工艺选择提供了试验依据。

1 试验材料与方法

1.1 试验材料

试验材料选用了含铌量为0.055%和0.075%的X80直缝埋弧焊钢管,钢管直径为1219mm,壁厚为22mm。两种铌含量钢管分别标记为N055和N075,具体化学成分见表1。N055和N075钢管的焊接冷裂纹敏感系数(CE_{Pcm})分别为0.165%和0.167%,管体纵向的拉伸屈服强度分别为567MPa和565MPa,抗拉强度分别为645MPa和689MPa,断后伸长率分别为21%和25%;N055和N075管体纵向平均冲击吸收能量在0℃时分别为401J和375J,在-20℃时分别为391J和340J。

表1　试验钢管的化学成分　　　　　　　　　　　　单位:%(质量分数)

编号	C	Mn	Si	P	S	Cr	Mo	Ni	Nb	V	Ti	Cu	B	Al	N	Fe	CE_{Pcm}
N055	0.049	1.74	0.15	0.010	0.0025	0.25	0.093	0.16	0.058	0.0043	0.011	0.022	0.0003	0.027	0.0031	余量	0.165
N075	0.051	1.72	0.16	0.013	0.0026	0.26	0.088	0.16	0.082	0.0045	0.012	0.026	0.0003	0.025	0.0033	余量	0.167

1.2　环焊缝焊接与试验方法

由于实际的环焊接头熔合区的形状极不规则,对熔合线处的夏比冲击离散性影响较大,因此,为了更准确地研究在熔化极气体保护自动焊(Auto-GMAW)和SMAW两种典型热输入下,不同Nb含量X80环焊接头热影响区的韧性,参照API RECOMMENDED PRACTICE 2Z《海上结构用钢板预生产评定推荐作法》标准,采用如图1所示的单侧V形坡口。GMAW环焊采用直径为1.0mm的ER80S-G(BOHLER SG 8-P)的实心焊丝;SMAW采用直径为3.2mm的E9018-G焊条进行根焊,填充和盖面焊采用直径为4.0mm的E11018-G焊条。GMAW和SMAW的焊接工艺参数分别见表2和表3。

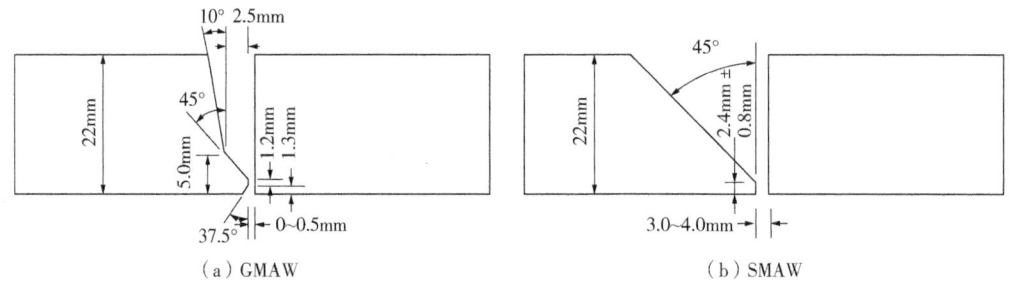

图1　单侧V形坡口示意图

表2　单侧双V形坡口GMAW焊接工艺参数

焊接层	焊接电流 I/A	焊接电压 U/V	送丝速度 $v_f/(m/min)$	焊接速度 $v_w/(mm/min)$	保护气体 混合比例($Ar:CO_2$)	保护气体 流量 $Q/(L/min)$	平均热输入 $\overline{Q}/(kJ/mm)$
根焊	150~200	22~26	8.13~9.65	460~660	4:1	20~25	0.48
热焊	180~240	24~27	11.43~12.92	560~760	4:1	30~35	0.52
填充焊	160~220	24~26	8.13~10.92	330~460	4:1	25~35	0.75
盖面焊	150~200	22~26	7.62~9.14	330~460	4:1	20~30	0.65

表3　单侧V形坡口SMAW焊接工艺参数

焊接层数	焊接电流 I/A	焊接电压 U/V	焊接速度 $v_w/(mm/min)$	平均热输入 $\overline{Q}/(kJ/mm)$
根焊	130~150	24~27	90~100	2.10
热焊	130~150	24~27	100~120	1.91
填充焊	170~210	24~27	90~120	2.84
盖面焊	170~200	24~27	90~120	2.70

焊接完成后,从环焊接头3点钟位置截取试块制备成金相试样,经2%硝酸酒精溶液浸蚀后,在OLS 4100激光共聚焦显微镜下观察直边侧坡口热影响区的微观组织。并从立焊位置(环焊缝2点钟至5点钟)截取试块,在壁厚中心制备V形缺口夏比冲击试样,以保证试样位于填充焊道热影响区。冲击试样尺寸为10mm×10mm×55mm,V形缺口轴向位于直边侧

HAZ不同位置处。$FL_{0.5}$，$FL_{1.0}$和$FL_{2.0}$（FL为缺口位置位于50%WM+50%HAZ）分别代表缺口轴线距FL处0.5mm，1.0mm，2.0mm。图2为缺口位于$FL_{0.5}$的示意图。并利用PSW750冲击试验机，依据GB/T 229—2020《金属材料夏比摆锤冲击试验方法》进行夏比冲击试验。Boltzmann函数具有S形曲线形状，与金属材料冲击吸收能量—温度关系曲线的形状非常吻合，满足下平台区、转变温度区和上平台区3阶段分布特征，物理意义明确、相关性高，是较为认可的数学模型[10-12]。因此，采用Boltzmann函数模型对系列冲击试验测试数据进行拟合，获得热影响区的韧脆转变温度（ductile-brittle transition temperature，DBTT）。为了进一步研究管体母材的性能和原始微观组织状态对环焊接头热影响区微观组织和韧性的影响，采用显微硬度计测试了GMAW环焊接头的硬度分布，再用Matlab软件绘制了硬度云图。借助扫描电镜观察了N055和N075管体的微观组织。同时，采用高温激光共聚焦显微镜在相同的条件下将N055和N075分别加热至1350℃保温1s，冷却至150℃后，二次加热至780℃保温1s，再冷却至200℃，以此模拟X80管道环焊接头粗晶区二次热循环，进一步对比不同Nb含量对X80环焊热影响区组织转变的影响。

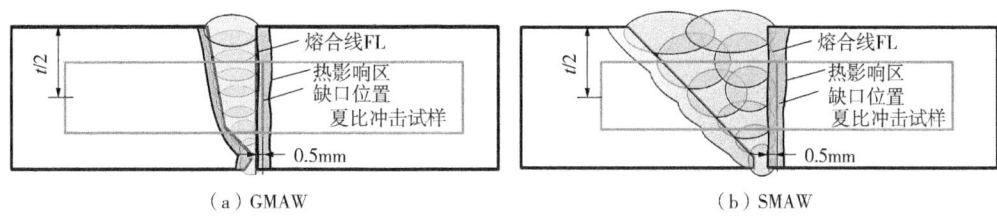

图2　冲击试样缺口位于$FL_{0.5}$处的示意图

2 试验结果与分析

2.1 不同热输入下热影响区的冲击韧性

图3为N055和N075钢管GMAW环焊接头热影响区不同位置的夏比冲击吸收能量与管体纵向夏比冲击吸收能量的对比。图3表明，N075和N055热影响区均具有较高的冲击韧性，其平均值高于150J。N055与N075环焊接头热影响区相比，采用较低热输入的环焊工艺时，N055焊接接头热影响区FL，$FL_{0.5}$，$FL_{1.0}$和$FL_{2.0}$在0℃和-20℃时夏比冲击吸收能量均略高于N075。

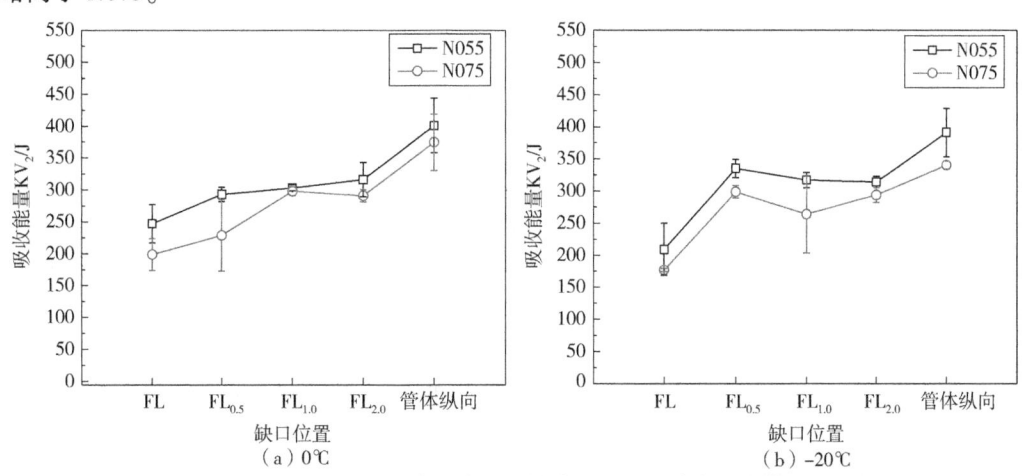

图3　GMAW环焊热影响区不同位置的夏比冲击吸收能量

图4为不同Nb含量X80单边V形坡口GMAW和SMAW环焊接头粗晶区的韧脆转变曲线。图4(a)表明，低热输入焊接时，N055和N075在粗晶区$FL_{0.5}$处韧脆转变的上平台温度分别达到-80℃和-60℃，且二者的上平台吸收能量相当，均约为300J。经Boltzmann函数拟合计算，N055的韧脆转变温度(DBTT)约为-104℃，N075的韧脆转变温度为-85℃。因此，采用低热输入的GMAW环焊时，N055和N075均具有优良的低温韧性。

图4(b)为N055和N075采用较高热输入的SMAW环焊接头$FL_{0.5}$处的韧脆转变曲线。由图4(b)可知，在较高热输入下，N055和N075在粗晶区$FL_{0.5}$处韧脆转变的上平台温度分别达到-30℃和-50℃，且N075的上平台能约为275J，而N055的上平台能约为230J。同时，从图4(b)可以看出，N055和N075较高热输入的SMAW环焊接头粗晶区的韧脆转变温度分别为-56℃和-77℃。因此，与N055相比，较大热输入环焊时，N075具有更高上平台冲击吸收能量和更低的上平台温度，其低温韧性更优异。

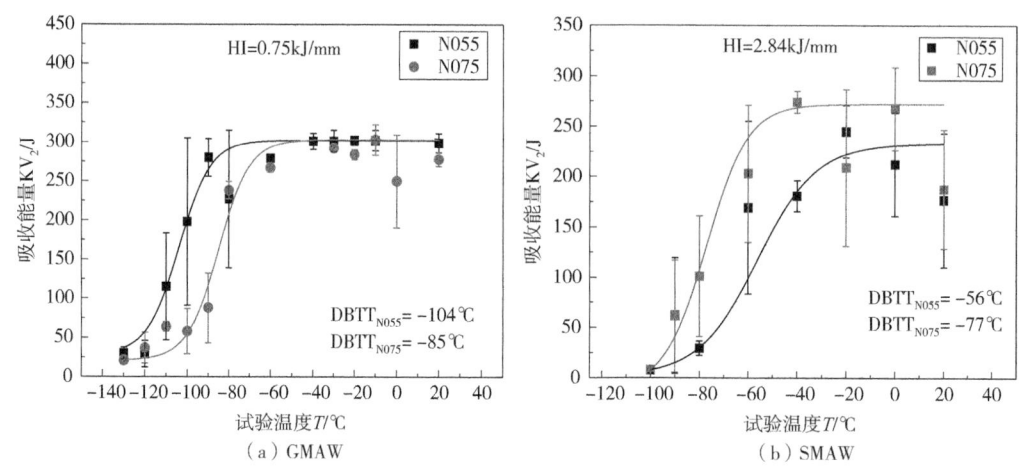

图4 环焊接头粗晶区的韧脆转变曲线

2.2 微观组织

焊接接头粗晶区的微观组织分布不均匀，且在临界二次加热粗晶区(Intercritically reheated coarse grained HAZ, IC CGHAZ)因存在链状分布的M-A组元而具有高的局部化特征被认为是其韧脆变差[13-14]的主要原因。图5和图6分别为不同铌含量X80在较低热输入的GMAW和较高热输入的SMAW焊接下粗晶区的微观组织。结果表明，在较低热输入和较高热输入下，N055和N075粗晶区(CGHAZ)的组织均以粒状贝氏体($B_{粒}$)为主。在较低热输入GMAW环焊时，N055和N075临界二次加热粗晶区(IC CGHAZ)的原始奥氏体晶界上分布有M-A链状组织，但是N075的IC CGHAZ内晶粒相对较小且更为均匀，M-A组织占比更高。当采用热输入较大的SMAW环焊时，N055和N075粗晶区晶界上无粗大的M-A链状组织分布，且N075粗晶区晶界清晰，其晶粒尺寸也比N055粗晶区的更细小。因此，N075 SMAW环焊粗晶区的韧性好于N055。

2.3 讨论与分析

焊接热影响区粗晶区的韧性恶化或波动影响因素极其复杂。通常认为，主要受热影响区中粗大的M-A组元、取样位置、缺口尖端组织状态和缺陷、试验温度、合金元素含量和碳当量的影响[4-8]。同时，管体母材的强度、韧性和原始组织状态对热影响区韧性的降低也有明显的遗传性影响。

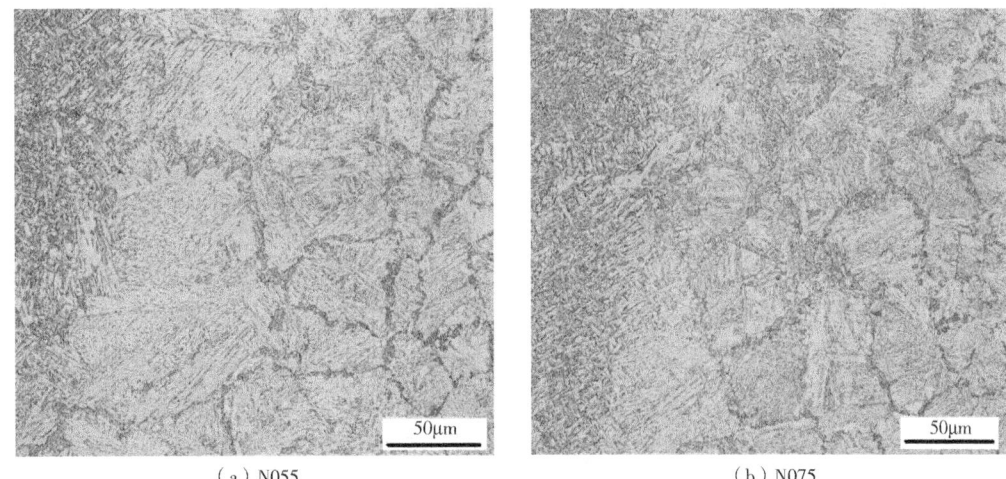

(a) N055　　　　　　　　　　　　　(b) N075

图 5　较低热输入的 GMAW 环焊接头粗晶区微观组织

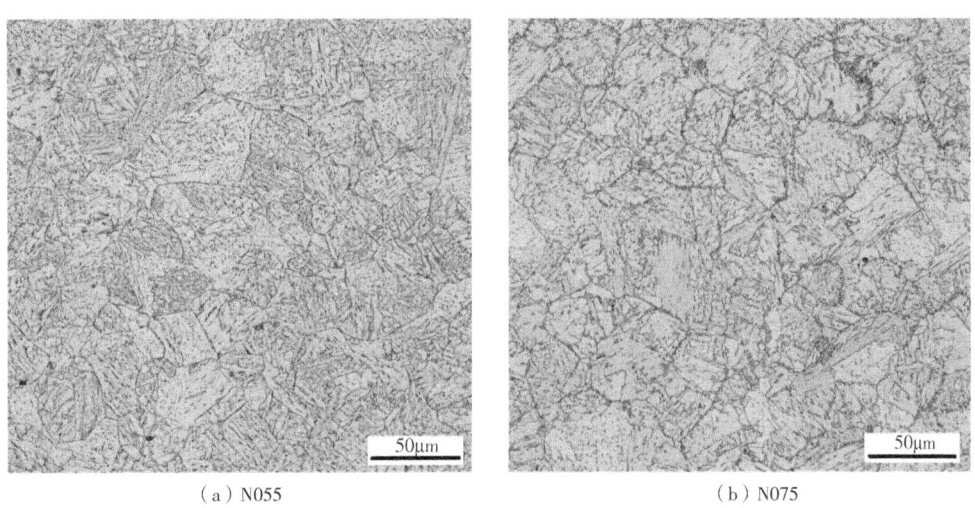

(a) N055　　　　　　　　　　　　　(b) N075

图 6　较高热输入的 SMAW 环焊接头粗晶区微观组织

图 7 为 GMAW 环焊接头的硬度云图。图 7 表明，N075 接头两侧管体硬度高于 N055，壁厚中心附近 N075 热影响区的平均硬度约为 205HV0.5，N055 热影响区的平均硬度约为 198HV0.5。因此，N075 热影响区的硬度略高于 N055，即表明 N075 热影响区的强度也高于 N055。同时，图 3 的对比结果表明，N055 管体的纵向冲击韧性高于 N075。图 8 为扫描电镜下 N055 和 N075 管体壁厚中心的微观组织。从图 8 可以看出，与 N055 相比，N075 管体具有更多的粒状贝氏体组织，且在晶粒内分布有较多的板条亚结构。因此，N075 管体纵向强度高于 N055，而平均冲击韧性低于 N055。管体原始的组织和性能导致 0.055%Nb 的 X80 钢管 GMAW 环焊接头热影响区不同位置的冲击韧性高于 0.075%Nb 的 X80 钢管热影响区韧性。

CGHAZ 的韧性受贝氏体相变后其晶体学结构影响，而 IC CGHAZ 则主要受沿原奥氏体晶界形成的链状 M-A 影响。在试样受冲击过程中，裂纹在二次热循环产物 M-A 处形成核。当遇到岛型 M-A 组分时，裂纹发生偏转；而当遇到 M-A 型纤细组分时，则呈直线传播。裂纹偏转越小，传播路径消耗的能量越小，从而降低韧性[15]。通过优化母材合金成分（比如加入 Nb 元素），细化 CGHAZ 的奥氏体晶粒尺寸，获得最佳晶体学结构匹配的组织，有利于改善焊接热影响区韧性[16]。

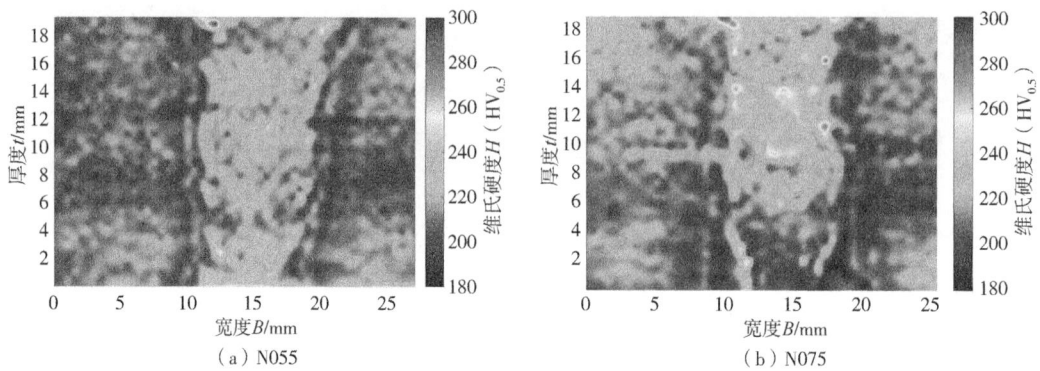

(a) N055　　　　　　　　　(b) N075

图 7　GMAW 环焊接头的硬度云图

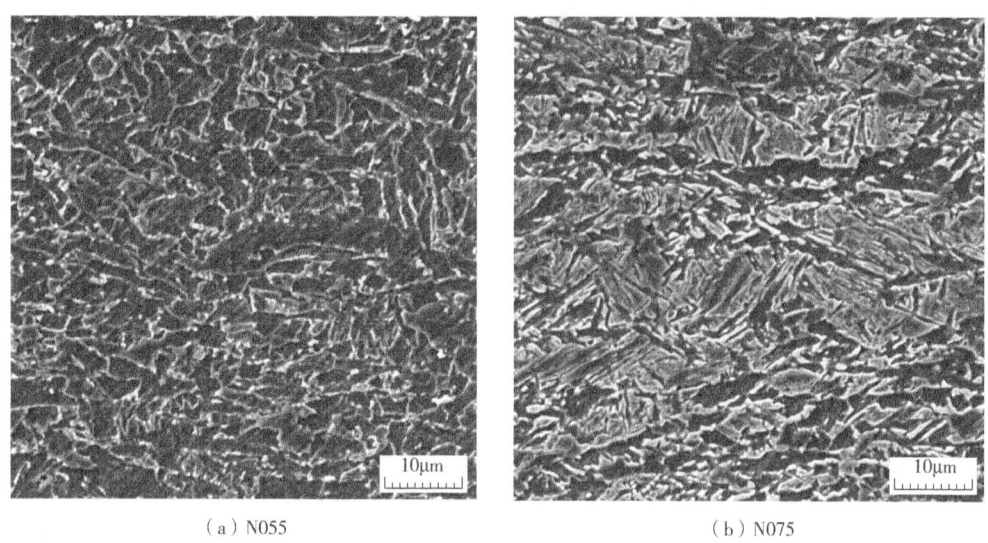

(a) N055　　　　　　　　　(b) N075

图 8　X80 管体 SEM 微观组织

图 9 为激光共聚焦观察到 N055 和 N075 二次热循环后的微观组织。从图 9 可以看出，N075 在二次热循环后热影响区的晶粒尺寸明显小于 N055。GMAW 环焊时，由于热输入较低，热影响区很窄，而 N075 在 IC CGHAZ 形成的 M-A 较粗，但其晶粒相对较细。因此，M-A 所占比例较高，在管体原始冲击韧性遗传影响下，N075 热影响区的韧性低于 N055。当较高热输入的 SMAW 环焊时，N075 热影响区的奥氏体晶粒尺寸细化，M-A 也随之细化，M-A 分布趋于不连续或消失，IC CGHAZ 表现出更好的韧性。因此，较高热输入的 SMAW 环焊时，与 N055 相比，N075 粗晶区具有更高韧脆转变上平台能和更低韧脆转变温度。

3　结论

（1）在 0℃和-20℃时，0.075%Nb 和 0.055%Nb 的 X80 钢管 GMAW 环焊接头热影响区均具有较高的冲击韧性，其平均冲击吸收能量高于 150J。但是，0.055%Nb 略高于 0.075%Nb 的 X80 钢管 GMAW 环焊接头热影响区的夏比冲击吸收能量。

（2）当采用低热输入焊接时，0.055%Nb 的 X80 环焊接头粗晶区比 0.075%Nb 的 X80 的粗晶区的韧脆转变温度更低，具有更好的低温韧性；当采用较高热输入焊接时，0.075%Nb 的 X80 环焊热影响区的上平台冲击吸收能量更高，且具有更低的上平台温度和韧脆转变温度，其低温韧性更优异。

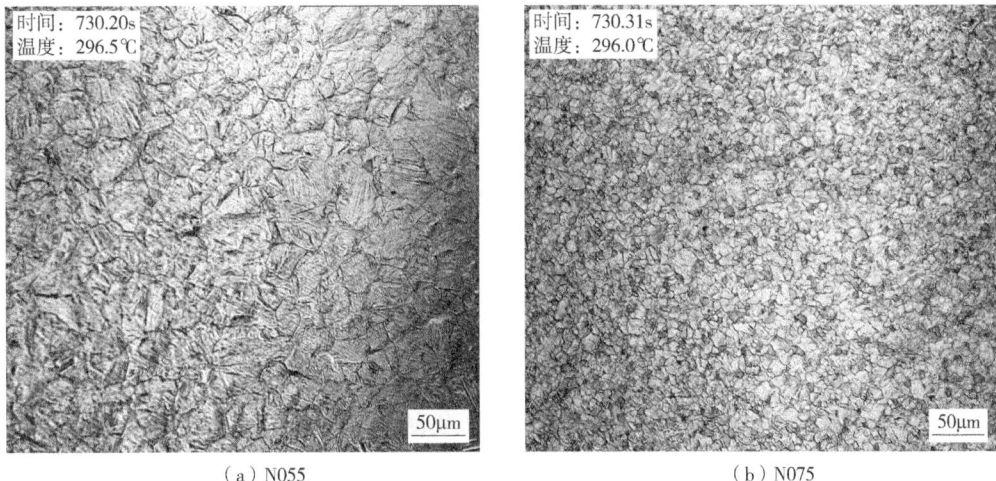

(a) N055　　　　　　　　　　　　(b) N075

图9　X80二次热循环冷却至约296℃的微观组织

（3）X80环焊接头热影响区的冲击韧性不仅与焊接热输入量和热影响中M-A形状、大小、分布有关，而且还受管体的化学成分、原始的强韧性和微观组织状态的遗传影响。

参 考 文 献

[1] MORRISON W B. Microalloy steels-the beginning[J]. Materials Science and Technology, 2009, 25(9): 1066-1073.

[2] DEARDO A, HUA M, CHO K, et al. On strength of microalloyed steels: an interpretive Review[J]. Materials Science and Technology, 2009, 25(9): 1074-1082.

[3] FATEH F, MATTHIAS M. Effect of Nb on austenite formation and decomposition in an X80 linepipesteel[J]. Journal of Iron and Steel Research, International, 2011, 18: 658-663.

[4] 缪成亮, 尚成嘉, 王学敏, 等. 高Nb X80管线钢焊接热影响区显微组织与韧性[J]. 金属学报, 2010, 46(5): 541-546.

[5] TEIXEIRA P, LOUREIRO A, RODRIGUES D, et al. Effect of the HAZ microstructural gradient on the unstable fracture behaviour of welds in a high strength steel[J]. Materials Science Forum, 2006, 514-516 (5): 539-543.

[6] 徐杰, 卓小敏, 李朋朋. 温度对X80管线钢韧/脆转变区断裂韧性的影响[J]. 工程力学, 2016, 33(S): 56-61.

[7] 袁军军, 禅志善, 曹睿, 等. 多层多道平焊接头冲击性能不稳定的原因分析[J]. 焊接学报, 2017, 38(5): 100-103.

[8] MOHAMMADIJOO M, COLLINS L, RASHID M F, et al. Influence of steel chemistry and field girth welding procedure on performance of API X70 line pipe steels[C]. International Pipeline Conference. ASME, 2020, Virtual, Online.

[9] WANG H, WANG J, TONG Z, et al. Characterization of Nb interface segregation during welding thermal cycle in microalloyed steel by atom probe tomography[J]. Metallurgical and Materials Transactions A, 2018 (12): 6224-6230.

[10] 赵建平, 张秀敏, 沈士明. 材料韧脆转变温度数据处理方法探讨[J]. 石油化工设备, 2004, 33(4): 29-32.

[11] 周昌玉, 夏翔鸣. CrMo钢材料韧脆转变温度曲线的回归分析[J]. 压力容器, 2003, 20(6): 13-18.

[12] 周腾飞, 关凯书. 不同缺口形式小冲杆试样测试3Cr1MoV钢韧脆转变温度的对比[J]. 机械工程材料,

2018, 42(12): 21-26.

[13] 李学达, 尚诚嘉, 韩昌柴, 等. X100 管线钢焊接热影响区中链状 M-A 组元对冲击韧性和断裂机制的影响[J]. 金属学报, 2016, 52(9): 1025-1035.

[14] 李学达, 李春雨, 曹宁, 等. 高强管线钢焊接临界再热粗晶区中逆转奥氏体的逆相变晶体学[J]. 金属学报, 2021, 57(8): 967-976.

[15] QI X N, HUAN P C, Wang X N, et al. Effect of root welding heat input on microstructure evolution and fracture mechanism in intercritically reheat–coarse grained heat–affected zone of X80 pipeline steel[J]. Materials Today Communications, 2022, 31: 1-9.

[16] 王学林, 李学达, 尚成嘉. 高强度管线钢焊接热影响区显微组织精细表征[J]. 焊管, 2019, 42(7): 27-37.

本论文原发表于《焊接学报》2024 年第 45 卷第 3 期。

油气管道工程用大口径 TE555 增材制造三通的开发

吉玲康[1]　胡美娟[1]　田　野[2]　王　俊[1]　陈越峰[1]
杨耀彬[1]　刘　琰[1]　李胜男[1]　李　鑫[1]

(1. 中国石油集团工程材料研究院有限公司，油气钻采及输送全国重点实验室；
2. 国家管网集团西部管道有限责任公司)

摘　要：本文从材料、制造工艺、力学性能、焊接性能、实物爆破试验性能，以及无损检测、残余应力等多个方面，介绍了采用电弧增材制造技术开发的-60℃和-40℃的低温环境用油气管道工程用 φ1219mm 大口径 TE555 增材制造三通。开发的增材制造三通：(1) 各位置强韧性均匀，无明显的方向性，无厚度效应，特别是具有较好的低温韧性；(2) 焊接性能良好，无软化现象；(3) 肩部加厚后爆破压力达到 57MPa；(4) 同时经过去应力处理后残余应力大幅度降低；(5) 无损检测结果表明增材制造工艺可行，无超标缺陷。增材制造三通在定制化、特殊环境和特殊用途工况情况下，具有极大的推广和应用前景。

关键词：电弧增材制造；埋弧；大口径三通；低温韧性；拉伸性能；焊接性能；爆破试验；残余应力

三通是油气管道工程中压力管道元件的典型产品，直接关系到整个管道系统的完整性及安全运行。近年来油气管道建设已进入了发展的高峰期，随着管道业的发展，三通管件也趋于向高强度、大口径、厚壁、高性能的方向发展[1]。例如，中俄东线天然气管道工程北段站场用 TE555 级 φ1400mm 三通首次设计采用裸露服役方式，取消了传统的保温伴热等措施，三通最低服役环境温度为-45℃，设计壁厚达 57mm。由于在裸露环境中服役，因此对三通的耐低温性能提出了更高的要求[2]。

热挤压三通为现今我国制造高强度、大口径油气输送管道三通管件的主要技术。这种工艺技术采用的坯料管一般是由钢板卷制、焊接形成的。坯料管加热到 A_{c3} 温度以上后放入模具中进行热挤压，金属在模具内腔内流动，沿模腔几何形状产生塑性变形而形成三通支管，然后再进行支管开口、翻边、扩径，最后进行淬火加回火热处理获得与干线钢管相匹配的强韧性。按此工艺生产的三通，其主体结构带一条纵向焊缝，关键技术为材料(包括钢板母材

基金项目：中国石油天然气集团有限公司科学研究与技术开发项目《石油管/管件/装备构件增材制造技术研究》(编号：2021DJ2702)；中国石油天然气集团公司基础研究和战略储备技术研究基金项目《特种管件开发中的增材构建技术研究》(编号：2017Z-02)。

作者简介：吉玲康，男，1966年生，教授级高级工程师，1989年毕业于西安交通大学金属材料及热处理专业，工学博士，现主要从事油气输送管的应用技术研究。E-mail：jilk@cnpc.com.cn

和与之工艺匹配的焊材)选用、结构尺寸成型及最终热处理工艺等[3-4]。由于原理性的制约,热挤压三通制造工艺技术主要存在以下问题:(1)满足现行设计标准中面积补强原理对壁厚的要求难度较大;(2)由于壁厚较厚,壁厚中心和表面存在组织和性能不均匀现象;(3)肩部在生产过程经历多次高温热循环,组织粗大,存在脆性开裂风险;(4)母管的焊缝是薄弱部位;(5)支管较短,不利于现场环焊质量保证和修复。上述问题给油气管道工程,特别是大口径高压油气管道带来了一定的安全隐患。因此,多年来,热挤压三通的质量提升一直是管道工作者关心和研究的热点问题[5-6]。

增材制造技术是一种颠覆性的金属零部件智能制造工艺方法,它可以改变构件结构、改变材料、改变重大装备的制造模式。它通过 CAD 设计数据,采用材料逐层累加的方法制造实体构件。金属增材制造作为增材制造领域的研究重点,广泛应用于航空航天、汽车及生物医学等各个领域[7-10]。它给解决传统热挤压三通存在的问题带来了可能。

笔者团队在国际上首次采用电弧作为热源,也就是电弧增材制造技术,针对高钢级大口径三通产品进行了多年的研究,取得了多项技术成果[11-14]。其中开发了两种成分和性能的油气管道工程用大口径 TE555 增材制造三通,分别适用于 -60℃和 -45℃的低温环境。三通产品的主、支管直径均为 φ1219mm,壁厚为 60mm,主管长度为 2m(图1)。通过多年的科学研究、技术攻关、产品试制,目前已经探索出了增材制造三通的完整制造工艺路线和检验方法;制定发布了产品企业标准 QSY-TGRC 201—2022《油气管道工程用 TE555 增材制造三通》。产品通过了国家石油管材质量检验检测中心的评价,以及中国石油和石油化工设备工业协会组织的专家鉴定,认为开发的增材制造三通产品为国际首创,总体处于国际先进水平。

图1 φ1219mm 大口径 TE555 电弧增材制造三通

以下将从材料、制造工艺、力学性能、焊接性能、实物爆破试验性能,以及无损检测、残余应力等多个方面对该产品进行介绍。

1 TE555 增材制造三通的丝材成分设计

TE555 增材制造三通用丝材成分设计充分考虑了其不仅需要具备较好的强韧性匹配,同时具有较好的焊接性能的要求,参考传统热挤压 TE555 三通的低碳合金钢的成分体系,结合增材制造工艺的低热输入、高冷却速度的特点,进行了两种低碳微合金钢焊丝成分体系的设计,分别适用于 -60℃和 -40℃的低温环境。最终确定的成分体系见表1。

表1 TE555 增材制造三通用丝材化学成分

试样	C	Si	Mn	P	S	Cr	Mo	Ni	Nb+V+Ti	其他	Pcm
丝材-1	≤0.12	≤0.5	≤1.8	≤0.02	≤0.01	≤0.1	0.50~0.75	≥1.2	≤0.015	痕量	≤0.25
丝材-2	≤0.12	≤0.5	≤1.8	≤0.02	≤0.01	≤0.1	0.40~0.55	0.8~1.2	≤0.015	痕量	≤0.25

可见,两种材料的微合金元素含量基本一致,相对于原热挤压三通,增加了 Mo 和 Ni 等合金元素。其中,适当提高 Mo 元素含量,可促进贝氏体的形成,细化晶粒并提高淬透性

和回火稳定性,且可以使零件在较高温度下回火,从而有效地消除或降低残余应力,提高塑性和韧性。由于 Ni 可以降低临界点并增加奥氏体的稳定性,因此提高 Ni 元素含量可以提高淬透性,一方面可以通过固溶强化强烈提高钢的强度,另一方面又保持较好的低温韧性水平,有效降低钢的韧脆转变温度,减小钢对缺口的敏感性。

2 ϕ1219mm TE555 增材制造三通的制造工艺

在工程材料院的电弧增材技术研究系统(图2),采用表1中两种成分的焊材进行 ϕ1219mm TE555 三通的制造工作。该设备可进行碳钢、合金钢、铝合金、不锈钢、镍基合金等多种材料的增材制造,承载重量达到 20t;成型尺寸最大 2m×2m×1.8m;效率最高 8kg/h。

制造采用埋弧工艺,两种焊丝的规格均为 ϕ4.0mm。焊丝在使用前需检查焊丝表面质量,应无锈蚀、油污、水迹等影响增材质量的物质。配套焊剂为 GXL-125,规格为 10~60 目。焊剂在使用前应烘干,根据焊剂生产厂家提供的烘焙温度、时间进行烘焙;增材过程中未参与烧结反应的焊剂可回收重复使用,若暴露在空气中的时间大于 4h,需重新烘焙。

图 2 电弧增材技术研究系统

三通增材制造工作分为两个阶段进行(图3),即从三通支管中心点将其分成两个部分,第一阶段完成后进行机加工后进行第二阶段。采用自主研发的 3DAM 切片软件编辑与生成增材制造三通整体打印轨迹,以层高 2.3mm 逐层进行轨迹布置(图4),并生成数控系统执行文件。增材制造三通的打印工艺主要参数为:电流 475A,电压 26V,层间温度小于 300℃。

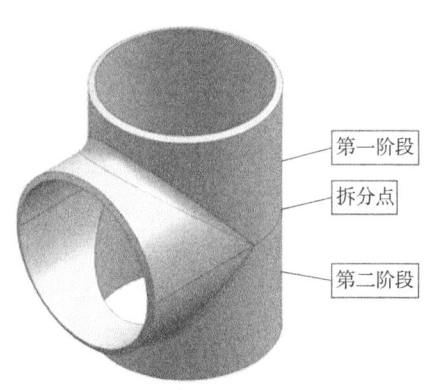

图 3 增材制造三通打印分解图

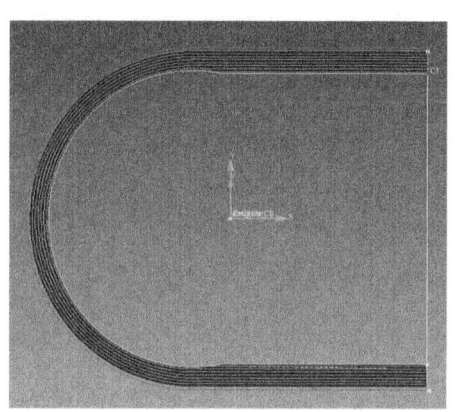

图 4 增材制造三通打印轨迹布置

在三通的增材制造过程中,其制造工序如下:(1)在设备工作平台上安装基体工装和辅助工装;(2)将基体上表面打印的区域预热到 100~150℃ 之间;(3)按模型轨迹逐层打印,在每打印高度增长 200mm 左右,上移辅助工装;(4)待构件高度增长至 300mm 时,将支管工装与构件焊接成为一体;(5)打印第一阶段完成后,机加去除基体工装及过渡层,将构件翻转重新固定到设备平台上;(6)开始第二阶段打印,完成整个三通打印;(7)进行消应力处理,热处理温度为650℃,保温2h后空冷至室温;(8)根据最终尺寸要求,进行二次加工。

3 TE555增材制造三通的性能

采用表1中的两种打印焊丝进行φ1219mm TE555三通的增材制造,每种丝材制造2个三通,分别用于理化性能的分析和水压爆破试验(表2),另外增材制造一个和Tee2相同规格和材料的管圈用于焊接试验。试验结果介绍如下:

表2 增材制造三通样品及试验项目

三通编号	打印丝材编号	适用环境温度/℃	试验目的
Tee1	丝材-1	−60	理化性能
Tee2	丝材-2	−45	理化性能、焊接性能
Tee3	丝材-1	−60	水压爆破试验
Tee4	丝材-2	−45	水压爆破试验

3.1 理化性能

分别在三通主管、支管,以及肩部、主管180°位置的壁厚中心,沿不同方向取φ12.7mm×50mm圆棒试样,其中试样方向中的横向、纵向分别为垂直于、平行于三通管的轴线方向。依据GB/T 228.1—2021《金属材料 拉伸试验 第1部分:室温试验方法》进行拉伸试验,试验结果见表3。可见,两件三通的拉伸性能均满足,且在不同的位置,各向拉伸性能均匀,不同位置的屈服强度差异在30MPa以内,抗拉强度在20MPa以内,屈强比0.88~0.91;并且拉伸性能无明显的方向性,表现出各向同性的性能特征。

表3 增材制造三通的拉伸性能

样品编号	取样位置	试样方向	$R_{t0.5}$/MPa	R_m/MPa	A/%	$R_{t0.5}/R_m$
Tee1	主管	横向	612	690	26.0	0.89
		纵向	612	689	24.0	0.89
	支管	横向	606	681	26.0	0.89
		纵向	609	680	29.0	0.90
Tee2	主管	横向	615	684	30.0	0.90
		纵向	602	680	24.0	0.88
	主管180°	横向	632	696	28.5	0.91
	支管	横向	634	696	26.0	0.90
		纵向	630	699	28.0	0.90
	肩部	横向	631	692	28.0	0.91
标准要求			555~690	625~825	≥20.0	—

分别在三通主管、支管,以及肩部、主管180°位置的壁厚中心,沿不同方向取10mm×10mm×55mm夏比冲击试样,其中试样方向中的横向、纵向分别为垂直于、平行于三通管的轴线方向。依据GB/T 229—2020《金属材料 夏比摆锤冲击试验方法》进行冲击试验,试验结果见表4。可见,两件三通在不同的部位、不同的方向上均呈现出较好的低温韧性,其

中，Tee1样品在-60℃的低温冲击功最小平均值为69J，平均值为97J；-45℃的低温冲击功最小平均值达到了102J，平均值为138J。而Tee2样品在-45℃的低温冲击功最小平均值也在64J以上，三通整体平均冲击功为88J。但是各部位的冲击韧性在不同的方向上表现出一定的差异，这和增材的打印路径和工艺有关，主要是由于垂直打印方向的大角度晶界占比高于平行打印方向，使垂直打印方向具有较高的冲击韧性[15]。

表4 夏比冲击试验结果

试样			试验温度/℃	CVN/J				SA/%				FATT50/℃
样品编号	取样位置	方向		单个值			平均值	单个值			平均值	
Tee1	主管	横向	-60	53	105	69	76	15	35	20	23	-48
			-45	173	140	116	143	80	60	50	63	
		纵向	-60	101	127	148	125	35	50	60	48	-57
			-45	167	176	166	170	65	70	60	65	
	支管	横向	-60	118	128	132	126	50	55	65	57	<-60
			-45	116	149	157	141	60	70	75	68	
		纵向	-60	90	112	68	90	30	35	25	30	-45
			-45	154	111	102	122	70	40	40	50	
	肩部	横向	-60	90	62	55	69	30	25	20	25	-40
			-45	115	110	115	113	45	40	45	43	
Tee2	主管	横向	-45	68	86	75	76	50	55	50	52	-46
		纵向	-45	169	152	129	150	85	75	65	75	<-60
	主管180°	横向	-45	76	104	61	80	50	65	45	53	—
	支管	横向	-45	57	72	62	64	35	45	40	40	-41
		纵向	-45	89	84	78	84	45	40	40	42	-39
	肩部	横向	-45	59	64	94	72	50	50	60	53	—

分别在三通主管、支管，以及肩部等多个位置取全壁厚试样，并在其纵、横向界面上，距试样上、下表面1.5mm及壁厚中心处，依据标准GB/T 4340.1—2009《金属材料 维氏硬度试验 第1部分：试验方法》进行10kg载荷维氏硬度的试验，试验结果见表5。可见，主管、支管、肩部的硬度较为均匀。

表5 硬度试验结果　　　　　　　　单位：HV

样品编号	位置	最大值	最小值	平均值	
Tee1	主管	230	211	217	219
	支管	231	209	222	
Tee2	主管	245	207	227	229
	支管	247	215	233	
	肩部	237	218	228	

分别在主管和支管上取纵、横向试样，并依据标准ASTM A370-17a进行导向弯曲试验。弯轴直径为100mm。弯曲角度为180°，试样均未出现裂纹。

3.2 焊接性能

将口径为φ1219mm的X80钢管和采用丝材-2增材制造的试验环安装在环焊对口机上进行组对，采用实心焊丝自动焊接系统进行环焊，以研究增材制造的TE555三通的焊接性能。焊接材料采用φ1.0mm的ER80S-G实芯焊丝，保护气体为20%CO_2+80%Ar。焊接采用GMAW全自动下向焊接，电流200~260A，电压25~29V，焊接速度15~25in/min。焊接完成后环焊缝形貌及微观组织情况如图5所示。

对环焊缝进行金相和力学性能分析：

（1）焊接接头拉伸试验：拉伸试样为板状全壁厚试样，试样宽度为19.1mm。抗拉强度为669~679MPa，断裂位于增材侧。

（2）焊缝区夏比冲击试验：在靠近样品外表面、内表面取夏比冲击试样，尺寸为10mm×10mm×55mm，V形缺口位于焊缝中心和热影响区。试验结果如图6所示。

（3）硬度试验：在焊缝上取全壁厚横截面试样，进行10kg载荷维氏硬度试验，试验标准为GB/T 4340.1—2009。硬度测试点及试验结果如图7所示。

（4）导向弯曲试验：垂直焊缝取焊接接头侧弯试样，试样宽度为13mm，厚度为原始壁厚，去除焊缝余高。弯轴直径为90mm，弯曲角度为180°，按照标准GB/T 2653—2008《焊接接头弯曲试验方法》进行导向弯曲试验，试样未出现裂纹。

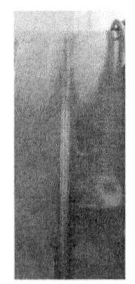 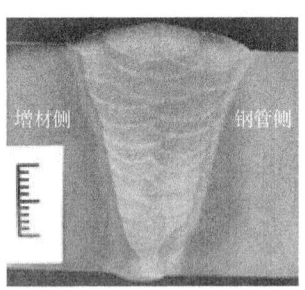  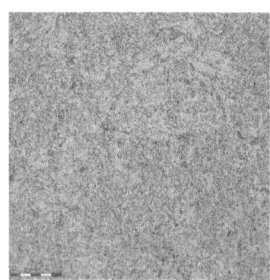

（a）环焊缝形貌　　（b）焊缝横截面　　（c）增材侧细晶区微观组织　　（d）增材区母材微观组织

图5　环焊缝宏观形貌及微观组织

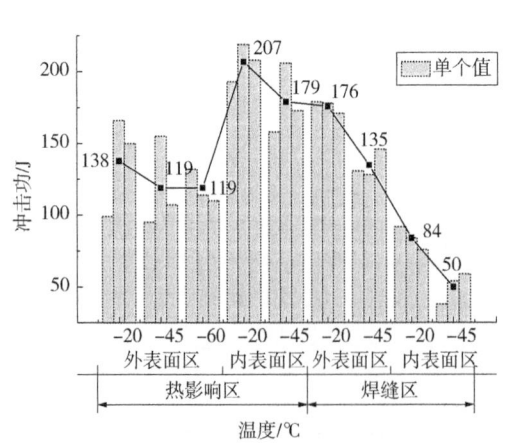

图6　环焊缝区夏比冲击试验结果

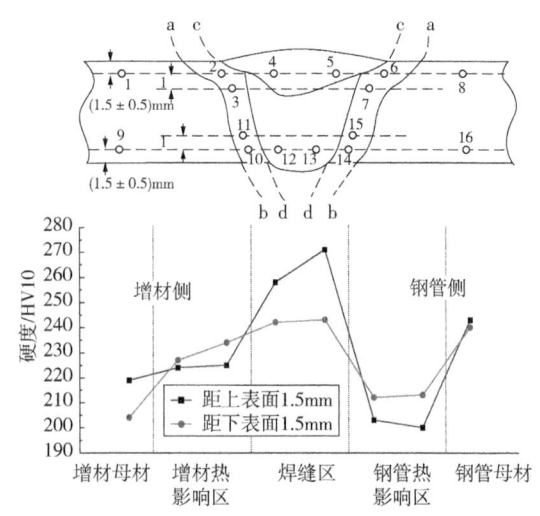

图7　环焊缝区硬度试验结果

环焊试验结果表明，增材三通和 X80 钢管对焊形成的焊接接头的强度、韧性等各项性能良好，特别是焊接后热影响区仍能保持较高的冲击韧性，而且，增材侧热影响区硬度相对于母材无下降，未出现软化现象，主要与较高的合金含量造成其细晶区不仅晶粒细小且仍能保持相当数量的粒状贝氏体有关。这一点明显区别于钢管侧。

3.3 实物爆破性能

按照三通相关标准的要求，管件承受内部压力的能力应不低于匹配钢管的耐压能力。耐压能力验证应通过计算和(或)验证试验方法验证。因此，分别采用表1中的两种丝材打印出 Tee3 和 Tee4 三通样品。其中 Tee4 的肩部较 Tee3 进行了加厚，最大厚度 164mm。

对 Tee3 和 Tee4 三通样品进行水压爆破试验。试验前在三通样品的主管和支管端部进行钢管短节、堵头的焊接。

通过注水口对试验样品加压，直至样品发生失效。其中：(1)Tee3 样品加压至 52MPa 时，压力开始下降，观察后为三通样品泄漏失效，水从失效处喷出。断裂开口在肩部，共两处。一处裂纹沿着支管方向，长为 180mm，完全张开；另一处裂纹沿着主管和支管相贯线方向，长度为 220mm，未完全张开[图8(a)]。两处断口均为完全韧性剪切断口[图8(b)]。(2)Tee4 样品加压过程中，48mm 厚短节首先发生变形，然后在 57.5MPa 发生爆破失效，随后压力快速下降。爆破裂纹沿短节轴向两边快速扩展，向左至堵头，向右则扩展至增材三通肩部，转而沿着圆周方向扩展 100mm 停止。短节和堵头的断口为韧脆混合断口。增材三通样品上断口为完全韧性剪切断口，局部位置在断裂时经历断口摩擦，呈现黑色，断面有微小的撕裂棱(图9)。

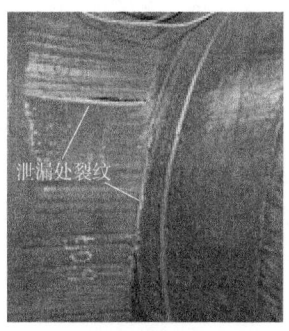 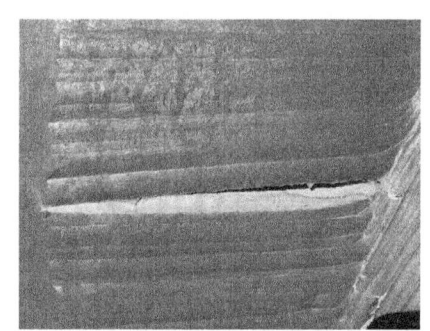

(a) 肩部泄漏处宏观形貌　　　　　(b) 第一处裂纹断口形貌

图8　Tee3 水压试验肩部泄漏处宏观形貌和断口形貌

 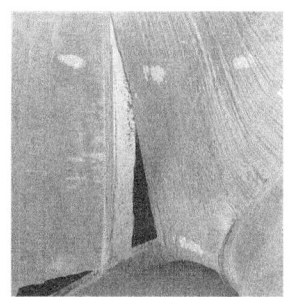

(a) 爆破后三通宏观形貌　　　　　(b) 三通环向裂纹断口形貌

图9　Tee4 水压爆破试验后样品宏观形貌和断口形貌

3.4 TE555增材制造三通的无损检测

对Tee3和Tee4三通样品的内表面进行机械加工。随后在机加工面进行100%磁粉检测和采用直探头和斜探头的超声波检测。检测结果均未发现相关缺陷显示或者超标信号。其中：

（1）磁粉检测按照ASTM E709或NB/T 47013.4进行，验收标准参照AWS D20.1/D20.1M：2019第8.4条标准或NB/T 47013.4标准执行。

（2）设计并制作了增材专用对比试块（Z-1），试块规格340mm×150mm×58mm，包括直径ϕ2mm、ϕ3mm、ϕ4mm、ϕ5mm，深度为5mm、10mm、20mm、30mm的平底孔缺陷9个和竖通孔缺陷10个。超声检测采用直探头、斜探头两种探头形式，并按照NB/T 47013.3进行检测和验收。其中，对于平行检测面的平面型缺陷采用纵波直入射法进行检测，采用增材专用对比试块（Z-1）ϕ2mm平底孔进行校准；其他内部缺陷采用横波斜入射扫查的方式进行检测，采用增材专用对比试块（Z-1）ϕ2mm竖通孔校准。

3.5 TE555增材制造三通的残余应力

采用盲孔法对Tee3样品去应力热处理前、后的残余应力进行检测。测试采用BE120-2CA-K型应变花和DH5921动态应力应变测试仪完成，其中盲孔钻削通过高速钻孔装置完成。检测位置分别在主管和支管的侧面与端面，以及相贯线5个位置，每个位置检测4点，计算并获得每个位置的最大残余应力σ_{max}。图10为最大残余应力测试及分析结果。可见，经去应力热处理后，各位置残余应力最大值从498~546MPa降低至139~233MPa，平均降低幅度达到62%。

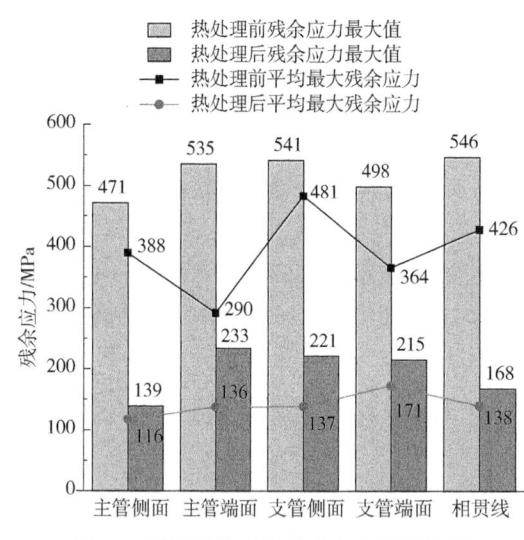

图10 增材制造三通残余应力测试结果

4 结语

各种试验和检验结果证明，开发的-60℃和-40℃的低温环境用TE555增材制造三通完全满足油气管道工程、压力容器的设计要求。(1)不同位置拉伸性能均匀，且无明显的方向性，无厚度效应；(2)具有较好的低温韧性，-60℃和-45℃产品低温冲击韧性平均值达到97J和88J；(3)增材三通为低碳微合金化钢，焊接性好，和X80钢管对焊形成的焊接接头的强度、韧性等各项性能良好，特别是焊接后热影响区仍能保持较高的冲击韧性，并无软化现象；(4)实物水压爆破试验表明，增材制造三通具有较高的爆破压力，且断口为完全韧性，具有较高的安全性；(5)无损检测结果表明，三通产品未发现相关缺陷显示或者超标信号；(6)经去应力热处理后，残余应力得到大幅度降低。

增材制造三通生产所用的材料、技术和工艺与现有热挤压三通产品存在较大差异，具有较多传统热挤压三通无法比拟的优点。如，增材制造三通：(1)克服热拔三通制造方法的工艺局限性，支管的设计壁厚可以和主管一致，长度亦可增加，这可减少连接管的使用和环焊缝的数量，可按照现有运行标准中面积补强原理的壁厚设计要求进行设计，可在局部受力承载区域进行厚度补强；(2)根据三通的应用环境可灵活选择打印材料，且材料利用率高；

(3)将产品数字化设计、制造、高度一体化,制造工序流程少、生产周期短;(4)增材制造三通不同部位、不同方向和不同壁厚位置的力学性能均匀一致,强韧性配合较好。因此,增材制造三通可以解决传统热拔三通制造和安装过程中的安全隐患,在定制化、特殊环境和特殊用途工况方面具有极大的推广和应用前景,可和传统热挤压制造三通并行,对其形成较好的补充。

参 考 文 献

[1] 胡美娟,刘迎来,吉玲康,等.油气管道用三通管件的研究[J].石油管材与仪器,2018,4(5):1-4.
[2] 赵志伟,吴亚军,安晓军.中俄东线-45℃低温环境用X80M钢级ϕ1400mm三通的研发[J].焊管,2021,44(3):1-6.
[3] 刘金生,祝鹏,李玉卓,等.X80钢级DN1200低温三通[J].石油科技论坛,2017(S1):76-79.
[4] 刘迎来,吴宏,井懿平,等.高强度油气输送管道三通验证试验研究[J].焊管,2014,37(3):28-33.
[5] 李昱坤,刘迎来,刘鹏,等.油气管道用管件制造存在问题分析及质量提升措施[J].石油管材与仪器,2023,9(2):91-94.
[6] 邓波,刘迎来,王高峰,等.油气管道工程用弯管和管件的生产现状及研究热点[J].石油管材与仪器,2019,5(6):7-10.
[7] 卢秉恒.我国增材制造技术的应用方向及未来发展趋势[J].表面工程与再制造,2019,19(1):11-13.
[8] 王华明.金属增材制造技术及其对重大装备制造业的影响[J].中国工业和信息化,2019(12):54-56.
[9] 宋文清,李晓光,曲伸.金属增材制造技术在航空发动机中的应用展望[J].金属加工(热加工),2016(2):44-46.
[10] 张宏亮.金属增材制造技术在船舶制造中的应用[J].世界有色金属,2017(21):240-241.
[11] 胡美娟,吉玲康,卓炎,等.一种电熔增材制造的三通管件:ZL201921316723.7[P].2020-10-13.
[12] 吉玲康,胡美娟,马秋荣,等.一种电熔增材制造X100钢级三通管件材料及使用方法:ZL201910913041.2[P].2021-07-02.
[13] 胡美娟,马秋荣,吉玲康,等.一种油气输送用厚壁大口径三通管件的电弧增材制造方法:202110092101.6[P].2022-10-04.
[14] 胡美娟,吉玲康,池强,等.一种控制晶粒尺寸的低温增材制造用丝材及制备和应用:202110475939.3[P].2023-02-21.
[15] 陈越峰,高琦,吉玲康,等.油气管道用埋弧增材三通的性能研究[J].西安石油大学学报(自然科学版),2023,38(6):109-117.

本论文原发表于《石油管材与仪器》2023年第9卷第5期。

L485 高应变海洋管道的焊接技术研究

李为卫[1]　栾陈杰[2]　邸新杰[3]　李箕福[4]　牛爱军[5]　陈　亮[2]

(1. 中国石油集团工程材料研究院有限公司；2. 海洋石油工程股份有限公司；
3. 天津大学；4. 安泰科技股份有限公司；5. 宝鸡石油钢管有限责任公司)

摘　要：深海油气输送管道建设和运行过程中需要承受纵向载荷，需要采用高钢级、大厚径比钢管和基于应变的设计方法，对管道环焊缝的性能和质量提出较高的要求。为了掌握 L485/X70 高钢级厚壁管道的焊接技术，采用理论计算、试验分析和实物试验等方法，开展管道环焊接头力学及断裂变形行为、环焊材料、环焊工艺及性能等三方面研究，形成了具有高强匹配、高韧性和良好变形能力的环焊缝工艺技术，为深海油气管道建设提供了参考。

关键词：海洋管道；基于应变设计；环焊缝；高强匹配；高韧性

海洋油气资源的开发是国家的重大战略需求，与陆上管道有所不同，海洋油气管道由于铺设过程的弯曲及运行过程的洋流冲击、地震等影响，普遍采用基于应变的设计。高强度管线钢虽然有很多优点，但要承受基于应变的载荷，对管道环焊缝的性能提出了新的挑战[1]。要保证管道的变形能力，必须提高环焊接头的抗变形能力，对焊缝的强度、韧性，以及质量提出很高的要求。针对国产化开发的 L485/X70 钢级、31.8mm 大壁厚高应变海洋钢管，开展了管道环焊接头力学及断裂变形行为研究、环焊材料研发、环焊工艺及性能研究等三方面内容的研究，本文介绍了三方面的主要研究成果，以期为从事相关工作的技术人员提供参考。

1　L485 厚壁高应变海洋管道环焊接头力学及断裂变形行为研究

1.1　高应变管道环焊接头强度匹配技术

焊接接头各区域的成分、热历史和显微组织存在很大差异，其力学性能必然存在很大差异，进而导致承载过程中变形能力的差异。焊接结构应采用等强、高强还是低强匹配，国内外的认识不一致[2-4]。针对 $\phi 559 mm \times 31.8 mm$ L485 高应变海洋管道环缝，采用有限元计算方法，用壁厚减薄率和纵向应变量指标，研究并掌握了环焊接头不同区域的强度、尺寸及缺陷对变形与断裂能力的影响规律[5]。结果表明，焊缝的高强匹配有利于提高接头的变形能力，正常尺寸范围内焊缝的余高和宽度对变形能力影响较小，高应变海洋管道环焊缝应采用超过母材屈服强度标准规定值上限的高强匹配；缺陷的高度比长度对变形能力的影响更大，

基金项目：国家重点研发计划课题"L485 高应变海洋管道环焊材料及工艺技术"(2018YFC0310305)；中油股份有限公司课题"高应变海洋管道关键服役性能评估及环焊技术研究"(2019E-23-0501)。

作者简介：李为卫，男，1965 年生，正高级工程师，1988 年毕业于西安交通大学焊接专业，获学士学位，现主要从事油气输送管道材料研究及标准化工作。E-mail：liweiwei001@cnpc.com.cn。

焊接工艺应尽量减少缺陷的产生,无损检测应采用灵敏度高的方法。

1.2 L485高应变管线钢及焊材裂纹敏感性

采用小铁研试验和插销试验法,研究两种成分与制造工艺的L485管线钢及研发焊丝的冷裂纹敏感性。研发的L485管线钢,配合开发的气保实心焊丝,焊缝扩散氢低,最高硬度小,无有害显微组织,焊接冷裂纹敏感性低,焊接性良好,主要试验结果如图1和表1所示。

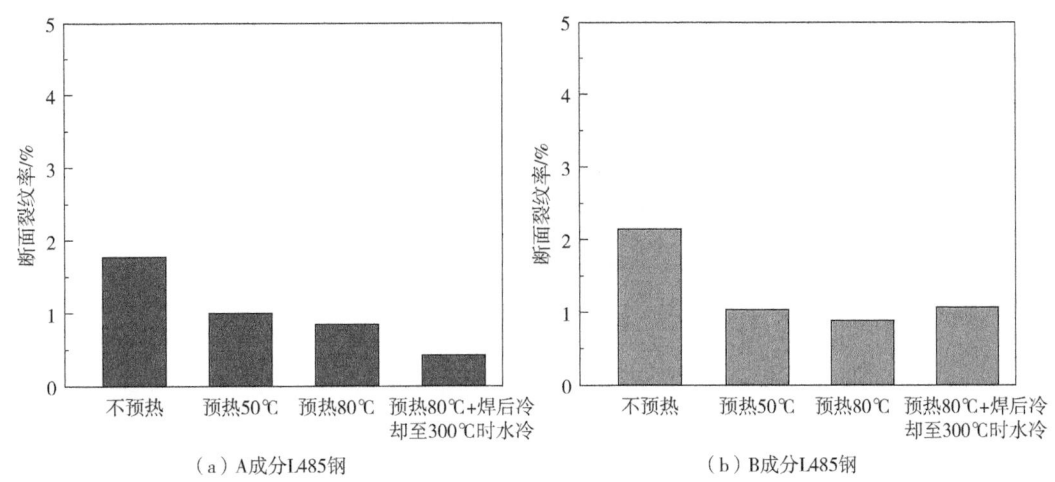

图1 小铁研试验试件平均断面裂纹率

表1 焊缝金属扩散氢测量结果

焊丝编号	熔敷金属质量/g	扩散氢体积/mL	扩散氢含量HD/(mL/100g)
HL565-A2	19.3	0.32	1.5
HL565-A3	23.7	0.76	2.9

1.3 L485高应变管道环焊接头的应力—应变行为

采用DIC(数字图像技术)拉伸、微区压痕法、微剪切、硬度云图等多种测试手段,研究了L485环焊接头的各区域的强度特征及应力—应变行为。图2为典型高强环焊接头(管号POJL20002,焊材HL565-A2)横向DIC拉伸应变分布,图3为典型环焊接头微剪切强度分布曲线。结果表明,开发的焊丝和工艺形成的焊缝达到高强匹配,拉伸断于母材,接头拉伸过程根焊强度稍低、有少量的变形,填充焊道强度高、变形小,远离焊缝的母材处变形量大,最终断于母材,表明焊缝的强度高于母材,焊接接头拉伸过程具有良好的抗变形能力。

1.4 L485高应变管道环焊接头脆化与软化行为

采用热模拟和实际焊接方法,研究了环焊接头热影响区的脆化和软化行为。结果表明,热输入对热影响区组织、韧性和硬度有较大影响(图4和图5);拘束度和热循环协同作用,对热影响粗晶区的组织和强度有较大影响(图6和图7),采用较少的热输入焊接有利于提高HAZ性能;研发的两种管线钢、配合开发的焊丝,接头各部位的韧性良好,韧脆转变温度低,HAZ软化不明显(图8)。

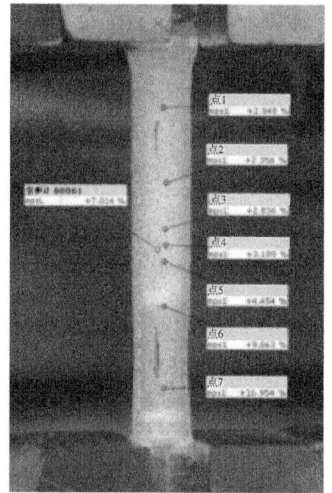
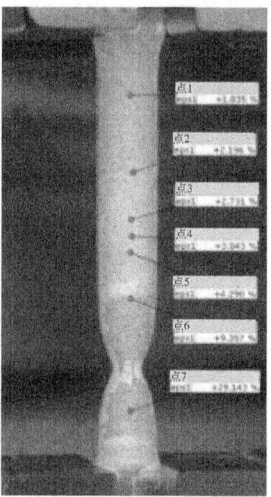
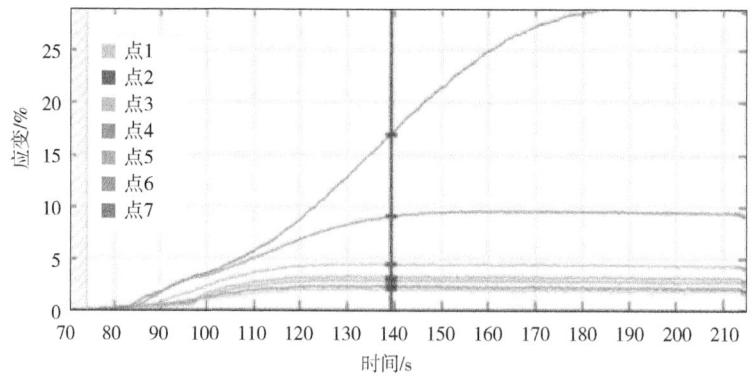

图 2 高强匹配环焊接头横向 DIC 拉伸应变分布

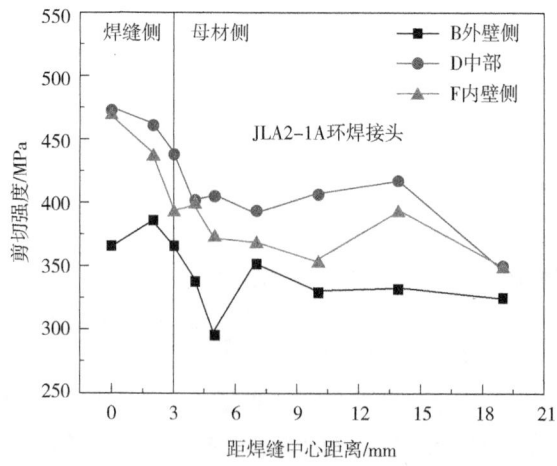

图 3 典型环焊接头微剪切强度分布曲线

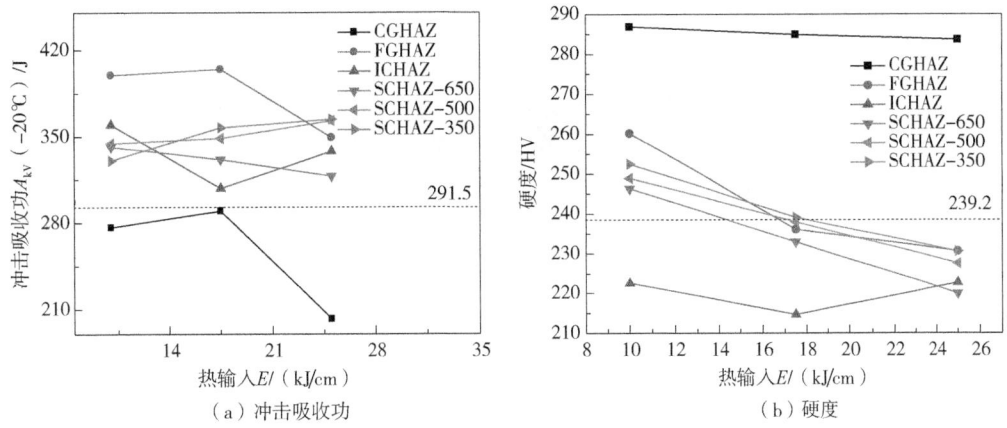

图 4 不同热输入下 HAZ 各亚区的冲击吸收功和硬度

图 5 粗晶区 EBSD IPF(晶粒取向)图及大小角度晶界图

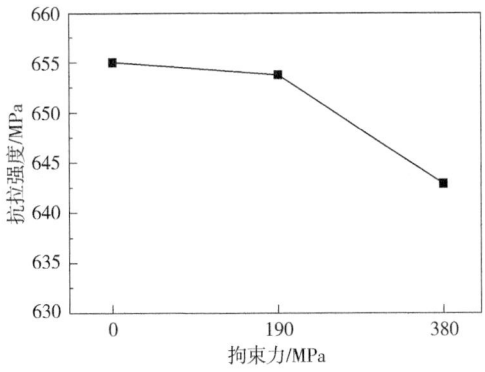

图 6 不同拘束度下焊接模拟 CGHAZ 的强度

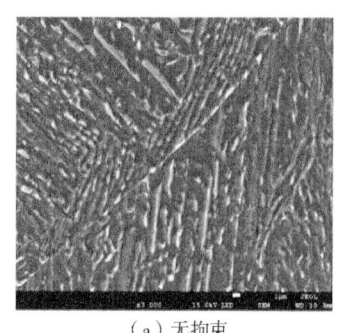  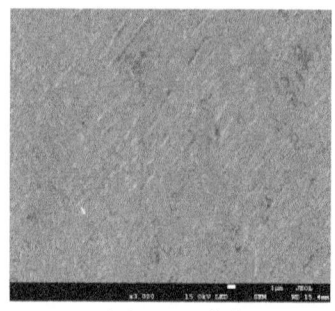

（a）无拘束　　　　　　　　（b）380MPa拘束　　　　　　　（c）190MPa拘束

图7　不同拘束度下模拟CGHAZ的SEM

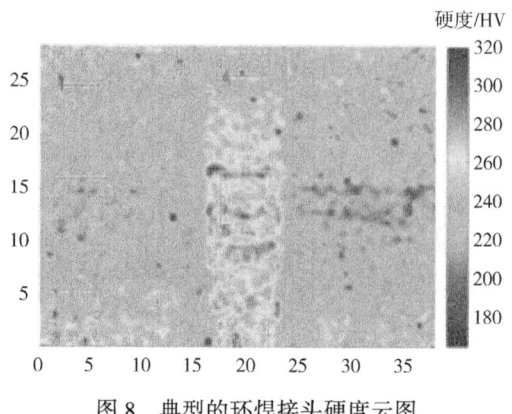

图8　典型的环焊接头硬度云图

2　L485高应变海洋管道环焊材料研究开发

2.1　焊丝的开发内控目标和成分设计

由于焊丝与母材熔合后共同形成的焊接接头，焊缝强度受母材、工艺参数等因素影响，实心焊丝气保焊焊缝金属的强度比焊丝自身的熔敷金属强度高出120~150MPa[6]，合理选择和使用焊接材料对保证焊缝金属的优良力学性能至关重要。分析了高应变海洋管道的环焊的技术需求，以及国内外相关资料和实物性能，研究了合金元素及纯净度对焊缝组织和性能的影响机理，提出了高强度、高韧性熔化极气体保护焊（GMAW）实心焊丝的研制内控目标，设计并冶炼了6种不同合金成分的焊丝用坯料（表2），拉拔试制出6种焊丝。焊丝的C含量，杂质元素S、P含量低。

表2　设计试制的6种焊丝的化学成分　　　　　单位:%（质量分数）

编号	C	Mn	Si	S	P	其他
HL565-A1	0.07	1.50	0.80	0.0100	0.015	Cu
HL565-A2	0.07	1.50	0.70	0.0100	0.015	Ni、Ti
HL565-A3	0.07	1.70	0.70	0.0050	0.015	Cr、Ni、Ti、Cu
HL565-B1	0.03	1.56	0.56	0.0039	0.012	Cr、Ni、Mo
HL565-B2	0.06	1.41	0.53	0.0040	0.005	Cr、Ni、Mo
HL565-B3	0.08	1.38	0.50	0.0040	0.005	Cr、Ni、Mo

2.2 焊丝熔敷金属性能测试

对试制的 6 种焊丝进行了熔敷金属的成分和性能测试。焊接过程工艺性能良好,焊接电弧稳定、熔池流动性良好、飞溅小、焊道成形美观,焊后经 X 射线拍片,焊缝未出现气孔、夹渣等缺陷。几种研制焊丝熔敷金属具有不同的强度水平,塑性指标和冲击韧性基本满足预期目标,具体测试结果见表3。由于熔敷金属试件焊接的热输入比实际海洋管道环缝焊接时的高很多,因此,熔敷金属测试的强度比实际焊缝的强度低很多,其值仅供参考。

表 3 焊丝熔敷金属拉伸和冲击韧性试验结果

试样编号	屈服强度 σ_s/MPa	抗拉强度 σ_b/MPa	延伸率 A/%	A_{kV}/J	保护气体
HL565-A1	498	585	28	−30℃:96, 88, 98	80%Ar+20%CO_2
HL565-A2	500	590	26	−30℃:135, 120, 118	80%Ar+20%CO_2
HL565-A3	576	672	26	−40℃:173, 156, 146, 150, 157, 161	50%Ar+50%CO_2
HL565-B1	530	585	24	−40℃:204, 200, 186, 174, 172	97%Ar+3%O_2
HL565-B2	598	678	22	−20℃:101, 121, 85	50%Ar+50%CO_2
HL565-B3	593	654	24	−20℃:67, 89, 73	50%Ar+50%CO_2

2.3 焊丝形成焊缝的性能试验

对试制焊丝在 31.8mm 厚、L485M 管线钢板上进行了适用性试验,研究了不同焊丝、不同工艺条件下焊缝的显微组织及与性能的关系,并与国外进口焊丝的性能进行了对比[7-8]。总体来看,研制焊丝得到的焊接接头的强度、塑性、韧性、硬度、弯曲性能在无缺陷情况下满足相关标准要求,焊缝的强度满足高强匹配,冲击韧性和塑性满足课题任务书要求,并优于或相当于国外著名企业同类焊材。在海洋管道较小焊接热输入工艺参数下,形成焊缝的一次凝固组织均为针状铁素体组织,整体组织均匀细化,虽然晶界处出现了先共析铁素体量,但总体数量可控。由于焊缝组织以细小、均匀和多位相分布的针状铁素体为主,因而保证了焊缝具有高强度的同时,具有良好的塑性和韧性。

3 L485 高应变海洋管道环焊工艺及性能研究

3.1 L485 高应变厚壁海洋管道环焊方法及设备选择

针对 ϕ559mm×31.8mm L485 高应变海洋管道环缝高质量、高效率的焊接需求,通过对国内外焊接方法和焊接设备的调研,结合我国海上石油工程实际情况和经验,开展了自动熔化极气体保护焊(AUTO GMAW)工艺参数试验研究,确定了 L485 高应变海洋管道环缝焊接的方法、设备型号、工艺方案及焊接参数,拟定了预焊工艺规程和工艺评定试验方案。

3.2 焊接工艺及接头性能试验研究

采用国产化第一轮试制的两种 L485 高应变海洋管、项目研发的三种气保实心焊丝和一种国外进口焊丝,开展的首次 6 件环焊缝焊接工艺试验(图9)和性能测试,按课题任务书评价 2 件有不合格项,4 件全部合格,按照 DNVGL-ST-F101-2017 标准评价 4 件有不合格项,2 件全部合格。产生性能不合格项的主要原因是因焊接缺陷导致的接头拉伸断于焊缝、弯曲出现超标开裂、宏观金相发现存在铜裂纹和未熔合缺陷，后续的焊接试验应加强对缺陷的控制，提高焊缝的质量。

图 9　工艺评定试验件焊接

汲取首次焊接工艺试验的经验，改进操作工艺和技术，主要是防止缺陷的产生，采用3种试制焊丝对国产化第二轮试制的两种 L485 高应变海洋管进行环焊缝焊接及性能评价试验，5件环焊缝按项目任务书和 DNVGL-ST-F101-2017 标准评价全部合格。表 4 和表 5 为环焊接头焊缝金属拉伸性能和不同部位韧性的统计分析。

表 4　环焊接头焊缝金属拉伸性能(圆棒试样)统计分析

项　　目		屈服强度 $R_{t0.5}$/MPa	抗拉强度 R_m/MPa	伸长率 A/%
试制焊丝 （10 道焊缝）	最小值	639	696	18.0
	最大值	718	765	27.0
	平均值	679	723	24.0
进口焊丝 Lincoln Pipeliner ER70S-6(1 道焊缝)	试验值 1	685	726	25.0
	试验值 2	676	725	26.0
	平均值	681	726	25.5
项目任务书要求		≥565	—	—

表 5　环焊接头不同部位的夏比冲击吸收能统计分析

焊丝	缺口位置	吸收能量 KV_2(-20℃)/J					
		单个值			平均值		
		最小值	最大值	平均值	最小值	最大值	平均值
研发焊丝 （10 道焊缝）	上焊缝中心	95	183	135	102	167	135
	下焊缝中心	86	180	138	107	169	138
	上熔合线	120	373	228	151	324	228
	下熔合线	127	324	259	165	305	259
	上熔合线+2mm	256	394	299	262	338	299
	上熔合线+5mm	245	361	299	259	343	299

续表

焊丝	缺口位置	吸收能量 KV$_2$(-20℃)/J					
		单个值			平均值		
		最小值	最大值	平均值	最小值	最大值	平均值
国外焊丝 （1道焊缝）	上焊缝中心	129	144	139	—	—	—
	下焊缝中心	128	146	137	—	—	—
	上熔合线	145	166	155	—	—	—
	下熔合线	125	141	135	—	—	—
	上熔合线+2mm	305	340	321	—	—	—
	上熔合线+5mm	308	319	313	—	—	—
项目任务书要求		平均值≥50J			—		

在0℃下的断裂韧性CTOD值：焊缝中心最小为0.181mm，最大为0.934mm，平均为0.380mm；熔合线最小为0.238mm，最大为0.905mm，平均为0.630mm。与国外焊丝（林肯PIPELINER 70S-6，焊缝中心最小为0.276mm，最大为0.378mm，平均为0.340mm；熔合线CTOD最小为0.435mm，最大为0.917mm，平均为0.615mm）相比，开发焊丝HL565-A1、HL565-A2焊缝中心的CTOD值相对较低，而开发焊丝HL565-A3焊缝中心的CTOD值相对较高。

含预制缺陷的环焊缝宽板拉伸和全尺寸弯曲试验表明，变形、断裂位置位于母材，环焊接头未产生明显的变形或断裂，应变能力满足海工设计要求（图10，图11和表6）。综合分析可以看出，研发焊丝与配套焊接工艺形成的ϕ559mm×31.8mm L485钢级环焊接头的强度与母材相比达到高强匹配，且具有良好的塑性、韧性和抗断裂、变形能力，环焊接头的力学性能满足项目任务书和DNV标准的要求。

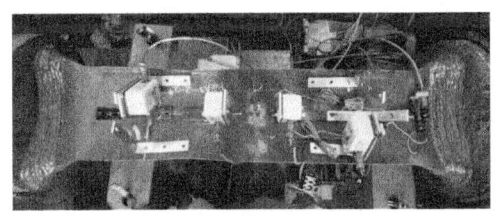

图10 含预制缺陷的环焊接头宽板拉伸试验

图11 含预制缺陷环焊接头的钢管全尺寸弯曲试验

表6 含预制缺陷的环焊接头宽板拉伸试验结果

试样编号	失效形式	拉伸应变/%	
		A侧	B侧
JLA2-2-1#	远端屈服	4.02	5.98
JLA2-2-2#	远端屈服	3.69	6.26
BGA1-2-1#	远端屈服	0.52	2.76
BGA1-2-2#	远端屈服	3.09	5.55
BG-A2-2#	远端屈服	3.59	5.79
BGA3-2-1#	远端屈服	3.49	6.31
BGA3-2-2#	远端屈服	4.39	2.04

4 结论

针对国产化开发的两种制管形式（JCOE 和 UOE）的 L485 高钢级大厚径比高应变海洋钢管，开展了管道环焊接头力学及断裂变形行为研究、环焊材料研发、环焊工艺及性能研究等三方面的研究，结论如下：

（1）高应变管道环焊接头力学及断裂变形行为表征技术，针对 L485 钢级、31.8mm 大壁厚高应变海洋管道的环焊技术开展机理研究，掌握了高应变管道高强匹配环焊缝的断裂变形机理及多元素微合金控轧高应变管线钢的焊接性特征，解决了拘束度和热循环协同作用对环焊接头显微组织和力学性能的影响机制科学问题，为高强、高韧焊接材料和工艺研究提供理论支撑。

（2）L485 高应变管道环焊用高强韧气保焊丝技术，通过合金强化、组织细化、杂质控制等技术，突破了高强韧塑性焊丝成分的设计，开发了高强度、高韧性熔化极气体保护焊（GMAW）实心焊丝，配合低热输入多层多道 GMAW 工艺，形成的 L485 高应变管道环焊缝经过第三方权威机构检测，其性能达到或优于国外先进产品的水平。

（3）L485 厚壁高应变管道环焊缝工艺技术突破 L485 高钢级、31.8mm 大壁厚管道 GMAW 自动焊工艺关键技术，形成的环焊接头的性能经第三方权威机构检测，实现了焊缝的高强匹配，且具有高韧性和良好的抗变形能力，满足项目任务书指标及 DNV 标准要求，焊接工艺技术及环焊接头性能达到国外先进水平。

开发出 L485 厚壁高应变海洋管道环缝用焊接材料和工艺技术，提升了我国高性能海洋管道工程建设的能力，降低了对国外材料和技术的依赖，对促进中国石油走向深海、推动我国海洋管道自主建设的技术进步和推广应用具有良好的社会和经济效益前景。

参 考 文 献

[1] VERMA N, FAIRCHILD D P, NOECKER F F Ⅱ. Advanced Strain-based Design Pipeline Welding Technologies[C]. Proceedings of the 2014 10th International Pipeline Conference. September 29－October 3, 2014, Calgary, Alberta, Canada：1-10.

[2] 李为卫，何小东，葛加林．油气管道环缝焊接国外先进标准的启示和借鉴[J]．石油管材与仪器，2020, 6(2)：1-7.

[3] 曹雷，孙谦，宗培．等强匹配焊接接头的特征及界定方法[J]．焊接学报，2006，27(7)：81-84.

[4] 庄传晶，冯耀荣，霍春勇．西气东输管道环焊缝强度匹配工艺探讨[J]．机械工程材料，2005，29(8)：32-34.

[5] 何小东，李为卫，吉玲康，等．环焊接头强度对高应变海洋管道轴向承载能力的影响[J]．焊接，2022(6)：1-7.

[6] 陆阳，邵强，隋永莉，等．大管径、高钢级天然气管道环焊缝焊接技术[J]．天然气工业，2020，40(9)：114-122.

[7] 李为卫，杨耀彬，何小东，等，国内外几种 GMAW 焊丝强韧性对比试验研究[J]．石油管材与仪器，2021，7(2)：62-64.

[8] 李为卫，李嘉良，梁明华，等．热输入量对熔化极气体保护焊焊缝强韧性影响[J]．焊管，2021，44(4)：1-4.

本论文原发表于《石油管材与仪器》2023 年第 9 卷第 2 期。

无缝钢管在线测厚系统的测量原理与应用实践

王久刚　岳世斌　孙建安

(宝山钢铁股份有限公司)

摘　要：介绍了同位素在线测厚系统的组成、测量原理和功能，分析了影响测厚精度的因素，以及其在无缝钢管生产现场的应用实践。该在线测厚系统的测量具有及时性和全面性，不但可以判断钢管壁厚精度，也可以反映出各生产环节存在的问题；但该系统是以采样点壁厚展示壁厚精度，放射源之间的位置的壁厚存在超差风险，需避免用显示壁厚代替管体全壁厚，造成对实际壁厚的误判。

关键词：无缝钢管；壁厚；在线测量；同位素；应用

在无缝钢管的生产过程中，壁厚是最重要的技术指标之一，直接影响了管材的使用性能。在无缝钢管轧制过程中，非接触的在线检测系统可以及时、全面地了解每支钢管所测位置壁厚情况，有利于及时有效地判断壁厚情况、排查影响因素和指导调整方向。通过测厚样本数据判断每支钢管的壁厚情况，判断各工序设备、工艺和轧制工具状态对壁厚产生的影响，以便进行针对性的判断和调整，从而提高壁厚精度和成材率，避免了人工测量的滞后性、样本单一性，显著减轻了劳动强度。现介绍了无缝钢管同位素在线测厚系统的系统组成、测量原理、功能及其在生产应用中的实践。

1 同位素在线测厚的基本原理

目前无缝钢管热轧生产线使用的在线壁厚测量系统主要有两种形式：一种是同位素测量方式，以德国 IMS 公司的壁厚测量系统为代表，如烟台鲁宝钢管有限责任公司 $\phi 460mm$ 机组、天津钢管制造有限公司 $\phi 258mm$ 机组等；另一种是激光测量方式，以加拿大泰克纳(Tecnar)公司为代表，如宝山钢铁股份有限公司无缝钢管厂等。由于同位素测量方式应用较多，所以主要介绍同位素在线测厚系统的原理和应用实践。

同位素在线测厚系统主要由辐射系统、测量系统、气动系统、冷却水系统和计算机系统组成[1]。辐射系统由辐射源、电离室、前置放大器组成。辐射源为放射性核元素铯-137，固体硫酸盐形式，半衰期 30.1 年，工作期 15 年，放射性活度 185GBq。测量系统除测厚仪外，还安装有电荷耦合器件(CCD)外径测量仪、热辐射测温仪、激光测长仪器[2]。

1.1 测厚原理

无缝钢管同位素测厚仪采用多条经过准直的、强度已知的放射线束穿过钢管中心，测量穿透钢管的射线剂量当量率来测出钢管的壁厚。钢管同位素测厚系统的原理[3]如图1所示。

一个已知放射源剂量当量率为 I_0 的放射线穿过钢管，测得的钢管壁厚关系式为：

$$S = (\ln I_0 - \ln I)/\mu\rho \tag{1}$$

作者简介：王久刚(1981—)，男，高级工程师，从事无缝钢管产品研究、开发和轧管工艺研究工作。

$$I/I_0 = e^{-\mu\rho s} \tag{2}$$

式中：S 为钢管被测位置厚度，cm；I 为穿过钢管的剂量当量率，Sv/h；μ 为质量吸收系数，cm^2/g；ρ 为钢管密度，g/cm^3。

1.2 测量精度

1.2.1 壁厚的影响

由公式(2)可见，在放射源发射的射线剂量当量率、被测钢管密度和质量吸收系数固定的条件下，穿过钢管的射线剂量当量率与钢管壁厚之间关系如图2所示。测量厚度越小，I/I_0 越大，测量精度越高；反之，壁厚越厚测量精度越小[4-12]。

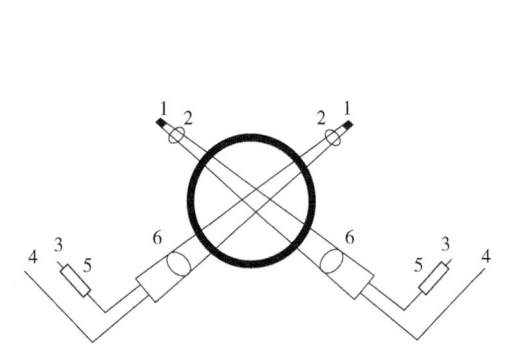

图1 钢管同位素测厚系统原理示意
1—放射源；2—准直孔；3—信号输出；
4—高压；5—放大器；6—电离室

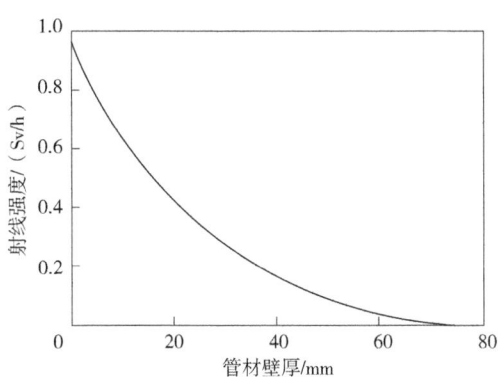

图2 穿过钢管的射线剂量当量率
与钢管壁厚的关系

1.2.2 其他因素的影响

钢管壁厚测量精度还受到钢管材质、外径、位置偏移和异物影响。不同牌号的钢管材质由于含有不同质量分数的组成元素，其密度存在差异。由公式(1)可见，钢管的密度直接影响了所测壁厚的数值。因此在生产中需要合理设定钢管密度，以保证所测壁厚的准确性。

测量系统根据不同的荒管外径设定值计算偏移值，外径设定值与电荷耦合器件(CCD)相机测量的荒管外径相差较大时，会造成壁厚测量计算结果失真，从而导致壁厚测量精度变差。其他导致荒管中心线偏移的因素同样会导致壁厚测量精度降低，如荒管输送辊道中心线不正、荒管在辊道上跳动等。

放射源发射的射线剂量当量率数值与正常偏差较大或放射源附近存在异物(如毛管尾部的铁耳子)时，放射源发射的射线穿过荒管和异物，剂量当量率明显降低，会造成测量壁厚精度降低。

1.3 测厚功能

在线检测系统放射源安装在 O 形架(连轧工序脱管机后)或 C 形架(定径或者张力减径机后)上，沿钢管周向分布。在荒管行进时，测厚系统通过一定的采样频率测量荒管壁厚，其采样点在整个钢管表面上呈格栅状分布。格栅点壁厚测量值进行处理和显示可以展示出壁厚的变化趋势，但是格栅点之间的部分壁厚无法采集到，且未被检测到的面积远远超过检测部分的面积，实际生产中存在壁厚超差点漏测的风险。

提取测量到的壁厚数值，同位素在线测厚系统经计算后可以展示出以下内容(图3)。

(1) 实测数值：各个通道壁厚测量曲线、壁厚色阶图、单截面壁厚趋势、外径、温度和长度；

(2) 壁厚统计值：所有通道壁厚平均值、最大值、最小值和偏心率，针对单截面快速傅里叶分析后提取的影响壁厚的偏心值(频率和振幅，如图 4 所示)；

(3) 平均壁厚的历史数据曲线；

(4) 根据炉号、批号等信息对壁厚历史数据进行查询及展示；

(5) 报警等其他相关信息。

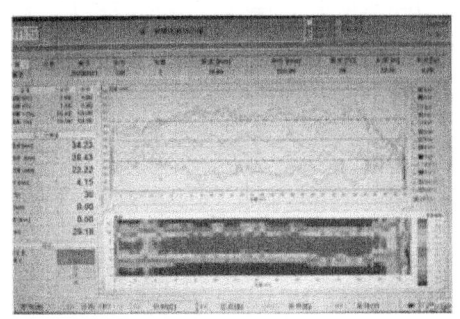

图 3　在线测厚系统显示实测数值
曲线和图形示意

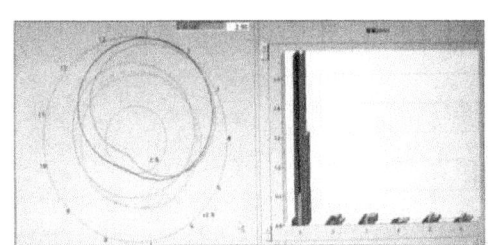

图 4　在线测厚系统显示单截面壁厚
数值和偏心值示意

2　同位素在线测厚仪应用实践

同位素在线测厚系统测量的荒管壁厚采样点在钢管表面上呈格栅状分布。通过格栅点的壁厚测量值可以反馈壁厚的变化趋势，但是格栅点之间的部分壁厚无法采集到，实际生产中存在壁厚超差点漏测的风险。因此在生产中既要充分利用在线测厚系统反馈的壁厚趋势估计实际壁厚情况，以便指导生产和调整；又要避免以点代面、以显示壁厚代替管体全部壁厚造成对实际壁厚的误判，尤其是对潜在的壁厚超差点的误判。

2.1　横截面壁厚监控

环形炉加热、穿孔毛管穿制和连轧延伸均容易产生壁厚不均，根据各工序不同的工艺特征可以有针对性地判断壁厚的影响因素，从而确定调整方向。

2.1.1　管坯断面均匀性判断

如图 4 左侧所示壁厚截面形状，周向壁厚过渡平缓，形状圆润、椭圆度很小、壁厚极值点处于对称位置，壁厚趋势朝向单方向严重偏心；如图 4 右侧偏心值显示，位置 1 的振幅极高，而其他位置振幅极低。这种情况主要原因为管坯加热均匀性较差，断面温差大、变形抗力差异大。在穿孔过程中，顶头仅在轴向有顶杆支撑，在孔型横截面方向没有支撑，处于浮动状态。受到不同的变形抗力影响，顶头向变形抗力低的方向移动，从而出现壁厚偏差。这种情况一般出现在出料节奏过快、停机开轧、长短坯切换、薄壁厚壁规格切换和钢种切换阶段，需要调整生产节奏或加热制度。

在环形炉内管坯采用品字形布料时，由于环形炉内外环加热条件存在差异，有时也会出现钢管一端壁厚偏大、另一端壁厚偏小的情况，需要对布料形式进行优化。

2.1.2　穿孔机调整和轧制工具异常判断

穿孔机斜轧过程中毛管螺旋前进，由于顶头偏心、顶杆弯曲、轧机中心线和三辊定心中心线不正等原因会产生一定程度的规律的对称性壁厚不均，如"糖葫芦"状壁厚不均，如图 5 所示。通过矫正中心线、剔除异常工器具、调整轧制参数等措施有利于减小壁厚偏差。

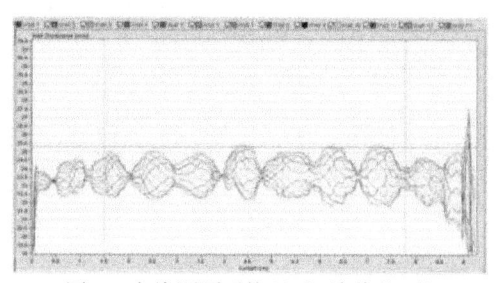

图5 在线测厚系统显示"糖葫芦"状规律性壁厚不均示意

2.1.3 连轧工序壁厚不均判断

连轧工序在无缝钢管生产流程中几乎是控制壁厚精度的最后一道工序。虽然张力减径机可以对平均壁厚和增厚端进行微调,但是对已经形成的壁厚偏差难以明显改善。因此控制连轧工序的壁厚偏差是保证无缝钢管壁厚精度的最重要一环。在线测厚系统在平均壁厚控制、上下限控制和局部壁厚判断、壁厚趋势判断等方面应用较多,在提高连轧荒管壁厚精度方面起着关键性作用。

(1) 平均壁厚控制。

连轧工序后变形机组为定径机或者张力减径机。定径机由于延伸系数小,无长度调整能力,连轧工序的平均壁厚决定了定径机后成品管长度。在不同管坯投料长度、质检检验要求和壁厚状态条件下,平均壁厚控制具有一定的浮动空间。平均壁厚的控制既要避免轧制长度过短导致弯曲、壁厚超差、外径超差、内折等缺陷切不掉造成短尺,又要避免长度过长导致切头尾损失大。因此借助于在线测厚系统反馈的平均壁厚和实际长度,可以进行综合判断、调整,有利于辅助平衡质量控制和成材率的关系,提高成材率。张力减径机的来料为连轧荒管,在张力减径工序通过调整参数微调成品管壁厚和长度时,连轧工序的平均壁厚控制直接影响了张力减径机的调整,因此连轧平均壁厚对张力减径的参数调整影响较大。

(2) 局部壁厚控制和判断。

在加热炉工况、设备状态、工器具装配精度和磨损程度,以及轧管机参数调整的综合作用下,生产中壁厚时常会有波动。对于壁厚公差带窄的管子,局部壁厚有可能会出现超差的情况,因此通过在线测厚设备观察局部壁厚情况,并根据图形展示判断壁厚超差情况,可以预估壁厚实际情况,即时确定调整方向。

① 采用轧态和调质态不同工艺的钢管,荒管壁厚上下线控制略有区别,调质管要留出烧损余量;

② 荒管壁厚极差较大时,壁厚偏厚的位置比壁厚偏薄的位置增厚更多,因此不同位置增厚存在差异,不同(张力减径/定径)减径量的管子增厚量也有差异,需要及时根据在线检测壁厚进行调整;

③ 在线测厚曲线出现鼓肚现象时,说明局部截面壁厚极差较大,由于沿荒管周向放射源数量有限,壁厚极值点会大概率出现在放射源中间位置,尤其是出现单点壁厚超差时最危险,由于壁厚变化幅度较大,实际壁厚极薄点很有可能漏测,出现单点壁厚超差。出现鼓肚壁厚曲线如图6所示,局部截面壁厚极差较大,存在壁厚极值点漏测的风险,因此需要及时进行调整。从生产经验来看,图6所示壁厚偏差与环形炉内外环加热有较大相关性,出现超差需及时采取控制措施,如调整出炉温度、增加穿孔前管坯等待时间促进表面温度降低、调整穿孔导距、咬入角、毛管壁厚或更换导板

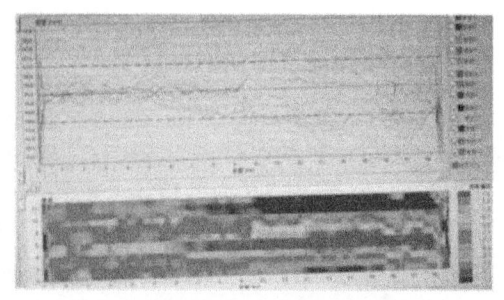

图6 在线测厚系统显示鼓肚壁厚曲线示意

等,甚至停机保温。

(3)管体截面壁厚异常。

在管体上沿纵向全长的壁厚偏差通常在连轧工序或定(减)径工序形成,为纵向延伸的形式决定的。如图7(a)和图7(b)所示,壁厚趋势具有规律性。图7(a)荒管横截面上壁厚分布呈现规律的三角(PQF,三辊)形状,图7(b)壁厚分布呈现规律的椭圆(MPM,两辊)形状,说明连轧工序单双架减壁量匹配不当,需检查和调整单双架的匹配情况。如果某个机架出现严重的辊缝异常,将会出现如图7(c)的截面壁厚单点超差现象。出现以上壁厚情况都需对连轧辊缝进行调整,如果大幅度调整后,某一方向仍然存在不可接受的壁厚偏差,则需归零重新调整甚至更换部分或全套机架。

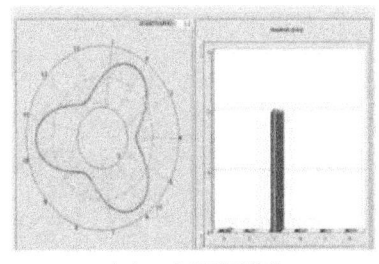 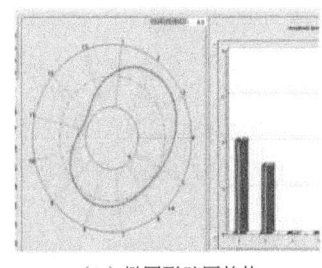 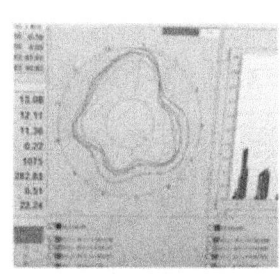

(a)三角形壁厚趋势　　　　(b)椭圆形壁厚趋势　　　　(c)单点壁厚超差

图7　在线测厚系统显示管体截面特殊特征的壁厚情况示意

2.2　纵向壁厚控制

2.2.1　拉钢状态判断

限动芯棒连轧管机组连轧工序机架间拉钢严重的情况下,在线测厚曲线呈现出中间偏薄、形如微笑曲线的形状,原因为张力导致管体中间壁厚被拉薄。严重的拉钢将导致荒管出现严重的拉凹(管壁收缩)缺陷,需要从毛管几何尺寸、连轧管机轧制参数、除氧化剂喷涂和芯棒温度方面进行调整,避免拉钢。

2.2.2　壁厚超差点位置确定

在发现在线测厚壁厚曲线显示单点壁厚超差或者有超差危险时,可以根据激光测长仪显示的位置,考虑定径延伸系数后确定成品管上壁厚异常点位置,通过人工进行更详细的实测,便于确定显示壁厚超差点的实际壁厚情况,为调整确定方向。壁厚单点超差曲线如图8所示。

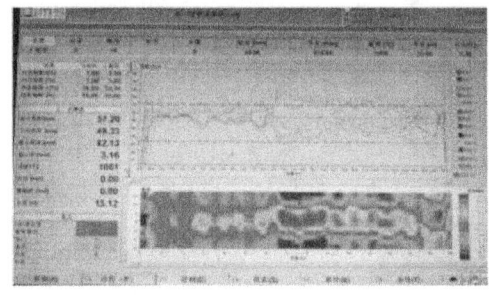

图8　在线测厚系统显示壁厚单点超差曲线示意

2.2.3　管壁收缩和孔洞的监控

在部分机组,在线测厚系统可以监控到管壁收缩、孔洞和单点壁厚超标的情况,因此依据图形显示进行及时和便利地相应连续调整。

2.2.4　使用削尖后壁厚监控

在连轧工序使用削尖功能后,管端壁厚偏薄。实际的壁厚情况需要根据在线测厚系统显示的壁厚情况判断削尖是否启用及其参数是否合理。合理使用削尖功能对于提高成材率有重要意义。

2.3 历史数据查询和对比

2.3.1 轧制状态复盘

出现壁厚质量事故后,需要对当时的轧制状态进行复盘分析。而在线测厚系统存储的壁厚曲线就是最好的参照,便于分析当时各工序调整和控制情况。通过壁厚情况的查询、分析和判断,结合各工序监控曲线和数据,对于确定产生质量问题的要因帮助非常大。成功的复盘可以形成典型案例和控制要点,有助于质量控制能力的提升。

2.3.2 指导难轧合同生产前生产预案的制定

在生产壁厚公差带窄和通径要求严时,壁厚精度是成品质量和通径控制的重要指标。在生产前对于壁厚控制措施的制定和壁厚状态的预估,往往需要参考历史上相近的合同,所以调取历史数据进行查询和分析就显得尤为重要。

3 结语

(1) 同位素在线测厚系统为无缝钢管壁厚的在线测量带来巨大的便利,解决了人工测量的弊端,减少了调整的盲目性和滞后性,提高了钢管质量控制能力,提升了生产质量与效率,并且便于产品质量跟踪。

(2) 同位素在线测厚具有及时性和全面性,不但可以判断出钢管壁厚精度,还可以反映出各个生产环节存在的问题,有利于对相应工序和设备进行调整、更换。

(3) 由于在线测厚系统以采样点壁厚展示壁厚趋势,因此位于放射源之间位置的钢管壁厚有超差的风险,尤其是壁厚偏差大时,需要避免以点代面、以显示壁厚代替管体全部壁厚造成对实际壁厚的误判,避免产生大批量壁厚超差缺陷。

参 考 文 献

[1] 王波. 无缝钢管壁厚测量系统[J]. 冶金自动化,2008(S1):252-254.

[2] 张朔共. 钢管在线测厚和应用[J]. 宝钢技术,1996(3):19-24.

[3] 胡儒卓,郑贵英,庄钢. 钢管热测壁厚的应用与实践[J]. 钢管,2000,29(2):31-33.

[4] 田党,修雪峰,侯伟,等. 钢管超声波测厚示值减小的原因分析[J]. 钢管,2011,40(5):57-61.

[5] 徐亮,刘世泽. 海洋管线管焊缝全壁厚试样与减薄试样的 CTOD 值分析[J]. 钢管,2012,41(1):72-74.

[6] 张进,朱宝禄,韩建新. 连轧钢管的壁厚不均原因分析[J]. 钢管,2013,42(1):55-58.

[7] 陆卫中. 核电用 NS3105 合金 U 型管壁厚偏差工艺设计及控制[J]. 钢管,2013,42(6):50-53.

[8] 陈今良,白磊,王军,等. 单机架减径率对无缝钢管周向壁厚影响的研究[J]. 钢管,2015,44(1):33-37.

[9] 张卫华,项春萍,陈小伟,等. 厚壁焊管强度在壁厚方向分布规律的研究[J]. 钢管,2015,44(3):25-30.

[10] 汉斯·约阿希姆·佩勒. 无缝钢管壁厚偏心率的测量分析及降低方法[J]. 钢管,2016,45(1):41-44.

[11] 欧阳建,王锐,覃宣,等. 大中型多辊连轧管机组产品壁厚精度的模拟对比[J]. 钢管,2016,45(5):52-55.

[12] 肖永忠,王文强,刘国庆,等. 张力减径机头尾壁厚控制系统的开发与实践[J]. 钢管,2017,46(6):36-38.

本论文原发表于《钢管》2023 年第 52 卷第 1 期。

燕尾形超高抗扭特殊螺纹接头研究

詹先觉 高 展

(宝山钢铁股份有限公司)

摘 要：对比了传统的 API 圆螺纹和偏梯形螺纹的优缺点，设计了一种新型的"倒梯形的楔子"式的燕尾形螺纹齿型，以及一种经济型的超高抗扭特殊螺纹接头。详细介绍了该接头的设计、加工、检验过程，并采用有限元分析程序计算了接头上扣过程中螺纹的应力和应变。分析结果表明：该接头上扣到位后，螺纹所有侧面完全啮合，并且具有很高的接触应力，使接头在复杂的环境载荷下能始终保证螺纹间配合的稳定性。

关键词：油套管；特殊螺纹接头；抗扭；经济型；有限元分析

石油开采中使用量最大的油套管产品一直是 API 标准中的圆螺纹和偏梯形螺纹接头[1-5]。圆螺纹易加工，且拧接后只在齿顶、齿底圆角处有间隙，在被螺纹脂填充后密封性能较好。但该螺纹上扣过盈量大，易发生批量螺纹黏结；且圆螺纹抗滑脱性能差，接头抗拉性能只有管体的 60%~80%，当拉伸力增大，接头将发生脱扣，使管柱掉井，引发生产事故。偏梯形螺纹抗拉性能高，但是该螺纹 4 个圆角(0~0.05mm)和轴向齿宽(0.03~0.20mm)存在的间隙使得偏梯形螺纹密封性能不佳，虽然可以使用螺纹脂在一定程度上填充内、外齿之间的空隙，但是接头在交变的拉伸、压缩载荷下，螺纹导向面和承载面之间的间隙交替出现，无法保证接头轴向的稳定性，因此 API 螺纹在井况恶劣的深井、高压井中难以使用。因此，国内外钢管厂家都开发了高气密封特殊螺纹接头[6-9]。该类接头通常采用改进的偏梯形螺纹，同时在端部增加一个金属对金属过盈密封结构和止扭台肩。该结构可弥补偏梯形螺纹的不足，提高了密封性能，但是对产品加工公差和上扣操作要求高，因此价格高于 API 接头。

1 问题背景

随着易开采石油逐渐地开采殆尽，石油开采难度越来越大，而且随着新开采工艺的应用，对接头的抗扭、抗压缩、密封、经济性提出了很高要求。

通常，在接头端部增加一个扭矩台肩可以提高接头的抗扭和抗压缩性能，但是由于内通径、管体壁厚等限制，抗扭性能提高有限；而且该方式增加了产品加工和使用成本，提高了开采成本。为了提高接头的抗扭性能，同时仍然保证产品的经济性，设计了一种新型的"倒梯形的楔子(Wedge of Inverse Trapezium)"式燕尾形螺纹，简称 WIT 螺纹。

2 WIT 螺纹与接头开发

WIT 螺纹的展开似一个截面为倒梯形的螺旋的楔子，外螺纹接头沿-X 方向宽度逐渐增

作者简介：詹先觉(1986—)，男，高级工程师，从事特殊螺纹接头产品的研发工作。

大的螺纹齿,与内螺纹接头沿 X 方向宽度逐渐减小的螺纹槽正好匹配,内、外螺纹在相互旋紧配合过程中形似一个楔子打入楔槽。该接头内、外螺纹尺寸被设计成最合适的匹配值,当接头旋转到位,螺纹齿顶、齿底和齿侧达到全啮合的状态。因此接头能够提供较高压力的内外气体密封能力。又因为接头螺纹为螺旋楔形结构,越拧接配合越紧,过盈越大,所需的拧接扭矩也越大。该接头螺纹为粗牙螺纹,且螺纹接触面积大,能承受的最大上扣扭矩比常规接头大得多,能够承受大扭矩。由于该螺纹齿形为上大下小的倒梯形,螺纹导向面和承载面为负角度,内外螺纹相互咬合在一起,只能通过反向旋转分离,无论承受多大的拉伸、压缩、弯曲、内压、外压,除非螺纹整体破坏,接头不会相互脱离。

2.1 WIT 螺纹结构设计与加工

WIT 螺纹配合如图 1 所示。为了方便螺纹加工时编程,考虑到 $\arctan(1/10)=5.71°$,导向面角 α 和承载面角 β 设计为 $-5.71°$,螺纹导向面齿高为 h_S,承载面齿高为 h_L。因螺纹导向面螺距 p_L 小于承载面螺距 p_L,螺纹齿宽沿轴线呈线性变化[10]。螺纹齿顶和齿底可以平行于轴线,也可以平行于锥母线,现设计 WIT 螺纹的螺纹齿顶和齿底为平行于轴线,则导向面螺纹锥度 T_S 为:

$$T_S = 2(h_S - h_L)/p_S \tag{1}$$

选取整条螺纹的中段长度设置于钢管端部,螺纹加工刀片为单齿刀,齿刀宽度 B_T 至少比螺纹最小齿宽 B_{min} 小 $(\tan\alpha + \tan\beta)h_L$,即 $B_T \leq B_{min} - (\tan\alpha + \tan\beta)h_L$。为了提高螺纹加工效率,所设计的内、外螺纹最大齿宽 B_{max} 与最小齿宽 B_{min} 之比小于 2,即 $B_{max}/B_T \leq 2$,如图 2 所示,车削加工时,只需要变换 2~3 次螺距就可以加工出螺纹[11]。

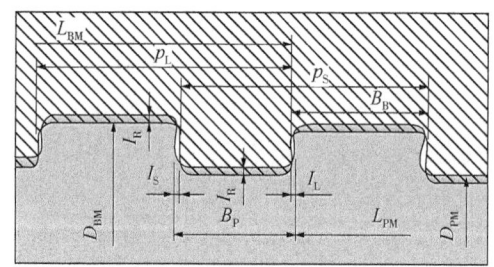

图 1 WIT 螺纹配合示意

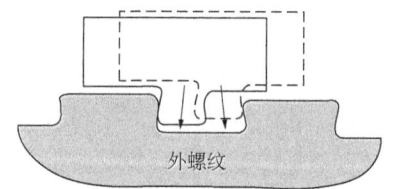

图 2 WIT 外螺纹加工示意

I_R—径向过盈量(半径);I_S—导向面轴向间隙量;
I_L—承载面轴向间隙量;B_P—测量位置外螺纹齿宽;
B_B—测量位置内螺纹齿宽;L_{PM}—B_P 对应的轴向测点;
D_{PM}—B_P 对应的前一螺纹齿底直径;
L_{BM}—B_B 对应的轴向测点;
D_{BM}—B_B 对应的前一螺纹齿底直径

2.2 WIT 接头开发

设计 WIT 接头的关键在于除了选定合适的基本参数 p_S、p_L、h_S、h_L、α 和 β 外,还需要选定机械拧紧状态最合适的 I_R、I_S 和 I_L,随着机械拧紧时上扣扭矩的不断上升,导向面轴向间隙量 I_S 和承载面轴向间隙量 I_L 越来越逼近更小值,从而封堵住轴向间隙,又因为螺纹两侧均为负角度,所以无论在内压、外压、拉伸和压缩的复合载荷下,都始终保持内、外螺纹间的配合稳定性,并且螺纹所有侧面同时承载高扭矩。

WIT 螺纹将满足关系式式(2)至式(4)：

$$B_P + B_B = p_S + h_L \tan\beta + h_S \tan\alpha + I_L - I_R \tan\beta + I_S - I_R \tan\alpha \tag{2}$$

$$D_{BM} - D_{PM} = 4h_S - 2h_L - 2I_R \tag{3}$$

$$L_S = L_{BM} + L_{PM} + h_L \tan\beta + I_L - I_R \tan\beta \tag{4}$$

其中，L_S 为上紧后外螺纹接头端部至接箍端面长度，即上扣损失长度。

当 $B_P = B_B$，由公式(2)可得：

$$B_P = (p_S + h_L \tan\beta + h_S \tan\alpha + I_L - I_R \tan\beta + I_S - I_R \tan\alpha)/2 \tag{5}$$

为外螺纹选定 L_{PM}、D_{PM} 后，则与外螺纹配合的内螺纹结构参数 L_{BM}、D_{BM} 可由公式(3)和公式(4)得出。

2.3 WIT 螺纹测量

通常用于加工特殊螺纹接头的数控车床在加工前都会对钢管车平端面进行零位对刀，因此轴向偏差较小，但是径向偏差较大。因此在确保加工程序正确的情况下，轴向尺寸不需要逐个测量，仅需要对螺纹的径向尺寸进行测量。但是为了确保加工的首件轴向尺寸正确，抽检时仍需要对螺纹轴向尺寸进行测量[12]。

外螺纹宽度为 B_P' 的位置对应的轴向与径向测量长度为：

$$L_{PM}' = L_{PM} - p_L(B_P' - B_P)/(p_L - p_S) \tag{6}$$

$$D_{PM}' = D_{PM} - 2(h_S - h_L)(B_P' - B_P)/(p_L - p_S) \tag{7}$$

内螺纹宽度为 B_B' 的位置对应的轴向与径向测量长度为：

$$L_{BM}' = L_{BM} - p_L(B_B' - B_B)/(p_L - p_S) \tag{8}$$

$$D_{BM}' = D_{BM} + 2(h_S - h_L)(B_B' - B_B)/(p_L - p_S) \tag{9}$$

3 有限元分析

现基于 Python 的 Abaqus 有限元分析程序[13]构建二维参数化 WIT 螺纹接头，相比于偏梯形螺纹接头，WIT 螺纹的上扣止扭位置与齿形密切相关，当外螺纹在齿宽为 B_P 的载面再拧进 θ 角度，该载面的螺纹齿宽 B_P' 和尺寸 D_{PM}'、L_{PM}' 分别为：

$$B_P' = B_P - \theta(p_L - p_S)/360 \tag{10}$$

$$D_{PM}' = D_{PM} + 2(h_S - h_L)\theta/360 \tag{11}$$

$$L_{PM}' = L_{PM} - \theta p_L/360 \tag{12}$$

通过所构建的有限元分析程序，模拟分析了 110 钢级 ϕ88.9mm×6.45mm 规格油套管 WIT 螺纹接头在拧进角度 θ 分别为 0°、20°、40°、60°时的螺纹应力情况，结果如图3所示。当拧进角度逐渐增大，齿侧间隙由间隙转为过盈配合。局部螺纹应力分布如图4所示，螺纹表面有较大的应力使所有侧面完全啮合，因此上紧后，无论内外压力、拉伸力和压缩力如何交替变化，始终保证螺纹间配合的稳定性。

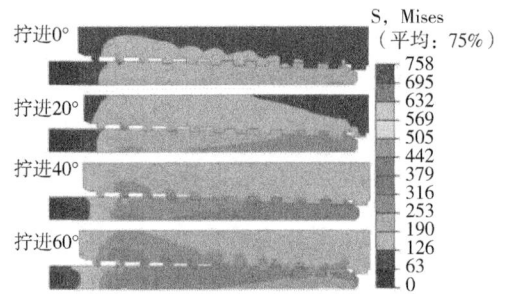
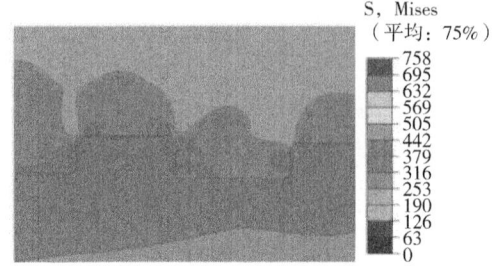

图3 WIT螺纹上扣应力变化　　　　图4 WIT螺纹局部应力分布(拧进角度60°)

4 结语

WIT接头弥补了API偏梯形螺纹轴向齿宽存在0.03~0.20mm间隙的缺陷，具有稳定的气密封能力。为了提高WIT螺纹的加工效率，所设计的结构只需变换2~3次螺距就可以加工出螺纹。采用Abaqus有限元分析计算了接头在上扣过程中的螺纹应力/应变，当接头上扣到位后，螺纹所有侧面完全啮合，具有很高的接触应力与配合的稳定性。

参 考 文 献

[1] 孙建安,卢小庆,吕庆刚.油套管双头螺纹接头开发试验研究[J].钢管,2022,51(1):61-64.
[2] 李小兵,胡志立.油套管特殊螺纹的发展[J].钢管,2020,49(5):15-22.
[3] 曹杉,高进伟,孔祥领,等.油井管特殊螺纹设计技术分析[J].钢管,2015,44(1):15-23.
[4] 马刘宝,朱靖,赖兴涛.油套管螺纹粘扣原因分析及研究现状[J].钢管,2011,40(3):27-30.
[5] 张居勤,高连新,李飞键,等.油井管螺纹粘扣类型及其原因分析[J].钢管,2004,33(4):16-20.
[6] 王建东,李玉飞,汪传磊,等.油套管用特殊螺纹连接密封完整性探讨[J].钢管,2021,50(3):19-25.
[7] 梅丽,史彬,张军,等.TP-JC特殊螺纹接头的开发及应用[J].钢管,2014,43(2):58-62.
[8] 王怡,张建兵,聂艳,等.高温高压井用特殊螺纹接头的设计与评价现状[J].钢管,2020,49(1):72-76.
[9] 廖凌,崔顺贤,叶顶鹏,等.汉廷特殊螺纹接头油套管的技术特点与应用分析[J].钢管,2009,38(4):44-47.
[10] 詹先觉,王珂.气密封楔形燕尾锥螺纹接头设计及试验研究[J].宝钢技术,2018(1):20-23.
[11] 崔顺贤.燕尾螺纹及其加工方法[J].石油机械,1990,18(3):12-15.
[12] 吴维新,陈文治.用于管路连接的新型燕尾牙型螺纹及其测量[J].石油钻采机械,1984(6):79-84.
[13] 詹先觉,左宏志.基于Python的特殊螺纹接头参数化有限元分析[J].钢管,2022,51(1):70-74.

本论文原发表于《钢管》2023年第52卷第1期。

X80 管线钢与氢环境相容性试验研究

张伟卫[1,2] 封 辉[2] 陈越峰[2] 高雄雄[2]

(1. 中国石油集团工程材料研究院有限公司石油管材及装备材料服役行为与结构安全国家重点实验室；2. 西安建筑科技大学)

摘 要：本文针对 X80 管线钢材料在高压气相氢环境下开展了慢应变速率拉伸试验、断裂韧度试验、疲劳裂纹扩展速率试验，分析了不同氢分压对 X80 管线钢力学性能、疲劳性能及断口形貌的影响规律。研究结果表明，氢分压是影响材料氢脆和疲劳性能的重要因素，随着氢分压的增加，X80 管线钢的氢脆敏感性显著增大，缺口疲劳试样的疲劳循环次数显著降低，断口韧窝间出现典型的具有小平面和撕裂棱组成的准解理特征脆性断裂形貌，疲劳裂纹扩展速率急剧增加，提高了管道失效风险。本研究成果可为高强度管道用于输送掺氢天然气的工程临界评估提供参考。

关键词：X80 管线钢；断裂；疲劳；氢分压；氢脆

管道输送是大规模、长距离输送流体介质的重要方式，也是实现大范围资源调配、能源互联互通的重要途径[1-3]。近年来，随着与氢能利用相关产业的高速发展，国际上利用管道输送纯氢、掺氢天然气和合成天然气等含氢气体已得到较为广泛的认可并开始实施[4-7]。由于纯氢管道建设成本较高，同时氢气利用领域具有一定的局限性，纯氢管道的发展远远落后于天然气管道，难以满足氢能产业迅速发展的需求。因此将氢气以一定比例掺入天然气中，通过成熟、完善的天然气管网输送至下游终端用户，成为氢气输送的最优途径[8]。

在临氢环境中，管道中的氢气分子吸附于钢材表面，随后以原子形式渗入金属内部，使管线钢发生氢致开裂、氢鼓泡等损伤现象，材料的韧性、塑性、强度显著下降，甚至使材料的断裂行为从韧性断裂转变为脆性断裂，加剧管道失效的突发性，这成为利用金属管道输送氢气的关键制约因素。为保障天然气管道输送氢气的安全性，管线钢的环境氢脆研究备受重视，国内外学者已针对金属管道中的氢脆现象开展了大量研究[9-13]，研究对象多为无缝钢管或低强度焊管，且鲜见管线钢与氢环境的相容性系统研究的报道。

我国是 X80 管线钢应用总里程最长的国家，掌握掺氢天然气对 X80 管线钢力学行为的影响规律，对服役管道的工程临界评估具有重要的现实意义。本工作依据 GB/T 34542.2—2018《氢气储存输送系统 第 2 部分：金属材料与氢相容性试验方法》，针对 X80 管线钢开展了慢应变速率拉伸试验、断裂韧度试验、疲劳裂纹扩展速率试验，探究 X80 管线钢在氢环境下原位力学性能和断口形貌变化规律，为高强度管道用于输送掺氢天然气的工程临界评估提供技术支持。

基金项目：中国石油天然气集团公司科学研究与技术开发项目"中长距离管道纯氢/掺氢输送关键技术研究"（项目编号：2021DJ5002）。

作者简介：张伟卫(1981—)，男，高级工程师，主要从事输送管与管线材料研发应用及标准化方面的研究工作。

1 试验

1.1 试验材料及试样

试验材料为某公司生产的 API 5L PSL2 等级的 X80 钢管，沿钢管周向制取试样。钢管外径为 1219mm，壁厚为 18.4mm，其化学成分见表 1。拉伸试验测得钢管的抗拉强度 R_m 为 720MPa，屈服强度 $R_{t0.5}$ 为 641MPa，断后伸长率为 19.4%。

制备两组（每组 8 个）完全相同的缺口圆棒试样，用于慢应变速率拉伸试验和断裂韧度试验。制备一组（7 个）紧凑拉伸试样（CT 试样），机械加工 2mm 深的缺口，并预制 3mm 长的疲劳裂纹，用于疲劳裂纹扩展速率试验。试样缺口方向与管道轴向平行，试样形状如图 1 和图 2 所示，试样尺寸见表 2 和表 3。

表 1 试验材料的化学成分

元素	C	Si	Mn	P	S	Cr+Mo+Ni+Cu	Nb+V+Ti	CE_{Pcm}
质量百分数/%	0.06	0.21	1.77	0.006	0.002	0.54	0.116	0.19

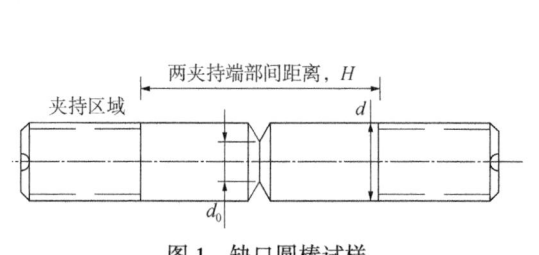

图 1 缺口圆棒试样

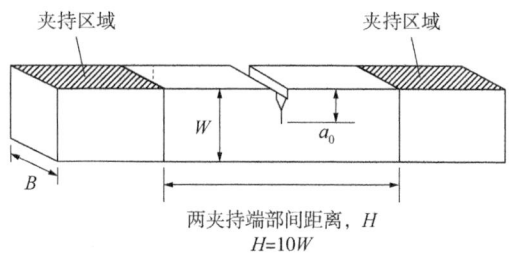

图 2 紧凑拉伸试样

表 2 缺口圆棒试样尺寸

试样长度 H/mm	试样直径 d/mm	缺口处试样直径 d_0/mm	夹持端长度/mm
38.2	12.0	6.0	19.0

表 3 紧凑拉伸试样尺寸

试样长度 H/mm	试样厚度 B/mm	试样宽度 W/mm	初始裂纹 a_0/mm
100	10	10	5

1.2 试验方法

试验时试样放置于不同气质的环境中，根据氢气体积分数的不同，分为四个气质环境，环境箱试验压力 12MPa。试验环境气质组分见表 4。

表 4 试验环境气质组分

序号	气质组分体积百分数/%		压力/MPa
	氮气	氢气	
1	100.0	0	12
2	99.0	1.0	12
3	97.8	2.2	12
4	95.0	5.0	12

（1）慢应变速率拉伸试验。

将缺口应力集中系数 $K_t \geq 3$ 的标准带缺口圆棒试样置于试验气质环境中，以位移控制方式（拉伸位移速率选取 0.01mm/min，对应的应变速率为 $6.56×10^{-6}s^{-1}$），对其进行单轴拉伸，直至试样断裂，得到 X80 管线钢在高压不同氢含量环境下的抗拉强度、屈服强度、缺口抗拉强度、断后伸长率及断面收缩率等力学性能数据。

（2）低周疲劳试验。

采用缺口应力集中系数 $K_t \geq 3$ 的标准带缺口圆棒拉伸试样，将其置于试验气质环境中，采用力值比 R 为 0.1 的正弦加载波形，对其施加单轴拉—拉循环载荷，最小和最大力分别采用 2kN 和 20kN，加载频率为 1Hz，直至试样断裂。获取试验过程中管材 X80 钢在高压含氢环境下的循环次数，即疲劳寿命。

（3）疲劳裂纹扩展试验。

采用 CT 试样，首先在空气中预制裂纹，采用力值比 R 为 0.1 的正弦加载波形，对试样施加循环载荷，加载频率取 5Hz，初始力值范围取 25kN，阶段性降载过程见表5，当裂纹长度扩展到 3.0mm 附近以后，将试样置于试验气质环境中，继续采用应力比为 0.1 的正弦波形对试样循环加载，力值保持 19kN 不变，加载频率变为 1Hz，获取试验过程中裂纹扩展长度及应力强度因子随载荷循环次数的变化关系，最终获得该材料在高压含氢环境下的疲劳裂纹扩展速率，定量反映管材 X80 钢在高压含氢环境下的缺陷敏感性。

表5　空气环境下 CT 试样预制裂纹应力设置

力值范围/kN	25	23	21	19
裂纹长度/mm	0	1	2	3

2　结果与讨论

2.1　氢对拉伸性能的影响

表6给出了缺口圆棒的慢拉伸力学性能试验数据。试验测得氢气分压由低到高环境下 X80 钢平均缺口抗拉强度分别为 1458.4MPa、1448.3MPa、1441.3MPa 和 1435.7MPa，断后试样伸长量分别为 1.24mm、1.13mm、1.10mm 和 0.98mm，断面收缩率分别为 35.8%、34.9%、32.5% 和 31.0%。

表6　缺口圆棒的慢拉伸力学性能

氢气体积分数/%	抗拉强度/MPa		伸长量/mm		断面收缩率/%	
	试验值	平均值	试验值	平均值	试验值	平均值
0	1443.0	1458.4	1.25	1.24	37.1	35.8
	1473.7		1.23		34.5	
1.0	1450.6	1448.3	1.10	1.13	34.2	34.9
	1445.9		1.15		35.6	
2.2	1430.0	1441.3	1.12	1.10	32.9	32.5
	1452.5		1.08		32.1	
5.0	1427.9	1435.7	0.98	0.98	30.5	31.0
	1443.4		0.98		31.4	

断后伸长率和断面收缩率是反映材料塑性的重要指标。通过计算不同氢分压条件下试样与无氢环境试样的断后伸长率、断面收缩率的降低率，定义为氢脆敏感指数，以此反映试样的氢脆程度。试样氢脆敏感指数越大，其氢脆程度越高，力学性能劣化越显著。

氢分压对缺口圆棒试样抗拉强度的影响如图3所示，抗拉强度随着氢分压的增加略有降低。与纯氮气环境相比，氢含量1%时，抗拉强度降低0.69%，氢含量2.2%时，抗拉强度降低1.17%，氢含量5%时，抗拉强度降低1.56%。氢分压对材料强度的影响并不显著。

氢分压对试样断后伸长量(位移)和断面收缩率的影响如图4和图5所示。随着氢分压增大，试样的断后伸长率和断面收缩率降低较明显，氢脆敏感指数相应增大。氢含量1%时，断后伸长率降低8.87%，断面收缩率降低2.51%。氢含量2.2%时，断后伸长率降低11.29%，断面收缩率降低9.21%。氢含量5%时，断后伸长率降低20.97%，断面收缩率降低13.41%。

可见随着氢分压的增加，X80管线钢的缺口抗拉强度无显著变化，但材料的氢脆指数迅速增大，导致材料发生较大的塑性损失。在氢含量5%时，试样的断后伸长率降低率高达20.97%，试样在该环境中可能会因塑性损失而发生危险，给管道实际运行带来隐患。

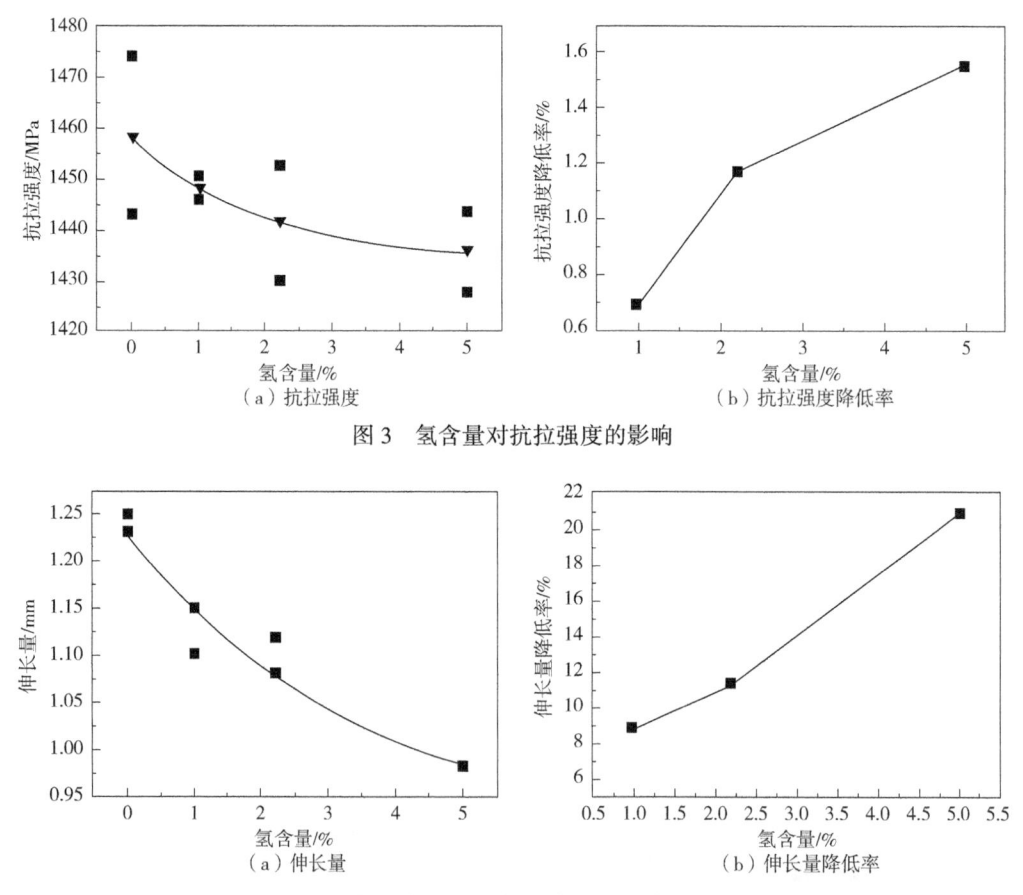

图3 氢含量对抗拉强度的影响

图4 氢含量对断后伸长量的影响

图6和图7分别为无氢气和氢气含量5%时缺口拉伸试样的断口形貌，由图6和图7可以看出，无论是否存在氢，裂纹都会在断面的边缘萌生，随后扩展并导致断裂。两种气质环境下缺口拉伸试样的中心都可以观察到韧窝为主的韧性断裂形貌。无氢环境下，断口呈明显

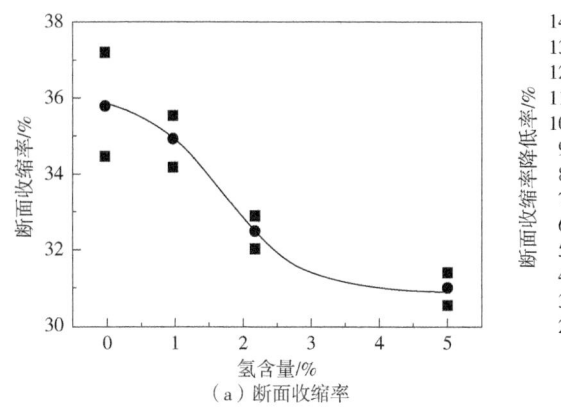 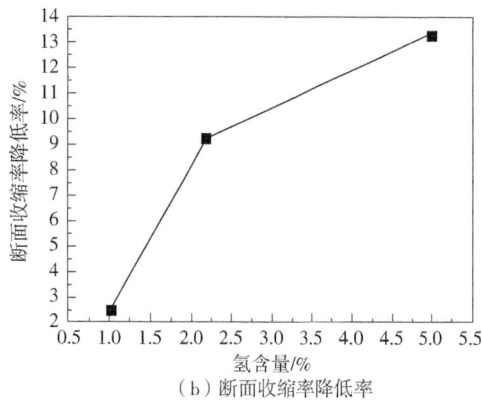

（a）断面收缩率　　　　　　　　　　（b）断面收缩率降低率

图5　氢含量对断面收缩率的影响

的韧性断裂模式，可以观察到明显的纤维区与剪切唇，断口上分布众多微孔（图6a）。在高倍扫面电镜下观察发现，试样中心呈等轴状韧窝形貌（图6b），韧窝尺寸约为 $50\mu m \times 50\mu m$，且韧窝底部存在第二相粒子。

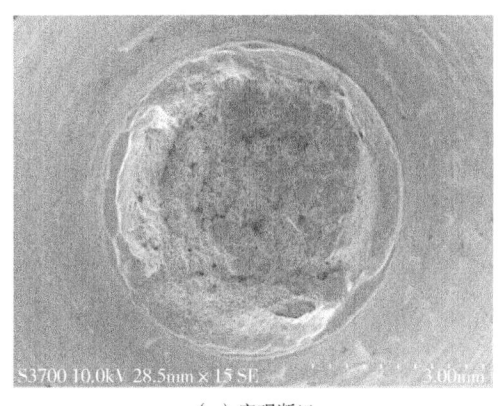 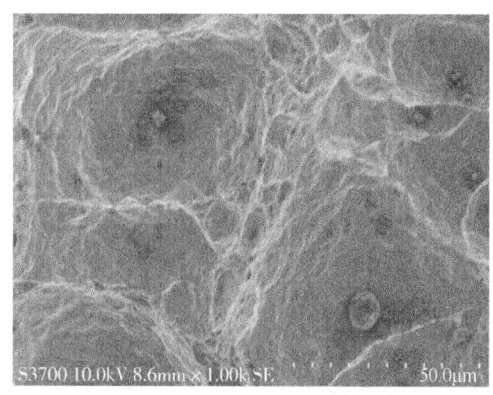

（a）宏观断口　　　　　　　　　　（b）试样中心断口形貌

图6　缺口拉伸试样断口形貌(无氢气环境)

5%氢气环境明显改变了试样的断口形貌，在断口周围外表面产生了明显的二次裂纹[图7（a）]，而无氢环境试样表面并没有观测到这种裂纹。这可能是因为进入金属材料内部的氢原子在晶界或内部缺陷的位置聚集结合形成氢分子产生局部高压，促进了微裂纹的形成与扩展，破坏了金属内部结构的连续性，最终导致材料的破坏。在高倍扫描电镜下观察发现，试样断口形貌的韧窝相较无氢气环境更小，韧窝间出现典型的具有小平面和撕裂棱组成的准解理特征脆性断裂形貌[图7（b）]。

有研究指出，随着氢分压的增大，吸附在金属表面的氢原子浓度升高，导致材料内部的氢原子浓度升高，材料发生塑性变形时，内部生成大量的空位、位错等缺陷，氢原子进入缺陷后产生钉扎效应，抑制了缺陷的移动，氢原子浓度越高这种抑制作用越明显，最终金属塑性变形能力下降，表现为金属的断后伸长率和断面收缩率随氢浓度升高而降低[14-18]。

2.2　氢对金属疲劳的影响

不同氢气含量条件下疲劳试验得到的断后循环次数如图8（a）所示，可见随着氢分压增加，断后疲劳循环次数急剧下降，含氢量0、1%、2.2%、5%时X80钢的平均断后循环次数为4274次、1461次、1198次和398.5次，氢气的掺入对X80管线钢的疲劳性能造成了极大

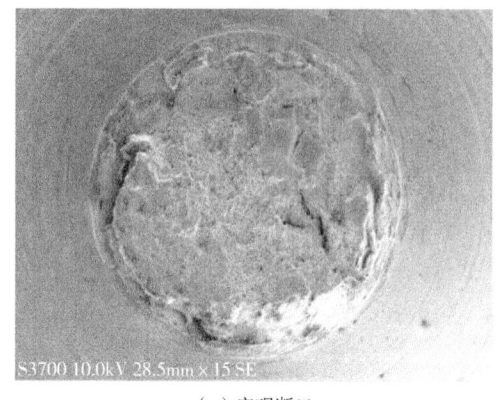
（a）宏观断口

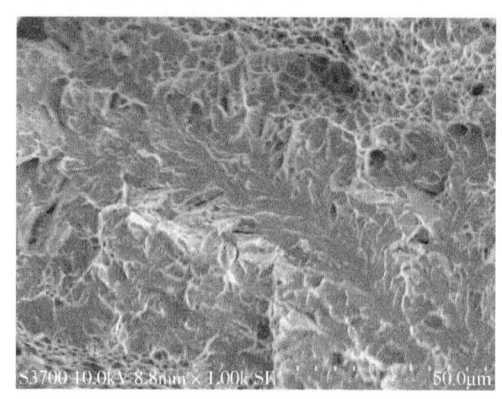
（b）试样中心断口形貌

图 7　缺口拉伸试样断口形貌(5%氢气含量)

损伤。与无氢气环境相比，氢含量 1%时，断后循环次数降低 65.82%，氢含量 2.2%时，断后循环次数降低 71.79%，氢含量 5%时，断后循环次数降低 90.68%，如图 8(b)所示。

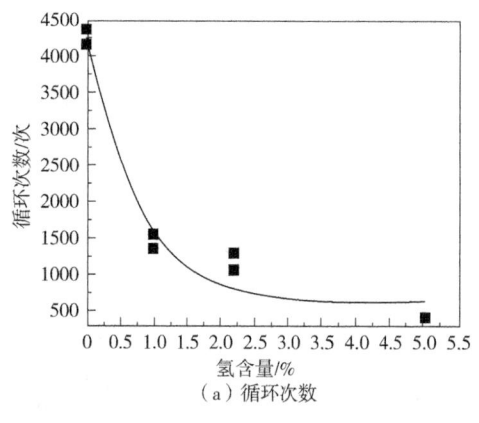

（a）循环次数　　　　　　　　　　　　（b）循环次数降低率

图 8　氢含量对试样疲劳循环的影响

试验得到的 X80 管线钢疲劳裂纹扩展曲线如图 9 所示。可见相对于无氢气环境，在含氢量 1%、2.2%、5%时，疲劳裂纹扩展速率明显增加，其中在氢含量为 1%时疲劳裂纹扩展速率约为氮气环境中的 7.2 倍，在氢含量为 2.2%时疲劳裂纹扩展速率约为氮气环境中的 11.2 倍，在氢含量为 5%时疲劳裂纹扩展速率约为氮气环境中的 13.5 倍。疲劳裂纹扩展速率随氢气分压的增加显著升高，增加了管线钢服役过程中的失效风险。

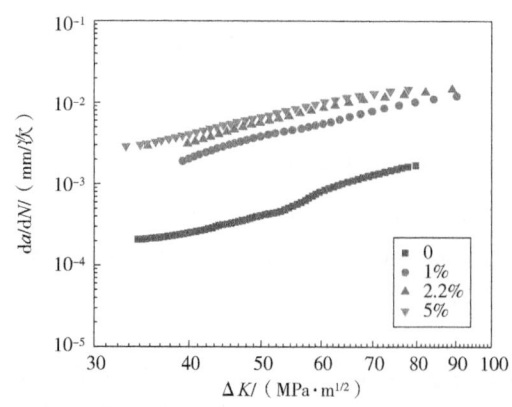

图 9　母材 CT 试样的 da/dN—ΔK 曲线

图 10 和图 11 分别为无氢气和氢气含量 5%环境中缺口疲劳试样的断口形貌。可以看出，两种气质环境中，疲劳裂纹都在断面的边缘缺口根部萌生，两种环境下缺口拉伸试样断口都已没有韧窝特征。

放大后发现，无氢气环境中的断口显微形貌具有显著的塑性变形特征，在表面可以观察到分布较为均匀的驻留滑移带，当试样受到循环载荷时,驻留滑移带内部大量无规则和不可逆的循环滑移变形使得断口表面形成滑移台阶、挤出片等结构[图10(b)]。

在氢气含量5%环境中试样的断面显示出准解理特征[图11(b)],滑移台阶和挤出片结构数量明显小于无氢气环境,氢气的掺入改变了材料微结构的变形模式,明显加快了缺口试样的疲劳裂纹扩展速度,少量驻留滑移带出现以后,快速的裂纹形核萌生及扩展,使滑移台阶还未完全形成之前,试样就已经发生断裂,从而导致了宏观循环塑性变形行为的改变。

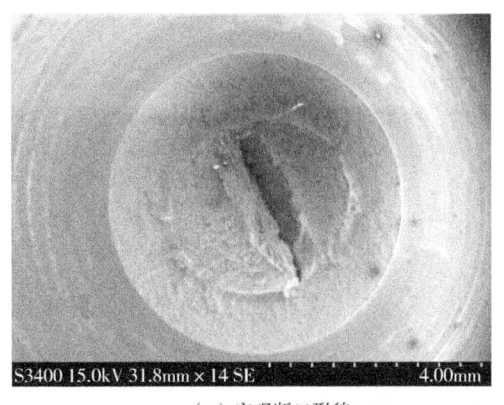

(a)宏观断口形貌　　　　　　　(b)微观断口形貌

图10　缺口疲劳试样断口形貌(无氢气环境)

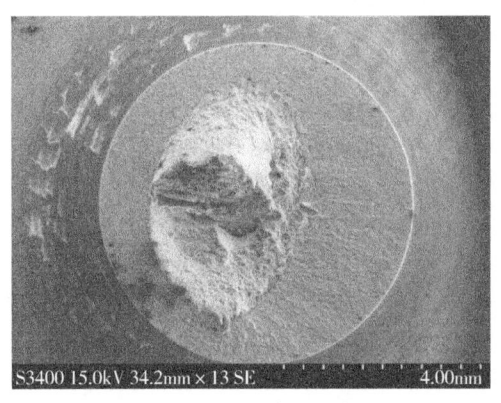 

(a)宏观断口形貌　　　　　　　(b)微观断口形貌

图11　缺口疲劳试样断口形貌(5%氢气含量)

3　结论

(1)氢气环境中试样的抗拉强度、断后伸长率、断面收缩率发生不同程度的损失。氢分压是影响材料氢脆的重要因素,在本研究试验气质环境中,随着氢分压的增加,X80管线钢的氢脆敏感性显著增加。

(2)氢气的加入显著降低了X80管线钢缺口疲劳试样的疲劳循环次数,使X80管线钢的疲劳裂纹扩展速率增加了一个数量级左右,增加了管道的失效风险。

(3)临氢环境改变了X80管线钢材料微结构的变形模式,使试样断口韧窝间出现典型的具有小平面和撕裂棱组成的准解理特征脆性断裂形貌,滑移台阶和挤出片结构数量减少,加快了缺口试样的疲劳裂纹扩展速度,导致宏观循环塑性变形行为发生改变。

参 考 文 献

[1] BUDNY C, MADLENER R, HILGERS C. Economic feasibility of pipe storage and underground reservoir storage options for power-to-gas load balancing[J]. Energy Conversion and Management, 2015, 102: 258-266.

[2] HAWKINS A S. Technological characterisation of hydrogen storage and distribution technologies[J]. UKSHEC Social Working Paper, 2006, 21(1): 1-36.

[3] 程玉峰, 孙颖昊, 张引弟. 氢气管道发展与管线钢氢脆挑战[J]. 长江大学学报(自然科学版), 2022, 19(1): 54-69.

[4] CERNIAUSKAS S, JUNCO A J C, GRUBE T, et al. Options of natural gas pipeline reassignment for hydrogen: Cost assessment for a Germany case study[J]. International Journal of Hydrogen Energy, 2020, 45(21): 10295-12107.

[5] OGDEN J, JAFFE A M, SCHEITRUM D, et al. Natural gas as a bridge to hydrogen transportation fuel: Insights from the literature[J]. Energy Policy, 2018, 115: 217-329.

[6] REDDI K. Hydrogen delivery infrastructure options analysis[R]. 2016-1-19. http://www.doc88.com/p-9965242465504.html.

[7] DODDS P E, STAFFELL I, HAWKES A D, et al. Hydrogen and fuel cell technologies for heating: A review[J]. International Journal of Hydrogen Energy, 2015, 40(5): 2065-2083.

[8] HANLEY E S, DEANE J P, GALLACHÓIR B Ó. The role of hydrogen in low carbon energy futures-a review of existing perspectives[J]. Renewable and Sustainable Energy Reviews, 2018, 82(Part 3): 3027-3045.

[9] LAN L, KONG X, QIU C, et al. A review of recent advance on hydrogen embrittlement phenomenon based on multiscale mechanical experiments[J]. Jinshu Xuebao/Acta Metallurgica Sinica, 2021, 57(7): 845-859.

[10] DU Y, GAO X, LAN L, et al. Hydrogen embrittlement behavior of high strength low carbon medium manganese steel under different heat treatments[J]. International Journal of Hydrogen Energy, 2019, 44(60): 32292-32306.

[11] ZHENG Y, ZHANG L, SHI Q, et al. Effects of hydrogen on the mechanical response of X80 pipeline stees subject to high strain rate tensile tests[J]. Fatigue and Fracture of Engineering Materials and Structures, 2020, 43(4): 684-697.

[12] 兰亮云, 孔祥伟, 邱春林, 等. 基于多尺度力学实验的氢脆现象的最新研究进展[J]. 金属学报, 2021, 57(7): 845-859.

[13] 李玉星, 张睿, 刘翠伟, 等. 掺氢天然气管道典型管线钢氢脆行为[J]. 油气储运, 2022, 41(6): 732-742.

[14] 张家轩, 王财林, 刘翠伟, 等. 掺氢天然气环境下管道钢氢脆行为研究进展[J]. 表面技术, 2022, 51(10): 76-88.

[15] 张体明. 高压煤制气管线 X80 钢焊接接头的氢致脆化研究[D]. 青岛: 中国石油大学(华东), 2016.

[16] 赵颖, 王荣. X70 管线钢电化学充氢后的力学行为研究[J]. 中国腐蚀与防护学报, 2004, 24(5): 293-296.

[17] 刘玉, 李焰, 李强. 阴极极化对 X80 管线钢在模拟深海条件下氢脆敏感性的影响[J]. 金属学报, 2013, 49(9): 1089-1097.

[18] GADALA I M, ALFANTAZI A. Low alloy X100 pipeline steel corrosion and passivation behavior in bicarbonate-based solutions of pH 6.7 to 8.9 with groundwater anions: an electrochemical study[J]. Metallurgical and Materials Transactions A, 2015, 46(7): 3104-3116.

本论文原发表于《焊管》2023 年第 46 卷第 3 期。

热处理工艺对膨胀管组织和力学性能的影响

朱兴华[1]　董晓明[2]　高　展[2]

（1. 宝山钢铁股份有限公司钢管条钢事业部；
2. 宝山钢铁股份有限公司中央研究院）

摘　要：采用两相区热处理工艺研究了膨胀管用低碳中锰钢组织演变规律和力学性能。结果表明：采用两相区热处理工艺的低碳中锰钢组织为回火索氏体+富碳马氏体/贝氏体+少量铁素体的复相组织+残余奥氏体，残余奥氏体分布在原奥氏体晶界上和马氏体/贝氏体板条界上，残余奥氏体主要通过临界淬火富集 C 和 Mn 元素达到稳定，室温下稳定的残余奥氏体含量最高可达 12%。由于残余奥氏体的应变诱导塑性（TRIP）效应，低碳中锰钢具有良好的塑性，断后总延伸率高于 40%，均匀延伸率高于 20%。

关键词：残余奥氏体；两相区热处理；中锰钢

1　概述

可膨胀套管技术是当前油气井工程领域内的重大高新技术之一，具有优化井深结构、降低钻采成本、节约作业时间和提高单井产量等优势，可广泛用于钻完井、采油和修井作业中，被称为"21 世纪石油钻采领域中的一次技术革命"[1-2]。可膨胀套管是一种由特殊材料制成的金属钢管，具有良好的塑性，在井下可通过机械或液压的方法使可膨胀套管在直径方向膨胀 10%~30%，同时在冷作硬化效应下提高自身刚性。壳牌公司和哈里伯顿公司联合成立 Enventure 公司，研发和推广可膨胀套管技术，具有较好的发展前景。Enventure 公司在国内胜利油田进行了两口井的膨胀套管固井完井试验，一次性膨胀长度为 398m，成功完成了膨胀套管作业，成为世界第 200 口和第 201 口膨胀套管井。至今为止，Enventure 公司已在全球建设了 209 口膨胀套管作业井，累计膨胀管长度为 62862m，膨胀接头共计 5417 个。

可膨胀套管在服役的过程中要承受巨大的外压力，因此用作膨胀管管体的材料首先要具备较高的强度及较强的抗挤压能力，同时为了满足管体膨胀过程中的大变形要求，膨胀管材料自身还应具备优良的延伸性能。为此，许多学者对钢的微观组织结构，对膨胀管的力学性能进行了大量的研究，Caballero 等[3]指出，硬相+软相+过渡相的多相组织结构在提高钢材综合力学性能方面的作用十分显著，同时获得了高强度和高塑性。

本文在国内外研究成果的基础上，研究了可用于膨胀管的新钢种和热处理工艺，并对其微观组织进行了表征，以求揭示材料的微观组织结构与力学性能的关系，从而为高性能膨胀管技术的研究和发展提供组织调控方向的借鉴和参考。

作者简介：朱兴华，工程师，1986 年生，2008 年毕业于东北大学，现从事高等级油井管制造工艺技术研究。电话：15154519436。E-mail：zhuxinghua@baosteel.com。

2 试验材料与方法

试验用膨胀管的材料采用真空感应炉冶炼并浇铸成钢锭,化学成分见表1。试验钢经真空冶炼之后浇铸到模具中,铸锭在1250℃保温2h后在900~1200℃范围内锻造成180mm×120mm×200mm的锻坯。锻坯经过轧制及热处理工艺(将试样V1、V2、V3分别在900℃、790℃、810℃保温30min后淬火至室温,然后再加热到690℃保温50min后空冷)处理后进行力学性能检测。

表1 化学成分　　　　　　　单位:%(质量分数)

$w(C)$	$w(Si)$	$w(Mn)$	$w(P)_{max}$	$w(S)_{max}$	$w(Ni)$
0.100~0.160	0.200~0.500	2.500~3.500	0.012	0.003	0.200~0.400

试样经过抛光后用4%硝酸酒精溶液腐蚀,再用光学显微镜观察金相组织。采用光学显微镜、电子探针、EVO MA25扫描电镜和JEM 2100F透射电镜对试样进行微观组织分析。拉伸性能采用圆棒试样通过MTS 810-15试验机测试,冲击性能采用JBN-300B设备,根据ASTM A370—2010标准测试材料的夏比V形缺口冲击功,试样尺寸为10mm×10mm×55mm。

3 试验结果及分析

3.1 力学性能

表2为试验钢经不同温度热处理后的拉伸性能(未预变形)。由表2可知,V1的屈服强度、屈强比较高,冲击韧性也略高于V2和V3,而延伸率和均匀延伸率均低于采用两相区淬火的V2和V3。V2和V3屈服强度随着临界热处理温度的升高而降低,而抗拉强度则随着临界热处理温度的升高出现了增大的现象。

膨胀套管膨胀施工时需要保证更低的屈服强度和更好的变形能力,膨胀后需要保证较高的强度以获得良好的抗挤毁性能,因此对V1~V3试验钢采用预变形来模拟套管膨胀施工,研究材料变形后的力学性能。膨胀管一般膨胀量在10%以上,因此V1~V3试验钢的预变形量为10%,表3为V1~V3试验钢预变形后的力学性能。

表2 V1~V3试验钢预变形前的力学性能

编号	屈服强度 $R_{p0.2}$/MPa	抗拉强度 R_m/MPa	屈强比/%	延伸率 $A_{50.8}$/%	均匀延伸率/%	0℃横向全尺寸夏比冲击功/J
V1	439	630	0.69	33.5	14	115,105,108
V2	351	581	0.60	43.0	21	101,95,90
V3	300	642	0.47	40.0	21	96,92,85

表3 V1~V3试验钢预变形后的力学性能

编号	屈服强度 $R_{p0.2}$/MPa	抗拉强度 R_m/MPa	延伸率 $A_{50.8}$/%	均匀延伸率/%	0℃横向全尺寸夏比冲击功/J
V1	737	777	26.5	6	52,50,55
V2	636	668	44.0	11	45,50,58
V3	699	738	40.0	12	55,60,52

从表3可知,预变形10%的试样屈服强度显著提升,均达到90ksi(1ksi=6.895MPa)钢级以上,其中V1钢的屈服强度和抗拉强度最高,但是其延伸率较V2钢种和V3钢种显著降

低。在冲击韧性指标上3种钢未有显著差异。

3.2 金相组织分析

图1为试验钢经不同热处理后的金相组织。可以看出，V1试样在900℃完全奥氏体化淬火+回火后的组织主要为回火索氏体，而经过两相区热处理后的V2和V3组织为回火索氏体+铁素体+残余奥氏体+少量粒状贝氏体。随着两相区淬火温度的提高，V3贝氏体的板条结构更加清晰。临界区淬火是一个逆相变过程，随着淬火温度的升高，逆转变的奥氏体逐渐增多，淬火后所形成的新生马氏体/贝氏体也逐渐增多，板条结构越来越清晰。

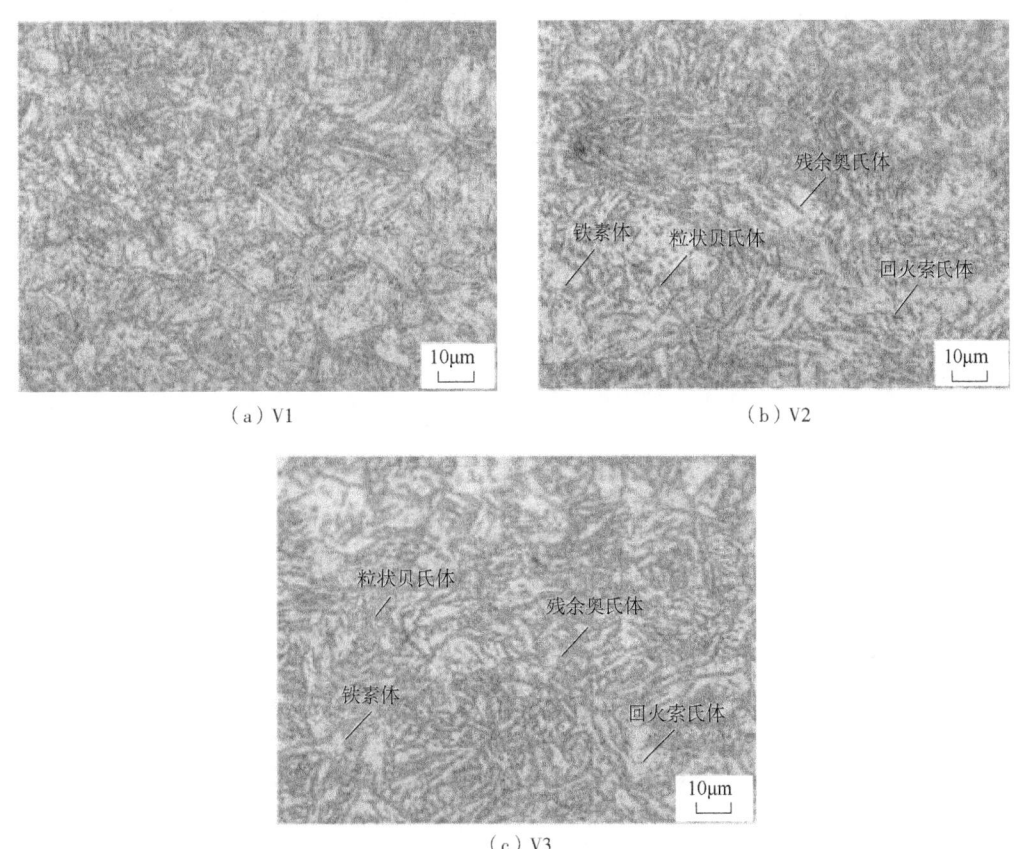

图1 不同热处理试样金相组织

对试验钢采用热膨胀法检测其热轧和两相区淬火后试样的相变点。热轧态试样马氏体向奥氏体转变的开始温度A_{c1}为710℃，而经过两相区淬火之后试样的A_{c1}降低至680℃。A_{c1}降低的原因是在两相区淬火过程中，发生了合金元素的再分配，合金元素富集的区域相变温度降低，同时有利于提高逆相变奥氏体的稳定性。因此V2和V3在进行回火热处理时回火温度高于A_{c1}，因此发生了马氏体向奥氏体的逆转变。

V2试样的临界淬火温度较低，形成的奥氏体含量较少，奥氏体淬火后形成马氏体，在回火过程中马氏体形成逆变奥氏体；V3试样的临界淬火温度较高，形成的奥氏体含量较多，奥氏体淬火后形成马氏体，因此在回火过程中形成的逆变奥氏体也就更多。此种逆变奥氏体在后续的冷却过程中部分转化为贝氏体，部分保留下来成为残余奥氏体，有利于提高材料的塑性。

3.3 残余奥氏体组织分析

图2为在不同临界温度下淬火后试样的EBSD图像,图2中红色区域代表残余奥氏体。由图2可知,随着临界淬火温度升高,V1~V3试样中残余奥氏体的含量逐渐增多,采用两相区淬火的试样残余奥氏体含量高于完全奥氏体化淬火的试样。图2的黑色线代表取向差大于15°的大角度晶界,可以看到残余奥氏体主要分布在取向差较大的大角度晶界上,这是因为大角度晶界上能量较高,逆转变奥氏体容易在此形核长大[4]。

对两相区热处理后的V3试样进行精细结构观察,相应的TEM组织表征结果如图3所示。可以看出,组织中铁素体基体占很大比例,同时条状残余奥氏体与基体相间分布,厚度约为几十纳米,如图3(a)和图3(b)所示;同时,也观察到少量部分尺寸较大的粗大板条状残余奥氏体,如图3(c)所示,其板条宽度约460nm。此外,还观察到块状奥氏体,可确定块状残余奥氏体主要分布在原奥氏体晶界及贝氏体板条群边界处,通常,晶界处是热处理过程中逆转变奥氏体形核的位置。

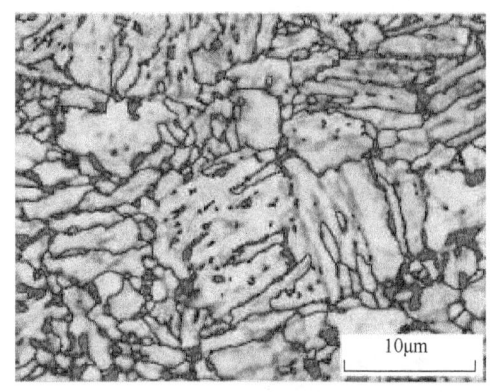

(a) V1残余奥氏体8.85%

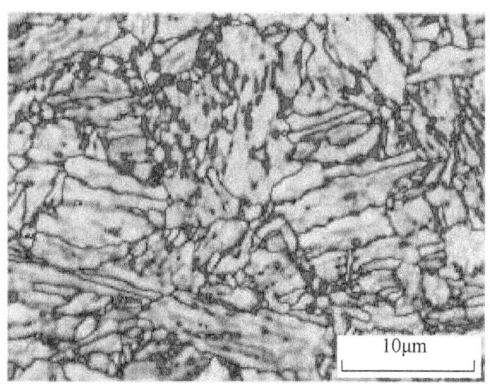

(b) V2残余奥氏体10.60%

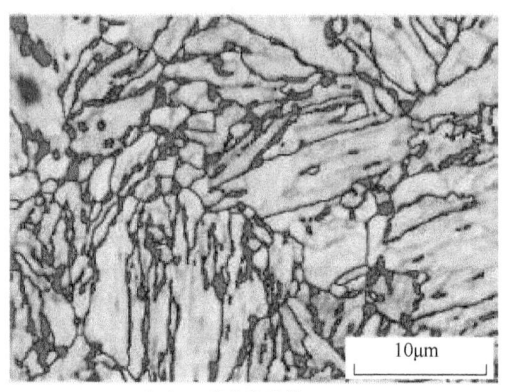

(c) V3残余奥氏体12.10%

图2 不同热处理后试样的EBSD图像

用TEM-EDS对V3两相区热处理后的组织进行微区成分分析,微区组织形貌和Mn元素面扫描结果如图4(a)和图4(b)所示,采用EDS测定薄膜状残余奥氏体(001位置)与周围铁素体基体中(002位置)Mn的含量分别为4.47%和0.71%,与设计成分中的Mn含量相比,奥氏体相中的Mn含量大大增加,Mn元素含量在奥氏体与铁素体中比例为6∶1。证实了经过临界区热处理后,逆转变形成的残余奥氏体从邻近的铁素体、奥氏体中富集了大量的Mn元素,极大地提高了其稳定性,因此可以稳定至室温。

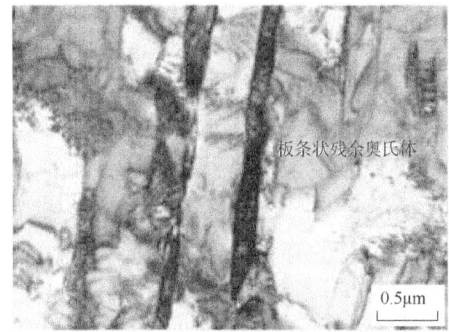
(a) 板条状残余奥氏体，明场像

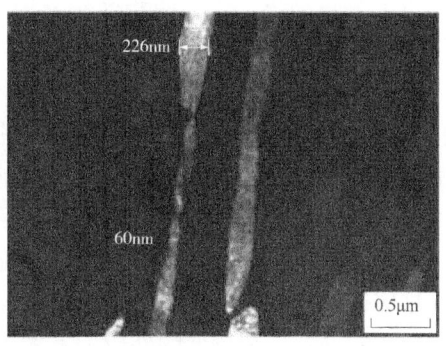
(b) 板条状残余奥氏体，暗场像

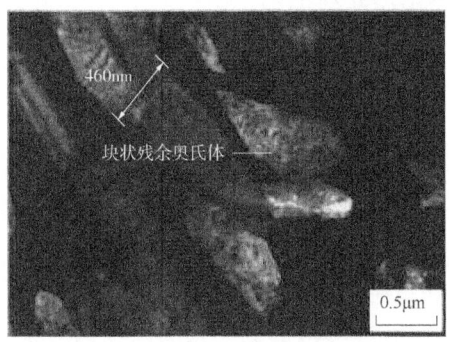
(c) 粗大板条状和块状残余奥氏体，暗场像

图3 热处理后的V3试样TEM图像

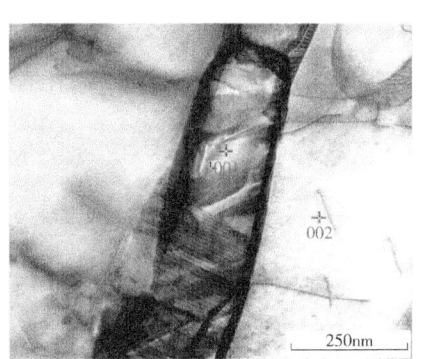
(a) 微区组织与EDS成分分析位置

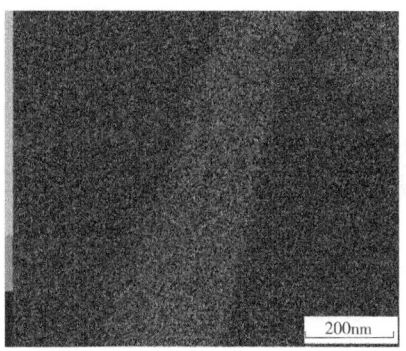
(b) 微区Mn元素EDS面扫描图像

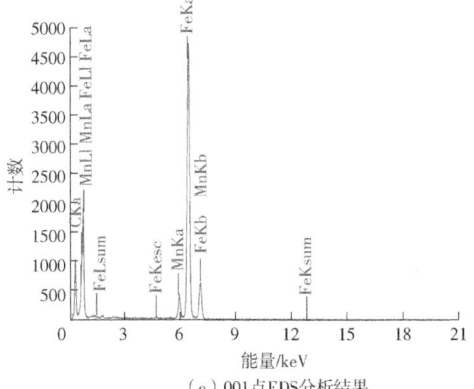

(c) 001点EDS分析结果　　　　　　　　(d) 002点EDS分析结果

图4 热处理后的V3试样TEM-EDS表征结果

根据以上分析结果，采用两相区热处理工艺的 V2 和 V3 试样呈现出由回火索氏体、残余奥氏体、铁素体所组成的复相组织结构，这种多相组织结构能够细化材料的有效晶粒尺寸，从而有利于改善材料的塑韧性，其中回火索氏体能够改善材料的韧性，碳化物同时起到析出强化作用，可以提高材料的抗拉强度，使膨胀管具备较好的抗挤毁、抗内压等实体力学性能；由于回火索氏体和相变诱发马氏体（形变过程中由残余奥氏体转变而来）的共同作用，V2 和 V3 钢具备了 600MPa 以上的高屈服强度；大量弥散分布的小尺寸铁素体晶粒的存在，则保证了该类钢具有较低的屈服强度，并同弥散分布的过渡相残余奥氏体共同提高了钢的延伸性能，这应该就是该类材料具备优良的均匀延伸率的一个重要原因，同时铁素体组织可以降低材料的屈服强度，在较低的外应力下发生塑性形变，有利于降低膨胀管作业压力要求。而 V1 钢采用的是完全奥氏体化热处理工艺，回火之后的金相组织为回火索氏体单一组织，残余奥氏体组织含量较少，相变诱发马氏体的作用较小，导致其强度较高，而均匀延伸率较低，不适合作为膨胀管的材料。

4 结论

（1）试验钢采用两相区淬火时合金元素富集在马氏体中，在随后的高温回火过程中，富集合金元素的马氏体发生逆转变，同时在奥氏体中富集更多的合金元素，提高了逆转变奥氏体的稳定性，从而使其可以在室温或者更低的温度稳定存在。

（2）室温条件下，采用两相区淬火+高温回火的试验钢中获得回火索氏体+富碳马氏体/贝氏体+少量铁素体的复相组织+残余奥氏体，较高的残余奥氏体使得试验钢可以获得较高的塑性，有利于膨胀管在膨胀施工过程中的均匀变形。

（3）采用两相区淬火+高温回火的试验钢的均匀延伸率达到 20% 以上，延伸率达到 40% 以上，同时具有良好的冲击韧性，预变形 10% 后屈服强度达到了 90ksi，适合用于高膨胀量的膨胀套管产品。

参 考 文 献

[1] WILLIAM F. Expandable casing program helps operator hit TD with larger tubulars[J]. Offshore：Incorporating the Oilman, 2000, 60(1)：48-50.

[2] 贺飞, 尚成嘉, 袁胜福, 等. 膨胀管用钢的热处理工艺与扩径性能[J]. 材料热处理学报, 2012, 33(S)：37-41.

[3] CABALLERO F G, BHADESHIA H K D H. Very strong bainite[J]. Current opinion in solid state & materials science, 2004, 8(3)：251-257.

[4] KUZMINA M, PONGE D, RAABE D. Grain boundary segregation engineering and austenite reversion turn embrittlement into toughness：example of a 9 wt.% medium Mn steel[J]. Acta Materialia, 2015, 86：182-192.

本论文原发表于《宝钢技术》2023 年第 5 期。

Effect of Nb Content and Second Heat Cycle Peak Temperatures on Toughness of X80 Pipeline Steel

Chen Yuefeng[1,2], Yang Yaobin[1,2], He Xiaodong[1,2]
Chi Qiang[1,2], Qi Lihua[1,2], Li Weiwei[1,2], Li Xin[1,2]

(1. National Key Laboratory of Oil and Gas Drilling and Production Transportation Equipment, Tubular Goods Research Institute CNPC;
2. International Welding Technology Center)

Abstract: The microstructure evolution and variation of impact toughness in the heat affected zone(HAZ) of X80 pipeline steel with different Nb content under different peak temperatures in secondary thermal cycle were studied through welding thermal simulation, Charpy impact test, EBSD analysis, SEM observation, and TEM observation in this study. The results indicate that when the peak temperatures of the second pass were lower than A_{c1}, both X80 pipeline steels had high impact toughness. For the secondary peak temperatures were in the range of A_{c1} to A_{c3}, both X80 pipeline steels had the worst impact toughness, mainly due to the formation of massive blocky M-A constituents in chain-like on grain boundaries. When the second peak temperatures were higher than A_{c3}, both X80 pipeline steels had excellent impact toughness. Smaller grain size and higher proportions of HAGBs can effectively improve the impact toughness. Meanwhile, high Nb X80 pipeline steel had higher impact absorption energy and smaller dispersion. Adding an appropriate amount of Nb to X80 pipeline steel can ensure the impact toughness of SCCGHAZ and SCGHAZ in welded joints.

Keywords: X80 pipeline steel; Nb; Thermal simulation; HAZ; Toughness

1 Introduction

As a representative of high strength low alloy(HSLA) steel, X80 pipeline steel with excellent strength and toughness is widely used in the field of oil and gas transportation. With the construction of long-distance oil and gas pipelines in China, the manufacturing technology of X80 pipeline steel has been intensely developed[4]. Welding is an important process in the production and laying of oil and gas pipeline steel pipes. Multi-pass welding is widely used in pipeline manufacturing and field girth welding[5-9]. Compared to the base material, the thermal cycle experienced a series of peak temperatures during welding that can alter the microstructure and properties in the heat affected zone

---

Corresponding author: Chen Yuefeng, chenyf@cnpc.com.cn.

(HAZ). An important issue currently limiting the development of X80 pipeline steel is the significant difference in microstructure and mechanics performance between welded joints and base materials, with welds becoming the weakest part of X80 long-distance oil and gas pipelines[10-12].

According to the different peak temperatures in secondary thermal cycle, the coarse grain heat affected zone (CGHAZ) can be divided into super-critical coarse grain heat affected zone (SCCGHAZ), inter-critical coarse grain heat affected zone (ICCGHAZ), and sub-critical coarse grain heat affected zone (SCGHAZ)[13-15]. The peak temperatures in secondary thermal cycle corresponding to SCCGHAZ and SCGHAZ is higher than A_{c3} and lower than A_{c1}, respectively. ICCGHAZ is the pre-existing CGHAZ that was reheated to the temperature between A_{c1} and A_{c3} by a subsequent weld pass. Extensive research results showed that the worst impact toughness in the HAZ of multi-pass welding is ICCGHAZ, while the impact toughness of SCCGHAZ and SCGHAZ in the HAZ is relatively high[16]. Due to the existence of martensite-austenite (M-A) islands, the impact toughness of HAZ is prone to fluctuations after secondary thermal cycle[17-19]. The main reason for the formation of embrittlement zone is the M-A islands of ICCGHAZ with chain form distributed at the grain boundary, and the crack is easy to spread along or through the M-A[10,20-21]. Some investigations have shown that the volume fraction, size, and distribution of M-A islands can affect the low-temperature toughness of welding joints of pipeline steel[22-23]. A higher volume fraction, larger size, and smaller spacing between M-A islands distributed on the grain boundary reduce the low-temperature toughness of the welded joints. The influence of the second different peak temperatures on the HAZ of welded joint is mainly the change of hardness, strength and impact toughness of ICCGHAZ in welded joint[24]. The results indicated that under low heat input, the impact toughness of ICCGHAZ decreases slightly and can be improved[25-26]. However, under the large heat input, the impact toughness decreases obviously, and the impact toughness of ICCGHAZ deteriorates seriously. At the same time, some scholars have studied the effect of Nb content on the impact toughness of the welding HAZ of HSLA steels. For example, some scholars believe that high Nb can narrow the welded CGHAZ and make the grain size small, which is conducive to improving the impact toughness of the CGHAZ[27]. Zhang et al. believe that Nb element can promote the formation and coarsening of coarse bainite, expand the width of CGHAZ, and reduce the toughness of HAZ of HSLA steel[28]. Although some studies have been carried out on the impact toughness of Nb element on the HAZ of primary welding, there are few reports on the effect of different Nb content on the impact toughness of ICCGHAZ of X80 pipeline steel.

In order to clarify the influence of different Nb content in X80 pipeline steels on the impact toughness of HAZ of secondary thermal cycle welding, two X80 pipeline steels with different Nb content were selected as research objects in this study. The microstructure, high angle grain boundary distribution and toughness of X80 pipeline steel with different Nb content in the HAZ of secondary thermal cycle welding were investigated by welding thermal simulation test, which provided theoretical support for composition optimization of X80 pipeline steel with high weldability.

2 Experimental procedures

2.1 Materials

Two X80 SMAL (longitudinal submerged arc welding) steel pipes with different Nb contents were selected in this study, with dimensions of OD 1219mm×22mm. Both pipes were produced by the same pipe factory, and the steel plates used were from the same steel mill in China. The chemical composition of the two X80 pipeline steels used as experimental steels was listed in Table 1. In addition to the slight difference in Nb content, there was no significant difference in the content of other alloying elements in the two X80 pipeline steels.

Table 1 Chemical composition of two pipeline steels (%, mass fraction)

Sample	C	Si	Mn	Cr	Mo	Ni	Nb	V	Ti	Cu	Al
A	0.049	0.15	1.73	0.25	0.093	0.16	0.058	0.0043	0.011	0.022	0.027
B	0.049	0.15	1.72	0.26	0.090	0.16	0.084	0.0045	0.011	0.026	0.023

The microstructures in the center of wall thickness of A and B pipeline steels base materials were shown in Figure 1. As depicted in Figure 1, the metallographic structures of the two pipeline steels exhibit a typical acicular ferrite morphology, consisting of granular bainite (GB) and quasi-polygonal ferrite (QF). By comparing the metallographic structures of the two pipeline steels, it was found that the content of QF in A pipeline steel was higher than that in B pipeline steel, while the content of GB was lower than that in B pipeline steel. The grain size of the two pipeline steels were measured in optical images via the intercept method according to ASTM E112. The results indicated that the grain sizes of the two pipeline steels were both grade 11.5 and both steels exhibited fine grain sizes. Despite the similar grain sizes of the two pipeline steels, the grain size of B pipeline steel was smaller and more uniform than that of A pipeline steel.

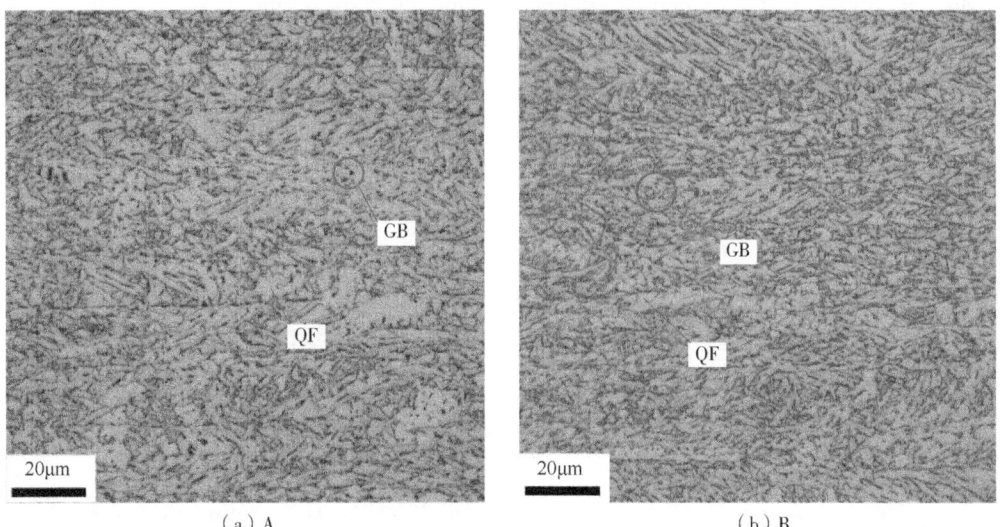

(a) A (b) B

Figure 1 Optical micrographs of two X80 pipe steels

A Gleeble-3500 thermomechanical simulator equipped with a thermal dilatometer was used to test the A_{c1} and A_{c3} of two samples. Round rod specimens with dimensions of ϕ6mm×70mm, taken from the center of wall thickness along longitudinal direction of steel pipe, were used to test A_{c1} and A_{c3}. The heating rate used for A_{c1} and A_{c3} tests was 0.05℃/s, from room temperature to 1000℃. The variation curve of expansion of two kinds of pipeline steel with temperature were shown in Figure 2. As the temperature increased, the crystal structure of pipeline steel changes from body centered cube(BCC)to face centered cube(FCC), accompanied by a sudden change in expansion amount. A_{c1} and A_{c3} of the two pipeline steels were determined by the change in the slope of the expansion curve. The A_{c1} and A_{c3} of sample A were 680℃ and 821℃, respectively. While the A_{c1} and A_{c3} of sample B were 685℃ and 840℃, respectively.

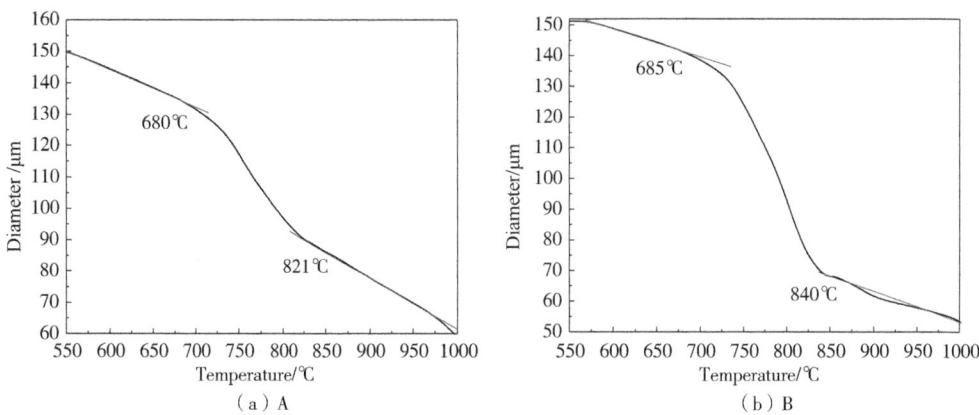

Figure 2　A_{c1} and A_{c3} of two samples

2.2　Welding Thermal Simulation

In order to simulate the changes in microstructure and impact toughness of different areas in the HAZ of X80 pipeline steel during multi-pass welding process, samples with dimensions of 10.5mm ×10.5mm×70mm, taken from the center of wall thickness along longitudinal direction of steel pipe, were used in this study. Subsequently, thermal simulation tests of the HAZ with different peak temperatures in secondary thermal cycle were conducted on a Gleeble-3500 thermal simulator. To precisely control the temperature, K-type thermo-couples were spot welded in the middle of the sample. Welding thermal simulation processes and its parameters were shown in Figure 3. The specimens were first heated to 1350℃ at a rate of 200℃/s and cooled to 150℃, to simulate the CGHAZ of the first welding pass. The peak temperatures selected for second thermal cycle were as follows: 450℃, 500℃, 550℃, 600℃, 650℃, 700℃, 725℃, 750℃, 775℃, 800℃, 825℃, 850℃, 900℃, and 1000℃, with a heat rate of 200℃/s. The cooling time from 800℃ to 500℃ ($t_{8/5}$) was 15.9s for cooling regimes of all thermal cycles, corresponding to the heat input of 2.5kJ/mm. This heat input was simulated the

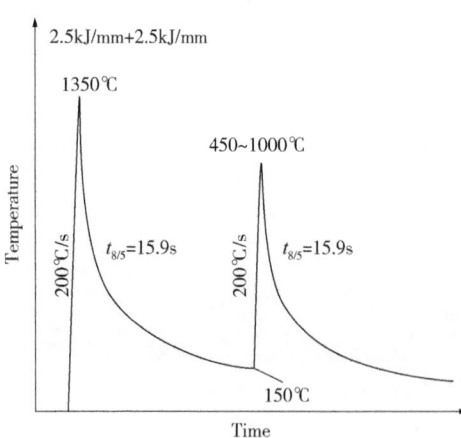

Figure 3　Welding thermal simulation processes diagram

shielded metal arc welding(SMAW) process.

2.3 Mechanical Properties

The specimens were machined into Charpy impact specimens with sizes of 10mm×10mm×55mm after the thermal simulation tests. The notch positions of V-type Charpy impact specimens were consistent with that of thermocouple. The impact toughness of the thermal simulated samples was tested at-10℃.

2.4 Microstructure Analysis

The surfaces of thermal simulation samples used for microstructure observation and analysis were the cross sections of thermocouples location of thermal simulation sample. Samples used for metallographic and scanning electron microscopy(SEM) observation were ground with a series of SiC papers, mechanically polished using a 1.5μm diamond paste, and etched with 4% nitrate alcohol solution, sequentially. Electron back-scattered diffraction(EBSD) specimens were electrochemically polished in 5% perchloric acid solution after mechanical polishing. AztecCrystal software was used to post-process EBSD data. Specimens with diameter of 3mm for the transmission electron microscopy (TEM) observation were sliced into 50μm through the abrasion on variant grit silicon carbide papers and electropolished using 90%(volume fraction) ethanol and 10%(volume fraction) perchloric acid. Axiovert 405M metallographic microscope, TESCAN CLARA scanning electron microscope equipped with an EBSD, and JEOL JEM-200CX electron microscope were used to characterize the microstructure and impact fracture morphology.

The samples used for microstructure and fracture analysis were numbered based on the different peak temperatures of second pass. When the peak temperatures in secondary thermal cycle were 550℃, samples A and B were named A-550℃ and B-550℃. When the peak temperatures in secondary thermal cycle were 700℃, samples A and B were named A-700℃ and B-700℃. When the peak temperatures in secondary thermal cycle were 1000℃, samples A and B were marked out A-1000℃ and B-1000℃.

3 Results and discussion

3.1 Impact Toughness

The impact absorption energy of A and B X80 pipeline steel base material at-10℃ were 352J and 335J respectively. The impact absorption energy of two X80 pipeline steels in the weld HAZ at different peak temperature of secondary heat cycle was shown Figure 4. According to the corresponding temperatures of A_{c1} and A_{c3} of the two pipeline steels in Figure 2, the second peak temperature corresponding to SCGHAZ was lower than 680℃, the peak temperature of secondary thermal cycle corresponding to ICCGHAZ was 680℃~840℃, and the secondary peak temperature corresponding to SCCGHAZ was higher than 840℃.

By fitting the impact absorption energy corresponding to different peak temperature of secondary heat cycle, the trend with the variation of secondary peak temperature of two pipeline steels was basically same. With the change of second peak temperature, there were three platforms in the absorption energy curve, corresponding to SCGHAZ, ICCGHAZ and SCCGHAZ respectively. According to the change of the curve of impact absorption energy, it was obvious that SCGHAZ and

SCCGHAZ had higher impact absorption energy than ICCGHAZ.

When the peak temperature of second pass was below 600℃, it corresponded to the SCGHAZ in multi-pass welded joints. The impact absorption energy values of SCGHAZ samples corresponding to the same peak temperature of samples A and B were basically same. For the second peak temperature of 450℃, the average absorption energy values of samples A and B had no obvious difference, and both of them were about 200J. There was a slight advantage in impact toughness of sample B for secondary peak temperature of 500℃, while sample A had a slight advantage for second peak temperature of 550℃. The values of average absorption energy of samples A and B for secondary peak temperature of 500℃ were 223J and 236J, respectively. At a second peak temperature of 550℃, the values of average absorption energy of samples A and B were 237J and 214J, respectively. For the thermal simulation sample with the secondary peak temperature of 600℃, the average and dispersion of absorption energy values of sample A were higher than that of sample B, but the difference in the average absorption energy values of the two samples was about 50J. When the second peak temperature was 650℃, the impact toughness values of samples A and B both significantly decreased. The average impact absorption energy value of sample A was 7J, while the average absorption energy value of sample B was 33J.

When the peak temperature of second pass were in the range of 680~840℃, it corresponded to the ICCGHAZ in multi-pass welded joints. A low valley appeared in the impact absorption energy curve, indicated that the toughness of ICCGHAZ sample was the lowest among the two X80 pipeline steels. The average absorption energy values of ICCGHAZ sample were basically same, with minimum impact absorption energy values of 6J. For sample A, when the secondary peak temperatures were within the range of 700~800℃, the values of average absorption energy were less than 20J. At reheated temperatures of sample B in the range of 700~775℃, the average values of absorption energy were less than 20J. The average value of impact absorption energy of sample B at the secondary peak temperature of 800℃ was 26J. For the secondary peak temperature of 825℃, the average absorption energy values of samples A and B were 35J and 53J, respectively.

When the peak temperature of second pass were higher than 840℃, it corresponded to the SCCGHAZ in multi-pass welded joints. For the second peak temperatures were higher than 840℃, the impact toughness of samples A and B quickly recovered to a higher level. Compared to ICCGHAZ, it can found that the samples of SCCGHAZ had higher impact toughness from Figure 4. But sample B had higher impact absorption energy and smaller dispersion. The values of average impact absorption energy of sample A at the second peak temperature of 850℃, 900℃, and 1000℃ were 275J, 215J, and 194J, respectively. The average values of impact absorption energy of sample B at the second peak temperature of 850℃, 900℃, and 1000℃ were 256J, 296J, and 301J, respectively.

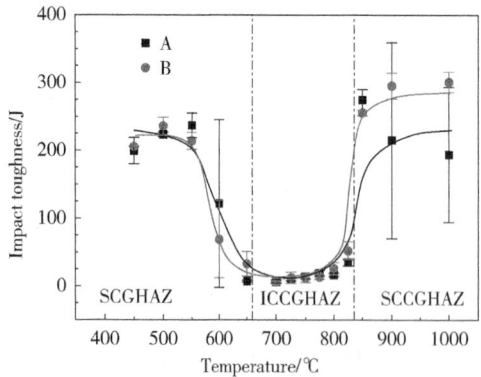

Figure 4 Impact toughness of two X80 steels with different peak temperatures of secondary thermal cycle

3.2 Microstructure Analysis

Opticaland SEM micrographs of the thermal simulated samples with different peak temperature of second cycle were presented in Figure 5 and Figure 6, respectively. When the secondary peak temperature was lower than A_{c1}, both A-550℃ and B-550℃ samples maintained the characteristics of primary CGHAZ, and the grain size was relatively coarse. The microstructures of the two thermal simulation samples were mainly dominated by GB, with a small number of M-A components distributed within the grains. By statistically measuring the grain size in ten metallographic micrographs, the average sizes of grain in specimens of A-550℃ and B-550℃ were 74.9μm and 72.6μm, respectively.

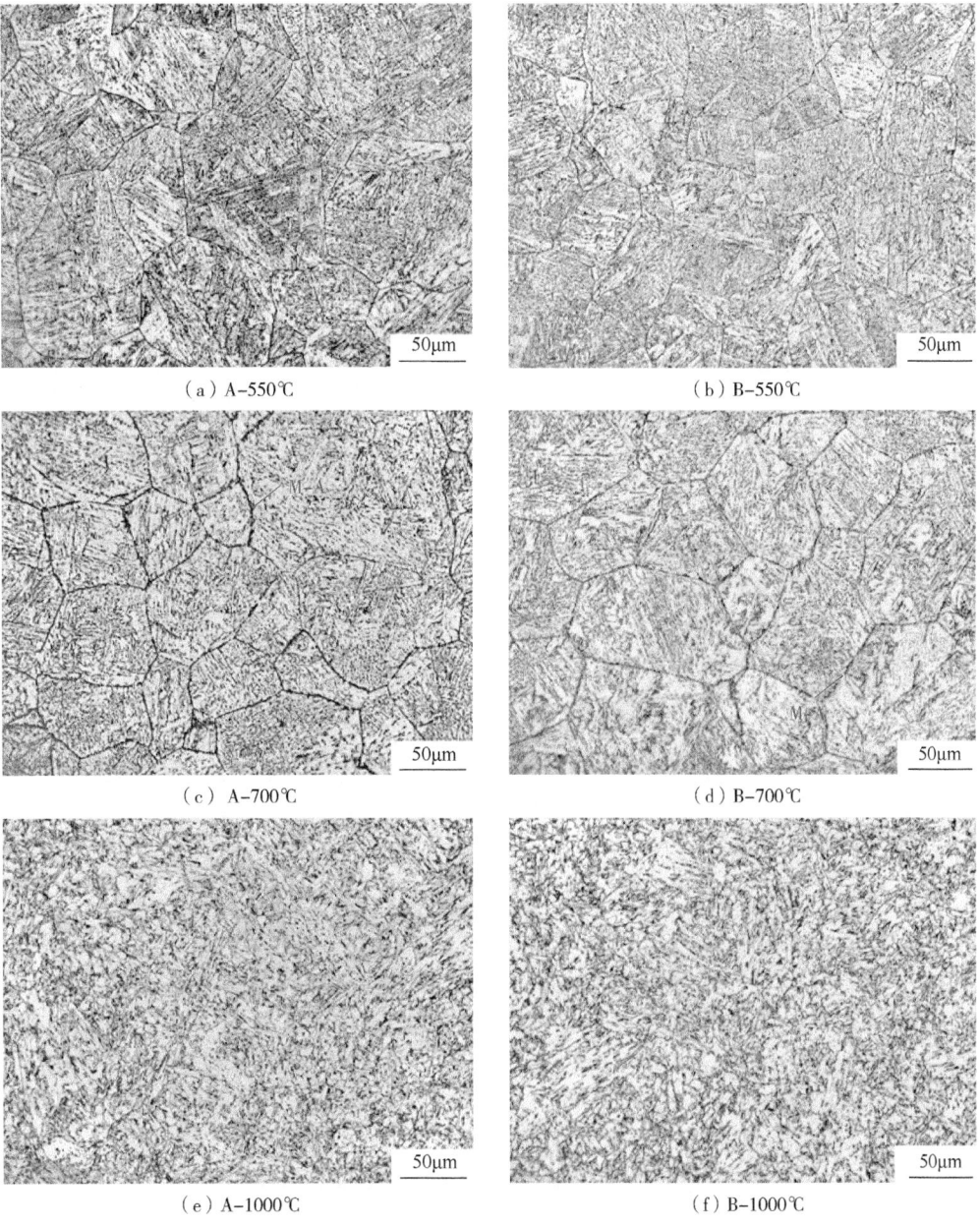

(a) A-550℃ (b) B-550℃
(c) A-700℃ (d) B-700℃
(e) A-1000℃ (f) B-1000℃

Figure 5 Optical micrographs of thermal simulation samples

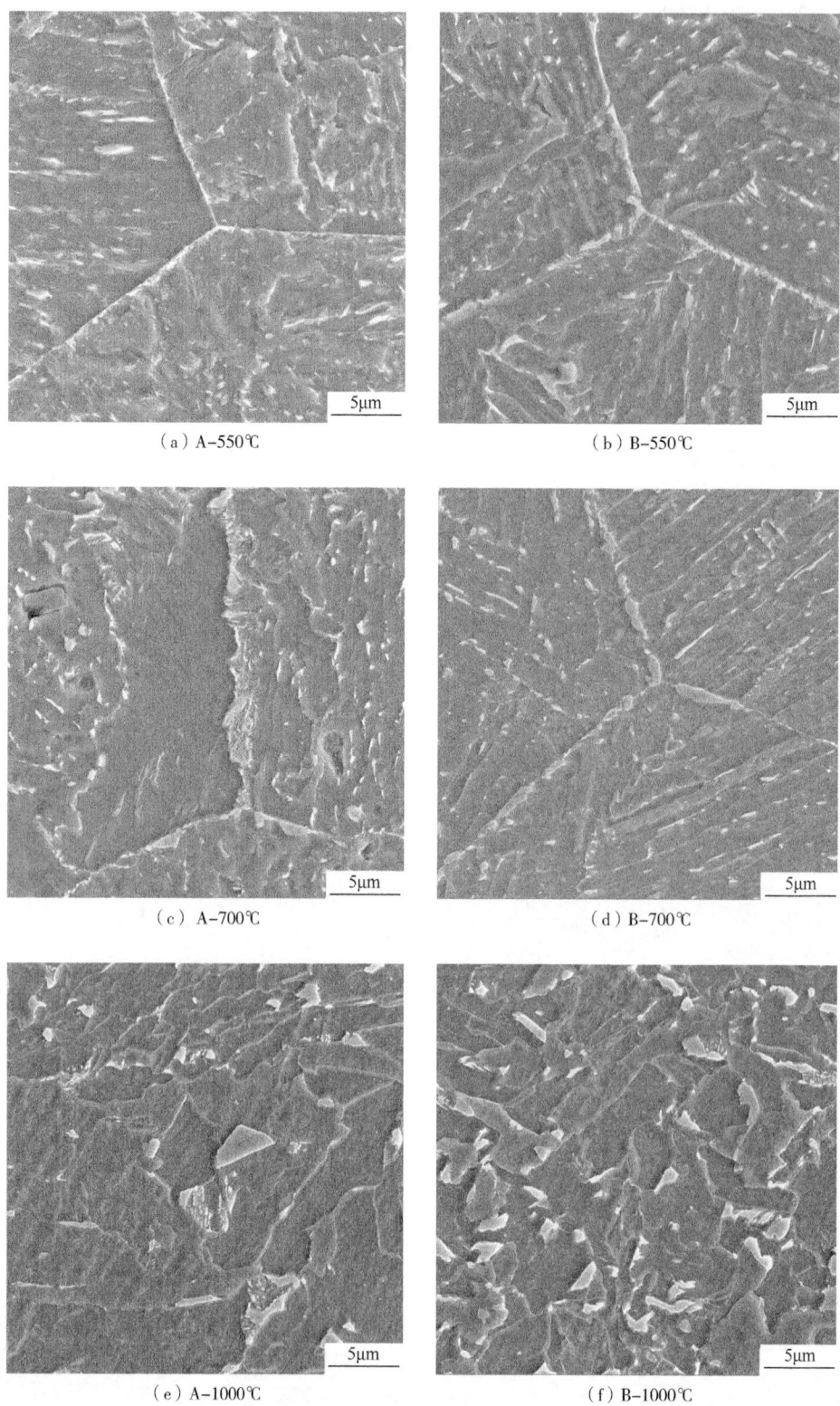

Figure 6 SEM micrographs of thermal simulation samples

At a reheated temperature of the second thermal cycle between A_{c1} and A_{c3}, the coarse grain microstructure formed during the first thermal cycle could not completely undergo austenite phase transformation, and only a portion of the microstructure could undergo austenite phase transformation. In the samples of A-700℃ and B-700℃, M-A component was generated in blocky form at the original austenite grain boundaries, with a large number of blocky M-A components formed in necklace-type. According to the statistics of grain size in specimens of A-700℃ and B-700℃, the average sizes of grain were 69.4μm and 68.2μm, respectively. Through the SEM analysis of the samples with the second peak temperatures of 550℃ and 700℃, it was found that the width of grain boundary increased obviously with the increase of the second pass peak temperature. This was mainly because when the secondary temperature was 700℃, a large number of massive M-A components were formed on the grain boundaries, which made the grain boundary width larger.

For the samples of second peak temperature higher than A_{c3}, the CGHAZ microstructure occurred in the first pass was completely austenitized, but the peak temperature of second pass had not yet reached the temperature of rapid grain growth and coarsening, resulted in smaller austenite grains that form polygonal ferrite(PF) and QF during the cooling process. At the same time, a small amount of fine and regularly shaped M-A of second pass was also formed at the grain boundaries. The grain size of samples with a secondary peak temperature of 1000℃ were measured by EBSD, with the average sizes of grain were 26.5μm and 24.7μm for samples A-1000℃ and B-1000℃, respectively.

TEM characterization of microstructures of thermal simulation samples were depicted in Figure 7. In TEM morphology of all thermal simulated samples, layer-by-layer structures appeared in the form of dark-light-dark. This kind of structures was also found in the reference studied by Li et al.[20,29], which was considered to be typical M-A constituent. When the second peak temperature was 550℃, the microstructure of A-550℃ and B-550℃ samples was dominated by parallel lath ferrite with high dislocation density, and thin film M-A components distributed between the paralleled lath ferrite. These nearly parallel laths were considered to have almost the same crystallographic orientation. According to the statistics of the width of ferrite laths in the samples, the width of ferrite laths in A-550℃ specimen was about 2.2μm, while the width of ferrite laths in B-550℃ specimen was about 2.7μm. For the secondary temperature of 700℃, a large number of blocky M-A island components appeared on the grain boundaries of A-700℃ and B-700℃ samples, which was in chain form. At a reheated temperature of 1000℃, the ferrite matrix was in the form of lath or quasi-polygon, and the dislocation density inside the ferrite laths was significantly reduced. M-A components were distributed between ferrite laths in the form of film, or distributed at the boundary of QF in the form of particle.

3.3 Fracture morphology analysis

Macroscopic and microscopic fracture morphology of thermal simulation sample for pipeline steel A was presented in Figure 8. The macroscopic fracture morphology of samples was shown on the left, which was divided into fiber zone(F), radiation zone(R), and shear lip zone(S). The microscopic morphology of the marked circle area in the macroscopic fracture morphology was shown on the right. For the secondary peak temperature of 550℃, the size and depth of the dimples in the fracture

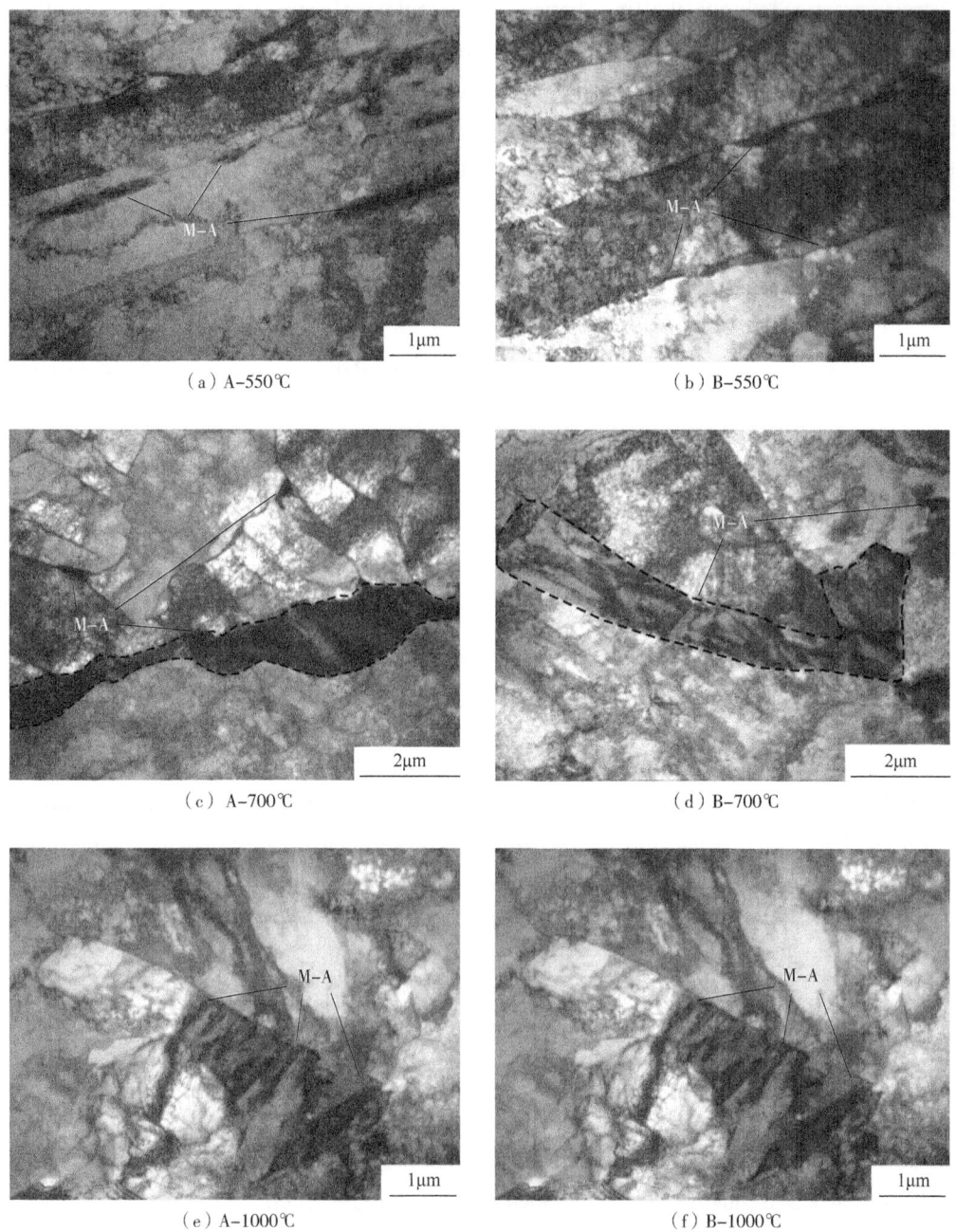

Figure 7 TEM micrographs of thermal simulation samples

surface of the sample were larger, and there was a certain proportion of cleavage zones. When the second peak temperature was 700℃, the fracture surface was almost in the radiation zone. The entire fracture surface was almost entirely cleavage fracture, with a river like pattern and no dimples presented. The reheated peak temperature of 1000℃, the fiber zone was mainly composed of dimples of varying sizes. Compared with the sample at A-550℃, the number of dimples decreased and they were shallower, while the number of cracks increased. The area of the cleavage zone, size of the cleavage step and number of dimples at the tearing edge were significantly decreased.

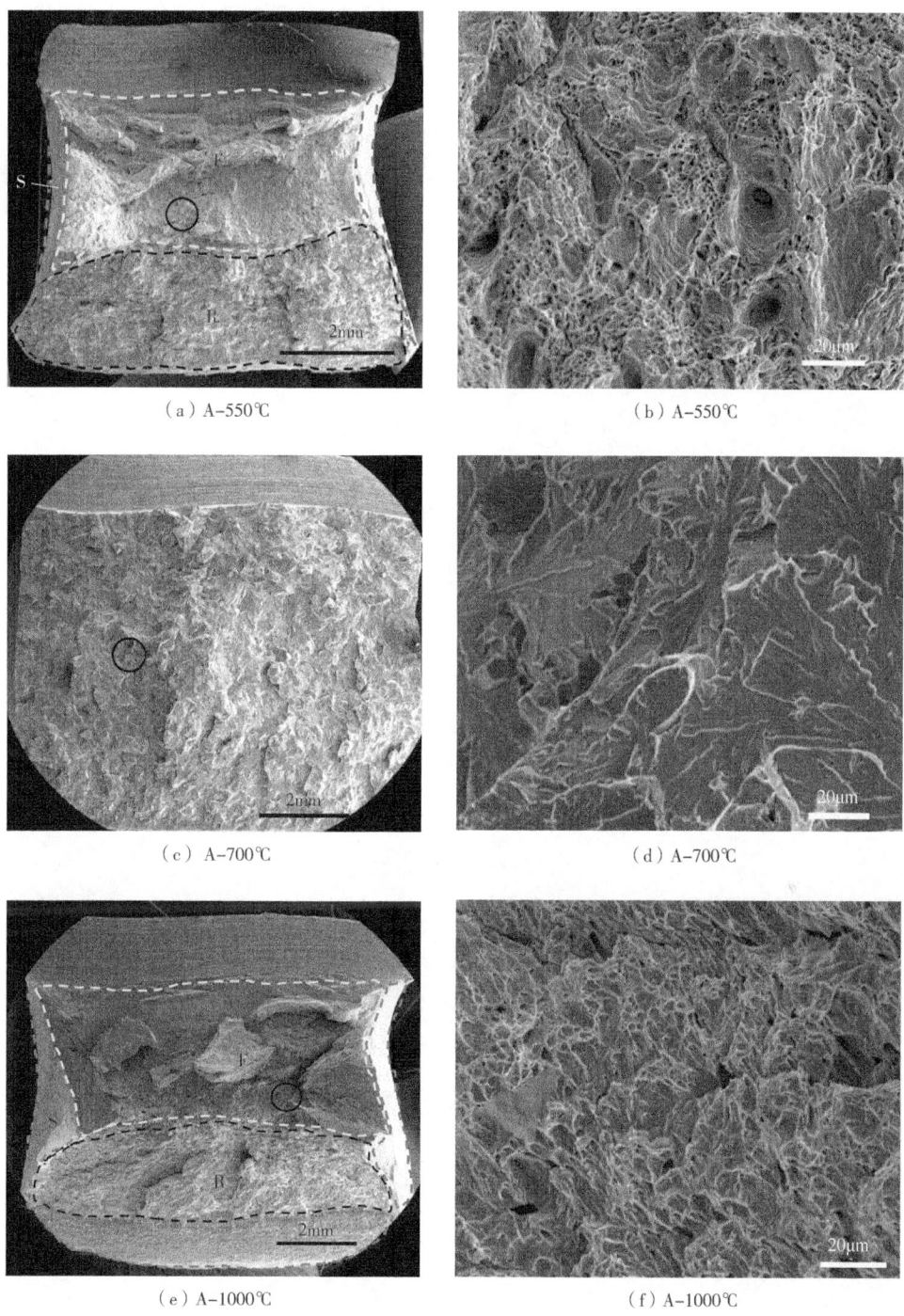

(a) A-550℃   (b) A-550℃

(c) A-700℃   (d) A-700℃

(e) A-1000℃   (f) A-1000℃

Figure 8 Macroscopic and microscopic fracture morphology of thermal simulation samples for specimens

Macroscopic and microscopic fracture morphology of thermal simulation samples for pipeline steel B was presented in Figure 9. Compared with the A-550℃ sample, the depth of dimples in the fracture surface of the B-550℃ sample was shallower, the size was smaller, and the cleavage zone area was slightly larger. When the secondary peak temperature was 700℃, similar to the A-700℃ sample, the fracture surface was almost in the radiation zone. The entire fracture surface was almost entirely of cleavage character, with a river like pattern and no dimples were present. For the second peak temperature of 1000℃, the shear lip zone ratio was the highest among all thermal simulation samples, indicated that this sample had experienced obvious plastic deformation under the impact load. Therefore, the sample B-1000℃ had high plastic toughness, and corresponded excellent impact toughness. For the microscopic fracture morphology of sample B-1000℃, there was a large number of deep and large-sized dimples in the fracture surface, and the cleavage zone area was relatively small.

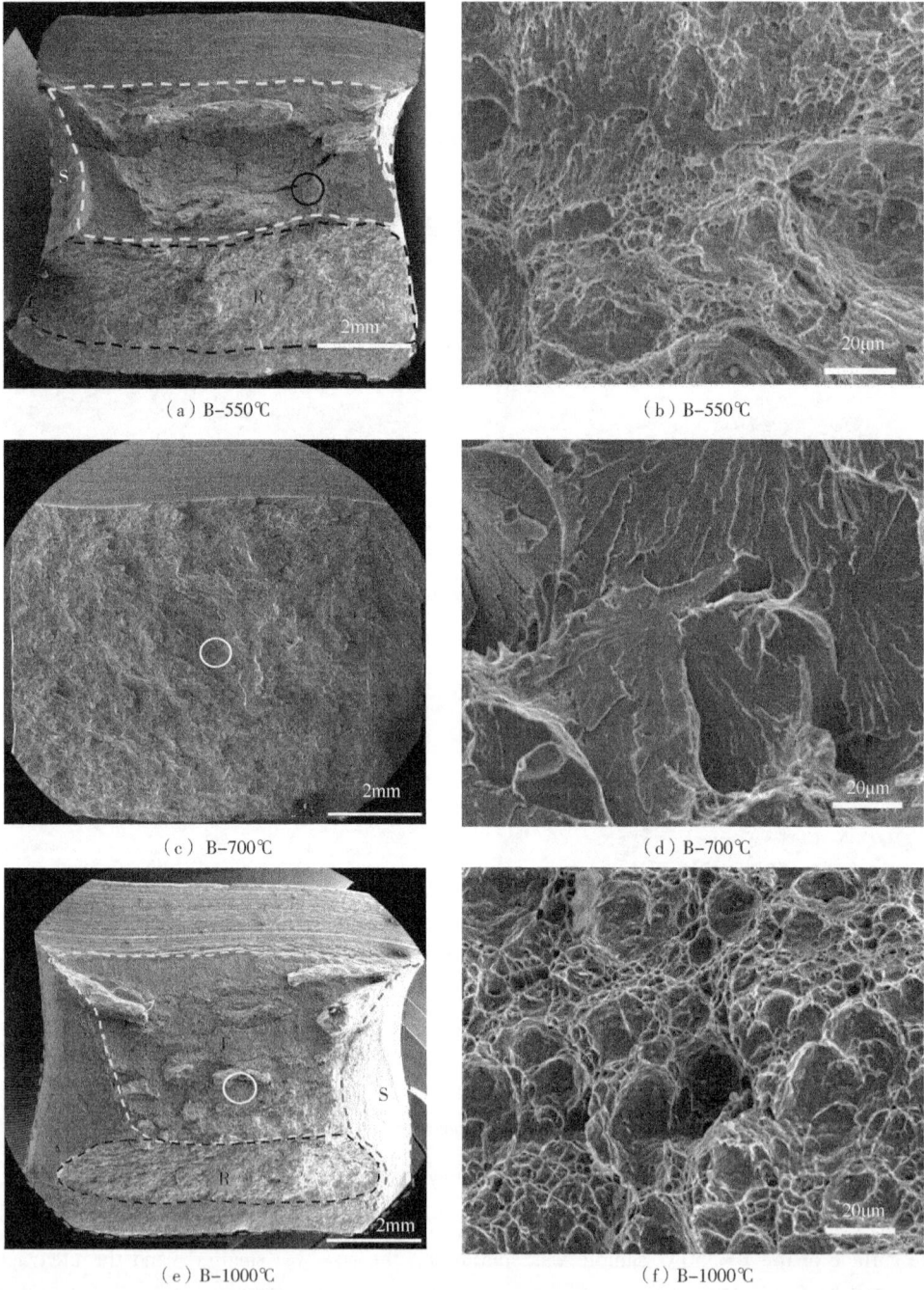

(a) B-550℃ (b) B-550℃
(c) B-700℃ (d) B-700℃
(e) B-1000℃ (f) B-1000℃

Figure 9 Macroscopic and microscopic fracture morphology of thermal simulation samples for specimens

3.4 Discussion

The EBSD maps of the secondary crack close to the main crack of the Charpy impact specimens with different reheated peak temperatures were presented in Figure 10. When the peak temperatures of secondary cycle were 550℃ and 700℃, the cracks propagate directly in the grain without deflection, and turn when encountering grain boundaries. While the peak temperature of secondary cycle was 1000℃, the crack did not extend in a straight line, but bended forward.

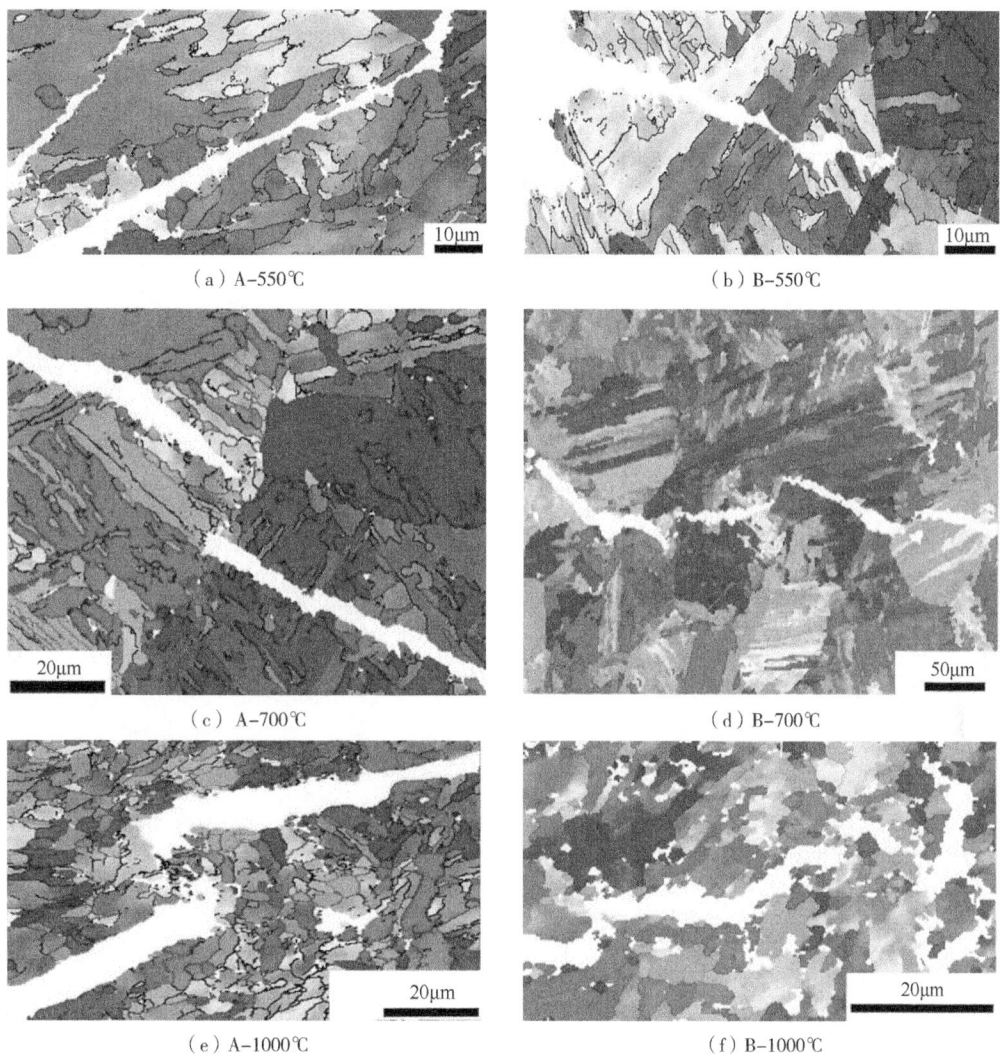

Figure 10　Propagation path of secondary cracks of simulation samples

The grain boundary maps and misorientation angle maps of thermal simulation samples at different peak temperatures of second thermal cycle were shown in Figure 11 and Figure 12, respectively. As presented in Figure 11, the misorientation angle within the range of 2°~15° was regarded as low-angle grain boundaries(LAGBs) marked in green lines, the black line indicated as the medium-angle grain boundaries(MAGBs) with the misorientation angle between 15° and 45°, while misorientation angle higher than 45° was characterized as high-angle grain boundaries (HAGBs) marked in red lines. It can be seen from Figures 11 and 12 that only a small amount of

MAGBs existed in all the thermal simulation samples. As illustrated from Figure 11(a) and Figure 11(b), there were HAGBs between the lath of bainite structure and grain boundaries in the A-550℃ and B-550℃ samples, while the LAGBs only distributed inside the laths. As presented in Figure 11(e) and Figure 11(f), For the A-1000℃ and B-1000℃ samples, the grain boundaries of PF and QF were HAGBs, and the interior of the QF was LAGBs.

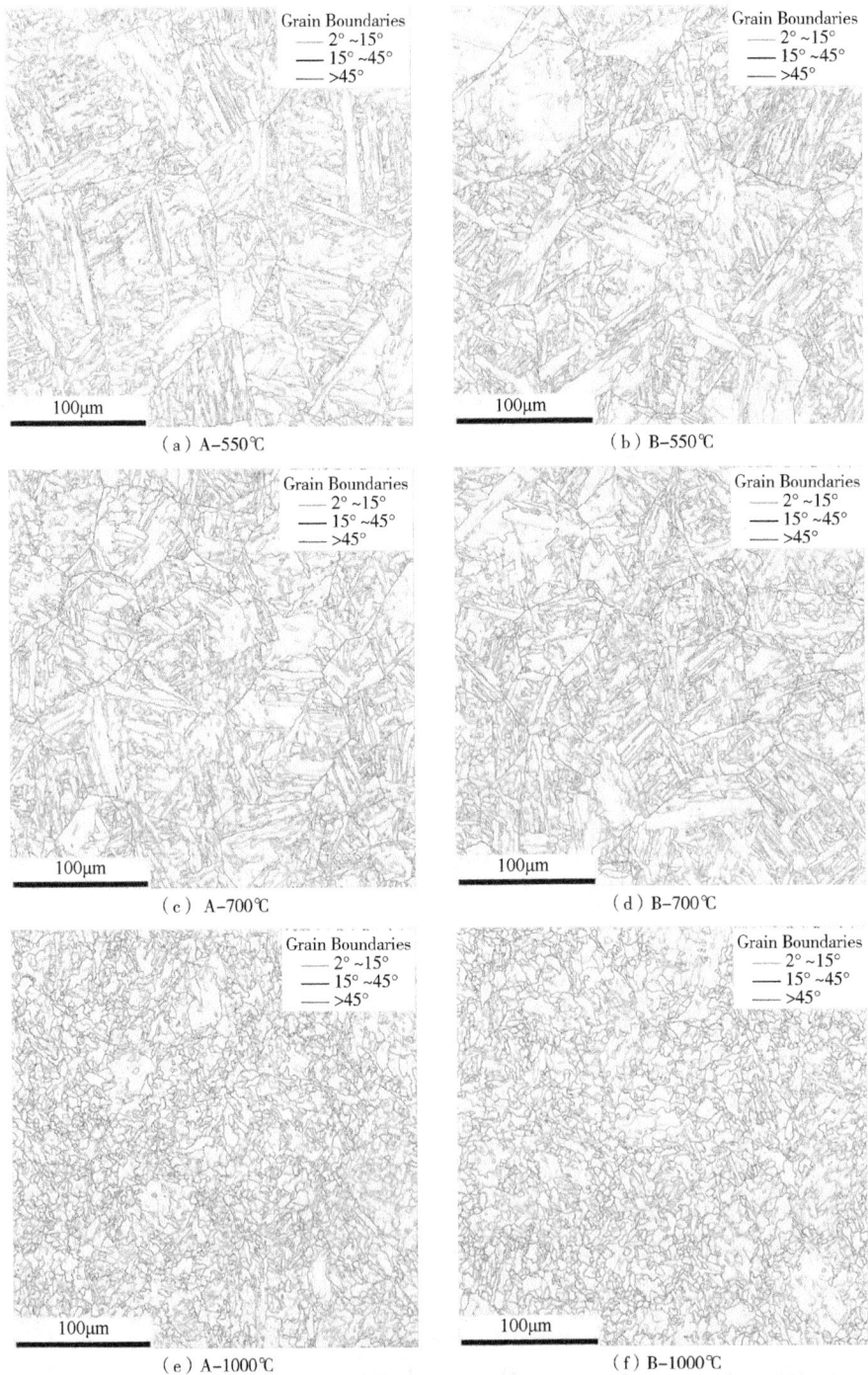

Figure 11 Grain boundary distribution of thermal simulation specimens

The statistical results of grain boundary lengths classified from different misorientation angles were shown in Table 2. According to the results of Table 2 and Figure 12, quantitative statistical analysis was conducted on LAGBs, MAGBs and HAGBs, and the statistical results were displayed in Table 3. There was no significant difference in length and proportions of HAGBs, MAGBs and LAGBs in samples A and B for secondary peak temperatures of 500℃ and 700℃, which were mainly related to the very close grain size thermal simulated samples at the two peak temperatures. However, the length and proportions of MAGBs and HAGBs in the thermal simulation sample with a secondary peak temperature of 1000℃ increased significantly, which corresponded to the refinement of grain size in the samples.

Table 2 Grain boundary length of thermal simulation specimens for different misorientation angles

mm

Sample	A-550℃	A-700℃	A-1000℃	B-550℃	B-700℃	B-1000℃
2°~15°	26.83	27.15	20.74	26.84	27.29	17.17
15°~45°	2.29	2.76	7.28	2.15	2.99	8.44
>45°	13.68	13.83	17.65	13.90	13.54	19.57

Table 3 Ratio of grain boundary for different misorientation angles

%

Sample	A-550℃	A-700℃	A-1000℃	B-550℃	B-700℃	B-1000℃
2°~15°	62.69	62.08	45.42	62.58	62.28	38.00
15°~45°	5.36	6.30	15.93	5.00	6.83	18.69
>45°	31.96	31.63	38.65	32.42	30.89	43.31

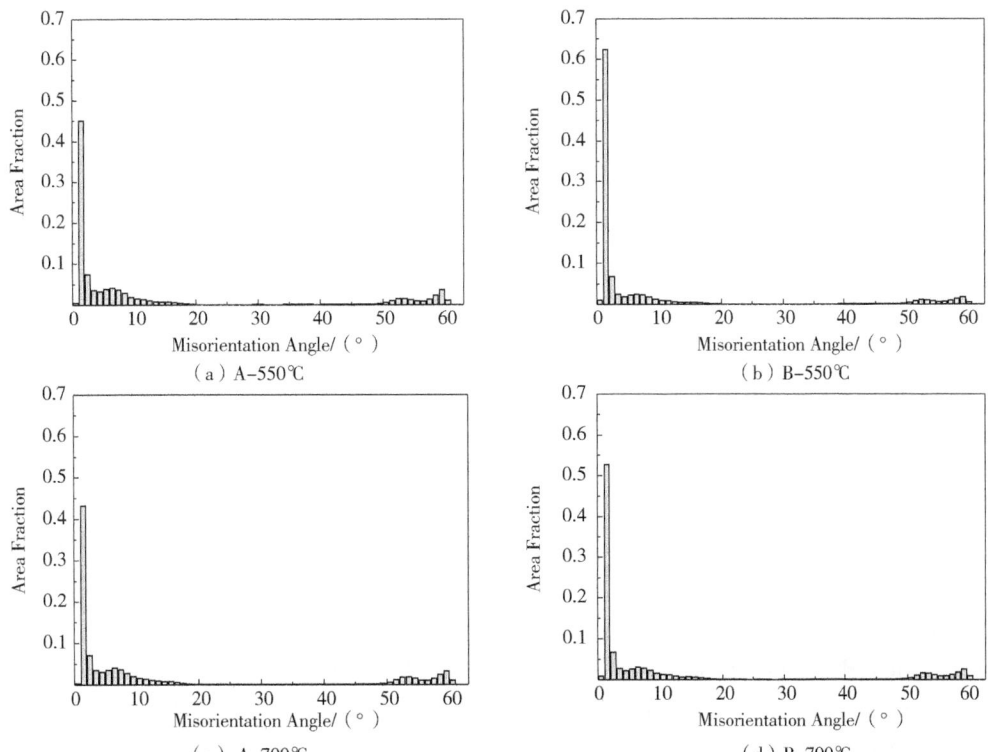

Figure 12 Grain orientation angles of thermal simulation specimens

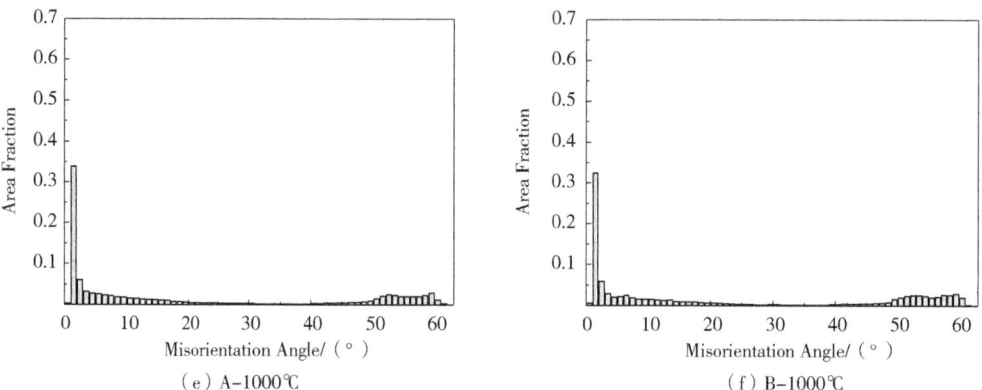

Figure 12 Grain orientation angles of thermal simulation specimens (continued)

Although the fractions of HAGBs, MAGBs and LAGBs in the thermal simulation samples for secondary peak temperatures of 550℃ and 700℃ was basically same, the corresponding impact toughness was quite different. According to the metallographic, SEM and TEM morphology analysis, a large number of M-A components were distributed on the grain boundaries of A-700℃ and B-700℃ samples compared with A-550℃ and B-550℃ samples. The results of Li et al. revealed that when the M-A component length was greater than 2μm and 1μm in thickness, this kind of M-A constituents can promote nucleation of brittle fracture either through cracking or debonding from the matrix. For the second peak temperature of 1000℃, the grains in A-1000℃ and B-1000℃ samples were obviously refined. Therefore, the length and ratio of MAGBs and HAGBs in B-1000℃ sample was the highest among the thermal simulation samples, corresponded to the highest impact absorption energy, and lower dispersion degree of the impact absorption energy values. The increase of the proportions of HAGBs helped to prevent the propagation of cracks, so that more energy was consumed during the propagation process, thus improving the impact toughness of the material[27-28].

4 Conclusions

The effect of different peak temperatures in second pass on the impact toughness of X80 pipeline steel with different Nb contents, microstructure and fracture morphology of the thermal simulation samples were investigated in this study. The main conclusions are derived as follows.

(1) The secondary peak temperatures corresponding to the toughness valley of the two X80 pipeline steels were located at A_{c1} to A_{c3}. For the peak temperature of second pass were lower than A_{c1}, both X80 pipeline steels had high impact toughness, and the impact absorption energy was close for both steels. When the second peak temperatures were higher than A_{c3}, X80 pipeline steel with higher Nb content had excellent impact toughness, lower impact dispersion, and higher impact absorbed energy than X80 pipeline steel with lower Nb content.

(2) When M-A components were distributed at the ferrite lath interface in a film-like form or in smaller sizes, they had less impact on the impact toughness of the HAZ of X80 pipeline steel. When chain-like distributed massive M-A constituents appeared on the grain boundaries, the impact toughness of the HAZ of X80 pipeline steel was seriously deteriorated.

(3) Adding an appropriate amount of Nb to X80 pipeline steel can improve the impact toughness of the SCCGHAZ and reduce thedispersion degree of impact absorption energy values.

Acknowledgements

The authors are grateful for the financial supports from the China National Petroleum Corporation for the research project (2023ZZ11 - 04), Companhia Brasileira de Metalurgia e Mineração(CBMM), and CITIC Metal.

Author Contributions

Conceptualization, Y. C., X. H., Q. C., Q. L. and W. L.; methodology, Y. C., Y. Y., X. H., X. L. and L. Q.; writing—original draft, Y. C. and Y. Y.; data curation, Y. Y. and X. H.; investigation, Y. C. and X. L.; formal analysis, W. L.; resources, Q. C.; writing - review & editing, Y. Y., Y. C., and X. H.; project administration, Q. C.; supervision, L. Q.; funding acquisition, Q. C. All authors have read and agreed to the published version of the manuscript.

Conflicts of Interest

The authors declare that they have no known competing financial interest or personal relationship that could have appeared to influence the work reported in this paper.

References

[1] Qi X, Huan P, Wang X, et al. Effect of root welding heat input on microstructure evolution and fracture mechanism in intercritically reheat-coarse grained heat-affected zone of X80 pipeline steel[J]. Materials Today Communications, 2022, 31, 103413.

[2] Sharma S K, Maheshwari S. A review on welding of high strength oil and gas pipeline steels[J]. Journal of Natural Gas Science and Engineering, 2017, 38, 203-217.

[3] Wang X, Wang D. Microstructure and mechanical properties of welded joint of X80 pipeline steel before and after ultrasonic impact treatment[J]. Journal of Materials Engineering and Performance, 2022, 31(2): 1465-1477.

[4] Ohaeri, E. G.; Szpunar, J. A., An overview on pipeline steel development for cold climate applications [J]. Journal of Pipeline Science and Engineering 2022, 2, (1), 1-17.

[5] Li L, Fu J, Wang X, et al. Comparative study of microstructure and toughness of automatic welded joints of X70 pipeline steel with no slope and 25° slope[J]. International Journal of Pressure Vessels and Piping, 2023, 206, 105054.

[6] Li B, Liu Q, Jia S, et al. Effect of V content and heat input on HAZ softening of deep-sea pipeline steel [J]. Materials, 2022, 15(3): 794.

[7] Vafaei, M.; Mashhuriazar, A.; Omidvar, H., et al., In-service welding of X70 steel gas pipeline: Numerical and experimental investigations [J]. Journal of Materials Research and Technology 2023, 26, 6907-6918.

[8] Sharma, S. K.; Maheshwari, S.; Singh, R. K. R., Modeling and optimization of HAZ characteristics for submerged arc welded high strength pipeline steel[J]. Transactions of the Indian Institute of Metals 2019, 72, (2), 439-454.

[9] Amori, K. E.; Hussain, M. N.; Hilal, H. B., Thermal analysis of in-service welding process for pipeline [J]. Journal of Petroleum Research and Studies 2019, 9, (1), 1-20.

[10] Qi, X.; Huan, P.; Wang, X., et al., Study on the mechanism of heat input on the grain boundary distribution and impact toughness in CGHAZ of X100 pipeline steel from the aspect of variant[J]. Materials Characterization 2021, 179, 111344.

[11] Sudin, V.; Stepanov, P.; Bozhenov, V., et al., Microstructural features of low-alloy pipeline steels that determine impact strength of welded joint heat-affected zone[J]. 2021, 65, 500-516.

[12] Costa, P.; Reyes-Valdés, F.; Saldaña-Garcés, R., et al., Thermal behavior of an HSLA steel and the impact in phase transformation: Submerged arc welding (SAW) process approach to pipelines [J]. Characterization of Metals and Alloys 2017, 85-98.

[13] Singh, M. P.; Shukla, D. K.; Kumar, R., et al., The structural integrity of high-strength welded pipeline steels: a review[J]. International Journal of Structural Integrity 2020, 12, (3), 470-496.

[14] Wang, X.; Wang, D.; Dai, L., et al., Effect of post-weld heat treatment on microstructure and fracture toughness of X80 pipeline steel welded joint[J]. Materials 2022, 15, (19), 6646.

[15] Efron, L.; Stepanov, P.; Zharkov, S., et al., Investigation of Low-Carbon Pipeline Steel Weldability by Welding Thermal Cycle Simulation[J]. Metallurgist 2022, 66, (7-8), 909-921.

[16] Wang, X.; Zhao, Y.; Guo, P., et al., Effect of heat input on MA constituent and toughness of coarse grain heat-affected zone in an X100 pipeline steel[J]. Journal of Materials Engineering and Performance 2019, 28, 1810-1821.

[17] Huda, N.; Lazor, R.; Gerlich, A. P., Study of MA effect on yield strength and ductility of X80 linepipe steels weld[J]. Metallurgical and Materials Transactions A 2017, 48, (9), 4166-4179.

[18] Bayraktar, E.; Kaplan, D., Mechanical and metallurgical investigation of martensite-austenite constituents in simulated welding conditions[J]. Journal of Materials Processing Technology 2004, s 153-154, 87-92.

[19] Singh, M. P.; Arora, K. S.; Gupta, A., et al., Experimental characterization of dynamic fracture toughness behavior of X80 pipeline steel welded joints for different heat inputs[J]. Welding in the World 2023, 67, (3), 617-636.

[20] Ramachandran, D. C.; Kim, S.; Moon, J., et al., Classification of martensite-austenite constituents according to its internal morphology in high-strength low alloy steel [J]. Materials Letters 2020, 278, 128422.

[21] Di, X.; Tong, M.; Li, C., et al., Microstructural evolution and its influence on toughness in simulated inter-critical heat affected zone of large thickness bainitic steel[J]. Materials Science and Engineering: A 2019, 743, 67-76.

[22] Di Schino, A.; Di Nunzio, P. E., Niobium effect on base metal and heat affected zone microstructure of girth welded joints[J]. Acta Metallurgica Slovaca 2017, 23, (1), 55-61.

[23] Singh, M. P.; Arora, K. S.; Kumar, R., et al., Influence of heat input on microstructure and fracture toughness property in different zones of X80 pipeline steel weldments[J]. Fatigue and Fracture of Engineering Materials and Structures 2021, 44, (1), 85-100.

[24] Fu, C.; Li, X.; Li, H., et al., Influence of ICCGHAZ on the low-temperature toughness in HAZ of Heavy-wall X80 pipeline steel[J]. Metals 2022, 12, (6), 907.

[25] Moeinifar, S.; Kokabi, A. H.; Hosseini, H. R. M., Effect of tandem submerged arc welding process and parameters of Gleeble simulator thermal cycles on properties of the intercritically reheated heat affected zone [J]. Materials and Design 2011, 32, (2), 869-876.

[26] Andia, J. L. M.; de Souza, L. F. G.; Bott, I. d. S., Microstructural and mechanical properties of the intercritically reheated coarse grained heat affected zone (ICCGHAZ) of an API 5L X80 pipeline steel[J].

Materials Science Forum 2014, 783-786, 657-662.

[27] Frantov I I; Velichko A A; Bortsov A N, et al., Weldability of niobium-containing high-strength steel for pipelines[J]. Welding Journal 2014, 93, (1), 23-29.

[28] Zhang, Y.; Zhang, H.; Zhao, S., et al., Effects of Nb on microstructure and toughness of high-strength structural steels heat affected zone at high heat input[J]. Transactions of the China Welding Institution 2008, 29, (9), 96-100.

[29] Li, X.; Ma, X.; Subramanian, S. V., et al., Structure-property-fracture mechanism correlation in heat-affected zone of X100 ferrite-bainite pipeline steel[J]. Metallurgical and Materials Transactions E 2015, 2, (1), 1-11.

本论文原发表于《Materials》2021年第14期。

Safety Analysis of Casing String after Sidetracking in Existing Production-resumed Wells

Zhou Sheng[1,2] Wang Jianjun[1,2] Hao Xuelei[3]
Shen Zhaoxi[2] Du Xu[4] Zhao Huanzhen[1,2]

(1. School of Mechanical Engineering, Xi'an Shiyou University;
2. CNPC Tubular Goods Research Institute, State Key Laboratory of Performance and Structural Safety for Petroleum Tubular Goods and Equipment Materials;
3. The First Gas Plant of PetroChina Changqing Oilfield Company;
4. CNPC Logging Corporation Changqing)

Abstract: In order to study the remaining service performance of the casing string in the sidetracking development of the existent well in Changqing Oilfield, and ensure the safe production in the later stage of the oilfield. The MIT-MTT logging technology is used to obtain the actual geometric size of the old well casing string, and the theoretical internal pressure resistance and collapse resistance strength of the existent well casing string are calculated. The experimental data of internal pressure resistance and collapse of the pipe are used to study the influence of the change of pipe string size on the strength of the pipe string. According to the service environment of the casing string and the process parameters during the sidetracking window opening process, the structural strength of the casing string in the existent sidetracking well is evaluated and analyzed. The experimental results show that: (1) the internal pressure resistance and external collapse resistance of the casing attenuate with the increase of radial deformation; (2) the theoretical value of the internal pressure resistance and external collapse resistance of the casing is close to the experimental value, and the theoretical value is generally lower than that of the experimental value; (3) The safety factor of casing string in old wells against internal pressure>1.1, and the safety factor against external extrusion>1.0. It is concluded that the service performance of the old well casing string is attenuated after sidetracking development, but it can still meet the design production requirements.

Keywords: Existent well; Sidetracking; Klever – Tamano collapse pressure formulae; Well integrity

1 Introduction

Sidetracking of old wells has become the main measure for the development of remaining

Corresponding author: Wang Jianjun, wangjianjun005@cnpc.com.cn.

reservoirs in old oilfield blocks, while sidetracking operations will weaken the service performance of existent well casing strings[1], casing columns as an important part of oil and gas production, with the role of isolating water layers, protecting well walls, and sealing formations[2-3]. In the process of oil and gas development, there are many factors that cause casing damage, such as corrosion, external pressure extrusion, annular air belt pressure, et al. [4-5]. In the case of production wells, external pressure extrusion is a major factor in casing damage[6]. The reasons for the external pressure extrusion are poor cementing quality, salt–creep formation, high–pressure water injection and so on. TakingChangqing Oilfield as an example, in 2015, there were 10787 medium and low production wells and long-stopped wells in Changqing Oilfield, accounting for 20.7% of the total number of oil wells, while 5292 medium and low-production wells and 475 long-stopped wells in Changqing Oilfield, accounting for 50% and 4.55% of the total number of gas wells, respectively[2]. The damaged parts of the casing are mainly in the production layer. The other well sections of the existent well except for the production well section have deformation, but they can still be used.

Therefore, the extrusion test and internal pressure resistance test of the deformation casing are adopted in this paper, and the method of formula calculation is used to explore the influence of geometric size change on the residual strength of the casing. In this way, the service safety after the sidetracking of the existent well casing string is evaluated.

2 Examples

An oil well and a gas well are selected as a case for the theoretical analysis of safety, and the schematic diagrams of the well structure are shown in Fig. 1 and Fig. 2. The same casing specifications were used for the two sidetracks and the geological conditions were similar, but Well SU14 was deeper than Well YC59. The cement sheath above the hanger position is well consolidated.

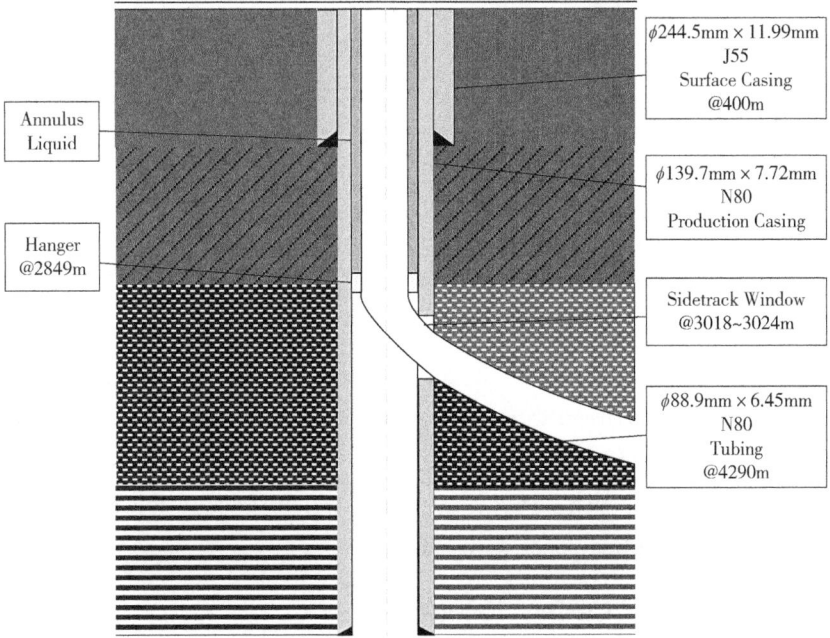

Fig. 1 Schematic of Well SU14

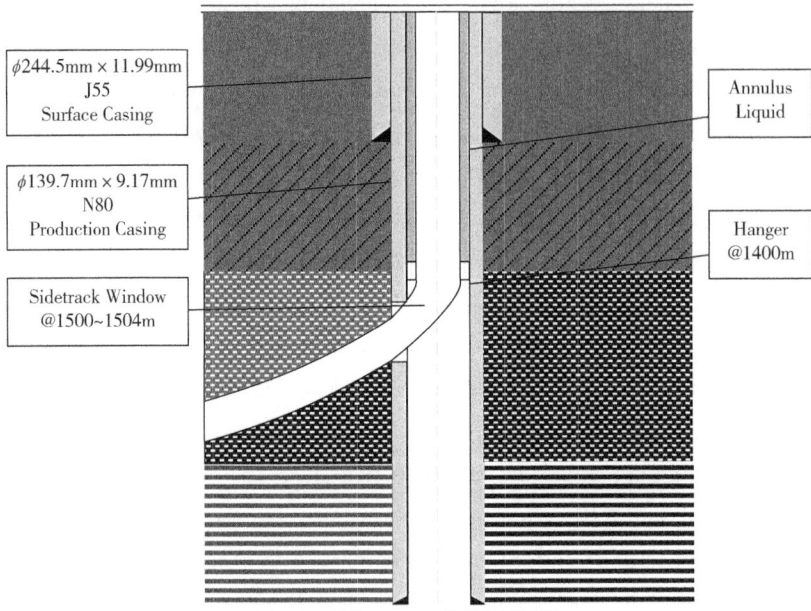

Fig. 2　Schematic of Well YC59

The MIT-MTT (Multifinger Imaging Tool and Magnetic Thickness Tool) logging tool portfolio was used to detect flaws in these two wells, and the parameters such as wall thickness, inner diameter and outer diameter of the production casing strings were measured (see Fig. 3).

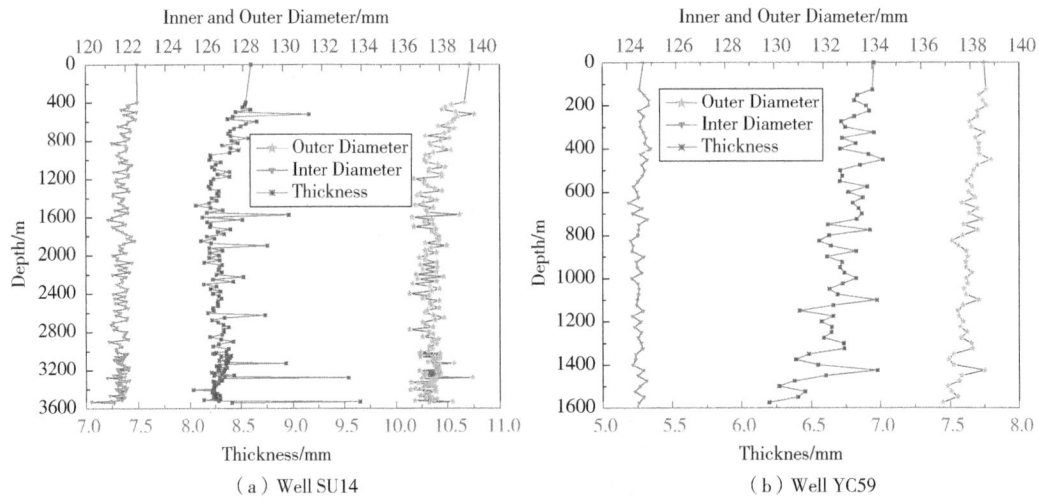

Fig. 3　Diagram of the change of casing geometry versus well depth

According to the logging information, it can be found that the wall thickness and outer diameter of the casing show a decreasing trend with the increase of the well depth. Wall thicknesses and outer diameters are mostly lower than nominal.

In places with a large dog-leg degree, there is a certain error in the measurement results of logging tools. the data used in this paper were corrected.

3　Safety Analysis of Casing Strings

This section mainly analyzes the tensile, external and internal pressure resistance of the existent well casing string in three stages: pressure test, fracturing and production.

3.1 Calculations for Loads on Casing String

3.1.1 Axial Load

Casing strings in existent wells generally have a cement sheath in the external annulus for cementing. If the cement sheath returns to the wellhead, it can be considered that during the cement sheath curing, the axial load in the casing (excluding the floating weight) is basically evenly distributed on the cement sheath, and the casing floating weight load is basically linearly distributed from the cement sheath. bear. After the casing string is in service, due to the large thickness of the cement sheath and the close connection between the casing and the formation, at this time, the change of the casing internal pressure will cause little deformation in the casing circumferential and axial directions. The effect of casing axial load is also small. At this time, the main influence on the axial load of the casing is the temperature change.

(1) Cement sheath outside the casing.

When there is a cement ring outside the casing, when the temperature changes at $\Delta T(\text{℃})$, the casing is constrained by the cement ring, and the additional axial load generated in the casing at a specific location can be directly calculated as:

$$F_T = \alpha \cdot \Delta T \cdot E \cdot A \tag{1}$$

Use the Eq. (1) to calculate at each position, and then superimpose it with the original axial load of the casing at this point, that is, the axial load distribution in the casing after the temperature changes.

(2) No cement sheath outside the casing.

When the casing is filled with completion fluid or formation water, the additional axial load caused by temperature change and internal and external pressure changes is:

$$F_p = \frac{\mu A}{L} \int_0^L \frac{2p_{ix}d^2 - p_{ox}D^2}{D^2 - d^2} dx + AE \int_0^L \frac{T_x \alpha}{L} dx \tag{2}$$

The calculation result of Eq. (2) is the additional axial load evenly distributed on the entire casing string, and the additional axial load is superimposed with the original axial load of the casing string, which is the current axial load distribution of the casing string.

3.1.2 Internal Load

(1) Gas well.

Internal load of casing strings is calculated by using Eq. (3) [7].

$$p_{ix} = p_{smin} / e^{1.11548 \times 10^{-4} G(H - h_v)} \tag{3}$$

(2) Oil well.

Assuming that the density of the internal liquid is close to the density of the formation water, the internal pressure of the casing string is approximated by the hydrostatic pressure. Under fracturing conditions, the density of the fluid in the casing is replaced by the density of the fracturing fluid.

If $h_v \leq h_y$, internal load is calculated by Eq. (4).

$$p_{ix}=0 \tag{4}$$

If $h_v > h_y$, internal load is calculated by Eq. (5).

$$p_{ix}=0.00981\rho_o(h_v-h_y) \tag{5}$$

3.1.3 External Load

It should be noted that the cement return height outside the casing, and the suspended section is considered to be full of drilling fluid in harsh conditions.

If $h_v \leq h_y$, external load is calculated by Eq. (6).

$$p_{ox}=p_{ohB}+0.00981\rho_c h_d \tag{6}$$

If $h_v > h_y$, external load is calculated by Eq. (7).

$$p_{ox}=p_{ohB}+0.00981\rho_c h_d+0.00981\rho_w(h_v-h_d) \tag{7}$$

3.2 Calculations for the Strength of Casing Strings

3.2.1 Tensile Strength

Tensile strength is calculated by Eq. (8) ~ (14)[7].

The onset of yield is defined as:

$$\sigma_e = f_{ymn} \tag{8}$$

The equivalent stress is defined as:

$$\sigma_e = \sqrt{\sigma_r^2+\sigma_c^2+(\sigma_a+\sigma_b)^2-\sigma_r\sigma_c-\sigma_r(\sigma_a+\sigma_b)-\sigma_c(\sigma_a+\sigma_b)+3\tau_{ac}^2} \tag{9}$$

With:

$$\sigma_r = \frac{p_{ix}t^2-p_{ox}^2 D^2}{D^2-t^2} - \frac{(p_{ix}-p_{ox})t^2 D^2}{4(D^2-t^2)r^2} \tag{10}$$

$$\sigma_c = \frac{p_{ix}t^2-p_{ox}^2 D^2}{D^2-t^2} + \frac{(p_{ix}-p_{ox})t^2 D^2}{4(D^2-t^2)r^2} \tag{11}$$

$$\sigma_a = \frac{F_a}{A} \tag{12}$$

$$\sigma_b = \pm \frac{M_b r}{I} \tag{13}$$

$$\tau_{ac} = \frac{Tr}{J_p} \tag{14}$$

3.2.2 Collapse Strength

Due to the influence of the measurement error of the logging instrument, when the casing ovality calculated by equation 4 is greater than 5%[8], the influence of the casing ovality on the casing collapse strength should be considered. Strength should be calculated using the Eq. (15) ~ (19)[7,9]. Due to the curvature of the tubing string in directional wells, this affects the distribution of the axial force of the tubing string. Therefore, the equivalent yield strength is used for calculation.

$$P_{des} = \frac{k_{e\,des}P_{e\,des}+k_{y\,des}P_{y\,des}-\sqrt{(k_{e\,des}P_{e\,des}-k_{y\,des}P_{y\,des})^2+4k_{e\,des}P_{e\,des}k_{y\,des}P_{y\,des}Ht_{des}}}{2(1-Ht_{des})} \quad (15)$$

Where,

$$P_{e\,des} = \frac{2E}{(1-\mu^2)(D/t)(D/t-1)^2} \quad (16)$$

$$P_{y\,des} = \frac{2f_{yax}}{D/t}\left[1+\frac{1}{2(D/t)}\right] \quad (17)$$

$$Ht_{des} = 0.127ov + 0.0039ec - 0.440(\sigma_{rs}/f_{ymn}) + h_n \quad (18)$$

$$f_{yax} = \left[\sqrt{1-\frac{3}{4}\left(\frac{\sigma_a}{f_{ymn}}\right)^2} - \frac{\sigma_a}{2f_{ymn}}\right]f_{ymn} \quad (19)$$

3.2.3 Burst Strength

The burst strength is calculated by Eq. (20)[10].

$$P_b = 0.875 \times \frac{f_{yax}}{2(D/t)} \quad (20)$$

The loads on the existent well casing string after sidetracking are:

(1) The effect of drilling pressure, window opening, vibration and so on the casing in the process of sidetracking window opening;

(2) Wear on the internal wall of the casing during sidetracking;

(3) In the early service process, the existent well casing corrosion damage;

(4) Existent wellbore casing pressure test. Oil wells 21MPa, gas wells 25MPa;

(5) If theexistent well casing is used for fracturing operations, the effect of pressure on the casing.

(6) Oil Well: Casing specification ϕ139.7mm × 7.72mm, steel grade N80, lamination fracture pressure 25.6 ~ 38.8MPa, working pressure 13.2 ~ 26MPa, injection time 30 ~ 80 minutes.

(7) Gas well: Casing specification ϕ139.7mm × 7.72mm, steel grade N80, lamination fracture pressure 40~58.8MPa, working pressure 39.2~60MPa, injection time 60~190 minutes.

According to the different loads on the casing string under the three working conditions, the safety factors of tensile, internal pressure and external pressure resistance of oil and gas wells under the three working conditions are calculated respectively.

According to the measured data in Section 2 and the theoretical basis of this section, the safety factor of the production well section of the old well under the three operating conditions is calculated as shown inFig. 4 to Fig. 9. It can be seen from the figure that the safety factor of the calculated well sections is greater than the standard value.

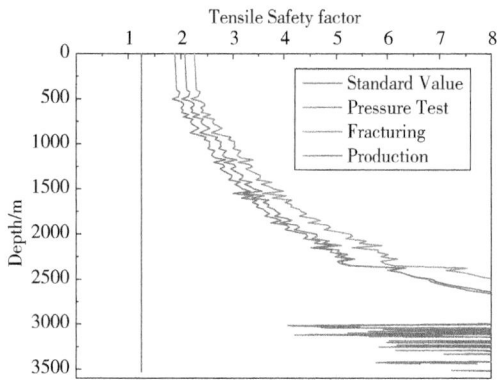

Fig. 4 Tensile Safety Factor Curve of Well SU14

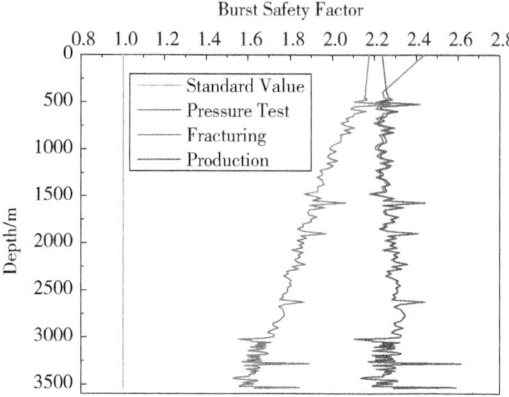

Fig. 5 Burst Safety Factor Curve of Well SU14

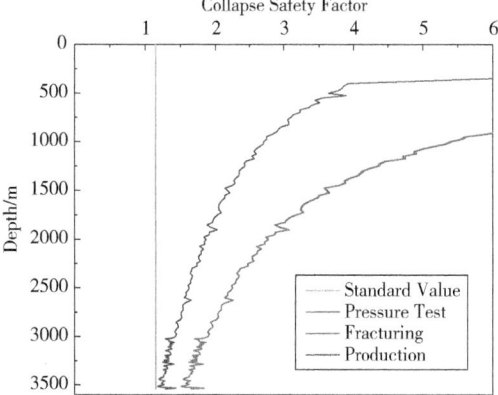

Fig. 6 Collapse Safety Factor Curve of Well SU14

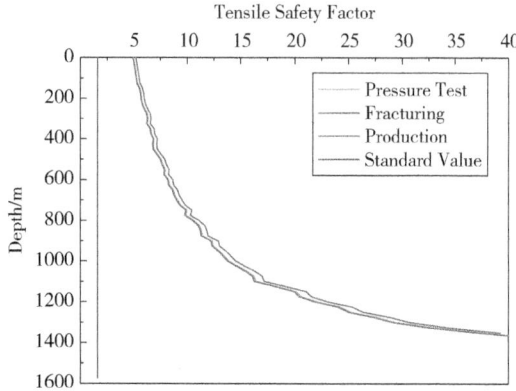

Fig. 7 Tensile Safety Factor Curve of Well YC59

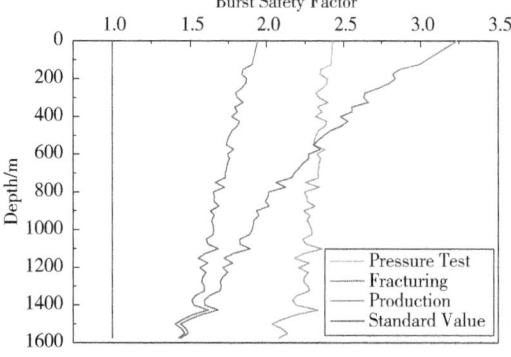

Fig. 8 Burst Safety Factor Curve of Well YC59

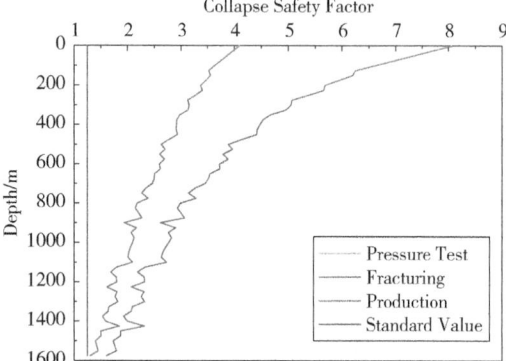

Fig. 9 Collapse Safety Factor Curve of Well YC59

The standard value still retains a certain margin compared with the actual working conditions, so if the safety factor of a well is 95% of the standard value, it cannot be considered that the well is unsafe. However, in the design of the wellbore structure, try to ensure that the safety factor of the whole well section is greater than the standard value.

4 Deformed Casing Collapse and Burst Test

The purpose of this section is to compare the difference between the internal pressure resistance and external collapse resistance of the deformed casing and the complete casing, and to analyze the pressure bearing capacity of the casing string under the condition of repeated fracturing after long-term production. According to the maximum external diameter of the bridge plug required by the site, the radial deformation amount of 3.5mm+3.5mm is achieved for ϕ139.7mm×7.72mm N80 casings by mechanical loading, and the loading method is shown in Fig. 10. A total of 2 casings were radial deformation, and the internal pressure resistance test and the squeeze test were carried out, and the test results were compared with the test results of the non-deformed casing.

The sample and test system used in the test are shown in Fig. 11.

Fig. 10 Diagram of mechanical loading deformation of casing

Fig. 11 Composite Collapse Test System

After the ϕ139.7mm×7.72mm N80 casing is deformed by 7mm (considering the diameter of 121.08mm, it cannot exceed 10mm in practice), the short axis is 132.7mm, and the long axis can be calculated to be 147.07mm. According to the ISO/TR 10400 standard strength theory calculation[7], the theoretical value of the internal compressive strength after deformation of the casing is 43.8MPa, which is 12.02% lower than the standard value of 49.8MPa; The theoretical value value of 42.7MPa decreased by 14.25%.

Comparing the calculated and experimental values of the internal pressure resistance and collapse resistance, it is found that there is not much difference between the calculated value of the deformation casing strength and the experimental value, and the theoretical calculation value is higher than the experimental value. It is safe. As shown in Fig. 12 and Fig. 13, the internal compressive strength and collapse strength of ϕ139.7mm×7.72mm N80 casing decrease with the increase of radial deformation, and the decrease of the test value is lower than the calculated value, but testing to determine casing strength changes is still recommended.

5 Conclusions

(1) According to the results of the physical test, the method of safety analysis of casing string in old oil and gas wells adopted in this paper is feasible. Under fracturing conditions, it is necessary

to ensure that the casing has sufficient tensile strength, under production conditions, more attention should be paid to the collapse strength of the casing.

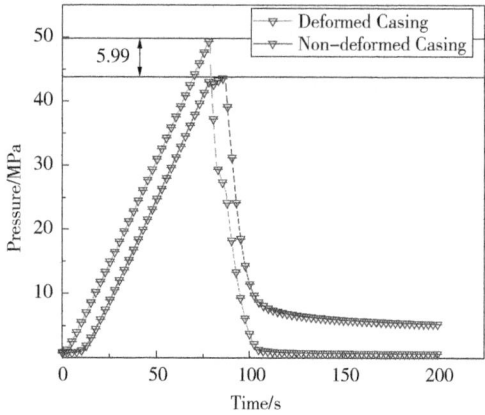
Fig. 12　Comparison of test results of casing collapse performance

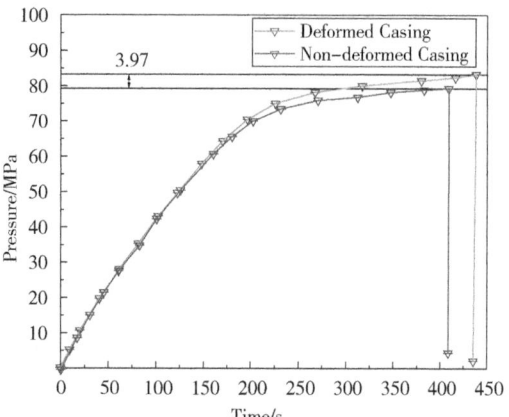
Fig. 13　Comparison of test results of casing burst performance

(2) The change of the geometric parameters of the casing string will weaken its structural strength and mechanical properties. It is mainly manifested in the weakening of the collapse strength, this also causes fluctuations in the safety factor value.

(3) If the inner diameter of the casing can make the downhole tool run smoothly, and the deformation of the casing does not exceed 5%, its mechanical properties can still meet the actual production requirements.

Acknowledgments

The project is supported by "Research on key technologies for evaluation and repair of downhole casing damge" (Number 2021DJ2705).

Nomenclature

A　the area of the casing cross section
D　the average external diameter measured
E　Young's modulus
ec　eccentricity
F_T　additional axial loads generated by temperature changes in casing, N
F_p　additional axial loads generated by pressure changes in casing, N
f_{yax}　equivalent yield strength in the presence of axial stress
f_{ymn}　specified minimum yield strength
H　total depth of well
Ht_{des}　decrement factor for design collapse strength
h_d　cement back-up depth outside casing
h_n　stress-strain curve shape factor
h_v　depth of the calculation point

h_y depth of liquid level in the casing

I the moment of inertia of the casing cross section

J_p polar moment of inertia of the casing cross section

$k_{e\,des}$ down-rating factor for design elastic collapse

$k_{y\,des}$ down-rating factor for design yield collapse

M_b bending moment

ov ovality

P_{des} design collapse pressure

P_y yield collapse term

P_e pressure for elastic collapse

$p_{s\,min}$ minimum wellhead pressure (gas well)

p_{ix} internal pressure at depth x

p_{ox} external pressure at depth x

P_{ohB} production casing annulus pressure

r the radial coordinate, $(d/2) \leqslant r \leqslant (D/2)$

T the applied torque

T_x temperature change of casing at depth x, ℃

t specified pipe wall thickness

α the coefficient of thermal expansion of the casing material

ΔT temperature change

μ possion's ratio

ρ_a anulus protective fluid density

ρ_c cement density

ρ_o well production fluid density

ρ_w formation water density outside the casing

σ_a component of axial stress not due to bending

σ_b component of axial stress due to bending

σ_c circumferential stress

σ_e equivalent stress

σ_r radial stress

σ_{rs} residual stress

τ_{ac} shear stress

References

[1] Chen Xinyong, F, Fu Xiao, S, Li Dongjie, T. Research Status and Prospect of Sidetracking Technology for Old Wells in China. Oil Field Equipment 48(6), 73-76(2019).

[2] Zhang Jinwu, F, Wang Guoyong, S, He Kai, T. Practice and understanding of sidetracking horizontal drilling in old wells in Sulige Gas Field. Petroleum Exploration and Development 46(2), 370-377(2019).

[3] Sun Ning, F. Sidetracking Horizontal Well to Become Important Technological Scheme for Stimulation of Low-yield Oil and Gas Wells in Old Areas. Oil Forum 37(2), 28-35(2018).

[4] Tan Chengjin, F, Xu Bingye, S, Gao Deli, T. Numerical Analysis On The Remaining Collapsing Strength of Worn Casing. Oil Drilling & Production Technology 22(1), 6-8+13(2000).

[5] Lin Yuanhua, F, Deng Kuanhai, S, Zeng Dezhi, T. The Deformation Law in the Collapsing Process of C110 Casing Under Non-uniform Loading. Mechanical Science and Technology for Aerospace Engineering 34(2), 315-319(2015).

[6] Han Zhiyong, F. Down-hole Tubular Mechanics in Hydraulic Environment. Petroleum Industry Press, Beijing (2011).

[7] International Organization For Standardization. ISO/TR 10400: 2018. Petroleum and Natural Gas Industries-Formulae and Calculations for the Properties of Casing, Tubing, Drill Pipe and Line Pipe Used As Casingor Tubing. Switzerland(2018).

[8] Wang Jianjun, F, Yan Xiangzhen, S, Lin Kai, T. Experiment on collapsing deformation of high-grade steel casing with inside ovality. Journal of China University of Petroleum(Edition of Natural Science) 35(2), 123-126(2011).

[9] F. J. Klever, F, T. Tamano. S. A New OCTG Strength Equation for Collapse under Combined Loads. SPE Drilling & Completion, 21(3), 164-179(2006).

[10] Huang Xiaoguang, F, Chen Yanyun, S, Lin Kai, T. Burst Strength Analysis of Casing With Geometrical Imperfections. J. Pressure Vessel Technol 129(4), 763-770(2007).

本论文原发表于《IFEDC 2022》2022 年。

Creep Crack Opening Tip Displacement (CCTOD) of X80 Pipeline Steel at Room Temperature

Wang Peng[1,2] Yang Sen[1] Chen Fuxing[1] Che Qi[1]
Xie Jiamiao[3] Wang Fenghui[1] Hao Wenqian[3,4]

(1. Tubular Goods Research Institute, China National Petroleum Corporation;
2. Department of Engineering Mechanics, Bi-Inspired and Advanced Energy Research Center, Northwestern Polytechnical University;
3. School of Aerospace Engineering, North University of China;
4. Underground Target Damage Technology National Defense Key Discipline Laboratory, North University of China)

Abstract: The room temperature creep behavior of X80 pipeline steel has attracted increasing attention because it is considered to be the main cause oftime-delayed failure in pipeline steel. The crack opening tip displacement (CTOD) is a general criterion to characterize the time-delayed failure and assessment safety of pipeline steel. However, there are few methods to demonstrate crack behavior of the pipeline steel caused by room temperature creep. Based on the traditional CTOD theoretical model, a modified CCTOD theoretical model is presented to calculate the creep CTOD from the crack mouth opening displacement(CMOD). The room temperature creep experiment with a single edge notched beam subjected to three-point bending creep load for 36 hours at room temperature is performed to know more about the creep CTOD of X80 pipeline steel. Based on a power-law time delayed relationship among creep strain, stress and time fitted from experiment results of tensile creep tests, the numerical approach is also taken placed for evaluation the creep CTOD of pipeline steel at room temperature. It is found that there is about 10% CTOD increment caused by room temperature creep added to that of the traditional elastic and plastic crack tip opening displacement, which indicates that room temperature creep is also further blunted the crack tip of X80 pipeline steel. Meanwhile, the loads effects and load history on the CCTOD at room temperature creep are also tested and discussed. The results will provide a approach for failure evaluation and an integrity guideline for X80 pipeline steel.

Keywords: Creep crack opening tip displacement (CCTOD); Room temperature creep; Modified model; X80 pipeline steel; Numerical approach

Corresponding author: Wang Fenghui, fhwang@nwpu.edu.cn; Hao Wenqian, wqhao@nuc.edu.cn.

1 Introduction

As the high demand for clean gas energy sources such as natural gas, the blended hydrogen-natural gas increases, they need to be transported and stored more efficiently, securely and economically via high-grade pipeline steel[1-3]. These targets have motivated the continuous development of pipeline steel technology in the direction of high strength and high toughness. High-grade X80 pipeline steel is one of the most popular and durable pipeline steels with approximately 30000km constructed worldwide[4-5]. However, in the practical engineering applications, the service reliability of X80 pipeline steel is affected by various factors, including creep, hydrogen embrittlement, fatigue, corrosion and so on[2,6-10]. The serious leakage problem of transported high-pressure gas pipeline steel is also observed at room temperature due to the fracture behavior after the stable operation. It has been believed that a defect in a pipeline steel, such as a dent and gouge, can fail at an operation pressure after some period of time, which is called as a time-delayed failure.

As we known, creep is a deformation time-delayed failure after the long-term operation of material under constants load. Irreversible deformation occurs when the material is loaded at the temperature above 40% of the melting point for material ($T/T_m > 0.4$), which is known as high-temperature creep. However, the room temperature creep is also observed in X-series pipeline steel and other kinds of steels at the low temperatures ($T/T_m < 0.2$) and the deformation is relatively small[11]. Room temperature creep behavior of X80 pipeline steel has attracted increasing attention because some failure in fields believed to cause by creep. It is caused by stresses that facilitate dislocations to cross barriers and substantial dislocation slip over time, which is manifested as room temperature creep deformation of the material macroscopically[12-13]. Based on previous experimental results, the factors that affect room temperature creep behavior are loading stress, loading rate, loading history and fatigue[14-16]. Some constitutive models, including time-hardening model[17], strain-hardening model[18-23], Johnson-Cook[24-25] and Cowper-Symonds[26], are used to describe creep behavior of metallic material[27-28]. The influences of stress and temperature on creep behavior of material was analyzed by Cadek et al.[29]. The relationship among stress, temperature and creep rate of alloy material was also deduced, and the control mechanism of creep rate was clarified by theorical and experimental methods[10,16]. The room temperature creep occurs at the condition of $1/3R_{p0.2}$ ($R_{p0.2}$ is the yield strength corresponding to the 0.2% non-proportional extension rate)[30] and it only includes the first and second creep stages, which investigated by Liu et al.[31]. The room temperature creeps of X70 pipeline and 304 stainless steels were investigated by Nie et al.[32], it was found that the room temperature creep in the first stage satisfied the logarithmic law. By studying the room temperature creep of 304 steel and 304 heat treated steel, Kassner et al.[33] found that the room temperature creep plasticity of stainless steel is driven by thermally activated dislocation movement, dislocation interaction and interaction between dislocation and solute atom.

Many investigations have been conducted to study in depth the cracking arrest and fracture evaluation behaviors of various grades of pipeline steels under the room temperature creep. The fracture toughness in different area of pipeline steel under the room temperature creep, especially in the stress concentration areas, such as the pipeline circumferential welds, the scratches, the weld joints, the dent and so on, can be effectively evaluated by fracture toughness experiment[34-40]. The J-integral, CTOA and CTOD are usually selected to represent fracture toughness in predicting the parameters of steady crack propagation in pipeline steel[41-42]. The CTOD of the composite lamellae was observed by in-situ SEM method by Poursartip et al.[43]. The multi-specimen method is usually adopted to determine the fracture toughness parameters, and the fracture resistance and R-resistance curve are calculated by measuring crack mouth opening displacement(CMOD). The crack opening tip displacement(CTOD) is a general criterion to characterize the failure of pipeline steel. So the calculated CTOD and measured CMOD are significant related to permitted pressure for design and service in pipeline steel. Although creep does not lead to large deformations at room temperature, it has a significant effect on the stress concentration areas (crack tips, corrosion defects) and changes in properties in the stress relief area[27-28]. The schematic diagram of yield zone for the cracked specimen in uniaxial creep tension experiment is shown in Fig. 1. The yield zone and CTOD of specimen at crack tip are observed in Fig. 1(b). The room temperature creep occurred when specimen loaded around yield stresses and CTOD also increased by the blunt crack tip correspondently, which was demonstrated by Wang et al.[16].

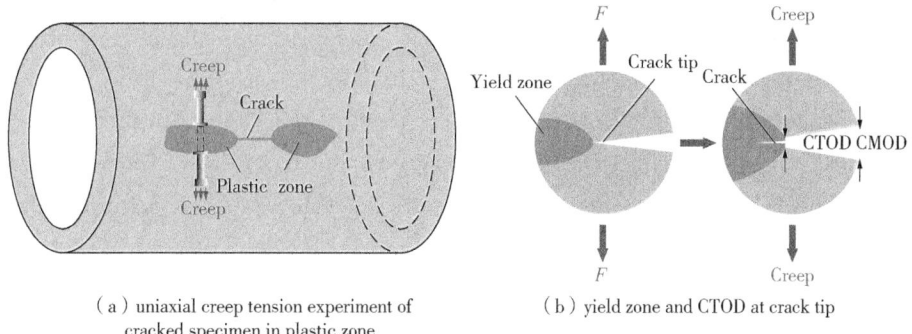

(a) uniaxial creep tension experiment of cracked specimen in plastic zone

(b) yield zone and CTOD at crack tip

Fig. 1 Schematic diagram of yield zone for the cracked specimen in uniaxial creep tension experiment

Although a number of research works are available in the literature, which report the creep behavior of various grades of pipeline steels, but determining the creep CTOD and understanding the influence of different loads and load history on the room temperature creep of X80 pipeline steel are still the problems. In order to investigate the creep CTOD of X80 pipeline steel, the room temperature creep experiments of the single edge notched beam subjected to three-point bending retention load for 36 hours at room temperature are performed in this study. Based on the traditional CTOD theoretical model, a modified CTOD theoretical model is firstly presented to calculate the creep CTOD from the CMOD measurement results. A power-law function fitted from experiment results is also taken forward into the numerical simulation model. The numerical simulation results and experiment results are compared. In addition, the effects of different loads and load history on the room temperature creep of the crack tip are also discussed. The results of this research will provide a reference to the study of the fracture toughness evaluation and safe application of X80 pipeline steel under room temperature creep.

2 Material and methods

2.1 Room temperature creep experiment

The base metal specimens machined from X80 longitudinal submerged arc welded pipe in service at the China-Russia East-Route Natural Gas Pipeline (Heihe-Changling) were selected in the experiment. The diameter and wall thickness of X80 pipeline manufactured according to American Petroleum Institute Specification 5L (API SPEC 5L) standard[44] were 1422mm and 21.4mm, respectively. The chemical composition (mass fraction) of base metal specimens for X80 pipeline steel are shown in Table 1.

Table1 Chemical composition base metal specimens for X80 pipeline steel

Chemical composition	C	Si	Mn	P	S	Cr	Ni	Ti	Nb	V	Mo
Mass fraction/%	0.063	0.280	1.830	0.011	0.0006	0.03	0.03	0.016	0.061	0.059	0.220

According to ISO 204: 2018 standard[45], The stress-strain curves of base metal specimens for X80 pipeline steel were obtained by the uniaxial tensile experiment at room temperature. The tensile rate was set to 0.1mm/min in the uniaxial tensile experiment of X80 pipeline steel. The geometric parameters of base metal specimens and the uniaxial tensile experiment results were given in previous researches investigated by Wang et al.[16,42]. Based on the uniaxial creep experiment in tension results, the room temperature creep curve of base metal specimens for X80 pipeline steel under three loading stresses of $1.03R_{p0.2}$(619MPa), $1.00R_{p0.2}$(601MPa) and $0.97R_{p0.2}$(582MPa) was obtained. $R_{p0.2}$ is the yield strength corresponding to the 0.2% non-proportional extension rate. The parameters of creep time-hardening theoretical model were fitted based on the uniaxial creep experiment in tension results in Section 2.2.

2.2 Creep time-hardening model

In order to calculate the crack tip deformation, a time hardening constitutive relationship of X80 pipeline steel is obtained as necessary. According to previous work[16], the creep time-hardening model is often used to represent the relationship of creep deformation with time for X80 pipeline steel. The creep time-hardening model is considered following Bailey-Norton creep law, and the creep strain rate of a material is related to the loading stress, temperature and time[46]. The creep rate can be expressed in terms of stress σ and time t, which can be expressed as

$$\dot{\varepsilon}_c = A\sigma^n t^m \quad (1)$$

where $\dot{\varepsilon}_c$ is the creep rate, A (power law multiplier) and m (time exponent) are the material-related constants and $-1 < m \leq 0$, n is the stress exponent (positive number). Integrating Eq. (1) over time and then taking the logarithm to obtain Eq. (2), as shown by

$$\lg\varepsilon_c = \lg\left(\frac{A}{m+1}\right) + n\lg\sigma + (m+1)\lg t \quad (2)$$

According to Eq. (2), the creep can be formulated as a binary function with respect to stress σ and time t. Based on the creep experiment results, the room temperature creep parameters (A, m and n) of base metal specimens for X80 pipeline steel under three loading stresses can be obtained by nonlinear fitting method. Based on Norton power function, the nonlinear fitting curves of room temperature creep of base metal specimens under the different loading stresses was divided into two stages, including the primary stage and the second stage, which was investigated by Wang et al.[16]. Let $D=\lg\left(\dfrac{A}{m+1}\right)$, $B=n$ and $C=m+1$, Eq. (2) is reduced to

$$\lg\varepsilon_c = D + B\lg\sigma + C\lg t \tag{3}$$

For calculation simply, here a general regression of stresses and time model including both the primary stage and the second stage data is obtained. The Gauss-Newton iteration method[47-48], logarithmic theory and multiple nonlinear regression methods are adopted to fit the room temperature creep parameters of base metal specimens under multiple loading stresses, as listed in Table 2. Where R^2 is the determination coefficient of correction, which is used to describe the correlation degree between regression function and experiment data. The closer R^2 is to 1, the higher the reference value of fitting results; the closer R^2 is to 0, the lower the reference value.

Table 2 Time-hardening model parameters

Time-hardening model parameters	Power law multiplier D	Stress exponent B	Time exponent C	R^2
Values	0	28.98	0.131	0.987

The comparison between multiple nonlinear fitting results and experiment results of room temperature curves of base metal under three loading stresses of $1.03R_{p0.2}$ (619MPa), $1.00R_{p0.2}$ (601MPa) and $0.97R_{p0.2}$ (582MPa) are shown in Fig. 2. As can be seen from Fig. 2, the multiple nonlinear fitting results at $0.97R_{p0.2}$ loading stress agree well with the experiment results. On the whole, the multiple nonlinear fitting results under multiple loading stresses have a good coincidence degree with the experiment results. The fitting time-hardening model parameters of base metal are reasonable and acceptable, which can be used for finite element analysis in Section 2.5.

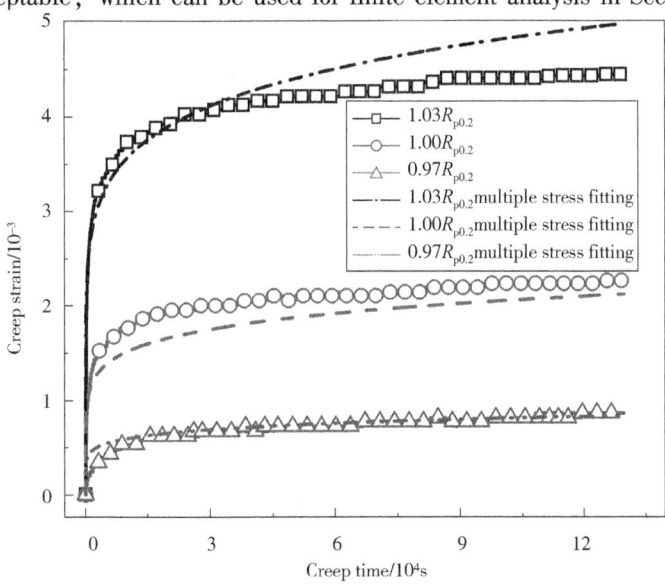

Fig. 2 Comparison between multiple nonlinear fitting results and experiment results of room temperature curves of base metal under three loading stresses of $1.03R_{p0.2}$(619MPa), $1.00R_{p0.2}$(601MPa) and $0.97R_{p0.2}$(582MPa)

2.3 Crack growth resistance curve and critical crack growth load

The specimens from Base part used in fracture toughness experiment were machined as single edge notched beam (SENB), as shown in Fig. 3(a). The three-point bending experiment was proposed, and three identical specimens (Specimen S1, Specimen S2 and Specimen S3) were tested in this experiment based on standard[44]. The geometric model of SENB specimen with side grooves are shown in Fig. 3(b). The length L, height W, thickness B and gauge length S of the SENB specimen were $L = 138$mm, $W = 30$mm, $B = 15$mm, and $S = 120$mm. The detailed fracture toughness experiment setting is shown in Appendix A.

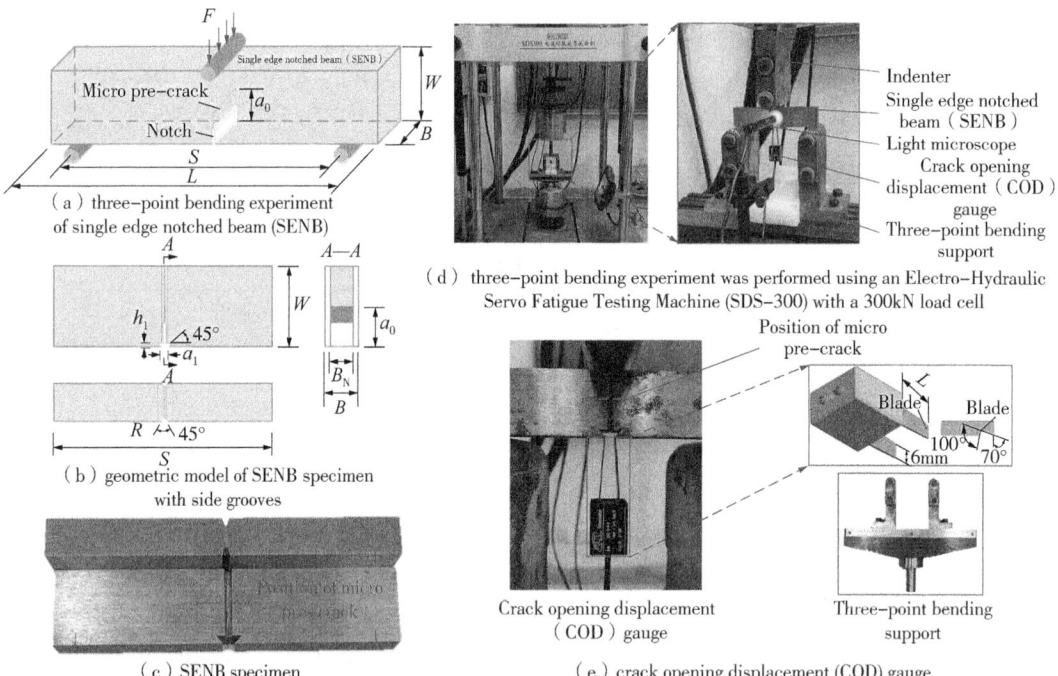

Fig. 3 Fracture toughness experiment

Determination of maximum fatigue crack prefabricated force are given in Appendix B. The determination of three-point bending load in the creep experiment is closely related to the critical CTOD. The experiment results of the crack extension resistance and the creep CTOD for the SENB specimen subjected to three-point bending load of 15kN are shown in Fig. 4. The crack extension resistance curves (R-curves) are plotted, and the critical CTOD $\delta_{0.2}$ is estimated according to the intersection of fitted curve and construction curve. Where Δa is the stable crack extension, Δa_{max} is the crack extension limit for δ-controlled crack extension. δ is the CTOD evaluated using the rotation point formula, δ_g is

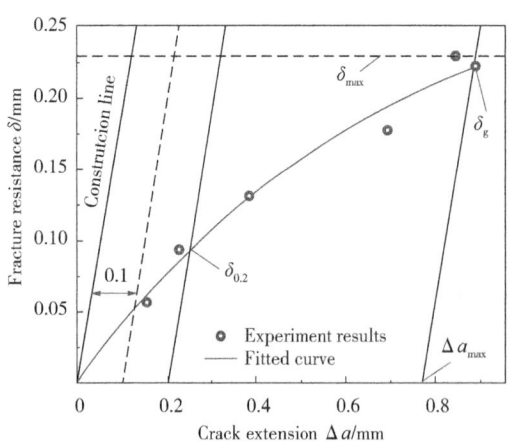

Fig. 4 Crack extension resistance curves (R-curves) and the critical CTOD $\delta_{0.2}$

the δ at the limit of δ-controlled crack extension, $\delta_{0.2}$ is the size insensitive fracture resistance at 0.2mm crack extension offset from construction curve.

2.4 Creep crack opening tip displacement

The CMOD of pre-crack SENB specimens was recorded by COD gauge. The testing principle of CTOD is shown in Fig. 5(a). The relationship between the CMOD and the load is shown in Fig. 5(b). $V_e(t)$, $V_p(t)$ and $V_c(t)$ denoted the elastic CMOD in linear elastic loading, the plastic CMOD in plastic loading and the creep CMOD in creep loading, respectively.

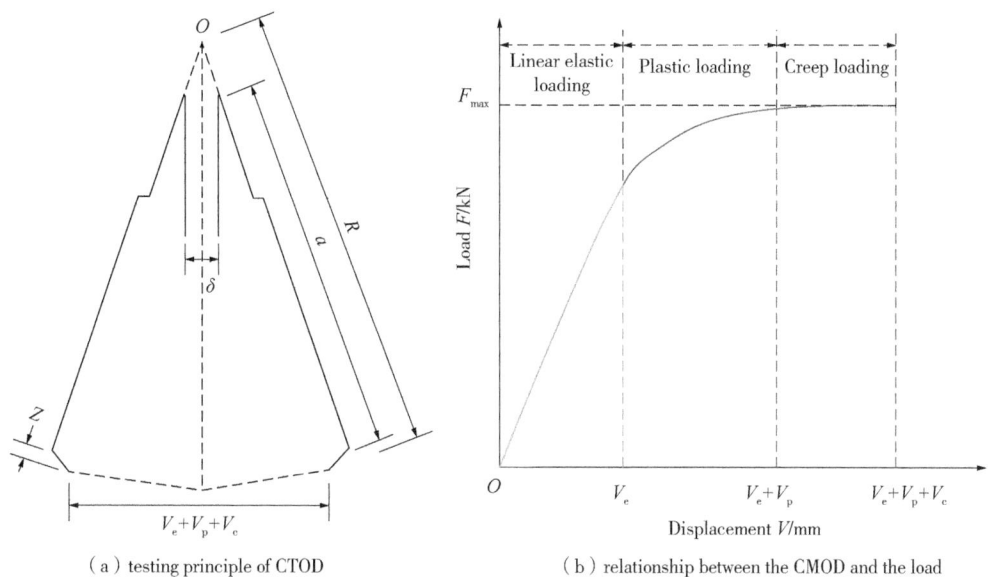

(a) testing principle of CTOD   (b) relationship between the CMOD and the load

Fig. 5 Testing principle of CTOD and relationship between the CMOD and the load

The composed of the CTOD considering the effect of room temperature creep can be stated as follows

$$\delta = \delta_e + \delta_p + \delta_c \tag{4}$$

where δ_e is the elastic CTOD in linear elastic loading, δ_p is the plastic CTOD in plastic loading and δ_c is the room temperature creep CTOD in creep loading. The elastic CTOD in linear elastic loading δ_e and the mode I stress intensity factor K_I are given by

$$\delta_e = K_I^2 \cdot \frac{1-v^2}{2R_{p0.2}E} \tag{5}$$

$$K_I = \left(\frac{S}{W}\right) \frac{F}{(BB_N W)^{0.5}} \times g_1\left(\frac{a_0}{W}\right) \tag{6}$$

where B is the specimen thickness, W is the specimen width, S is the specimen gauge length, B_N is effective thickness of the specimen with side grooves, $R_{p0.2}$ is the specified non-proportional elongation strength of 0.2% perpendicular to the crack plane at the test temperature. F is the three-point bending load, E is the elastic modulus, v is the Poisson's ratio, $g_1(a_0/W)$ is the stress intensity factor coefficient, as given in Eq. (B.3).

The plastic CTOD in plastic loading δ_p and the room temperature creep CTOD in creep loading

δ_c in Eq. (4) can be obtained from the geometrical relationship of specimen after plastic deformation, as shown in Fig. 5(a). When the SENB specimen is subjected to three-point bending load, the crack rotates around the point O at a certain angle. Point O is the rotation center. According to Fig. 5 (a), the geometrical relationship between of the sum of the plastic CTOD term δ_p and the room temperature creep CTOD term δ_c and the sum of the plastic CMOD term V_p and the creep CMOD term V_p is given by

$$\frac{\delta_p+\delta_c}{V_p+V_c}=\frac{R-a-Z}{R} \tag{7}$$

where a is the crack length, R is the rotation radius, Z is the knife edge thickness (Z is positive when the measuring position of CMOD is outside the notch edge, while Z is negative when measuring position of CMOD inside the notch edge). Through the rotation correction of SENB specimen, the rotation radius R can be expressed as[49]

$$\frac{R}{W} = \sum_{i=0}^{6} k_i \left(\frac{a}{W}\right)^i \left[\frac{\left(\frac{a}{W}\right)^2}{1-\frac{a}{W}}\right] \tag{8}$$

where k_i ($i = 0, 1, \cdots, 6$) is the correction coefficient, which are $k_0 = 10.511$, $k_1 = -18.236$, $k_2 = -92.964$, $k_3 = 406.186$, $k_4 = -648.184$, $k_5 = 482.261$, $k_6 = -139.817$ and $a/W = 0.45 \sim 0.55$.

In Eq. (7), $V_c(t)$ denoted the room temperature creep CMOD in creep loading, which is regarded as a power-law function of creep time. The room temperature creep CMOD in creep loading is given by

$$V_c(t)=\beta t^{\alpha+1} \tag{9}$$

where β is the constant related to three-point bending load, α is the time hardening constant. α and β can be obtained by fitting the CMOD-time curve obtained by three-point bending experiment in the load retention process.

Therefore, combining Eqs. (4) ~ (7) and Eq. (9), the CTOD including elastic CTOD, plastic CTOD and creep CTOD of SENB specimens can be expressed as

$$\delta(t)=\left[\left(\frac{S}{W}\right)\frac{F}{(BB_NW)^{0.5}}\times g_1\left(\frac{a_0}{W}\right)\right]^2\left[\frac{(1-v^2)}{2R_{p0.2}E}\right]+\frac{(R-a-Z)(V_p+\beta t^{\alpha+1})}{R} \tag{10}$$

From Eq. (10), the time-dependent CTOD considered the effect of room temperature creep of SENB specimens can be calculated from CMOD measured by COD gauge.

2.5 Numerical calculation of creep CTOD

Based on the three-point bending experiment, the numerical simulation model of SENB specimens with a crack tip is established using the ABAQUS nonlinear finite element software. The numerical simulation model material is assumed to be an isotropic and ideal elastic-plastic material. The creep time-hardening model is adopted in the creep process, which is shown in Eq. (3). The time-hardening model parameters are shown in Table 2. The deformation of the fixture is not considered during the finite element analysis, so the crosshead(indenter) and support fixture in the three-point bending experiment are simplified to three rigid thin-walled cylinders, as shown in Fig. 6(a). The crack tip deformation is mainly composed of three parts: elastic deformation, plastic deformation and creep deformation during creep loading. The true stress and true strain used in calculation and can be obtained from the uniaxial tensile experiment measured the nominal stress and nominal strain, which are depicted as

$$\varepsilon = \ln(1+\varepsilon_{nom}) \tag{11}$$

$$\sigma = \sigma_{nom}(1+\varepsilon_{nom}) \tag{12}$$

where ε_{nom} is the nominal strain and σ_{nom} is the nominal stress. The true plastic strain of the base metal specimens is the difference between the total strain measured in the uniaxial tensile experiment and the elastic strain. The true plastic strain can be indicated by

$$\varepsilon^{pl} = \varepsilon^{t} - \varepsilon^{el} = \varepsilon^{t} - \frac{\sigma}{E} \tag{13}$$

where ε^{pl} is the true plastic strain, ε^{t} is the total strain, ε^{el} is the true elastic strain, σ is the true stress and E is the elastic modulus of the base metal specimens. The eight-node brick element with reduced integration element (C3D8R) is selected and the meshing division of finite element model is shown in Fig 6(b). It is assumed that tangential slip is not allowed at contact surfaces between the indenter and the SENB specimens.

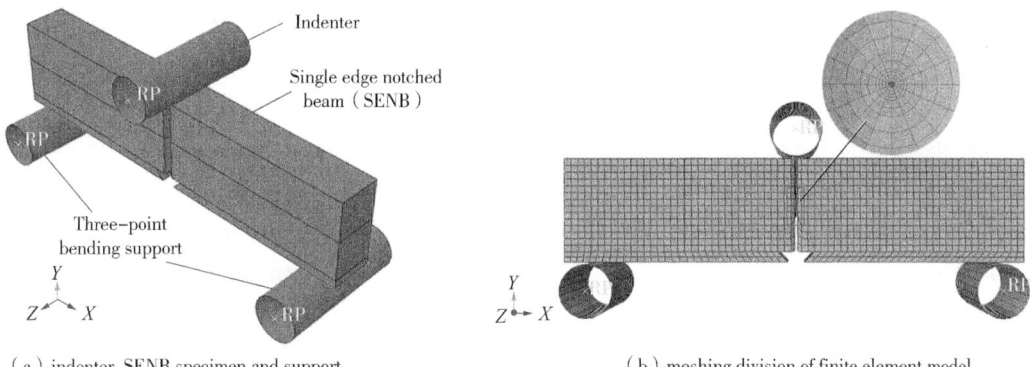

(a) indenter, SENB specimen and support (b) meshing division of finite element model

Fig. 6　Numerical simulation model of SENB specimen

3　Results and discussions

3.1　Creep CTOD of experiments results

The SENB specimens S1, S2 and S3 are loaded by three-point bending experiment and the time-dependent deformation of the crack tip is observed to measure the CMOD of the SENB specimens. The CMOD during the load retention process is obtained by subtracting the CMOD at the beginning of the load retention process from the CMOD at the whole loading process. The CMOD, CTOD, creep CMOD and creep CTOD of SENB specimens S1, S2 and S3 under three-point bending load of 15kN are shown in Fig. 7. The change of the CMOD with load of SENB specimens under the three-point bending load F from 0kN to 15kN is obtained by the room temperature creep experiment [Fig. 7(a)]. The change of the CTOD with load ($F = 0 \sim 15\text{kN}$) is calculated by the relationship between CMOD and CTOD in Eq. (10), which is shown in Fig. 7(b). According to Eq. (9), the fitted power-law function $V_c(t) = 0.01423 \cdot t^{0.10349}$ with constants $\beta = 0.01423$, $\alpha = -0.89651$ and $R^2 = 0.93561$ are obtained by fitting the creep CMOD-time curve obtained by three-point bending experiment in the load retention process [Fig. 7(c)]. The results show that there is a good agreement among the SENB specimens S1, S2 and S3 in room temperature creep experiment. In the load retention process ($F = 15\text{kN}$), the change of the creep CMOD with time is measured by COD gauge. According to Eq. (10), the change of the creep CTOD with time is calculated.

The change of creep CMOD and creep CTOD curves of SENB specimens demonstrates that the room temperature creep at the crack tip is in the creep stage I with the first 30000 seconds. The significant creep rate exists in this creep stage (creep time < 30000s). As the creep progresses, the creep rate decreases and the creep deformation increases slowly by a low rate. The room temperature creep is in the creep stage II (30000s < creep time < 129600s). After 129600s of room temperature creep, the creep CMODs of the three specimens are 0.047mm, 0.049mm and 0.045mm, respectively [Fig. 7(c)]. The deformation in the creep stage I accounts for more than 90% of the whole experiment. Based on the creep CMOD measurement results, the calculated creep CTODs of the SENB specimens in the 15kN load retention are 0.00843mm, 0.00874mm and 0.00783mm after 129600s of room temperature creep, as shown in Fig. 7(d). Based on Eq. (10), the CTOD caused by room temperature creep of the SENB specimens in the 15kN load retention is 0.0098mm. According to previous research[50-51], the crack initiation toughness of X80 pipeline steel at room temperature is 0.094mm, which indicates that the increment of CTOD under 15kN load exceeds 10% of the crack initiation toughness. Therefore, the effect of room temperature creep on the passivation of crack tip for the X80 pipeline steel is not negligible. As the creep time at room temperature increases, the passivation zone of crack tip tends to be saturated. When the CTOD exceeds the crack initiation toughness, the service performance of X80 pipeline steel is affected due to crack propagation.

In order to determine whether the CTOD is due to creep or crack propagation, fatigue crack tip propagation of the SENB specimens is observed by a light microscope, as shown in Fig. 8. It can be founded that the crack tip length remains the same after 129600s room temperature creep, and the crack tip edge remains at $0.55W$. Therefore, it is obvious that the increasing crack opening displacement during creep process is caused by the time-dependent deformation at the crack tip.

3.2 Creep CTOD of numerical results

The stress distribution of the SENB specimen is observed after the room temperature creep numerical simulation. The numerical simulation model of the SENB specimen in creep stage is divided along the pre-crack surfaces to capture the stress nephogram at 15kN for 36 hours (129600 seconds), as shown in Fig. 9. The plastic zone of SENB specimen is fishtailed when subjected to three-point bending load. The stress near the crack surface increases with loading time, and the maximum stress exists at the crack tip. The plastic deformation occurs at the edge of the side groove firstly, as shown in Fig.9(b).

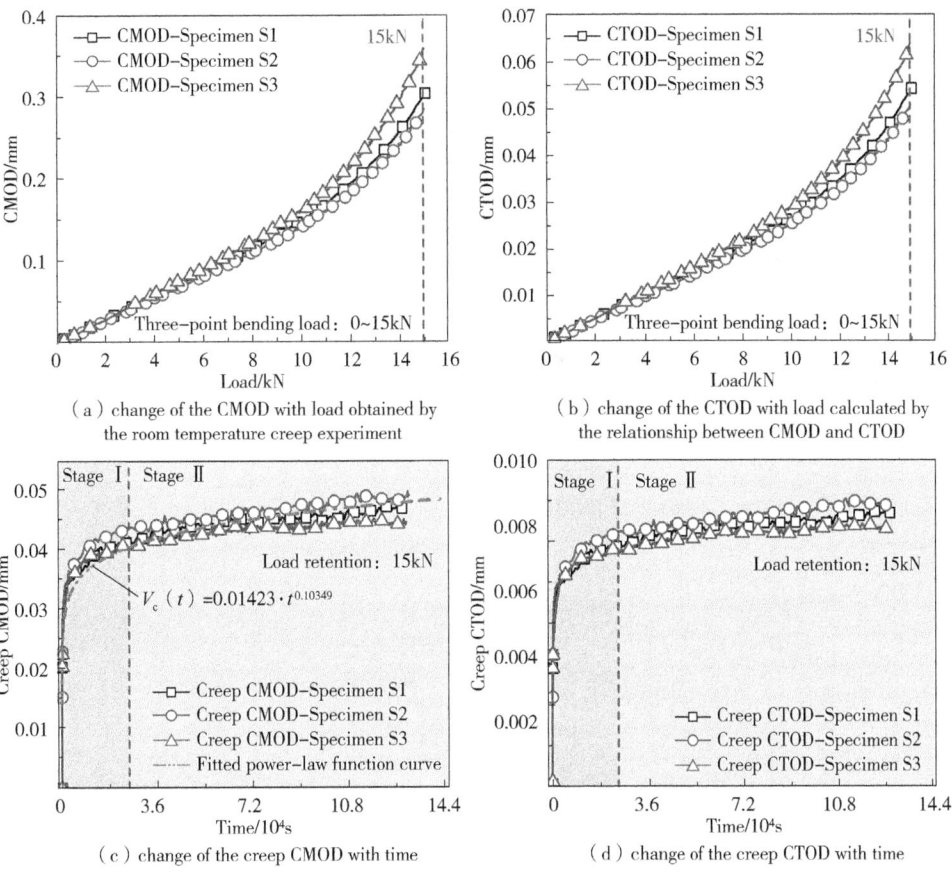

Fig.7 CMOD, CTOD, creep CMOD and creep CTOD of SENB specimens S1, S2 and S3 under three-point bending load of 15kN

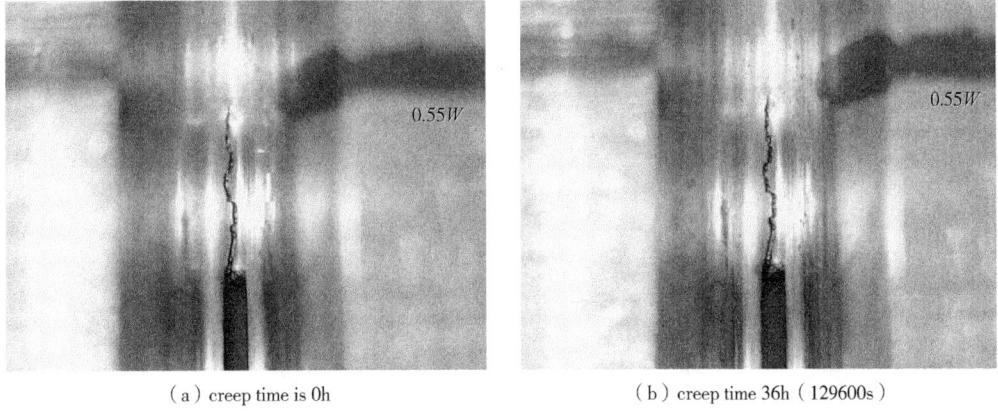

(a) creep time is 0h (b) creep time 36h (129600s)

Fig.8 Variation of fatigue crack tip propagation in the room temperature creep process

The variations of the CMOD (X-direction) of the SENB specimen subjected to three-point bending load of 15kN and loading time of 0s and 29600s at the room temperature creep stage are demonstrated in Fig.10. The numerical simulation results indicate that the room temperature creep behavior occurs at the crack tip. The X-directional displacement of the SENB specimen is symmetric about the crack face. When the crack tip is taken as the center of rotation, the X-directional displacement gradually increases with loading time along the SENB height direction. The X-directional displacement of SENB specimen reaches 0.21mm when the loading time reaches 29600s, which increases 0.03mm compared with the initial displacement.

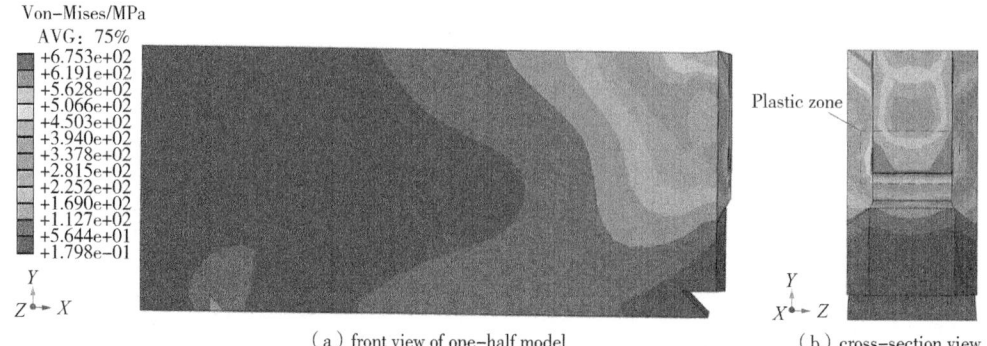

Fig. 9 Stress nephogram of the SENB specimen in creep stage at 15kN for 36 hours

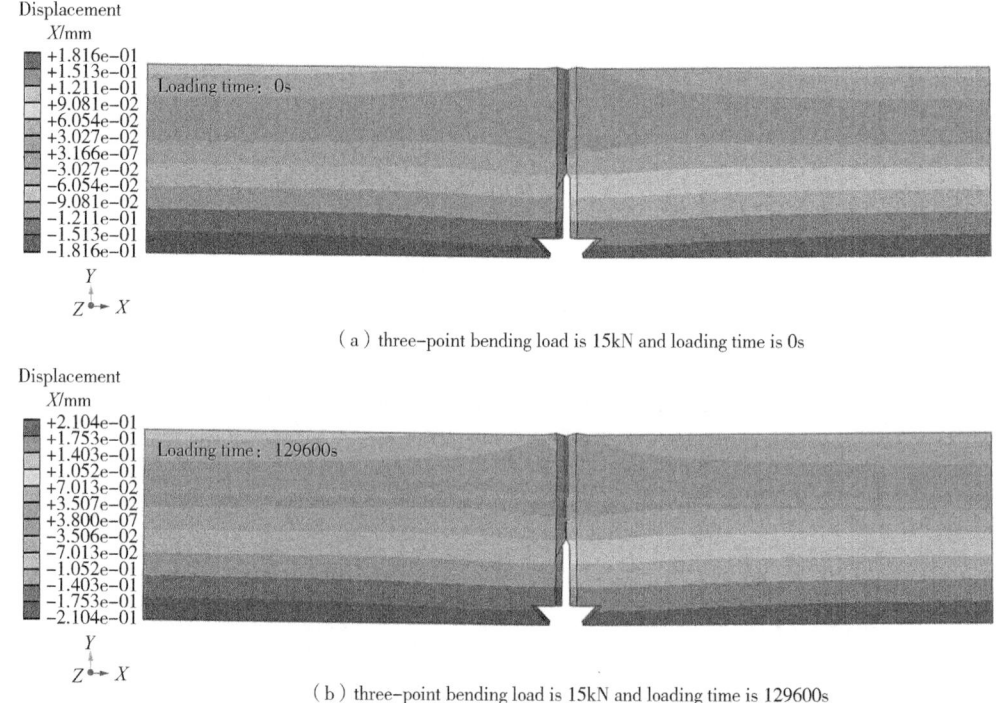

Fig. 10 Variation of CMOD(X-direction) of the SENB specimen at the room temperature creep stage

The comparisons between numerical simulation results and experiment results of the crack extension resistance and the creep CTOD for the SENB specimen subjected to three-point bending load of 15kN are shown in Fig. 4 and Fig. 11. It can be found that the numerical simulation results are in good agreement with the experiment results. According to previous research, the determination of three-point bending load in the creep experiment is closely related to the critical CTOD[52]. The creep CTOD of the numerical simulation result is 0.0098mm when loading time is 29600 s, which is greater than that of experiment result of 0.0084mm, as shown in Fig. 11. The slight load fluctuation under the sustained experiment could lead to a deviation.

3.3 Effect of load level on creep CTOD

The SENB specimens with same dimensions are loaded to 14.5kN, 15kN and 15.5kN at the same rate under room temperature condition, and loads are maintained for 36 hours. The effect of different loads on creep CTOD is illustrated in Fig. 12. The obtained corresponding creep CTODs of SENB specimens loaded to 14.5kN, 15kN and 15.5kN are 0.0065mm, 0.0086mm and 0.01195mm, respectively. The increment of creep CTOD at 15.5kN is twice as large as that at 14.5kN, which exhibits the strong load sensitivity. As the stress level increases, the rate of room temperature also increases and the deceleration creep stage continues longer.

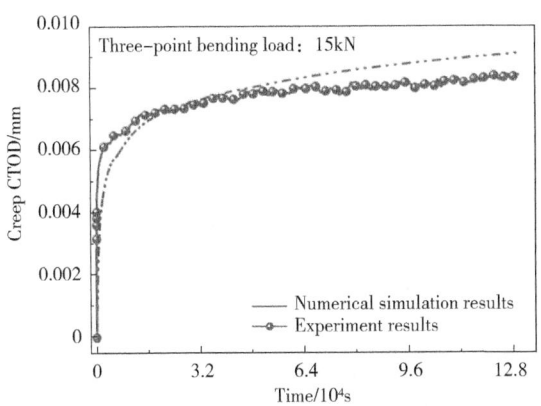

Fig. 11 Comparison between numerical simulation and experiment results of creep CTOD curve for the SENB specimen subjected to three-point bending load of 15kN

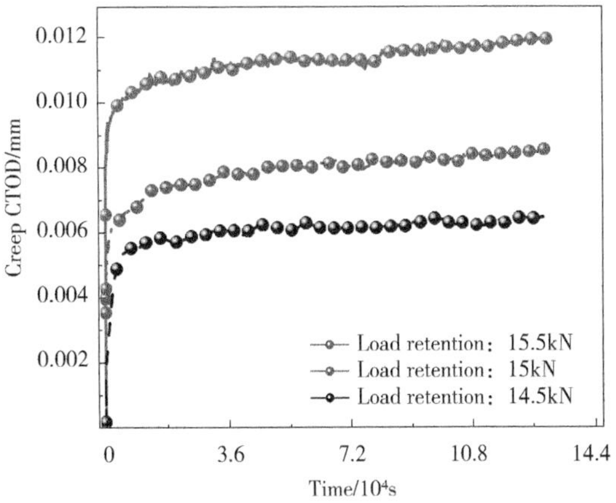

Fig. 12 Effect of different loads on creep CTOD of SENB specimens with same dimensions

3.4 Effect of load history on creep CTOD

After the SENB specimens are loaded to 15kN firstly, room temperature creep data is recorded. Then, the same SENB specimens are unload and reloaded to 15kN again for 36 hours in the room temperature creep experiment. The creep CTOD of the same SENB specimens after the second loaded to 15kN in the room temperature creep experiment is shown in Fig. 13. The effect of load history on creep CTOD is investigated based on the room temperature creep experiment. The creep CTOD of the SENB specimens after the second loading is 0.0040mm, which is one-third of the first loading of 0.0121mm. This indicates that the room temperature creep deformation of SENB specimens is significantly weakened when it is subjected to constant loading again. The creep time of the reloaded specimens from the creep stage I to the creep stage II is significantly shortened and the creep rate is reduced, as shown in Fig. 13.

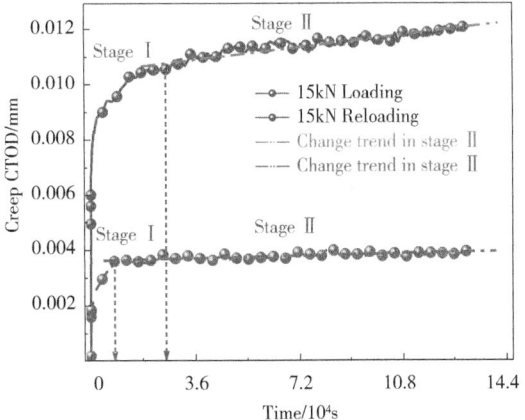

Fig. 13 Creep CTOD of the same SENB specimens after the second loaded to 15kN in the room temperature creep experiment

According to the movable dislocation theory of creep[10,53-54], the main reason for plastic deformation of metal materials is due to the slippage of movable dislocations within the material, the density of movable dislocations within the pipeline steel decreases after the first loading, and the immobile dislocations increase during the second loading creep, thus showing a significant reduction in the amount of room temperature creep.

4 Conclusion

The room temperature creep behavior of X80 pipeline steel has attracted increasing attention because it is considered to be the main cause of failure in pipeline steel. In order to investigate the creep CTOD of X80 pipeline steel, the room temperature creep experiments of the SENB specimens subjected to three-point bending retention load for 36 hours at room temperature are performed in this study. The room temperature creep of the crack tip of X80 pipeline steel is investigated by observing the variation of the CMOD at the bottom of the SENB specimens with time. The change of creep CTOD at the retention load of 15kN is analyzed and the effects of different loads and load history on the room temperature creep of the crack tip for X80 pipeline steel are analyzed by changing the load and loading process. The main conclusions in this study are as follows:

(1) A typic creep curve consists the primary creep stage and stable creep stage and the creep deformation is mainly reflected in the primary creep stage. The creep deformation in the deceleration stage accounts for more than 90% of the whole creep in the room temperature creep experiment.

(2) Based on the traditional CTOD theoretical model, a modified CTOD theoretical model is presented to calculate the creep CTOD from the CMOD measurement results. A power-law function fitted from experiment results is also taken forward into the numerical simulation model. The numerical simulation results are in good agreement with the experiment results.

(3) The increment of room temperature creep CTOD is closed to 10% of the crack initiation toughness, which indicates that it affects the passivation of the crack tip of X80 pipeline steel. As the creep time at room temperature increases, the passivation zone of crack tip tends to be saturated. When the CTOD exceeds the crack initiation toughness, the service performance of X80 pipeline steel is affected due to crack propagation.

(4) The creep CTOD of X80 pipeline steel has the strong load sensitivity. The creep deformation and CTOD increase with load. The room temperature creep deformation of is significantly weakened when it is subjected to constant loading again. The room temperature creep deformation and creep rates of SENB specimens subjected to second constant load are significantly reduced compared with that of the first constant load. The mechanism of these behaviors is related to the yield zone at crack tip. The yield zone becomes larger with the increase of load and the creep CTOD also becomes larger. The room temperature creep mainly depends on movable dislocation at yield zone. Therefore, the creep deformation deeply reduces with the decrease of the movable dislocation of SENB specimens subjected to second constant load.

Declaration of competing interest

The authors declare that they have no known competing financial interests or personal relationships that could have appeared to influence the work reported in this paper.

Acknowledgements

This work is supported by the National Natural Science Foundation of China (grant numbers: 11572253, 11972302, 12102399 and 12202407), Fundamental Research Program of Shanxi Province (grant numbers: 20210302124263 and 20210302124383), Open Research Fund for Underground Target Damage Technology National Defense Key Discipline Laboratory, North University of China (grant numbers: DXMBJJ2021 − 03). The authors would like to gratefully acknowledge these supports.

Appendix A. Detailed fracture toughness experiment setting

The width of the lower edge for the notch and the angle of the side grooves on both sides were $a_1 = 5mm$ and $45°$, respectively. The thickness and depth of the side grooves on both sides of specimen were $0.1B$ ($B_N = 0.8B$) and $h_1 = 3mm$. B_N is effective thickness of the specimen with side grooves, and $B_N = B$ when the specimen without side grooves[49]. The fillet radius at the side groove bottom was $R = 0.5mm$. The angle between the notch cut surface and the horizontal plane was $45°$ [Fig. 3(b)]. The initial crack length of the notch was $13.5mm$ ($0.45W$). The SENB specimen and the position of micro pre−crack are shown in Fig. 3(c). The micro pre−crack ($0.10W$) was then further prefabricated in the location of notch tip by an Electro − Hydraulic Servo Fatigue Testing Machine (SDS−300, Sinotest Equipment Co., Ltd., Changchun, China) with a 300kN load cell (accuracy: ±0.5%), as shown in Fig. 3(d). So the length of original fatigue crack is $a_0 = 0.45W + 0.10W$[49]. The equipment of three − point bending experiment mainly included indenter, SENB specimen, light microscope, crack opening displacement (COD) gauge and support. The fatigue crack tip propagation of the SENB specimens was observed by a light microscope (B011, SuperEyes Electronics Technology Co., Ltd., Guangzhou, China) so that the leading crack front did not exceed $16.5mm$ ($0.55W$). The COD gauge (Sinotest Equipment Co., Ltd., Changchun, China) with gauge length of 5mm and measuring range of [−1mm, 4mm] was used to measure and record the opening displacement of notch during the dwell loading. The dimensions of the COD gauge were shown in Fig. 3(e). Since the loading time was short compared to the creep time, the acquisition frequency of the loading stage was 0.2Hz, and the data was collected each 60s during the sustaining load stage. The fatigue stress ratio was set as 0.1, the frequency of fatigue pre−cracking was 5Hz.

Appendix B. Determination of maximum fatigue crack prefabricated force

Since the fatigue crack of the specimen is prefabricated, the minimum prefabricated fatigue crack growth shall be greater than 1.3mm or 2.5%W, whichever is greater. When the crack growth is 1.3mm, the corresponding maximum fatigue crack prefabricated force is[42,49]

$$F_f^{1.3} = 0.8 \times \frac{B(W-a_0)^2}{S} \times R_{p0.2} \tag{B.1}$$

where B is the specimen thickness, W is the specimen width, S is the specimen gauge length, a_0 is the initial crack length, $R_{p0.2}$ is the specified non−proportional elongation strength of 0.2%

perpendicular to the crack plane at the test temperature.

While when the crack growth is 2.5% W, the corresponding maximum fatigue crack prefabricated force is[49]

$$F_f^{2.5\%W} = \xi \times E \left[\frac{(W \times B \times B_N)^{0.5}}{g_1\left(\frac{a_0}{W}\right)} \right] \left(\frac{W}{S}\right) \quad (B.2)$$

where $\xi = 1.6 \times 10^{-4} \, m^{1/2}$, E is the Young's modulus of specimen, B_N is effective thickness of the specimen with side grooves, and $g_1(a_0/W)$ is the stress intensity factor coefficient and its expression is[49]

$$g_1\left(\frac{a_0}{W}\right) = \frac{3\left(\frac{a_0}{W}\right)^{0.5} \left[1.99 - \left(\frac{a_0}{W}\right)\left(1 - \frac{a_0}{W}\right)\left(2.15 - \frac{3.93a_0}{W} + \frac{2.7a_0^2}{W^2}\right)\right]}{2\left(1 + \frac{2a_0}{W}\right)\left(1 - \frac{a_0}{W}\right)^{1.5}} \quad (B.3)$$

The minimum value between Eq. (B.1) and Eq. (B.2) can be taken as the maximum fatigue crack prefabricated force.

References

[1] E. Ohaeri, U. Eduok, J. Szpunar. Hydrogen related degradation in pipeline steel: A review. International Journal of Hydrogen Energy, 2018, 43(31): 14584-14617.

[2] R. Pourazizi, M. A. Mohtadi-Bonab, J. A. Szpunar. Investigation of different failure modes in oil and natural gas pipeline steels. Engineering Failure Analysis, 2020, 109: 104400.

[3] D. Zhou, T. Li, D. Huang, Y. Wu, Z. Huang, W. Xiao, Q. Wang, X. Wang. The experiment study to assess the impact of hydrogen blended natural gas on the tensile properties and damage mechanism of X80 pipeline steel. International Journal of Hydrogen Energy, 2021, 46(10): 7402-7414.

[4] Y. Shuai, X. H. Wang, C. Feng, Y. Zhu, C. L. Wang, T. Sun, J. Han, Y. F. Cheng. A novel strain-based assessment method of compressive buckling of X80 corroded pipelines subjected to bending moment load. Thin-Walled Structures, 2021, 167: 108172.

[5] B. Liu, Z. Li, X. Yang, C. Du, X. Li. Microbiologically influenced corrosion of X80 pipeline steel by nitrate reducing bacteria in artificial Beijing soil. Bioelectrochemistry, 2020, 135: 107551.

[6] T. An, H. Peng, P. Bai, S. Zheng, X. Wen, L. Zhang. Influence of hydrogen pressure on fatigue properties of X80 pipeline steel, International Journal of Hydrogen Energy, 2017, 42(23): 15669-15678.

[7] B. B. Zand, A. Steiner. Effect of room temperature creep on hydrostatic leak test. Proceedings of the 2018 12th International Pipeline Conference. Volume 3: Operations, Monitoring, and Maintenance; Materials and Joining. Calgary, Alberta, Canada, 2018: V003T04A047.

[8] Y. Chen, H. Zhang, J. Zhang, X. Liu, X. Li, J. Zhou. Failure assessment of X80 pipeline with interacting corrosion defects. Engineering Failure Analysis, 2015, 47: 67-76.

[9] S. Toyoda, S. Goto, T. Okabe, Y. Kato, S. Igi, T. Inoue, M. Egi. Mechanical properties of newly developed API X80 grade HFW linepipe for long-term exposure at elevated temperature. Proceedings of the ASME 2014 33rd International Conference on Ocean, Offshore and Arctic Engineering. Volume 5: Materials Technology; Petroleum Technology. San Francisco, California, USA, 2014: V005T03A022.

[10] P. Wang, W. Hao, J. Xie, J. Ding, F. Wang, C. Huo. Primary creep X80 pipeline steel at room temperature

using molecular dynamics simulation. Applied Physics A: Materials Science & Processing, 2022, 128(3): 204.

[11] D. F. Nie, J. Zhao. Room temperature creep and its effect on fatigue crack growth in a X70 steel with various microstructures. Key Engineering Materials, 2007, 353-358: 138-141.

[12] O. Nassif, T. J. Truster, R. Ma, K. B. Cochran, D. M. Parks, M. C. Messner, T. L. Sham. Combined crystal plasticity and grain boundary modeling of creep in ferritic - martensitic steels: I. Theory and implementation. Modelling and Simulation in Materials Science and Engineering, 2019, 27(7): 075009.

[13] A. A. Ayubali, A. Singh, B. P. Shanmugavel, K. A. Padmanabhan. A phenomenological model for predicting long-term high temperature creep life of materials from short-term high temperature creep test data. International Journal of Mechanical Sciences, 2021, 202-203: 106505.

[14] S. H. Wang, Y. Zhang, W. Chen. Room temperature creep and strain-rate-dependent stress-strain behavior of pipeline steels. Journal of Materials Science, 2001, 36: 1931-1938.

[15] J. Cao, K. Wang, W. Ma, J. Ren, H. Nie, W. Dang, X. Liang, T. Yao, X. Zhao. Indentation creep deformation behavior of local zones for X70 girth weld. International Journal of Pressure Vessels and Piping, 2022, 199: 104776.

[16] P. Wang, J. R. Zhi, W. Q. Hao, J. M. Xie, F. H. Wang, C. Y. Huo. Room temperature creep behaviors of base metal and welding materials for X80 pipeline steel. Materials Science and Engineering A, 2022, 856: 144038.

[17] Y. Du, J. Y. Richard Liew, J. Jiang, G. Q. Li. Improved time-hardening creep model for investigation on behaviour of pre-tensioned steel strands subject to localised fire. Fire Safety Journal, 2020, 116: 103191.

[18] W. Q. Hao, J. M. Xie, F. H. Wang. Theoretical prediction for large deflection with local indentation of sandwich beam under quasi-static lateral loading. Composite Structures, 2018, 192: 206-216.

[19] W. Q. Hao, J. M. Xie, F. H. Wang, Z. F. Liu, Z. H. Wang. Analytical model of thin-walled corrugated tubes with sinusoidal patterns under axial impacting. International Journal of Mechanical Sciences, 2017, 128-129: 1-16.

[20] W. Q. Hao, J. M. Xie, F. H. Wang. Theoretical prediction of the progressive buckling and energy absorption of the sinusoidal corrugated tube subjected to axial crushing. Computers & Structures, 2017, 191: 12-21.

[21] W. Q. Hao, J. M. Xie. Reducing diffusion-induced stress of bilayer electrode system by introducing pre-strain in lithium-ion battery. Journal of Electrochemical Energy Conversion and Storage, 2021, 18(2), 20909.

[22] W. Q. Hao, J. M. Xie, F. H. Wang. The indentation analysis triggering internal short circuit of lithium-ion pouch battery based on shape function theory. International Journal of Energy Research, 2018, 42(11): 3696-3703.

[23] W. Q. Hao, J. M. Xie, X. Q. Bo, F. H. Wang. Resistance exterior force property of lithium-ion pouch batteries with different positive materials. International Journal of Energy Research, 2019, 43(9): 4976-4986.

[24] Y. G. Cao, Y. Zhen, M. Song, H. J. Yi, F. G. Li, X. Y. Li. Determination of Johnson-Cook parameters and evaluation of Charpy impact test performance for X80 pipeline steel. International Journal of Mechanical Sciences, 2020, 179: 105627.

[25] W. Q. Hao, P. Zhang, J. M. Xie, M. Y. Hou, Z. J. Wang, X. F. Bai. Investigation of impact performance of perforated plates and effects of the perforation arrangement and shape on failure mode. Engineering Failure Analysis, 2022, 140: 106638.

[26] X. D. Gao, Y. B. Shao, C. Chen, H. M. Zhu, K. S. Li. Experimental and numerical investigation on transverse impact resistance behaviour of pipe-in-pipe submarine pipelines after service time. Ocean Engineering, 2022, 248: 110868.

[27] P. Cui, W. Guo. Crack-tip-opening-displacement-based description of three-dimensional elastic-plastic

crack border fields. Engineering Fracture Mechanics, 2020, 231: 107008.

[28] O. Hembara, O. Y. Chepil. Modeling of the deformation of structural elements under the conditions of creep, corrosion cracking, and hydrogenation. Materials Science, 2022, 57: 557-561.

[29] J. Čadek, V. Šustek, M. Pahutová. An analysis of a set of creep data for a 9Cr-1Mo-0.2V (P91 type) steel. Materials Science & Engineering A, 1997, 225(1-2): 22-28.

[30] ISO 6892-1: 2019 Metallic materials-Tensile testing-Part 1: Method of test at room temperature. International Organization for Standardization: London, UK, 2019.

[31] C. Liu, P. Liu, Z. B. Zhao, D. O. Northwood. Room temperature creep of a high strength steel. Materials & Design, 2001, 22(4): 325-328.

[32] J. Zhao, T. Mo, D. F. Nie. The occurrence of room-temperature creep in cracked 304 stainless steel specimens and its effect on crack growth behavior. Materials Science and Engineering: A, 2008, 483-484: 572-575.

[33] M. E. Kassner, P. Geantil, R. S. Rosen. Ambient temperature creep of type 304 stainless steel. Journal of Engineering Materials and Technology, 2011, 133(2): 021012.

[34] Y. Yang, L. Shi, Z. Xu, H. Lu, X. Chen, X. Wang. Fracture toughness of the materials in welded joint of X80 pipeline steel. Engineering Fracture Mechanics, 2015, 148: 337-349.

[35] T. An, S. Zhang, M. Feng, B. Luo, S. Zheng, L. Chen, L. Zhang. Synergistic action of hydrogen gas and weld defects on fracture toughness of X80 pipeline steel. International Journal of Fatigue, 2019, 120: 23-32.

[36] R. Ghajar, G. Mirone, A. Keshavarz. Ductile failure of X100 pipeline steel – experiments and fractography. Materials & Design, 2013, 43: 513-525.

[37] S. Xu, W. R. Tyson. Effects of strain rate on strength, and of orientation on toughness, of modern high-strength pipe steels. Journal of Pipeline Engineering 2015, 14(3): 211-224.

[38] L. Lu, S. Wang, G. Tong. Relationship between incremental J integral and crack tip opening angle in elastic plastic materials. European Journal of Mechanics-A/Solids, 2019, 75: 399-409.

[39] J. Y. Yoo, S. S. Ahn, D. H. Seo, W. H. Song, K. B. Kang. New development of high grade X80 to X120 pipeline steels. Materials and Manufacturing Processes, 2011, 26(1): 154-160.

[40] Y. J. Chao, J. D. Ward, R. G. Sands. Charpy impact energy, fracture toughness and ductile-brittle transition temperature of dual-phase 590 Steel. Materials & Design, 2007, 28(2): 551-557.

[41] N. Blanco, D. Trias, S. T. Pinho, P. Robinson. Intralaminar fracture toughness characterisation of woven composite laminates. Part I: Design and analysis of a compact tension (CT) specimen. Engineering Fracture Mechanics, 2014, 131: 349-360.

[42] P. Wang, W. Hao, J. Xie, F. He, F. Wang, C. Huo. Stress triaxial constraint and fracture toughness properties of X90 pipeline steel. Metals, 2022, 12(1): 72.

[43] A. Poursartip, A. Gambone, S. Ferguson, G. Fernlund. In-situ SEM measurements of crack tip displacements in composite laminates to determine local G in mode I and II. Engineering Fracture Mechanics, 1998, 60(2): 173-185.

[44] API SPEC 5L Specification for Line Pipe (45th edition). American Petroleum Institute, Washington DC, 2012.

[45] ISO 204: 2018. Metallic materials-Uniaxial creep testing in tension-Method of test. International Organization for Standardization: London, UK, 2018.

[46] M. L. Zheng, L. L. Han, Z. P. Qiu, H. Y. Li, Q. L. Ma, F. Che. Simulation of permanent deformation in high-modulus asphalt pavement using the Bailey-Norton creep law. Journal of Materials in Civil Engineering, 2016, 28(7): 4016020.

[47] J. Bonet, R. D. Wood. Nonlinear Continuum Mechanics for Finite Element Analysis (2nd edition). Cambridge: Cambridge University Press, 2008.

[48] R. S. Motta, H. L. D. Cabral, S. M. B. Afonso, R. B. Willmersdorf, N. Bouchonneau, P. R. M. Lyra, E. Q. de Andrade. Comparative studies for failure pressure prediction of corroded pipelines. Engineering Failure Analysis, 2017, 81: 178-192.

[49] ISO Standard 12135: 2016(E). Metallic Materials-Unified Method of Test for the Determination of Quasistatic Fracture Toughness: International Organization for Standardization: London, UK, 2016.

[50] X. Li, Z. Ding, C. Liu, S. Bao, Z. Gao. Evaluation and comparison of fracture toughness for metallic materials in different conditions by ASTM and ISO standards. International Journal of Pressure Vessels and Piping, 2020, 187: 104189.

[51] ASTM E1820-17(a). Standard Test Method for Measurement of Fracture Toughness. ASTM International, West Conshohocken, PA, 2017.

[52] L. K. Lu, S. N. Wang. Relationship between crack growth resistance curves and critical CTOA. Engineering Fracture Mechanics, 2017, 173: 146-156.

[53] J. Jiang, Z. Liu, Q. Gao, H. Zhang, A. Hao, F. Qu, X. Lin, H. Li. The effect of isothermal aging on creep behavior of modified 2.5Al alumina-forming austenitic steel. Materials Science and Engineering: A, 2020, 797: 140219.

[54] W. Chen, H. Zhu, S. Wang. Low temperature creep behaviour of pipeline steels. Canadian Metallurgical Quarterly, 2009, 48(3): 271-283.

本论文原发表于《International Journal of Fatigue》2024 年第 178 卷。

Prediction of Collapsing Strength of High Collapse-resistant Casing Based on Machine Learning

Wang Peng[1]   Zhong Chengxu[2]   Fan Shuai[1]   Li Dongfeng[1]
Zhang Shengyue[3]   Liu Peihang[3]   Ji Yu[3]

(1. State Key Laboratory for Performance and Structure Safety of Petroleum Tubular Goods and Equipment Materials, CNPC Tubular Goods Research institute;
2. Shale Gas Research Institute, PetroChina Southwest Oil & Gasfield Company;
3. School of Electronic Engineering, Xi'an Shiyou University)

Abstract: With the increasing complexity of shale gas extraction conditions, a large number of high collapse-resistant casing is applied to the extraction of unconventional oil and gas resources. There are errors in the traditional API collapse strength formula. A high precision and low computational cost model is needed for predicting the strength of high collapsible casing. The key influencing factors of casing anti-collapse strength was determined as outer diameter, wall thickness, yield strength, ovality, wall thickness unevenness, and residual stress by analyzing the casing collapse mechanism. In response to the key factors mentioned above, a dataset was formed by measuring the geometric parameters of the full size casing and collecting data on the results of the anti collapse strength experiment, which divided into a training set (70%) and a testing set (30%). Three machine learning algorithms, neural network, random forest, and support vector machine, were trained to predict the anti-extrusion strength. By used the correlation coefficient R^2, root mean square error RMSE, and average relative MRE to evaluate indexes for model preference evaluation. The results show that machine learning algorithms has unique advantages in casing anti-collapsing strength prediction. In which the neural network prediction model has the best prediction effect, and its characteristics of high precision, low cost and high efficiency are more suitable for the prediction of casing extrusion strength. It's testing set R^2 is 0.9733, RMSE is 0.0267 and MRE is 0.0782, and the prediction accuracy can reach 92.2% which much higher than the API calculation result (63.3%). The network prediction model is suitable for casing anti-collapsing strength prediction and meets the actual prediction requirements.

Keywords: Prediction of collapsing strength; high-strength collapse-resistant casing; Machine learning

Corresponding author: Wang Peng, wangpeng008@cnpc.com.cn, 86-29-81887665.

1 Introduction

With the increasingly harsh mining conditions, the casing deformation problem has seriously constrained the shale gas extraction process[1-3]. To avoid a series of economic losses caused by casing damage due to extrusion, high collapse-resistant casing is beginning to be used in large quantities in unconventional oil and gas extraction. At present, the strength prediction for high-grade steel casing is based on the API formula. Meanwhile, there are also related experts who use the combination of theoretical analysis and numerical simulation to make the prediction, but its reliability and accuracy need to be further improved. API 5C3 specification of the strength formula is mainly based on experimental data and elastic mechanics and other related theories derived from the column, so in practical application is vulnerable to the pipe manufacturing processing deviations and other factors constraints[4-5]. The strength design of high collapse-resistant casing in complex geological environments is over or under-designed due to deviations in existing specifications, which cannot effectively guide the practical application of high collapse-resistant casing. In order to avoid under or over design due to such problems, there is an urgent need for a comprehensive and accurate prediction model to realize accurate prediction of the extrusion strength of high collapse-resistant casing.

In recent years, driven by advancements in big data and artificial intelligence technologies, the concept of data-driven modeling for intelligent decision-making has been increasingly embraced across various fields[6]. From the early days of mathematically-driven technologies based on statistics to later advanced machine learning algorithms such as neural networks and deep learning, data-driven technologies have found successful applications in various industries[7-9]. Machine learning algorithms can train models based on historical data and circumstances without knowing the underlying physical mechanisms, enabling accurate predictions of models[10-12]. Domestically, scholars have successively integrated it with oilfield big data for relevant research, and the technology has been applied in multiple fields including the oil and gas industry[13-14]. Kriti Yadav et al.[15] provided an overview of the applications of machine learning and artificial intelligence in the upstream sector of the oil and gas industry and other fields, highlighting the significant importance of machine learning-based intelligent systems in mitigating risks and reducing maintenance costs in the oil and gas industry. Martirosyan et al.[16] utilized a supercomputer to analyze the temperature field behavior of reservoirs and illustrated the differences in modeling quality and implementation costs between supercomputers and conventional personal computers. Auwalu I. Mohammed et al.[17] introduced a novel approach for studying casing structural integrity by combining Finite Element Analysis(FEA) and machine learning techniques. By employing multi-parameter combined loading, they established relationships among various parameters, uncovering the effects of these parameters on stress, displacement, and casing safety factor. Yan et al[18] explored the casing life prediction method through the casing damage factors, and finally achieved the casing service life prediction with the help of support vector machine modeling after comparing 32 influencing parameters. Wang[19] combined the key parameters influencing the comprehensive quality of oil casing, built an assessment model with BP neural network to effectively utilize the original island data to predict the possible failure of oil casing. Zhang et al[20] contrasted BP neural network with Bayesian neural network in the set damage prediction study, and after comparing the prediction results, they found that the Bayesian neural network has certain advantages in handling the overfitting issues, and has higher prediction accuracy compared to the traditional model. Zhang J et al[21] proposed a data-based method for predicting the risk of casing failure at well and reservoir granularity, The predictive model can achieve an accuracy of 92.45%, which is much higher than the traditional model prediction accuracy. Qin Feng Di et al[22] introduced an artificial intelligence method based on Support Vector Machine(SVM) to predict the maximum stress of eccentric casing under non-uniform in-situ stress, this method can be effectively used to predict the maximum stresses in eccentric casing under complex downhole conditions with lower cost and higher accuracy than finite element simulations. In the study of oil casing performance and casing damage, considering the diversity and randomness of casing damage factors, it is therefore feasible to use the machine learning method to realize the prediction of casing collapse strength by combining the measured data.

By analyzing the factors influencing the crush resistance, the key parameters were determined. A full-size physical test is designed to obtain the collapse strength of casing of each specification, from which a data set is established, and then the model training is carried out using a machine learning approach. Thus, the best prediction model is optimally evaluated to achieve accurate prediction of casing strength against external collapse.

2 Analysis of influencing factors and acquisition of experimental data

2.1 Analysis of factors influencing the strength of collapsing resistance

The API 5C3 standard developed by API has guided many international oil companies to conduct casing collapse strength research since the mid-1970s. The calculation of API anti-collapse is different from the general anti-collapse calculation formula, which is based on the difference in the limits of the sample diameter-thickness ratio to determine the anti-collapse model under different conditions, to calculate the anti-collapse strength of the casing. Due to the occurrence of anti-collapse, there are four different collapse cases, yield collapse, plastic collapse, transition collapse, and elastic collapse. To calculate the collapse strength under the four different cases, API has determined the dividing line and calculation formula between different collapse models based on the casing yield strength and diameter-thickness ratio. However, the formula has low calculation accuracy and incomplete consideration of the influencing factors of casing collapsing strength performance[23]. For this reason, the API/ISO working group compared and validated 11 models proposed by scholars around the world for calculating collapse strength, and issued a new standard ISO/TR 10400-2007, recommending the KT formula[24-25] proposed by Klever and Tamano in 2004 as the new model for calculating crush strength of the casing. The KT formula, which takes into account ovality, wall thickness non-uniformity, and residual stress, has proven to be highly accurate. In response to the inherent defects of the pipe, the API organization rectified the problems in the 5C3 specification and issued the API 5CT standard in July 2011 to revise the calculation formula for crush strength taking into account manufacturing defects (ovality, wall thickness unevenness, residual stress). At the same time, related studies have shown that the outer diameter out-of-roundness, wall thickness non-uniformity, material strength, and residual stress have a certain influence on the collapse strength of casing[26-28].

In summary, the outer diameter, wall thickness, yield strength, ovality, wall thickness non-uniformity, and residual stress are taken as the main influencing factors of the collapsing strength. The relevant data were collected by full-size measurement, residual stress detection, and casing collapse test, and organized to form the collapse strength prediction data set for machine learning prediction model training.

2.2 Experimental protocol and procedures

The data was obtained by physical crush test according to ISO 11960/API RP 5C3 standard. For the tubing specimens with a nominal outer diameter (D) less than $9\frac{5}{8}$ in, the minimum length was 8 times the nominal outer diameter; for the tubing with a nominal outer diameter (D) greater than $9\frac{5}{8}$ in, the minimum length was 7 times the nominal outer diameter. Due to the limited experimental equipment in the field, it was not possible to simulate the complex mechanical environment in the actual well, so the collapse test was completed by applying a uniform external load to the pipe specimen to be tested at room temperature only. The overall experimental program flow is shown in Figure 1.

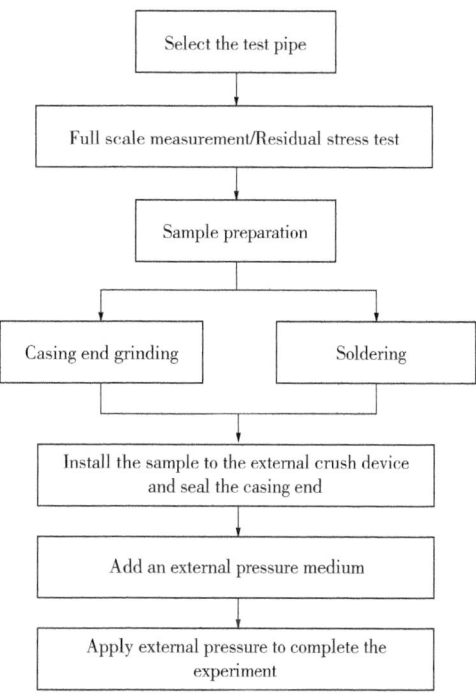

Figure 1　Experimental flow

To make full use of the historical data and complete the subsequent data set establishment. The selection of the experimental pipe should be combined with the information of the existing historical data of the casing to complete the selection, the outer diameter range between 4.5 ~ 13.4in pipe grouped as follows: < 5.5in, 5.5 ~ 7.8in, 8.6in, 9.7in, 10.82 ~ 11.84in and >13in a total of six groups. Based on the historical data, three tubes were selected from each of the above groups for the squeezing and destruction experiments. Before conducting the experiments, full dimensional measurements and residual stresses were measured on all specimens without radial or axial loads, after which the specimens were pressurized and slowly decompressed under the composite collapse test system to ensure that the specimens were crushed.

2.3 Full-size measurement and residual stress detection

The specimens for the collapse experimental study were subjected to geometric measurements before the experiments. The geometric measurement locations are shown in Figure 2. Each specimen is measured in 5 sections with 8 points per section. The measured geometry was used to calculate the average outside diameter, average wall thickness, ovality, and wall thickness unevenness of the pipe.

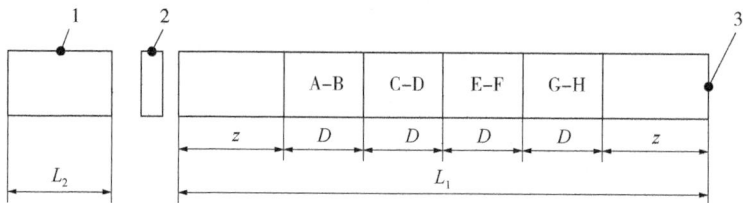

Figure 2 Measurement sample before collapsing test

1—Residual stress test specimen; 2—Tensile specimen; 3—Squeezed specimen; D—Outer diameter;
L_1—Minimum length of the squeezed specimen; L_2—Minimum length of residual stress specimen.

Eight points (A~H) were selected at five equally spaced locations to measure the average outer diameter, average wall thickness, ovality, and wall thickness non-uniformity

Table 1 shows the results of the full-scale measurement of the tube with nominal outer diameter $D = 139.7$ mm and nominal wall thickness $t = 12.7$ mm. The average outer diameter and average wall thickness of different sections of the tube are obtained after measuring the calibration points A to H. The manufacturing defect parameters of the tube can also be obtained after calculation: ovality and wall thickness non-uniformity. In full-size measurement, the formulas for the calculation of ovality and wall thickness non-uniformity are as follows:

(1) Ovality calculation formula:

$$\frac{2(D_{max} - D_{min})}{D_{max} + D_{min}} \times 100\% \qquad (1)$$

Where: D_{max} is the maximum measured outer diameter value on the same cross-section; D_{min} is the minimum measured outer diameter value on the same cross-section.

(2) Wall thickness non-uniformity calculation formula:

$$\frac{2(t_{max} - t_{min})}{t_{max} - t_{min}} \times 100\% \qquad (2)$$

Where: t_{max} is the maximum measured wall thickness value in the same section; t_{min} is the minimum measured wall thickness value in the same section.

Table 1 Specimen geometry inspection results (nominal outer diameter $D = 139.7$ mm pipe fitting as an example)

(a) Outside diameter test results (unit: mm)					
A-B	C-D	E-F	G-H	Average outer diameter	Ovality
141.58	141.18	141.07	141.17	141.25	0.36
141.00	141.22	141.21	141.09	141.13	0.16
141.10	141.56	141.14	141.33	141.28	0.33
141.01	141.16	141.19	141.01	141.09	0.13
141.17	141.21	141.09	141.09	141.14	0.09

continued

(b) Wall thickness test results (unit: mm)									
A	B	C	D	E	F	G	H	Average wall thickness	Unevenness of wall thickness
13.09	13.43	12.77	13.46	12.96	12.84	13.16	13.15	13.11	5.26
13.08	13.40	12.94	13.21	12.75	12.92	12.96	13.31	13.06	4.97
13.00	13.34	12.71	13.22	12.60	12.86	13.15	13.03	12.99	5.71
13.10	13.19	13.02	12.71	12.94	12.59	13.32	12.96	12.98	5.63
13.33	13.20	12.96	12.96	13.15	12.65	13.10	13.00	13.02	5.23

After the full-size measurement, the residual stress of each pipe will be tested. To ensure the integrity of the surface of the pipe to be tested, the residual stress of each piece to be tested in the experiment will be measured by ultrasonic nondestructive testing. The ultrasonic stress tester (developed by Jianwei Technology Co., Ltd.) is used as the testing device. The layout of the test points is shown in Figure 3(a~b), and Figure 3(c) shows the polishing treatment of the test points.

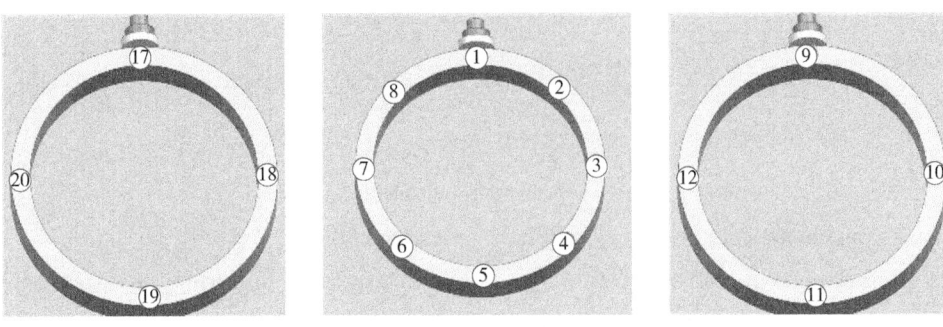

(a) Cross-sectional view of specimen layout

(b) Test piece tube body layout diagram

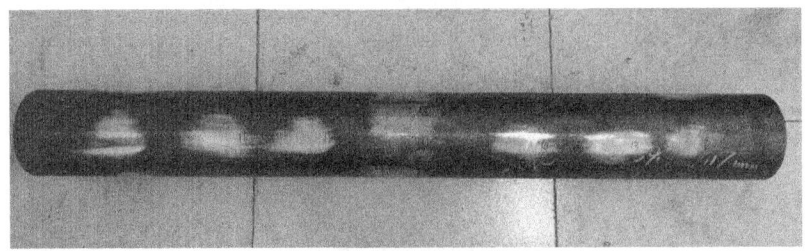

(c) Specimen layout grinding

Figure 3 Steel pipe stress detection layout

The test results of point 1 is shown in Figure 4, on the left side of the interface is the pre-collected reference waveform graph, while on the right side is the stress value of the point under test. It should be pointed out that prior to conducting measurements, the area on the pipe fittings where the test will be performed needs to be polished to ensure there is no rust. Subsequently, apply a coupling agent to the area and test one by one using an ultrasound probe. The same point can measured 2~5 times, and the residual stress at the final measurement point is averaged.

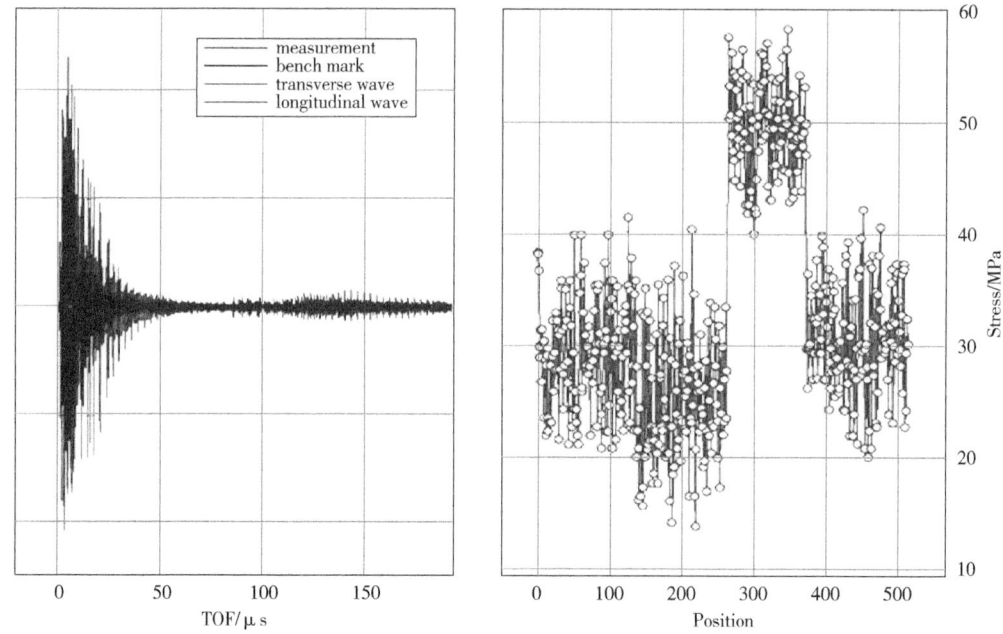

Figure 4 Point-by-point test result interface (test point 1)

2.4 Experiments and results of external pressure collapse

The external pressure collapse test system is shown in Figure 5. During the experiment, the collapse specimen is first put into the external pressure collapse cylinder, and the two ends of the specimen are sealed by installing gaskets and sealing flanges, which play a high-pressure sealing role during the experiment. At both ends of the sealing flange outside the installation of threaded connection flange, the connection flange directly with the external pressure extrusion cylinder on the thread for connection, through the connection flange to complete the specimen and external pressure extrusion experimental system connection. After the specimen shown in Figure 6 is installed, the pressure detection sensor is installed, and the pressurized medium is injected into the external pressure crush cylinder, and the pressure is applied until the specimen is destroyed by external pressure, and the shape of the specimen after the collapse failure is shown in Figure 7. The experimental process is monitored in real time by the pressure sensor, and the data of the external pressure extrusion failure value is recorded, and the extrusion experimental data is recorded after the experiment is completed.

The data from the external pressure collapse test are organized in Table 2.

Table 2 External pressure crush test data (partial)

Outer diameter/ mm	Wall thickness/ mm	Yield strength/ MPa	Ovality/ %	Unevenness of wall thickness/%	Residual stress/ MPa	Collapsing strength/ MPa
115.062	6.960	603.519	0.189	1.521	29.248	131.702
114.808	6.858	570.975	0.196	2.337	22.110	128.421

continued

Outer diameter/ mm	Wall thickness/ mm	Yield strength/ MPa	Ovality/ %	Unevenness of wall thickness/%	Residual stress/ MPa	Collapsing strength/ MPa
115.062	6.756	600.968	0.177	1.882	30.952	126.901
...	...	...	...	...	...	...
127.508	7.671	665.368	0.168	1.910	90.538	121.373
127.508	7.976	670.539	0.168	2.203	75.634	132.328
127.508	7.722	655.025	0.139	3.711	105.097	117.830
...	...	...	...	...	...	...
140.462	7.823	618.482	0.701	0.813	152.476	86.734
139.954	7.645	801.889	0.354	2.615	114.145	116.866
139.954	9.220	648.957	0.145	6.573	65.255	152.808
...	...	...	...	...	...	...
178.816	10.312	596.969	0.145	2.189	41.821	114.465
179.070	10.211	604.967	0.169	2.611	30.441	113.789
178.562	10.795	609.518	0.126	0.843	25.062	114.350
...	...	...	...	...	...	...

Figure 5 Experimental system of external pressure crush destruction

Figure 6 The shape of the specimen before crush failure

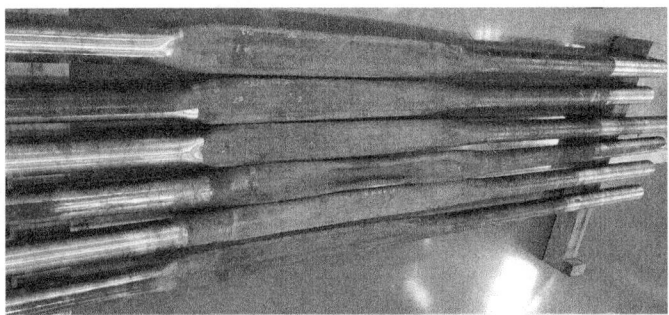

Figure 7 Shape of specimen after crush failure

3 Machine learning prediction model for casing collapse strength

3.1 Neural network prediction model

The neural network regression prediction model is a method of regression analysis and prediction using artificial neural networks. Its basic principle is to transform the input data nonlinearly by a certain number of neurons and optimize the connection weights of the neurons using a backpropagation algorithm to fit and predict the output data. Specifically, the neural network regression prediction model can be represented as a directed acyclic graph, in which the input layer receives the data, the output layer outputs the prediction results, and the middle hidden layer is responsible for the nonlinear transformation of the input data. Each neuron receives a certain number of input signals, and after weighting and calculation, the result is nonlinearly transformed by the activation function and output to the next layer of neurons[29].

Assuming that there are L layers of neurons, $x^{(i)}$ denotes the input vector of the i th sample, $y^{(i)}$ denotes the corresponding output, and $\hat{y}^{(i)}$ denotes the predicted value of the model for $y^{(i)}$, the neural network regression prediction model can be expressed as Formula(3):

$$\hat{y}^{(i)} = f(x^{(i)}; \theta) = f_L(f_{L-1}(\cdots f_2(f_1(x^{(i)}; \theta_1); \theta_2)\cdots); \theta_L) \tag{3}$$

Where, f_1 is the output of the 1th layer neuron and θ_1 is the parameters of the 1st layer neuron.

Determine the outer diameter, wall thickness, yield strength, ovality, wall thickness unevenness, and residual stress as the input data and the ultimate deformation load as the output data, and divide the normalized data into a training set (70%) for model training and parameter seeking, and testing set (30%) for model training effect evaluation.

Considering that the study data do not involve serial or temporal correlation and the dimensionality of the features to be processed is not high, it is a regression task with known input to predict the output. The feedforward neural network structure with a single hidden layer is chosen. The model structure is simple and easy to explain and understand, and the training time is short while ensuring the prediction accuracy. Combining the existing data samples, the outer diameter, wall thickness, yield strength, ovality, wall thickness unevenness, and residual stress are determined as the input data, and the ultimate deformation load as the output data. The network structure diagram(input layer-hidden layer-output layer) is shown in Figure 8.

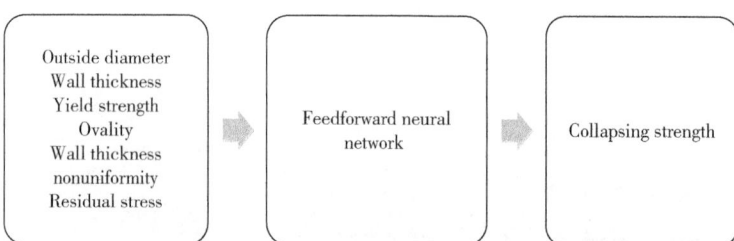

Figure 8 Network structure diagram

The LM algorithm(Levenberg-Marquardt algorithm) is an optimization algorithm mainly used to train feedforward neural networks. The LM algorithm uses a combination of the second-order Newton method and the first-order gradient descent method, which has an efficient convergence speed and stability and can adjust the learning rate adaptively during the training process, avoiding the oscillation and scattering problems of the gradient descent algorithm are avoided.

In the neural network super parametric tuning, mainly includes the determination of the number of neurons, learning rate, and the number of iterations. In terms of the learning rate, the LM algorithm is chosen as the training algorithm to adjust the learning rate adaptively. In terms of iteration number, the number of iterations can be set to a larger amount(1000), and the number of iterations can be constrained by setting an error threshold to stop training when the training error is less than the error threshold. Determining the number of neurons is a critical issue. The number of neurons directly affects the complexity and learning ability of the model. In general, the number of neurons should be large enough so that the model can fully learn the features and patterns in the dataset. However, it is also important to avoid too many neurons, which can lead to overfitting of the model or too long training time.

The number of neurons in the hidden layer is usually determined based on the number of samples. A commonly used formula is:

$$N=\sqrt{m+l}+\alpha \tag{4}$$

Where N is number of hidden neurons, m is number of nodes in the input layer, l is number of nodes in the output layer, and α is a constant between 1 and 10.

The trial-and-error range (4 ~ 13 shown in Table 3) of the trial-and-error method is determined by the empirical formula, and the neural network model is trained according to the range, and the root mean square error RMSE and correlation coefficient R^2 are recorded for each training. Comparing the data performance of each training set, the number of hidden layer neurons corresponding to the time when the root mean square error RMSE is the smallest and the correlation coefficient R^2 is the closest to 1 is determined as the optimal parameter of the model.

Table 3 Neuron number preference table

Neuron count	RMSE	R^2
4	0.0338	0.97132
5	0.0164	0.99323
6	0.0342	0.97358
7	0.0184	0.99148
8	0.0236	0.98600
9	0.0163	0.99328
10	0.0197	0.99022
11	0.0197	0.99024
12	0.0346	0.96996
13	0.0178	0.99200

The number of neurons was determined to be 9. The RMSE: 0.0163 is the minimum value when the number of neurons is 9 and R^2: 0.99328 is closest to 1. Therefore, the number of neurons was determined to be 9.

The topology of the finalized neural network prediction model is shown in Figure 9.

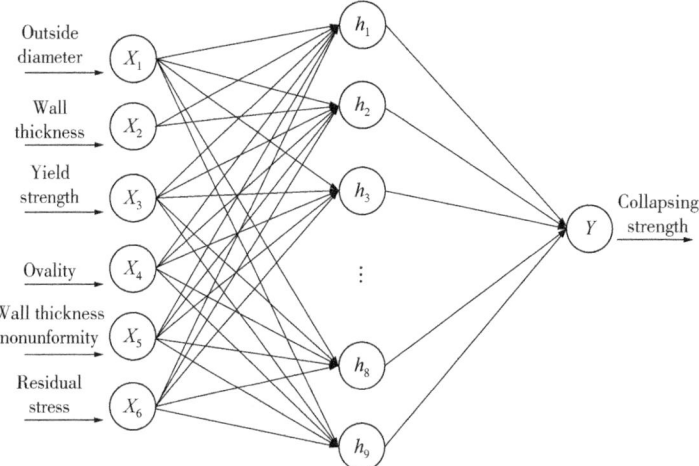

Figure 9 Neural network prediction model topology

The prediction results of the training set and test set of the neural network prediction model are shown in Figure 10.

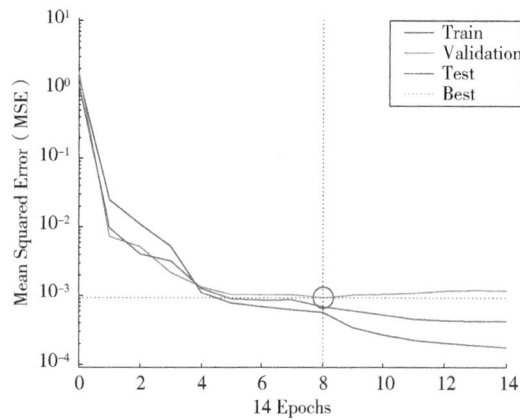

Figure 10 Number of model iterations
Best Validation Performace is 0.00092815 at epoch 8

The model reaches the best training effect after 8 iterations. After the training of the model, the training set, the test set, and the overall data set are predicted respectively, and the data are denormalized to scale the training data distributed between [0, 1] to the size of the actual values. The comparison curve between actual value and predicted value is shown in Figure 11.

From the prediction result curves, the high degree of fit between the predicted and actual value curves indicates the high accuracy of the current neural network model.

The evaluation metrics for assessing the model prediction results include Root Mean Square Error(RMSE), Coefficient of Determination(R^2), and Mean Relative Error.

Root Mean Square Error(RMSE) is a indicator used to assess the magnitude of prediction errors in regression models. It represents the square root of the mean of the squared differences between predicted values and actual values. The formula (5) is calculating of RMSE.

$$\text{RMSE} = \sqrt{\frac{1}{n}\sum_{i=1}^{n}(y_i - f_i)^2} \tag{5}$$

Where y_i is the true value, f_i is the predicted value, and n is the number of samples. The smaller the RMSE, the better the predictive ability of the model.

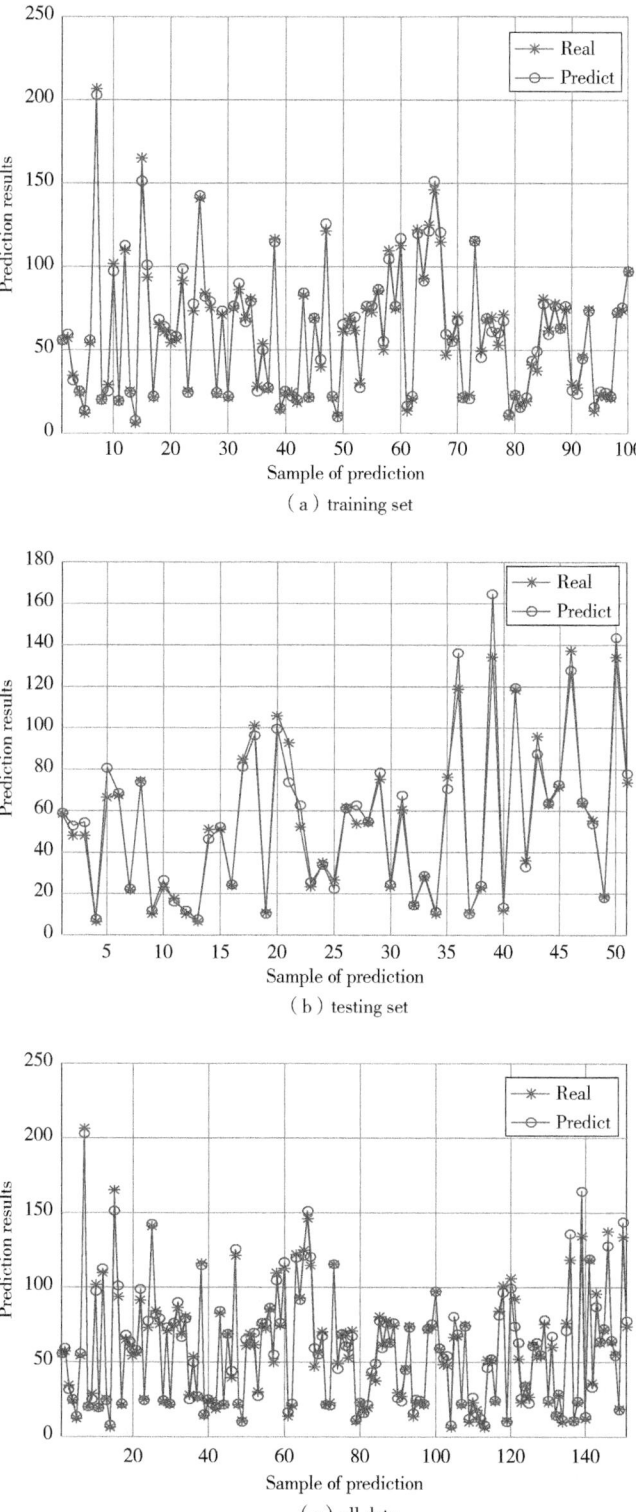

Figure 11　Neural network prediction comparison curve

R^2 is an indicator used to evaluate the goodness of fitting of a regression model, it indicates how much variation in the target variable can be explained by the model. It is calculated using the formula(6).

$$R^2 = 1 - \frac{SS_{res}}{SS_{tot}} \tag{6}$$

Where SS_{res} is the residual sum of squares and SS_{tot} is the total sum of squares. It can be calculated using the following formula:

$$SS_{res} = \sum_{i=1}^{n} (y_i - f_i)^2 \tag{7}$$

$$SS_{tot} = \sum_{i=1}^{n} (y_i - \bar{y})^2 \tag{8}$$

Where y_i is the true value, f_i is the predicted value, $\bar{y}$ is the mean value of y, and n represents the number of samples.

The value of R^2 ranges between 0 and 1. The closer to 1 means that the model fits the data better, indicating that the model is can explain most of the changes in the target variable. When R^2 is equal to 1, it means that the model fits the data completely, and when R^2 is equal to 0, it means that the model is unable to explain the changes in the target variable, and the model is unable to fit the data.

Mean Relative Error(MRE) is a common indicator of the difference between the predicted value and the true value used formula(9).

$$MRE = \frac{1}{n} \sum_{i=1}^{n} \left| \frac{y_i - y_i'}{y_i} \right| \tag{9}$$

Where n is the number of samples, y_i is the true value of the i th sample, and y_i' represents the corresponding predicted value.

The results of neural network model prediction are shown in Table 4.

Table 4 Neural network model prediction

Parameters	Training set	Testing set	All data
Correlation coefficient R^2	0.9947	0.9733	0.9893
Root mean square error RMSE	0.0146	0.0267	0.0195
Mean relative error MRE	0.0374	0.0782	0.0512

3.2 Random forest prediction models

Random forest is an integrated learning algorithm based on decision trees, which is usually used in classification and regression problems. In regression problems, the random forest model consists of multiple decision trees, each constructed by bootstrap sampling and random feature selection of the training data[30]. For each decision tree, a subset of samples and a subset of features are generated by randomly sampling the original data and randomly selecting the features. Based on this sample subset and feature subset, a decision tree model is trained. Here the process of random sampling and random feature selection reduces the variance of the decision tree.

The prediction function for each decision tree in the random forest regression prediction model is formula(10).

$$f(x) = \sum_{i=1}^{m} c_i I(x \in R_i) \quad (10)$$

Where m denotes the number of decision trees, c_i denotes the constant term in the i th decision tree, R_i denotes the leaf node region in the i th decision tree, and $I(x \in R_i)$ denotes whether sample x falls within leaf node R_i.

$$f(x) = \frac{1}{m} \sum_{j=1}^{m} f_j(x) = \frac{1}{m} \sum_{j=1}^{m} \sum_{i=1}^{T} c_{i,j} I(x \in R_{i,j}) \quad (11)$$

Where T denotes the number of leaf nodes in the j th decision tree and $c_{i,j}$ denotes the constant term of the i th leaf node in the j th decision tree.

Decision tree is the basic unit of random forest construction. Determine the outer diameter, wall thickness, yield strength, ovality, wall thickness unevenness, and residual stress as decision tree features, and the ultimate deformation load as prediction samples to build the decision tree. Then randomly selected features and sample data are used to construct multiple decision trees to build a random forest.

Given the number of trees in the random forest(n_estimators), the maximum depth of each tree(max_depth), and the range and interval of the minimum number of leaf node samples(min_samples_leaf). Within the given range of parameters, different combinations of parameters are generated to form a "grid". A random forest regression model is constructed using each combination of parameters, and the performance of the model on the validation set is evaluated, usually using metrics such as root mean square error (RMSE) or correlation coefficient (R^2). The optimal combination of parameters is selected based on the model performance.

The number of trees(n_estimators) is determined to be 100, the maximum depth of each tree (max_depth) is 5, and the minimum number of leaf node samples (min_samples_leaf) is 5 according to the grid search algorithm.

Training of the model using optimal parameters. The error variation curve with the number of decision trees is shown in Figure 12.

After the training of the model, the training set, the test set, and the overall data set are predicted respectively, and the data are denormalized to scale the training data distributed between [0, 1] to the size of the actual values. The comparison curve between actual value and predicted value is shown in Figure 13.

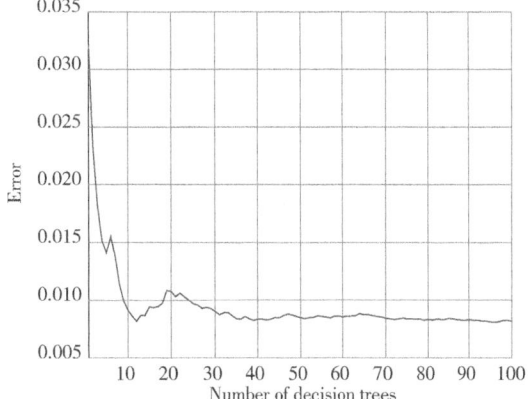

Figure 12　Error variation curve with the number of decision trees

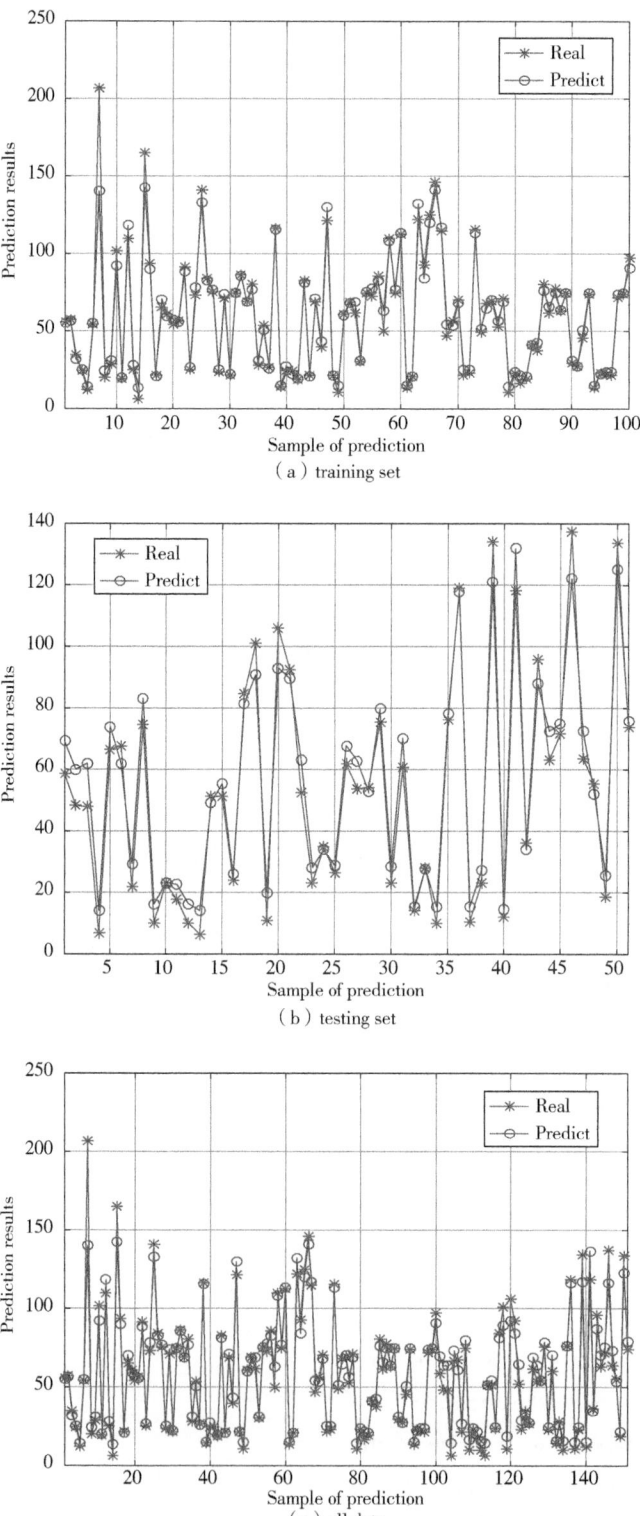

Figure 13 Random forest prediction comparison curve

From the comparison curves of the prediction results, which listed in Table 5, the random forest prediction model has higher prediction accuracy for sample points with more concentrated data, but there is a larger error in the prediction of outlier points with larger sample values.

Table 5 Prediction results of random forest model

Parameters	Training set	Testing set	All data
Correlation coefficient R^2	0.9032	0.8870	0.8992
Root mean square error RMSE	0.0621	0.0549	0.0597
Mean relative error MRE	0.1915	0.2221	0.2018

3.3 Support vector machine prediction model

Support Vector Machine (SVM) is a machine learning algorithm widely used in the fields of classification, regression, and anomaly detection. In regression problems, SVM can be used to fit a nonlinear function to describe the relationship between input variables (independent variables) and output variables (dependent variables).

The regression problem of SVM can be transformed into solving a minimizing convex quadratic programming problem to minimize the model complexity while maximizing the prediction error. Specifically, given a training data set x_i, $y_{i=1}^m$, where $x_i \in R^n$ is the independent variable and $y_i \in R$ is the dependent variable. The goal of the SVM regression model is to find a function $f(x) \leqslant w$, $x>+b$, where $w \in R^n$ is the weight vector and $b \in R$ is the bias, such that for all $i \in [1, m]$, the error $\varepsilon_i = |y_i - f(x_i)|$ is less than a given tolerance ε, while minimizing the complexity of the model[31].

The objective function of the SVM regression model can be expressed as formula(10).

$$\min_{w, b, \xi, \xi^*} \frac{1}{2}w^2 + C\sum_{i=1}^m (\xi_i + \xi_i^*) \tag{12}$$

Where ξ_i and ξ_i^* are relaxation variables for the non-separable case and C is a regularization parameter to control the complexity of the model. Also, this objective function needs to satisfy the following constraints:

$$\begin{cases} y_i - w \cdot x_i - b \leqslant \varepsilon + \xi_i \\ w \cdot x_i + b_i - y \leqslant \varepsilon + \xi_i^* \\ \xi_i, \xi_i^* \geqslant 0 \end{cases} \tag{13}$$

Where the first constraint indicates that $f(x_i) + \varepsilon$ is greater than or equal to y_i, and the second constraint indicates that $f(x_i) - \varepsilon$ is less than or equal to y_i. These constraints indicate that the training sample points must satisfy within the range of ε.

After solving the above convex quadratic programming problem to obtain the weight vectors and biases, the predicted values of the SVM regression model are:

$$\hat{y} = w \cdot x + b \tag{14}$$

The advantages of SVM regression models are their ability to handle high-dimensional and nonlinear data, as well as their robustness to noise and outliers. The disadvantages are the large amount of computation and storage space required, and the challenges for parameter selection and tuning.

Support vector machine regression prediction model hyperparameter determination usually includes the selection of the kernel function and the determination of the penalty factor. A kernel function is a function that maps the original data to a high-dimensional space and is used to deal with nonlinear problems. In Support Vector Machine (SVM), the kernel function is often used to construct classifiers or regressors to transform nonlinear problems into linear ones.

Radial Basis Function Kernel (RBF Kernel): It is the most commonly used kernel function with smooth nonlinear characteristics, which can better handle nonlinear problems, and has better robustness and adaptability. Its expression is $K(x_i, x_j) = e^{-\gamma \|x^i - x^j\|^2}$: where γ is a parameter that determines the rate of change of the function also known as bandwidth.

After determining the RBF as the kernel function, it is necessary to further determine the parameters γ and the penalty factor C of the radial basis kernel function. The penalty factor C is a hyperparameter that is used to control the complexity of the model. The larger C is, the larger the penalty on misclassified points, and the more complex the model is; the smaller C is, the smaller the penalty on misclassified points, and the simpler the model is. Therefore, the value of C needs to be optimized in the model selection. The parameter γ of the radial basis kernel function controls the rate of change of the radial basis kernel function. γ is larger, the value of the kernel function decreases rapidly with the increase of the distance between points, and the decision boundary becomes more complicated; γ is smaller, and the value of the kernel function decreases more slowly, and the decision boundary becomes smoother. In the model training process, the optimal γ and C values need to be selected by cross-validation. The $\gamma = 0.25$ and $C = 11.3137$ are determined by double-loop 5-fold cross-validation.

After the training of the model, the training set, the test set, and the overall data set are predicted respectively, and the data are denormalized to scale the training data distributed between [0, 1] to the size of the actual values. The comparison curve between actual value and predicted value is shown in Figure 14.

The comparison curves of the prediction results, which listed in Table 6, show that the support vector machine regression model has high accuracy in the training set, test set, and overall data prediction.

Table 6 Support vector machine model prediction results

Parameters	Training set	Testing set	All data
Correlation coefficient R^2	0.9840	0.9581	0.9775
Root mean square error RMSE	0.0252	0.0334	0.0282
Mean relative error MRE	0.8010	0.1003	0.8690

3.4 Comparative analysis of three prediction models and API formulas

The comparison curves of the three model prediction results, the calculated values by API formula (ISO10400) and the measured values of casing collapse strength are plotted as follows. Due to the relatively small sample size used for model training, the training set, the test set and the overall data results are now shown in their entirety. The model optimization is mainly based on the test set samples that are not involved in modeling training.

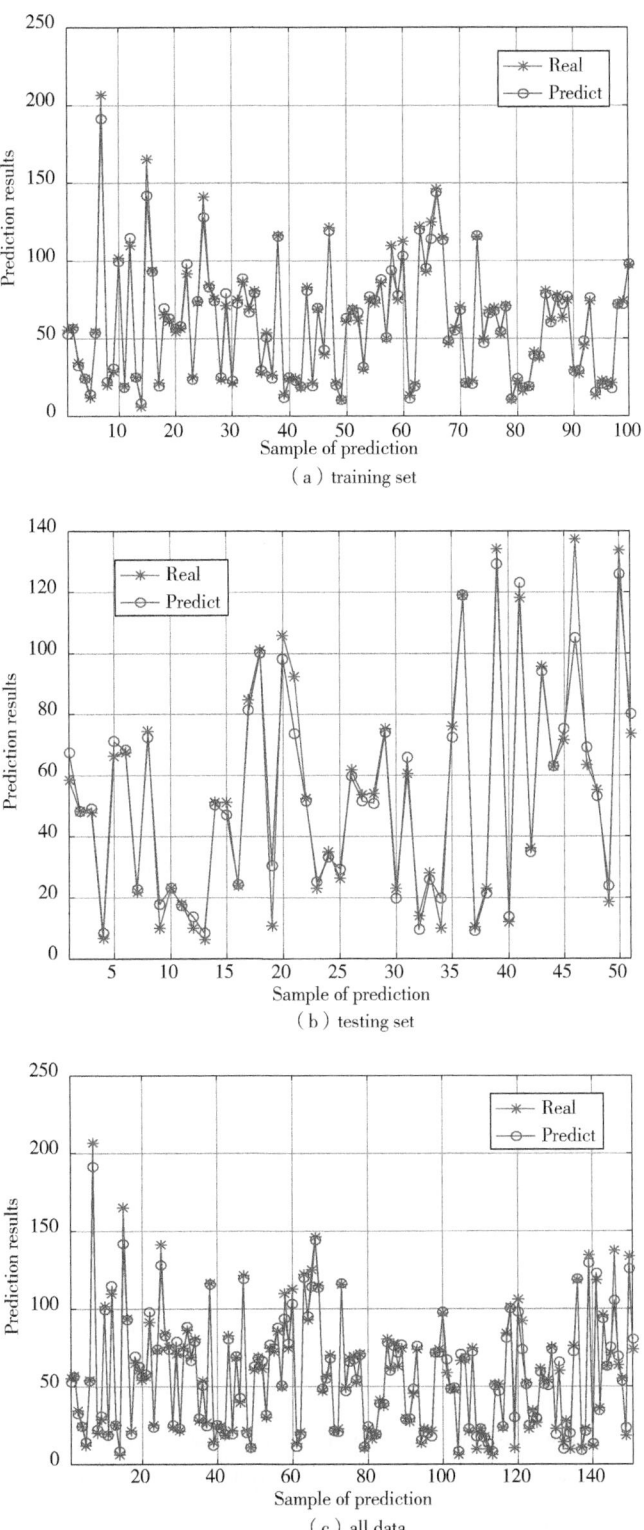

Figure 14 Support vector machine prediction comparison curve

Three casing limit deformation load regression prediction models, neural network, random forest, and support vector machine, were constructed based on the measured data, and all three machine learning prediction models have high prediction accuracy. From Figure 15, the best prediction models were selected based on the root mean square error RMSE, correlation coefficient R^2 and mean relative error MRE as the evaluation indexes.

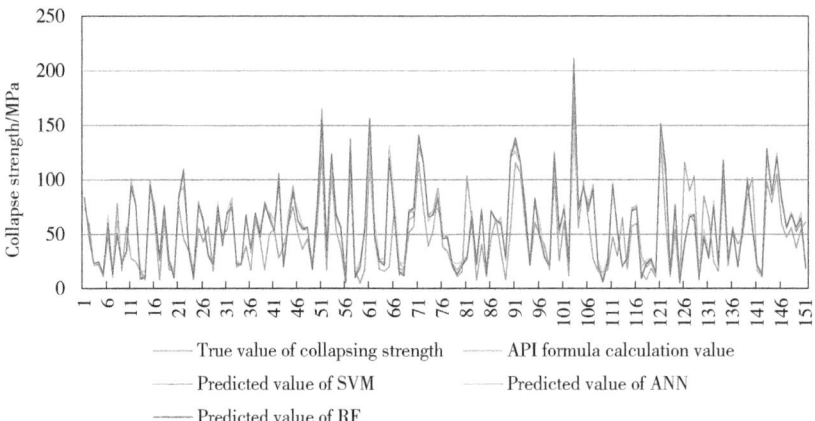

Figure 15 Comparison curves of actual value, formula value and predicted value of collapsing strength

The comparison curves of the actual, predicted, and calculated values from the formula show that all three machine learning prediction models have better prediction results, while the API calculation formula has a larger error. This indicates that the machine learning algorithm has some advantages over the traditional API formula in predicting the casingcollapse strength. The differences between the three prediction models and calculation formulas are further determined by specific evaluation indexes to determine the best prediction model. To exclude the training data interference, the prediction model evaluation metrics were used from the test set data without training.

The prediction results of the three models, neural network, support vector machine, and random forest, are summarized and compared, which shown in Table 7. The accuracy (ACC) is calculate by formula(15).

$$ACC = (1-MRE) \times 100\% \qquad (15)$$

Table 7 Summary of prediction results of three models

Parameters	Neural network	Random forest	Support vector machine	API formula
R^2	0.9733	0.8870	0.9581	—
RMSE	0.0267	0.0549	0.0334	—
MRE	0.0782	0.2221	0.1003	0.3660
Prediction accuracy/%	92.20	77.80	89.97	63.30

Among the three machine learning prediction models, the neural network prediction model has the best prediction effect, the correlation coefficient is 0.9733, which is closest to 1, and the root mean square error (0.0267) is the smallest. By comparing the actual value of the collapse strength with the predicted value, the average prediction accuracy of the traditional API calculation formula is only 63.3%, while the three machine learning prediction models have higher accuracy, and the average prediction accuracy of the neural network prediction model can reach 92.2%.

4 Conclusion

Through the analysis of the casing strength prediction model, the main factors affecting the casing collapse strength are outer diameter, wall thickness, yield strength, ovality, wall thickness unevenness, and residual stress.

By training the neural network prediction model through the trial-and-error method, it was recorded that when the number of neurons was 9, RMSE: 0.0163 was the minimum value, and R^2: 0.99328 was the closest to 1. The maximum prediction accuracy could reach 92.2%; by training the random forest prediction model according to the grid search algorithm. The number of trees(n_estimators) was determined to be 100, the maximum depth of each tree(max_depth) is 5, and the number of minimum leaf node samples(min_samples_leaf) is 5, and the highest prediction accuracy can be obtained up to 77.8%. The support vector machine prediction model $\gamma=0.25$ and $C=11.3137$ is determined by double-loop 5-fold cross-validation, and the highest prediction accuracy can be obtained up to 89.97%.

The machine learning algorithm has unique advantages in the prediction of casing collapse strength. The three prediction algorithms are higher than the API calculation formula(63.3%). The neural network prediction model is more suitable for the prediction of casing collapse strength, and the prediction accuracy can reach 92.2%, which can be used to guide the prediction of high collapse casing strength.

Acknowledgments

The authors would like to acknowledge the financial support of CNPC research and development project(2020B-4020). The Project Supported by Natural Science Basic Research Plan in Shaanxi Province of China(Program No. 2023-JC-QN-0554).

Data Availability Statement

Data Availability Statements are available in section "MDPI Research Data Policies" at https://www.mdpi.com/ethics.

Conflicts of Interest

The authors declare no conflict of interest.

References

[1] Yan X, Jun L, Chunqing Z, et al. A new investigation on casing shear deformation during multistage fracturing in shale gas wells based on microseism data and calliper surveys[J]. Journal of Petroleum Science and Engineering, 2019, 180: 1034-1045.

[2] Zeng B, Zhou X, Cao J, et al. A Casing Deformation Prediction Model Considering the Properties of Cement [J]. Processes, 2023, 11(3): 695.

［3］ Dong K, Liu N, Chen Z, et al. Geomechanical analysis on casing deformation in Longmaxi shale formation [J]. Journal of Petroleum Science and Engineering, 2019, 177: 724-733.

［4］ Zhang Z, Zheng Y, Hou D, et al. The influence of hydrogen sulfide on internal pressure strength of carbon steel production casing in the gas well [J]. Journal of Petroleum Science and Engineering, 2020, 191: 107113.

［5］ Deng K H. Theoretical and experimental study on the casing collapse under non-uniform load and working mechanics of casing repair[D]. Southwest Petroleum University, 2018.

［6］ Montáns F J, Chinesta F, Gómez-Bombarelli R, et al. Data-driven modeling and learning in science and engineering[J]. Comptes Rendus Mécanique, 2019, 347(11): 845-855.

［7］ Zhao Y H, Jiang H Q, Li H Q, et al. Research on predictions of casing damage based on machine learning [J]. Journal of China University of Petroleum(Edition of Natural Science), 2020, 44(4): 11.

［8］ Fisher O J, Watson N J, Escrig J E, et al. Considerations, challenges and opportunities when developing data-driven models for process manufacturing systems [J]. Computers & Chemical Engineering, 2020, 140: 106881.

［9］ Bahramian M, Dereli R K, Zhao W, et al. Data to intelligence: The role of data-driven models in wastewater treatment[J]. Expert Systems with Applications, 2022: 119453.

［10］ Li C L. Under the background of big data review of machine learning algorithms[J]. Information Recording Materials, 2018, 5: 4-5.

［11］ Niu C C, Li S B, Hu J J, et al. Application of machine learning in material informatics: a Survey[J]. Materials Reports, 2020, 23: 23100-23108.

［12］ Martirosyan, A V, Ilyushin, Y V. Modeling of the Natural Objects' Temperature Field Distribution Using a Supercomputer[J]. Informatics 2022, 9(3), 62.

［13］ Sabah M, Mehrad M, Ashrafi S B, et al. Hybrid machine learning algorithms to enhance lost-circulation prediction and management in the Marun oil field[J]. Journal of Petroleum Science and Engineering, 2021, 198: 108125.

［14］ Dixit N, McColgan P, Kusler K. Machine learning-based probabilistic lithofacies prediction from conventional well Logs: A case from the Umiat Oil Field of Alaska[J]. Energies, 2020, 13(18): 4862.

［15］ Sircar A, Yadav K, Rayavarapu K, et al. Application of machine learning and artificial intelligence in oil and gas industry[J]. Petroleum Research, 2021, 6(4): 379-391.

［16］ Ilyushin Y V, Kapostey E I. Developing a Comprehensive Mathematical Model for Aluminum Production in a Soderberg Electrolyser. Energies 2023, 16, 6313.

［17］ Mohammed A I, Bartlett M, Oyeneyin B, et al. An application of FEA and machine learning for the prediction and optimisation of casing buckling and deformation responses in shale gas wells in an in-situ operation[J]. Journal of natural gas science and engineering, 2021, 95: 104221.

［18］ Yan X Z, Zhang D F, Yang X J, et al. Study on strength reduction of casing with surface defects[J]. Oil Field Equipment. 2009, 38(11): 1-4.

［19］ Wang C Y. Research and development of tubing and casing quality evaluation system in intelligent manufacturing environment[D]. Xi'an University Of Technology, 2018.

［20］ Zhang X, Wang L, Meng F S, et al. Bayesian neural network approach to casing damage forecasting [J]. Progress in Geophysics, 2018, 33(3): 1319-1324.

［21］ Zhang J, Wu L, Jia D, et al. A Machine Learning Method for the Risk Prediction of Casing Damage and Its Application in Waterflooding[J]. Sustainability, 2022, 14(22): 14733.

［22］ Di Q F, Wu Z H, Chen T, et al. Artificial intelligence method for predicting the maximum stress of an off-center casing under non-uniform ground stress with support vector machine[J]. Science China Technological

Sciences, 2020, 63(12): 2553-2561.

[23] Fan S, Chen D F, Feng L, et al. Effect of ovality on casing collapse strength based on KT model[J]. Fault-Block Oil and Gas Field, 2017, 24(3): 426-429.

[24] Klever F J, Tamano T. A new OCTG strength equation for collapse under combined loads[R]. SPE 90904, 2004.

[25] Stewart G, Klever F J. Accounting for flaws in the burst strength of OCTG[R]. SPE 48330, 1998.

[26] Wang J, Tian X L, Fan Z X. Research on resistance properties against external collapse of SEW high collapse-resistant casing[J]. Steel Pipe, 2014, 43(2): 16-21.

[27] Bai Y, Igland R T, Moan T. Tube collapse under combined external pressure, tension and bending[J]. Marine Structures, 1997, 10(5): 389-410.

[28] Huang X, Mihsein M, Kibble K, et al. Collapse strength analysis of casing design using finite element method[J]. International Journal of Pressure Vessels and Piping, 2002, 77(7): 359-367.

[29] Jain A K, Mao J, Mohiuddin K M. Artificial neural networks: A tutorial[J]. Computer, 1996, 29(3): 31-44.

[30] Aljameel S S, Alomari D M, Alismail S, et al. An anomaly detection model for oil and gas pipelines using machine learning[J]. Computation, 2022, 10(8): 138.

[31] Hearst M A, Dumais S T, Osuna E, et al. Support vector machines[J]. IEEE Intelligent Systems and their applications, 1998, 13(4): 18-28.

本论文原发表于《Processes》2023 年第 11 期。

Effect of Deviation of Welding Parameters on Mechanical Properties of X80 Steel Girth Weld

Zhu Lixia[1]   Luo Jinheng[1]   Jia Haidong[2]   Li Lifeng[1]
Yu Wenchang[2]   Chen Yongnan[3]

(1. CNPC Tubular Goods Research Institute;
2. Pipeline Network Group(Xinjiang)United Pipeline Co., Ltd.; 3. Chang'an University)

Abstract: Tensile tests were carried out on X80 steel girth welds with four different parameters(no deviation of welding parameters, inadequate number of welding layers, inadequate preheating temperature and inadequate interpass temperature) under automatic welding technology. The strain distribution behavior, strain hardening law and the influence of welding parameter deviation on mechanical properties were studied during tensile deformation. The results show that the strain localization occurs at the interface of the weld filler layer when the preheating temperature of the weld is inadequate and the interpass temperature of the weld is inadequate, and the strain localization occurs at the maximum thickness of the filler layer when the number of welding layers is inadequate. The deviation of welding parameters significantly reduces the strength of girth weld. The yield strength and tensile strength are the lowest when the interpass temperature is inadequate. When the interpass temperature is inadequate, the strain hardening ability deteriorates rapidly. When there is no deviation of welding parameters, the fracture is dimple shaped, which is typical ductile fracture. When the welding parameters deviate, the fracture of the girth weld is a mixed fracture of toughness and brittleness; When the number of welding layers is inadequate, it is mainly brittle fracture; When preheating temperature and interpass temperature are inadequate, ductile fracture is the main mechanism.

Keywords: X80 Steel Girth weld; Welding process parameters; Mechanical properties; Strain hardening behavior

1 Introduction

In recent years, with the increasing consumption of natural gas & oil and the expansion of transportation facilities, the research on pipeline steel with large diameter, large thickness, high

---

Corresponding author: Zhu Lixia: zhulx@cnpc.com.cn; Luo Jinheng, luojh@cnpc.com.cn; Chen Yongnan, frank_cyn@163.com.

strength and toughness which suit for high-pressure transportation has been developed rapidly. X80 steel has been widely used in major projects such as the West East Gas Transmission Project in China[1-2] due to its high strength & toughness, good weldability and other characteristics. As a long-distance and trans regional transportation pipeline, X80 pipeline steel will inevitably suffer from the effects of environment and external loads during service, resulting in various deformations. Due to the non-uniformity microstructure and various welding defects, the weld joint often becomes the weakest position of long-distance transportation pipeline[3].

The number of welding layers, preheating temperature and interpass temperature in the welding processof girth weld will directly affect the microstructure of the weld joint, and then affect the mechanical properties. Tang et al.[4] studied the influence of welding layers on residual stress under three different welding process conditions with the help of the birth and death element module in the finite element analysis, and found that under the same boundary conditions, with the increase of the number of welding layers, the residual stress generated in the welding process is relatively reduced. This phenomenon is related to the heat treatment effect of multi-layer welding. The latter layer normalizes the former layer, improves the weld microstructure and reduce the residual stress caused by welding. Wang et al.[5] found that during the welding process, the slag and other defects will cause a significant decrease in the tensile strength of the welded joint, and have an impact on the toughness and fracture strain. The content and size of the slag and other defects will decrease with the increase of heat input, then the strength and plasticity will increase. This phenomenon may be related to the accumulation of defects such as slag between weld layers. During multi-layer welding, the cold crack sensitivity at the weld can be reduced by proper preheating temperature and interpass temperature[6]. Shi et al.[7] found that when the preheating temperature is inadequate, there will be incomplete fusion defects in the weld, and stress concentration will occur at the defects, which significantly reduce the mechanical properties of the weld.

On the basis of the automatic welding process of X80 steel girth weld in the Third West to East Gas Pipeline, four kinds of girth welds were welded with LINCOLN S500 automatic welding system by changing the number of welding layers, preheating temperature and interpass temperature (no deviation of welding parameters, inadequate number of welding layers, inadequate preheating temperature, inadequate interpass temperature). Tensile tests are carried out at constant strain rate, and the strain transfer behavior in the tensile deformation process is studied with strain digital image correlation(DIC)equipment, microstructure and fracture mechanism of tensile fracture were studied by electron backscatter diffraction (EBSD) and scanning electron microscope (SEM). The strain hardening behavior of four kinds of welds during tensile deformation was analyzed by Kocks Mecking (K-M) strain hardening model. The purpose of this paper is to reveal the relationship between welding parameter deviation and grain hardening, strain transfer, and discuss the influence of welding parameter deviation on weld mechanical properties. This work will provide theoretical support and scientific basis for the improvement of welding process and mechanical properties of X80 pipeline steel.

2 Experimental

2.1 Experimental materials and tensile test

The GMAW(root welding) + FCAW-S(filling and covering) process is used to prepare four kinds of X80 steel girth welds, as shown in Fig. 1. The main chemical composition of the welding wire is shown in Table 1. 80% Ar and 20% CO_2 protective atmosphere shall be used during welding. Specific welding parameters are shown in Table 2.

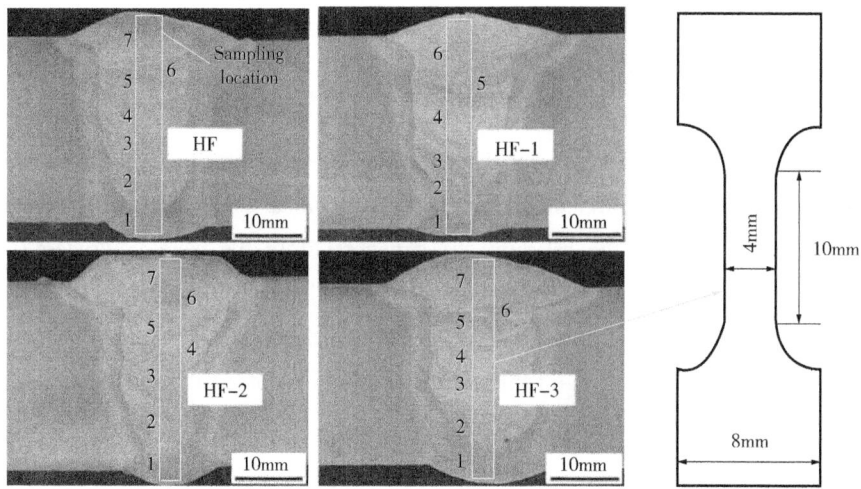

Fig. 1 Weld sampling diagram and tensile specimens

Table 1 Chemical compositions of the weld metal(%, mass fraction)

Elements	C	Mn	Si	S	P	Ni	Nb	Ti	V
Composition	0.07	1.89	0.34	0.002	0.011	0.030	0.045	0.003	0.048

Table 2 Welding parameters

No.	Voltage/V	Current/A	Welding speed/ (mm/min)	Gas flow rate/ (L/min)	Number of weld layers	Inter-pass temperature/℃	Preheat temperature/℃
HF	23~27	170~230	14~23	30~35	7	100	150
HF-1					6	100	150
HF-2					7	100	50
HF-3					7	40	150

Fig. 1 shows the weld morphology of four welding parameters (no parameter deviation: HF, insufficient number of layers: HF-1, insufficient preheating temperature: HF-2, insufficient interpass temperature: HF-3). The four kinds of welds are composed of weld layers and interfaces between layers. The fourth layer of HF-1 has the largest thickness and is located in the filling layer. The room temperature tensile test was carried out on QX-W750 tensile testing machine, and the loading rate was 0.1mm/min. In order to ensure the reliability of the test, the tensile test is repeated three times respectively. The tension process is equipped with DIC system, which is used to record the strain during the tension deformation process and analyze the strain response characteristics of the weld metal. At the beginning of the tensile test, the DIC camera is opened synchronously to correlate the flow stress and strain in the tensile process.

2.2 Characterization of properties and microstructure

Cut the sample from the weld metal fracture after tensile deformation, clean it with absolute ethanol and ultrasonic, dry it, and observe the fracture morphology with scanning electron microscope(SEM, Hitachi S-4800). After plastic deformation of weld metal, cut a sample 2~3mm away from its fracture surface and inlay it along the tensile direction. After the resin is completely cured, use different types of sandpaper for coarse grinding and fine grinding, and then use ethanol solution containing 5% perchloric acid for double jet electropolishing. The polishing time is 20~24s. Use electron backscatter diffraction technology(EBSD, JSM-6700F) to analyze the section and fracture mechanism. During the test, the accelerating voltage is 20kV, the sample inclination is 70°, the working distance is 12mm, and the step distance is 0.15μm.

3 Experimental results and discussion

3.1 Strain response characteristics and microstructure evolution of weld metal

Fig. 2 shows the DIC strain distribution cloud diagram of the weld specimen under tensile stress. As shown in Fig. 2a, this process can be divided into four stages, namely, elastic stage(Ⅰ), yield stage(Ⅱ), uniform deformation stage(Ⅲ), and fracture stage(Ⅳ). The four stages of HF sample are completely listed in Fig. 2a. Fig. 2b shows the strain distribution cloud diagram of HF-1, HF-2 and HF-3 when the flow stress is 690MPa. The results show that the deviation of welding parameters will cause the difference of strain distribution in the tensile process. In the elastic stage, the strain distribution of the four kinds of welds is relatively uniform, and there is almost no deformation concentration area. With the increase of flow stress, HF-2 and HF-3 samples yield at the interface between weld layers, and HF-1 samples yield in the weld layer, both of which have slight strain localization. When the tensile stress continues to increase, the strain transfer will be accelerated continuously, and the strain of HF-2 and HF-3 samples will continue to transfer to each weld layer; When the flow stress reaches the tensile strength, the interface between the layers of the weld will shrink and eventually fracture due to severe strain localization. For HF-1 sample, after the yield stage, the flow stress continues to increase, and the necking occurs in the filling layer with the largest thickness (the fourth layer). When the flow stress reaches 690MPa, the instantaneous strain of HF-1, HF-2, and HF-3 samples is significantly greater than HF, of which HF-3 is the largest, 32.64% higher than HF(Fig. 2b).

The mechanical properties of each layer and the interface between layersin the multi-layer welded are different. The reason is that the normalizing effect of the latter layer on the former layer during welding refines the grains, thus improving the mechanical properties. The grain refinement effect produced by normalizing in the fourth layer of HF-1 sample is not obvious, which will lead to the difference of grain size. In the elastic stage, when the flow stress is small, the strain is basically evenly distributed. In the plastic stage, the fine grain area in the fourth layer of weld continuously hardens, and the strain capacity is poor. Therefore, with the increase of the flow stress, the strain will transfer to the uneven grain size area with low hardening degree and strong deformation capacity. In other words, the strain will transfer to the weak area and fracture will occur eventually. When the preheating temperature and interpass temperature are insufficient, the normalizing effect of HF-2 and HF-3 samples is weakened, resulting in relatively coarse grains in the layer, while the grain size at the interpass interface is small, resulting in the non-uniformity of microstructure. As a typical non-uniform material, the complex and diverse internal microstructure of weld will cause serious strain incompatibility in the plastic deformation stage[8-9]. For HF-2 and HF-3 samples, the flow stress increases continuously, and the resulting strain incongruity occurs at the interface between the inner layers of the weld, and finally necking occurs, and the failure occurs at the interface.

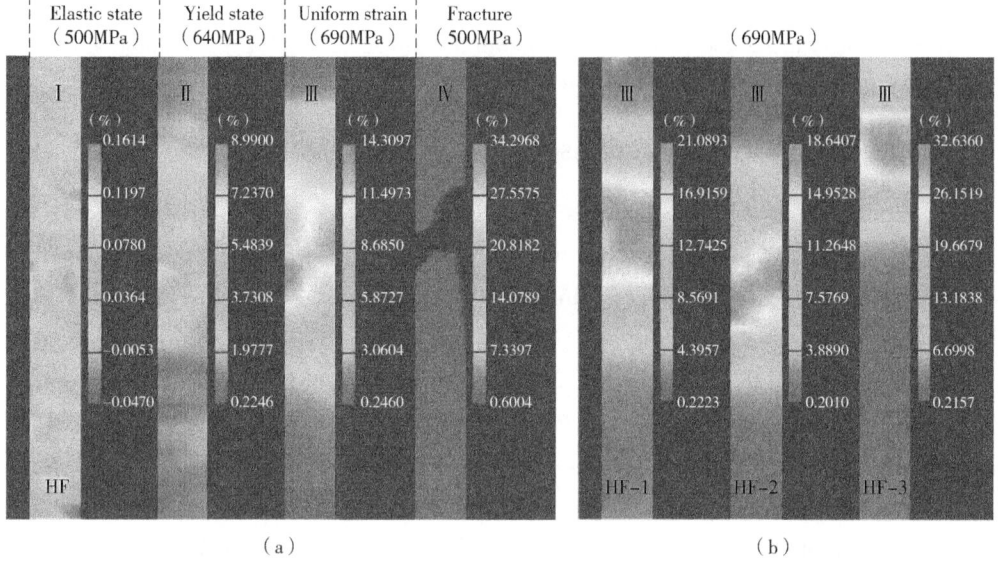

Fig. 2 (a) DIC strain distribution of weld metal during deformation of HF sample,
(b) DIC strain distribution of HF-1, HF-2 and HF-3 samples when flow stress is 690MPa

3.2 Mechanical properties and hardening behavior of welds with different welding process defects

Fig. 3 and Table 3 show the stress-strain curves and mechanical properties of X80 steel welds with different welding parameters. HF sample has high yield strength and tensile strength, which are 640MPa and 749MPa respectively, and the uniform elongation is 5.79%. In this experiment, when the welding parameters deviates such as insufficient preheating temperature and insufficient interpass temperature, the residual stress increases linearly, resulting in a significant decrease in the strength of HF-2 and HF-3 specimens, as well as a decrease in the uniform elongation. Stress reduction is due to stress release during normalizing process. Because normalizing can make grain refinement, material distribution uniform. The cooling process after normalizing can not only release the stress, but also improve the toughness and plasticity of the material. Deng[10] took the flat plate butt welding as the research object, and simulated the welding process by using the finite element analysis method. It was found that the residual stresses in the middle section of the welding were tensile stresses. When the temperature of the welding heat source increases, the longitudinal residual stresses will decrease linearly. In multi pass welding, the latter layer of welding has normalizing effect on the previous layer of welding, which can improve the morphology of the previous layer of welding microstructure and reduce the welding residual stress, and with the increase of the number of welding layers, the welding residual stress is relatively reduced[11]. The distribution of welding residual stress in joints is generally symmetrical along the weld center line, mainly concentrated in the vicinity of weld and heat-affected zone. During welding, the heat affected zone and the first completed weld undergo multiple heating and cooling processes. After the formation of the previous weld, it turns into the

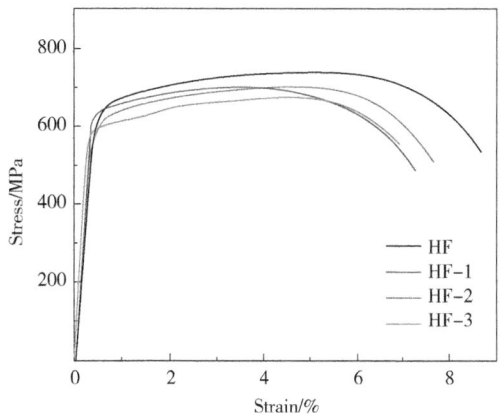

Fig. 3 Stress strain curves of welds with different welding parameters

heat affected zone of the next adjacent weld, and its residual stress is close to the welding heat affected zone, which is also a typical feature of multi-pass welding. That is to say, the first weld has experienced the most welding heating and cooling cycles, and the residual stress gradually decreases under the welding normalizing effect. In this experiment, the HF-1 sample has reduced the number of welding layers, resulting in the transverse and longitudinal shrinkage of the previous layer after cooling, in which the longitudinal shrinkage deformation will produce large tensile residual stress [12]. The thickness of the fourth layer of HF-1 welding is large, and the normalizing effect generated by the latter welding is not enough to eliminate the residual stress generated before in a short time. The less the number of welding layers, the more obvious this phenomenon is. In the tensile process of HF-1, the combined effect of residual tensile stress and external stress in the weld seam acts on the fourth layer with large thickness in the filler layer of the weld seam. The failure occurs after stress concentration in the weak part of the layer, leading to the decline of mechanical properties, especially the sharp decline of toughness. The yield strength and tensile strength are reduced to 621MPa and 697MPa respectively, and the uniform elongation is reduced to 3.96%.

Table 3 Mechanical properties of welds with different welding parameters

Sample	Yield stress/MPa	Tensile stress/MPa	Yield ratio	Uniform elongation/%
HF	640	749	0.85	5.79
HF-1	621	697	0.89	3.96
HF-2	610	701	0.87	5.08
HF-3	594	672	0.88	4.81

Kocks Mecking(K-M) strain hardening model is used to describe the relationship between flow stress and strain hardening rate of materials[13]. The five stages of strain hardening and the corresponding hardening mechanism have been confirmed[14]. The hardening process of weld materials mainly includes three stages. The corresponding third-order strain hardening behavior of K-M curve can be divided into stage III, stage IV and stage V, as shown in Fig. 4. The remarkable

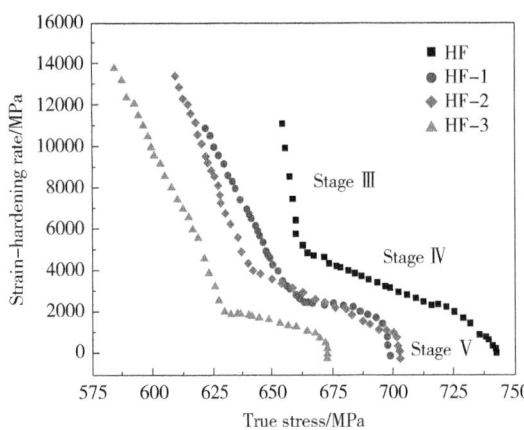

Fig. 4 Stress strain curves of welds with different welding parameters

feature of stage Ⅲ is that the strain hardening rate decreases with the increase of flow stress, which leads to the loss of hardening rate due to the dynamic recovery of micro structures such as dislocations. Under the large flow stress in the stage Ⅳ, the hardening rate caused by incomplete dynamic recovery and dislocation stacking is improved again. Finally, a sudden drop in the hardening rate can be observed at stage Ⅴ, which indicates the formation of internal micropores and other defects and the end of strain hardening behavior[15].

The flow stress value and strain hardening rate of transition between different stages in K-M curve are shown in Table 4. Under different welding parameters deviation, the flow stress and strain hardening rate of the transition from stage Ⅲ to stage Ⅳ and from stage Ⅳ to stage Ⅴ are significantly reduced, especially for the HF-3 specimen, the strain hardening ability is deteriorated rapidly. Inadequate number of layers, preheating temperature and interpass temperature will cause stress concentration in the weld. Under the combined effect of stress concentration and external stress, multiple slip systems in the weld grain will reach the critical cutting stress in advance, forcing the activation and proliferation of internal dislocation sources, increasing dislocation density and deformation capacity[16].

Table 4 Flow stress and strain hardening rate obtained from K-M analysis

Sample	$\sigma_{Ⅲ-Ⅳ}$/MPa	$\sigma_{Ⅳ-Ⅴ}$/MPa	$\theta_{Ⅲ-Ⅳ}$/MPa	$\theta_{Ⅳ-Ⅴ}$/MPa
HF	662	719	4985	2193
HF-1	661	691	2492	1543
HF-2	645	698	3865	1058
HF-3	629	669	1084	825

3.3 Tensile strength of weld metal under different welding processes

In order to explore the fracture behavior of weld caused by deviation of welding parameters, four kinds of weld tensile fractures were analyzed by SEM. A large number of dimples are distributed in the fracture surface of HF sample. These dimples are flat, densely distributed and uniform, indicating that ductile fracture is dominant(Fig. 5a). For HF-1 sample, the fracture dimple is flat, containing a large number of cleavage surfaces and dissociation steps(Fig. 5b). When the number of welding layers decreases, welding residual stress will be generated. When the external stress increases during tensile deformation, both of them will act on the weak part in the layer to produce stress concentration. Due to the large degree of hardening in the non weak area, the strain is difficult to continue to occur in this area, so it is easier to move to the weak area with weak hardening, and necking will occur. This shows that ductile fracture occurs in the normal area of HF-1, accompanied by brittle cleavage in the weak areas such as holes at the grain boundary, which is the direct reason for the toughness reduction of HF-1. The flat dimples at HF-2 and HF-3 fractures are uniformly arranged, and there are a few cleavage surfaces in local areas (Fig. 5c~d). This is related to the inadequate preheating temperature and interpass temperature during welding. The inadequate heat source temperature causes the difference of interpass structure and the accumulation of defects such as slag between the layers [17], resulting in poor deformation coordination ability. When the plastic deformation degree increases, the stress concentration is easily formed near the interface without effective release, and finally cracks along the grain boundary to form a brittle fracture zone. Bastola et al. [18] found that at the necking position of X80 pipeline steel weld, it is believed that there are inclusions, second phase particles etc. at the necking position of pipeline steel. These impurities will form defects such as micropores, and then grow through, eventually causing fracture. A dimple with pit characteristics appears at the fracture. For weld metal, the fracture mode is greatly affected by welding process.

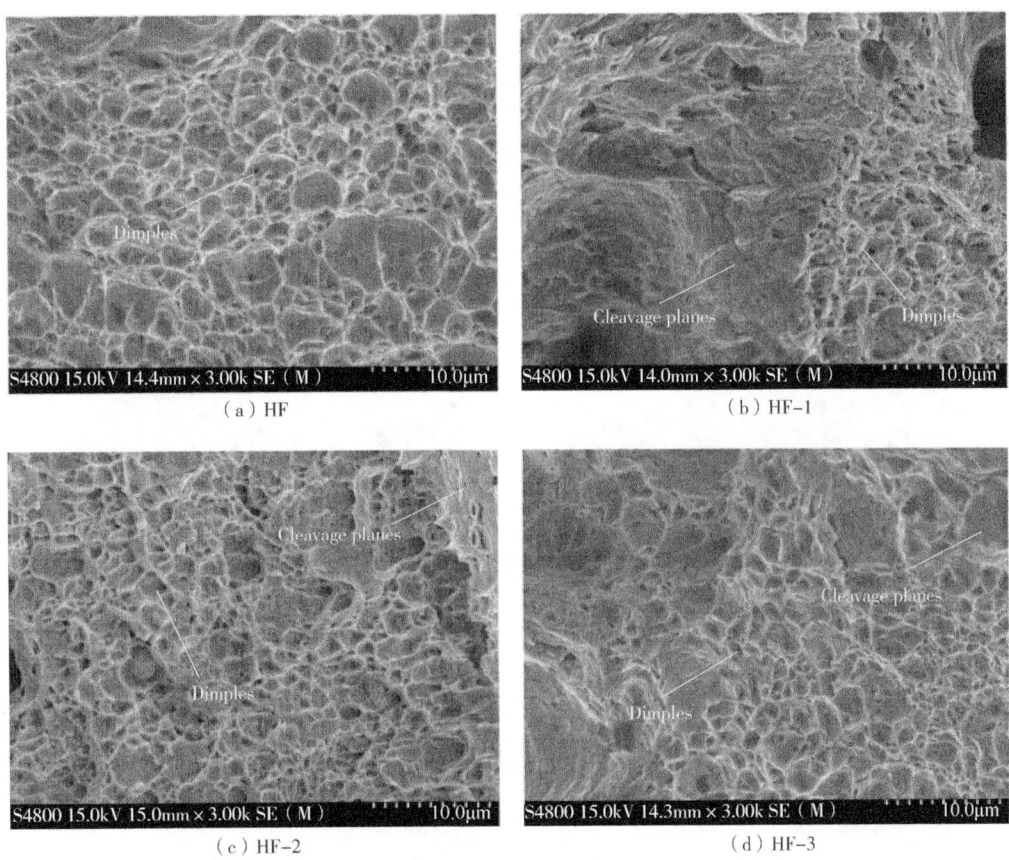

Fig. 5 Fracture morphologies of welds with different welding parameters

The internal mechanism of fracture was further studied and clarified. As shown in Fig. 6, the local average orientation difference (KAM) diagram of the section along the tensile direction of four kinds of welds was obtained by EBSD technology. The average orientation difference between the pixel and other adjacent pixels is defined as the KAM value of the pixel. The KAM diagram shows the defect density, which can reflect the strain history near the weld fracture [19]. Almost every grain in HF sample has experienced plastic deformation, and there is a certain strain gradient between grains(Fig. 6a), which is related to the grain size difference caused by welding and the strain distribution behavior of hard and soft phases[20]. At the same time, the strain gradient phenomenon here is weaker than that of HF-1, HF-2 and HF-3 samples, and residual broken grains can be seen at the edge of the fracture(Fig. 6), which is the result of the original grains being torn during deformation. The plastic deformation of HF-1 sample is obviously concentrated at the grain boundary (Fig. 6b), which leads to severe strain gradient in the grain interior and at the interface. At the same time, micropores appear at the grain boundary, indicating that intergranular fracture occurs at the grain boundary, while transgranular fracture occurs at the edge of the fracture. For HF-2 and HF-3 samples, severe plastic strain regionalization occurs in the grain, and the strain degree near the grain boundary is greater in the grain interior, so the grain boundary is a weak area, and intergranular fracture occurs. In addition, some torn broken grains were observed at the fracture edge, where transgranular fracture occurred(Fig. 6c~d). To sum up, the fracture mechanism of weld specimen depends to some extent on the welding parameters. HF specimen is a ductile fracture mechanism, and HF-1, HF-2, and HF-3 specimens are all a mixed fracture mechanism of toughness and brittleness. For HF-1 specimen, the brittleness mechanism is dominant. In addition, for welded samples, welding residual stress will be produced due to thermal expansion during welding. Although the existence of normalizing during multi-layer welding can eliminate part of the stress, negative expansion in the cooling process will make part of the stress remain in the sample, leading to mild brittle fracture.

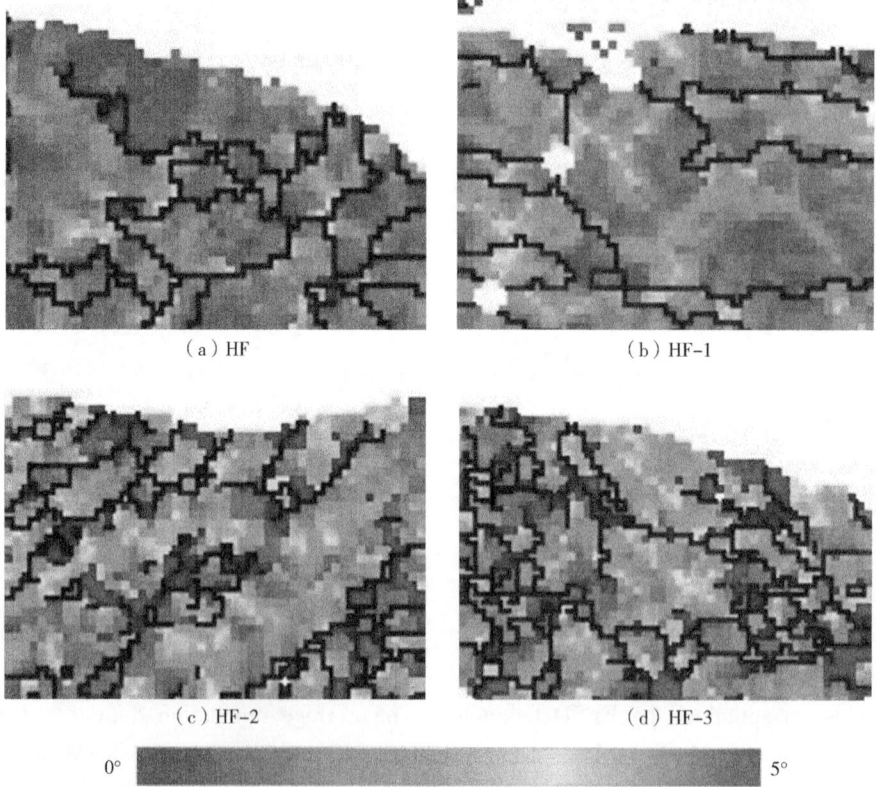

Fig. 6 Kernel average orientation(KAM) maps of weld section for different welding parameters

4 Conclusion

(1) When the welding parameters deviate, the strain distribution will be uneven, and the necking position will also change. When preheating temperature and interpass temperature are insufficient, necking occurs at the interface between weld layers, and when the number of welding layers is insufficient, necking occurs at the weak area in the filler layer.

(2) When the welding parameters deviate, the welding residual tensile stress will be generated in the weld, and the hardening rate will decrease, leading to the decrease of the weld strength. Especially when the interpass temperature is insufficient, the yield strength will decrease more significantly.

(3) Under normal welding parameters, the weld is ductile fracture mechanism; When the welding parameters deviate, the fracture mechanism is a mixture of toughness and brittleness; When the number of welding layers is insufficient, the brittleness mechanism is dominant; When preheating temperature and interpass temperature are insufficient, ductile fracture is the main mechanism.

Declaration of Competing interests

The authors declare that they have no known competing financial interests or personal relationships that could have appeared to influence the work reported in this paper.

Acknowledgments

This work was supported by the Science & Technology Project of CNPC(2021DJ2804).

References

[1] Wang N, Chen Y, Wu G, et al. Investigation on micromechanism involved in ferrite hardening after prestraining of dual-phase steel[J]. Materials Science and Engineering: A, 2021, 800: 140387.

[2] Qiao G, Chen X, Zhang Z, et al. Mechanical Properties of High-Nb X80 Steel Weld Pipes for the Second West-to-East Gas Transmission Pipeline Project[J]. Advances in Materials Science and Engineering, 2017, 2017: 1-13.

[3] Kong D, Ye C, Guo W, et al. Tensile fracture behaviors and fracture morphologies of X80 pipeline steel welded joint[J]. Journal of Building Materials, 2015, 18(1): 139-144.

[4] Tang X, Wang L, Chen Y, et al. Influence of welding layers on welding residual stress and deformation [J]. Welding Technique, 2014, 43(4): 5-8.

[5] Wang X, Liu H, Tang X, et al. Effect of heat input on defects, microstructure and mechanical properties of welded joint of Q235 steel[J]. Transactions of Materials and Heat Treatment, 2022, 43(4): 162-169.

[6] Wen H. Welding of X80 steel pipeline for long-distance pipeline[J]. Equipment manufacturing technology, 2013(2): 114-115.

[7] Shi X, Tian W, Chen L. Effects of Preheating Temperature on Microstructure and Mechanical Properties of Red-Copper MIG Welding Joints[J]. Hot Working Technology, 2023(7): 117-121.

[8] Wu X, Zhu Y. Heterogeneous materials: a new class of materials with unprecedented mechanical properties [J]. Materials Research Letters, 2017, 5(8): 527-32.

［9］ Ashby M. The deformation of plastically non-homogeneous materials［J］. The Philosophical Magazine: A Journal of Theoretical Experimental and Applied Physics, 2006, 21(170): 399-424.

［10］ Deng X. Influence of welding parameters on welding deformation and residual stress［J］. China Ship Repair, 2018, 31(2): 28-31.

［11］ Wen Q, Wu Z. Analysis of Multi-layer and Multi-pass Welding［J］. Equipment Manufacturing Technology, 2011(11): 145-146.

［12］ You M, Zheng X, Yu H. Discussion and investigation on mechanism of welding residual stress in mild steel ［J］. Transactions of The China Welding Institution, 2003, 2.

［13］ Mecking H, Kocks U. Kinetics of flow and strain-hardening［J］. Acta Metallurgica, 1981, 29(11): 1865-1875.

［14］ Rollett A D, Kocks U. A review of the stages of work hardening［J］. Solid state phenomena, 1993, 35: 1-18.

［15］ Galindo E, Sietsma J, Rivera P. Dislocation annihilation in plastic deformation: II. Kocks-Mecking Analysis ［J］. Acta materialia, 2012, 60(6-7): 2615-2624.

［16］ Huang T, Gou R, Dan W, et al. Strain-hardening behaviors of dual phase steels with microstructure features ［J］. Materials Science and Engineering: A, 2016, 672: 88-97.

［17］ Wang X, Liu H, Tang X. Effect of welding method on defect, microstructure and mechanical properties of welded joints of Q235 steel［J］. Transactions of Materials and Heat Treatment, 2021, 42(7): 150-159.

［18］ Bastola A, Wang J, Shitamoto H, et al. Investigation on the strain capacity of girth welds of X80 seamless pipes with defects［J］. Engineering Fracture Mechanics, 2017, 180: 348-365.

［19］ Calcagnotto M, Ponge D, Demir E, et al. Orientation gradients and geometrically necessary dislocations in ultrafine grained dual-phase steels studied by 2D and 3D EBSD［J］. Materials Science and Engineering: A, 2010, 527(10): 2738-2746.

［20］ Wang N, Chen Y, Wu G, et al. Influence of microstructural heterogeneity on strain partitioning behavior of metals containing columnar grains［J］. Materials Science and Engineering: A, 2021, 814: 141247.

本论文原发表于《Journal of Materials Research and Technology》2023 年第 22 期。

三、装备质量基础与服役安全

从技术视角看北溪管道泄漏事件

刘亚旭　李为卫　霍春勇　赵新伟　池　强　付安庆

(中国石油集团工程材料研究院有限公司，石油管材及装备材料服役行为与结构安全国家重点实验室)

摘　要：穿越波罗的海从俄罗斯输送大量天然气到德国的北溪管道，是目前世界上输送能力最大、长度最长、管径最大的海底天然气管道，技术含量高，对我国天然气资源的引进及管道建设与运行具有重要参考价值。本文介绍了北溪管道的概况，着重从技术角度分析了其断裂和泄漏的原因，以及修复的可行性，讨论了其对中俄天然气合作的影响，并就中俄油气管线建设关键技术研究提出了建议。

关键词：北溪管道；泄漏；断裂；修复；中俄天然气管线

世界上最长、输送能力最高、花费巨大、技术含量高，号称欧洲能源长期、安全、可靠大动脉的北溪天然气海底管道，随着俄乌冲突的升级陷入断输状态[1]。两年前，即将完工的北溪管道二期就陷入大国角逐的政治漩涡，成为2023年引起俄乌冲突的焦点之一，今天的停输状态及未来能否恢复运行的不确定性，使其极有可能沦为政治和军事冲突的牺牲品，不免让人惋惜，尤其是对从事能源行业的人们。北溪管道的真实断裂原因和实际损坏情况有待于现场调查和失效分析，因政治、军事、经济、技术等各种复杂因素，北溪管道未来的命运难以预测。本文介绍了北溪管道的技术概况，并根据现有资料和多年科研积淀，从技术视角对北溪管道的断裂原因、修复可行性，以及对中俄天然气合作的影响进行分析，对即将建设的中俄蒙天然气管道提出建议。

1　北溪管道概况

俄罗斯有丰富的天然气资源，欧洲天然气消费量巨大，苏联和俄罗斯迄今共修建通往欧洲的8大天然气管道，输气能力总计$3090×10^8 m^3/a$，北溪管道(Nord Stream)是其中输气能力($1100×10^8 m^3/a$)最大的项目。北溪管道包括北溪-1和北溪-2两条管道，线路走向和参数基本相同，从俄罗斯波托瓦亚湾的维堡、乌斯季卢加出发至德国格雷夫斯瓦尔德附近的卢布明，穿过波罗的海俄罗斯、芬兰、瑞典、丹麦和德国的专属经济区，以及俄罗斯、丹麦和德意志的领海，管道由注册于瑞士的公司Nord Stream AG建造和运营[1]。

北溪项目两期管道的技术参数几乎完全相同。北溪-1管道系统是由2条平行的海底管道组成，每条管道长1224km，沿途最大水深为210m，年输气能力均为$275×10^8 m^3$，设计寿命50年。由于该项目加强了欧盟能源市场并加强了供应安全，该项目曾被欧洲议会和理事

作者简介：刘亚旭，男，1968年生，正高级工程师，2007年毕业于西安交通大学管理科学与工程专业，获博士学位，现任中国石油集团工程材料研究院有限公司党委书记、执行董事，主要从事石油管工程技术和科技管理等工作。E-mail：liuyaxu@cnpc.com.cn。

会指定为"欧洲利益"项目[1]。

表1为北溪-1管道与世界上另一条著名的兰格里德(Langeled)海底管道(始于挪威海域,止于英格兰东海岸的Easington,据称是世界上最长的海底管道)技术参数的对比,可见,北溪管道的技术参数总体上高于Langeled海底管道,是目前世界上距离最长、综合技术参数最高、输送能力最大的海底天然气管道。

表1 北溪-1管道与Langeled管道技术参数对比

管道名称	年输气量/$10^8 m^3$	长度/km	管道外径/mm	设计压力/MPa	钢级	壁厚/mm	钢管量/10^4t	混凝土涂层量/10^4t	总投资/亿欧元
Langeled管道	200	1171	1067,1118	25.00,21.50,15.68	X70/L485	23.3~35.9	100	100	25
北溪-1管道	550	2×1224	1219	22.00,20.00,17.75	X70/L485	26.8~41.0	220	240	>74

北溪管道的独特之处在于,其采用了一泵到底的高压输送方式,中间没有加压站,从而节省了大量建设和运行成本[1-3]。管道在俄罗斯进气口的压力为22.0MPa,在德国出气口的压力为10.6MPa。为了保证环缝焊接的质量,以及便于清管器和内检测器通过,该管道全程采用内对齐方式,公称外径为1219mm(随壁厚变化),整条管线内径设计为1153mm。为节约管材总需求量、同时减少焊接量从而加快建设速度,该管道全线采用分段设计方法,从俄罗斯入海口到德国的登陆点整个长度共分为3段,设计工作压力分别为22.00MPa、20.00MPa、17.75MPa,前三分之一段钢管壁厚为34.6mm,中段壁厚为30.9mm,后三分之一段壁厚为26.8mm,在某些特殊地段的壁厚达41.0mm。

北溪-1管道采用的钢管钢级为L485/X70,单根管长12m,钢管内部有减阻涂层,外部有聚乙烯防腐(防止海水的侵蚀),最外部有60~100mm厚的钢丝混凝土配重层。钢管的主要制造商是位于德国穆勒海姆德鲁尔的EUROPIPE Gmbh,从2008年4月至2011年9月,该公司为北溪-1项目交付了15万根钢管(工程总用量近20万根)[1]。北溪-1还使用了38根壁厚分别为35.0mm、35.4mm、38.0mm,采用感应加热弯制、热处理炉整体加热淬火+回火工艺生产,具有良好低温(-38~-25℃)韧性的L450/X65和L485/X70感应弯管[4]。

为了提高施工效率,在铺管船上首先采用高效率的自动埋弧焊方法将两根钢管先焊在一起,形成双联管,双联管环焊缝超声检测合格后,进行管道环缝安装焊接,焊接方法为熔化极气体保护自动焊。环焊缝检验合格后进行防腐补口和装填保护材料,最后沿铺管架放入海水中。整个工程分三段施工,每段焊接工作完成后,分段充水进行压力测试,试验压力超过最高工作压力25%,测试时间为24h。测试合格后三段管段在水深80~110m的水下由专业的公司进行连头焊接。

为了保证管道的安全运行,管道建设完成采用内检测器(PIG)进行基线扫查,并定期进行缺陷检测。北溪-1管道自2011年运行以来,截至本次泄漏爆炸事件前尚未出现失效事故,北溪管道在水下情况如图1所示[1]。

北溪-1的两条管道已分别于2011年11月和2012年10月投入使用。自北溪-1开通和北溪-2开工建设,美国就始终反对这两条直接连接德国和俄罗斯的基础设施项目,反

图 1 北溪管道在水下示意图

复以它们是"地缘政治"工程为名向德国施加压力，要求德国放弃，并不断以各种手段从中阻挠。2021 年 9 月，在北溪-2 完成管线铺设即将运营之际，德国审批机构暂停管道项目的资格认证程序。乌克兰危机爆发后，欧俄天然气争端愈演愈烈，不仅北溪-2 无法投入使用，连北溪-1 在 2022 年 8 月底开始也进入停运状态[5]。

2 北溪管道泄漏事件

根据管道运营商 Nord Stream AG 官方网站消息[1]，2022 年 9 月 26 日控制中心发现北溪-1 两条管道压力下降，9 月 27 日推测为物理损坏所致，11 月 2 日发布了北溪-1 管道损坏现场检查的初步结果：在相距约 248m 的海床上发现了深度 3～5m 的技术成因坑，坑之间的管道部分被破坏，管道碎片散布半径至少为 250m。北溪-1 两条管道三处受损，北溪-2 管道 A 线一处受损，B 线完好。已发现四处管道的泄漏中，瑞典测量站 2022 年 9 月 26 日探测到两次强烈的水下爆炸，瑞典《快报》10 月 18 日公布了 50 多米的管道断裂并落入海床的画面。

国际上关于油气管道失效概率统计数据因口径不同而有差异，综合分析美国 DOT、加拿大 NEB 等权威数据库统计，欧洲、加拿大、英国、美国的天然气管道的失效概率分别约为 0.16 次/(10^3km·a)、0.10 次/(10^3km·a)、0.23 次/(10^3km·a)、0.11 次/(10^3km·a)，我国天然气管道失效概率约为 0.16 次/(10^3km·a)。海底管道从设计、管材制造到施工等各环节质量控制更加严格，失效概率一般更低。Nord Stream AG 官网甚至声称其管道故障或泄漏的概率低至 100000 年发生一次[1]。北溪-1 管道从 2011 年投运至本次事故前，未发生泄漏事故。从失效概率分析，管道 4 处一天内几乎同时断裂泄漏，管道本身的技术、质量或环境因素导致的可能性非常小，人为破坏可能性大。

李鹤林院士[6]将油气管道的失效原因划分为：外部干扰、腐蚀、焊接和材料缺陷、设备和操作等几类，北溪管道的失效原因属于外部干扰。什么破坏方式能造成如此厚的高强度钢管断裂？据丹麦和瑞典 2022 年 9 月 30 日发布的调查报告称，这是一系列的"人为蓄意破坏"所致，两条管线共发生四起水下爆炸，爆炸当量达到数百千克 TNT 炸药[5]。根据瑞典公布的管道断裂视频和图片(图 2 和图 3)，分析认为有可能采用线性聚能切割器实施爆炸破坏至管道断裂。

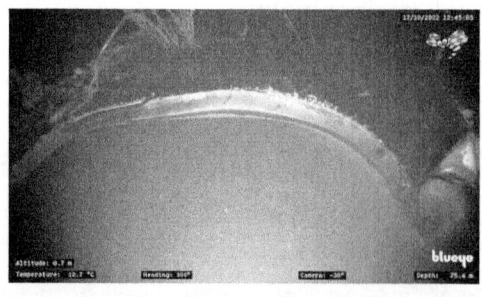

图 2 北溪-1 管道爆炸断口照片 1

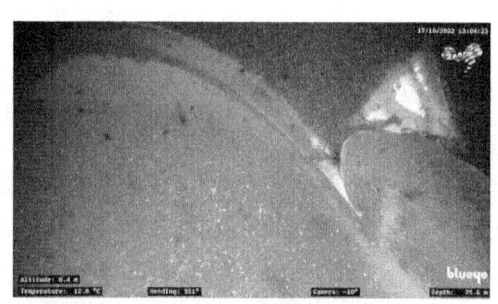

图 3 北溪-1 管道爆炸断口照片 2

工程材料院曾多次开展天然气管道止裂韧性爆破试验，委托具有爆破资质的专业公司实施爆破[7]。试验采用线性聚能切割器在外径1422mm、壁厚21.4mm、内压12MPa的X80管道上瞬间切出500mm长、宽度约10mm的纵向穿透型裂纹，如图4所示，然后在管道内高压气体的作用下，裂纹进一步扩展，最后止裂，根据测试的数据评估高强度钢管的止裂能力，管道最后断裂的断口形貌与图2和图3非常相似。该试验的线性聚能切割器采用了约0.5kg的TNT和RDX的混合炸药制作的聚能药包，据此推测，要将北溪管道直径1219mm、厚度26.8mm的钢管沿环向完全爆破切割，不考虑钢管外60~100mm厚的水泥配重层，约6kg的聚能炸药就可以实现。

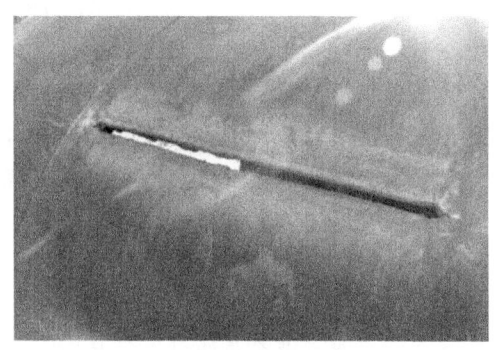

图4 天然管道止裂韧性试验聚能切割器在钢管上产生的裂纹

3 北溪管道泄漏后续预测与中俄天然气合作机遇

3.1 北溪管道泄漏后续预测

北溪管道泄漏事件在调查程序上疑点重重，整个调查工作呈现出碎片化的特点。2022年10月初，瑞典拒绝了俄罗斯对北溪天然气管道泄漏事件进行调查的要求，瑞典检察院封锁了事发海域，不让俄罗斯调查组进驻事故现场进行实地考察。目前的调查工作只能由瑞典、丹麦、德国和俄罗斯分别展开。俄罗斯天然气工业股份公司2022年10月13日称，从泄漏的速度来看，包括通往俄罗斯的数百千米管道已经充满海水，如想全面恢复北溪管道的运营，需要拆除更换很大一部分管道，相当于铺设一条全新管道[5]。

海水已从断口进入管道内部，部分管道内表面已暴露于海水中，面临海水的腐蚀风险。海水是含盐量相当高的腐蚀性介质，其中的氯化物、微生物等将增加腐蚀活性，破坏金属表面的钝化膜，引起点蚀、溃疡等局部腐蚀，导致管道管壁减薄直至穿孔[8]。某X65海底输油管线（φ323.9mm×12.7mm）因施工不当，海水进入管道内部，浸泡6个月后经吹扫投运16个月即发生腐蚀穿孔，工程材料院解剖分析发现管道内表面发生大面积点蚀，局部最大腐蚀速率达6.9mm/a，如图5所示。某海底双金属复合管线[316L/X65，φ457mm×(3+15.9)mm]因第三方破坏致海水进入管道，6个月后管道内检测发现3mm厚的316L不锈钢内衬层发生腐蚀穿孔，经工程材料院分析是由于氯离子和微生物协同腐蚀所致，如图6所示。

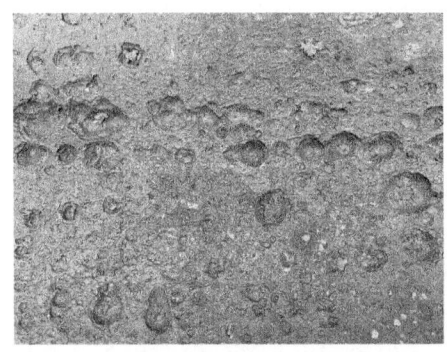

图5 某海底X65输油管线内壁腐蚀

图6 某海底316L双金属复合输油管线内壁腐蚀

尽管北溪管道内表面有涂层，但主要起减阻作用，防腐性能较差，而且遭海水浸泡后会发生脱落、鼓泡等，易造成局部腐蚀，管道在海水中暴露时间越长，腐蚀越严重，根据经验，北溪管道应在6个月之内修复，否则一旦发生大范围严重的局部腐蚀，修复极其困难，甚至要重新敷设部分管段。

北溪管道运营商宣称其具备管道维修的资源和技术能力，但维修成本高、周期长。目前，北溪管道控股方俄气已获准参加事故调查，但截至2022年12月12日，Nord Stream AG官网仍没有管道泄漏事件调查及维修的进一步信息。

3.2 中俄天然气合作机遇

从俄乌冲突导致德国等欧洲国家努力摆脱对俄罗斯能源依赖的情势看，无论北溪管道能否恢复使用，或者俄罗斯通过土耳其向欧洲供气的目的能否达到(现有"蓝溪"和"土耳其溪"两条管道，供气能力 $680×10^8 m^3/a$)，俄罗斯都将努力扩大东方的天然气出口。这将为我国加强与俄罗斯的天然气合作、扩大四大能源战略通道的进口天然气能力、进一步保障国家能源安全带来机遇。

早在1995年，中国石油就安排规划总院、管材研究所等单位针对中俄天然气管道开展科研和技术储备，2015年中俄东线动工修建，2019年12月北段通气[9]，截至2021年12月供气量累计仅 $136×10^8 m^3$。俄乌冲突后，俄罗斯加快了与我国能源合作步伐。2022年9月15日，中俄蒙元首第六次会晤商定积极推进中俄天然气管道过境蒙古国建设项目。同日，俄罗斯副总理诺瓦克表示将于2024年动工修建中俄天然气管道"西伯利亚力量-2"，起于亚马尔半岛、途经蒙古国，设计输气量 $500×10^8 m^3/a$。

2022年11月中旬，俄罗斯副总理阿布拉姆琴科透露，途经蒙古国接往中国的"东方联盟"天然气管道项目的可行性研究目前已经获批，现场勘察和研究的相关工作正在稳步推进，设计工作将会在2023年全部完成。这条管道如能实施，加上中俄东线 $380×10^8 m^3/a$ 的输量，以及2022年2月4日冬奥会期间中国石油与俄气签署的《远东天然气购销协议》中新增 $100×10^8 m^3/a$ 的输量[10]，我国来自俄罗斯的天然气将有望达到 $980×10^8 m^3/a$，这对保障我国天然气进口多元化和供应安全将起到重要作用。

4 结语与建议

北溪管道泄漏是国际能源行业的一次重大事件，其命运如何，有待持续观察。乌克兰危机导致俄罗斯被西方孤立并一定程度上陷入困境，俄罗斯加快了与我国的能源合作。为促进中俄蒙天然气管道顺利建设，从技术角度提出以下建议：

(1) 针对中俄蒙管道建设开展技术对接。中国石油在管道建设方面(管材研制、标准制定、管道设计、钢管及管件制造、施工等)具有显著技术优势，与俄气(Gazprom)有着较密切的合作关系，"十三五"以来开展了多项科研合作。建议与俄气和蒙古国的实施企业进行技术对接，争取采用我国具有国际先进水平的标准、管材和施工服务。

(2) 重视对超大输量天然气管道建设关键技术的研发，确保陆上管道科技持续领跑。根据工程材料院前期科研成果，2000年西气东输规划建设时，李鹤林院士、黄志潜总工建议采用1016mm管径、X70钢级、10MPa输送压力代替1118mm管径、X65钢级、8.4MPa的原设计方案，设计年输量 $120×10^8 m^3$，技术上实现了巨大跨越；2006年西二线规划建设时，工程材料院建议采用X80钢级、12MPa输送压力代替X70钢级、10MPa双线方案，不仅节约大量投资，而且年输量达 $300×10^8 m^3$，推动我国高强度管道技术进入国际领跑行列，我国

X80高钢级管道占全世界50%以上。受2017年、2018年中缅管道两次断裂事故影响，国内更高钢级的大输量管道研究与应用陷入停滞，我国2015年研制的X90钢管至今难以落实试验段。中俄蒙天然气管道是超大输量管道，从输送效率和经济性综合考虑，如果采用X90或更高钢级管材、更高输送压力，一条管道有望实现年输量$500×10^8 m^3$。建议在国家科技计划中设立相应项目，在保证中俄蒙管道如期高质量建成的同时，使我国陆上管道技术保持国际领先水平。

（3）重视海底管道的科研，储备海洋油气资源输送技术。我国在海洋管道建设运行方面有了技术基础，工程材料院也牵头完成了"高应变海洋管的研制"国家重点研发计划项目，但我国海底管道整体上相较北溪管道等国际先进水平存在较大差距。建议设立国家重点研发计划和国际合作项目，提升我国的海洋管道技术水平，为南海油气资源的开发做好技术储备。

参 考 文 献

[1] Nord Stream AG. Incident on the Nord Stream Pipeline[EB/OL].[2022-11-02]. https://www.nord-stream.com/press-info/press-releases/.

[2] HILLENBRAND H G，KALWAC，SCHROEDER J. Meeting highest requirements for the challenge of the Nord Stream project[C]. Proceedings of the Pipeline Technology Conference，Ostend，Belgium，October 12-14，2009.

[3] 董瑾. 北溪管道用管线钢管的质量控制述评[J]. 西安石油大学学报(自然科学版)，2022，37(4)：112-119.

[4] 吉玲康，董瑾. 北溪管道用感应加热弯管的生产和质量水平[J]. 石油管材与仪器，2021，7(5)：86-89.

[5] 赵晨. "北溪"管线泄漏事故：加速欧洲能源地缘政治重构[J]. 世界知识，2022(11)：39-43.

[6] 李鹤林. 油气管道失效控制技术[J]. 油气储运，2011，30(6)：401-410.

[7] 池强，杨坤，李鹤，等. 高钢级管道全尺寸气体爆破试验技术研究[J]. 焊管，2019，42(7)：78-82.

[8] 金伟良，张恩勇，邵剑文，等. 海底管道失效原因分析及其对策[J]. 科技通报，2004，20(6)：529-533.

[9] 新华网. 俄罗斯天然气通过中俄东线天然气管道正式进入中国[EB/OL].（2019-12-2）[2022-12-10]. http://www.xinhuanet.com/world/2019-12/02/c_1125299431.html.

[10] 祁治业. 俄罗斯天然气出口重心正在东移[J]. 世界知识，2022(17)：52-53.

本论文原发表于《石油管材与仪器》2023年第9卷第1期。

基于 FCE-SEW 模型的某工程项目风险评估研究

梁晓斌[1]　马卫锋[1]　谭朝成[2]　陈　丹[2]　贺雅丽[3]

(1. 中国石油集团工程材料研究院有限公司石油管材及服役行为与结构安全国家重点实验室；
2. 长庆油田分公司第十采油厂；3. 华陆工程科技有限责任公司)

摘　要：充分识别、科学评价并控制工程项目现场的安全风险可以将事故隐患消灭在萌芽期，对预防事故尤其是重特大事故的发生具有重要意义。针对目前工程项目现场常用风险分析方法存在的工作程序复杂、风险评估主观性大、风险评估结果不够准确等缺陷，本文建立了基于模糊综合评价法(FCE)和结构熵权法(SEW)的工程项目现场安全风险评估模型。根据某工程项目现场的实际情况出发，从现场制度因素、现场执行因素、现场监管因素、现场配置因素、安全教育因素5个方面构建工程项目现场风险评估指标体系，包含5个二级指标和32个三级指标。结构熵权法可以兼顾主观性和客观性，解决了各个指标因素存在的不确定性和认知误差问题。模糊综合评价方法可以将专家定性的评价结果定量化，充分考虑专家评价过程中的模糊性问题，实现对工程项目现场系统全面的评估。现场实例分析表明，工程项目现场的安全风险评估等级为中等风险，得出的结果贴合实际情况，为现场安全风险的分级管控和隐患排查治理提供技术支撑。

关键词：工程项目现场；结构熵权法；模糊综合评价；安全风险评估

建筑施工行业具有劳动密集，作业人员素质相对薄弱，作业环境复杂，安全风险因素众多等特点，多因素的共同作用使得生产安全事故频繁发生，对人民的生命财产安全造成严重威胁，已被国家列入高危行业[1]。特别是近年来，建设工程作业面由低层向高层、超高层发展，施工现场由较为广阔的场地向狭窄的场地变化、由单一施工向交叉施工等趋势带来了新的安全风险。施工现场的安全管理由以安全检查防范事故发生为主要手段，逐渐转化为提前预防生产过程中的各类安全风险。开展项目现场安全风险评估是深度预防事故发生的根本原因，从本质上提高安全管理能力，是实现事故预防企业主体责任的有效落实，对有效预防工程项目现场安全事故意义重大。

1　研究现状

2016年国务院安委办相继出台了关于构建双重预防机制的相关文件要求[2]，2021年新

作者简介：梁晓斌，男，1992年生，山西人，工程师，工学博士，2020年毕业于中国石油大学(北京)安全与科学工程专业，现主要从事石油管及装备完整性管理工作。E-mail：1963593432@qq.com；电话：15210012963，工作于陕西省西安市中国石油集团工程材料研究院有限公司。

修订的《安全生产法》提出要构建并落实安全风险分级管控和隐患排查治理双重预防机制[3-4]。而现场风险评估作为双重预防机制的前提条件和核心内容,可为风险分级管控和隐患排查治理指明方向。

早在20世纪50年代以前,针对工程项目现场的风险评估技术基本采用基于专家经验的定性评价法,例如安全检查表法、工作危害分析法等,此类风险评价主要依据专家多年经验,缺乏相关现场设计资料、运行资料等客观知识的支撑。在20世纪50年代之后,国外学者开始采用基于概率论和数理统计的"蒙特卡罗模拟"方法,直至现在,该方法已成为目前工程项目中最常用的风险评估方法之一。在20世纪70年代,由T. L. Saaty提出的层次分析法逐渐被广泛应用于工程项目风险评估。我国在工程项目风险评估方面研究起步较晚,李海文结合粗糙集和专家主观经验,得出了更为全面的工程项目风险指标权重值[5]。刘敬辉采用故障树分析和层次分析法分析了高速铁路存在的安全风险[6]。李素英利用模糊层次分析法评估了石济高速铁路项目的风险等级[7]。成威引入LEC法设计了一种隧道施工安全风险评估方法[8]。

总之,目前工程现场主要采用工作危险性分析、安全检查表法、LEC法等风险分析法,虽然能一定程度上反映现场安全状况,但是针对工程项目施工的风险评估研究仍未成熟,尚未建立全面完善的施工现场风险指标体系,且风险因素指标评估过程存在极大主观性,导致现场评估结果有时与现场实际情况偏差较大,甚至出现错误的评估结果。

针对现场存在的一系列问题,本文提出基于结构熵权法和模糊综合评价法的工程项目现场安全风险评估模型,以实际存在问题为导向,从现场制度因素、现场执行因素、现场监管因素、现场配置因素、安全教育因素5个方面建立工程项目现场的风险评估指标体系;利用结构熵权法充分兼顾指标的主观性和客观性,考虑了专家打分过程中可能的认知误差和不确定问题;最后利用模糊综合评价法对影响因素进行综合评价,同时以某工程项目现场为例,挖掘关键风险因素,为工程项目现场安全风险的双重预防机制提供技术支撑。

2 常用的风险评估方法

2.1 工作危害分析法

工作危害分析法分别对每一步作业过程进行深入分析,通过分析和对比发现有危害性的作业过程,并对其进行管控,是辨识危害因素及其风险的最普遍的方法之一。该方法是一种定性分析方法,作业步骤划分缺少相关准则和依据,可能存在作业过程分解不到位,风险辨识的结果与分析质量有直接关系,受作业风险分析人员的经验影响较大。

2.2 安全检查表法

安全检查表法是依据相应的标准,对已知危险源进行逐个检查。检查的对象必须要分为多个体系,这样做的目的是为了防止检查过程中的遗漏,通过专家打分法,将待检查项目按照相应规定,列出一张或几张表,这种方法称为安全检查表法[9]。检查表法可以做到不漏项,避免检查项出现遗漏问题。该方法也是最基础的评价方法。但是对于不同的检查对象,有时需要预先制定多张检查表,检查表的质量和大量的预先准备工作是最大的限制因素。

2.3 LEC法

LEC评价法是通过确定事故发生可能性L、人员暴露于危险环境内的频率E和发生事故可能后果C三种风险因素的取值并计算其三者的乘积D(危险性)来评价系统风险大小的一种半定量分析方法[10]。该方法操作简单,风险等级划分相对明确,缺点是风险因素受评估人员经验影响较大，最终得到三者乘积的危险性受主观因素影响更大，局限性较为明显。

3 SEW-FCE风险评估模型

3.1 模糊综合评价法

模糊综合评价法(Fuzzy comprehensive evaluation method, FCE)是建立在模糊集合上的一种方法，能够通过精确数字手段处理模糊的评价对象，可深入挖掘呈现模糊性的资料，并对其做出贴合实际的定量化评价[11]。具体实现方式是将指标集 U 和评语集 Y 之间建立映射关系。该方法首先需要建立 U 与 Y 的集合关系，然后建立起 U 到 Y 的 n 个模糊评价矩阵 R，最后再算出目标的综合评价值。

计算步骤如下：

第一步：建立评估指标集合 $U=(U_1, U_2, U_3, \cdots, U_n)$。

第二步：建立评估指标权重集合 $W=(W_1, W_2, W_3, W_4, \cdots, W_n)$。

第三步：建立评语集合：

$Y=(Y_1, Y_2, Y_3, Y_4, Y_5)=$（可忽略风险，可允风险，中度风险，高风险，极高风险）$=((0,1],(1,2],(2,3],(3,4],(4,5])$

设置评价等级矩阵为：

$$Z=(1, 2, 3, 4, 5)^{\mathrm{T}} \tag{1}$$

第四步：建立隶属矩阵：由各位专家对各个指标分别进行打分，由此形成某指标在5个风险区间内的评语集合，所有指标的评分结果的汇总称为隶属度矩阵，即：

$$R_i = (r_{ij})_{n \times 5} \tag{2}$$

第五步：建立评估矩阵：根据权重计算法获取的权重集合与隶属度矩阵的结合，可得到评估矩阵为：

$$A = WR \tag{3}$$

第六步：判定风险等级。

风险等级结果为：

$$E = AZ \tag{4}$$

通过将 E 与风险级别划分标准对应，得出评估对象所处的风险级别。

3.2 结构熵权法

结构熵权法(Structure entropy weight method, SEW)是由程启月在2010年首次提出，基本原理是根据各个基本事件的贡献大小排序，采用熵值法定量化分析潜在的不确定性，通过熵值和盲度分析方法实现对偏差数据的统计分析，最终得到各事件权值[12-13]。本文采用SEW法确定某工程项目安全风险的评估指标权值。具体说明如下：

步骤1：采集专家意见，形成典型排序。

选择熟悉该工程项目主要风险隐患及对应指标的专业技术人员组建评价小组，评价小组成员应具有多年经验或者相关专业知识。通常评价小组人员设定为10~50人，此数量可用于对事件权重问卷进行定性排序。根据德尔菲方法的相关规定，选定的评价小组人员被邀请进行无记名填写问卷。汇总所有评价人员意见，达成共识，形成评价意见的典型排序。表1表示一份关于事件重要度排序的问卷。评价人员根据各自的经验或者专业知识，用对号或者数字对事件进行排序。最重要的索引按 1 排序，依此类推。

表 1 事件重要程度排序的问卷

事件	专家	第一选择	第二选择
事件 A_1	E_1	√	
	E_2	√	
	E_3		√
事件 A_2	E_1		√
	E_2		√
	E_3	√	

步骤 2：典型排序的不确定分析。

不同的专家对工程项目安全风险都有不同的认知局限性，对典型排序的评价人员意见通常也存在一定的不确定性和认知误差。其中采用熵理论有效解决了典型排序的不确定性，区间评分有效避免了评价小组人员的认知误差。通过对表 1 中各事件的定性结果统计分析，采用熵理论求解熵值，减少了评价人员对事件排序的不确定性。具体方法如下：

假设 M 位评价人员被邀请进行问卷调查，得到 m 份反馈。每一份问卷都对应一组事件，可以表示为 $U=(u_1, u_2, \cdots, u_n)$，对应典型排序为 $(a_{i1}, a_{i2}, \cdots, a_{in})$。则 m 组问卷的典型排序可以用矩阵表示为 $A(A=(a_{ij})_{m \times n}, i=1, 2, \cdots, m, j=1, 2, \cdots, n)$。其中矩阵 A 中的元素 a_{ij} 表示第 i 个专家对第 j 个指标分析得出的序号。

将典型排序的定性值变换为定量值，则序号 x 的信息熵计算公式为：

$$H(x) = -kP(x)\ln P(x), \quad 1 \leq x \leq n \tag{5}$$

其中 $P(x) = \eta \dfrac{\lambda - x}{\lambda - 1}$，$k = \dfrac{1}{\ln(\lambda - 1)}$，$\lambda = n+2$，$\lambda$ 为信息熵转换参数。

事件的重要性强度计算公式如下：

$$\gamma_x = \lambda - x \tag{6}$$

排序为 x 的事件相对重要性系数可以表示为：

$$\eta = \dfrac{\gamma_x}{\gamma_1} \tag{7}$$

根据计算推理，定量化排序结果，从而能够反映评价人员对各事件的认知程度。

将 $P(x) = \eta \dfrac{\lambda - x}{\lambda - 1}$，$k = \dfrac{1}{\ln(\lambda - 1)}$ 代入 $H(x)$，得到：

$$H(x) = -\dfrac{(\lambda - x)\ln(\lambda - x)}{(\lambda - 1)\ln(\lambda - 1)} \tag{8}$$

当 $\lambda - x \geq 1$ 时：

$$e_x = \dfrac{H(x)}{\dfrac{\lambda - x}{\lambda - 1}} = 1 - \dfrac{\ln(\lambda - x)}{\ln(\lambda - 1)} \tag{9}$$

根据熵权法得到事件排序的隶属度函数为：

$$\psi(x) = 1 - e_x = \frac{\ln(\lambda - x)}{\ln(\lambda - 1)} \quad (10)$$

式中：x 为所有评价人员给出事件 u_j 的定性排序数。如表 1 所示，如果事件 A_1 是第一选择，那么 x 等于 1，依此类推。$\psi(x)$ 是对应于 x 的隶属度函数值。

将排序数 $x = a_{ij}$ 代入方程（10），定性值 a_{ij} 实现量化，获取定量变换值 b_{ij}，即（$b_{ij} = \psi(a_{ij})$）。b_{ij} 表示排序数 x 的隶属度，则 $\boldsymbol{B} = (b_{ij})_{m \times n}$ 表示隶属度矩阵。

假设 m 位评价人员话语权相同。将 m 位评价人员对事件 u_j 的一致看法称为平均识别度。计算过程如下：

$$b_j = (b_{1j} + b_{2j} + \cdots + b_{mj})/m \quad (11)$$

由评价人员知识产生的不确定性称为盲度 Q_j。计算方法如下：

$$Q_j = |\{[\max(b_{1j}, b_{2j}, \cdots, b_{mj}) - b_j] + [\min(b_{1j}, b_{2j}, \cdots, b_{mj}) - b_j]\}/2| \quad (12)$$

显然，$Q_j \geq 0$。

步骤 3：归一化处理。

定义所有被邀请评价人员对每个事件 u_j 均有一个整体认识 c_j，即：

$$c_j = b_j(1 - Q_j), \quad c_j > 0 \quad (13)$$

则 m 位评价人员对所有事件的整体评估向量为：

$$\boldsymbol{C} = (c_1, c_2, \cdots, c_n) \quad (14)$$

为获取事件 u_j 的权重，将 $c_j = b_j(1 - Q_j)$ 归一化，则：

$$w_j = c_j \bigg/ \sum_{j=1}^{\lambda} c_j \quad (15)$$

其中 $j = 1, 2, \cdots, n$，$\sum_{j=1}^{n} w_j = 1$。$\boldsymbol{W} = \{w_1, w_2, \cdots, w_n\}$ 为事件集合 $\boldsymbol{U} = \{u_1, u_2, \cdots, u_n\}$ 的加权向量，即 m 位评价人员对事件重要度的总体判断。它满足所有 m 位评价人员的意愿和认知。

4 某工程项目现场指标体系构建

4.1 危险源辨识

危险源指的是工程项目系统中有可能释放危险或造成环境破坏的设施设备等。危险源识别就是对可能出现的上述情况和条件进行识别的过程，危险源识别需要主动结合预测性的安全数据收集方法进行识别，其中包括一些事件的调查结果。项目现场的安全管理机构需要制定相应的危险源识别程序，并随着对事件的总结，对程序进行不断的规划与完善，以确保所制定出地程序能够快速地识别出危险源[14]。

4.2 评估指标提取

按照各指标之间的隶属关系，将施工现场安全风险评估指标划分为三层。通过对工程项目现场可能存在的安全风险分析，提出项目现场安全风险涉及 5 个方面内容，包括现场制度风险、现场执行风险、现场监管风险、现场配置风险、安全教育风险。综合分析现场安全调研结果和安全风险管理状况，构建某工程项目现场指标体系，如图1所示[15-16]。

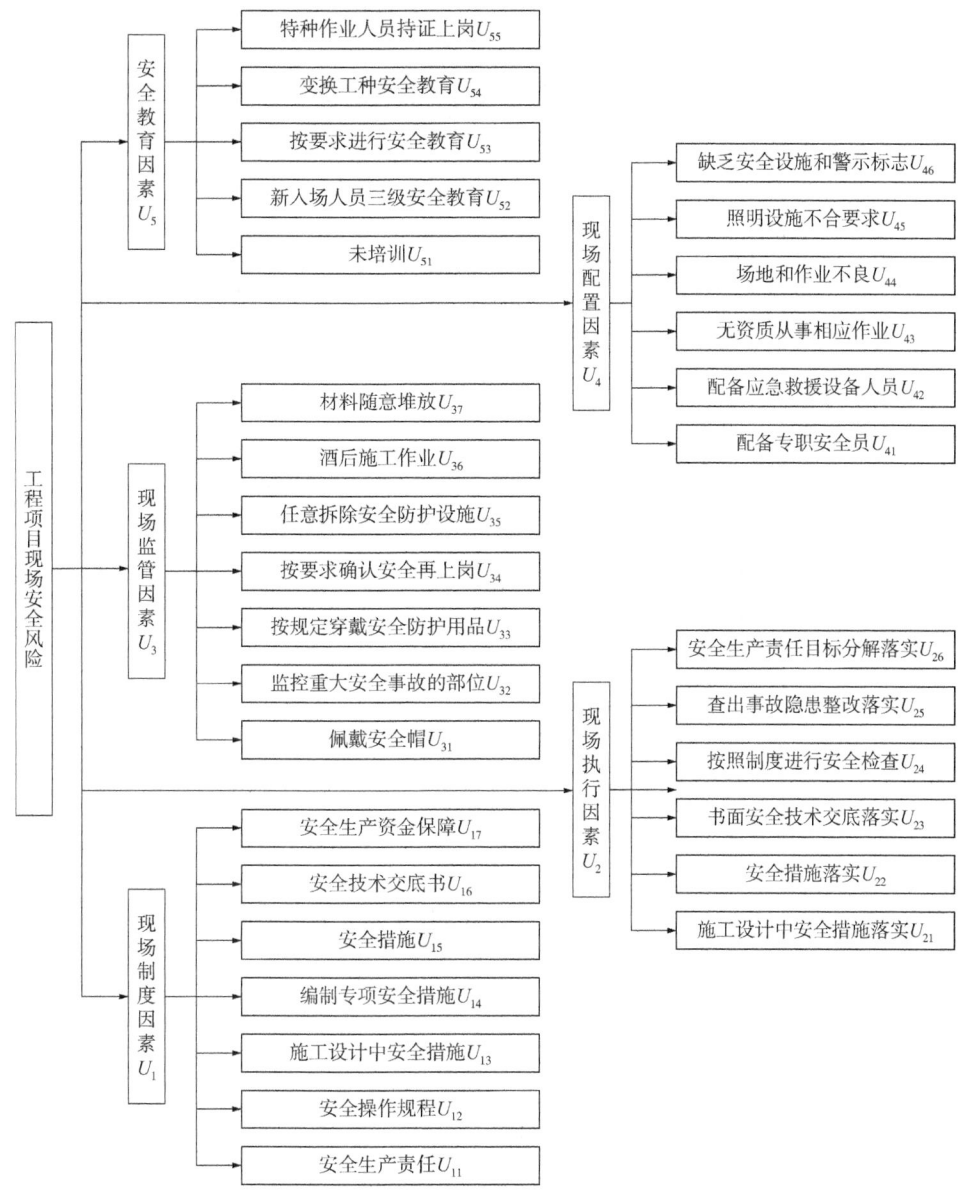

图1 工程项目现场安全风险评估指标体系

5 某工程项目现场安全风险评估

5.1 项目概况

某工程项目负责建设 10×10^4t/a 1,4-丁二醇装置、4.6×10^4t/a 聚四氢呋喃、10×10^4t/a 乙二醇装置、22×10^4t/a 甲醛装置、490×10^8m³/h 气体分离装置、新建 22000m³/h 空分装置、40000m³/d 净水厂、770t/h 脱盐水装置、30000m³/h 循环水厂等设备及装置。现场有管理人员 28 人，其中专职安全管理人员 4 人。

5.2 指标权重计算

5.2.1 二级指标权重的确定

选择3组现场评估人员分别为分组1、分组2、分组3。对二级指标的重要性进行排序，通过重要程度的排序结果，采用结构熵权法对其权重进行确定，主要采用公式(11)至公式(15)分别对专家的平均识别度、盲度、整体认识及归一化权重进行求解，相关参数计算结果见表2[17]。

表2 事件排序的 SEW 计算表

计算值	U_1	U_2	U_3	U_4	U_5
分组1	3	5	1	4	2
分组2	2	5	4	1	3
分组3	2	3	5	1	4
b_j	0.857	0.516	0.667	0.871	0.762
$1-Q_j$	0.979	0.724	0.974	0.936	0.994
c_j	0.839	0.374	0.650	0.815	0.757
w_j	0.244	0.109	0.189	0.237	0.220

5.2.2 三级指标权重的确定

与表2求解过程相同，对每个二级指标下隶属的三级指标按照结构熵权法的过程分别进行计算，计算结果见表3至表7。

表3 U_1 的下属指标权重表

指标	U_{11}	U_{12}	U_{13}	U_{14}
权重	0.067	0.190	0.195	0.146
指标	U_{15}	U_{16}	U_{17}	
权重	0.139	0.120	0.143	

表4 U_2 的下属指标权重表

指标	U_{21}	U_{22}	U_{23}
权重	0.207	0.134	0.165
指标	U_{24}	U_{25}	U_{26}
权重	0.121	0.150	0.223

表5 U_3 的下属指标权重表

指标	U_{31}	U_{32}	U_{33}	U_{34}
权重	0.199	0.137	0.144	0.089
指标	U_{35}	U_{36}	U_{37}	
权重	0.175	0.179	0.077	

表6 U_4 的下属指标权重表

指标	U_{41}	U_{42}	U_{43}
权重	0.241	0.242	0.124
指标	U_{44}	U_{45}	U_{46}
权重	0.123	0.146	0.125

表7 U_5 的下属指标权重表

指标	U_{51}	U_{52}	U_{53}
权重	0.124	0.264	0.143
指标	U_{54}	U_{55}	
权重	0.254	0.215	

5.3 模糊综合评价

5.3.1 建立评估指标集

根据评估指标体系，建立一级评估指标集 $U = \{U_1, U_2, U_3, U_4, U_5\}$，其次，二级评估指标集分别为 $U_i = \{U_{i1}, U_{i2}, U_{i3}, \cdots, U_{in}\}$，其中，$i$ 为二级指标的编号，$i = 1, 2, 3, 4, 5$，n 为每个二级指标隶属的三级指标的编号。

5.3.2 建立评估指标权重集合

依据结构熵权法的计算结果可知，一级指标的权重集合为：

$$W = \{0.244, 0.109, 0.189, 0.237, 0.220\}$$

二级指标的权重集合为：

$$W_1 = \{0.067, 0.190, 0.195, 0.146, 0.139, 0.120, 0.143\}$$

$$W_2 = \{0.207, 0.134, 0.165, 0.121, 0.150, 0.223\}$$

$$W_3 = \{0.199, 0.137, 0.144, 0.089, 0.175, 0.179, 0.077\}$$

$$W_4 = \{0.241, 0.242, 0.124, 0.123, 0.146, 0.125\}$$

$$W_5 = \{0.124, 0.264, 0.143, 0.254, 0.215\}$$

5.3.3 建立评语集合

根据现场工程项目安全管理的实际情况，将现场安全管理的三级指标的评语分为5个等级，即：

$Y = \{Y_1, Y_2, Y_3, Y_4, Y_5\} = \{可忽略风险，可允风险，中等风险，重大风险，极高风险\}$

评价等级矩阵为：

$$Z = \{1, 2, 3, 4, 5\}^T$$

风险等级划分标准见表8。

表8 风险等级划分标准

等级	可忽略风险	可允许风险	中等风险	重大风险	极高风险
分值	(0, 1]	(1, 2]	(2, 3]	(3, 4]	(4, 5]

5.3.4 建立隶属度矩阵

通过实地调查访谈，得到20名项目管理人员对三级指标的评语打分结果，见表9。

表9 三级指标评语一览表

编号	评价因子	评语集的样本数				
		Y_1	Y_2	Y_3	Y_4	Y_5
1	U_{11}	2	5	3	6	4

续表

编号	评价因子	评语集的样本数				
		Y_1	Y_2	Y_3	Y_4	Y_5
2	U_{12}	3	5	4	3	5
3	U_{13}	4	2	5	7	2
4	U_{14}	1	8	5	4	2
5	U_{15}	2	7	4	4	3
6	U_{16}	1	6	6	5	2
7	U_{17}	3	7	7	2	1
8	U_{21}	4	6	4	3	3
9	U_{22}	6	4	5	3	2
10	U_{23}	4	3	9	3	1
11	U_{24}	10	2	3	4	1
12	U_{25}	5	3	6	5	1
13	U_{26}	1	7	7	4	1
14	U_{31}	2	6	5	4	3
15	U_{32}	1	2	8	3	6
16	U_{33}	4	2	4	5	5
17	U_{34}	3	2	3	4	8
18	U_{35}	1	9	6	3	1
19	U_{36}	2	8	4	5	1
20	U_{37}	1	3	6	8	2
21	U_{41}	10	6	2	2	0
22	U_{42}	9	5	3	2	1
23	U_{43}	1	5	4	9	1
24	U_{44}	12	1	1	5	1
25	U_{45}	4	2	5	4	5
26	U_{46}	4	3	3	8	2
27	U_{51}	4	2	4	6	4
28	U_{52}	1	5	6	6	2
29	U_{53}	3	4	8	3	2
30	U_{54}	8	3	1	3	5
31	U_{55}	6	3	5	5	1

根据评估打分结果，利用公式(2)建立各二级指标的隶属度矩阵，如下所示：

$$R_1 = \begin{bmatrix} 0.10 & 0.25 & 0.15 & 0.30 & 0.20 \\ 0.15 & 0.25 & 0.20 & 0.15 & 0.25 \\ 0.20 & 0.10 & 0.25 & 0.35 & 0.10 \\ 0.05 & 0.40 & 0.25 & 0.20 & 0.10 \\ 0.10 & 0.35 & 0.20 & 0.20 & 0.15 \\ 0.05 & 0.30 & 0.30 & 0.25 & 0.10 \\ 0.15 & 0.35 & 0.35 & 0.10 & 0.05 \end{bmatrix}$$

$$R_2 = \begin{bmatrix} 0.20 & 0.30 & 0.20 & 0.15 & 0.15 \\ 0.30 & 0.20 & 0.25 & 0.15 & 0.10 \\ 0.20 & 0.15 & 0.45 & 0.15 & 0.05 \\ 0.50 & 0.10 & 0.15 & 0.20 & 0.05 \\ 0.25 & 0.15 & 0.30 & 0.25 & 0.05 \\ 0.05 & 0.35 & 0.35 & 0.20 & 0.05 \end{bmatrix}$$

$$R_3 = \begin{bmatrix} 0.10 & 0.30 & 0.25 & 0.20 & 0.15 \\ 0.05 & 0.10 & 0.40 & 0.15 & 0.30 \\ 0.20 & 0.10 & 0.20 & 0.25 & 0.25 \\ 0.15 & 0.10 & 0.15 & 0.20 & 0.40 \\ 0.05 & 0.45 & 0.30 & 0.15 & 0.05 \\ 0.10 & 0.40 & 0.20 & 0.25 & 0.05 \\ 0.05 & 0.15 & 0.30 & 0.40 & 0.10 \end{bmatrix}$$

$$R_4 = \begin{bmatrix} 0.50 & 0.30 & 0.10 & 0.10 & 0 \\ 0.45 & 0.25 & 0.15 & 0.10 & 0.05 \\ 0.05 & 0.25 & 0.20 & 0.45 & 0.05 \\ 0.60 & 0.05 & 0.05 & 0.25 & 0.05 \\ 0.20 & 0.10 & 0.25 & 0.20 & 0.25 \\ 0.20 & 0.15 & 0.15 & 0.40 & 0.10 \end{bmatrix}$$

$$R_5 = \begin{bmatrix} 0.20 & 0.10 & 0.20 & 0.30 & 0.20 \\ 0.05 & 0.25 & 0.30 & 0.30 & 0.10 \\ 0.15 & 0.20 & 0.40 & 0.15 & 0.10 \\ 0.40 & 0.15 & 0.05 & 0.15 & 0.25 \\ 0.30 & 0.15 & 0.25 & 0.25 & 0.05 \end{bmatrix}$$

5.3.5 计算评估矩阵

根据各二级指标的隶属度矩阵和权重集合，利用公式(3)建立各二级指标 U_i 的评估矩阵 $B_i = W_i R_i$ 为：

$$B_1 = [0.12 \quad 0.28 \quad 0.25 \quad 0.22 \quad 0.14]$$
$$B_2 = [0.22 \quad 0.23 \quad 0.29 \quad 0.18 \quad 0.08]$$
$$B_3 = [0.10 \quad 0.26 \quad 0.26 \quad 0.22 \quad 0.17]$$
$$B_4 = [0.36 \quad 0.20 \quad 0.15 \quad 0.21 \quad 0.07]$$
$$B_5 = [0.23 \quad 0.18 \quad 0.23 \quad 0.23 \quad 0.14]$$

由此可知，一级指标 U 的评估矩阵 $A = W[B_1 \quad B_2 \quad B_3 \quad B_4 \quad B_5]^T$ 为：

$$A = [0.19 \quad 0.19 \quad 0.19 \quad 0.20 \quad 0.20]$$

5.3.6 确定风险等级

基于评语集合，则各个二级指标的风险等级 $E_i = B_i Z$ 为：

$$E_1 = 3.01; \quad E_2 = 2.67; \quad E_3 = 3.13; \quad E_4 = 2.40; \quad E_5 = 2.90$$

则一级指标的风险等级 $E = AZ$ 为：

$$E = 2.94$$

由此可知，工程项目安全处于中等风险水平，该工程项目负责人及安全管理人员应该保持高度警惕，重点对现场监管因素、现场制度因素、安全教育因素等三方面加强建设和定期检查，尤其是在现场监管过程中如发现酒后施工、任意拆除安全防护设施等特殊危险行为，应立即要求工程项目施工现场停止作业，对其进行相应整改，整改完成方可继续施工。同时对酒后施工人员进行警告和教育培训，培训合格后方可继续上岗工作，确保安全风险降低到可允许风险范围，进而防止人员伤亡事故的发生。

6 总结

（1）工程项目现场安全风险指标众多，各评估指标存在的不确定性和认知偏差往往导致评估结果与实际情况差别很大，开展科学、有效的先进风险评估方法可以有效地解决此方面的问题。本文通过制度因素、执行因素、监管因素、配置因素和安全教育因素五大方面建立了工程项目安全风险评价指标体系，提出采用基于主观权重和客观权重优势的结构熵权法，减少了指标权重的不确定性和认知误差，赋予了更为合理的指标权重。

（2）采用基于结构熵权法和模糊综合评价法的风险评估模型为工程项目现场提供了新的

风险管控管理办法,并将其应用于现场分析。实例评估表明:某现场工程项目的安全风险等级为中等风险,与实际情况吻合,说明具有一定的实用性和合理性。针对中风险评估等级,管理者日后应重点关注现场监管因素、现场制度因素等重要指标参数,为双重预防机制的实施提供技术支持。

参 考 文 献

[1] 尹成龙,翟成凯.风险分级管控和隐患排查治理双控体系研究与讨论[J].内江科技,2019,40(6):52-70.

[2] 赵连池,张桂英,孟圆圆,等.危化品企业如何做好双重预防体系与安全生产标准化[J].山东化工,2019,48(17):132-135.

[3] 张畅.企业构建双重预防机制的管理理论分析[J].中国商论,2018(2):104-105.

[4] 李忠财.关于我省重点行业领域安全生产风险分级管控和隐患排查治理"双重"预防机制建设工作的思考[J].吉林劳动保护,2018(2):25-27.

[5] 李海文.基于粗糙集理论的工程项目风险评估研究[J].江苏科技信息,2016(3):43-45.

[6] 刘敬辉.基于FTA-AHP的铁路安全风险综合评估方法[J].中国铁道科学,2017,38(2):138-144.

[7] 李素英,田崖,吴永立.基于FAHP模型的铁路工程项目风险评估研究[J].铁道工程学报,2019,36(7):92-99.

[8] 成威.基于LEC法的隧道施工安全风险评估研究[J].交通世界,2021(36):11-13.

[9] 吴济民.国外化工企业工艺安全技术管理概述[J].中国安全生产科学技术,2011(7):192-198.

[10] 王永祥,吴滔,李亮,等.基于突变级数法的地铁盾构施工安全风险评价[J].安全与环境工程,2021,28(1):95-102.

[11] LI B, WANG E Y, SHANG Z, et al. Quantification study of working fatigue state affected by coal mine noise exposure based on fuzzy comprehensive evaluation[J]. Safety Science, 2022, 146: 105577.

[12] 程启月.评测指标权重确定的结构熵权法[J].系统工程理论与实践,2010,30(7):1225-1228.

[13] LIANG X B, LIANG W, ZHANG L B, et al. Risk assessment for long-distance gas pipelines in coal mine gobs based on structure entropy weight method and multi-step backward cloud transformation algorithm based on sampling with replacement[J]. Journal of Cleaner Production, 2019, 227: 218-228.

[14] 许兴武,吴浩,刘永强,等.基于改进LEC法的堤防施工危险源辨识及评价[J].水电能源科学,2022,40(5):131-134.

[15] ZHONG C H, YANG Q C, LIANG J, et al. Fuzzy comprehensive evaluation with AHP and entropy methods and health risk assessment of groundwater in Yinchuan Basin, northwest China[J]. Environmental Research, 2022, 204: 111956.

[16] 杨斯玲,黄和平,刘伟,等.基于结构熵权和修正证据理论的装配式建筑施工安全风险评价[J].安全与环境工程,2019,26(6):143-149.

[17] 黄萍,林杰钦.基于层次分析法的DFT城市综合管廊盾构施工风险评价研究[J].安全与环境工程,2020,27(5):116-121.

本论文原发表于《油气田地面工程》2023年第42卷第12期。

Corrosion Behavior and Failure Mechanism of V150 Drill Pipe in HPHT and Ultra-deep Drilling Process

Chen Zihan[1] Sun Shixuan[2] Huang Hao[3] He Jibo[3] Ouyang Zhiying[4]
Yu Shijie[4] Li Xuanpeng[1] Fu Anqing[1] Feng Yaorong[1]

(1. CNPC Tubular Goods Research Institute; 2. PetroChina Coalbed Methane Company Limited;
3. PetroChina Changqing Oilfield Company; 4. Shanghai Hilong Oil Tubular Goods Research Institute)

Abstract: The simulation test of the corrosion resistance of V150 drill pipe samples with different surface conditions in the field drilling fluid showed that different drilling fluid types and different temperature conditions had a great influence on the corrosion resistance of drill pipe. After immersion in one particular type of drilling fluid at 150℃, scaling on the surface of the drill pipe sample caused serious pitting corrosion, and the average corrosion rate reached 0.85mm/a. When the temperature was lowered to 120℃, the scaling phenomenon disappeared and the average corrosion rate decreased to 0.2mm/a. The corrosion behavior of V150 drill pipe was dominated by oxygen corrosion. When there was a scale layer on the surface of the drill pipe, a macroscopic oxygen concentration cell was formed inside and outside the scale layer, which promoted the acidification and autocatalysis process under the scale, eventually accelerating the under-deposits corrosion.

Keywords: V150 drill pipe; Drilling fluid; Pitting corrosion; HTHP; Oxygen corrosion

1 Introduction

With the continuous development of the oil and gas industry, various complex oil and gas fields have appeared one after another and made the oil and gas extraction conditions increasingly harsh and severe. Especially high temperature and high pressure(HTHP) as well as high sulfur oil and gas fields are important areas for drill pipe failure[1-2]. Drill pipe failure will not only cause substantial economic losses, but also serious casualties[3-7]. The traditional G105 and S135 drill pipes do not meet the operational needs of ultra-deep wells due to low safety factors and tensile strength[8-10]. Therefore, V150 drill pipe as a new type of high-strength drill pipe has gained much attention in oil and gas field development[11-15].

Several research has been conducted on V150 drill pipe in recent years. Shu et al.[16] analyzed

---

Corresponding author: Chen Zihan, chenzihan@cnpc.com.cn.

the interaction between tensile stress-strain and torsional stress-strain of V150 drill pipe under tensile-torsional composite load conditions using the tensile-torsional composite load test method of pre-torsional followed by tensile and pre-torsional followed by torsional. The results showed that the design of V150 drill pipe against tensile-torsional composite load capacity according to the Von Mises strength criterion was conservative. In contrast, the tensile – torsional elliptical strength criterion contained two benchmark parameters of material tensile yield strength and torsional yield strength, and was in good agreement with the test data, which was more valuable for engineering applications. Liu et al.[17] investigated the material properties of V150 drill pipe with different heat treatments using experimental tests and microstructure analysis. The study showed that inclusions and defects during the heat treatment of V150 were the main cause of the drill pipe limited properties. Ouyang et al.[18] analyzed the causes of puncture failure of V150 drill pipe, and the results showed that corrosion fatigue is the main cause of drill pipe puncture. From the research, it can be seen that there are few studies related to V150 drill pipe. These studies mainly focus on the performance of V150 drill pipe tubing and material properties, and few in-depth studies on the actual use performance of V150 drill pipe in ultra-deep wells.

Since 2019, the V150 drill pipe produced by a Chinese oil service industry group has experienced early localized corrosion during the drilling of ultra-deep wells in an oil field in northwest China, mainly manifesting as outer wall corrosion. A detailed study of the 75 V150 drill pipe from five wells on site revealed that the corroded drill pipe mainly concentrated in the 3000~6000m section of the well, and the drilling fluid system used was KCl-polysulfide. The corrosion behavior of the drill pipe of these five wells has some similarities, mainly intensive pitting, and there is a certain pattern in the distribution of the pit location. It is assumed that after the straightening process, there might be some changes in the surface condition of the drill pipe. The location of the straightening marks is susceptible to the formation of rust bands in a corrosive environment and there is where pits tend to form. In addition, most corroded or failed drill pipe is scratched in service, therefore, pits tend also to form in such locations. Finally, most of the pitting pits were mainly concentrated on the side of the drill pipe. Among them, the average corrosion depth difference between the front and back sides is about 1mm, and the pits on the side with severe corrosion are distributed along the axial direction. It is assumed that this phenomenon may occur downhole, not in time to clean the mud cover, and in the well site due to the long time placed by gravity and other factors led to the lower part of the drill pipe pit propagation. The local macro corrosion morphology of the drill pipe is shown in Figure 1.

Drill pipe corrosion is a common problem in drilling engineering, which is more prominent in deep wells, ultra-deep wells, deep-sea wells and highly corrosive wells[19-21]. In the drilling operation, to meet the needs of various drilling processes, drilling fluid systems such as brine and potassium-based polymers are used and contain a variety of additives, which are strongly corrosive under the action of high temperature and pressure downhole[22]. The corrosiveness of drilling fluids is primarily related to dissolved oxygen. As drilling fluid passes through mud tanks on the surface and is cleaned and treated on shakers, it comes into contact with air, causing oxygen from the air to enter the drilling fluid. Some of the oxygen is dissolved in the drilling fluid until it is saturated, so

Figure 1 Macro corrosion morphology of drill pipe surface

dissolved oxygen corrosion is the most important reason for the decrease in drill pipe life[23-25].

This paper mainly focuses on the corrosion analysis of V150 drill pipe corrosion phenomenon, and investigates the influence of different drill pipe surface conditions and different mud compositions as well as different temperatures on the corrosion behavior of V150 drill pipe.

2 Experimental method

2.1 Microscopic corrosion morphology and corrosion product composition analysis

A cross-sectional specimen was taken from the surface of a corroded drill pipe (No. K8), and the microscopic morphology of the corrosion pit and the surrounding substrate was investigated using a metallurgical microscope and image analysis system. In addition, the X-ray diffraction (XRD) analysis and EDS spectroscopy were performed on the corrosion products in the pits formed at the outer wall of the drill pipe.

2.2 Chemical composition and metallographic structure analysis

The chemical compositions of one corroded drill pipe (No. K8) and one uncorroded drill pipe (No. D1) were analyzed using the ThermoFisher ARL 4460 direct reading spectrometer. In addition, the microscopy and image analysis after metallographic preparation were used to examine the microstructure, grain size and cleanliness of the two drill pipes.

2.3 High temperature and high pressure corrosion test

Corrosion tests were carried out considering two aspects. Firstly, the corrosion resistance of the drill pipe samples with different surface conditions was studied. The uncorroded drill pipe (No. D1) was used to take specimens with a smooth surface, with the original surface, and to simulate roll marks and scratches. Subsequently, the corrosion resistance of these specimens with different surface conditions was evaluated in two drilling fluids. The second is an environmental media sensitivity study. Smooth surface specimens were taken from the uncorroded drill pipe (No. D1) to evaluate its corrosion resistance under different temperature conditions. High temperature and high pressure corrosion tests were carried out in an autoclave under static conditions according to standard ASTM G111-21[26]. The specific test parameters and conditions were shown in Table 1, and the shape and dimensions of the samples were shown in Figure 2. After the immersion testing, the sample surfaces were cleaned, following the NACE RP0775‑2005 standard, in order to remove the corrosion scales, and the corrosion rate (CR) was calculated based on Eq. (1):

$$CR = \frac{3.65 \times 10^5 \times W}{ATD} \quad (1)$$

Where: CR is corrosion rate, mm/a; W is weight loss, g; A is surface area of the metal specimen, mm^2; T is corrosion test time, d; D is the density of the specimen, g/cm^3.

Table 1 Corrosion experimental parameters

Temperature/℃	80, 120, 150
Time/h	168
Total pressure/MPa	10
O_2 partial pressure/MPa	2.1
Solution	Two drilling fluids of the KCl‑water‑based polysulfide system with a Cl^- concentration of approximately 20mg/L

3 Results and Discussion

3.1 Microscopic corrosion morphology and corrosion product composition analysis

The cross-sectional morphology of the corrosion pits of K8 drill pipe is shown in Figure 3. It can be seen that the corrosion degree of K8 drill pipe is very high, and the maximum depth of corrosion pits reaches 2.45mm, with a width of 7.72mm. The microstructure around the pitting pits is full of tempered martensite, which is not abnormal.

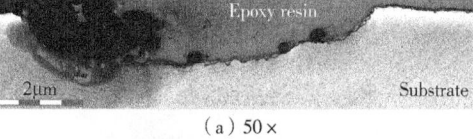
(a) 50×

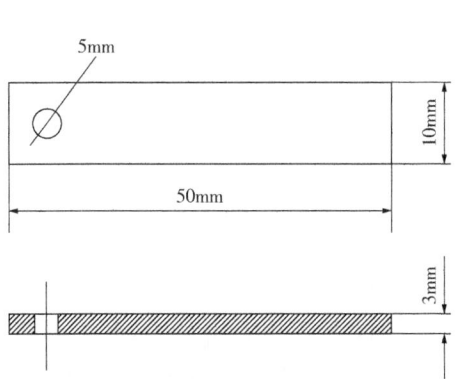

Figure 2 Sample size specifications for corrosion testing

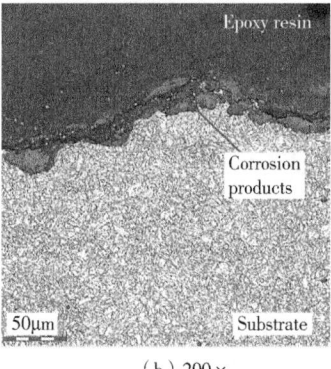
(b) 200×

Figure 3 The cross-sectional morphology of the corrosion pits of K8 drill pipe

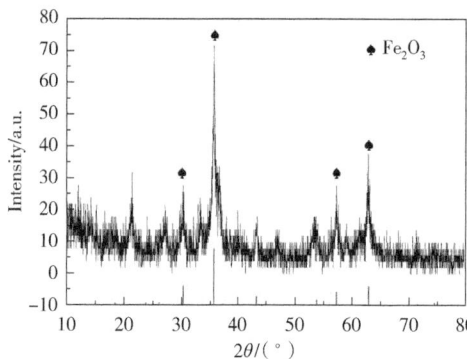

Figure 4 XRD pattern of the product inside the corrosion pit on the surface of K8 drill pipe

The corrosion products in the corrosion pit of the outer wall of K8 drill pipe were analyzed by XRD, and the XRD spectra is shown in Figure 4. The material phase obtained by matching is mainly Fe_2O_3.

The cross-sectional samples of K8 drill pipe were investigated using SEM and the corrosion products inside the corrosion pits were analyzed for elemental composition using EDS as shown in Figure 5. The results show that there is a large accumulation of corrosion products inside the corrosion pit. The main constituent elements of the corrosion products are Fe and O, which is consistent with the results of Fe_2O_3 in the XRD pattern, and it is presumed that oxygen corrosion has mainly occurred in the drill pipe. In addition, small amounts of C, Cr, and Cl were detected, of which Cr originated from the matrix and Cl probably originated from the service environment. The distribution of O is more uniform, while for Fe and Cl there is an apparent downward trend from the bottom of the corrosion pit to the outer surface, indicating that the surface corrosion product film is not dense enough, external Cl^- and other corrosion ions are easy to penetrate the corrosion product layer, thereby supporting pit propagation.

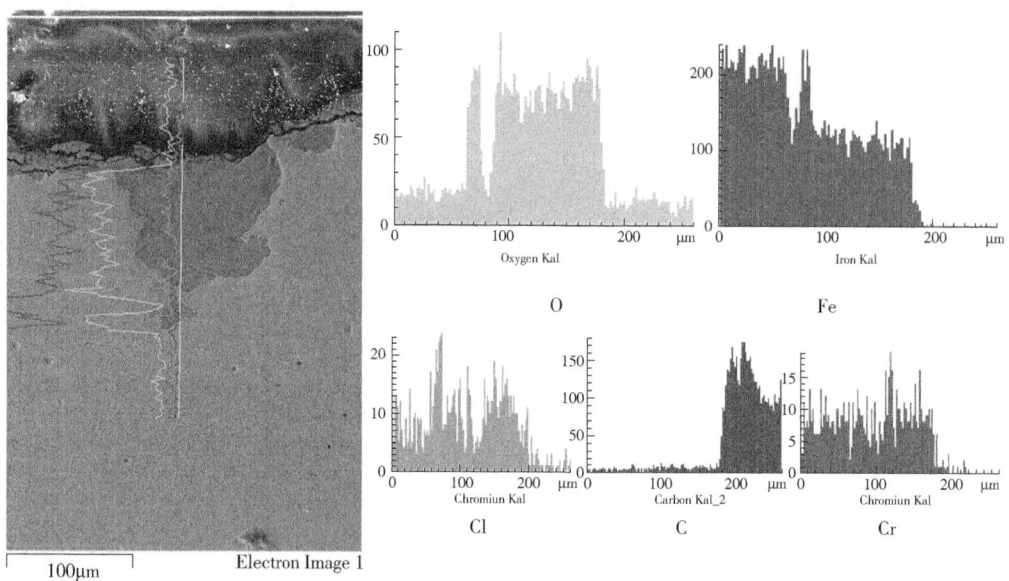

Figure 5 Line scan EDS spectrum of corrosion products inside the K8 drill pipe corrosion pit

3.2 Chemical composition and metallographic structure analysis

The chemical compositions of K8 drill pipe samples and D1 drill pipe samples were tested separately, and the results are shown in Table 2. It can be seen that there is no significant difference in the chemical composition of the two drill pipe samples, in which the content of P and S elements are also at a low level.

Table 2 Chemical composition analysis results(%, mass fraction)

Samples	C	Si	Mn	P	S	Cr	Mo	Ni
D1	0.24	0.23	0.46	0.0066	<0.002	1.03	0.70	0.59
K8	0.25	0.23	0.47	0.0061	<0.002	1.06	0.70	0.58

The metallographic structure, grain size, and non-metallic inclusions of K8 drill pipe specimens and D1 drill pipe specimens were examined and analyzed, respectively. The test results are shown in Table 3, and the metallographic structure is shown in Figure 6. It can be seen that the structure of both samples is tempered martensite, the grain size has reached 10.0 grade, and no excessive non-metallic inclusions are found. Therefore, from the results of chemical composition analysis and metallographic analysis, it can be concluded that the local corrosion behavior of the drill pipe occurred and its tissue composition does not have a significant correlation.

Table 3 Results of metallographic structure analysis

Samples	Non-metallic inclusions								Metallographic structure	Grain size/Grade
	A		B		C		D			
	Thin	Thick	Thin	Thick	Thin	Thick	Thin	Thick		
D1	0.5	0	0.5	0	0	0	0.5	0	Tempered martensite	10.0
K8	0.5	0	0.5	0	0	0	0.5	0	Tempered martensite	10.0

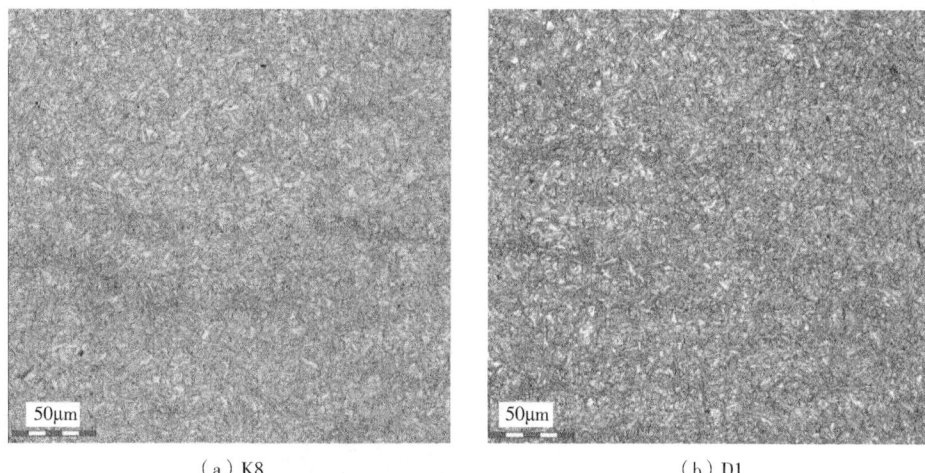

(a) K8   (b) D1

Figure 6 Metallographic morphology of the drill pipe samples

3.3 High temperature and high pressure corrosion test

Since the drill pipe is typically straightened by a roller press during the manufacturing process to ensure straightness, it is possible to find surface defects or marks due to contact with the rollers. In addition, the drill pipeis prone to be scratched during the drilling process by rocks. In order to study the influence of the surface condition of the drill pipe on its corrosion resistance, specimens with different surface conditions were taken from Drill pipe D1: smooth surface, original surface, original surface with scratches and roller marks. The specimens were subjected to 168h immersion tests in two drilling fluids obtained from the oilfield site. The average corrosion rate of three specimens from each surface condition after the immersion test is shown in Figure 7, and the macroscopic corrosion morphology is shown in Figure 8 and Figure 9.

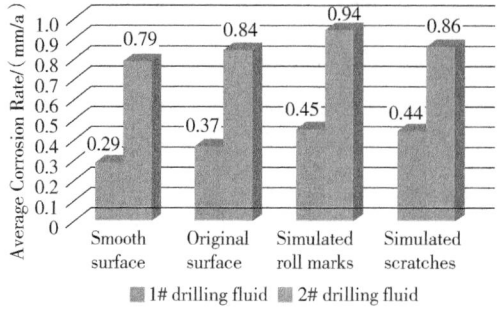

Figure 7　The corrosion rates of the samples with four different surface conditions in two different drilling fluids at 150℃

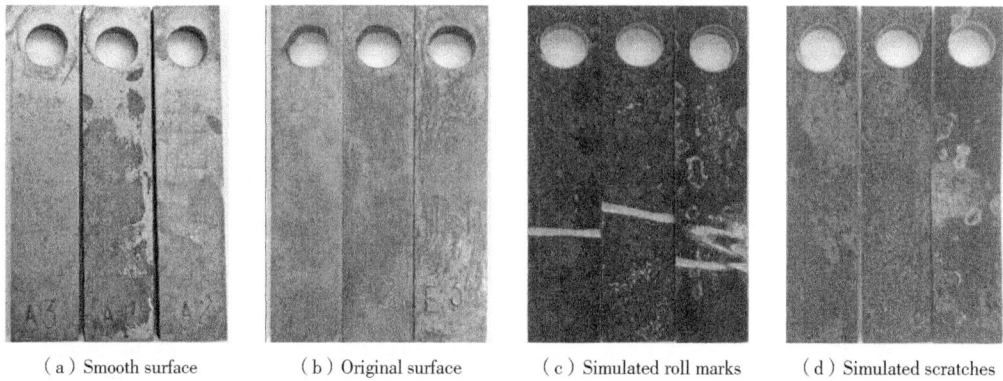

(a) Smooth surface　　(b) Original surface　　(c) Simulated roll marks　　(d) Simulated scratches

Figure 8　Macroscopic corrosion morphology of specimens after immersion test in drilling fluid 1

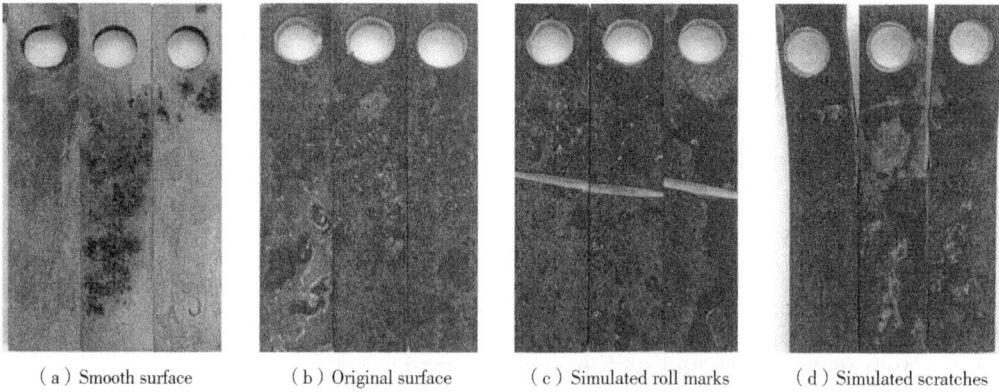

(a) Smooth surface　　(b) Original surface　　(c) Simulated roll marks　　(d) Simulated scratches

Figure 9　Macroscopic corrosion morphology of specimens after immersion test in drilling fluid 2

It was found that the corrosion rate of the four surface conditions in drilling fluid 2 was consistently almost twice as high as that in drilling fluid 1. A certain degree of scaling was observed on the surface of all the specimens immersed in drilling fluid 2 before the corrosion products were removed from the samples. The scale layer was black and brown, as shown in Figure 10. After removing the scale layer on the surface of the samples by the mechanical method, pits appeared under the scale to varying degrees. The surface of the specimens immersed in drilling fluid 1 had only a small amount of corrosion products adhering to the surface, and the surface was uniformly corroded after removal. The smooth surface specimens immersed and de-filmed in the two different drilling fluids were observed under a laser confocal microscope to determine the corrosion morphology and the pit depth. As shown in Figure 11, the surface of the samples in drilling fluid 1 was relatively flat with local pits, and the maximum pit depth was about 27μm. The surface of the specimen in drilling fluid 2 was densely distributed with a maximum pit depth of 290μm.

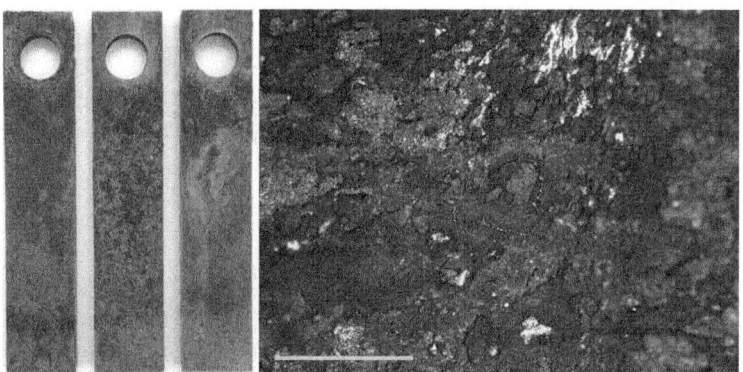

Figure 10　Macroscopic and microscopic morphology of the scale layer on the surface of the specimens

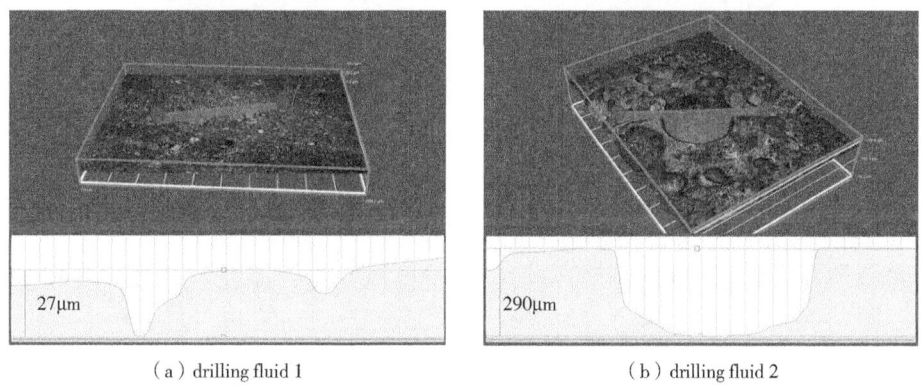

(a) drilling fluid 1　　　　　　　　　　　　(b) drilling fluid 2

Figure 11　Surface pitting pit depth of smooth specimens after immersion in different drilling fluids at 150℃

In order to investigate the effect of temperature on the corrosion rate further, smooth surface specimens of drill pipe D1 were taken for immersion tests in drilling fluid 2 at 150℃, 120℃ and 80℃, respectively. The test results are shown in Figure 12. It can be seen from the figure that the corrosion rate of all the specimens show a significant downward trend as the test temperature decreases. The average corrosion rate of specimens taken from drill pipe D1 at 150℃ was as high as 0.79mm/a, while at 120℃ the average corrosion rate dropped to 0.18mm/a,

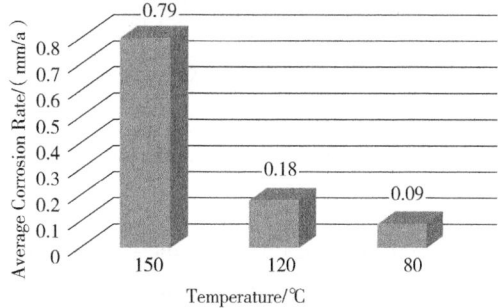

Figure 12　Corrosion rate of drill pipe D1 specimens in drilling fluid 2 at different temperatures

and at 80℃ the average corrosion rate was further reduced to 0.09mm/a.

The macroscopic corrosion morphologies of the drill pipe specimens in drilling fluid 2 under the test conditions of 120℃ and 80℃ are shown in Figure 13. Neither fouling nor pitting was observed at the surface of either group of specimens, which is quite different from the corrosion morphology of the samples after immersion at 150℃. Therefore, the temperature greatly affected on the scaling and under-deposit corrosion on the surface of the drill pipe. The microscopic corrosion morphology of the specimens' surface after removing the corrosion products was further investigated under the laser confocal microscope. The results, shown in Figure 14, confirmed that there were small corrosion pits on the surface of the samples, and the overall corrosion was uniform.

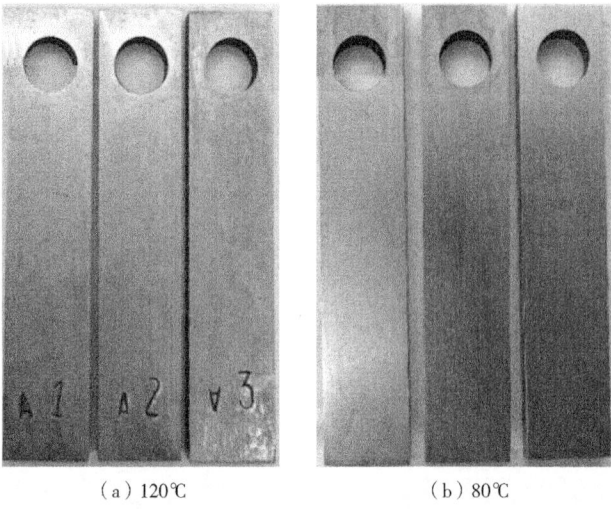

(a) 120℃   (b) 80℃

Figure 13   The macroscopic corrosion morphologies of the drill pipe specimens in drilling fluid 2 at different temperatures

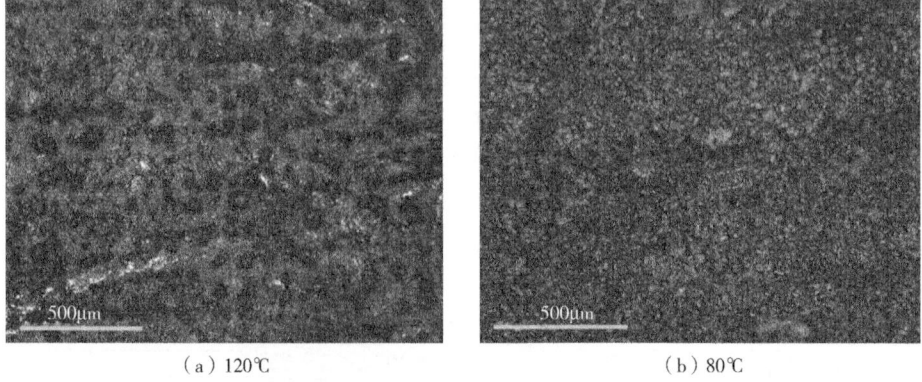

(a) 120℃   (b) 80℃

Figure 14   Microscopic corrosion morphology of the surface of the samples after the removal of corrosion products

The main components and major elemental compositions of the two drilling fluids are shown in Table 4. Both drilling fluids are close in composition, consisting of potassium polyacrylate, sodium lignosulfonate, dimethyldiallylammonium chloride and water. Potassium polyacrylate for drilling fluids is a non-toxic well wall stabilizer, which is easily soluble in water and has the effect of inhibiting the dispersion of mud shale and drill cuttings, as well as having the properties of lowering water loss, improving flow pattern and increasing lubrication. Sodium lignosulfonate is a polymer substance produced by lignin sulfonation, which is an excellent dispersant, flocculant and corrosion and scale inhibitor. Dimethyldiallylammonium chloride is a drilling fluid treatment agent, oil well cement additive, acid fracturing additive and oil repellent agent used for temperature and salt resistance. A comparison of the drilling fluid compositions revealed that the contents of Ca, Ba, and Mg in drilling fluid 2 were higher than those in drilling fluid 1. In addition, drilling fluid 2 contained lower levels of slow-release scale inhibitor additives than drilling fluid 1 and had a significantly lower pH value. The drilling fluid composition analysis and corrosion test results show that the drilling fluid pH and additive components affect the corrosion rate of drill pipe. The corrosion rate of the V150 drill pipe can be significantly reduced when the pH is increased and the content of sodium lignosulfonate is increased.

Table 4 Results of drilling fluids composition analysis

Content of testing		Drilling Fluid 1	Drilling Fluid 2
The Main Components	Potassium polyacrylate	3%~4% (mass fraction)	1%~2% (mass fraction)
	Sodium lignosulfonate	15%~16% (mass fraction)	10%~11% (mass fraction)
	Dimethyldiallylammonium chloride	2%~3% (mass fraction)	1%~2% (mass fraction)
	Water	78%~79% (mass fraction)	86%~87% (mass fraction)
Major Elemental Compositions	Ca	64.20 μg/g	157.00 μg/g
	Fe	155.77 μg/g	132.10 μg/g
	Al	73.01 μg/g	76.47 μg/g
	Ba	103.88 μg/g	165.31 μg/g
	Mg	24.61 μg/g	69.13 μg/g
pH		8.92	7.70

Since the drilling fluid circulation system is not airtight, atmospheric oxygen is mixed into the drilling fluid as it circulates through the mud sumps, high pressure mud guns, mud pumps, and other equipment. Some of the oxygen is dissolved in the drilling fluid until it becomes saturated, so the corrosion process that occurs at the drill pipe is typically assumed to be supported by the dissolved oxygen in the drilling fluid[27].

The dissolution of Fe, on the one hand, occurs at the anode to generate Fe^{2+}, and the reduction of O_2 and H_2O at the cathode generate OH^-. Fe^{2+} is hydrolyzed to generate $FeO(OH)$, and then dehydrated to further oxidation to form Fe_2O_3. In general, $FeO(OH)$ and Fe_2O_3 are loose and porous, have poor adhesion to the substrate and cannot play a protective role. The reactions can be described as follows:

$$Fe - 2e^- \longrightarrow Fe^{2+} \tag{2}$$

$$O_2 + 2H_2O + 4e^- \longrightarrow 4OH^- \tag{3}$$

$$Fe^{2+} + 2OH^- \longrightarrow Fe(OH)_2 \tag{4}$$

$$2Fe(OH)_2 + \frac{1}{2}O_2 \longrightarrow 2FeO(OH) + H_2O \tag{5}$$

$$2FeO(OH) \longrightarrow Fe_2O_3 + H_2O \qquad (6)$$

Dissolved oxygen corrosion usually has general corrosion and localized corrosion in two forms, and the corrosion surface has brown corrosion products. Total corrosion causes little change in wall thickness, but localized corrosion produces pitting pits scattered on the outer surface of the drill pipe. The severe corrosion will form pits, as well as grooves in an annular distribution, which is consistent with the surface state of the failed drill pipe.

Consequently, oxygen corrosion can generally continue without obstacles. During the experimental simulation, a distinct scale layer exists on the sample surface at 150℃ and a relatively closed microenvironment is formed under the scale. Due to the obstructive effect of the scale layer, it is tough for oxygen to reach the metal interface under the scale layer through the crevices or scale layer micro-pores. Therefore, as corrosion proceeds, an oxygen-poor zone is formed under the scale, forming a macroscopic oxygen concentration cell with the rest of the external part of the scale. Due to the presence of the scale layer, it is also difficult for the metal cations to diffuse to the outside of the scale layer, which will cause the accumulation of Fe^{2+} and the formation of excess electrons. In order to maintain charge conservation, external corrosion ions such as Cl^- will migrate under the scale layer. In addition, the excess Fe^{2+} hydrolysis forms $Fe(OH)_2$, which causes H^+ ions to accumulate under the scale layer, thus forming an acidification autocatalytic process, eventually leading to accelerated corrosion under the scale. The mechanistic model is shown in Figure 15. In the corrosion process under the scale, the excess electrons generated by the dissolution of the pit are transferred through the metal, making the metal around the scale layer appear similar to the "cathodic protection" effect, thus inhibiting the corrosion of the metal around the scale layer[28].

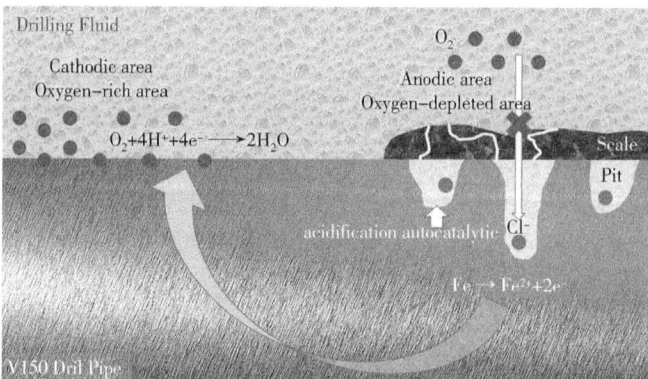

Figure 15 A mechanistic model illustrating the galvanic corrosion effect between bare steel and the steel under the scale layer

4 Conclusion and Recommendation

(1) The localized corrosion mechanism of drill pipe grade V150 during drilling in an oil field in northwest China is mainly attributed to an oxygen concentration cell and under-deposit corrosion. When a scale layer is formed at the drill pipe surface, a macroscopic oxygen concentration difference cell is formed inside and outside the scale layer, which promotes the acidification autocatalytic process under the scale, leading to an increase in corrosion.

(2) Different drilling fluid types and temperatures significantly affected the corrosion behavior of drill pipe grade V150. Drill pipe specimens immersed in drilling fluid 2 at 150℃ showed scaling, resulting in severely localized corrosion under deposits. When the temperature was reduced to 120℃ and below, the scaling phenomenon disappeared and the average corrosion rate decreased from 0.79mm/a to 0.18mm/a.

(3) The corrosion rate of drill pipe grade V150 is also affected by the condition of the metallic surface. The experimental results showed that the corrosion rates of the original surface and those with scratches and roll marks were slightly higher compared to the ones determined on smooth surfaces.

(4) It is suggested to increase the pH of the drilling fluid and add an appropriate amount of corrosion inhibitor when the oxygen corrosion rate is moderate. When pitting occurs, or the corrosion rate of oxygen corrosion is severe, it is recommended that a deoxidizer be added to control the degree of corrosion. In addition, after the drill pipe is lifted out, it is necessary to wash the surface with water to clean the residual drilling fluid mud, and store it after sufficient drying.

Acknowledgments

The authors wish to acknowledge the Scientific Research and Technology Development Project of Petrochina(2021ZZ01-04), CNPC Scientific and Technical Project(No. 2021DJ2703), CNPC Basic Research and Strategic Reserve Technology Research Fund Project[No. 2022DQ03(2022Z-01)].

References

[1] M Liu, et al. Corrosion fatigue crack propagation behavior of S135 high-strength drill pipe steel in H_2S environment, Engineering Failure Analysis. 97(2019): 493-505.

[2] C H Peng, Z Y Liu, X Z Zhao. Failure analysis of a steel tube joint perforated by corrosion in a well-drilling pipe, Engineering Failure Analysis. 25(2012): 13-28.

[3] L Elliott, V Buchoud, T Krepp. High-strength, thin-wall-steel drill pipe may provide solution for ultra-extended-reach wells, Drilling Contractor. 2(2009): 65.

[4] Z Peng. On hydrogen induced cracking(HIC) and hydrogen stress cracking(HSC), Journal of Chinese Society of Corrosion and Protection 2.2(1982): 42-47.

[5] X Huang, G Lu. Overall performance study on P110SS sulfur-resistant casing in the high sulfide gas well annulus protection fluid, Chemical Engineering of Oil and Gas. 42(2011): 45-48.

[6] H Wang, L Hong. Effect of tempering temperature on precipitate and mechanical properties of an anti-sulfur drill pipe steel, Transactions of Materials and Heat Treatment. 33(2012): 88-93.

[7] D Xia, W Feng. Stress corrosion test and study of high strength drilling rod in sour medium, Natural Gas and Oil. 15(1997): 25-32.

[8] S Dan, K Tong, et al. Failure analysis of S135 drill pipe body fracture in a well, Engineering Failure Analysis. 145(2023): 106998.

[9] Z M Yu, D Z Zeng, et al. The failure patterns and analysis process of drill pipes in oil and gas well: A case study of fracture S135 drill pipe, Engineering Failure Analysis. 138(2022): 106171.

[10] S J Luo, M Liu, X X Zheng. Characteristics and life expression of fatigue fracture of G105 and S135 drill pipe steels for API grade, Engineering Failure Analysis. 116(2020): 104705.

[11] H Wang, et al. Deep and ultra-deep oil and gas well drilling technologies: Progress and prospect, Natural Gas Industry B. 9(2022): 141-157.

[12] M Jellison, B Foster, et al. Light weight ultra-high strength drill pipe for extended reach and critical deep drilling, NACE International, Houston, TX, 2010.

[13] L Zhu, et al. Corrosion Failure Reason of a G105 Steel Drill Pipe, Corrosion & Protection. 37(2016): 775-780.

[14] H Bi, Y Jie, P Zhao. Corrosion failure analysis of G105 drill pipe in deposited, Physical Testing and Chemical Analysis Part A Physical Testing. 41(2005): 304-306.

[15] M Chen, Z Ouyang, S Yu. Reasons for corrosion failure of S135 steel grade drill pipe in a certain well, Materials for Mechanical Engineering. 45(2021): 93-97.

[16] Z Shu, Z Ouyang, P Yuan. The Mechanical Performance of V150 Drill Pipe under Combined Tension-Torsion Loading, Petroleum Drilling Techniques. 47(2019): 68-73.

[17] G Liu, et al. Effect of heat treatment on microstructure and properties of V150 drill pipe steel, Transactions of Materials and Heat Treatment. 37(2016): 149-155.

[18] Z Ouyang, M Chen, X Xie. Piercing Cause of V150 High Strength Drill Pipe, Materials for Mechanical Engineering. 46(2022): 64-70.

[19] P Zhao, J Yu, J Guo. Analysis of Influencing Factors on Dissolved Oxygen Corrosion of Drill Pipe, Steel Pipe. 39(2010): 29-33.

[20] C Wang. Cause analysis of corrosion holes on S135 drill pipe, Surface Technology. 45, 3(2016): 58-63.

[21] W Tian, Z Yang, et al. Study on property of corrosion & failure for S135 drill rod joints, Physical Testing and Chemical Analysis(Part A: Physical Testing). 44(2008): 575-578.

[22] M Bai, J Luo, et al. Effects of Cl^- on corrosion of S135 steel for drill pipe, Materials Protection. 54(2021): 140-144.

[23] J Wang, X Pu, et al. Corrosion deposit of drill pipe steel in high density saturated saline drilling fluid, Chemical Engineering & Equipment. 11(2010): 51-53.

[24] W Liu, T Shi, et al. Corrosion mechanism of a drill pipe, Corrosion & Protection. 32(2011): 806-809.

[25] L Wang, R Hu, et al. The Oxygen corrosion behavior of S135 drill pipe steel in drill fluid, China Petroleum Machinery. 34(2006): 1-4.

[26] ASTM. Standard G 111-21, Standard Guide for Corrosion Tests in High Temperature or High Pressure Environment, or Both, ASTM International, West Conshohocken, PA, 2021.

[27] NACE Standard SP0775-2013, Preparation, Installation, Analysis, and Interpretation of Corrosion Coupons in Oilfield Operations, NACE International, Houston, TX, USA, 2013.

[28] Q Xiong, J Hu, et al. The study of under deposit corrosion of carbon steel in the flowback water during shale gas production, Applied Surface Science. 523(2020): 146534.

本论文原发表于《Engineering Failure Analysis》2024 年第 158 卷。

Failure Analysis of Corrosion and Fracture of P110 Tubing in a Development Well

Han Yan[1]   Ren Hong[1]   Zhang Wenbin[2]
Fu Anqing[1]   Ma Qingwei[1]

(1. State Key Laboratory for Performance and Structure Safety of Petroleum Tubular Goods and Equipment Materials, Tubular Goods Research Institute of CNPC;

2. Well Testing and Workover Company, CNPC Chuanqing Drilling Engineering Co., Ltd., Chengdu)

Abstract: In this paper, a fractured and corroded tubing string was analyzed through macroscopic observation, chemical composition analysis, metallographic and mechanical test. The corrosion products were studied by SEM, EDS, and XRD method. The results showed that the corrosion of tubing in this well was caused by CO_2 corrosion, the well contains CO_2 medium and increasing water cut in the late stage of development were the main reasons of serious corrosion, the scaling on the tubing wall aggravated the corrosion. Finally, some suggestions were proposed for avoiding or slowdown this kind of corrosion failures.

Keywords: P110 tubing; Wall thickness thinning; CO_2 corrosion; Fracture; Failure analysis

1 Introduction

With the increasing demand for energy in the production development, oil and gas well shift to the harsh environment, the service environment of tubing and casing are more and more severe, and the corrosion fracture of tubing and casing frequently[1-4]. Tubing is the lifeline of oil and gas exploitation. Once the corrosion fracture occurs, it will seriously threaten the underground safety of wells, affect the normal production of oilfield, and bring huge economic losses [5-7]. Therefore, it is necessary to analyze the causes of corrosion fracture of tubing and put forward targeted preventive measures to reduce the occurrence of similar accidents.

In this paper, the reason of a fractured tubing in a development well in the west of China was

---

Corresponding author: Han Yan, hanyan003@cnpc.com.cn.

studied. The depth of the well was 5700m, and the specification of tubing was $\phi 88.9mm \times 6.45mm$ P110EUE. It was found that the pressure dropped sharply during the gas lift operation, and the 471# tubing was found fractured when it was taken out from the well. There was obvious corrosion occurred in the inner wall of the 372#~471# tubing, and accompanied by a certain degree of scaling phenomenon. Among them, the 467#~471# tubing scaling heavily, and the 471# tubing is the most corroded. No obvious outer wall corrosion was found in other tubing except the 471# and 521# tubing. A certain degree of scaling was found on the outer wall of the tubing, and there was no significant change in scaling degree on the outer wall.

2 Experimental methods

To study the reason of thecorrosion and fracture of tubing, the following tests were conducted: (1) macroscopic observation, (2) chemical composition analysis, (3) metallographic structural characterization, (4) mechanical properties, (5) scanning electron microscopy (SEM) and energy-dispersive spectrometry (EDS) of corrosion products, (6) X-Ray Diffraction (XRD) of corrosion products.

3 Results and discussion

3.1 Macroscopic observation

The macroscopic morphology of the failed tubing is shown in Figure 1 to Figure 3. The 471# tubing is a fractured tubing. The inner and outer walls of the tubing are pit-like corrosion morphology, and the wall thickness is reduced obviously. The remaining wall thickness is between 1.5mm to 4mm. Due to the thin wall thickness, it has been flattened after salvage, and the fracture is also damaged severely, as shown in Figure 1. It can be seen from the fracture at the lower part of 471# that there were more corrosion products on the internal wall of the 471# tubing and the thinning was serious. The wall thickness thinning is serious, resulting in a sharp decline in the tensile strength of the tube, and eventually the tube fracture. The internal and external walls of the 472# tubing are corroded lightly compare to 471# tubing, there was no obvious thickness thinning, and the remaining wall thickness is between 5.5mm to 6mm, as shown in Figure 2. The internal and external walls of 521# tubing were attached with heavy oil, the inner and outer walls are pit-like corrosion morphology, and the remaining wall thickness is between 2.5mm to 4.5mm, as shown in Figure 3. The inner wall corrosion is relatively serious than that of the outer wall on the other tubing, and a large number of corrosion products and scaling are attached to both the inner and outer walls.

The original wall thickness of the tubing in this well was 6.45mm and the service time was 7.5 years. Based on this, theaverage corrosion rate of tubing was calculated and the qualitative categorization of tubing corrosion was determined according to NACE SP0775-2023[8] standard (Table 1). The results are shown in Table 2. Tubing 471# and 521# experienced severe corrosion, while tubing 472# experienced moderate corrosion.

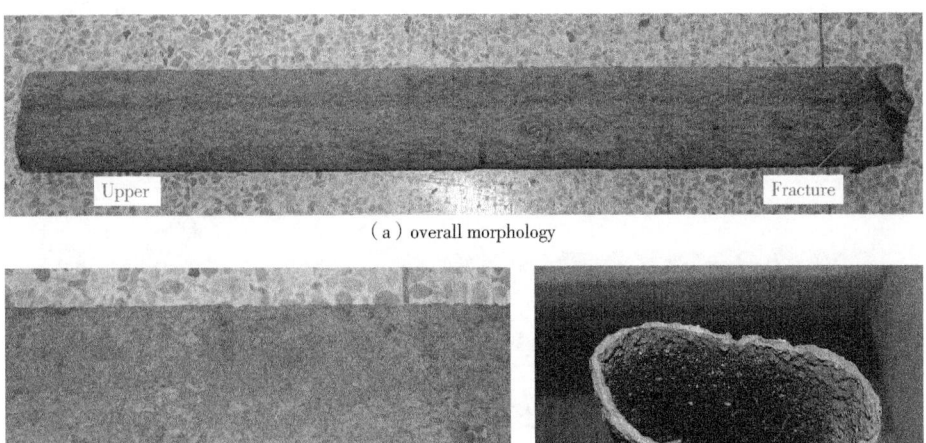

(a) overall morphology

(b) external corrosion

(c) internal corrosion and distortion

Figure 1 Macroscopic morphology of the fractured tubing (471#)

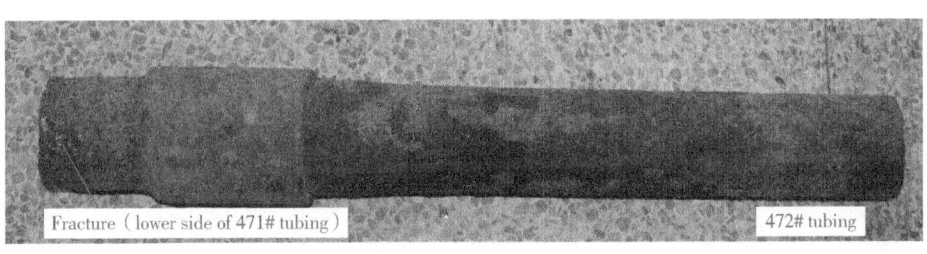

(a) overall morphology

(b) fracture surface (471#)

(c) internal corrosion (472#)

Figure 2 Macroscopic morphology of the fractured tubing (471# and 472#)

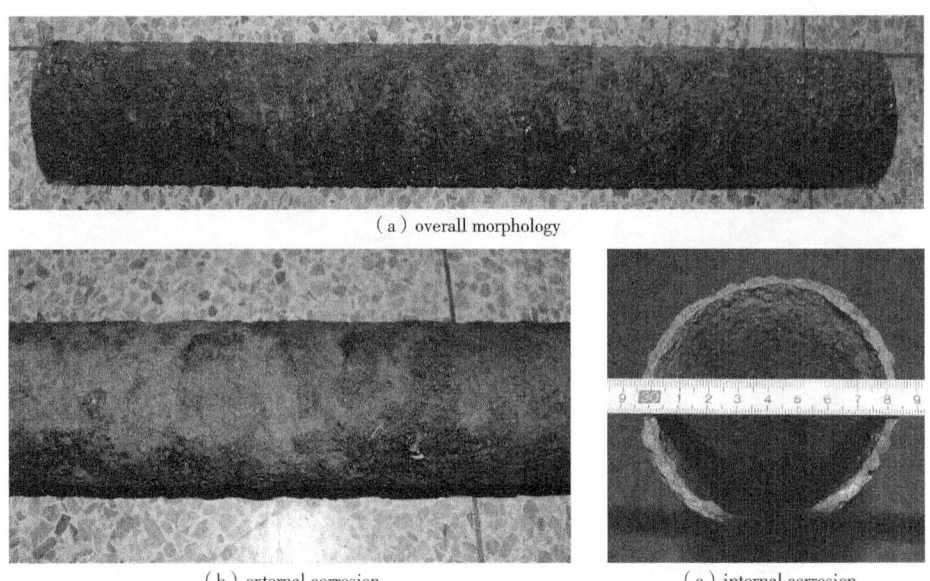

Figure 3 Macroscopic morphology of the 521# tubing

Table 1 Qualitative categorization of carbon steel corrosion rates

Qualitative categorization	Average general corrosion rate/(mm/a)	Maximum pitting rate/(mm/a)
Low	<0.05	<0.13
Moderate	0.05~0.20	0.13~0.30
High	>0.20	>0.30

Table 2 Qualitative categorization of tubing

NO.	471#	472#	521#
Average general corrosion rate/(mm/a)	0.33~0.66	0.06~0.13	0.26~0.53
Qualitative categorization	High	Moderate	High

In addition to the remaining wall thickness of 471# tubing, which is too thin to be analyzed, the 472# and 521# corroded tubing were tested and analyzed.

3.2 Chemical composition analysis

The chemical composition of the P110 tubing is shown in Table 3. The elements content accorded with the API Spec 5CT [9] standard requirement, the chemical composition is qualified, but the content of P, S, Cr and Mo elements in 472# and 521# tubing is different significantly. Compared with 521# tubing, the 472# tubing with higher corrosion resistance elements Cr, Mo and less impurities P and S elements showed less corrosion.

Table 3 Chemical composition of the tubing (%, mas fraction)

Elements	C	Si	Mn	P	S	Cr	Mo	Ni	V	Cu
472#	0.23	0.21	1.21	0.0075	0.0042	0.016	0.023	<0.001	0.0037	0.0036
521#	0.22	0.20	1.22	0.014	0.0066	0.021	0.0025	<0.001	0.0036	0.0036
API Spec 5CT	—	—	—	≤0.03	≤0.03	—	—	—	—	—

3.3 Metallographic structural characterization

The metallographic structural of the tubing is show in Figure 4, the microstructure is tempered sorbite. No oversized nonmetallic inclusions were found, and the grain size of tubing is ASTM 9.0 grade.

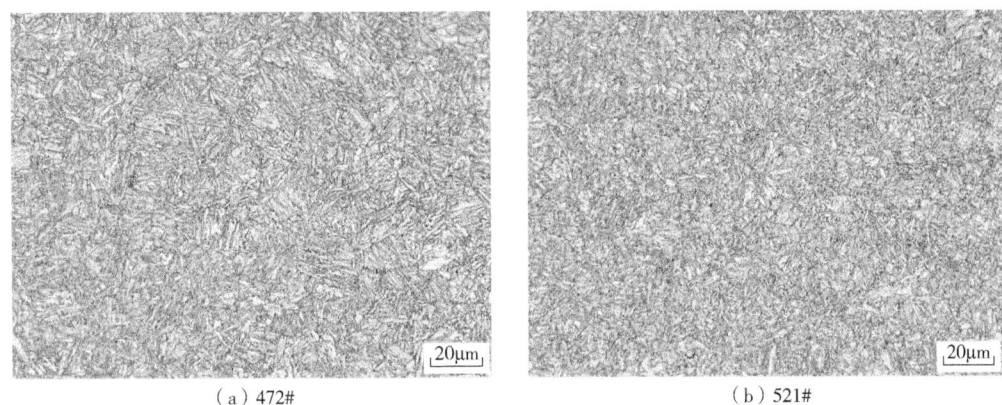

(a) 472#   (b) 521#

Figure 4　Microstructure of tubing

3.4 Mechanical properties

The results of tensile test showed that the tensile strength of 521# tubing is far below the requirement in API Spec 5CT standard, one of the 472# sample is below the 862MPa, as shown in Table 4. Since the test samples are all corroded samples, the exact cross-sectional area cannot be measured, so the elongation value A cannot be obtained; The wall thickness of the 521# tubing is thinned seriously, and the yield strength and impact test results cannot be measured. Both the internal and external surfaces of the samples were corroded, and the residual wall thickness is not enough to process the standard impact sample. The 2.5mm×10mm×55mm sample is a non-standard sample, and the measured values are for reference only, as shown in Table 4.

Table 4　Results of tensile test and charpy impact test

Sample No.		Results of tensile test			Results of charpy impact test (0℃)	
		R_m/MPa	$R_{t0.6}$/MPa	A/%	Size of sample/ mm×mm×mm	CVN/J
472#	Sample 1	860	795	—	2.5×10×55	24
	Sample 2	885	820	—		24
	Sample 3	880	820	—		25
521#	Sample 1	565	—	—	2.5×10×55	—
	Sample 2	390	—	—		—
	Sample 3	460	—	—		—
API Spec 5CT		≥862	758~965	≥12	10×10×55	≥27

The results of hardness value were shown in Table 5. The average hardness of 472# and 521# tubing is 34.2 and 29.3, respectively.

Table 5 Results of hardness

Sample No.	HRC		
472#	33.0	34.5	35.0
521#	28.5	29.0	30.5
API Spec 5CT	—		

3.5 SEM observation and EDS analysis

The micro-morphology and energy spectrum analysisspecimen were taken from internal and external of the tubing. Electron microscope observation showed that thick corrosion product films were attached to both the inner and outer walls of the sample, and the corrosion films were uneven, especially for external wall, the morphology of which was shown in Figure 5 and Figure 6.

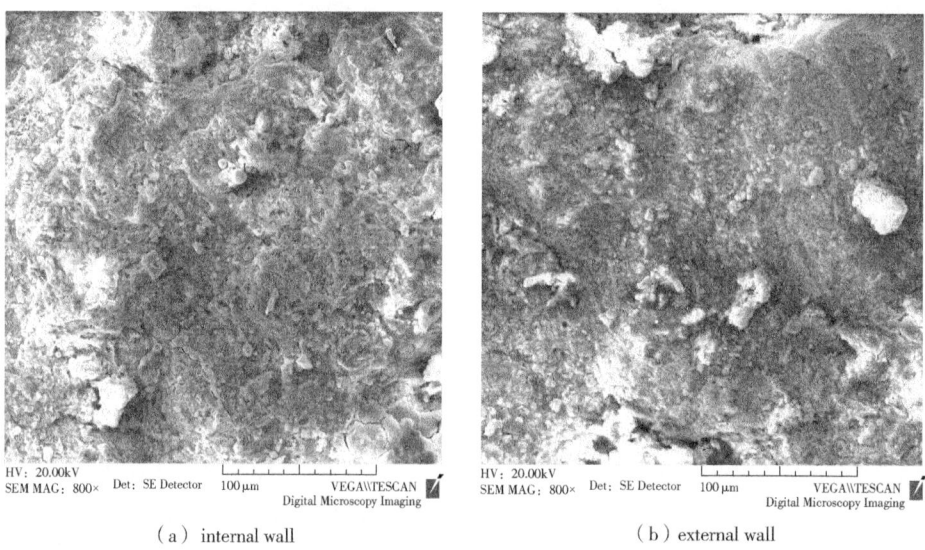

(a) internal wall   (b) external wall

Figure 5 Micro-morphology of the 472# tubing

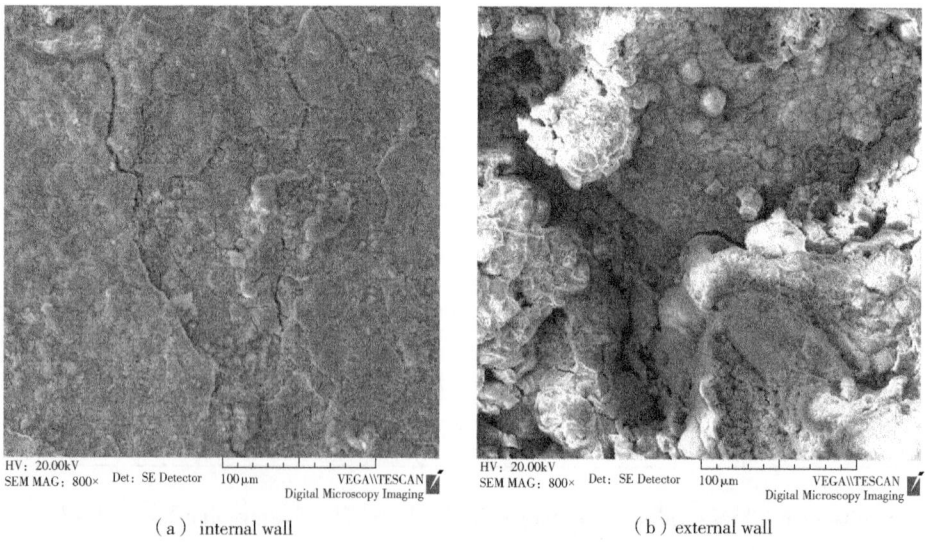

(a) internal wall   (b) external wall

Figure 6 Micro-morphology of the 521# tubing

The EDS results show that the corrosion products mainly contain of Fe, C, O, S and Cl, Si, Ca, Al elements, as show in Figure 7 and Figure 8. The C, O and Fe elements were the main components of corrosion products, while Cl, Si, Ca and Al elements were the main components of well fluid. It suggested that the corrosion film on the surface is mainly CO_2 corrosion product.

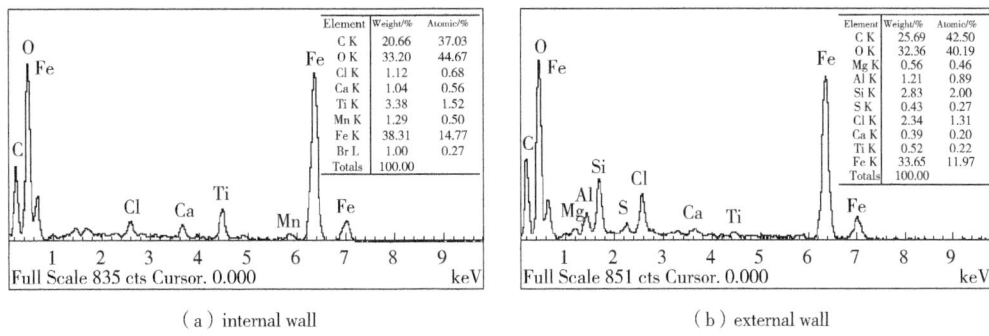

Figure 7   Energy spectrum of the corrosion products on 472# tubing

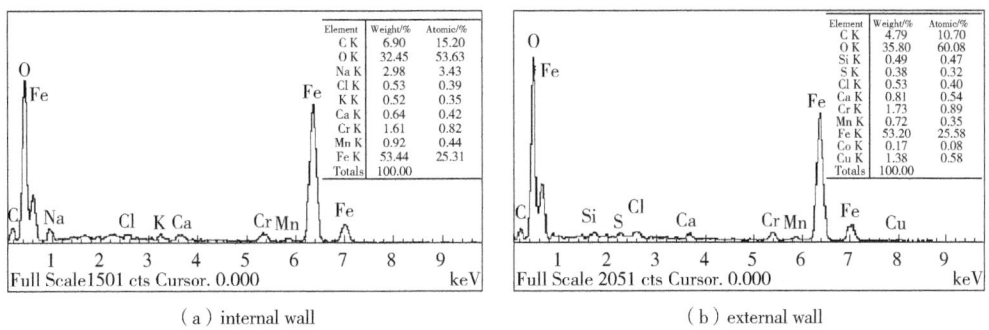

Figure 8   Energy spectrum of the corrosion products on 521# tubing

3.6   XRD analysis

The corrosion products were FeO(OH) and $FeCO_3$, shown in Figure 9 and Figure 10. The $CaCO_3$ shown in Figure 10 was the scale deposit. The corrosion products are mainly iron oxides and carbon dioxide corrosion products.

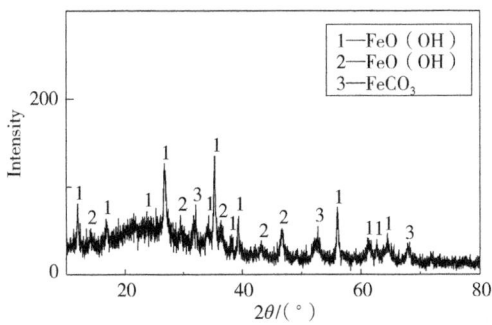

Figure 9   XRD result of corrosion products on 472# tubing

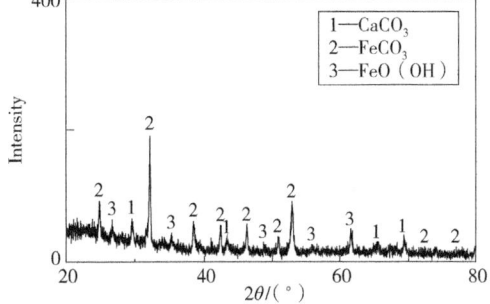

Figure 10   XRD result of corrosion products on 521# tubing

4 Discussions

According to the field data, the average water content of the crude oil in the well is greater than 30% (mass fraction), the average salt content is greater than 10000mg/L, and it contains a certain amount of sulfur and wax formation, and the associated gas in the well contains 2% (volume fraction). CO_2 and trace H_2S, which belongs to the gas reservoir with low sulfur content and medium CO_2 content, and the pipe string was worked in a harsh corrosion environment. The formation water contained HCO_3^-, Ca^{2+}, Mg^{2+}, and Cl^- plasmas accelerated tubing corrosion, as shown in Table 6.

Table 6 The composition of formation water

Composition	SO_4^{2-}	HCO_3^-	Ca^{2+}	Mg^{2+}	Cl^-	pH
Content/(mg/L)	300.00	3.35	501.40	76.30	3437.90	5.8

In oil and gas production systems, CO_2 is not corrosive without water, but it is corrosive when dissolved in water [7]. In the late stage of the well, the moisture content is as high as 48%, and CO_2 is very soluble in water, then the carbonic acid is obtained after it is dissolved in water, and hydrogen ions are released. Hydrogen ions are strong depolarizing agents, which are easy to seize electron reduction, promote the dissolution of anode iron and lead to corrosion. The reaction formula is as follows [10-11]:

Step 1: CO_2 dissolve.

$$CO_2 + H_2O = H_2CO_3 \tag{1}$$

Step 2: Carbonic hydrolysis.

$$H_2CO_3 = H^+ + HCO_3^- \tag{2}$$

$$HCO_3^- = H^+ + CO_3^{2-} \tag{3}$$

Anode process:

$$Fe + H_2O = FeOH + H^+ + e^- \tag{4}$$

$$FeOH = FeOH^+ + e^- \tag{5}$$

$$FeOH^+ + H^+ = Fe^{2+} + H_2O \tag{6}$$

Total reaction:

$$Fe = Fe^{2+} + 2e^- \tag{7}$$

Cathode process:

$$H^+ + e^- = H, pH \leq 4 \tag{8}$$

$$H_2CO_3 + e^- = H + HCO_3^-, 4 < pH \leq 6 \tag{9}$$

$$H_2O + e^- = H + OH^-, pH > 6 \tag{10}$$

Then:

$$2H = H_2 \qquad (11)$$

Result:

$$CO_2 + H_2O + Fe = FeCO_3 + H_2 \qquad (12)$$

In the process of corrosion reaction, the following reactions may also exist:

$$Fe^{2+} + 2H_2O = FeO(OH) + 3H^+ + e^- \qquad (13)$$

This is consistent with the results of EDS and XRD analysis of corroded tubing. The corrosion products are mainly $FeCO_3$ and $FeO(OH)$ which are CO_2 corrosion products, indicated that CO_2 corrosion mainly occurs in tubing. $FeO(OH)$ in corrosion products can also be considered as a hydrated oxide of Fe^{2+} in the presence of oxygen, with the following formula:

$$4Fe^{2+} + 6H_2O + O_2 = 4FeO(OH) + 8H^+ \qquad (14)$$

In the process of CO_2 corrosion, temperature is one of the most important factors, and the influence of temperature on the corrosion rate and corrosion form is largely reflected by the influence on the corrosion product film [12-13].

The results[14] show that at about 60℃, a small amount of soft and undense $FeCO_3$ is formed on the surface of carbon steel, and the corrosion rate is determined by the rate of CO_2 hydrolysis to carbonic acid and the diffusion of carbonic acid to the metal surface. In this case, the corrosion is uniform corrosion. At about 100℃, the corrosion products are thick but very loose, and the corrosion rate is determined by the mass transfer process of the corrosive medium through the product film. At this time, the corrosion rate is the largest, forming a deep pit or ring corrosion. Under the temperature condition of about 150℃, the formation of dense and strong adhesion $FeCO_3$, the corrosion process is basically prevented, and the corrosion rate is low. It is because of the strong influence of temperature on corrosion that local corrosion often occurs selectively at a certain depth in oil and gas wells.

For this failure well, the corrosion degree of tubing gradually increased from 372#~471# when the temperature of the corroded pipe section was in the range of 110~120℃, but below the 521# tubing, the corrosion degree of tubing decreased, indicated that the corrosion degree tended to stabilize as the temperature exceeded 125℃.

In addition, from the macroscopic morphology of the corroded tubing and the XRD results of the corrosion products, there is obvious scaling phenomenon in the tubing, and the deposit of scale on the tubing will accelerate the corrosion of metal. Corrosion products such as FeS, $FeCO_3$, FeO and other iron compounds and $CaCO_3$, $MgCO_3$, $CaSO_4$, $BaSO_4$ and silicon dirt oil deposited on the surface of the tubing to form scaling. Scale is mostly formed in the lower part of the wellbore and near material defects or joints. Due to the high salinity in the produced liquid and the strong penetrating active anion Cl^-, the corrosion film generated on the metal surface is destroyed, and it is difficult to form a dense scale layer or corrosion product protection layer. Microscopic corrosion galvanic cells are formed in local locations, forming pitting pits and rapidly developing in depth. The local corrosion rate is much higher than the average corrosion rate, which accelerates the corrosion. The severity of scaling in the well should be closely related to the quality of its produced water.

According to the formation water analysis data of the oilfield, the concentration of Ca^{2+} and Mg^{2+} ions in the formation water is not very high, but the water cut of the well has increased significantly since 2008, and the water quality in the later stage of development may have changed. Therefore, the formation water quality of the well should be further analyzed.

According to the above macroscopic morphology analysis and corrosion product analysis of corroded tubing, combined with the service conditions of tubing and the mechanism of CO_2 corrosion, it can be seen that tubing corrosion meets the characteristics of CO_2 corrosion, the main cause of corrosion is CO_2 corrosion, and the high water content in the late stage of development is the key reason for corrosion. The reason why the corrosion of the inner wall is more serious than that of the outer wall may be due to the fact that the outer wall of the tubing is in a static protective fluid environmental, and the inner wall is constantly in contact with the produced fluid containing corrosive media.

5 Conclusions and suggestions

(1) The corrosion of tubing in this well belongs to CO_2 corrosion;

(2) The oil well contains CO_2 corrosive medium and the increasing water cut in the late stage of development are the main reasons of serious corrosion, and the corrosion is aggravated by scaling on the surface of the tubing.

It is suggested that targeted corrosion inhibitors and scale removal agents should be added continuously in the late stage of oil well development to reduce the influence of corrosion medium on pipe string corrosion and scale formation.

Acknowledgments

This work wassupported by the National Natural Science Foundation of China (No. 52071338) and Scientific Research and Technology Development Project of CNPC (2021ZG09).

References

[1] Z.Y. Liu, H. Li, Z.J. Jia, et al., Failure analysis of P110 steel tubing in low-temperature annular environment of CO_2 flooding wells. Engineering Failure Analysis 60 (2016) 296-306.

[2] Zhang Zhi, Zheng Yushan, Li Jing, et al., Stress corrosion crack evaluation of super 13Cr tubing in high temperature and high-pressure gas wells. Engineering Failure Analysis 95 (2019) 263-272.

[3] Yang Xinyong, Yan Zhilun, Li Fang, et al., Analysis of the corrosion and fracture of P110S tubing in a well of Tahe oilfield, Materials protection 55 (2022) 222-226.

[4] Zhang Jiangjiang, Zeng Dezhi, Peng Zhengde, et al., Failure analysis of corrosion and fracture of P110 Steel tubing in a condensate well in west China, Materials protection 53 (2020) 122-129.

[5] ZHU S. D., WEI J. F., BAI Z. Q., et al. Failure analysis of P110 tubing string in the ultra-deep oil well[J]. Engineering Failure Analysis, 2011,18:950-962.

[6] QIU Zhichao, XIONG Chunming, CHANG Zeliang, et al. Major corrosion factors in the CO_2 and H_2S coexistent environment and the anti-corrosion method: Taking Tazhong I gas field, Tarim Basin, as an example[J]. Petroleum exploration and development, 2012,39:238-242.

[7] LIN Hai, XU Jie, FAN Baitao, et al., Review on CO_2 corrosion rule of down-hole strings in Bohai oil field and

current status of anticorrosion material selection[J]. Surface technology,2016,45:97-103.
[8] NACE SP0775-2023, Preparation, installation, analysis, and interpretation of corrosion coupons in hydrocarbon operations. 2023. Houston: AMPP.
[9] API Spec 5CT, Casing and Tubing. 10th edition, 2018. Washington DC: API.
[10] ZHOU Jiming, Corrosion behavior of tubing steel in high temperature and high pressure water medium containing CO_2/H_2S and the role of protection technology[D]. Northwestern university,2002:2-6.
[11] ZHANG Haibao, Carbon dioxide corrosion research of tube and casing[D]. China university of petroleum, 2008:27-28.
[12] Hua Y, Xu S S, Wang Y, et al. The formation of $FeCO_3$ and Fe_3O_4 on carbon steel and their protective capabilities against CO_2 corrosion at elevated temperature and pressure [J]. Corrosion Science., 2019, 157: 392.
[13] HUANG Jiahe, YUAN Xi, CHEN Wen, et al. Effect of Temperature on Corrosion Behavior of Pipeline Steels N80 and TP125V in Artificial CO_2-saturated Fracturing Fluid of Shale Gas [J]. Journal of Chinese Society for Corrosion and protection, 2023,43: 251-260.
[14] ZHANG Xueyuan, DU Yuanlong. The rule and research trend of CO_2 corrosion in oil and gas industry [J]. Materials protection,1997,30:21-23.

本论文原发表于《Journal of Physics: Conference Series》2023 年第 2639 卷。

Collapse Failure Analysis of S13Cr-110 Tubing in a High-pressure and High-temperature Gas Well

Ji Nan[1,2] Zhao Mifeng[3] Wu Zhenjiang[3] Wang Peng[1]
Feng Chun[1] Xie Junfeng[3] Long Yan[1,2]
Song Wenwen[3] Xiong Maoxian[3]

(1. State Key Laboratory of Performance and Structure Safety of Petroleum Tubular Goods and Equipment Materials, CNPC Tubular Goods Research Institute;
2. China University of Petroleum (East China);
3. Tarim Oilfield Company, PetroChina Company Limited)

Abstract: Due to excellent CO_2 corrosion resistance and low cost, martensitic stainless steels such as super 13Cr have been widely used as tubing strings in high-pressure and high-temperature wells. However, in recent years, with the further development of the oilfield, super 13Cr tubing failure accidents keep happening in high-pressure and high-temperature gas well, resulting in the loss of well integrity and huge economic losses. So a systematic research on the failure mechanism of the super 13Cr tubing in HPHT gas well under different service conditions is vital to the reduction of the failure accidents. In this paper, a collapse failure accident of S13Cr-110 tubing in a high-pressure and high-temperature gas well was investigated by metallographic microscope, scan-ning electron microscope, energy dispersive spectrometer, X-ray diffraction and finite element analysis, aiming to find out the root causes and put forward useful suggestion for the establish-ment of the preventive measures. The results showed that the chemical composition and me-chanical properties of the tubing were all in accordance with the corresponding parameter requirements, and the collapse failure was attributed to the combined effect of formation sand production, high temperature and stress corrosion cracking on the outer surface. The tubing plugging induced by the formation sand production caused the increase of pressure difference between the annulus and tubing, which was the main reason for the collapse failure. In addition, high temperature in the downhole would result in a decreasing of the tubing's anti-collapseability. Meanwhile, the annular environment containing Cl^-, PO_4^{3-}, further induced the stress corrosion cracking under service stress. The presence of the cracks on the outer surface could cause a significant decrease of the pressure-bearing capacity of tubing strings, which would finally

Corresponding author: Ji Nan, jinan003@cnpc.com.cn.

have an obvious influence on the acceleration of the failure process.

Keywords: S13Cr-110 tubing; Collapse failure; Formation sand production; Stress corrosion cracking; Annular environment

1 Introduction

With the ever-growing energy demand of human socio-economic development and continuous improvement of the drilling and completion technics, high-pressure and high-temperature ultra-deep oil and gas reservoir has gradually become a new growth point of the oil and gas recoverable reserve[1]. Gas wells in Kuche foreland basin thrust belt of Traim oilfield in western China is a repre-sentative of the typical high-pressure and high-temperature gas well, where tubing and casing strings are subjected to harsh service conditions. For example, the downhole temperature and pressure are as high as 180℃ and 138MPa. What's more is the production fluids contain high content of corrosive substance, such as the maximum CO_2 partial pressure could be 4MPa, and the content of Cl^- in formation water could be as high as 160000mg/L and high volume of the corrosive residual acid flowback which injected in the acid fracturing process. In addition to this, permanent load in the normal production process, alternating load and vibration load caused by frequent opening and shutting in well also formed complex load condition. Therefore, tubular string failure problems induced by complex service conditions have become big issues that threat the integrity of the high pressure and high temperature wells, and also put forward high service property requirements to the material of tubing and casing.

Super 13Cr martensitic stainless steel has been widely used for corrosion control in sweet and weak acidity service conditions for its high strength, medium level corrosion resistance and relatively economic cost compared with other corrosion resistant alloys such as 22Cr[2-11]. In recent years, as the continuous expansion of the scale and quantity of super 13Cr martensitic stainless steel that used in HPHT wells as tubing strings, the failure problems of super 13Cr tubing have also gradually emerged and increased. Common failure modes of the tubing could be classified as corrosion, perforation, gluing, fracture, cracking, and plastic deformation[12-17]. Till now, researchers have conducted many works on the failure analysis of super 13Cr tubing in HPHT wells[18-22], and revealed that most of the failure mode were stress corrosion cracking or fracture, CO_2 and packer fluid were the main environmental reasons for the failure, and in addition to this, other failure mode is rare. With the further development of the ultra-deep reservoir, some new failure risk factors that never existed before have gradually come to light. In some high gas rate wells, ultra-high natural gas production brings an excessive large production pressure difference, which would result in a serve formation sand production[23-24]. At present, most high pressure and high temperature wells use slotted screen or perforated screen for sand prevention, but whether slotted screen or perorated screen, the sand prevention effects are all limited[25-28], so this would lead to the formation sand go into the tubing string easily and result in a tubing plugging or sand erosion of the tubing wall. Both the tubing plugging and sand erosion of the tubing would bring failure risks to the tubing strings. Sand erosion would result in a reduction of the tubing's wall thickness and severely would lead to a perforation of the tubing, meanwhile, tubing plugging would lead to a decreasing of the oil pressure, that means an increasing of the tubing's inner and outer pressure difference, and this would result in a rising risk of the collapse failure. Especially when combined with environmental factor in the downhole, would further enhance the failure risks.

In this work, collapse failure accidents of an S13Cr-110 tubing in a high-pressure and high-temperature gas well was investigated by magnetic particle inspection, chemical and mechanical properties tests, energy dispersive spectrometer and X-ray diffraction analysis, collapse stress finite element analysis. Aiming to make out the collapse failure causes of the tubing and put forward correspondingly suggestions to avoid similar failure accidents from happening again.

2 Background of the failure

2.1 Failure process

An HTHP gas well which has a depth of 7160m, finished drilling and well completion on December 15, 2012, and February 3, 2013 respectively. The maximum reservoir pressure and temperature is 118MPa and 180℃. The type of the packer fluid is weight4 phosphate, which has a density of 1.4g/cm³. After well testing on December 25, 2013, the HTHP gas well turned to production on October 17, 2014. The produced medium contain natural gas and formation water, the produced gas contained CO_2 concentration 1.13% ~ 1.74% (mole fraction) without H_2S, the type of the formation water is calcium chloride, and has a density of 1.12 g/cm³, and the con- centration of various ions in the formation water is shown in Table 1.

Table 1 Chemical composition of the formation water (mg/L)

Component	Cl^-	SO_4^{2-}	Ca^{2+}	Mg^{2+}	HCO_3^{2-}	$Na^+ + K^+$	Ba^{2+}	Salinity
Content	71557	151	9670	787	219	71800	15.3	234000

Fig.1 Sand in the positive choke

The type of the HPHT gas reservoir in this area is fractured tight sandstone gas reservoir, and about 55% of the HPHT gas well here suffer from severe sand production. The screen pipe completion method was used in this well, and the type of the screen is slotted screen which made from the tubing used in the well. But because of the low sand control accuracy and erosion resistance of the slotted tubing screen, the effect of the sand control measure is limited, massive sand can be found in the positive choke, as shown in Fig. 1. Fig. 2 illustrated the production and pressure condition of the HPHT gas well during production. There has been sustained casing pressure in the "A" annulus and "B" annulus since the beginning of the well production. The "A" annulus pressure ranged from 25MPa to 62MPa, and the "B" annulus pressure ranged from 0MPa to 53MPa. The sustained casing pressure indicated that there is a leakage in the thread joint of the tubing string. After production for 14 months, there had been sustained casing pressure in the "C" annulus for about 6 months until on September 30, 2016, the wellhead pressure suddenly decreased from 40.28MPa to13MPa, and then followed by the "A" annulus dropped from 39.44MPa to 0MPa. After that the choke and valve on the Christmas tree was closed to stop the production for inspection. The next day sampled the produced water for test, the result revealed that the density of the produced water was 1.38 kg/cm^3, which very close to the density of the packer fluid. When killing the well on October 17 for the subsequent well repair operation, the wellhead pressure was 40MPa, the "A" annulus pressure was 62.2MPa, the "B" annulus pressure was 13.17MPa and the "C" annulus pressure was 0MPa.

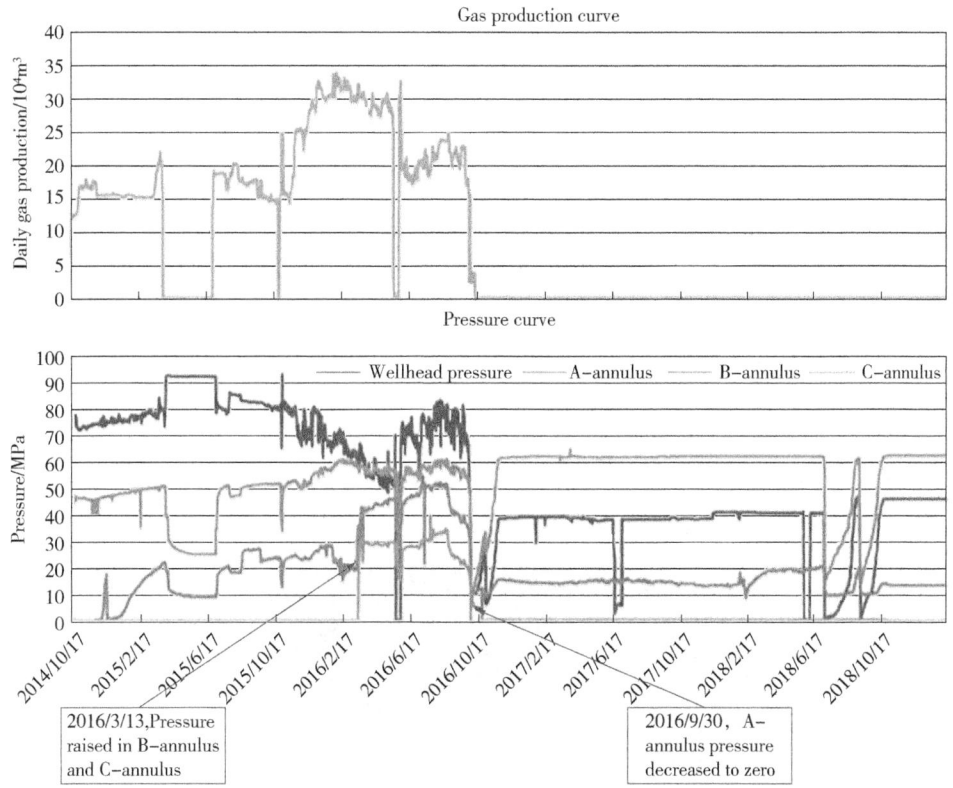

Fig.2 Gas production curve and pressure curve of the HPHT gas well

2.2 Tubing string dredging operation

For there was no gas production since the fluctuation of the wellhead pressure on September 30, 2016, coiled tubing dredging operation was conducted on the tubing string at September 15, 2018. Wellhead-pressure release operation was conducted before the tubing string dredge operation, the pressure monitoring curve was shown in Fig.3. During the oil pressure release stage, the casing pressure decreased obviously with the decreasing of the oil pressure, this was an evident that tubing string connected with the outside production casing string. During the well shut-in stage, there was only 0.7MPa increasing of oil pressure and casing pressure in 3h, this can illustrated that tubing string was blocked at a certain depth, and only a few gas could effuse from it.

After the release of wellhead pressure, coiled tubing dredging was applied inside the tubing

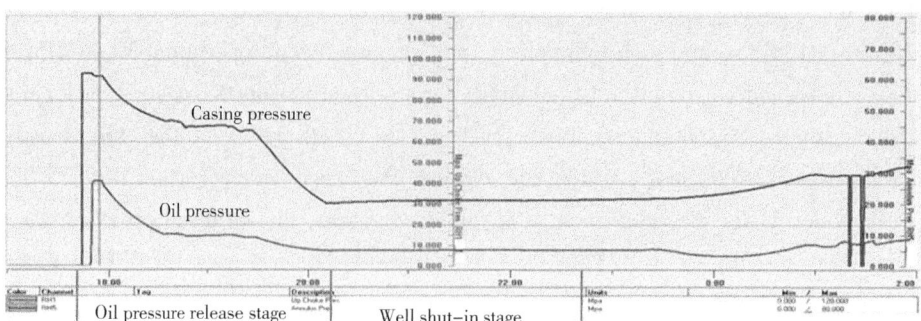

Fig.3 Wellhead pressure monitoring curve of the gas well

string and met resistance at the depth of 6491.37m, and failed to get through even at a maximum 2 tons load application. During this time, a total amount of 9.6m³ organic salt was injected in the downhole, but the metering equipment showed that the injected organic salt had been stable at the amount of 9.6m³, and there was no loss into the formation layer. That is to say, the tubing string which located above the blocked position and the "A" annulus had no connection with the formation layer, the tubing string was totally blocked at the depth of 6491.37m.

2.3 Failure sample information

When pulling the tubing string out from downhole, and found that the tubing in the depth of 6486.88m, No. 7 tubing counted from the packer was collapsed, as shown in Fig.4(a), the tubing string structure is shown in Fig.4(b). The whole tubing was totally collapsed from the external thread side to the coupling side, and an obvious collapse crack can be observed on the coupling side of the tubing, as shown in Fig.4(c) and Fig.4(d).

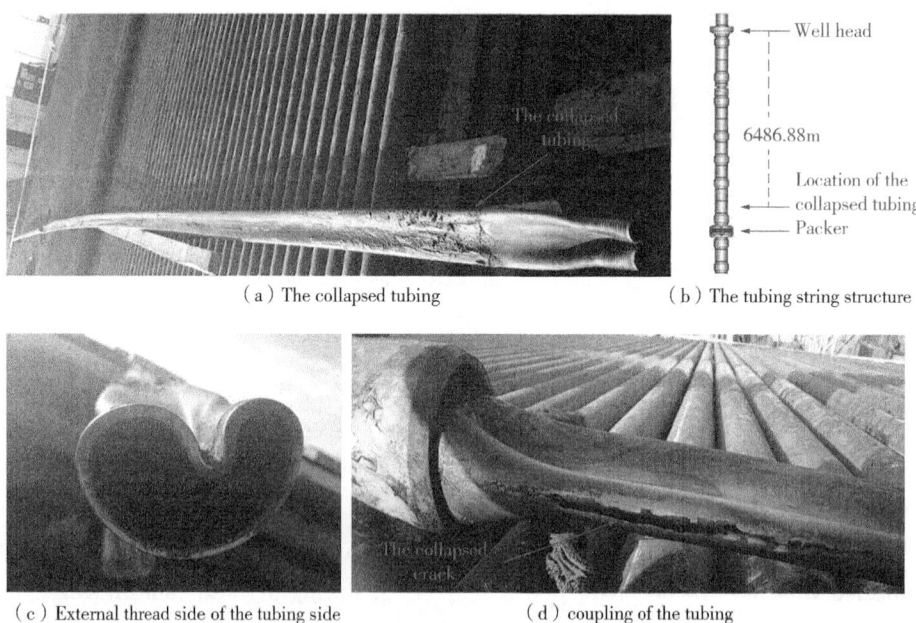

(a) The collapsed tubing   (b) The tubing string structure

(c) External thread side of the tubing side   (d) coupling of the tubing

Fig.4

The type and material of the collapsed tubing is $\phi 88.9mm \times 6.45mm$ S13Cr-110, and had been serviced in the well for 970 days. It is observed that the outer wall of the tubing was covered by thick scale layers, as shown in Fig.5(a). In addition to this, there also existed black bulkheads inside the tubing, as shown in Fig.5(b).

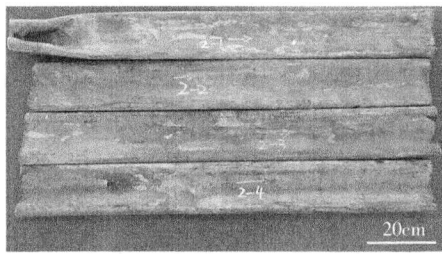
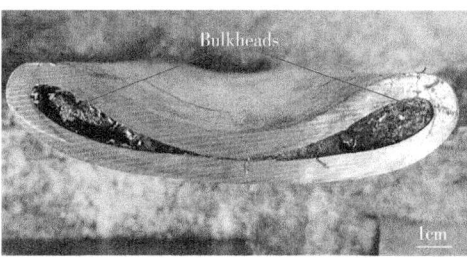

(a) Scale layer on the outer wall of the tubing　　　(b) The obstruction inside the tubing

Fig.5

3　Experiments

In order to make out the collapse failure causes of the tubing, a series tests were conducted during the failure analysis process. Magnetic particle inspection was performed on the outer surface of the collapsed tubing according to ASTM E709-2015 standard[29] to find out whether there were cracks or not, then wall thickness inspection was conducted using an ultrasonic thickness meter. Chemical composition of the collapsed tubing was analyzed using direct-reading spectrometer on the basis of ASTM A751-14a standard[30]. The metallographic examination, including the microstructure, grain size and non-metallic inclusions of the collapsed tubing were examined by microscopy according to the ASTM E3-11 (2017)[31], ASTM E112-13[32] and ASTM E45-18a[33] standards, respectively. Microscopy and scanning electron microscopy (SEM) were utilized to investigate the morphology of the cracks, Energy Dispersive Spectrometer (EDS) was used to analyze the relative chemical elements content of the cracks and corrosion products. Corrosion products on the outer surface and obstructions inside the tubing were analyzed by XRD diffraction analyzer.

Because of the severe deformation of the collapsed tubing, mechanical properties tests were performed on the same batch of the as-received super 13Cr tubing. The tensile properties were tested at room temperature according to the ASTM A370-20 standard[34]. The charpy V-notch impact properties were tested at -20℃ according to the ASTM A370-20 standard[34]. The Rockwell hardness tests were performed according to ASTM E18-2016[35]. The full-scale anti-collapse test was conducted according to API RP 5C5-2017[36] and API TR 5C3-2018[37]. At last, a finite element analysis was performed to simulate the collapse failure process of the tubing.

4　Results

4.1　Magnetic particle testing

Magnetic particle inspection results of the collapsed tubing are shown in Fig.6. The imaging agent is a black magnetic particle. Numerous transverse cracks were discovered on the outer surface of the collapsed tubing, and no cracks were observed on the inner surface. This revealed that the cracks may be in relationship with the annulus environment.

4.2 Diameter and wall thickness inspection

Considering the collapse strengthen is in big relationship with the diameter and wall thickness of the tubing, and the severe deformation of the collapsed tubing, diameter and wall thickness inspection were conducted on the same batch of the collapsed tubing. Fig.7 and Table 2 displayed the inspection position and results respectively. The diameter and wall thickness of the as-received super 13Cr tubing meet the requirements of API 5CT-2018 and the tubing purchasing technical agreement of Tarim Oilfield Company.

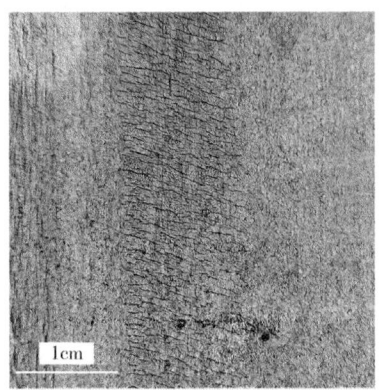

Fig.6 Inspection of collapsed super 13Cr tubing by magnetic particle testing

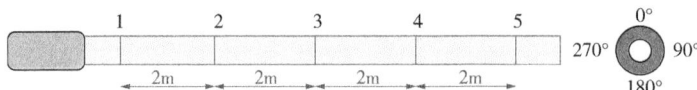

Fig.7 Schematic diagram of the diameter and wall thickness inspection

Table 2 Diameter and wall thickness inspection results of the same batch of collapsed super 13Cr tubing (mm)

Location	1	2	3	4	5
$t_{0°}$	6.48	6.63	6.56	6.53	6.52
$t_{90°}$	6.26	6.45	6.34	6.77	6.55
$t_{180°}$	6.30	6.67	6.58	6.60	6.60
$t_{270°}$	6.48	6.53	6.89	6.48	6.79
$D_{0°}$	89.02	89.12	89.16	89.04	89.12
$D_{90°}$	89.10	89.08	89.10	89.10	89.12

4.3 Chemical composition analysis

4.3.1 Chemical composition analysis of super 13Cr tubing

ARL-4460 direct-reading spectrometer is used for the chemical composition analysis of the collapsed tubing, and the results are listed in Table 3. All the elements contained in super 13Cr steel are in accordance with the tubing purchasing technical agreement of Tarim Oilfield Company.

Table 3 Chemical composition of the collapsed tubing (%, mass fraction)

Sample	C	Si	Mn	P	S	Cr	Mo	Ni
Collapsed tubing	0.036	0.17	0.40	0.010	0.0004	13.7	2.19	5.7
Tubing purchasing technical agreement of Tarim Oilfield Company	—	—	—	≤0.015	≤0.010	—	—	—

4.3.2 Chemical composition analysis of bulkheads

Fig. 4 revealed that the tubing was blocked by bulkheads, which present a black mud morphology, as shown in Fig.8. In order to make out the composition of the bulkheads, the following operations were conducted. First, the bulkheads were put into a drying oven at 105℃ for the separation of free water, and then burned the dried bulkheads at 550℃ for the separation of combined water. After that, the residuary bulkheads were dissolved into dilute nitric acid and filtered out the undissolved substance, formed the solution marked as solution A, the undissolved substance marked as residue B. The X-ray diffraction method was used for making out the composition of the residue B, the result is shown in Fig.9. The residue B mainly composed of $BaSO_4$ and SiO_2, considering SiO_2 can dissolve in the hydrofluoric acid, so when dissolved the residue B in the hydrofluoric acid, the loss in weight is the content of SiO_2, and the rest is $BaSO_4$, as shown in Table 4. The method of atomic absorption spectrometry and flame atomic absorption spectrometry were used to quantitative analysis the ion composition of solution A. The chemical composition of the solution A is shown in Table 4. The solution A mainly contains the ion of K^+, Na^+, Ca^{2+}, Mg^{2+}, Cl^-, Fe^{3+}, SO_4^{2-}.

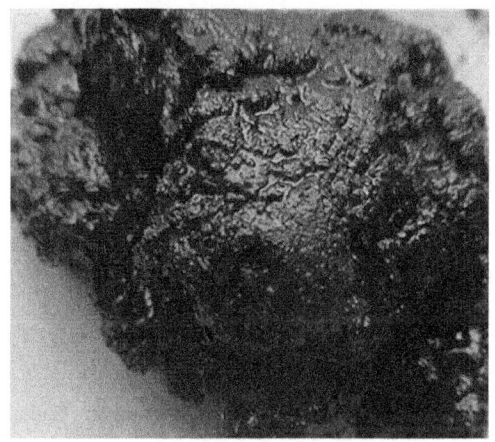

Fig.8 Macro-morphology of the bulkheads inside the collapsed tubing

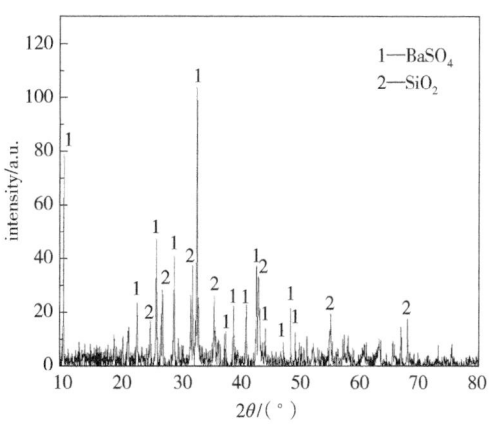

Fig.9 XRD analysis result of the residue B

Table 4 Chemical composition of the bulkheads inside the collapsed tubing (%, mass fraction)

Sample	Unsolvable in Nitric Acid		Solvable in Nitric Acid						
	Silicate	Barium sulfate	K^+	Na^+	Ca^{2+}	Mg^{2+}	Fe^{3+}	SO_4^{2-}	Cl^-
Bulkheads	12.0	31.8	2.3	1.4	1.7	0.5	4.0	0.8	2.2

As mentioned above, the effect of the sand control measure is limited, massive sand can be found in the positive choke, and considering SiO_2 is the main composition of sand, so the composition of SiO_2 in the bulkheads come from sand production from the formation layer. The chemical composition in Table 1 illustrated that the formation water mainly composed of Cl^-, SO_4^{2-}, Ca^{2+}, K^+, Na^+, Mg^{2+}, Ba^{2+}, HCO_3^{2-}, and the ion composition is very similar to that showed in Table 4. In addition to this, the salinity of the formation water is 1.45×10^5, which has a high scaling tendency, and the scale substance in such regions are mainly batium, so $BaSO_4$ in the bulkheads resulting from the scaling of the formation water. Considering the residue in the bulkheads has a big contribution on the obstruction of the tubing, it can be concluded that the bulkheads inside the tubing is a combination of sand production form the formation layer and scale formation from the formation water.

4.4 Metallographic examination

The microstructure of the collapsed tubing was tempered lath martensite, as shown in Fig.10, which was a normal metallographic structure of the martensite stainless steel. The grain size is grade 8.5 according to the ASTM E112-13 standard, while nonmetallic inclusion is grade A0.5, B0.5, D0.5 according to the ASTM E45-18a standard.

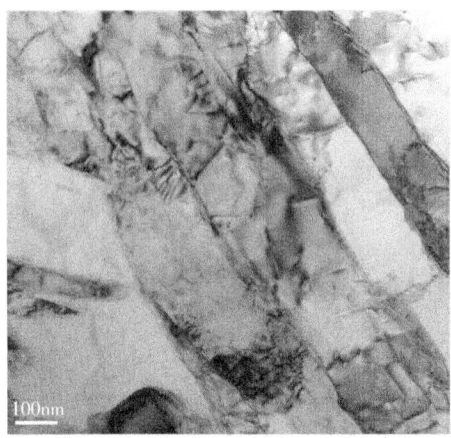

(a) Metallurgical structure of the collapsed super 13Cr tubing   (b) TEM observation image

Fig.10 Metallurgical structure of the collapsed super 13Cr tubing and TEM observation image

4.5 Mechanical properties

Tensile, chapry V-notch impact and Rockwell hardness tests results of the same batch of the as-received super 13Cr tubing were shown in Table 5 to Table 7, respectively. The results revealed that the mechanical properties test results were all satisfy the requirements of the tubing purchasing technical agreement of Tarim Oilfield Company.

Considering the high temperature in the downhole, tensile tests of the same batch of the as-received super 13Cr tubing were also conducted at different temperature, the results as shown in Table 8 and Fig.11 revealed that tensile properties at 180℃ were almost 86% of that at room temperature.

Table 5 Tensile test results of the same batch of collapsed super 13Cr tubing

Specimen	Width×Gauge/mm×mm	Tensile Strength/MPa	Yield Strength/MPa	Elogation/%
Super 13Cr samples	19.1 × 50	958	867	23
		949	858	22
		951	862	22
Tubing purchasing technical agreement of Tarim Oilfield Company		≥827	758~896	≥16

Table 6 Charpy V-notch impact test results of the same batch of collapsed super 13Cr tubing

Sample	Temperature/℃	A_{kv}/J	Average A_{kv}/J
Super 13Cr samples	−10	79	80
		81	
		81	
Tubing purchasing technical agreement of Tarim Oilfield Company		⩾77	

Table 7 Rockwell hardness test results of the same batch of collapsed super 13Cr tubing

Sample		HRC		
Super 13Cr samples	first quadrant	26.9	23.3	23.3
	second quadrant	28.8	29.3	29.2
	third quadrant	28.0	28.6	28.8
	fourth quadrant	29.3	29.6	29.4
Tubing purchasing technical agreement of Tarim Oilfield Company		⩽32.0		

Table 8 High temperature tensile test results of the same batch of collapsed super 13Cr tubing

Specimen	Temperature/℃	Tensile Strength/MPa	Yield Strength/MPa	Elogation/%
Super 13Cr samples	120	862	793	20
	160	846	771	19
	180	817	739	19
	200	811	719	19

4.6 Full scale anti-collapse test

A full scale anti-collapse test was conducted on the same batch of the as-received super 13Cr tubing. The schematic of the test apparatus is shown in Fig.12. The tubing body was placed in an external pressure chamber, fastened and sealed through flange, metal O ring and fastening bolt on both ends of the chamber. The pressurized medium in the test was water. According to the regulation of API TR 5C3, the test apparatus should not impose radial or axial restrains on the specimen, so there were no restrains on both end of the test tubing body. API TR 5C3 standard specifies that the minimum length of the tubing body should be

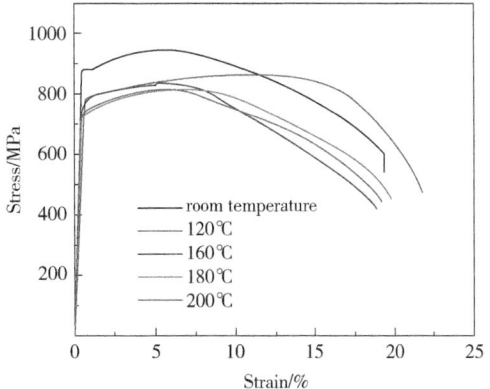

Fig.11 Stress-strain curve of the same batch of collapsed super 13Cr tubing at different temperature

eight times the specified diameter (≥0.71m), and considering the length of the external pressure chamber, the actual length of the testing tubing body is 2m. Pressurized water was injected into the chamber until the tubing body collapsed. Table 9 revealed that the anti-collapse strengthen of the tubing body at room temperature satisfies the requirements of the tubing purchasing technical agreement of Tarim Oilfield Company.

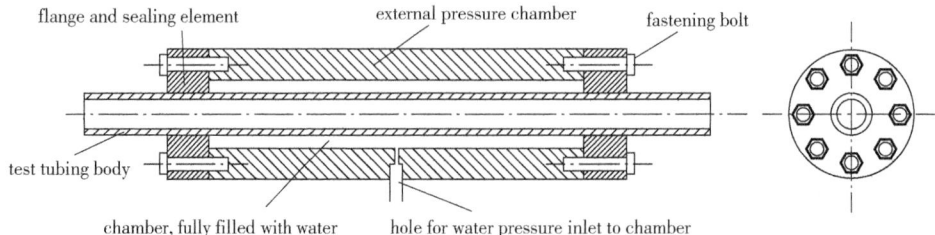

Fig.12 Schematic diagram of full scale anti-collapse test apparatus

Table 9 Full scale anti-collapse test result of the same batch of collapsed super 13Cr tubing

Specimen	Collapse strength/MPa
Super 13Cr sample	102.3
Tubing purchasing technical agreement of Tarim Oilfield Company	≥93.3

4.7 Characterization of cracks

4.7.1 Micro morphology of the cracks

The transverse cracks on the outer surface of the collapsed tubing, as depicted in Fig.6 were chosen for further analysis. Fig.13(a) shows the general micro morphology of the cracks. All cracks originated from the outer surface of the tubing body and propagated to the inner surface along the cross section. In addition, secondary cracks that propagate along paths that different from the main cracks are observed in the magnified views, as shown in Fig.13(b). On the whole, the cracks exist a trans-granular propagation feature, but inter-granular propagation feature can also be seen in some local area. When open the cracks, the micro morphology of the crack surfaces including inter-granular feature in the crack initiation zone as shown in Fig.14(a), and cleavage feature in the crack prop-agation zone can be seen, as shown in Fig.14(b).

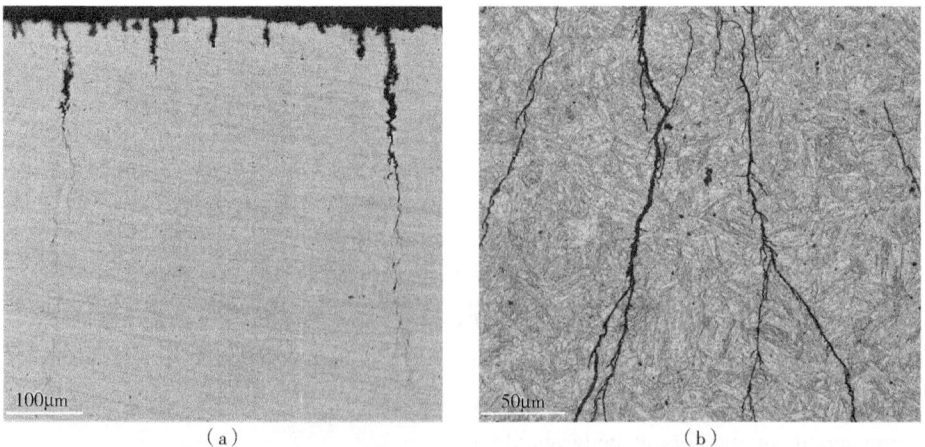

Fig.13 Microscopic morphologies of the cracks on the outer surface of the collapsed super 13Cr tubing

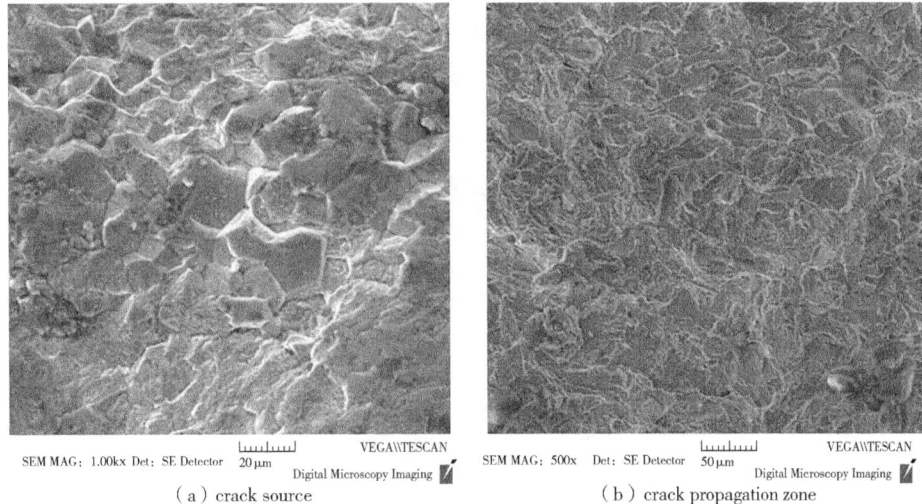

(a) crack source (b) crack propagation zone

Fig.14 Microscopic morphologies of crack surface

4.7.2 Chemical element distribution along the depth of cracks

As described in the preceding section, the cracks initiated from the outer surface of the tubing, which corresponds to the annulus environment in terms of phosphate packer fluid. In order to determine whether or not the cracks are related with the phosphate, SEM-EDS have been used to analyze the chemical element distribution at different position along the crack. As depicted in Fig.15, four different positons along the path of the crack were selected for evaluating the chemical element distribution, that is one crack initiation zone, two crack middle zones and one crack tip zone. The elements of C, O, Fe, P, S, Ca, Cr, Fe, Ni were detected at all the four positions, as listed in Table 10. In addition to this, one position contains numbers of cracks was selected to determine the principal chemical constituent of P, S, O, Cl, Fe, Cr, as depicted in Fig.16.

5 Discussion

Based on the experiments above, chemical composition, mechanical properties and collapse strengthen of the same batch of collapsed super 13Cr tubing all meet the requirements of the tubing purchasing technical agreement of Tarim Oilfield Company. Considering the bulkheads inside the tubing and the cracks on the outer surface of the tubing, it can be inferred that the collapse failure may in relationship with the service environment. The environmental factor that associate with the collapse failure are as follows:

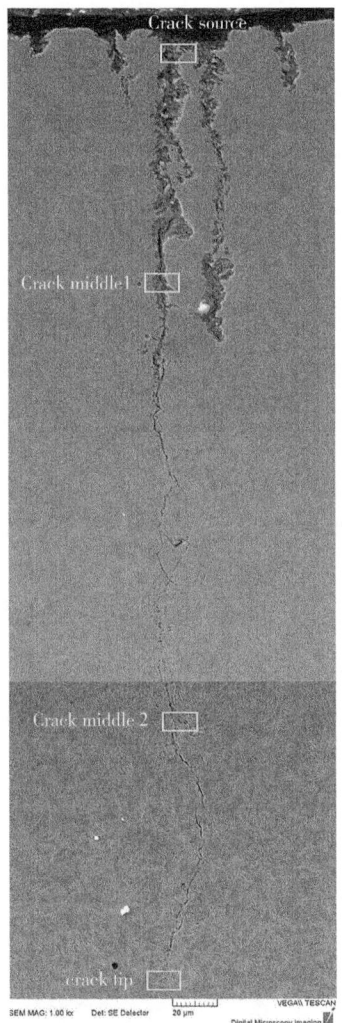

Fig.15 SEM observation of the crack illustrated in Fig.13

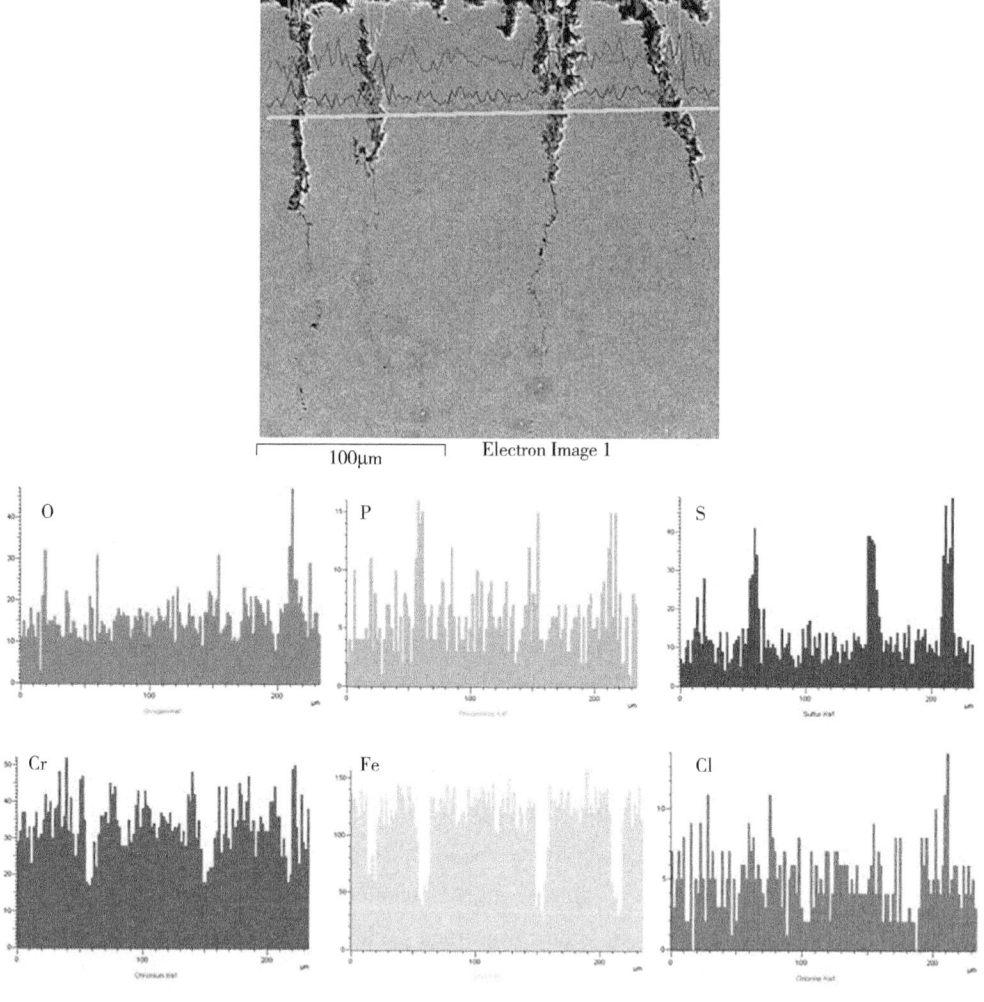

Fig.16 Chemical element distribution analysis of the crack illustrated in Fig.7 by using SEM-EDS line scanning

Table 10 Chemical element distribution along the crack in Fig.10 by using SEM-EDS point scanning (%, mass fraction)

Element	Crack source	Crack middle1	Crack middle2	Crack tip
C	16.6	21.4	19.5	11.0
O	38.6	14.5	30.6	6.6
P	3.4	5.5	3.6	0.6
S	0.6	1.2	2.4	3.3
Cl	—	0.7	1.1	—
K	1.2	1.2	1.3	—
Ca	6.3	1.9	3.5	0.9
Cr	5.1	10.9	9.4	13.1

continued

Element	Crack source	Crack middle1	Crack middle2	Crack tip
Fe	24.8	41.4	23.5	59.8
Ni	0.4	2.1	3.1	4.7
Na	0.9	—	2.0	—
Si	2.1	—	—	—

5.1 Tubing plugging

As mentioned above, the ineffective sand control measure result in a continuous sand production flow into the tubing, and finally formed as bulkheads when mixed with the scaling products $BaSO_4$ from formation water. The bulkheads inside the tubing would prevent the gas from getting through, which would lead to a fluctuation of the wellhead pressure. As the amount of the bulkheads increased up to completely plug the tubing at a certain depth, there would definitely an obvious drop of the wellhead pressure. Furthermore, there has been sustained casing pressure in the "A" annulus and "B" annulus since the beginning of the well production, which would result in a higher external pressure of the tubing string compared with those have no casing pressure. Considering the collapse pressure that the tubing suffer from is a differential pressure (Δp) between the external pressure (p_E) and internal pressure (p_I) of the tubing, as shown in Eqs. (1)

$$\Delta p = p_E - p_I \tag{1}$$

So a higher external pressure caused by casing pressure in "A" annulus and a drop of the internal pressure induced by tubing plugging would result in a increasing of the differential pressure, which would also increase the collapse failure risk of the tubing.

So the collapse failure process could be concluded as follows: tubing plugging caused by bulkheads result in a fluctuated of the wellhead pressure, until on September 30, 2016, one of the tubing in a certain depth was totally plugged, so the wellhead pressure suddenly decreased from 40.28MPa to 13MPa, and lead to an increasing of the differential pressure, which may result in a collapse failure of the tubing in the plugging depth. The collapse of the tubing could not only generated penetrate collapse cracks on the tubing body, but also could form obvious leakage path in the thread due to the deformation. So after that, the medium in the "A" annulus quickly flow into the tubing string through the collapsed tubing, so the pressure in "A" annulus suddenly released from 39.44MPa to 0MPa.

A differential pressure calculation carried out to verify the collapse failure conclusion. the external pressure could be calculated by the sum of A annulus pressure and packer fluid column pressure, as shown in Eqs. (2):

$$p_E = p_A + \frac{\rho g h}{1000} \tag{2}$$

Where ρ is the density of the packer fluid in A-annulus, g/cm^3; p_A is the A-annulus pressure, MPa; h is the depth of the collapsed tubing, m.

The internal pressure is mainly stem from the gas pressure inside the tubing string. As the natural gas flow from formation layer to the wellhead through tubing string, there could be a certain content of pressure drop, so the internal pressure could be calculated by the sum of the wellhead pressure and pressure drop in different depth, as shown in Eqs. (3):

$$p_1 = p_W + \frac{0.25h}{100} \tag{3}$$

Where p_W is the wellhead pressure, MPa; h is the depth of the collapsed tubing, m; 0.25/100 represent the pressure drop of the natural gas is 0.25MPa per 100m.

The differential pressure was calculated by Eqs. (1), Eqs. (2) and Eqs. (3). When the wellhead pressure was 40.28MPa, the collapse pressure of the tubing at 6486.88m was 56.49MPa. When the wellhead pressure dropped to 13MPa, the collapse pressure would increased to 99.22MPa. In this situation, the collapse pressure had already exceed the specified minimum collapse strengthen of the super 13Cr tubing (93.3MPa), and also very close to the actual collapse strengthen of the super 13Cr tubing in room temperature (102.3MPa) attained by experiment. So the tubing plugging would bring a high collapse failure risk to the super 13Cr tubing.

5.2 High temperature and combined load condition in the downhole

The collapse mode of the tubing could be divided into yield strength collapse, plastic collapse, transition collapse and elastic collapse based on yield strengthen and D/t value of the tubing. According to the range of yield strengthen and D/t value calculated from API TR 5C3 standard, the collapse mode of the collapsed super 13Cr tubing is plastic collapse. The plastic collapse strengthen of the tubing could be calculated by Eqs.(4) to Eqs.(7)[37]:

$$P_P = f_{ymn}[A_c/(D/t) - B_c] - C_c \tag{4}$$

$$A_c = 2.8762 + 0.15489 \times 10^{-3} f_{ymn} + 0.44809 \times 10^{-6} f_{ymn}^2 - 0.44809 \times 10^{-9} f_{ymn}^3 \tag{5}$$

$$B_c = 0.026233 + 0.50609 \times 10^{-6} f_{ymn} \tag{6}$$

$$C_c = -465.93 + 0.030867 f_{ymn} - 0.10483 \times 10^{-7} f_{ymn}^2 + 0.36989 \times 10^{-13} f_{ymn}^3 \tag{7}$$

Where P_P is the plastic collapse strengthen, MPa; f_{ymn} is the yield strength of the tubing material, MPa; D is the diameter, mm; t is the wall thickness, mm; A_c, B_c, C_c is the empirical coefficient that in connection with the yield strengthen of the tubing material. Eqs.(4) indicate that in addition to tubing diameter and wall thickness, yield strengthen of the tubing material also has a big influence on the collapse strengthen of the super 13Cr tubing. The collapsed tubing was in the depth of 6486.88m in the downhole, and the temperature gradient revealed that the temperature at that depth was 180℃. Based on the high temperature tensile properties tests above, the yield strengthen of the super 13Cr tubing at 180℃ was only 86% of that at room temperature, so the collapse strengthen at 180℃ was correspondingly 86% of that at room temperature, that is 87.98MPa. This is lower than the collapse pressure when the tubing plugged (99.22MPa).

When combined with the service loading conditions of axial stress, internal and external pressure in the downhole, the yield stress f_{ymn} used for calculating the collapse strengthen in Eqs.

(4) would be modified to a combined loading equivalent grade according to Eqs. (8) and Eqs. (9)[37]:

$$f_{yax} = \{[1-0.75((\sigma_a+\Delta p)/f_{ymn})^2]^{1/2} - 0.5(\sigma_a+\Delta p)/f_{ymn}\}f_{ymn} \quad (8)$$

$$\sigma_a = \frac{F_a}{A_p} \quad (9)$$

Where σ_a is the component of axial stress, MPa; f_{yax} is the combined loading equivalent grade, MPa. Eqs. (8) revealed that $f_{yax}/f_{ymn} < 1$, so the combined load condition of axial stress, internal and external pressure in the downhole would further reduce the collapse strengthen of the tubing.

Dynamic explicit finite element method in ABAQUS was used for the simulation of the tubing's service condition under different differential pressure. As shown in Fig. 17, the element type is C3D8R, the number of units is 12000, and non-uniform pressure was applied on the outer surface of the tubing.

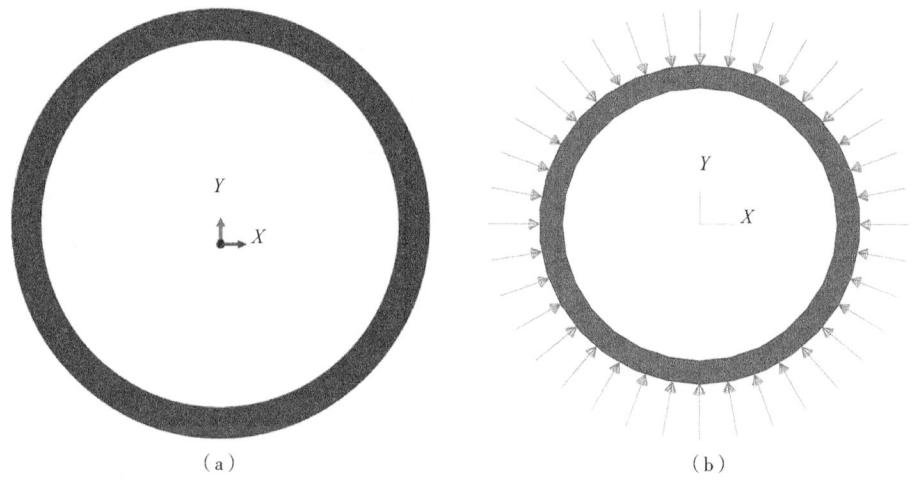

Fig.17 Mesh generation (a) and pressure apply of the tubing(b)

Fig.18 and Fig.19 shows the Von Mises stress distribution of the tubing at the collapse pressure of 56.49MPa and 99.22MPa. It can be seen that at normal production of the gas well, the collapse pressure of 56.49MPa, there is no evident change on the shape of the tubing. When the tubing plugging happened, the ovality of the tubing serviced at 180℃ has gradually increased, as this would lead to an instability of the tubing, and finally totally collapsed. But at such collapse pressure, there is only a little ovality changes on tubing service at room temperature.

5.3 SSC cracks on the outer surface of the tubing

As depicted inFig.6 and Fig.13, beside the final collapse crack showed in Fig.4(b), there are still numerous unpenetrated transverse cracks on the outer surface of the tubing, the existence of such unpenetrated cracks could definitely decrease the anti- collapse ability of the tubing. As the differential pressure of the tubing increased, the collapse failure risk is also higher compared with those no crack tubing.

From the former crack characterization results, the characteristic of the cracks on the outer surface of the tubing are discussed as follows. First, the cracks are all originated from the etch pits

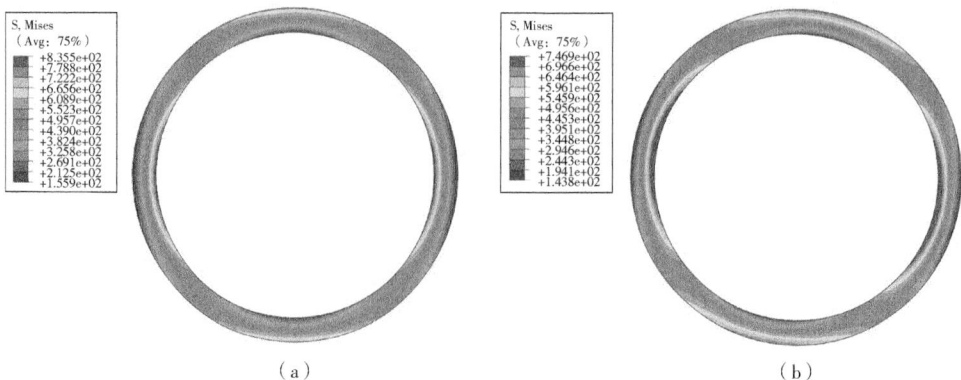

Fig.18 Von Mises stress distribution of super 13Cr tubing at 56.49MPa collapse pressure (a) room temperature and (b) 180℃

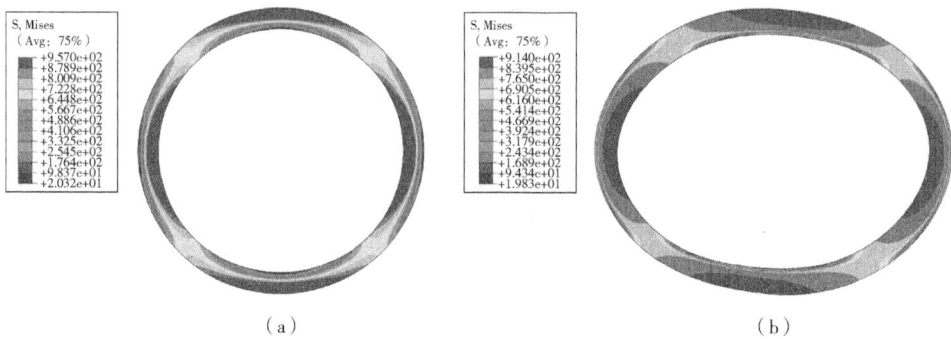

Fig.19 Von Mises stress distribution of super 13Cr tubing at 99.22MPa collapse pressure (a) room temperature and (b) 180℃

on the outer surface of the tubing. Second, cracks as a whole exist a trans-granular propagation feature, but in some local area exist an inter-granular extension feature. Third, the micro morphology of the crack initiation surface including inter-granular feature in the crack initiation zone and cleavage feature in the crack propagation zone. Based on the above characteristics, it can be inferred that the unpenetrated cracks on the outer surface of the tubing have the typical characteristic of the stress corrosion cracking (SCC) crack, which was normally caused by the combined effect of sensitive material under tensile stress and the prescribed corrosion environment on the a-annulus.

5.3.1 Sensitive material

At present, there are three main types of the high density water-based annulus protection fluid: halogen, phosphate and formate. Numerous research[38-40] have proved the material sensitivity of super 13Cr in halogen annulus protection fluid such as $CaCl_2$ annulus protection fluid, and the sensitivity is mainly in relationship with CO_2 gas invasion, pollution by sulfur oxides or oxidant pollution of the annulus protection fluid. In recently years, the SCC behavior of super 13Cr tubing in phosphate annulus protection fluid have gained more attention, numerous research[41-42] have proved the sensitivity of super 13Cr tubing in high temperature and insufficient oxidant phosphate annulus protection fluid.

5.3.2 Specified corrosive environment

As depicted in Fig.16 and Table 10, there exist the element of P, O, S, Cl along the crack propagation path. It is widely known that the occurrence of SCC is caused by the presence of H_2S or Cl^-, which is a well-known phenomena that has been observed with a wide range of metals, from carbon steel to stainless steel. In the HPHT gas well in western China, there are two main sources that could generate the gas of H_2S. One is the formation layer itself, the other is the use of polysulfonate drilling fluid in the drilling process, whose thermal degradation over 150℃ would generate the gas of H_2S. Because the produced gas contain CO_2 concentration 1.13%~1.74% (mole fraction) without H_2S, and there was also no polysulfonate drilling fluid used in the former drilling process, so there is no H_2S gas produced in any depth of the downhole. For only S from H_2S could contribute to the occurrence of SCC failure, it can be deduced that S in the cracks may come from the formation water which contain the substance of $BaSO_4$, and flow into the A-annulus and crack after the collapse of the tubing, and have nothing to do with the occurrence of SCC. Considering the outer surface of the tubing exposed to high density phosphate annulus protection fluid, and the element of P was abundant from the crack source to the tip, the corrosive environment should in relationship with the phosphate annulus protection fluid. In addition, Table 10 show that the content of C in the crack is far beyond that in the super 13Cr material as shown in Table 3, and also the above discussion in Fig.2 revealed that there has been sustained casing pressure in the "A" annulus and "B" annulus since the beginning of the well production, So there is the evidence that the produced gas contained CO_2 is definitely invaded into annulus[43]. As a result, protection fluids in a-annulus are exposed to a substantial amount of the production gas of CO_2. So the phosphate with contaminant of CO_2 exhibits an enhanced corrosion effect on super 13Cr martensitic stainless steels.

5.3.3 Tensile stress

The stress state of the completion tubing string including self-weight of the tubing, tension and compression load induced by temperature variation that lead to expansion and contraction of the tubing string, external and internal pressure, and vibration or any combination thereof. At production stage of the HPHT gas well, tubing string would be elongated along the axial direction because of the increasing of temperature during production stage. Because the top of the whole tubing string was constrained by Christmas tree in the wellhead, the bottom of its was constrained by packer, the elongation of the tubing was restricted, as a result, the tubing could only bending, and the bending degree increased from the wellhead to the packer. The location of the collapsed tubing is the seventh tubing above the packer, that is only 70m away, based on the mechanics of the tubing string, it can be concluded that the outer surface of the collapsed tubing was under bending-induced tensile stress. The tensile stress could not only damage the oxide scale/film, but it also weakens the metal atomic bonds, resulting in an increase in the steel's surface activity.

Therefore, based on the above issues, it can be concluded that the collapse failure of the super 13Cr tubing is mainly attributed to the tubing plugging caused by formation sand production and formation water scaling. In addition to this, high temperature in the downhole and the presence of the SSC cracks on the outer surface of the tubing which could cause a significant decrease of the pressure-bearing capacity of tubing strings would also give an acceleration of the collapse failure process.

6 Conclusions and suggestions

This research investigated the collapse failure causes of super 13Cr tubing in HPHT gas wells. The corresponding conclusions and suggestions are as follows:

(1) The main causes for the collapse failure is tubing plugging caused by formation sand production and formation water scaling, which would give an increasing of the differential pressure between the external and internal of the tubing.

(2) The higher service temperature in the depth of the collapse tubing would result in a reduction of the anti-collapse ability.

(3) Type of the cracks on the outer surface of the tubing is SSC crack, which induced by synergistic action of phosphate annulus protection fluid and CO_2 invasion. The presence of the cracks could be regarded as a reduction of the wall thickness which could cause a significant decrease of the pressure-bearing capacity of tubing strings.

(4) To prevent the collapse failure, the density of the annulus protection should be reduced in order to lower the pressure that induced by the liquid column. At the same time, the production differential should also be controlled to reduce the possibility of formation sand production.

Declaration of Competing Interest

The authors declare that they have no known competing financial interests or personal relationships that could have appeared to influence the work reported in this paper.

Data availability

The authors are unable or have chosen not to specify which data has been used.

Acknowledgement

This work is sponsored by Scientific Research and Technology Development Project of CNPC (Grant No. 2020B-4020).

References

[1] R.M. WilliamDurnie, BrianKinsella AlanJefferson, Harmonic analysis of carbon dioxide corrosion, Corros. Sci. 43 (2008) 1213–1221.

[2] X.Y. Zhang, F.P. Wang, Y.F. He, Y.L. Du, Study of the inhibition mechanism of imidazoline amide on CO_2 corrosion of armco iron, Corros. Sci. 43 (2001) 1417–1431.

[3] Z. Bai, C. Chen, M. Lu, J. Li, Analysis of EIS characteristics of CO_2 corrosion of well tube steels with corrosion scales, Appl. Surf. Sci. 252 (2006) 7578–7584.

[4] W. Li, L. Xu, L. Qiao, J. Li, Effect of free Cr content on corrosion behavior of 3Cr steels in a CO_2 environment, Appl. Surf. Sci. 425 (2017).

[5] M.N. Zafar, R. Rihan, L. Al-Hadhrami, Evaluation of the corrosion resistance of SA-543 and X65 steels in emulsions containing H_2S and CO_2 using a novel emulsion flow loop, Corros. Sci. 94 (2015) 275–287.

[6] X. Jiang, Y.G. Zheng, D.R. Qu, W. Ke, Effect of calcium ions on pitting corrosion and inhibition performance

in CO_2 corrosion of N80 steel, Corros. Sci. 48 (2006) 3091–3108.

[7] S. Guo, L. Xu, W. Chang, M. Lu, Experimental study of CO_2 corrosion of 3Cr pipe line steel, Acta Metall. Sin. 47 (2011) 1067–1074.

[8] M. Jian, Y. Juntao, X.u. Han Yan, L.L. Xiuqing, W. Ke, Corrosion Behavior of P110 Tubing Steel in the CO_2-saturated Simulated Oilfield Formation Water with Element Sulfur Addition, Rare Metal Mater. Eng. 47 (2018) 1965–1972.

[9] J.M. Zhao, Y. Lu, H.X. Liu, Corrosion and control of P110 oil tube steel in CO_2-saturated solution, Corros. Eng. Sci. Technol. 43 (2008) 313–319.

[10] Ikeda A Udea, Mukai S. CO_2 behavior of carbon and Cr Steels. In Hausler R H Giddard H P(Eds), Advances in CO_2 Corrosion.Vol. 1, NACE, Houston, Texas.1984: P39–52.

[11] Schmitt G. Fundamental aspects of CO_2 corrosion in Hausler R H. Giddard HP(EDS), Advances in CO_2 Corrosion. NACE(I), Houston Texas,1984.3:43–51.

[12] X.W. Lei, Y.R. Feng, A.Q. Fu, J.X. Zhang, Z.Q. Bai, C.X. Yin, C.H. Lu, Investigation of stress corrosion cracking behavior of super 13Cr tubing by full-scale tubular goods corrosion test system, Eng. Fail. Anal. 50 (2015) 62–70.

[13] Z. Zhi, Z. Yushan, L.i. Jing, L. Wanying, L. Mingqiu, G. Wenxiang, S. Taihe, Stress corrosion crack evaluation of super 13Cr tubing in hightemperature and high- pressure gas wells, Eng. Fail. Anal. 95 (2019) 263–272.

[14] Y. Long, J. Luo, M. Yue, W.u. Gang, M. Zhao, N. Ji, W. Song, Q. Jin, X. Kuang, Y. Fan, Investigation on leakage cause of 13Cr pipe flange used for a Christmas tree in a high-pressure and high-temperature gas well, Eng. Fail. Anal. 142 (2022), 106793.

[15] Y. Long, G. Wu, A.Q. Fu, J.F. Xie, M.F. Zhao, Z.Q. Bai, J.H. Luo, Y.R. Feng, Failure analysis of the 13Cr valve cage of tubing pump used in an oilfield, Eng. Fail. Anal. 93 (2018) 330–339.

[16] W. Liu, T. Shi, L.u. Qiang, Z. Zhang, C. Ming, J. Gong, J. Ren, Failure analysis on fracture of S13Cr–110 tubing, Eng. Fail. Anal. 90 (2018) 215–230.

[17] Hannah, I.M., Seymour D.A, Shearwater super duplex tubing failure investigation, Presented at the Corrosion 2006 conference, San Diego, California; Paper No. 06491.

[18] A.Q. Fu, Y. Long, H.T. Liu, M.F. Zhao, J.F. Xie, H. Su, X.P. Li, J.T. Yuan, X.W. Lei, C.X. Yin, Y.R. Feng, Stress corrosion cracking behavior of super 13Cr tubing in phosphate packer fluid of high pressure high temperature gas well, Eng. Fail. Anal. 139 (2022), 106478.

[19] S. Luo, F.u. Anqing, M. Liu, Y. Xue, N. Lv, Y. Han, Stress corrosion cracking behavior and mechanism of super 13Cr stainless steel in simulated O_2/CO_2 containing 3.5 wt% NaCl solution, Eng. Fail. Anal. 130 (2021), 105748.

[20] G.Y. Zhu, Y.Y. Li, B.S. Hou, Q.H. Zhang, G.A. Zhang, Corrosion behavior of 13Cr stainless steel under stress and crevice in high pressure CO_2/O_2 environment, J. Mater. Sci. Technol. 88 (2021) 79–89.

[21] Y.Z. Li, X. Wang, G.A. Zhang, Corrosion behaviour of 13Cr stainless steel under stress and crevice in 3.5 wt.% NaCl solution, Corros. Sci. 163 (2020), 108290.

[22] G. Xiao, Z.Y. SiZhou Tan, B. Dong, Y. Yi, G. Tian, H. Yu, S. Shi, CO_2 corrosion behaviors of 13Cr steel in the high-temperature steam environment, Petroleum. 6 (1) (2020) 106–113.

[23] X. Yue, L. Zhang, C. Sun, S. Xu, C. Wang, M. Lu, A. Neville, Y. Hua, A thermodynamic and kinetic study of the formation and evolution of corrosion product scales on 13Cr stainless steel in a geothermal environment, Corros. Sci. 169 (2020), 108640.

[24] A. Kozhagulova, Nguyen Hop Minh, Yong Zhao, Sai Cheong Fok, Experimental and Analytical Investigation of Sand Production in Weak formations for Multiple Well Shut-Ins, J. Pet. Sci. Eng. 195 (2020), 107628.

[25] D. Zivara, S. Shada, J. Foroozeshb, S. Salmanpour, Experimental study of sand production and permeability enhancement of unconsolidated rocks under different stress conditions, J. Pet. Sci. Eng. 181 (2019), 106238.

[26] Y. Jin, J. Chen, M. Chen, F. Zhang, L.u. Yunhu, J. Ding, Experimental study on the performance of sand control screens for gas wells, J. Petrol. Explor. Prod. Technol. 2 (2012) 37–47.

[27] C. Dong, K. Gao, S. Dong, X. Shang, W.u. Yanxin, Y. Zhong, A new integrated method for comprehensive performance of mechanical sand control screens testing and evaluation, J. Pet. Sci. Eng. 158 (2017) 775–783.

[28] Y. Guo, M. Roostaei, A. Nouri, V. Fattahpour, M. Mahmoudi, H. Jung, Effect of stress build-up around standalone screens on the screen performance in SAGD wells, J. Pet. Sci. Eng. 171 (2018) 325–339.

[29] ASTM E 709-2015 standard, Standard Guide for Magnetic Particle Testing.

[30] ASTM A751-14a standard, Standard Test Methods, Practices, and Terminology for Chemical Analysis of Steel Products.

[31] ASTM E3-11 (2017) standard, Standard Guide for Preparation of Metallographic Specimens.

[32] ASTM E112-13 standard, Standard Test Methods for Determining Average Grain Size.

[33] ASTM E45-18a standard, Standard Test Methods for Determining the Inclusion Content of Steel.

[34] ASTM A370-20 standard, Standard Test Methods and Definitions for Mechanical Testing of Steel Products.

[35] ASTM E18-2016 Standard Test Methods for Rockwell Hardness of Metallic Materials.

[36] API Recommended practice 5C5, Procedures for Testing Casing and Tubing Connections.

[37] API Technical Report 5C3 seventh edition, June 2018, Calculating Performance Properties of Pipe Used as Casing or Tubing.

[38] X. Lei, Y. Feng, J. Zhang, A. Fu, C. Yin, D.D. Macdonald, Impact of Reversed Austenite on the Pitting Corrosion Behavior of Super 13Cr Martensitic Stainless Steel, Electrochimicia Acta 191 (2016) 640–650.

[39] Y. Zhao, J. Xie, G. Zeng, T. Zhang, D. Xu, F. Wang, Pourbaix diagram for HP-13Cr stainless steel in the aggressive oilfield environment characterized by high temperature, high CO_2 partial pressure and high salinity, Electrochim. Acta. 293 (2019) 116–127.

[40] Jin Qiang, Jiang Wenchun, Gu Wenbin, et al. A primary plus secondary local PWHT method for mitigating weld residual stresses in pressure vessels. Int. J. Pressure Vessels Piping, 2021,192(3): 104431.

[41] S.D. Zhu, J.F. Wei, R. Cai, Z.Q. Bai, G.S. Zhou, Corrosion failure analysis of high strength grade super 13Cr-110 tubing string, Eng. Fail. Anal. 18 (8) (2011) 2222–2231.

[42] L. Bin Niu, K. Nakada, Effect of chloride and sulfate ions in simulated boiler water on pitting corrosion behavior of 13Cr steel, Corros. Sci. 96 (2015) 171–177.

[43] M. Mainguy, R. Innes, Explaining sustained A-annulus pressure in major North Sea high-pressure/high temperature fields, SPE Drilling Completions 34 (01) (2018) 71–80.

本论文原发表于《Engineering Failure Analysis》2023 年第 148 卷。

Investigation of Corrosion Behavior of 2205 Duplex Stainless Steel Coiled Tubing in Complex Operation Environments of Oil and Gas Wells

Luo Jinheng[1]   Yan Pai[2]   Fan Yujie[1]   Luo Sheji[2]   Long Yan[1]

(1. State Key Laboratory for Performance and Structure Safety of Petroleum Tubular Goods and Equipment Materials, CNPC Tubular Goods Research Institute; 2. Xi'an Shiyou University)

Abstract: In this paper, the corrosion behavior of 2205 duplex stainless steel coiled tubing in acidizing and production environments was studied by high-temperature and high-pressure corrosion weight loss tests, stress corrosion cracking test, electrochemical tests, microscopic morphological analysis and corrosion product analysis. The results show that in the acidizing environment, the duplex stainless steel has high corrosion rates especially in the acidizing solution with high HCl concentration, with the selective dissolution of ferrite phase. In the production environment containing CO_2 and H_2S, the duplex stainless steel has excellent resistances to corrosion and stress corrosion cracking. With the increase of pH value, the pitting potential and polarization resistance exhibit the increasing trend, indicating the better passivation performance. Therefore, 2205 duplex stainless steel coiled tubing can be recommended to be mainly used for the operations in production environment rather than acidizing operations.

Keywords: Duplex stainless steel; Coiled tubing; Acidizing environment; CO_2 and H_2S corrosion

1 Introduction

With the increasing demand for oil and gas resources, high-pressure and high-temperature wells have been developed in succession, accompanied by the increasingly prominent problem of tubing strings corrosion caused by harsh working conditions, such as high temperature, high pressure, highly mineralized formation water and CO_2/ H_2S-containing corrosive gas[1-2]. At present, most of oil and gas wells need to be acidified to increase production, but the high acidizing environment may seriously corrode and damage the tubing strings, which acutely threatens the

---

Corresponding author: Luo Jinheng, Tel:+8602981887989; Email:luojh@cnpc.com.cn.

production safety of oil and gas wells[2]. Coiled tubing is recognized for its high efficiency, economy and low contamination to the formation. Therefore, as a technology with broad development prospects, it has been widely used for well repair, completion and production enhancement[3-4]. For example, coiled tubing can be used in selective acidizing construction operations to avoid direct contact between acidizing working fluid and tubing strings. However, the carbon steel is susceptible to uniform or localized corrosion with high corrosion rates caused by CO_2 corrosion in service[5]. Furthermore, the coiled tubing will be subjected to self-weight tensile stress and periodic plastic strain when it is wound around guide arches and spools, and the synergistic effect of H_2S gas may cause material degradation and plasticity reduction, leading to the cracking of the coiled tubing[6]. Considering the severe corrosion of carbon steel and high cost of high corrosion resistant alloy, such as nickel-based alloy, the duplex stainless steel coiled tubing with excellent corrosion resistance and economy has been developed. Duplex stainless steel contains austenitic phase (γ) and ferritic phase (α), with Cr and Mo elements mainly presenting in the ferritic phase and Ni elements dominating in the austenite phase[7-9]. The 2205 duplex stainless steel has the balanced microstructure between austenite and ferritic phases, achieving the good mechanical properties and corrosion resistance[10-12].

However, in recent years, many researchers investigated the corrosion causes of theduplex stainless steel in service. It has been reported that the corrosion failures usually started with the action of a localized corrosion mechanism, like pitting corrosion or preferential corrosion[13]. Atxaga et al determined the reason for the failure of a group of duplex stainless steel valves after 15 years in service, revealing the failure was due to pitting corrosion initiated at the austenite - ferrite interfaces[14]. Kölblinger et al. studied the failure cause of superduplex stainless steel UNS S32760 flange for seawater service, observing the preferential corrosion of the ferrite phase[15]. Meanwhile, severe corrosive environment will significantly accelerate corrosion behavior of the duplex stainless steel. Bautista et al. reported that in sulfuric acid and Cl^--containing solutions, when the corrosion potential was approached, the ferrite in duplex stainless steel was selectively corroded; when the austenite was subjected to selective corrosion, the potential in the anodic activation zone increased[16]. Fu et al. observed a selective corrosion behavior of 2205 duplex stainless steel in the dual-phase system in nitric acid and Cl^- solutions, and the corrosion in austenite phase is more serious[17]. However, the selective corrosion of 2205 duplex stainless steel in which acidizing medium will occur and which main acid determines the understanding is still unclear. It is generally believed that the existence of CO_2 can promote corrosion[18], while H_2S can either accelerate the corrosion by acting as a promoter of anodic dissolution through sulfide adsorption and affecting the pH, or it can decrease the CO_2 corrosion rate by the formation of a protective sulfide scale[19]. In addition, H_2S can influence the type and properties of the corrosion scales formed, either improving or undermining them[19-20]. Therefore, the difference in the relative content of CO_2 and H_2S will determine the corrosion process controlled by CO_2 or H_2S. In addition, with the development of high-temperature and high-pressure wells, stress corrosion cracking (SCC) is the most common and serious failure under the action of stress and environments. Lv et al. reported that the presence of H_2S and the increase in Cl^- concentration significantly reduced the pitting corrosion potential and increased the SSC sensitivity of stainless steels[21]. However, the relatively complete theoretical system about the corrosion behavior of duplex stainless steel in the full life cycle operation environments (including acidizing operation and subsequent production environments) in the wells has not been clarified, which severely limits the wide application of duplex stainless steel coiled tubing.

The purpose of this paper is to reveal the environmental suitability of duplex stainless steel coiled tubing in oil and gas well environments. The corrosion behavior of 2205 duplex stainless steel in acidizing and production environments was investigated by electrochemical tests, high-temperature and high-pressure corrosion tests, stress corrosion cracking tests, observation of corrosion morphology and analysis of corrosion products. Some new viewpoints on the application conditions of 2205 duplex stainless steel coiled tubing has been presented.

2 Experimental

2.1 Material

The 2205 duplex stainless steel coiled pipe was produced by laser welding and then solid solution heat treatment at 1070℃. The sample material used in this paper was taken from 2205 duplex stainless steel coiled pipe and away from the weld position. The chemical composition of the coiled tubing is listed in Table 1. The microstructure of 2205 duplex stainless steel coiled tubing body is shown in Fig.1, consisting of ferrite (α phase, gray color) as well as austenite (γ phase, white color).

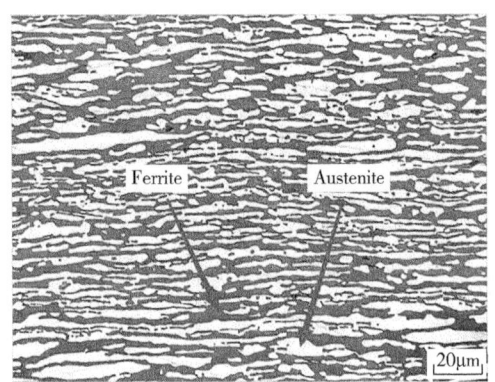

Fig.1 Microstructure of 2205 duplex stainless steel coiled tubing

Table 1 Chemical composition analysis of 2205 duplex stainless steel coiled tubing (%, mass fraction)

C	Si	Mn	P	S	Cr	Ni	Mo	Nb	V	Ti	Cu
0.018	0.53	1.52	0.025	0.0007	21.57	5.58	2.96	0.0081	0.10	0.0071	0.15

2.2 Corrosion weight loss tests

Considering that the service environments of the coiled pipe mainly involve two steps (i.e. live acid injection and production operations), the corrosion rates of 2205 duplex stainless steel in acidizing environment and production environment were measured by high-temperature and high-pressure corrosion weight loss tests. The samples for the corrosion weight loss tests were the corrosion coupons with the size of 40mm×10mm×2.7mm. All the surfaces of the coupons were polished sequentially with 240#~1200# metallographic sandpaper. The tests were conducted in an autoclave under the acidizing and production environment conditions, respectively. For the tests in the acidizing environment, the test solutions were four types of acid solutions with 5% corrosion inhibitors, the test pressure were 10MPa N_2 atmosphere, the test temperature was 160℃, and the test time was 4 h, as linsted in Table 2. For the production environment, the test was conducted in deaerated simulated formation water solution which chemical composition is listed in Table 4, with 4MPa CO_2 and 1MPa H_2S, and the test time extended to 720 h, as listed in Table 3.

Table 2 Test conditions for corrosion weight loss in acidizing environment

Test parameters	Parameter values
Test solutions	12% HCl +3% HF +5% corrosion inhibitor 9% HCl +3% HF +3% HAc +5% corrosion inhibitor 9% HCl +3% HF +2% HBF_4 +5% corrosion inhibitor 9% HCl +3% HF +8% HBF_4 +5% corrosion inhibitor
Test time/h	4
Total pressure/MPa	10
Test temperature/℃	160

Table 3 Test conditions for corrosion weight loss in production environment

Test parameters	Parameter values
Test solution	Simulated formation water which ions composition is listed in Table 4
Test temperature/℃	160
Test pressure/MPa	10
Partial pressure of CO_2/MPa	4
Partial pressure of H_2S/MPa	1
Test time/h	720

Table 4 Concentration of various ions of simulated formation water solution

Ion	Cl^-	SO_4^{2-}	HCO_3^-	Mg^{2+}	Ca^{2+}	$Na^+ + K^+$
Concentration/(mg/L)	118614	766	405	622	7240	71000

After the tests, the coupons were rinsed with distilled water and then absolute alcohol. The corrosion product on the surface of the coupon tested in the production environment was analyzed by energy dispersive spectroscopy (EDS) and X-ray diffraction (XRD). Then, all the coupons were cleaned using film stripper solution to remove the corrosion products or surface films, and then weighed to calculate the average corrosion rate. For the samples tested in the acidizing environments, the corrosion rate is calculated by Eq.1[22] and Eq.2[23]. For the samples tested in the production environment, the corrosion rate is calculated by Eq.2[23].

$$V_i = \frac{10^6 \Delta m_i}{A_i \times \Delta t} \quad (1)$$

Where V_i is the corrosion rate, g/(m² · h); Δm_i is the mass loss, g; Δt is the test time, h; A_i is the sample area, mm².

$$CR = \frac{3.65 \times 10^5 \times W}{ATD} \quad (2)$$

Where CR is the corrosion rate, mm/a; W is the mass loss, g; A is the initial exposed surface area of coupon, mm^2; T is the exposure time, d; D is the density of specimen, g/cm^3. The corrosion morphology was observed by optical microscope (OM) and scanning electron microscope (SEM).

2.3 Stress corrosion cracking test

In the production environment containing CO_2 and H_2S, the SCC sensitivity of 2205 duplex stainless steel need to be further evaluated high-temperature and high-pressure. The SCC test in production environment was conducted by four-point bending loading method with the loading stress of 496.8MPa (i.e. 90% of the measured minimum yield strength). Then the loaded samples were put into a high-temperature and high-pressure autoclave for testing. The test conditions including the test solution, CO_2 and H_2S partial pressures, total pressure and test time were the same as the corrosion weight loss test in production environment described in section 2.2. After the SCC test, the longitudinal-sectional morphology of maximum stress area of the sample was observed by OM.

2.4 Electrochemical tests

Considering that the variety of pH values may makean obvious effect on the corrosion behavior of the 2205 duplex stainless steel coiled tubing, the electrochemical tests were conducted under different pH value conditions. The size of electrochemical samples is 10mm × 10mm × 3mm. The samples were polished, rinsed and dried, and then sealed with epoxy resin to expose $1cm^2$ of the samples as the working surface. The electrochemical test solution was the artificial simulated formation water solution which chemical composition is listed in Table 4, and the pH value of the solution was adjusted by NaOH and HCl. A VersaSTAT-type electrochemical workstation with a three-electrode system was used, in which the working electrode was an electrochemical sample of 2205 duplex stainless steel studied in this work, the auxiliary electrode was a platinum plate, and the reference electrode was a Ag/AgCl electrode. The electrochemical test temperature was 60℃.

The test sequence of electrochemistry is generally as follows: pre-polarization open circuit potential-AC impedance spectrum-cyclic polarization curve. The pre-polarization time is 120s with a potential of −0.8V Ag/AgCl; the open circuit potential test time is 7200s; the test frequency of the electrochemical impedance spectrum is 10mHz ~ 100kHz with an applied sine wave amplitude of 10mV; the scan rate of the cyclic polarization curve is 0.333mV/s, and the scan potential range is −0.3 ~ 1.2V. After the measurement was completed, the cyclic polarization curve was subjected to potential analysis, the AC impedance spectrum was fitted, and the data was processed with ZSimp Win analysis software. At the same time, the corresponding electrochemical parameters were obtained, and the conclusion was drawn by analysis.

3 Results

3.1 High-temperature and high-pressure acidizing environment

By comparing the corrosion behaviors of 2205 duplex stainless steel in different acidizing environments, it can be seen from Table 5 that the average corrosion rate is the highest when the acidizing medium contains 12% HCl. With the increase of HBF_4 content, the corrosion rate of 2205 duplex stainless steel is slightly increased. Fig.2 shows the morphology of pitting pits on the surface of 2205 duplex stainless steel under different acidizing environments. It can be seen that deep pitting pits were formed on the surface of the samples, and the depth of the maximum pitting pits was measured, as linsted in Table 6, 12% HCl had the greatest influence on the corrosion behavior of 2205 duplex stainless steel, and 2% HBF_4 had the smallest influence on the corrosion behavior of 2205 duplex stainless steel. However, with the increase of HBF_4 concentration, the maximum pitting rate increased. The microscopic corrosion morphology of 2205 duplex stainless steel in different acidified environments is shown in Fig.3. There are lots of holes between the austenitic grains on the surface of the samples exposed in four acidizing environments, indicating the selective corrosion of the ferrite phase. Furthermore, Fig.4 shows the metallographic pictures of the cross-section of 2205 duplex stainless steel samples in different acidizing environments. As illustrated in the figure, the austenite phase remains white, and the ferrite phase is stained with brown color. Some brown ferrite near the surface of the sample is found to disappear, but that in the austenite phase still exists.

Table 5 Corrosion rates of 2205 duplex stainless steel in acidizing solutions

Acidified solutions	Corrosion rates/[g/(m² · h)]	Corrosion rates/(mm/a)
12% HCl +3% HF +5% corrosion inhibitor	104.60	116.13
9% HCl +3% HF +3% HAc +5% corrosion inhibitor	89.32	99.17
9% HCl +3% HF +2% HBF_4 +5% corrosion inhibitor	31.64	35.13
9% HCl +3% HF +8% HBF_4 +5% corrosion inhibitor	50.16	55.69

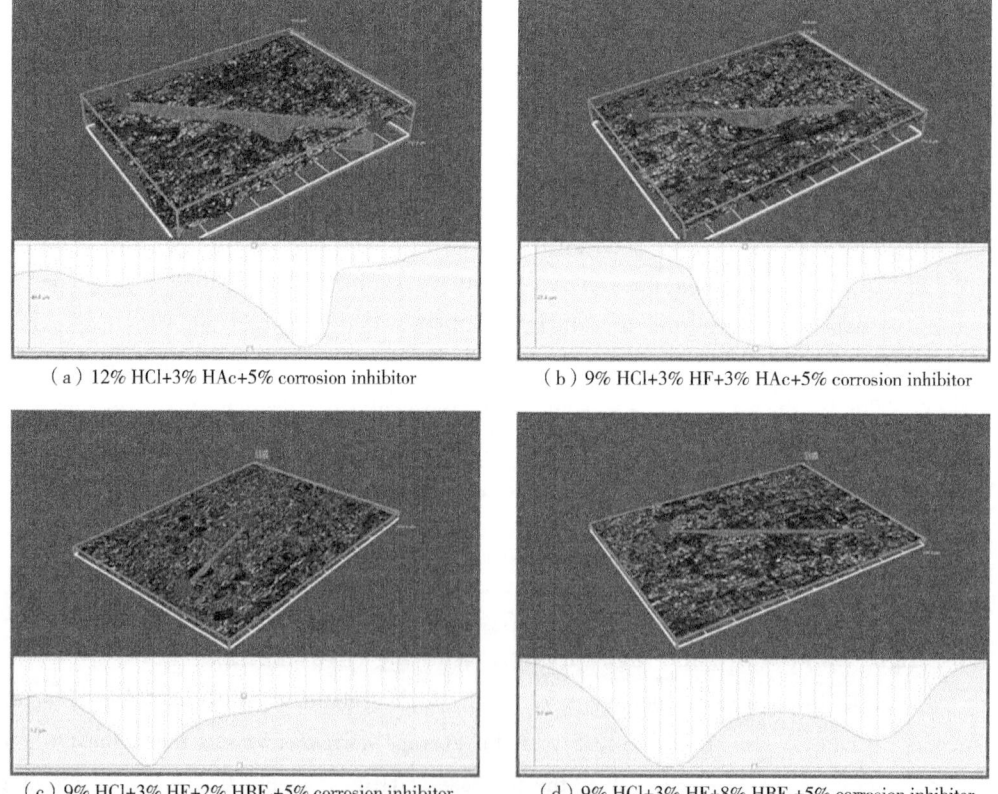

(a) 12% HCl+3% HAc+5% corrosion inhibitor (b) 9% HCl+3% HF+3% HAc+5% corrosion inhibitor

(c) 9% HCl+3% HF+2% HBF_4+5% corrosion inhibitor (d) 9% HCl+3% HF+8% HBF_4+5% corrosion inhibitor

Fig.2 Morphology of pits on the surface of 2205 duplex stainless steel under different acidizing environments

Table 6 Measurement results of maximum pitting depth

Acidified solutions	Maximum pitting depth/μm
12% HCl +3% HF +5% corrosion inhibitor	40.4
9% HCl +3% HF +3% HAc +5% corrosion inhibitor	27.4
9% HCl +3% HF +2% HBF_4 +5% corrosion inhibitor	1.2
9% HCl +3% HF +8% HBF_4 +5% corrosion inhibitor	5.2

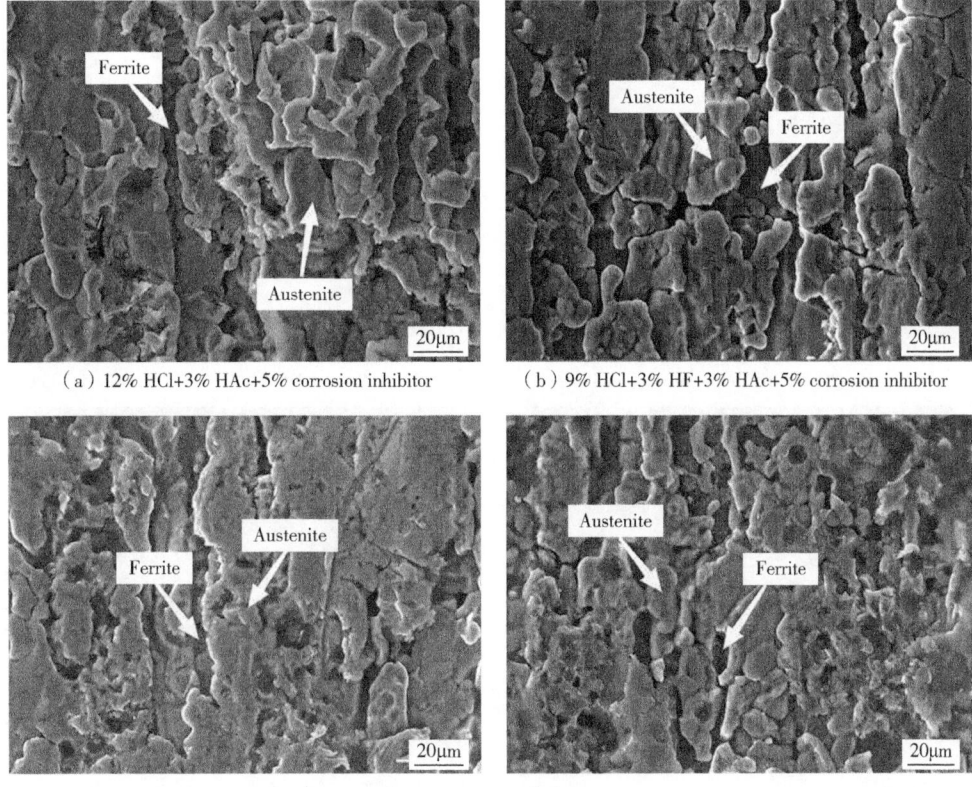

(a) 12% HCl+3% HAc+5% corrosion inhibitor (b) 9% HCl+3% HF+3% HAc+5% corrosion inhibitor

(c) 9% HCl+3% HF+2% HBF_4+5% corrosion inhibitor (d) 9% HCl+3% HF+8% HBF_4+5% corrosion inhibitor

Fig. 3 SEM images of 2205 duplex stainless steel under different corrosion inhibitor acidification environments

3.2 High-temperature and high-pressure production environment

The corrosion rate of 2205 duplex stainless steel was 0.0255 mm/a after 720 h of high-temperature and high-pressure production environment. It can be concluded that the corrosion rate of 2205 duplex stainless steel in production environment is much lower than that in high temperature acidification environment. Fig.5 shows the macro and micro corrosion morphology of 2205 duplex stainless steel after corrosion weight loss tests in simulated formation water containing CO_2 and H_2S. As shown in Fig.5 (b), after exposure at 160 ℃ for 720 h, there is no accumulation of a large number of corrosion products on the surface of the sample under low magnification, and the local corrosion pits can be observed on the surface of the sample. As shown in Fig. 5 (c), high magnification observation found that there are a small number of granular corrosion products attached to the surface of the sample. Fig.6 shows the element distribution of the corrosion products on the surface. It can be seen that the corrosion products contain mainly Fe, Cr, Ni and S elements, and a few Ca and Si elements (Table 7). The XRD result shows that the corrosion products formed on the surface are composed of FeS_2, as shown in Fig.7.

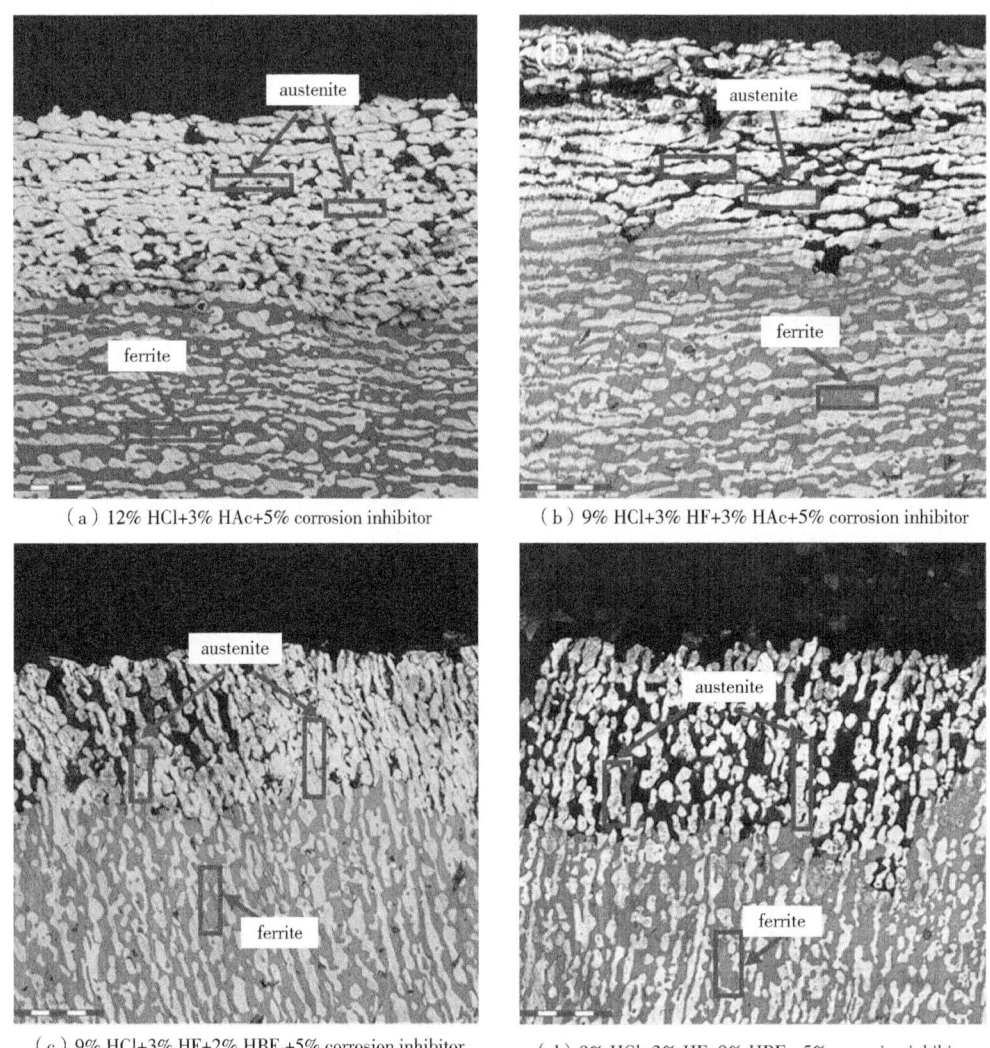

(a) 12% HCl+3% HAc+5% corrosion inhibitor

(b) 9% HCl+3% HF+3% HAc+5% corrosion inhibitor

(c) 9% HCl+3% HF+2% HBF_4+5% corrosion inhibitor

(d) 9% HCl+3% HF+8% HBF_4+5% corrosion inhibitor

Fig. 4 Cross-sectional morphology of samples after corrosion under different acidification environments

Table 7 Elemental compositions of corrosion products of 2205 duplex stainless steel in simulated formation water containing H_2S and CO_2

Element	O	Si	S	Ca	Cr	Fe	Ni
Area 1	21.39	0.75	23.43	2.52	34.93	2.31	14.67
Area 2	14.46	1.08	8.87	0.80	22.57	48.74	3.49
Area 3	20.23	1.06	25.91	2.52	28.86	9.17	12.23

Fig.5 (a) Macroscopic morphology of 2205 duplex stainless steel after corrosion weight loss tests in the simulated formation water containing CO_2 and H_2S; Microscopic corrosion morphology of 2205 duplex stainless steel in the simulated formation water containing CO_2 and H_2S at (b) low magnification and (c) high magnification, respectively

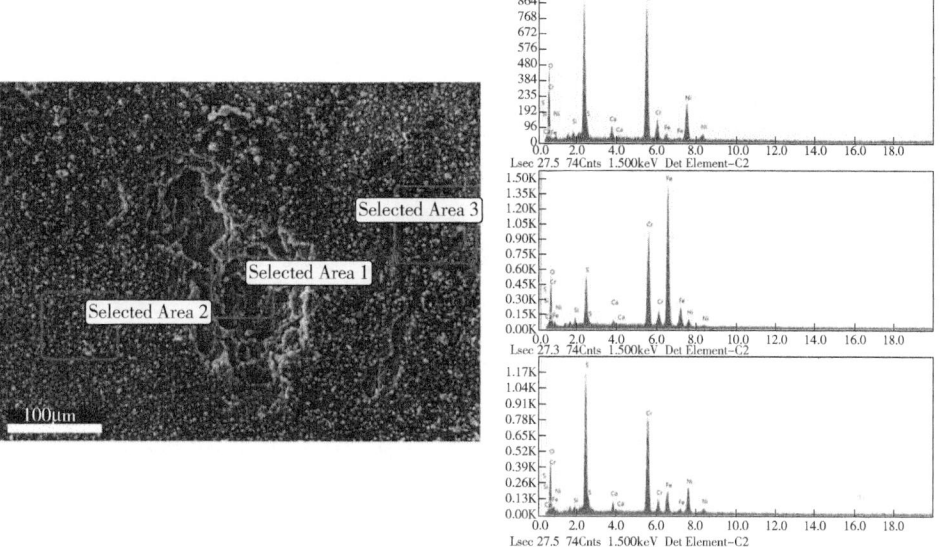

Fig.6 EDS spectrum of 2205 duplex stainless steel tested in simulated formation water containing CO_2 and H_2S

Through the stress corrosion cracking test of simulated formation water containing H_2S and CO_2, it is found that the samples have no macroscopic cracking characteristics, but obvious corrosion pits can be observed on the surface of the samples. The metallographic section analysis of the stress corrosion cracking sample shows that no cracking features are found, and the high-magnification SEM image further shows that no cracking has occurred at the bottom of the corrosion pit, as shown in Fig.8, which indicates that the 2205 duplex stainless steel has excellent stress corrosion cracking resistance in the simulated formation water containing H_2S and CO_2.

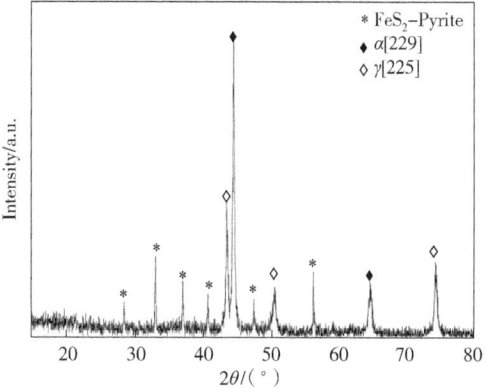

Fig.7 XRD result of corrosion product of 2205 duplex stainless steel in simulated formation water containing CO_2 and H_2S

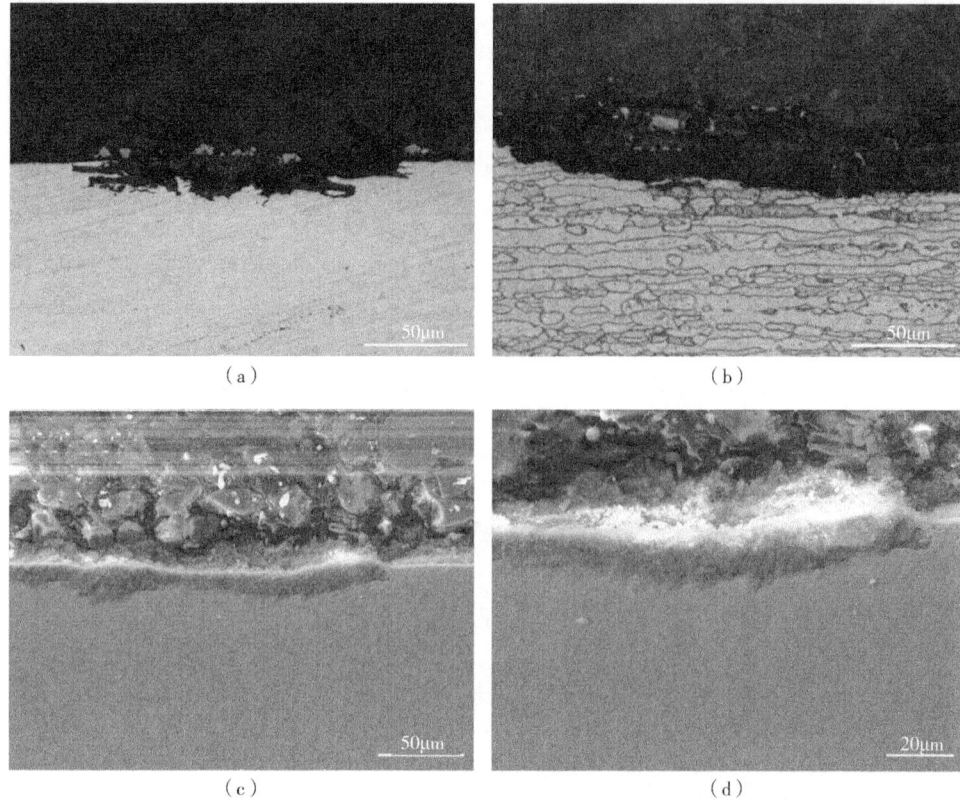

Fig.8 Longitudinal-sectional morphology of maximum stress area of the four point bending sample after the SCC test in production environment at high CO_2 and H_2S partial pressures. (a, b) Metallographic images of the corrosion pit. (c) SEM image and (d) corresponding high magnification image of the corrosion pit

3.3 Electrochemical test

3.3.1 Cyclic polarization curve

Fig.9 shows the potentiodynamic cyclic polarization curves of 2205 duplex stainless steel in different pH environments. It can be seen that despite the difference in pH values, the potentiodynamic cyclic polarization curves basically show the same trend and have the passivation characteristics in the anodic polarization zone. Under different pH conditions, the zero potential of the cyclic polarization curve of 2205 duplex stainless steel is within the ranges of $-800 \sim -200$ mV Ag/AgCl. With the increase of potential, the sample surface begins to activate and enters the anodic dissolution zone of Tafel. When the voltage increases to a certain value, the electrode is passivated and enters the passivation region. With the

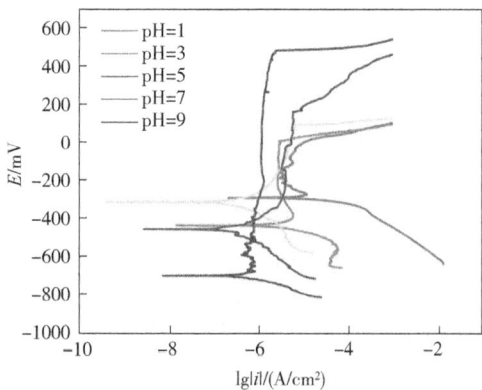

Fig.9 Polarization curves of 2205 duplex stainless steel in different pH environments

increase of the potential, the anodic current density has no obvious change. However, when the potential rises to a certain position, the current density increases rapidly, which indicates that the passivation film on the sample surface has begun to be punctured. With the increase of the pH value of the solution, a higher potential is required for the passivation to breakdown the film, that is, the pH value of the solution has a great effect on the passivation breakdown of sample surface. Table 8 shows the potential statistical data of 2205 duplex stainless steel in the cyclic polarization curve under different pH environments. The lower the pitting potential of metal material is, the easier it is to produce pitting corrosion[24]. With the increase of pH value, the pitting potential increases significantly. The lower the pitting potential of metal materials, the easier it is to produce pitting. It indicates that Cl⁻ accelerates the corrosion rate of the passive film of duplex stainless steel in low pH solution, causing serious pitting corrosion[25].

Table 8 Pitting and passivation values of the materials in the polarization curve under different pH environments

pH	E_{corr}/mV	E_b/mV	I_p/(mA/cm^2)
1	-287.158	-27.39	0.00514
3	-432.427	13.03	0.00339
5	-308.859	96.06	0.00432
7	-451.801	165.47	0.00438
9	-696.511	483.43	0.00117

3.3.2 Electrochemical impedance spectroscopy

Fig.10 shows the electrochemical impedance spectrum of 2205 duplex stainless steel in different pH environments. According to the Nyquist diagram in Fig.10 (a), the rules are observed based on the different radii of the capacitive arc appearing in different pH environments. In the acidic and neutral environments, the radius of the capacitive reactance arc increases with the increase of the pH value of the solution (the radius of the capacitive reactance arc is the largest in the environment of pH=7), and the radius of the capacitive reactance arc in the weak alkaline environment is larger than that in the acidic environment, indicating that the less prone to corrosion of the sample is in the neutral or weak alkaline environment, the stronger the corrosion resistance of the sample is. It can be seen on the Bode curve in Fig.10 (b) that the peak value of the measured phase angle becomes larger with the increase of the pH value of the solution in acidic and neutral environments, which indicates that the passivation ability of the passivation film on the sample surface is enhanced.

The equivalent circuit shown in Fig.11 was used to analyze the electrochemical impedance map in Fig.10 and fit the electrochemical parameters. The results are listed in Table 9. R_s, R_{ct} and Q_{dl} are solution resistance, electrochemical transfer resistance and double layer capacitive reactance, respectively. As can be seen from Table 9, R_{ct} increases with the increase of the pH value of the environmental solution, reaching the maximum at pH of 7, and then at pH of 9. It can be seen from Table 9 that the maximum R_f appears in the environment with pH of 7, followed by the environment with pH of 9, 5, 3 and 1. It illustrates that as the environmental pH value increases, the integrity and density of 2205 duplex stainless steel passivation membrane will increase, the resistance of ion passing through the passivation membrane will increase, the protection effect of the passivation film on 2205 duplex stainless steel will be enhanced, and the corrosion rate of sample will decrease. In a neutral environment, the corrosion resistance of passivation membrane is the strongest, but it is weak in acidic environment and even worse in alkaline environment. That is, the results obtained by impedance spectrum analysis are consistent with the data obtained by the above polarization curve fitting.

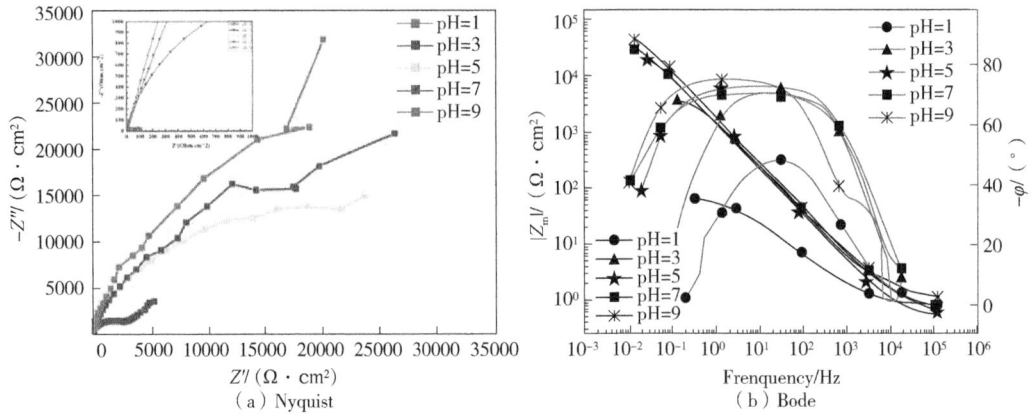

Fig.10 Electrochemical impedance spectrums of 2205 duplex stainless steel in different pH environments

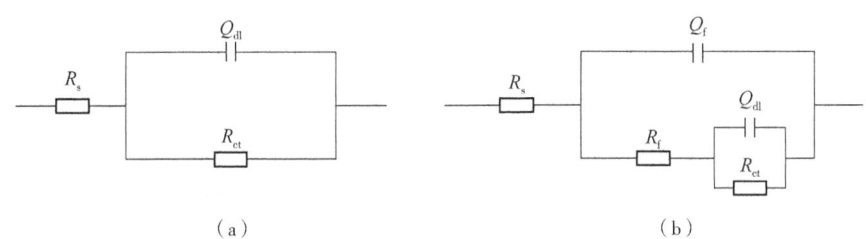

Fig.11 Equivalent circuit models for EIS fitting for (a) solutions with pH 1 and 3, (b) solutions with pH 5, 7 and 9

Table 9 Electrochemical impedance fitting results of 2205 duplex stainless steel in different pH environments

pH	$R_s/(\Omega \cdot cm^2)$	$Q_{dl}/(\Omega^{-1} \cdot cm^{-2} \cdot s^{-n})$	n_{dl}	$R_{ct}/(\Omega \cdot cm^2)$	$Q_f/(\Omega^{-1} \cdot cm^{-2} \cdot s^{-n})$	n_f	$R_f/(\Omega \cdot cm^2)$
1	0.7	0.00332960	0.60668	96.61	—	—	—
3	0.7	0.00147060	0.79432	5160	—	—	—
5	0.6	0.00013530	0.75288	38879	0.000032953	0.98100	31.630
7	0.7	0.00012760	0.78285	47394	0.000016387	0.97430	6.983
9	0.9	0.00010919	0.82889	78835	0.000022325	0.88349	2.512

4 Discussion

According to the issues mentioned above, it can be seen thatthe corrosion behaviors of 2205 duplex stainless steels significantly vary with the high – temperature and high – pressure service environments. In four types of acidizing operation environments, selective corrosion of duplex stainless steel occurs with the dissolution of ferrite phase. Selective dissolution corrosion is associated with the distribution of alloy elements and the ionization tendency of metal atoms in duplex stainless steel[26]. In duplex stainless steel, the alloying elements are unevenly distributed between phases. Generally, Cr, Mo, and Si are largely concentrated in ferrite, Ni, Mn, and Cu are more concentrated in austenite, and N is concentrated in austenite because of its limited solubility in ferrite[27]. Symniotis revealed that the weight loss of SAF2205DSS in 2M H_2SO_4+0.1M HCl solution consists of two contributions of the α phase and γ phase, which are potentially dependent other elements, the potential of ferrite is lower than that of austenite[28]. In the acidic solution, because that the Cr in ferrite is ionized more easily than other elements, the potential of ferrite is lower than that of austenite[26]. Eguchi et al. investigated the corrosion behavior of 17Cr and 22Cr steels in 15% HCl, observing that 17Cr consisting of martensite, ferrite, and retained austenite phases, exhibited higher corrosion resistance than the 22Cr duplex stainless steel[29]. Note that the increasing HCl concentration can significantly accelerate the average corrosion rate and the maximum pitting rate of duplex stainless steel in the present work. Lo et al. revealed that the extent of dissolution of both ferrite and austenite phases was enhanced as the concentration of HCl was increased[25]. In addition, under acid corrosion condition with extremely low pH value, the formation of passivation film will be inhibited, while acidification corrosion inhibitor is good choice to mitigate corrosion of stainless steels[30]. However, the present high temperature corrosion inhibitor can not avoid the selective corrosion behavior of duplex stainless steels. Comparison of corrosion behavior of duplex stainless steels by four types of acid solutions with 5% corrosion inhibitors, the descending order of corrosion are 12% HCl+3% HF, 9% HCl+3% HF+3% HAc, 9% HCl+3% HF+8% HBF_4, and 9% HCl+3% HF+2% HBF_4. Considering that HAc (CH_3COOH) as one of the most prevalent organic acids had lower corrosiveness to metals than that of HCl, HAc can partially replace HCl for acidizing operation solution[31]. Furthermore, HBF_4 can be hydrolyzed into H_3BO_3 and HF, which can accelerate the anodic dissolution, but H^+ ionized by HCl will inhibited the hydrolysis and ionization of HBF_4[32]. Therefore, the duplex stainless steel exhibits relatively low corrosion rate and pitting rate in acidizing environments containing a small quantity of HBF_4.

Compared withcorrosion behavior in acidizing environment, 2205 duplex stainless steel shows better corrosion resistance in high – temperature and high – pressure production environment. Macdonald et al. proposed that the thickness of the steady-state passivation film was not only related to the point defect reaction and potential, but also varied linearly with the pH value of the solution[33]. Proper increase of the pH value can reduce the dissolution rate of the passivation film, and further enhance the corrosion resistance of stainless steels[34]. In the present research, in the range of acidic solution to neutral solution, with the increase of pH value, the radius of the capacitive reactance arc increases and the E_b is more positive, indicating the better stability of the passivation and lower pitting sensitivity. By contrast, the lower pH value in acidizing solution will accelerate the dissolution of the protective passivation film on the surface, which further induce the metal loss.

In addition, the presence of CO_2 and H_2S can significantly influence the corrosion behavior due to the the formation of corrosion products which depend upon the various corrosion reactions[35]. It is well known that CO_2 dissolves in the water to form H_2CO_3, then ionize twice in solution to generate HCO_3^-, CO_3^{2-} and H^+, and the corresponding processes are shown in Eq.3 to Eq.5[36]. In CO_2 environment, the corrosion product of $FeCO_3$ could form via the solid state reaction (Eq. 6) or the precipitation reaction with these ions in solution (Eq. 7), while the H_2S dissolves in the water to generate HS^-, S^{2-} and H^+, as shown by Eq. 8 and Eq.9[37]. However, it has been reported that there are multiple iron sulfide corrosion product films on the surface, mainly including mackinawite (FeS_{1-x}, FeS_m), pyrrhotite (FeS_{1+x}) and pyrite (FeS_2), and their type plays an important role on corrosion behavior[38]. Mackinawite is the most common H_2S corrosion product at low temperature, and the widely accepted formation mechanism is mainly based on the solid-state reaction (Eq.10 to Eq. 12) and precipitation (Eq.13) kinetics[38-40]. It is noteworthy that mackinawite is thermodynamically metastable with standard molar Gibbs energy of −88.43kJ/mol and usually not very protective for metals[35,41].

$$CO_2 + H_2O \rightleftharpoons H_2CO_3 \tag{3}$$

$$H_2CO_3 \rightleftharpoons H^+ + HCO_3^- \tag{4}$$

$$HCO_3^- \rightleftharpoons H^+ + CO_3^{2-} \tag{5}$$

$$Fe + CO_2 + H_2O \rightleftharpoons FeCO_3 + 2H^+ + 2e^- \tag{6}$$

$$Fe^{2+} + CO_3^{2-} \rightleftharpoons FeCO_3 \tag{7}$$

$$H_2S \rightleftharpoons HS^- + H^+ \tag{8}$$

$$HS^- \rightleftharpoons S^{2-} + H^+ \tag{9}$$

$$Fe + H_2S + H_2O \rightleftharpoons FeHS_{ads}^- + H_3O^+ \tag{10}$$

$$FeHS_{ads}^- \rightleftharpoons FeHS_{ads}^+ + 2e^- \tag{11}$$

$$FeHS_{ads}^+ \rightleftharpoons FeS_{1-x} + xHS^- + (1-x)H^+ \tag{12}$$

$$Fe^{2+} + H_2S \rightleftharpoons FeS + 2H^+ \tag{13}$$

With increasing temperature and H_2S partial pressure, more stable pyrrhotite and pyrite with lower standard molar Gibbs energy of −114.5kJ/mol and −160.1kJ/mol[38], respectively, tend to be formed, and the possible reaction pathways are shown in Eq.14 and Eq.15[42].

$$FeS + xH_2S \rightleftharpoons FeS_{1+x} + xH_2 \tag{14}$$

$$FeS + H_2S \rightleftharpoons FeS_2 + H_2 \tag{15}$$

Taylor et al. reported that the macknawite underwent sequential conversion to pyrrhotite and pyrite in a aqueous H_2S at temperatures in excess of 100℃ [43]. In addition, only pyrite phase can be observed at very high H_2S concentrations and temperature [44]. Since pyrrhotite and pyrite are more than an order of magnitude less soluble than mackinawite, their continuous film will be more corrosion protective than that of mackinawite film in solution [38]. In the present research, no iron carbonate can be observed except for pyrite, indicating that the main corrosion mechanism might be controlled by H_2S corrosion. In a mixed CO_2 and H_2S system, the key influencing factor is the ratios of CO_2 and H_2S partial pressures, and the H_2S tend to govern the corrosion mechanism below a CO_2/H_2S ratio of 20 [35]. Therefore, the present production environment containing high H_2S partial pressure will favor the formation of dense and stable pyrite films which act as a barrier for mass transport to mitigate the pitting behaviors on the surface of stainless steels. Furthermore, no cringking characteristics could be observed at the bottom of pits under external stress condition, while pitting defects usually serve as nucleation sites for cracks.

Based on thediscussion above, it can be concluded in acidizing environment, duplex stainless steels have the lower passivation performance, and selective corrosion occurs in ferrite phase. Reasonable acid composition design still needs to be considered, such as the proper addition of HBF_4. On the other hand, due to the increasing passivation performance with the increase of pH value and the formation of stable corrosion products, duplex stainless steels achieve excellent resistances to corrosion and crack in production environment containing high H_2S partial pressure. Therefore, 2205 duplex stainless steel coiled tubing can be recommended to be mainly used for the operations in production environment rather than acidizing operations.

5 Conclusions

The corrosion behavior of 2205 duplex stainless steel in the full life cycle operation environments (including acidizing operation and subsequent production environments) was investigated. The following conclusions were obtained:

(1) In the acidizing operation environments with corrosion inhibitors, selective corrosion of 2205 duplex stainless steel occurs with the selective dissolution of ferrite phase. The higher concentration of HCl significantly accelerate corrosion behavior. However, the proper addition of HBF_4 to partially replace HCl can reduce corrosion rate and pitting rate.

(2) Inthe production environment containing high CO_2 and H_2S partial pressures, 2205 duplex stainless steel exhibits low corrosion rate, and only local corrosion occurs. Furthermore, under external stress condition, no cracking characteristics can be observed at the bottom of the pit, indicating the excellent resistance to stress corrosion cracking.

(3) With the increase of the pH value, the pitting potential shifts positively, while the polarization resistance increases and reachs the optimum in the neutral solution, indicating that 2205 duplex stainless steel have better passivation performance in the production environment. Therefore, it is concluded that the duplex stainless steel coiled tubing is more suitable for production operations in the oil and gas wells.

Acknowledgements

This work was supported by the Basic Research and Strategic Reserve Technology Research Fund Project of CNPC [Grant No. 2021DQ03(2022Z-07)].

References

[1] F. Zhang, X. Yang, J. Peng, N. Li, S. Lv, N. Zeng, R. Zhang, Well integrality technical practice of ultra deep ultrahigh pressure well in Tarim Oilfield, 6th International Petroleum Technology Conference, International Petroleum Technology Conference, Beijing, 2013, pp. IPTC-17126-MS.

[2] X.F. Yuan, Ultra high pressure well fracturing in KS Area, World Oil's 8th Annual HPHT Drilling and Conpletions Conference, World Oil, Houston, 2013.

[3] A.R. Crabtree, W. Gavin, Coiled tubing in sour environments – theory and practice, SPE/ICoTA Coiled Tubing Conference and Exhibition, Houston, TX, 2004, pp. SPE-89614-MS.

[4] A.R. Crabtree, CT – failure monitoring: a decade of experience, SPE/ICoTA Coiled Tubing and Well Intervention Conference and Exhibition, The Woodlands, TX, 2008, pp. SPE-113676-MS.

[5] Z.H. Zhang, J.B. Guo, Law of CO_2 corrosion of oil country tubular goods and the study of its progress at home and abroad, Baosteel Technol. (4) (2000) 54–58.

[6] R. Hampson, C. Moir, T. Freeney, Working with coiled tubing in H_2S and CO_2 wells: a global perspective, SPE/ICoTA Coiled Tubing & Well Intervention Conference and Exhibition, The Woodlands, TX, 2009, pp. SPE-121294-MS.

[7] X.F. Wang, W.Q. Chen, H.G. Zheng, Influence of isothermal aging on σ precipitation in super duplex stainless steel, Int. J. Miner. Metall. Mater. 17(4) (2010) 435–440.

[8] W.T. Tsai, M.S. Chen, Stress corrosion cracking behavior of 2205 duplex stainless steel in concentrated NaCl solution, Corros. Sci. 42(3) (2000) 545–559.

[9] H.M. Ezuber, A. El-Houd, F. El-Shawesh, Effects of sigma phase precipitation on seawater pitting of duplex stainless steel, Desalination 207(1–3) (2007) 268–275.

[10] H. Liu, X. Jin, Electrochemical corrosion behavior of the laser continuous heat treatment welded joints of 2205 duplex stainless steel, J. Wuhan Univ. Technol. Mater. Sci. Ed. 26(6) (2011) 1140–1147.

[11] H. Luo, C. Dong, X. Cheng, K. Xiao, X. Li, Electrochemical behavior of 2205 duplex stainless steel in NaCl solution with different chromate contents, J. Mater. Eng. Perform. 21(7) (2012) 1283–1291.

[12] A.I. Mourad, A. Khourshid, T. Sharef, Gas tungsten arc and laser beam welding processes effects on duplex stainless steel 2205 properties, Mater. Sci. Eng. A 549 (2012) 105–113.

[13] C.R.D.F. Azevedo, H.B. Pereira, S. Wolynec, A.F. Padilha, An overview of the recurrent failures of duplex stainless steels.Eng. Fail. Anal. 97 (2019) 161–188.

[14] G. Atxaga, A.M. Irisarri, Study of the failure of a duplex stainless steel valve. Eng. Fail. Anal. 16 (2009) 1412–1419.

[15] A.P. Kölblinger, S.S.M. Tavares, C.A. Della Rovere, A.R. Pimenta, Failure analysis of a flange of superduplex stainless steel by preferential corrosion of ferrite phase. Eng. Fail. Anal. 134 (2022) 106098.

[16] R. Merello, F. Botana, J. Botella, M. Marcos, Determination of the weaker phase in the pitting corrosion of non-standard low-Ni high-Mn-N duplex stainless steels, Mater. Corros. 55(2) (2004) 95–101.

[17] Y. Fu, C. Lin, W.T. Tsai, A study of the selective dissolution behavior of duplex stainless steel by micro-electrochemical technique, Acta Metall. Sin. 41(3) (2005) 302–306.

[18] C.D. Wang, M.L. Yan, X.W. Zhao, P.Q. Li, H. Wang, Research progress of H_2S/CO_2 corrosion in oil and

gas development, J. Xi'an Shiyou Univ. Nat. Sci. Ed. 20(5) (2005) 66-70.

[19] W. Li, Y. Zhou, Y. Xue, Corrosion behavior about tubing steel in environment with high H_2S and CO_2 content, J. Wuhan Univ. Technol. Mater. Sci. Ed. 28(5) (2013) 1038-1043.

[20] D. Jingen, Y. Wei, L. Xiaorong, D. Xiaoqin, Influence of H_2S content on CO_2 corrosion behaviors of N80 tubing steel, Pet. Sci. Technol. 29(13) (2011) 1387-1396.

[21] X.H. LÜ, G.X. Zhao, Y. Wang, J.B. Zhang, K.Y. Xie, SSC resistance of super 13Cr martensitic stainless steel, J. Mater. Eng. 1(2) (2011) 17-21.

[22] National Energy Administration, SY/T 5405-1996, The Method for Measurement and Evaluating Indicator of Corrosion Inhibitor for Acidizing, 1996.

[23] NACE, NACE Standard RP0775-2005, preparation, installation, analysis, and interpretation of corrosion coupons in oilfield operations, NACE, Houston, TX, 2005.

[24] X. Zhang, X.W. Zheng, H.Y. Zuo, F. Liu, M. Gong, Pitting Corrosion Behavior of 2205 Duplex Stainless Steel in Sodium Chloride Solution Under Vacuum Condition, J. Sichuan Univ. Sci. Eng. Nat. Sci. Ed. 33(1) (2020) 20-26.

[25] S. Jin, Study on corrosion and passivation behavior of super duplex stainless steel 2507, Tianjin University, 2015.

[26] G.X. Zhao, Y.Q. Wang, L. Ji, J.J. Liang, J.F. Xie, X.H. Lv, Corrosion behavior of 2205 duplex stainless steel in severe environment of oilfield, Mater. Mech. Eng. 42(2) (2018) 82-87.

[27] R. Merello, F. Botana, J. Botella, M. Marcos, Determination of the weaker phase in the pitting corrosion of non-standard low-Ni high-Mn-N duplex stainless steels, Mater. Corros. 55(2) (2004) 95-101.

[28] E. Symniotis, Galvanic effects on the active dissolution of duplex stainless steels, Corrosion 46(1) (1990) 2-12.

[29] K. Eguchi, Y. Ishiguro, H. Ota, Corrosion behavior of multi-phase stainless steel in 15% hydrochloric acid at a temperature of 80℃, Corros. 71 (2015) 1398-1405.

[30] X.H. Lv, F.X. Zhang, X.T. Yang, J.F. Xie, G.X. Zhao, Y. Xue. Corrosion performance of high strength 15Cr martensitic stainless steel in severe environments, J. Iron Steel Res. Int. 21 (2014) 774-780.

[31] L. Yu, Y.P. Shi, S.Y. Chen, X.Y. Yang, J.H. Cai, Experimental study on stimulation fluids by regular acidification in low permeable limestone wells, J. Water Resour. Water Eng. 31 (2020) 97-103.

[32] J. Du, Y.N. He, P.L. Liu, Y.G. Liu, X.H. Meng, L.Q. Zhao, Corrosion inhibition of N80 steel in 10% HCl+ 8% HBF_4 solution, Anti-Corros. Methods Mater. 66 (2019) 1-10.

[33] D.D. Macdonald. The History of the Point Defect Model for the Passive State: A Brief Review of Film Growth Aspects, Electrochim. Acta 56 (2011) 1761-1772.

[34] C.O.A. Olsson, D. Landolt. Passive Films on Stainless Steels-chemistry, Structure and Growth, Electrochim. Acta 48 (2003) 1093-1104.

[35] A.M. El-Sherik, Trends in Oil and Gas Corrosion Research and Technologies, Woodhead Publishing, 2017.

[36] X.P. Li, Y. Zhao, W.L. Qi, J.F. Xie, J.D. Wang, B. Liu, G.X. Zeng, T. Zhang, F. H. Wang, Effect of extremely aggressive environment on the nature of corrosion scales of HP-13Cr stainless steel, Appl. Surf. Sci. 469 (2019) 146-161.

[37] L. Wei, X.L. Pang, K.W. Gao, Corrosion of low alloy steel and stainless steel in supercritical $CO_2/H_2O/H_2S$ systems, Corros. Sci. 111 (2016) 637-648.

[38] F.W. Herbert, Mechanisms governing the growth, reactivity and stability of iron sulfides, Mater. Sci., Univ. Oxf. (2015).

[39] X.L. Wen, P.P. Bai, B.W. Luo, S.Q. Zheng, C.F. Chen, Review of recent progress in the study of corrosion products of steels in a hydrogen sulphide environment, Corros. Sci. 139 (2018) 124-140.

[40] Y. Long, W.W. Song, A.Q. Fu, J.F. Xie, Y.R. Feng, Z.Q. Bai, C.X. Yin, Q.W. Ma, N. Ji, X.R. Kuang, Combined effect of hydrogen embrittlement and corrosion on the cracking behaviour of C110 low alloy steel in O_2-contaminated H_2S environment, Corros. Sci. 194 (2022) 109926.

[41] L.G. Benning, R.T. Wilkin, H.L. Barnes, Reaction pathways in the Fe-S System below 100℃, Chem. Geol. 167 (2000) 25-51.

[42] Y. Li, R.A. van Santen, Th. Weber, High-temperature $FeS-FeS_2$ solid-state transitions: Reactions of solid mackinawite with gaseous H_2S, J. Solid State Chem. 181 (2008) 3151-3162.

[43] P. Taylor, T.E. Rummery, D.G. Owen, Reactions of Iron Monosulfide Solids with Aqueous Hydrogen Sulfide Up to 160℃, J. Inorg, Nucl. Chem. 41 (1979) 1683-1687.

[44] S.N. Smith, The Carbon Dioxide/Hydrogen Sulfide Ratio-Use and Relevance, Mater. Perform. 54 (2015) 64-67.

本论文原发表于《Engineering Failure Analysis》2023 年第 151 卷。

The Application of Terahertz Nondestructive Testing Technology in the Detection of Polyethylene Pipe Defects

Nie Hailiang[1] Hao Fengdan[2] Wang Litao[3] Guo Qiang[4]
Chen Hongda[1] Ren Junjie[1] Wang Ke[1] Dang Wei[1]
Liang Xiaobin[1] Ma Weifeng[1]

(1. Institute of Safety Assessment and Integrity, State Key Laboratory for Performance and Structure Safety of Petroleum Tubular Goods and Equipment Materials, Tubular Goods Research Center of CNPC;
2. China Promotion Association for Special Equipment Safety and Energy-saving;
3. The Third Oil Transportation Department of Changqing Oilfield Branch of China National Petroleum Corporation;
4. Shaanxi city gas industry development co.,LTD.)

Abstract: At present, polyethylene pipeline is widely used in urban gas projects, but a relatively mature and reliable nondestructive testing technology has not been formed. Therefore, it is urgent to develop a new nondestructive testing technology to meet the increasing demand for inspection of non-metallic pipes. The terahertz testing technology and related equipment has played an increasingly important role in the nondestructive testing of many nonmetallic structures, but it has not been applied to polyethylene (PE) pipes. In this work, terahertz time-domain spectroscopy was used to detect prefabricated defects inside the PE pipe specimens. The results show that the terahertz nondestructive testing technology can be used to detect common defects in nonblack PE pipes with a detection error of less than 10%. Higher power terahertz devices can detect defects in black polyethylene pipe, while lower power terahertz devices cannot. Because the black polyethylene pipe contains carbon and has strong absorption of terahertz waves. The penetration of lower power terahertz devices to the black polyethylene pipe is not enough, resulting in a low resolution of the imaging. The results of this work may promote the progress of the nondestructive testing technology of nonmetallic pipelines.

Keywords: Polyethylene pipe; Terahertz detection; Prefabricated defects; Characteristic signals; Nondestructive testing

---

Corresponding author: Nie Hailiang, niehailiang88@163.com.

1 Introduction

In recent years, the use of nonmetallic pipelines has increased in environmental protection projects, water supply projects, and urban gas projects, etc. High-density PE pipes have gradually replaced traditional steel pipes and cast-iron pipes in medium-pressure and low-pressure gas transportation because of their good corrosion resistance, absence of leakage, high strength and toughness, excellent flexibility, easy handling and installation, etc., thus, they have become the preferred pipes for urban gas transportation[1]. The safety of nonmetallic pipelines has also attracted increasing attention, and its evaluation will attract large economic interests and will undergo future development[2]. 163 cases of PE pipeline failure in 4 companies were investigated, and it was found that the third-party damage accounted for 59% of the cases. Third party damage refers to the man-made damage to the oil and gas pipeline except the owner, construction company and operation and management company. Third-party damage generally results in the loss of pipeline material. In recent years, the research field on damage detection in nonmetallic pipe structures has rapidly developed, mainly focusing on straight pipes and pipe joints for nonmetallic pipes[3]. At Dalhousie University (Canada), piezoelectric ceramics were used to study the damage at the joint of nonmetallic pipes[4]. The torsional-mode-guided wave probe designed by Ecole Centrale de Lyon (France) was used to study the guided wave response and damage characteristics of nonmetallic pipelines[5]. At the University of Rome (Italy), simulations and experimental studies were carried out on the groove damage on nonmetallic circular rods via Gaussian-pulse-guided wave excitation[6]. Theoretical studies achieved good results, and satisfactory results were obtained in laboratory experiments. However, due to the lack of testing equipment suitable for practical engineering, these methods can only be used in research laboratories and have not been used in practical engineering applications. In addition to the above conventional detection methods, some scholars have also developed unconventional detection methods, among which air coupled ultrasonic detection technology has made great progress. Arno[7] has proved that it is possible to detect artificial defects using air coupled ultrasonic inspection in thin polymer pipe walls, however, it is challenging to separate defects from normal variation in the measured signal.

Terahertz (THz) radiation refers to electromagnetic waves with frequencies ranging from 0.1 THz to 10THz (1THz = 10^{12} Hz), low energy (4.1meV), high signal-to-noise ratio, high resolution, and other characteristics, and is used in the field of nondestructive testing[8]. Since the 1990s, scientists have been working on devices that transmit and receive terahertz waves[9]. In 2008, Stoiks[10] applied terahertz Time-domain spectroscopy to nondestructive testing of aircraft fiberglass composites and evaluated the degree of thermal damage using simple amplitude two-dimensional images. Owing to its continuous development, the terahertz testing technology and related equipment will play an increasingly important role in the nondestructive testing of nonmetallic pipes.

Researchers have conducted extensive investigations in the field of terahertz nondestructive testing of nonmetallic materials and made some progress. Ole et al.[11] analyzed and calculated the terahertz time-domain spectrum of PE samples and obtained their stratified thickness without knowing the sample refractive index. Kwang-hee et al.[12] studied the terahertz transmission of thick fiber-wound composites, analyzed the obtained time-domain spectrum in detail, and verified the feasibility of terahertz time-domain spectroscopy (THz-TDS) and imaging for nondestructive testing of composite materials. Wietzke et al.[13] carried out a number of studies in the field of nonmetallic nondestructive testing and used the terahertz continuous imaging system to detect internal defects in nonmetallic materials, such as PE. The results show that the terahertz imaging technology has good application prospects for detecting defects in nonmetallic materials, such as delamination, inclusion, mechanical damage, and thermal damage. Albert et al.[14] used a portable terahertz system to study materials such as gypsum artwork and carbon fibers and found that terahertz spectroscopy has an excellent spectral resolution for plastic samples; they obtained accurate information on the bonding condition and sample of binder in the sample.

Many research institutes have also studiedthe terahertz nondestructive testing technology for pipelines. Wang et al.[15] used the terahertz reflection imaging technology and the THz time-domain spectral transmission technology to detect PE electric fusion joints (inclusions, delamination, and other defects); they were able to clearly identify the inclusions (wire) and burial defects in the samples. Jiang[16] measured the thickness of a PE pipe via THz-TDS and carried out defect detection tests. His research mainly focused on theoretical and feasibility studies. Chen[17] proposed a sample feature recognition algorithm based on THz-TDS and proposed a feature recognition model. This model overcame the issues of spectral feature similarity recognition, feature defect recognition, and signal denoising in terahertz detection, laying a theoretical foundation for the terahertz detection technology.

From the abovementioned studies, it can be seen that terahertz nondestructive testing has developed rapidly in recent years, and a large number of research results have been obtained. However, research institutions mainly focus on defect detection in flat materials and obtain good laboratory test results; however, in actual pipelines, the structure has a given curvature.

Whether the internal defects of such nonplanar structures can be detected and identified using the existing terahertz technology remains to be further verified. In this work, a method for the preparation of pipes with defects was mainly developed for PE pipes, which are widely used in urban gas transportation systems. Typical defects with minimum size of 2mm were prefabricated in PE pipes. Using the THz-TDS technology, the prefabricated defects were successfully detected. These results provide theoretical and technical support for the detection of defects in nonmetallic pipes.

2 Materials and samples

This study focuses on scratch damage defects on the surface of PE pipes as well as damage defects internal to the pipe material. Two types of pipeline materials were studied, namely yellow and black pipes.

The pipe samples used in thiswork are yellow and black PE pipes produced by Huida Pipe Industry with an outer diameter of 315mm and a wall thickness of 30mm. When manufacturing defects, 1/4 of the circumferential pipe section was considered, and different sizes and types of defects were fabricated at different locations on the pipe section using different drill bit sizes. Buried defects and inner surface defects were distributed at the two end faces of the pipe section. Outer surface defects were distributed in the middle axial direction of the pipe section, and the defects were elongated along the axial direction of the pipe section. A perpendicular drill was used to fabricate the buried and inner surface defects at the two end surfaces. When fabricating buried defects, the center of the drill bit was located in the middle of the wall; on the other hand, when fabricating inner surface defects, the center of the drill bit was at a depth of 1/2 of the diameter of the bit from the inner surface of the pipe segment. The diameter of the bit was used to control the diameter of the buried defects and the inner surface defects, and the depth of the drill was used to control the axial length of the defects. Outer surface defects were fabricated by drilling the bit perpendicular to the outer wall of the pipe segment. After drilling to the desired defect depth, the bit could move slowly along the axis of the pipe segment while cutting the material of the pipe body until the moving distance reached the desired defect length. The different defect types and sizes are listed in Table 1; the minimum defect diameter was 2mm. The yellow and black PE tube samples are shown in Fig.1.

Table 1 Defect types and sizes

Defect type	Defect diameter/mm	Defect length/mm
Outer surface defect	2, 3, 4, 5	25
Buried defect	2, 3, 4, 5	30
Inner surface defect	2, 3, 4, 5	15

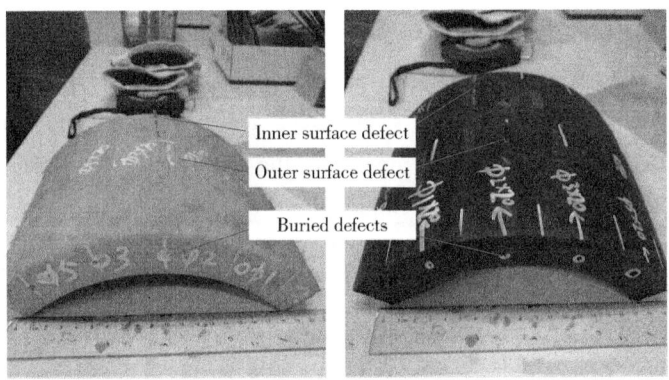

(a) Yellow pipe specimen   (b) Black pipe specimen

Fig.1 Photographs ofthe PE pipes with prefabricated defects

The defectswere fabricated directly on the nonmetallic pipe sections, and the samples were finally cut into tiles. This permitted not only to retain the curvature of the nonmetallic pipes themselves but also to reduce the space occupied by the samples, thereby easing their transportation, which lays a foundation for the indoor simulation of the field nondestructive testing of pipelines.

3 Test methods

3.1 Experimental equipment

In this work, the T-Ray 5000 smart TCU control unit produced by API Corporation and a manipulator are used. The T-Ray 5000 intelligent TCU control unit contains a laser source, and has many practical functions such as ultra-high speed optical delay scanning to obtain terahertz waveforms, display digital terahertz waveforms, numerical analysis and transmission of test reports over the network. The laser source is made of mode locked titanium sapphire, with pulse width of 100fs, average output power of 20mW, and detection spectrum width of 0.2~2.5THz. The sampling interval is 0.1ps. The focusing lens diameter is 75mm, and the spot diameter is about 2mm with a Rayleigh length of 5~50mm (the thickness of the test sample is 30mm). The sample surface was scanned in steps of 1mm during the detection process. The detection data was stored in a computer, and the defect image was obtained after processing using a self-developed software. The terahertz detection device is shown in Fig.2.

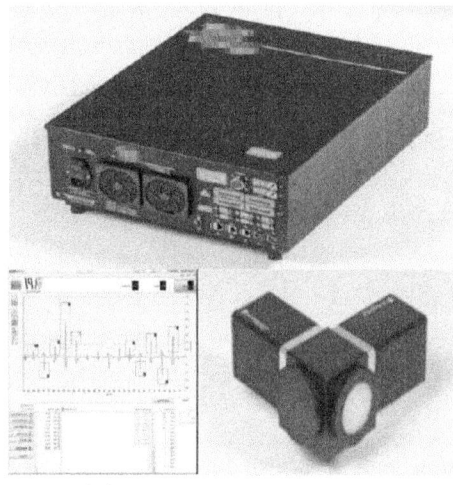
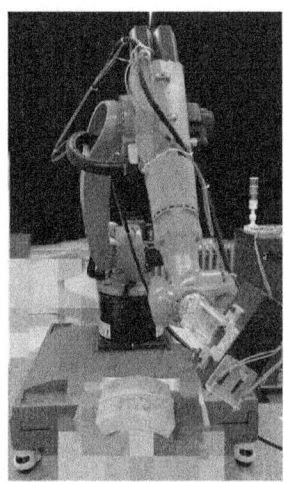

(a) Intelligent terahertz control unit　　　　(b) Automated imaging scanning system

Fig.2　Terahertz detection equipment

During the experiments, we found that defects in black PE pipe could not be detected with the T-ray 5000, so we chose TK-LAB from Terakalis for alternative, equipped with optical devices with an average output power of 2mW, frequency range of 100~600GHz. The focusing lens diameter is 60mm, and the spot diameter is about 2mm with a Rayleigh length of 3~40mm. This device is only used in the detection of black PE pipe samples.

3.2 Principle of defect detection

The main purpose of this paper is to verify that the terahertz nondestructive testing technology can be applied to the defect detection of polyethylene pipeline engineering. The specific signal processing process is not the focus of this paper, so this paper adopts the data processing software of the test equipment to process the terahertz signal. Here, we only describe the basic principle of signal processing, without describing the specific operation process in detail.

3.2.1 Terahertz imaging technique

The terahertz imaging system was used to process and analyze the reflection spectrum information (including the amplitude and phase information) of the imaged sample, and the terahertz image of the sample was thus obtained. The terahertz imaging system is equipped with an image processing device and a scanning control device.

The terahertz time-domain spectral imaging system acquires three-dimensional (3D) space and time data [i.e., two space (x, y) axes and one time axis]. Using this 3D dataset, terahertz images of a series of samples can be obtained. In addition, because the terahertz image at a time point contains very little information (only the waveform information from one detection point), it is usually necessary to obtain the whole 3D dataset for defect reconstruction. The reconstruction of terahertz images is usually based on specific parameters of the terahertz time-domain waveform or the delay time of the peak position. At present, there exist mainly five methods for sample reconstruction:

(1) Time-of-flight imaging: the time delay information of each pixel is used to image the THz signal. This imaging method reflects the refractive indexto THz radiation of the sample.

(2) Time-domain maximum value, minimum value, and peak value imaging: the maximum value, minimum value, or the difference between the maximum and minimum values of the THz time-domain signal in each pixel is used for imaging; the time-domain maximum-value imaging reflects the extinction coefficient of the sample to the terahertz wave.

(3) Specific frequency amplitude (phase) imaging: the amplitude (phase) value ofthe THz frequency-domain signal at a certain frequency in each pixel is used for imaging.

(4) Power spectral imaging: the image information is obtained by integrating the square of the amplitude of the THz frequency-domain signal of each pixel within a certain frequency range.

(5) Pulse-width imaging: using the pulse-width imaging of the main peak of the terahertz radiation, the imaging model mainly reflects the dispersion characteristics of an object, and it can clearly present the outline of an object.

The detection of internal defects in structures can be carried out by analyzing the two-dimensional scanning image in the results.

In terahertz imaging, two-dimensional scanning imaging is performed on the row and column of a pixel in the detection area, where the horizontal axis indicates the position of the row or column, while the vertical axis indicates the signal strength at the corresponding time of flight. Two-dimensional scanning imaging can be used to locate and analyze a defect, and the flight time corresponding to the defect can be used to analyze the depth direction of the defect. In addition, the defect characteristics in the terahertz time-domain waveform can be extracted, and the imaging analysis of the measured object can be carried out according to these characteristics.

3.2.2 Calculation principle of the defect height

When THz waves propagate in different media, both transmission and reflection occur at the interface, as shown in Fig. 3. Assuming that the sample is isotropic and the THz wave is incident at an angle θ_i, a THz echo E_{up} is generated at the air-medium interface due to the change in the refractive index. The transmitted THz wave is reflected on the upper surface of the defect. After the time difference ΔT, the detector detects the THz echo E_{down} successively.

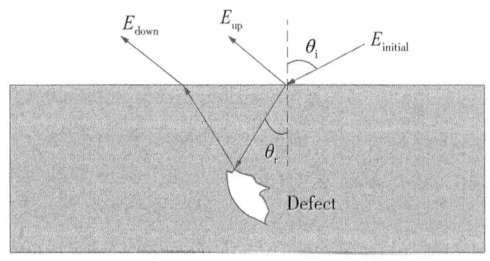

Fig.3 Reflection principle of THz waves

There is a linear relationship between the buried depth of the defect and the flight time difference of the THz echo. The reflected single-point thickness extraction model is as follows:

$$d = \frac{c\sqrt{n^2 - n_0^2 \sin^2 \theta_i}}{2n}(T_{up} - T_{down}) = \frac{c\sqrt{n^2 - n_0^2 \sin^2 \theta_i}}{2n} \Delta T \quad (1)$$

Where n and n_0 are the refractive index of air and the material to be measured respectively, T_{up} and T_{down} are the flight time of E_{up} and E_{down}, and c is the speed of light in air. When a THz wave is incident vertically, the single-point distance extraction model can be simplified as:

$$d = \frac{c}{2}(T_{up} - T_{down}) = \frac{c}{2} \Delta T \quad (2)$$

According to the time difference between the two peaks, that is, the flight time difference, the characteristics of the flight time difference of terahertz waves in different materials can be obtained using the peak extraction technology, and the buried depth of the upper surface of the defect can then be calculated. The peak extraction technology is to identify the defect characteristic waveform of the detection signal by comparing the actual detection signal with the standard test block detection signal, combined with the set threshold of frequency and amplitude. Using the same principle, the buried depth of the lower surface of the defect can be calculated. The height of the defect can be calculated by subtracting the buried depth of the lower surface of the defect from the buried depth of the upper surface of the defect.

4 Results

4.1 THz wave refractive index and absorption of materials

The refractive indices and absorption coefficients for the used materials of THz wave at different wavelength were measured, as shown in Table 2. It can be seen that the THz absorption coefficient of the black PE pipe is much higher than that of the yellow PE pipe. This is because the carbon black is added to the black PE pipe to prevent ultraviolet aging, and the carbon black absorbs a large amount of THz wave.

Table 2 Refractive index and absorption coefficient of tested materials to terahertz wave

Wavelength/μm	Refractive index		Absorption value/m^{-1}	
	Yellow PE pipes	Black PE pipes	Yellow PE pipes	Black PE pipes
70.5	1.554	1.561	81.52	99.54
96.5	1.549	1.551	53.79	74.79
118.8	1.546	1.548	51.63	71.63
122.4	1.542	1.546	48.87	68.87
158.5	1.536	1.542	42.78	62.78
184.3	1.530	1.538	23.80	43.80
214.6	1.523	1.528	22.96	42.96

4.2 Defect detection in yellow PE pipe samples

4.2.1 Two-dimensional scanning image of the specimens

The circular plane image of the prefabricated defect was obtained using the 2nd method (Time-domain maximum value, minimum value, and peak value imaging), as shown in Fig.4, where the color depth represents the reflected wave amplitude normalized by reflected wave/incident wave. As can be seen from the figure, there are four rows and three columns of detected defects, as expected. It can be seen that the first column represents buried defects, the second column represents outer surface defects, and the third column represents inner surface defects. The diameter of the defects decreases from the first row to the fourth row. The detected defects are ranked as follows: first row: 1-1#, 1-2#, 1-3#; second row: 2-1#, 2-2#, 2-3#; third row: 3-1#, 3-2#, 3-3#; fourth row: 4-1#, 4-2#, 4-3#.

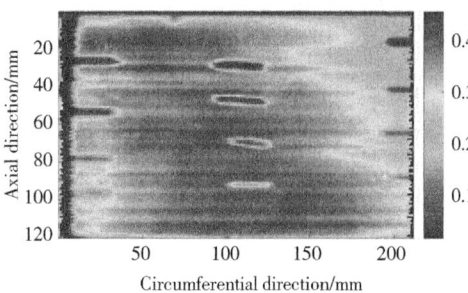

Fig.4 Toroidal two-dimensional scanning image of the defects

4.2.2 Tomographic images of the defects in the direction of the wall thickness

The location map of the defects in the direction of the thickness of the sample can be obtained using the tomographic imaging method, a technique to superimpose the reflected signals of different thickness on the same scan line to show the defect characteristics of different thickness layers. In fact, terahertz waves will generate reflection and transmission at the interface between solids and gases. In the experiment, we adopted the reflection imaging method, which is based on the amplitude characteristics of the reflected waves. Therefore, the image contour will be deeper in some parts. The depth of the imaging color has no practical significance, but the position features reflected by it are relatively accurate, which can help us identify and calculate the geometric features of the structure.

The tomographic images of the three defect positions in the first row were extracted, as shown in Fig.5. In the figure, the ordinate is the scanning time, and the abscissa is the axial scanning position. In the figure, the shallower contour on the uppermost layer is the result of the superposition of reflected waves on the upper surface of the sample, while the contour with the deepest color on the lower layer is the result of the superposition of reflected waves on the lower surface of the sample.

From Fig.5(a), it can be seen that the buried defect gives rise to a steep contour, which is the result of the superposition of reflected waves on the upper surface. This contour is located between the upper and lower surface contours; the contour is not clear due to the diffuse reflection of waves on the lower surface of the defect.

Fig.5(b) shows the existence of defects on the upper surface, which leads to discontinuous fuzzy segments in the contour after the reflection waves on the upper and lower surfaces are superimposed. This is because the diffuse reflected waves on the lower surface of the terahertz wave on the outer surface defect cause the received reflected wave amplitude to become smaller, and the image is not clear. Therefore, the main characteristics of the outer surface defects are discontinuous fuzzy segments in the image profile of the reflected wave on the inner and outer surfaces.

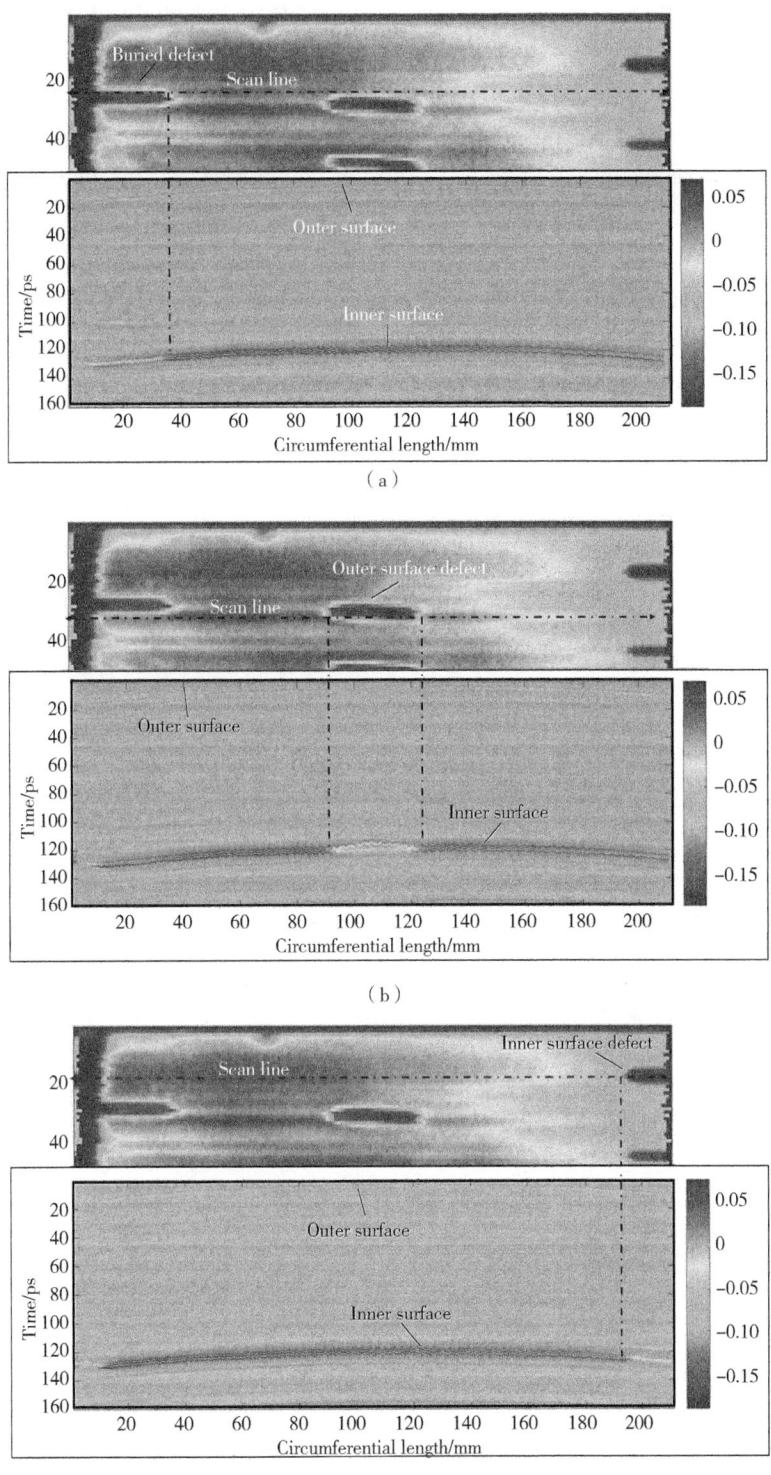

Fig.5 Tomographic image of the defects at the first row of Fig.4:
(a) 1-1#, the buried defect on the left, (b) 1-2#, the external surface defect in the middle, (c) 1-3#, the inner surface defect on the right

Fig.5(c) shows the existence of an inner surface defect (the profile of the surface reflection wave is a discontinuous fuzzy segment). At the same time, on the surface and under the surface imaging profile between a profile, the outline of the inner surface defect of reflected wave imaging results, thus, imaging characteristics of inner surface defects, the surface profile of imaging discrete fuzzy segment. On the other hand, there is an imaging profile between the inner and outer surface profiles.

The distribution of defects in the thickness direction of the sample can also be analyzed by extracting the single waveform of the defects. The single waveform of three defects in the first row is shown in Fig.6. It can be seen that the amplitude of the reflected wave in the area without defects is larger, with the amplitude changing from positive to negative and back to positive; on the other hand, for the area containing defects, the amplitude of the reflected signal is smaller.

As can be seen from Fig.6(a), the reflected signal of the buried defect presents two continuous peaks, one positive and one negative.

As can be seen from Fig.6(b), the reflected signal from the outer surface defect presents multiple positive and negative interlaced continuous peaks.

As can be seen from Fig.6(c), the reflected signal from the inner surface defect presents two discrete positive peaks.

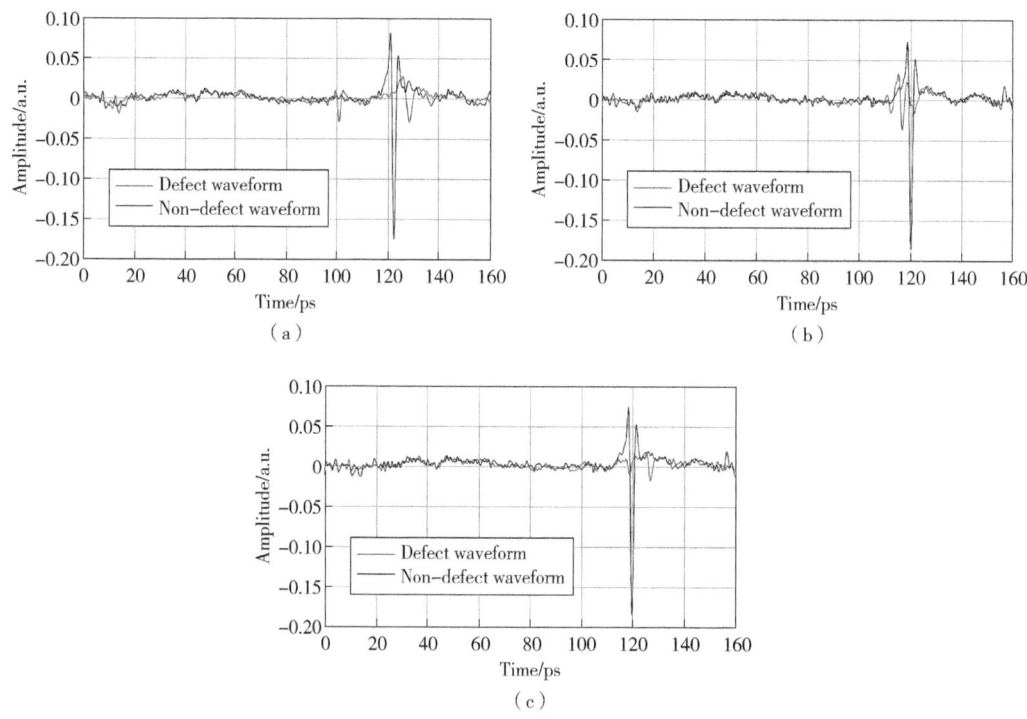

Fig.6 Comparison between the defect and non-defect waveforms at the first row:
(a) 1-1#, (b) 1-2#, and (c) 1-3#

4.2.3 Axial and circumferential dimensions of the defects

The axial dimension (L_a) and the circumferential dimension (L_w) of a defect can be calculated from the experimental results (as shown in Fig.7). The calculation method is as follows: the axial and circumferential lengths of the defect are both 1mm, so the pixel size is 1mm×1mm. L_a and L_w can be calculated by measuring the number of pixel points of defects in Fig.4. The measurement results are shown in Table 3. Compared with the actual defect size, the measurement error is about 10%.

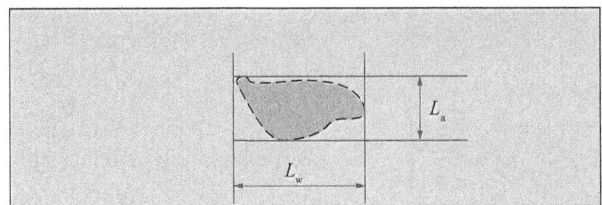

Fig.7 Schematic of the axial and circumferential dimensions of a defect

Table 3 Calculated defect dimensions

ID	L_a/mm	L_w/mm	ID	L_a/mm	L_w/mm	ID	L_a/mm	L_w/mm
1-1#	5	28	1-2#	5	31	1-3#	6	16
2-1#	5	27	2-2#	4	31	2-3#	4	16
3-1#	3	27	3-2#	3	29	3-3#	3	16
4-1#	2	29	4-2#	2	28	4-3#	2	16

4.2.4 Defect height calculation

The height of the defect is calculated as shown in Fig.8. The height of the defect can be calculated by determining the flight time of the upper and lower surfaces of the defect, as shown in Fig.9. Only buried defects can be used to calculate their height. Considering that the terahertz beam diameter is very small, only about 2mm, and the sample thickness is not large, the beam dispersion can be ignored when calculating the defect size. The calculation results are shown in Table 4, and the calculation error is less than 10%.

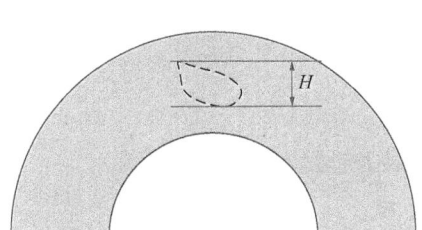

Fig.8 Diagram of the height of a defect

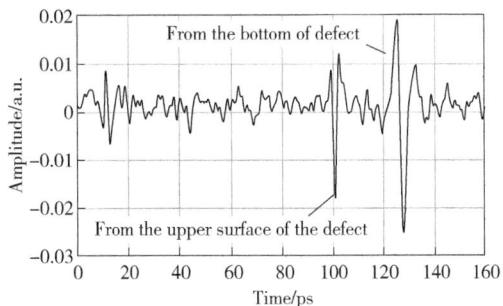

Fig.9 Defect height calculation using the flight time

Table 4 Calculated defect heights

ID	H/mm	Error/%
1-2#	4.8	4.0
2-2#	3.9	2.5
3-2#	2.9	3.3
4-2#	1.8	10.0

4.3 Defect detection in black PE pipe samples

T-ray 5000 series equipment was also used to detect the sample of black polyethylene pipe. The two-dimensional imaging results of the black PE pipe sample are shown in Fig.10. It can be seen from that when T-Ray 5000 series equipment is used to detect the black pipe, the terahertz waveform cannot detect the defect signal. After analysis, it is found that the T-Ray 5000 equipment we used is a time-domain spectral system with low energy (the power is 550W), while the black polyethylene tube contains carbon and has strong absorption of terahertz waves. The penetration of T-Ray 5000 to the black polyethylene tube is not enough, resulting in a low resolution of the imaging.

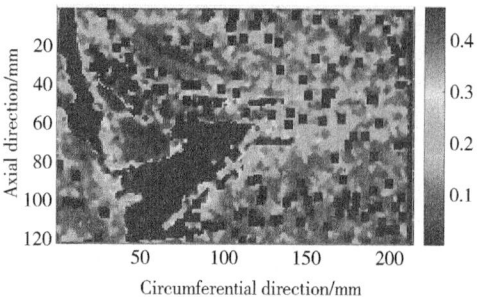

Fig.10 Two-dimensional imaging results for the black PE pipe sample

To verify the above analysis, we selected a higher power terahertz detection device Terakalis TK-Lab (the power is 4.4kW) to test the same black polyethylene tube sample. The disadvantage of this device is that the original detection platform only contains translational carrier platform, which cannot realize equidistance vertical scanning of test sample surfaces. The imaging results of black polyethylene pipe with Terakalis TK-Lab series equipment are shown in Fig.11. It can be seen that the defect signal can be detected by Terakalis TK-Lab. The reason is that the equipment has higher power and has a stronger penetration ability to carbon.

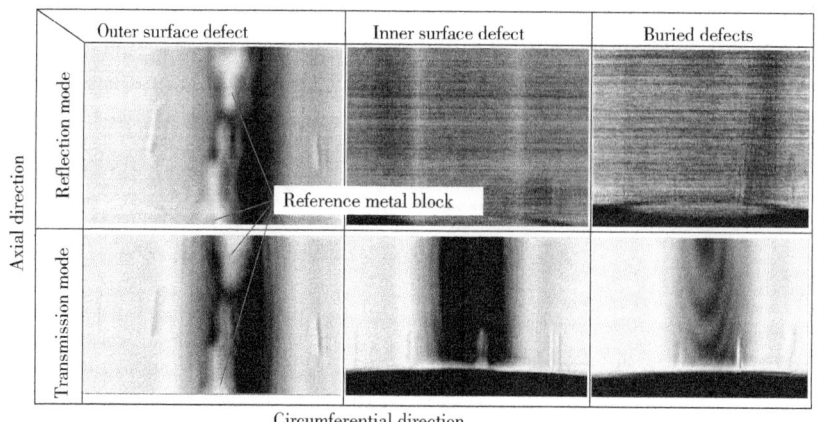

Fig.11 Imaging results of black polyethylenepipe by Terakalis TK-Lab series equipment in France

Through the above analysis, it can be seen that there are problems in the black tube defect detection as follows:

(1) Higher power terahertz devices can detect defects in black polyethylene pipe, while lower power terahertz devices cannot.

(2) In reflection mode, the detection effect is not ideal, and this is the large curvature artifacts existing equipment platform (Terakalis TK - Lab) can only along the planar scanning contradiction

has much to do, without a change in Angle probe and test face case, most of the reflected signal cannot be received, if the proper scanning mechanism (e.g., mechanical arm), The effect will be greatly improved.

(3) In addition to the defects, you can also see from the picture of other different distribution of light and shade, these phenomena still because there is a radian workpiece tested, there is no guarantee that position always in the same plane, for testing prototype (plane scan), detection results can reflect the change of position, can be used in the actual detection of mechanical arm programming scan way, The problem can be solved by keeping the distance between the probe and the workpiece always consistent.

(4) If the power increases, T-ray 5000 series equipment can also detect defects in black polyethylene pipe.

5 Discussion

5.1 Material requirements

Terahertz radiation penetrates into nonmetallic pipes but is completely reflected by metallic materials; thus, the terahertz detection technology cannot be used for metallic materials. In addition, carbon black strongly absorbs terahertz radiation, and black PE pipes often contain carbon black; thus, the terahertz technology cannot be used to detect defects in black PE pipes.

5.2 Structure requirements

The terahertz detection technology is a noncontact detection method that requires radiation perpendicularly incident onto the workpiece surface. When there is a bulge or depression with large curvature on the workpiece surface, the reflected wave may deviate from the path of the incident wave at a certain angle, resulting in the probe not being able to receive the reflected wave signal.

The time window of the existing terahertz data acquisition system is small, generally 160ps or 320ps, and the wave reflected beyond this window will not be collected. Therefore, the distance between the probe and the workpiece surface should be kept constant during the detection process to ensure that the waves reflected from the upper and lower surfaces can be received. Therefore, the terahertz detection technology requires a workpiece with smooth surface and uniform thickness. A great variation in the thickness may lead to signal loss from the thicker parts.

5.3 Requirements for PE pipe equipment

Terahertz detection of defects in PE pipes is feasible, but specific equipment requirements must be satisfied. A multidimensional transformation of the manipulator is necessary to obtain the 3D structural information of the sample to be tested. The manipulator can be guided to drive the probe for detection through system modeling.

5.4 Advantages and disadvantages of terahertz detection of defects in PE pipes

A comparison of the terahertz technology with other NDT technologies is provided in Table 5. The terahertz technology has the following technical advantages for defect detection in nonmetallic pipes: good penetration depth, low energy, transient, high resolution. Furthermore, it is a noncontact technique, permits good positioning to be achieved, provides qualitative and quantitative detection of defects, and is capable of 3D imaging of defects.

Table 5　Comparison of the terahertz technology with conventional detection techniques

NDT technology	Characteristics
Ultrasonic detection	Low efficiency, unable to penetrate multilayer structures, and high loss in nonmetallic materials. The technology is mature
Acoustic emission detection	The item under test must be loaded. One can only detect defects that are being generated and expanded. The technology is mature
Thermal imaging	Sensitive only to superficial defects. Immature technology
Ray detection	A quarantine area is required for this test. The technology is mature
Air coupled ultrasonic detection technology	Non-contact detection, the surface requirements are low, but the signal extraction and processing is more difficult
Terahertz detection	Good penetration depth, low energy, transient, and high resolution. This technique is noncontact; it can detect defects qualitatively and quantitatively and is capable of 3D imaging of defects

The terahertz detection technology also has certain limitations. The terahertz detection process must be perpendicular to the workpiece surface, which requires higher testing tooling, and a structure with a too large surface curvature cannot be tested with this technique. Additionally, it requires that the distance between the probe and the workpiece surface remains constant during detection to avoid data acquisition failure. Finally, the existing terahertz testing equipment is only suitable for indoor testing. Equipment and tooling for field testing have not been developed yet, and further research and development is needed.

6　Conclusions and prospects

In this study, defects with a minimum size of 2mm were prefabricated in typical PE pipe bodies. Using THz-TDS, the prefabricated defects inside the yellow PE pipe were successfully detected. At the same time, the circumferential and axial dimensions of the defects as well as the height of the buried defects were calculated. The results show that the terahertz nondestructive testing technology can be used to detect common defects in yellow PE pipes, and the detection error is less than 10%.

Higher power terahertz devices can detect defects in black polyethylene pipe, while lower power terahertz devices cannot. Because the black polyethylene tube contains carbon and has strong absorption of terahertz waves. The penetration of lower power terahertz devices to the black polyethylene pipe is not enough, resulting in a low resolution of the imaging.

The present study was conducted to verify the feasibility of indoor destructive testing. With further development in the terahertz detection technology, it is believed that it will be possible to use this technique in field detection. Real-time adjustment of the terahertz probe can be realized through real-time 3D optical scanning of the surface of the pipeline; this method not only exhibits high detection efficiency, but also greatly improves the detection accuracy. In addition, as this technique can be used to obtain 3D information, reconstructing the 3D shape of defects will become a research focus in the future. We have conducted some preliminary research work in this field and believe that 3D reconstruction of defects can be realized in the near future.

Acknowledgments

This work was supported by the Natural Science Fundation of Shannxi Province, China (grant number 2021JQ-947), and the Basic Research and Strategic Reserve Technology Research Fund of China National Petroleum Corporation [2019D-5008 (2019Z-01), 2022DQ03(2022Z-03)]. The authors also thank Entaike (Beijing) Technology Co., LTD for providing terahertz detection equipment and data processing software.

Data availability

The datasets generated and/or analysed during the current study are not publicly available due to the company confidentiality requirements but are available from the corresponding author on reasonable request.

References

[1] Schrock J. Underground Plastic Pipe. New York Ny American Society of Civil Engineers, 2015.

[2] Stepanova L N, Kabanov s I, Ramazanov I S, et al. Acoustic-emission testing of multiple-pass welding defects of large-size constructions. Russian Journal of Nondestructive Testing, 2015, 51(9):540-545.

[3] Nara T, Takanashi Y, Mizuide M. A sensor measuring the Fourier coefficients of the magnetic flux density for pipe crack detection using the magnetic flux leakage method. Journal of Applied Physics, 2011, 109(7): 07E305-07E305-3.

[4] Cheraghi N, Riley M J, Taheri F. A novel approach for detection of damage in adhesively bonded joints in plastic pipes based on vibration method using piezoelectric sensors.Systems, Man and Cybernetics, 2005 IEEE International Conference on. IEEE, 2005,4:3472-3478.

[5] Bareille O, Kharrat M, Zhou W, et al. Distributed piezoelectric guided-T-wave generator, design and analysis. Mechatronics, 2010, 22(5):544-551.

[6] Pau A, Vestroni F. Damage characterization in bars using guided waves.Proceedings of the XIX AIMETA Congress, Ancona. 2009: 65-75.

[7] Römmeler A, Furrer R, Sennhauser U, et al. Air coupled ultrasonic defect detection in polymer pipes. NDT and E International, 2019, 102:244-253.

[8] Stoik C D, Bohn M J, Blackshire J L. Nondestructive evaluation of aircraft composites using transmissive terahertz time domain spectroscopy. Optics Express, 2008, 16(21):17039-17051.

[9] Benicewicz P K, Roberts J P, Taylor A J. Scaling of terahertz radiation from large – aperture biased photoconductors. J. Opt. Soc. Am. B, 1994, 11(12): 2533-2546.

[10] Stoik C, Bohn M, Blackshire J. Nondestructive evaluation of aircraft composites using reflective terahertz time domain spectroscopy[J]. Ndt & E International, 2010, 43(2):106-115.

[11] Peters O, Wietzke S, Jansen C, et al. Nondestructive detection of delaminations in plasticweld joints.IEE. 2010, 978-14244-6657-3/10.

[12] Im K H, Hsu D K, Chiou C P, et al. Influence of Terahertz Waves on the Penetration in Thick FRP Composite Materials. AIP Publishing Proceedings, 2014, 1568-1575

[13] Jansen C, Wietzke S, Peters O, et al. Terahertz imaging: applications and perspectives. Applied optics, 2010, 49(19): E48-E57.

[14] Redo-Sanchez A, Norman L, Brian S, et al. Non-destructive Imaging with Compact and Portable Terahertz

Systems. AIP Publishing Proceedings, 2014, 1583-1586.

[15] Hong W, Liao X, Liu Y. Nondestructive Evaluation of Electrofusion Joint of Polyethylene Pipeline Using Terahertz Wave. China Special Equipment Safety, 2016.(in Chinese)

[16] Qiang J, Lingwie Y, He Y, et al. Deconvolution Algorithm of THz-TDS Polyethylene Thickness Measurement Based on Modified Autocorrelation Algorithm. Hongwai Jishu/Infrared Technology, 2020, 42(5):473-482.(in Chinese)

[17] Chen S, Han X, Lijuan L, et al. Research on Plastic Detection Based on Terahertz Time Domain Spectroscopy. Journal of Changchun University of Science and Technology(Natural Science Edition), 2017. (in Chinese)

本论文原发表于《ACS Omega》2023 年第 8 期。

Failure Analysis of the Crack and Leakage of a Crude Oil Pipeline under CO_2-Steam Flooding

Song Chengli[1] Li Yuanpeng[2] Wu Fan[2] Luo Jinheng[1]
Li Lifeng[1] Li Guangshan[1]

(1. State Key Laboratory for Performance and Structure Safety of Petroleum Tubular Goods and Equipment Materials & CNPC Tubular Goods Research Institute;
2. Research Institute of Experiment and Detection of Xinjiang Oil Field Company, CNPC)

Abstract: This paper presents the failure analysis of the crack and leakage accident of a crude oil pipeline under CO_2-steam flooding in the western oilfield of China. To analyze the failure behavior and cause, different testing, including nondestructive testing, chemical composition analysis, tensile property testing, metallographic analysis, and microanalysis of fracture and chloride stress corrosion cracking (SCC) testing, are applied in the present study. The obtained results showed that the pipeline under the insulation layer of high humidity, high oxygen content, and high Cl^- environment occurred pit corrosion, and the stress concentration area at the bottom of the corrosion pit sprouted cracks. Besides, it is demonstrated that the cracks were much branched, mostly through the crystal, and the fracture showed brittle, which is consistent with the typical characteristics of chloride SCC. Meanwhile, the insufficient Ni content of the pipeline material promoted the process of chloride SCC, and the high-temperature working conditions also aggravated the rate of chloride SCC. In addition, efficient precautions were provided to avoid fracture.

Keywords: Crude oil pipeline; 316L stainless steel; CO_2-steam flooding; Failure analysis; Chloride stress corrosion cracking

1 Introduction

Dissolving CO_2 gas in crude oil can improve the oil flow ratio and swell the crude oil to achieve the effect of enhanced recovery[1-2]. Moreover, in the global trend of carbon reduction, CO_2 can be effectively buried in this way. Therefore, CO_2 flooding has the economic benefits of recovery

---

Corresponding author: Song Chengli, clsong2022@163.com.

enhancement and social benefits of carbon reduction. The application of CO_2 flooding is gradually increasing globally, but it is still in the stage of industrial trials and enhances the application of benefits[3-4]. The literature has so far focused on the study of oil reservoirs applicability assessment, production parameter control and optimization, and miscible effect of CO_2 with crude oil[5-8], with few reported cases of failure of injection and recovery pipelines.

In the western oil field of China, there are many thick oil and super thick oil blocks. To improve recovery, in addition to using CO_2 to reduce the viscosity of the crude oil, steam is also injected to improve oil washing efficiency and wave area again[9-11]. Given the increased corrosion of the recovery media caused by the artificial injection of CO_2 and high-temperature steam, 316L stainless steel is used for the crude oil recovery pipeline to cope with internal corrosion. Moreover, this pipeline lay in the ground and was completely soaked in sanding water underground. The groundwater is mainly recharged by atmospheric precipitation, ground runoff and infiltration and discharged by underground runoff and evaporation. The annual variation of groundwater is about 0.5~1.0m in the region. But, the pipeline is not coated for corrosion protection, except for the external surface, which is covered with an insulation layer.

Fig. 1 Macroscopic morphology of the pipeline sample

Nevertheless, the 316L stainless steel pipeline suddenly leaked out after nine months of service. Figure 1 shows the failed pipeline sample after cleaning with water + paraffin. The material of the pipeline is manufactured according to ASTM-A312[12]. This pipeline sample is ϕ168mm×5mm and 600mm long. Table 1 presents the design and operating conditions of the pipeline which operating pressure, temperature and flow rate are within allowable limits. In this work, the pipeline's material properties and cracking characteristics will be analyzed by several tests to clarify the causes of leakage failure. And the study results will provide a scientific basis for the selection of materials and corrosion protection of the pipeline under the new conditions of CO_2-steam flooding.

Table 1 The design and operating conditions of the pipeline

Medium Type	Design Pressure/ MPa	Operating Pressure/ MPa	Design Temperature/ ℃	Operating Temperature/ ℃	Allowable Flow Rate/ (m³/s)	Operating Flow Rate/ (m³/s)
Crude oil containing water, associated gas	1.6	0.3	200	98	10	0.4

2 Materials and Methods

2.1 Nondestructive Test

Nondestructive tests can visually detect defects such as cracks and pitting. Given that the material of the failed pipeline is weakly magnetic, as well as large corrosion pits, the penetration testing method is more suitable for the nondestructive testing of this pipeline. To facilitate testing, the pipeline sample was cut into four equal pieces. The external and internal surfaces of the sample are then cleaned with a cleaning agent and sprayed with a uniform layer of white penetrant. After waiting for 10min, the penetrant is wiped off with a dry cloth, and then the sample surface is wiped with a paper towel soaked with cleaning agent until all the penetrant is wiped off. After the sample has dried naturally in the air, the surface is sprayed with a further layer of red developer. The resulting defect image is then observed and determined.

2.2 Physical and Chemical Performance Test

The chemical composition, mechanical properties and metallographic organization of metal pipes are the most basic physical and chemical properties, which are also the main basis for reflecting the corrosion resistance and strength of pipes. Firstly, an direct reading spectrometer (SPECTRO ARL 4460) was used to analyze the chemical composition of the pipeline body and corrosion pits area. Secondly, three parallel specimens (1#, 2#, 3#) were taken from the pipe's body to test tensile properties by material testing machine (MTS 810), including tensile strength, yield strength and elongation after a fracture. Moreover, the microstructure, grain size and nonmetallic inclusion of the pipeline body and cracks were analyzed by a metallographic microscope and image analysis system (LEICA MEF4M). The above tests determine whether there are any abnormalities in the material properties of the pipeline.

2.3 Microscopic Characterisation Analysis

This pipeline's corrosion morphology, products and cracking characteristics were characterized to analyze the mechanism of corrosion and cracking. The crack was mechanically opened, and the fracture morphology was analyzed using a scanning electron microscope (TESCAN VEGA 3). The surface products at the cracks were analyzed by an energy spectrum analyzer (XFORD INCA350) and X-ray diffractometer (D8 ADVANCE). In addition, the grains of the metallographic organization of the pipe body was subjected to energy spectroscopy by line scan to characterize whether there were changes in the elements within the grains and at the boundaries. Also, to determine if intergranular corrosion is a possibility. This leads to determining the possibility of intergranular corrosion.

2.4 Corrosion Test

Boiling magnesium chloride SCC standard test can determine the susceptibility of austenitic stainless steel to SCC. According to the standard of ASTM G36, three rectangular specimens (size: 75mm×15mm×2mm) were extracted from the pipeline sample. The specimens were bent U-shaped with an indenter with a radius of 8mm. Then the 42% $MgCl_2$ solution was added to the experimental vessel with a thermometer and condensation tube. The solution will be heated to a constant boiling point of (155 ± 1) ℃ and then put the specimens into it. And their appearance is monitored periodically at 1h intervals. The susceptibility of the pipeline to chloride SCC is determined by observing whether cracks develop on the surface of the specimens.

3 Results

3.1 Visual Inspection

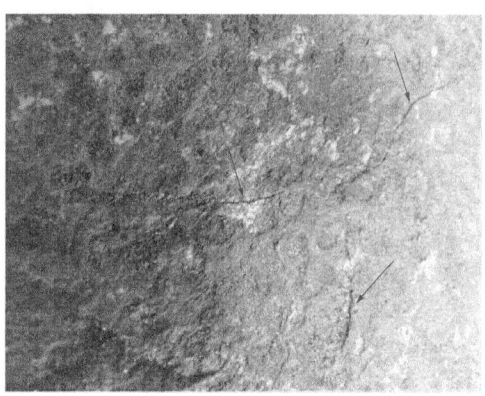

Fig. 2 Macroscopic appearance of cracks and corrosion pits on the external surface (The arrows point to the cracks)

Figure 2 shows the external wall of the pipeline, which had a large contiguous blackened area. It results from dense corrosion pits in which some oil has been deposited. The depth of these corrosion pits is generally in the range of 1 ~ 2mm. Furthermore, corrosion pits and cracks are in the same area. The multiple crack crosses are observed at the bottom of the pits 30 ~ 150mm in length. The main cracks are distributed along the pipeline transversely and longitudinally, as indicated by the arrows in Figure 2. The uncorroded area had a bright metallic luster, and no cracks were found. This suggests that uneven corrosion of the pipeline has occurred.

Figure 3 shows that the internal wall of the pipeline possessed multiple cracks at the same location as those on the external wall. Hence, penetration was suspected, but no corrosion pits were observed in the internal wall. Therefore, it should be indicated that the inner wall corrosion of the pipeline is not apparent under the fluid medium and working condition.

Figure 4 shows that the cracks originated from the external wall and expanded continuously to the internal wall according to the cross-sectional observation. It is more visualized and proved that the external soil environment of pipeline operation is the main influencing factor of cracking.

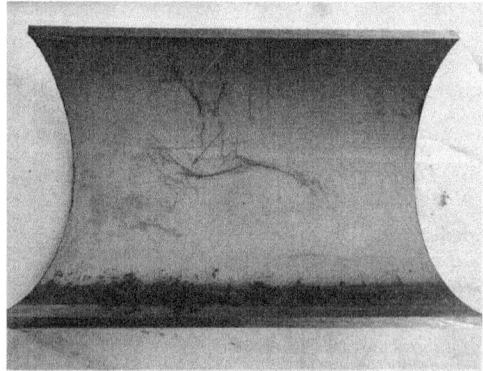

Fig. 3 Macroscopic appearance of cracks on the internal surface (The arrows point to the cracks)

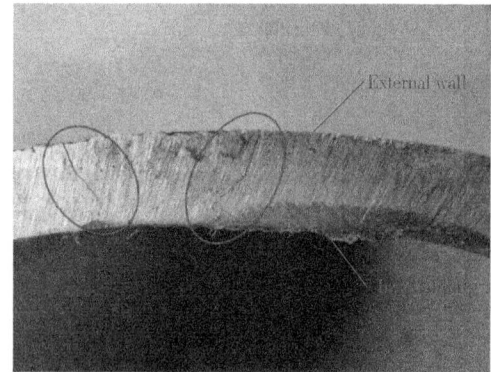

Fig. 4 Macroscopic appearance of cracks extension along the section

3.2 Nondestructive Test

Figure 5 shows the macroscopic morphology of the pipeline after the penetration test. Large defects are observed on the external surface of the pipeline sample, including corrosion pits and cracks. In addition, many obvious branching cracks are observed on the internal surface of the pipeline sample. The cracks on the external wall are concentrated in the corrosion pit area, where the majority are longitudinal, and the cracks are determined as penetrated.

 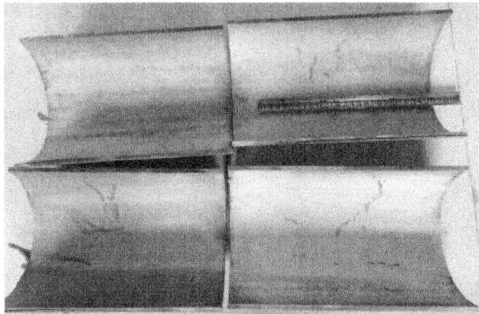

(a) external wall　　　　　　　　　　　(b) internal wall

Fig. 5　Macroscopic appearance of the pipeline sample after penetration testing

3.3　Chemical Composition

Table 2 presents the results of the chemical composition analysis of the pipe body and pitting area, in which all elements are within the standard requirements except for the nickel content. The nickel content is lower than the lower limit required by the standard of ASTM-A312, which is an unqualified product. Stainless steel materials are more corrosion resistant because of the addition of the alloy element chromium, molybdenum and nicke. Therefore, a reduction in nickel content will reduce the corrosion resistance of the material[13-14].

Table 2　Results of chemical composition analysis (%, mass fraction)

Element	C	Si	Mn	P	S	Ni	Cr	Mo	Nb	V	Cu	Al
Pipe body	0.019	0.40	0.91	0.030	0.0017	9.92	16.12	2.03	0.008	0.082	0.25	0.008
Pitting area	0.018	0.40	0.90	0.030	0.0017	9.89	16.14	2.01	0.007	0.083	0.25	0.009
ASTM-A312	≤0.035	≤1.00	≤2.00	≤0.045	≤0.030	10.00~14.00	16.00~18.00	2.00~3.00	—	—	—	—

3.4　Tensile Property

Table 3 illustrates the obtained results of the tensile properties of the pipe body. The tensile strength, yield strength and elongation after fracture are consistent with the requirements of ASTM-A312 for 316L steel. In addition, it indicates that the pressure-bearing performance of the pipe material without defects can meet the design operating conditions.

Table 3　Tensile characteristics test results

Sample	Original Gauge Length L_1/mm	Final Gauge Length L_2/mm	Yield Force F_m/kN	Maximal Force F_{eL}/kN	Original Cross-Sectional Area S/mm²	Tensile Strength R_m/MPa	Yield Strength R_{eL}/MPa	Elongation after Fracture A/%
1#	50	81.92	31.92	70.19	128.9	545	248	64
2#	50	81.81	29.40	66.91	121.9	549	241	64
3#	50	81.89	32.07	70.25	128.9	545	249	64
ASTM-A312	—	—	—	—	—	≥485	≥170	≥35

3.5 Metallographic Analysis

Figure 6 shows that the metallographic structure of the pipe body is austenite. No other abnormalities in the metallographic structure of nonmetallic inclusions and grain size. In addition, all cracks in the pipeline sample had similar characteristics, in which they all started from the external wall and extended to the internal wall. The cracks appear to be bifurcated, and no tissue distortion is seen. The main cracks are through crystal cracks, and part of the bifurcation is observed along the crystal fine cracks.

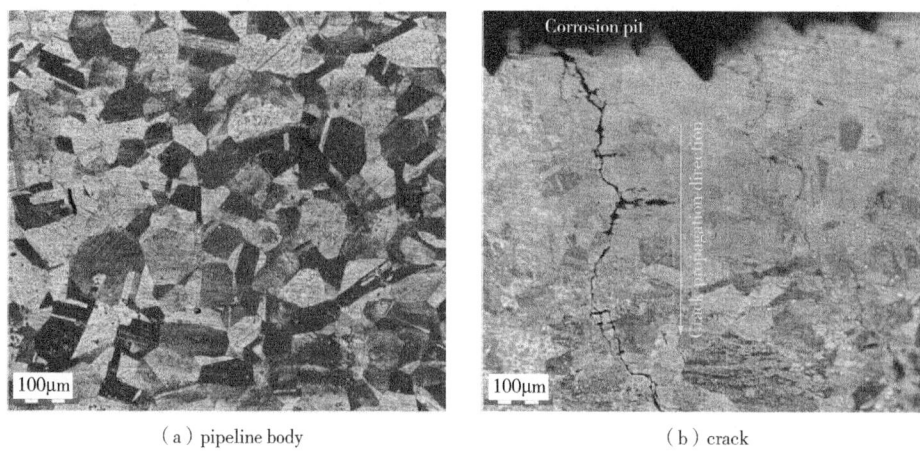

(a) pipeline body　　　　　　　(b) crack

Fig. 6　Metallographic structure of the pipeline

3.6 Microanalysis of Fracture

Figure 7 shows an SEM photo of the fracture at different magnifications. The fracture is flat, without necking, obvious deformation and thinning of wall thickness. As a result, this fracture indicates a clear, brittle fracture. Figure 8 and Table 4 show that the production elements within the cracks have O and Cl from external sources in addition to the metal matrix itself. The physical analysis of the product showed that it is mainly Fe_2O_3 which stems from oxygen corrosion, as shown in Figure 9.

Table 4　Elemental content based on EDS analysis

Elements	Area 1		Area 2	
	In mass fraction/%	In atom fraction/%	In mass fraction/%	In atom fraction/%
O	17.14	40.99	23.26	49.37
Si	0.70	0.95	1.33	1.60
Mo	5.21	2.08	5.02	1.78
Cl	3.04	3.28	9.47	9.07
Cr	43.75	32.19	28.47	18.60
Fe	25.82	17.69	27.11	16.49
Ni	4.34	2.83	5.34	3.09

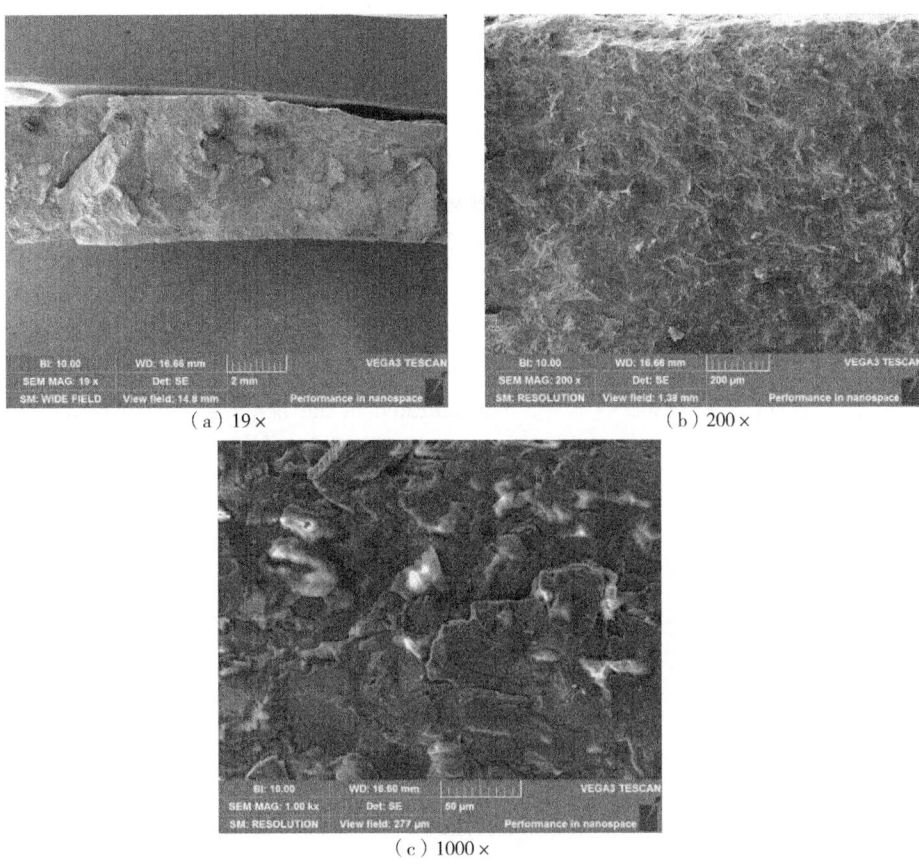

Fig. 7 Micromorphology of the fracture

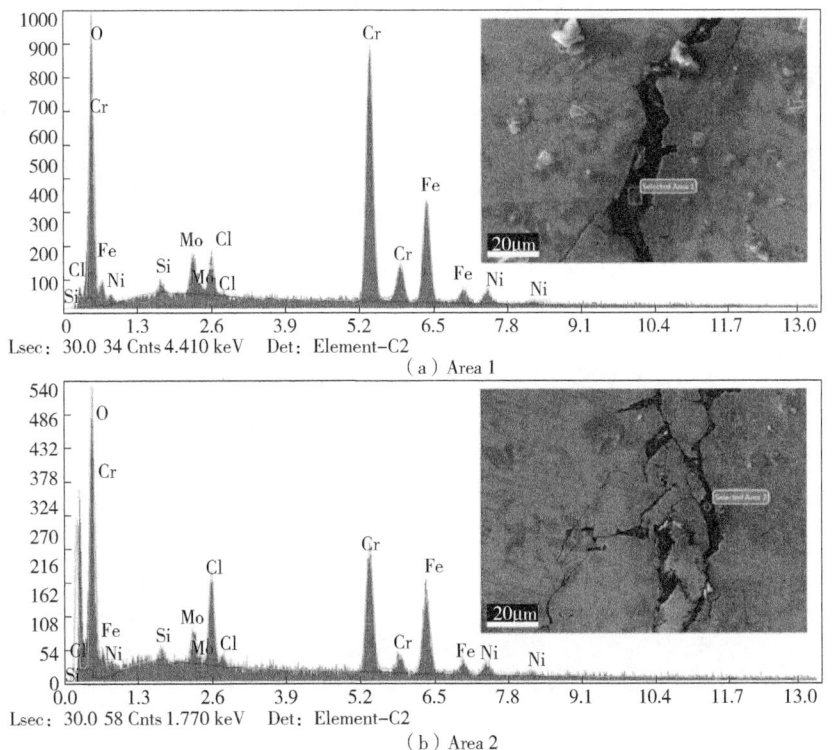

Fig. 8 Micromorphology and EDS analysis result of the cracks

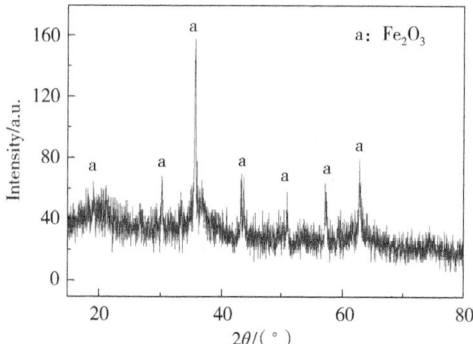

Fig. 9　XRD patterns of the corrosion products

Besides, Figure 10 shows the content of each element has no significant change along grain boundaries. Therefore, it shows no carbide precipitation at the grain boundary, excluding the possibility of intergranular corrosion.

3.7　Chloride SCC Test

After 20h of the chloride SCC test, cracks appear on the surface of specimens (1#, 2# and 3#), as shown in Figure 11. And the cracks are mainly concentrated in the bending section of the specimen, with fewer cracks in the straight edge section, as shown in Figure 12.

 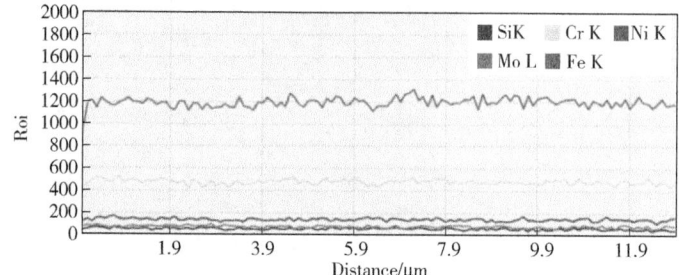

Fig. 10　Linear EDS analysis along grain boundaries

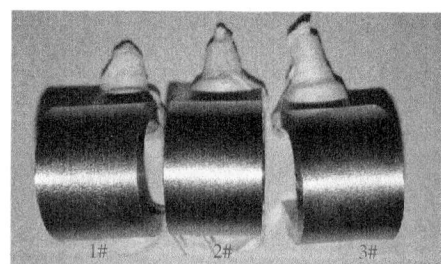 

(a) before the test　　　　　　　　　　　(b) after the test

Fig. 11　Chloride SCC test results

 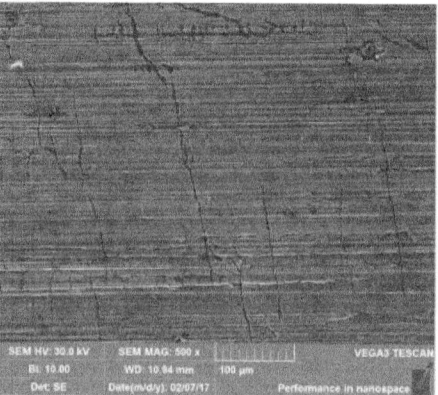

(a) bending section　　　　　　　　　　(b) straight edge

Fig. 12　Crack morphology

To further analyze the crack fracture, specimen 1# is separated along the longest crack in the center. Figure 13a presents the upper part of the fracture is corroded, indicating that the cracking starts from the outer wall and gradually extends to the interior. Figure 13b shows a tearing ling-like quasi-dissociative fracture, which was typical fracture morphology of crystal penetration.

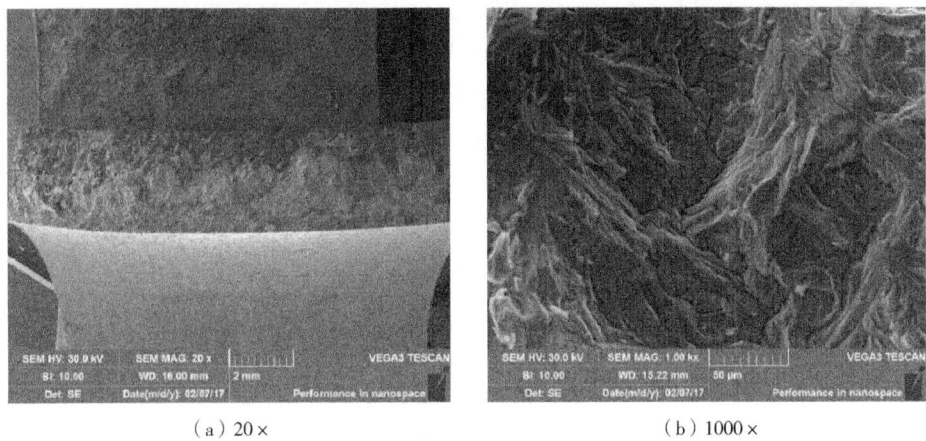

(a) 20×   (b) 1000×

Fig. 13 Fracture morphology of specimen 1#

4 Discussion

The failure type of the pipeline is cracking, with the cracks starting from the bottom of the corrosion pit on the external wall and extending to the internal wall. Moreover, the branching of the cracks, which are mainly crystalline, and the fracture show brittle characteristics. Coupled with the fact that the pipe material is 316L austenitic stainless steel, it is judged that the failure of the pipeline is consistent with the significant characteristics of chloride SCC. SSC is a localized corrosion damage in metal materials under the combined action of tensile stress and corrosive media. The following three specific analyses regarding environment, stress and material will be carried out.

4.1 Corrosion Environment Analysis

It uses a cyclic injection process of CO_2 and steam in the oilfield, resulting in a high recovery pipeline temperature of 98℃. The API RP 571 standard describes the starting temperature at which chloride SCC occurs as 60℃[15]. The API RP 581 standard states chloride cracking must be considered for environments above 38℃ under severe conditions[16]. Therefore, the operating temperature of the pipeline is sensitive to chloride SCC.

Macroscopic inspection reveals the existence of dense corrosion pits on the external wall of the pipeline, and XRD analysis further verifies the presence of Fe_2O_3, which is mainly a product of oxygen corrosion[17-18]. This is mainly because groundwater often contains dissolved oxygen from the air. Furthermore, it indicates that the insulation layer had broken down. Hence, localised pipeline corrosion is caused by underground water penetration into the insulation. Moreover, the higher the temperature, the more serious the corrosion will be, eventually resulting in large areas of pitting pits on the external wall of the pipeline.

In addition, underground water in the western region of China usually contains more Cl ions

(200~1000mg/L), constituting the basic condition for chloride SCC. At higher temperatures, the evaporation of water from the metal surface leads to a constant concentration and deposition of Cl ions. This will cause localised rupture of the stainless-steel passivation film. The formation of passivation-activation microcells at metal surfaces with and without passivation films will accelerate anodic dissolution and produce anodic polarisation[19-21]. As the corrosion pits deepen, small anodes and large cathodes will appear inside and outside, resulting in ever larger corrosion pits[22-24].

Moreover, it has been shown that the susceptibility to chloride SCC is significantly increased in the presence of both Cl ions and dissolved oxygen[25-27]. This is because the rate of oxygen consumption in the crack is greater than the rate of diffusion, leaving the crack tip still in a low oxygen state[28-29]. A corrosion potential gradient drives anions (e.g., chloride, sulphate, and hydroxide ions) deeper into the crack, while cations (hydrogen, sodium, and zinc ions) move outwards from the crack. This, in turn, causes the chloride ions to accumulate rapidly at the crack tip, creating very high concentrations. As a result, Cl ions destroy the passivation film more quickly and further reduce the rate of passivation film formation[30-31]. The aggressive Cl ions invaded the grain in multi-directions promoted by dislocation motion, facilitating the main crack to bifurcate[32].

4.2 Stress Analysis

Corrosion pits have formed after significantl ocalised corrosion has occurred on the external wall of this pipeline. Under operating pressure, soil pressure and other residual stress, there is a large stress concentration at the bottom of the corrosion pits. Residual stresses account for the largest proportion of several stresses because the pipeline is subjected to various processes, such as cooling, thermal processing, welding, etc., which can cause residual stresses. Besides, 316L austenitic stainless steel also has process hardening characteristics. The presence of stress makes the passivation film surface in the stress concentration area enriched with more chloride ions, which reduces the thickness, integrity of the passivation film, and pitting resistance[33-34]. This results in faster anodic dissolution in the stress concentration zone. In the low-stress area, the concentration of chloride ions is relatively low, and the passivation film thickness is greater and more complete, resulting in greater resistance to pitting corrosion. Figure 14 shows the model for stress corrosion of stainless steel.

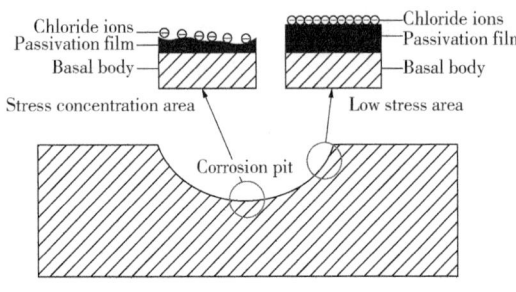

Fig. 14　Mechanism model of chloride SCC

The stress concentration area is prone to cracking[35-36], and the cracks in this pipeline all start at the bottom of the corrosion pits. As a result, once a crack has developed, the passivation film formed at the crack's tip differs from that away from the crack[37]. It is much looser and less stable, leading to cracking of the passivation film. The corrosion rate at the tip of the crack will be higher than at other locations[37], thus allowing the crack to expand in a direction perpendicular to the stress. Once the formation of micro cracks, its expansion rate is much faster than other types of localized corrosion, so SCC is the most destructive and damaging type of corrosion among all types of corrosion.

4.3 Material Analysis

The composition of alloy materials, organizational structure, and heat treatment will also affect its SCC resistance. For example, austenitic stainless steel is generally considered susceptible to chloride SCC. In contrast, ferritic stainless steel and duplex stainless steel have better resistance to chloride SCC performance when exposed to Cl⁻ environments for long periods. This pipeline insulation layer broke down, resulting in groundwater containing Cl ions and dissolved oxygen seeping into the insulation. Thus, the environment of high temperature, high humidity, and high Cl⁻ under the insulation provides favourable conditions for chloride SCC for 316L material.

In addition, Ni is the only important element to improve the stress corrosion resistance of austenitic stainless steel[38-39]. The standard API RP 571 also proposes that the Ni content of the alloy material is the most sensitive to chloride SCC at 8% ~ 12%[15], while the chemical composition analysis of the pipeline has a Ni content of about 9.9%, which is in the sensitive range of chloride SCC. At the same time, an EDS line scan of the pipeline grain structure revealed no significant change in alloying element content, ruling out intergranular corrosion cracking due to intergranular Cr depletion.

The accelerated test for chloride SCC confirmed that the pipeline cracked after 20h of testing, and it can be concluded that the pipeline does have a high susceptibility to chloride SCC.

Based on the above analysis, it can be seen that the pipeline leakage failure process is divided into three stages. In the first stage, due to the presence of groundwater containing dissolved oxygen and Cl ions, the corrosion process was localized. Under the action of high temperature (98℃), the local corrosion intensifies continuously. In the second stage, many cracks were generated at the bottom of the corrosion pits. In the third stage, due to the local concentration of stress and the continuous action of corrosion, the crack grew rapidly and led to the final failure of the material.

5 Methods for Chloride SCC Control of Stainless Steel

Three factors are required for SCC to occur: material, environment and stress. If any of these factors can be controlled, then it is possible to prevent or avoid the occurrence of chloride SCC in stainless steel.

Firstly, according to the specific environment in which the material is used, avoid using materials sensitive to chloride SCC. In general, in hot water and high-temperature water conditions, high chromium ferritic stainless steel, ferritic-austenitic duplex stainless steel, ultra-low carbon stainless steel and high nickel stainless steel can be considered to choose. However, in both, the need to resist SCC and require higher strength, ferritic-austenitic duplex stainless steel is more appropriate. On the other hand, in high concentrations of chloride media, ferritic stainless steel with low carbon and high chromium can be used, and high silicon chromium-nickel stainless steel is also a better choice.

Secondly, isolating the material from the corrosive environment is the most effective way to avoid SCC, such as using coatings or corrosion inhibitors. Reducing the concentration of Cl ions and operating temperature in the environment and preventing Cl ions adsorption and concentration are also the ways to slow down SCC. In addition, the mass fraction of oxygen should be reduced to a lower value when stainless steel is used to dissolve oxygen chloride.

Thirdly, measures should be taken during pipeline manufacturing to eliminate or reduce the residual stress in processing and welding. Surface treatment methods (such as shot peening, surface heat treatment, etc.) can also be used to reduce the residual compressive stress on the surface. Stress removal or elimination can also be performed using hydrostatic tests, temperature difference tensile and vibration.

6 Conclusions and Recommendations

(1) The pipeline experienced localized external corrosion in groundwater containing dissolved oxygen and Cl ions, and leakage failure occurred due to chloride SCC in the stress concentration area at the bottom of the corrosion pits.

(2) The Ni content of the pipeline material was lower than the standard product requirements and within the sensitive content range of chloride SCC, which accelerated the cracking of the pipeline. As well as the high temperature of the recovered medium under CO_2 and steam combined flooding promoted the progress of chloride SCC.

(3) Several Specific and practical recommendations are then proposed from both manufacturing and maintenance points of view. First, replacing the pipeline with 2205 duplex stainless steel pipe is feasible. Second, by reducing the operating temperature of the pipeline, the development of SCC will be slowed. Third, the buried pipelines should adopt an anticorrosion layer+cathodic protection to slow the occurrence of external corrosion damage under the new process of CO_2-steam combined flooding. Fourth, similar pipelines need to be excavated for defect inspection and safety evaluation, and the severely corroded external pipeline section needs to repair by B-type sleeve, carbon fiber (or glass fiber) reinforcement, etc.

Author Contributions

Methodology, C.S.; Data curation, Y.L. and F.W.; Writing—original draft J.L. and L.L.; Writing review & editing, G.L. All authors have read and agreed to the published version of the manuscript.

Acknowledgement

This research was funded by Shaanxi Province Innovation Capacity Support Program Project (2023KJXX-091) and CNPC research found(2020D-5008).

References

[1] Yao, J.; Han, H.; Yang, Y.; Song, Y.; Li, G. A review of recent progress of carbon capture, utilization, and storage (CCUS) in China. Appl. Sci. 2023, 13, 1169.

[2] Zhou,X.; Yuan, Q.; Peng, X.; Zeng, F.; Zhang, L. A critical review of the CO_2 huff 'n' puff process for enhanced heavy oil recovery. Fuel 2018, 215, 813-824.

[3] JLuo, S.; Hou, Z.M.; Feng, G.Q.; Liao, J.X.; Haris, M.; Xiong, Y. Effect of reservoir heterogeneity on CO_2 flooding in tight oil reservoirs. Energies 2022, 15, 3015-3022.

[4] Jiang, K.; Ashworth, P.; Zhang, S.; Liang, X.; Angus, D. China's carbon capture, utilization and storage (CCUS) policy: A critical review. Renew. Sustain. Energy Rev. 2019, 119, 109601.

[5] Makarova, A.A.; Mantorova, I.V.; Kovalev, D.A.; Kutovoy, I.N. The modeling of mineral water fields data structure. In Proceedings of the 2021 IEEE Conference of Russian Young Researchers in Electrical and Electronic Engineering (ElConRus), Moscow, Russia, 26-29 January 2021; pp. 517-521.

[6] Makarova, A.A.; Kaliberda, I.V.; Kovalev, D.A.; Pershin, I.M. Modeling a production well flow control system using the example of the Verkhneberezovskaya area. In Proceedings of the 2022 Conference of Russian Young Researchers in Electrical and Electronic Engineering (ElConRus), Saint Petersburg, Russia, 25-28 January 2022; pp. 760-764.

[7] Wang, T.; Wang, L.; Meng, X.; Chen, Y.; Song, W.; Yuan, C. Key parameters and dominant EOR mechanism of CO_2 miscible flooding applied in low-permeability oil reservoirs. Geoenergy Sci. Eng. 2023, 225, 211724.

[8] Morgan, A.; Ampomah, W.; Grigg, R.; Dai, Z.; You, J.; Wang, S. Techno-economic life cycle assessment of CO_2-EOR operations towards net negative emissions at Farnsworth field unit. Fuel 2023, 342, 127897.

[9] Heidari, P.; Kordestany, A.; Sepahvand, A. An experimental evaluation of oil recovery by steam alternative CO_2 injection in naturally fractured reservoirs. Energy Sources 2013, 35, 1498-1507.

[10] Yuan, R.; Yang, Z.D.; Guo, B.; Wang, X.W.; Zhang, L.Q.; Lin, R.Y. Potential analysis of enhanced oil recovery by superheated steam during steam-assisted gravity drainage. Energy Technol. Gener. Conversion. Storage Distrib. 2021, 9, 2100135.

[11] Liu, W.; Du, L.; Zou, X.; Liu, T.; Wu, X.; Wang, Y.; Dong, J. Experimental study on the enhanced ultra-heavy oil recovery using an oil-soluble viscosity reducer and CO_2 assisted steam flooding. Geoenergy Sci. Eng. 2023, 222, 211409.

[12] ASTM-A312-2018; Standard specification for seamless, welded, and heavily cold-worked austenitic stainless steel pipes. ASTM: West Conshohocken, PA, USA, 2018.

[13] Yu, H.; Luo, Z.; Zhang, X.; Feng, Y.; Xie, G. A comparative study of the microstructure and corrosion resistance of Fe-based/B4C composite coatings with Ni-added or Cu-added by vacuum cladding. Mater. Lett. 2023, 335, 133730.

[14] Zhang, S.; Bian, T.; Mou, L.; Yan, X.; Zhang, J.; Zhang, Y.; Liu, B. Alloy design employing Ni and Mo low alloying for 3Cr steel with enhanced corrosion resistance in CO_2 environments. J. Mater. Res. Technol. 2023, 24, 1304-1321.

[15] API RP 571-2020; Damage Mechanisms Affecting Fixed Equipment in the Refining Industry. API: Minnesota City, MN, USA, 2020.

[16] API RP 581-2016; Risk-based inspection methodology. API: Minnesota City, MN, USA, 2016.

[17] Luo, B.W.; Zhou, J.; Bai, P.P.; Zheng, S.Q.; An, T.; Wen, X.L. Comparative study on the corrosion behavior of X52, 3Cr, and 13Cr steel in an $O_2-H_2O-CO_2$ system: Products, reaction kinetics, and pitting sensitivity. Int. J. Miner. Metall. Mater. 2017, 24, 646-656.

[18] Song, C.L.; Liu, X.B.; Pan, X. Failure analysis of the leakage and ignition of an oil-gas mixture transportation pipeline. J. Fail. Anal. Preven. 2022, 22, 259-266.

[19] Zhang, S.H.; Hou, L.F.; Wei, Y.H.; Du, H.Y.; Wei, H.; Liu, B.S.; Chen, X.B. Dual functions of chloride ions on corrosion behavior of mild steel in CO_2 saturated aqueous solutions. Mater. Corros. 2019, 70, 888-896.

[20] Wang, Z.; Liu, Z.-X.; Jin, J.; Tang, D.-Z.; Zhang, L. Selective corrosion mechanism of CoCrFeMoNi high-entropy alloy in the transpassive region based on the passive film characterization by ToF-SIMS. Corros. Sci. 2023, 218, 111206.

[21] Cui, Y.-W.; Chen, L.-Y.; Chu, Y.-H.; Zhang, L.; Li, R.; Lu, S.; Wang, L.; Zhang, L.-C. Metastable pitting corrosion behavior and characteristics of passive film of laser powder bed fusion produced Ti-6Al-4V in NaCl solutions with different concentrations. Corros. Sci. 2023, 215, 111017.

[22] Cong, S.; Tong, K.; Li, D.F.; Chen, Z.X.; Cai, K. Leakage failure analysis of the ERW steel pipeline. Mater. Sci. Forum 2020, 993, 1224–1229.

[23] Hua, Y.; Shamsa, A.; Barker, R.; Neville, A. Protectiveness, morphology and composition of corrosion products formed on carbon steel in the presence of Cl^-, Ca^{2+} and Mg^{2+} in high-pressure CO_2 environments. Appl. Surf. Sci. 2018, 455, 667–682.

[24] Zhang, B.; Ma, X.L. A review—Pitting corrosion initiation investigated by TEM. J. Mater. Sci. Technol. 2019, 35, 1455–1465.

[25] Congleton, J.; Shih, H.C.; Shoji, T.; Parkins, R.N. The stress corrosion cracking of type 316 stainless steel in oxygenated and chlorinated high-temperature water. Corros. Sci. 1985, 25, 769–788.

[26] Andresen, P.L.; Morra, M.M. IGSCC of non-sensitized stainless steels in high-temperature water. J. Nucl. Mater. 2008, 383, 97–111.

[27] Du, D.H.; Chen, K.; Lu, H.; Zhang, L.F.; Shi, X.Q.; Xu, X.L.; Andresen, P.L. Effects of chloride and oxygen on stress corrosion cracking of cold worked 316/316L austenitic stainless steel in high-temperature water. Corros. Sci. 2016, 110, 134–142.

[28] Andresen, P.L.; Young, L.M. Characterization of the Roles of Electrochemistry, Convection and Crack Chemistry in Stress Corrosion Cracking; No. CONF-950816-; NACE International: Houston, TX, USA, 1995; pp. 579–596.

[29] Congleton, J.; Shoji, T.; Parkins, R.N. The stress corrosion cracking of reactor pressure vessel steel in high-temperature water. Corros. Sci. 1985, 25, 633–650.

[30] Xu, L.; Wu, P.; Zhu, X.; Zhao, G.; Ren, X.; Wei, Q.; Xie, L. Structural characteristics and chloride intrusion mechanism of the passive film. Corros. Sci. 2022, 207, 110563.

[31] Ibrahim, M.A.M.; Rehim, S.S.A.E.; Hamza, M.M. Corrosion behavior of some austenitic stainless steels in chloride environments. Mater. Chem. Phys. 2009, 115, 80–85.

[32] Chu, T.; Shao, C.; Wang, Y.; Ma, N.; Lu, F. Crack branching behavior and amorphous film formation mechanism during SCC expanding test for multi-layers weld metal of NiCrMoV steels. Mater. Des. 2022, 216, 110520.

[33] Hou, Q.; Liu, Z.Y.; Li, C.T.; Li, X.G.; Shao, J.M. Degradation of the oxide film formed on Alloy 690TT in a high-temperature chloride solution. Appl. Surf. Sci. 2019, 467, 1104–1112.

[34] Song, Z.; Zhang, Y.; Liu, L.; Pu, Q.; Jiang, L.; Chu, H.; Luo, Y.; Liu, Q.; Cai, H. Use of XPS for quantitative evaluation of tensile-stress-induced degradation of passive film on carbon steel in simulated concrete pore solution. Constr. Build. Mater. 2021, 274, 121779.

[35] Wenman, M.R.; Trethewey, K.R.; Jarman, S.E.; Chard-Tuckey, P.R. A finite-element computational model of chloride-induced transgranular stress-corrosion cracking of austenitic stainless steel. Acta Mater. 2008, 56, 4125–4136.

[36] Beber, V.C.; Schneider, B. Fatigue of structural adhesives under stress concentrations: Notch effect on fatigue strength, crack initiation and damage evolution. Int. J. Fatigue 2020, 140, 105824.

[37] Zhang, G.A.; Cheng, Y.F. Micro-electrochemical characterization of corrosion of pre-cracked X70 pipeline steel in a concentrated carbonate/bicarbonate solution. Corros. Sci. 2010, 52, 960–968.

[38] Yazdanpanah, A.; Pezzato, L.; Dabalà, M. Stress corrosion cracking of AISI 304 under chromium variation within the standard limits: Failure analysis implementing microcapillary method. Eng. Fail. Anal. 2022, 142, 106797.

[39] Das, N.K.; Suzuki, K.; Ogawa, K.; Shoji, T. Early stage SCC initiation analysis of fcc Fe-Cr-Ni ternary alloy at 288℃: A quantum chemical molecular dynamics approach. Corros. Sci. 2009, 51, 908–913.

本论文原发表于《Processes》2023 年第 11 期。

Failure Risk Prediction Model for Girth Welds in High-strength Steel Pipeline Based on Historical Data and Artificial Neural Network

Wang Ke[1,2]　Zhang Min[1]　Guo Qiang[3]
Ma Weifeng[2]　Zhang Yixin[4]

(1. Xi'an University of Technology; 2. Tubular Goods Research Institute of CNPC;
3. Shaanxi City Gas Industry Development Co., Ltd.; 4. Northwest University)

Abstract: Pipelines are the most economical and sensible way to transport oil and gas. Long-distance oil and gas pipelines consist of many steel pipes or pipe fittings joined by welded girth welds, so girth welds are an essential part of the pipelines. Due to the limitations of welding conditions and the complexity of controlling weld quality in the field, some defects are inevitably present in the girth welds and adjacent weld areas. These defects can lead to pipeline safety problems; therefore, it is necessary to perform the failure risk assessment of pipeline girth welds. In this study, an artificial neural network model was proposed to predict the failure risk of pipeline girth welds with defects. Firstly, many pipeline girth weld failure cases, pipeline excavation and inspection data were collected and analysed to determine the main factors influencing girth weld failure. Second, a spatial orthogonal optimization method was used to select training samples for the artificial neural network model to ensure that the training sample set could cover the feature space with a minimum number of samples. Third, a prediction model based on BP neural networks was established to predict the failure risk levels. The prediction accuracy is more than 83%. This study can provide a valuable reference for pipeline operators to prevent pipeline accidents.

Keywords: Pipeline; Girth welds; Sample selection; Failure risk

1 Introduction

Oil and natural gas are the most common energy sources. Compared with railway and road transportation, the pipeline is still the safest and most efficient oil and gas transportation mode. However, with the increasing energy demand, more and more long-distance oil and gas pipelines

---

Corresponding author: Zhang Min, zhmmn@ xaut. edu. cn.

are used (Biezma et al., 2020; Chen et al., 2021; Dai et al., 2017; Guo et al., 2016). Long-distance pipelines consist of a large number of steel pipes or fittings connected by girth welds formed by welding.

Due to the limitation of welding technical conditions and the complex control of field construction quality, various welding defects such as cracks, air holes, slag inclusion, and incomplete fusion are inevitable in the girth weld and the adjacent weld areas (Sun et al., 2019; Yang et al., 2021; Zapata et al., 2011). Under the service conditions, the girth weld may become one of the relatively weak parts in the pipeline structure, which is easy to crack or even break, causing a large amount of leakage of oil and natural gas and sudden and catastrophic accidents such as fire and explosion (Cao et al., 2022). The Myanmar-China pipeline once had leaked and exploded due to the weak quality of girth welds, resulting in 24 people injured. It is urgent to investigate the safety risks of pipeline girth welds, and the existing investigation is mainly based on the results of internal inspection of magnetic flux leakage, negative reassessment and directional sampling. However, the length of China's pipelines exceeds 40000km and there are more than 3 million welds, so the workload is large and high-risk welds cannot be effectively found.

At present, many studies have focused on the failure prediction of pipelines, including failure risk (Xie and Tian, 2018), failure probability and reliability (Oliveira et al., 2016), failure consequences (Parvizsedghy and Zayed, 2016), failure types (Davis et al., 2006), failure rate (Liao et al., 2012), failure pressure (Parvizsedghy and Zayed, 2015; Su et al., 2021), and others. These include qualitative, quantitative, and semi-quantitative assessment methods. For instance, Markovki and Mannan (2009) developed risk assessment methods for oil and gas pipelines through fuzzy logic and fuzzy rule-based systems. Shahria and Sadiq (2012) developed a sustainability assessment method based on fuzzy tie analysis for oil and gas pipeline risk analysis.

With the improvement of computer computing power, researchers are increasingly interested in computer simulation and intelligent methods, and machine learning methods are widely used in safety risk field. Kumari (2022) developed a comprehensive risk prediction model. They selected the influencing factors and established an artificial neural network model to predict the causes and consequences of accidents, according to the importance of corrosion-induced pipeline accidents. Ren (2012) developed a BPNN prediction model using the mileage, height difference, inclination angle, pressure, and Reynolds number of the natural gas pipeline as input parameters and the maximum average corrosion rate of the pipeline as output parameters. The results show that the model has good fitting accuracy and prediction results. Li (2023) established an improved SVR model to predict subsea crude oil pipeline corrosion effectively. The model will serve as a valuable online tool to support the safety and digitalization of process systems. However, all these researches focus on the whole pipes rather than welds.

For girth welds, Chang (2023) used a numerical simulation method based on GTN model to analyze the crack initiation and dynamic fracture behavior of welded pipe under pure bending load. Wu et al. (2023) investigated the effects of crack size, pipe diameter-thickness ratio, and material parameters on the fracture assessment accuracy of pipe girth welds based on the failure assessment diagram theory and the equivalent stress-strain relationship method. He (2022) established a numerical simulation model of the stress induced magnetic signal of a girth weld with unequal wall thickness and used the model to calculate and analyze the quantitative variation law of the magnetic gradient signal of the girth weld. The current researches on girth welds focuses on the failure mechanism, detection and evaluation, and repair technology, with less research on failure risk prediction. Fortunately, a large amount of data on girth weld construction, operation, failure and testing were accumulated during the construction and operation of pipelines, which provides a strong support for the study of the failure risk prediction of pipeline girth welds (Feng et al., 2019; Feng et al., 2021).

This study uses a large amount of collected girth weld failure data to conduct risk prediction model research. The focus is on (1) the identification of failure factors of oil and gas pipeline girth welds; (2) the use of a spatially orthogonal optimal method to purposefully select samples with extreme imbalance in data distribution; (3) the use of artificial neural networks to construct a pipeline girth weld failure risk prediction model.

The remainder of this paper is organized as follows: Section 2 focuses on the collection and preprocessing of pipeline girth weld failure data. Section 3 proposes a sample data selection method based on a spatially orthogonal optimal method. Section 4 describes the construction of the failure risk prediction model based on neural network. In section 5, the influence factor analysis, model prediction performance and sensitivity analysis are carried out. Section 6 concludes the paper.

2 Data collection and preprocessing

2.1 Data sources

According to pipeline girth weld failure cases, and field inspection data, a lot of the girth failure sample data in this study have been collected. These samples are classified into three risk levels: high, medium and low, based on the service conditions of the welds. The samples that failed or required replacement are defined as high risk welds, those that required repair rather than replacement are defined as medium risk welds and others as low risk. It was found that most of the available data samples focused on X70 and X80 steel pipes, therefore the prediction model will be applied to these two pipe types. Table 1 shows the basic information on the samples.

Table 1 The number and risk level of the collected raw girth weld samples

Risk level	Data sources	Number	X70 steel number	X70 steel proportion	X80 steel number	X80steel proportion	X70 and X80 proportion
High risk	Failed welds	86	15	17%	29	34%	51%
Medium risk	Excavation and repair of welded joints	2905	313	11%	2039	79%	90%
Low risk	Excavation without repair of welded joints	24773	4365	18%	17140	69%	87%

2.2 Influencing factors and normalization

Many factors affect the failure risk of girth welds in oil and gas transmission pipelines, but the industry generally agrees that the three key factors are pipe materials and properties, weld defects and loads(Wu et al., 2023; Xu et al., 2022). Based on the samples collected, the factors that cause girth weld failures can be divided into three categories: pipe type and performance related indicators, defect related indicators, and load related indicators. Table 2 shows the details of the indicators and their normalization. There are 20 specific indicators of girth weld failure risk. These indicators need to be normalized to make them valid training samples for machine learning models. Some of these factors, such as welding process, repair and defect type, are Boolean or enumerated types and require special normalization methods, as shown in Table 2.

Table 2 The indicators affecting failure risks of girth welds of pipelines

Level I indicators	Level II indicators	Symbol	Parameter	Normalization method
Pipe material and performance	Steel grade	X1	X70; X80	X70: 0 X80: 1
	Diameter	X2	D	$(D-1016)/(1219-1016)$
	Wall thickness	X3/X4	t_1; t_2	$(t-18.4)/(50-18.4)$
	Yield strength	X5	σ	$(\sigma-485)/(700-485)$
	Toughness	X6	CVN	$(CVN-20)/(300-20)$
	Welding process	X7	Semi-automatic welding; Manual welding; Full-automatic welding	Semi-automatic welding: 1 Manual welding: 0.5 Full-automatic welding: 0
	Construction in winter or not	X8	Yes or No	Yes: 1 No: 0
	Repaired or not	X9	Yes or No	Yes: 1 No: 0
	Joint or not	X10	Yes or No	Yes: 1 No: 0
	Fixed joint or not	X11	Yes or No	Yes: 1 No: 0
Welding defect	Defect Type	X12	Volumetric; Planar; Cracked	Cracked: 1 Planar: 0.5 Volumetric: 0
	Defect position along the depth direction	X13	Outer surface; Interlayer; Root	Root: 1 Outer surface: 0.5 Interlayer: 0

continued

Level I indicators	Level II indicators	Symbol	Parameter	Normalization method
Welding defect	Defect position along the circumference	X14	S	2-4 and 8-10:1 10-2:0.5 4-8:1
	Defect length	X15	L	$L/\pi D$
	Defect height	X16	a	a/t_2
	Weld radiographic grade	X17	H	$H/4$
Loading	Pressure	X18	p	$p/12$
	Axial stress	X19	Fixed joint; Elbow/Bend; Normal	Fixed joint:1 Elbow/Bend:0.5 Normal:0
	Geological area	X20	Yes or No	Yes:1 No:0

2.3 Sample extraction and screening

According to the field data sources and the definitions of high, medium and low risk samples, the format of each collected data was standardized. A total of 44 high risk, 1823 medium risk and 2950 low risk samples were extracted and sorted as shown in Table 3. Table 4 shows the examples of girth weld samples.

Table 3 The samples extracted from the raw data

Risk level	Source	Welds from X70 steel pipelines	Welds from X80 steel pipelines
High risk	failure analysis	13	7
	Cutting treatment	2	22
Medium risk	B-type sleeve	179	236
	Epoxy sleeve	14	464
	Composite material	30	866
	Polishing treatment	7	27
Low risk	Level 1 weld junction	53	590
	Level 2 weld junction	155	1863
	Level 3 welding junction	29	251
	Level 4 welding junction	0	9
Total	—	482	4335

Table 4 The examples of the girth weld samples

X1	X2	X3	X4	X5	X6	X7	X8	X9	X10	X11	X12	X13	X14	X15	X16	X17	X18	X19	X20	Risk level
X80	1219	18.4	555	60	Semi-automatic welding	No	18.4	No	No	No	Cracked	Root	6.33	23.5	0.90	4	12	Elbow/Bend	No	High
X80	1219	15.3	555	60	Semi-automatic welding	No	18.4	Yes	Yes	No	Cracked	Root	5.80	35.0	2.70	4	12	Elbow/Bend	No	High
X80	1219	18.4	555	60	Semi-automatic welding	Yes	22.0	No	No	No	Cracked	Root	6.00	64.0	3.50	4	12	Elbow/Bend	No	High
X80	1219	18.4	555	60	Semi-automatic welding	No	18.4	No	No	No	Cracked	Root	2.81	20.0	1.00	4	12	Elbow/Bend	No	High
X80	1219	18.4	555	60	Semi-automatic welding	Yes	18.4	No	No	No	Cracked	Root	6.00	38.0	9.30	4	12	Normal	No	High
X80	1219	16.5	555	60	Full-automatic welding	No	16.5	No	No	No	Planar	Root	0.75	70.0	10.00	4	12	Normal	No	High
X70	1016	15.3	555	60	Semi-automatic welding	No	15.3	No	No	No	Cracked	Root	4.00	48.0	0.20	4	10	Normal	No	Medium
X70	1016	15.3	555	60	Semi-automatic welding	No	15.3	No	No	No	Planar	Root	11.94	39.0	2.67	4	10	Normal	No	Medium
X70	1016	18.4	555	60	Semi-automatic welding	No	18.4	No	No	No	Planar	Root	4.78	65.0	2.80	4	10	Normal	No	Medium
X70	1016	15.3	555	60	Semi-automatic welding	No	17.5	Yes	No	No	Planar	Outer surface	6.96	220.0	2.94	4	10	Elbow/Bend	No	Medium
X80	1219	18.4	555	60	Manual welding	No	18.4	No	Yes	No	Planar	Outer surface	5.58	40.0	1.50	4	12	Normal	No	Medium
X80	1219	18.4	555	60	Manual welding	No	18.4	No	No	No	Volumetric	Root	2.78	60.0	2.00	4	12	Elbow/Bend	No	Medium
X80	1219	22.0	555	60	Manual welding	No	22.0	No	No	No	Planar	Root	8.46	17.0	2.86	4	12	Elbow/Bend	No	Medium
X80	1219	18.4	555	60	Semi-automatic welding	No	18.4	No	No	No	Planar	Root	0.05	8.0	2.40	4	12	Normal	No	Medium
X80	1219	18.4	555	60	Manual welding	No	18.4	No	Yes	No	Volumetric	Interlayer	3.95	4.0	1.20	2	12	Elbow/Bend	No	Low
X80	1219	18.4	555	60	Manual welding	No	18.4	No	No	No	Volumetric	Interlayer	6.27	0.9	1.10	1	12	Elbow/Bend	No	Low
X80	1219	18.4	555	60	Manual welding	No	18.4	No	No	No	Volumetric	Interlayer	0.97	8.0	1.70	2	12	Normal	No	Low
X80	1219	18.4	555	60	Manual welding	No	18.4	No	No	No	Volumetric	Interlayer	0	0	0	1	12	Normal	No	Low
X70	1016	18.4	555	60	Semi-automatic welding	No	18.4	No	No	No	Volumetric	Interlayer	9.47	7.0	1.20	2	10	Normal	No	Low
X70	1016	15.3	555	60	Semi-automatic welding	No	18.4	No	No	No	Volumetric	Interlayer	7.03	4.0	1.20	2	10	Normal	No	Low
X70	1016	15.3	555	60	Semi-automatic welding	No	15.3	No	No	No	Volumetric	Interlayer	11.75	3.0	1.10	1	10	Normal	No	Low

3 Selection of training samples

From the screened samples it can be seen that there is a serious imbalance between the various types of risk welds. The number of high-risk welds is only 44, while the number of low-risk welds is almost 3000, with huge differences between the numbers of welds of different risk levels. Whether linear regression or traditional neural network training with randomly selected samples is used, the prediction results are heavily biased towards the higher number of risk types. To address this challenge, this study proposed a spatially orthogonal optimal method for the targeted selection of training samples.

Mathematically, the minimum number of training samples can successfully train a neural network because the samples contain most of the feature information and the overall linear correlation of the training samples is minimal. Ideally, the training samples should be orthogonal to each other, so that the training sample set can contain the maximum number of features with the minimum number of samples. Therefore, the mathematical model for training sample set selection can be expressed as Eq. (1).

$$\min C = \sum_{i \in J; j \in J} V_i \cdot V_j \tag{1}$$

Where J is the set of sample vector numbers in the training set; V_i is the i-th sample vector. The solution objective of Eq. (1) is the J. Due to the large number of samples, it is difficult to determine the training set with the minimum overall linear correlation by conventional methods. Therefore, this study proposes a training set selection algorithm based on a heuristic method to achieve the overall correlation minimum. First, the sample vector correlation matrix C [Eq. (2)] can be constructed based on the sample vectors.

$$C = \begin{bmatrix} V_1 V_1 & V_1 V_2 & \cdots & V_1 V_n \\ V_2 V_1 & V_2 V_2 & \cdots & V_2 V_n \\ \cdots & \cdots & \cdots & \cdots \\ V_n V_1 & V_n V_2 & \cdots & V_n V_n \end{bmatrix} \tag{2}$$

Let $d_{ij} = V_i \cdot V_j$, then the algorithm for selecting the minimum correlation sample set can be described as follows:

(1) Construct a set $ASet = \{d_{ij} | 1 \leq i; i < j \leq N\}$ and let $J = \{\}$. Determine the number of training samples K;

(2) Search the smallest element $d_{i_{min} j_{min}}$ in $ASet$, and add its subscript (i_{min}, j_{min}) to J, then delete this element from $ASet$;

(3) Determine whether the number of elements in J (K_N) is greater than or equal to K, and if $K_N \geq K$, proceed to step (4), otherwise, proceed to step (2);

(4) Determine the corresponding V_{i_m} as the training sample based on the subscripts in the J.

4 Establishment of failure risk prediction model

4.1 Prediction model selection

The results of the girth weld failure risk prediction are divided into three categories: "high risk", "medium risk" and "low risk". This means that risk prediction is actually solving Eq. (3).

$$r = g(c_0, c_1, \cdots, c_{m-1}) \qquad (3)$$

Where $c_0, c_1, \cdots, c_{m-1}$ represent the corresponding indicators in Table 2. g is a mapping relationship. r is the predicted result.

Due to the complexity of the factors influencing the failure risk of the girth welds, it is extremely challenging to build a rigorous mathematical model to predict the weld failure risk. Artificial neural networks (ANNs) (El-Abbasy et al., 2014; Xu et al., 2017) have powerful nonlinear mapping capability to analyze any specific task such as classification, prediction and control, so it is a feasible approach to construct an ANN-based girth weld failure prediction model using the collected actual data. In this study, a fully-connected BP neural network was selected as the network architecture for girth weld failure risk prediction. The general framework of the research methodology is shown in Fig.1.

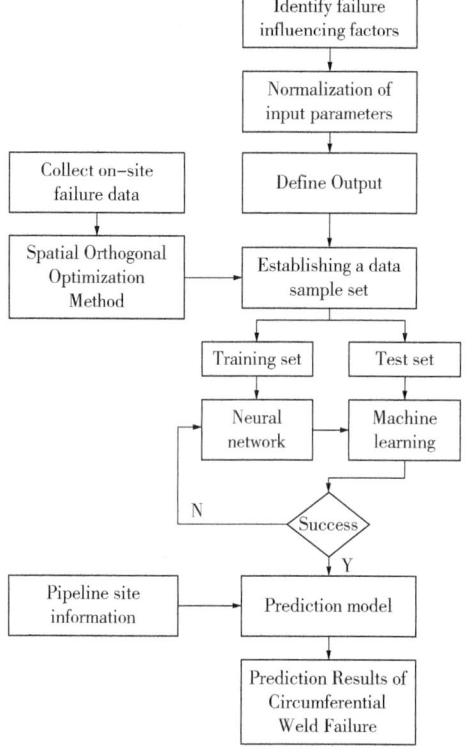

Fig.1 The framework of the research methodology

4.2 Prediction model establishment

BP (back propagation) network (Shaik et al., 2020) is a multilayer feed-forward neural network whose neurons are passed as Sigmoid function, which can achieve any nonlinear mapping from input to output. It is called a BP network because the weights are adjusted using a back propagation learning algorithm. In the practical application of ANNs, the majority of neural network models use BP networks and their variant forms (Kumari et al., 2022). A typical BP neural network consists of an input layer, at least one hidden layer, and an output layer. Each layer of the network is composed of neurons, and the structure of neurons is shown in Fig.2. In this figure, x_i represents the neuron input and b represents bias. $u = \sum x_i w_i + b$ and $y = f(u)$. The f function generally takes the Sigmoid function.

For the prediction of pipeline girth weld failure, the parameters involved are shown in Table 2. The number of neurons in the input layer is the same as the number of dimensions of the input parameters, which is 20. The number of nodes in the failure output layer is 3, which are "high risk", "medium risk" and "low risk" respectively.

The number of hidden layers of neural networks and the number of nodes in the hidden layers have a great impact on the prediction performance of networks(Shaik et al., 2021). If the number is too small, the neural network cannot obtain enough information to solve the prediction problem. If the number is too large, it will not only increase the learning time, but also may have the problem of "overfitting". In practice, the BP neural network is mostly 3 layers with only one hidden layer. That is, the structure of the neural network for girth weld failure risk prediction can be represented as Fig.3.

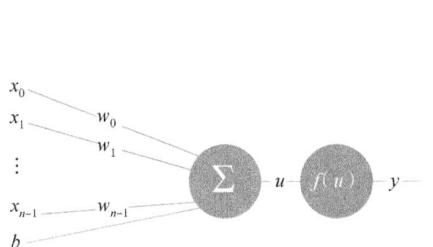

Fig.2 Schematic diagram of the structure of a neuron

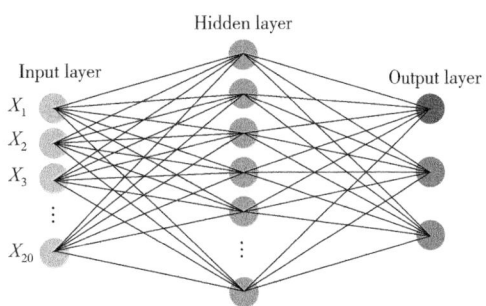

Fig.3 Schematic diagram of the neural network for girth weld failure risk prediction

By selecting the appropriate number of neurons in the hidden layer, the same mapping effect as multiple hidden layers can be achieved. This study adopted three-layer NN. Before determining the number of hidden layer neurons, the following speculation is given first:

Speculation 1: Since the number of inputs is 20, which is relatively small. Therefore, the actual network structure to achieve accurate defect prediction is more complex than the theoretical optimal network structure.

Speculation 2: For the three-classification pipeline failure risk prediction, the number of hidden layer neurons is in the same order of magnitude as $m \times n$, where m is the input quantity dimension and n is the output quantity dimension.

Also, in order to make the neural network practical, the number of nodes in the hidden layer should be less than $N-1$ (N is the number of training samples), otherwise the systematic error of the neural network will tend to 0, independent of the characteristics of the training samples, i.e., the neural network lacks generalization ability. In practice, the number of training samples must be more than the number of connection weights of the network 2 to 10 times. When the number of samples is not sufficient, the "rotational training" method is generally used to obtain a reliable neural network. Therefore, the number of neurons in the hidden layer is not only related to the prediction task, but also related to the number of training samples. In summary, the number of neurons in the hidden layer is recommended to satisfy the following equation.

$$\begin{cases} O(n_1) = O(m \cdot n) \\ n_1 \leq N-1 \end{cases} \quad (4)$$

Where n_1 is the number of neurons in the hidden layer. For the prediction task of this study, $m = 20$ and $n = 3$, According to speculations 1 and 2, it can be inferred that the minimum number of hidden layer neurons is 10 and the maximum is 100, i.e., $n_1 \in [10, 100]$. This interval range is still relatively large. From Eq. (4), it can be seen that the search range of n_1 can be further narrowed by determining the number of training samples.

From the perspective of statistics, it can be considered that the probability of selecting the appropriate number of training samples K obeys normal distribution,

$$f(K) = \frac{e^{-\frac{(K-\mu)^2}{2\sigma^2}}}{\sigma\sqrt{2\pi}} \tag{5}$$

This distribution is usually also denoted as $K \sim N(\mu, \sigma^2)$. From Eq. (3), we can derive $\mu \approx 50$. According to the "3σ criteria" in statistics, where almost all values of K are concentrated within the $[\mu-3\sigma, \mu+3\sigma]$, $\sigma \approx 10$ can be inferred, i.e., $K \sim N(50, 10^2)$.

Taking a confidence level of 0.9, i.e., the sample size satisfies the training requirement in 90% of cases, one has:

$$P(K < K_{max}) \geq 0.9 \tag{6}$$

Where P is the probability function, and K_{max} is the maximum number of samples sufficient to train the neural network.

Due to $K \sim N(50, 10^2)$, it follows that

$$\frac{K_{max}-50}{10} \sim N(0,1) \tag{7}$$

Knowing that $\varphi(1.29) = 0.9015$, then $\frac{K_{max}-50}{10} \approx 1.29$, i.e., $K_{max} \approx 63$. Since K obeys a normal distribution, we can get $K_{min} \approx \mu - (K_{max}-\mu) \approx 37$. According to Eq. (4), it can be inferred that the number of neurons in the hidden layer is $n_1 \in [10, 35]$. Considering that the output of the network is three categories, the n_1 can be roughly determined to be 30 by Speculation 1.

4.3 Training algorithm

The training process of BP neural network is mainly divided into two stages, the first stage is the forward propagation of the signal, which passes from the input layer to the hidden layer and finally reaches the output layer; the second stage is the backward propagation of the error, which goes from the output layer to the hidden layer and finally to the input layer, adjusting the weights and biases from the hidden layer to the output layer and the weights and biases from the input layer to the hidden layer in turn.

The overall output error of the neural network for failure risk prediction is

$$E = \frac{1}{2}\sum_{j_L=1}^{N_L}(T_{j_L} - O_{j_L}^L)^2 \tag{8}$$

Where L is the number of neural network layers; N_L is the number of neurons in the layer L; T_{j_L} is the standard output of the layer L; $O_{j_L}^L$ is the actual output of the layer L. For the neural network of girth weld failure risk prediction, $L=3$ and $N_L=3$.

Let $w_{j_L-\dot{v}_L}^L$ represent the weight of the connection between the j-th neuron of the layer $L-1$ and the j-th neuron of the layer L, then the partial derivative of the error to $w_{j_L-j_L}^L$ can be expressed as

$$\frac{\partial E}{\partial w_{j_L-j_L}^L} = \frac{\partial E}{\partial O_{j_L}^L} \cdot \frac{\partial O_{j_L}^L}{\partial z_{j_L}^L} \cdot \frac{\partial z_{j_L}^L}{\partial w_{j_L-j_L}^L} \tag{9}$$

where the output $O_{j_L}^L$ and the input $z_{j_L}^L$ are sigmoid functions [see Eq. (10)].

$$O_{j_L}^L = \frac{1}{1+e^{-z_{j_L}^L}} = s(z_{j_L}^L) \tag{10}$$

According to Eq. (10), it can be inferred that

$$\frac{\partial O_{j_L}^L}{\partial z_{j_L}^L} = s(z_{j_L}^L) \cdot [1-s(z_{j_L}^L)] \tag{11}$$

Then

$$\frac{\partial E}{\partial O_{j_L}^L} = -(T_{j_L} - O_{j_L}^L) \tag{12}$$

Due to the fact that the total input of neurons can be expressed as

$$z_{j_L}^L = \sum_{j_{L-1}=1}^{N_{L-1}} w_{j_L-j_L}^L \cdot O_{j_{L-1}}^{L-1} + b_{j_L}^L \tag{13}$$

Therefore, it can be introduced that

$$\frac{\partial z_j^L}{\partial w_{j_L-j_L}^L} = O_{j_{L-1}}^{L-1} \tag{14}$$

Take Eq. (11), Eq. (12) and Eq. (14) into Eq. (9) to get the partial derivative of output layer deviation and weight

$$\frac{\partial E}{\partial w_{j_L-j_L}^L} = -(T_{j_L} - O_{j_L}^L) \cdot s(z_{j_L}^L) \cdot [1-s(z_{j_L}^L)] \cdot O_{j_{L-1}}^{L-1} \tag{15}$$

The partial derivative of the bias can be easily obtained

$$\frac{\partial E}{\partial b_{j_L}^L} = -(T_{j_L} - O_{j_L}^L) \cdot s(z_{j_L}^L) \cdot [1-s(z_{j_L}^L)] \tag{16}$$

Let $\delta_{j_L}^L = -(T_{j_L} - O_{j_L}^L) \cdot s(z_{j_L}^L) \cdot [1-s(z_{j_L}^L)]$, then

$$\frac{\partial E}{\partial w_{j_L-j_L}^L} = \delta_{j_L}^L \cdot O_{j_{L-1}}^{L-1} \tag{17}$$

$$\frac{\partial E}{\partial b_{j_L}^L} = \delta_{j_L}^L \tag{18}$$

For the hidden layer $L-1$, the deviation of the error on the output of neurons in the $L-1$ layer can be expressed as

$$\frac{\partial E}{\partial O_{j_{L-1}}^{L-1}} = \sum_{j_L}^{N_L} \frac{\partial E_{j_L}}{\partial z_{j_L}^L} \cdot \frac{\partial z_{j_L}^L}{\partial O_{j_{L-1}}^{L-1}} = \sum_{j_L}^{N_L} \left\{ - (T_{j_L} - O_{j_L}^L) \cdot s(z_{j_L}^L) \cdot [1 - s(z_{j_L}^L)] \cdot \frac{\partial z_{j_L}^L}{\partial O_{j_{L-1}}^{L-1}} \right\} \quad (19)$$

From $z_{j_L}^L = \sum_{j_{L-1}}^{N_{L-1}} O_{j_{L-1}}^{L-1} w_{j_{L-1}j_L}^L + b_{j_L}^L$, it can be concluded that

$$\frac{\partial z_{j_L}^L}{\partial O_{j_{L-1}}^{L-1}} = w_{j_{L-1}j_L}^L \quad (20)$$

Taking equation (20) into equation (19) yields

$$\frac{\partial E}{\partial O_{j_{L-1}}^{L-1}} = \sum_{j_L}^{N_L} \frac{\partial E_{j_L}}{\partial z_{j_L}^L} \cdot \frac{\partial z_{j_L}^L}{\partial O_{j_{L-1}}^{L-1}} = \sum_{j_L}^{N_L} \{ - (T_{j_L} - O_{j_L}^L) \cdot s(z_{j_L}^L) \cdot [1 - s(z_{j_L}^L)] \cdot w_{j_{L-1}j_L}^L \}$$

$$= \sum_{j_L}^{N_L} (\delta_{j_L}^L \cdot w_{j_{L-1}j_L}^L) \quad (21)$$

Then

$$\frac{\partial E}{\partial w_{j_{L-2}j_{L-1}}^{L-1}} = \sum_{j_L}^{N_L} (\delta_{j_L}^L \cdot w_{j_{L-1}j_L}^L) \cdot \frac{\partial O_{j_{L-1}}^{L-1}}{\partial z_{j_{L-1}}^{L-1}} \cdot \frac{\partial z_{j_{L-1}}^{L-1}}{\partial w_{j_{L-2}j_{L-1}}^{L-1}} =$$

$$\sum_{j_L}^{N_L} (\delta_{j_L}^L \cdot w_{j_{L-1}j_L}^L) \cdot s(z_{j_{L-1}}^{L-1}) \cdot [1 - s(z_{j_{L-1}}^{L-1})] \cdot O_{j_{L-2}}^{L-2} \quad (22)$$

$$\frac{\partial E}{\partial b_{j_{L-1}}^{L-1}} = \sum_{j_L}^{N_L} (\delta_{j_L}^L \cdot w_{j_{L-1}j_L}^L) \cdot s(z_{j_{L-1}}^{L-1}) \cdot [1 - s(z_{j_{L-1}}^{L-1})] \quad (23)$$

Similarly, let $\delta_{j_{L-1}}^{L-1} = \sum_{j_L}^{N_L} (\delta_{j_L}^L \cdot w_{j_{L-1}j_L}^L) \cdot s(z_{j_{L-1}}^{L-1}) \cdot [1 - s(z_{j_{L-1}}^{L-1})]$, and it follows that

$$\frac{\partial E}{\partial w_{j_{L-2}j_{L-1}}^{L-1}} = \delta_{j_{L-1}}^{L-1} \cdot O_{j_{L-2}}^{L-2} \quad (24)$$

$$\frac{\partial E}{\partial b_{j_{L-1}}^{L-1}} = \delta_{j_{L-1}}^{L-1} \quad (25)$$

Due to the number of layers in the failure prediction neural network being 3, the back propagation of errors will reach the $L-2$ layer. The deviation of output error on the output of $L-2$ layer can be expressed as

$$\frac{\partial E}{\partial O_{j_{L-2}}^{L-2}} = \sum_{j_{L-1}}^{N_{L-1}} \frac{\partial E}{\partial O_{j_{L-1}}^{L-1}} \cdot \frac{\partial O_{j_{L-1}}^{L-1}}{\partial O_{j_{L-2}}^{L-2}} \quad (26)$$

Take $\dfrac{\partial E}{\partial O_{j_{L-1}}^{L-1}} = \sum\limits_{j_L}^{N_L}(\delta_{j_L}^L \cdot w_{j_L\text{-}j_L}^L)$ into Eq. (26), then

$$\dfrac{\partial E}{\partial O_{j_{L-2}}^{L-2}} = \sum_{j_{L-1}}^{N_{L-1}} \sum_{j_L}^{N_L}(\delta_{j_L}^L \cdot w_{j_L\text{-}j_L}^L) \cdot \dfrac{\partial O_{j_{L-1}}^{L-1}}{\partial O_{j_{L-2}}^{L-2}} \tag{27}$$

Considering $\dfrac{\partial O_{j_{L-1}}^{L-1}}{\partial O_{j_{L-2}}^{L-2}} = \dfrac{\partial O_{j_{L-1}}^{L-1}}{\partial z_{j_{L-1}}^{L-1}} \cdot \dfrac{\partial z_{j_{L-1}}^{L-1}}{\partial O_{j_{L-2}}^{L-2}} = s(z_{j_{L-1}}^{L-1}) \cdot [1 - s(z_{j_{L-1}}^{L-1})] \cdot \dfrac{\partial z_{j_{L-1}}^{L-1}}{\partial O_{j_{L-2}}^{L-2}}$, from $z_{j_{L-1}}^{L-1} = \sum\limits_{j_{L-2}}^{N_{L-2}} O_{j_{L-2}}^{L-2} w_{j_{L-2}\text{-}j_{L-1}}^{L-1} + b_{j_{L-1}}^{L-1}$, it can be concluded that

$$\dfrac{\partial z_{j_{L-1}}^{L-1}}{\partial O_{j_{L-2}}^{L-2}} = w_{j_{L-2}\text{-}j_{L-1}}^{L-1} \tag{28}$$

Then

$$\dfrac{\partial O_{j_{L-1}}^{L-1}}{\partial O_{j_{L-2}}^{L-2}} = s(z_{j_{L-1}}^{L-1}) \cdot [1 - s(z_{j_{L-1}}^{L-1})] \cdot w_{j_{L-2}\text{-}j_{L-1}}^{L-1} \tag{29}$$

Introducing Eq. (29) into Eq. (26) includes

$$\dfrac{\partial E}{\partial O_{j_{L-2}}^{L-2}} = \sum_{j_{L-1}}^{N_{L-1}} \dfrac{\partial E}{\partial O_{j_{L-1}}^{L-1}} \cdot \dfrac{\partial O_{j_{L-1}}^{L-1}}{\partial O_{j_{L-2}}^{L-2}} =$$

$$\sum_{j_{L-1}}^{N_{L-1}} \left\{ \sum_{j_L}^{N_L}(\delta_{j_L}^L \cdot w_{j_L\text{-}j_L}^L) \cdot s(z_{j_{L-1}}^{L-1}) \cdot [1 - s(z_{j_{L-1}}^{L-1})] \right\} \cdot w_{j_{L-2}\text{-}j_{L-1}}^{L-1} \tag{30}$$

By incorporating $\delta_{j_{L-1}}^{L-1} = \sum\limits_{j_L}^{N_L}(\delta_{j_L}^L \cdot w_{j_L\text{-}j_L}^L) \cdot s(z_{j_{L-1}}^{L-1}) \cdot [1 - s(z_{j_{L-1}}^{L-1})]$ into Eq. (30), the backward error propagation of the neural network can be obtained

$$\dfrac{\partial E}{\partial O_{j_{L-2}}^{L-2}} = \sum_{j_{L-1}}^{N_{L-1}} \delta_{j_{L-1}}^{L-1} \cdot w_{j_{L-2}\text{-}j_{L-1}}^{L-1} \tag{31}$$

Based on Eq. (30), there are

$$\dfrac{\partial E}{\partial w_{j_{L-3}\text{-}j_{L-2}}^{L-2}} = \dfrac{\partial E}{\partial O_{j_{L-2}}^{L-2}} \cdot \dfrac{\partial O_{j_{L-2}}^{L-2}}{\partial z_{j_{L-2}}^{L-2}} \cdot \dfrac{\partial z_{j_{L-2}}^{L-2}}{\partial w_{j_{L-3}\text{-}j_{L-2}}^{L-2}}$$

$$= \sum_{j_{L-1}}^{N_{L-1}}(\delta_{j_{L-1}}^{L-1} \cdot w_{j_{L-2}\text{-}j_{L-1}}^{L-1}) \cdot s(z_{j_{L-2}}^{L-2}) \cdot [1 - s(z_{j_{L-2}}^{L-2})] \cdot O_{j_{L-3}}^{L-3} \tag{32}$$

Similar to $\delta_{j_{L-1}}^{L-1} = \sum\limits_{j_L}^{N_L}(\delta_{j_L}^L \cdot w_{j_L\text{-}j_L}^L) \cdot s(z_{j_{L-1}}^{L-1}) \cdot [1 - s(z_{j_{L-1}}^{L-1})]$, let $\delta_{j_{L-2}}^{L-2} = \sum\limits_{j_{L-1}}^{N_{L-1}}(\delta_{j_{L-1}}^{L-1} \cdot w_{j_{L-2}\text{-}j_{L-1}}^{L-1}) \cdot s(z_{j_{L-2}}^{L-2}) \cdot [1 - s(z_{j_{L-2}}^{L-2})]$, then

$$\dfrac{\partial E}{\partial w_{j_{L-i}\text{-}j_{L-i}}^{L-i}} = \delta_{j_{L-i}}^{L-i} \cdot O_{j_{L-i-1}}^{L-i-1} \tag{33}$$

$$\frac{\partial E}{\partial b_{j_{L-2}}^{L-2}} = \delta_{j_{L-2}}^{L-2} \tag{34}$$

Eq. (33) and Eq. (34) are the error derivation of the backward iterative calculation of the neural network.

4.4 Sensitivity analysis

Sensitivity analysis is a method for studying and analyzing the sensitivity of changes in the state or output of a system (or model) to changes in system parameters or surrounding conditions. Sensitivity analysis is often used in optimization methods to study the stability of the optimal solution when the original data is inaccurate or changes occur. Sensitivity analysis can also be used to determine which parameters have a greater effect on the system or model.

The conventional methods of sensitivity analysis are used by varying a particular input on a fixed basis of all inputs and observing the change in output at that point. However, this method does not work for practical nonlinear mapping systems. In the case of girth weld failure risk prediction, for example, once the material of the pipe is changed, parameters such as pipe diameter, toughness and design pressure will also change. Obviously, the sensitivity analysis method that fixes other input values cannot obtain an accurate solution. In this study, the analytical solution of the degree of influence of each input on the risk prediction results is derived from the perspective of mathematical analysis.

Since BP neural network was used in the girth weld failure risk prediction, the sensitivity of the output of the second layer $o_{i_2}^2$ to the input of the input layer $z_{i_1}^1$ can be expressed as

$$\frac{\partial o_{i_2}^2}{\partial o_{i_1}^1} \cdot \frac{\partial o_{i_1}^1}{\partial z_{i_1}^1} = s(z_{i_2}^2)[1-s(z_{i_2}^2)]w_{i_1 i_2}^1 = \zeta_{i_2 i_1}^2 \tag{35}$$

The sensitivity of the output of the third layer $o_{i_3}^3$ to the input layer $z_{i_1}^1$ can be expressed as

$$\frac{\partial o_{i_3}^3}{\partial z_{i_1}^1} = \sum_{i_2=1}^{N_2} \frac{\partial o_{i_3}^3}{\partial o_{i_2}^2} \cdot \frac{\partial o_{i_2}^2}{\partial z_{i_1}^1} = \sum_{i_2=1}^{N_2} \frac{\partial o_{i_3}^3}{\partial o_{i_2}^2} \cdot \zeta_{i_2 i_1}^2 =$$
$$\sum_{i_2=1}^{N_2} s(z_{i_3}^3)[1-s(z_{i_3}^3)] \cdot w_{i_2 i_3}^2 \cdot \zeta_{i_2 i_1}^2 = \zeta_{i_3 i_1}^3 \tag{36}$$

The sensitivity of the output of the fourth layer $o_{i_4}^4$ to the input layer $z_{i_1}^1$ can be expressed as:

$$\frac{\partial o_{i_4}^4}{\partial z_{i_1}^1} = \sum_{i_3=1}^{N_3} \frac{\partial o_{i_4}^4}{\partial o_{i_3}^3} \cdot \frac{\partial o_{i_3}^3}{\partial z_{i_1}^1} = \sum_{i_3=1}^{N_3} \frac{\partial o_{i_4}^4}{\partial o_{i_3}^3} \cdot \zeta_{i_3 i_1}^3 =$$
$$\sum_{i_3=1}^{N_3} s(z_{i_4}^4)[1-s(z_{i_4}^4)] \cdot w_{i_3 i_4}^3 \cdot \zeta_{i_3 i_1}^3 = \zeta_{i_4 i_1}^4 \tag{37}$$

By analogy, it can be inferred that the sensitivity of the output of the layer L to input $z_{i_1}^1$ is:

$$\frac{\partial o_{i_L}^L}{\partial z_{i_1}^1} = \sum_{i_{L-1}=1}^{N_{L-1}} \frac{\partial o_{i_L}^L}{\partial o_{i_{L-1}}^{L-1}} \cdot \frac{\partial o_{i_{L-1}}^{L-1}}{\partial z_{i_1}^1} = \sum_{i_{L-1}=1}^{N_{L-1}} \frac{\partial o_{i_L}^L}{\partial o_{i_{L-1}}^{L-1}} \cdot \zeta_{i_{L-1}i_1}^{L-1} = $$
$$\sum_{i_{L-1}=1}^{N_{L-1}} s(z_{i_L}^L)[1 - s(z_{i_L}^L)] \cdot w_{i_{L-1}i_L}^{L-1} \cdot \zeta_{i_{L-1}i_1}^{L-1} = \zeta_{i_Li_1}^L \quad (38)$$

The sensitivity of the input parameters on the output results can be calculated by the above equation.

5 Results and discussion

5.1 Influencing factor analysis

Fig.4 shows the effects of some main factors on the failure of the pipeline girth welds. From Fig. 4(a), it can be found that the percentage of girth weld failures for pipe diameters above 900mm is 36%, 26% for pipe diameters from 600mm to 900mm and 300mm to 600mm respectively, and 12% for pipe diameters below 300mm. Fig.4(b) gives the relationship between failed girth welds and wall thickness, and it can be seen that in all failure cases, the percentage of pipes with wall thicknesses of 10mm and above is 65%, and the percentage of those with wall thicknesses below 10mm is 35%. 83% of the girth weld defects are located in the root of the welds [Fig.4(c)]. Fig.4(d) gives the circumferential distribution of defects, 48% of the defects are at the top of the pipe (10 o'clock-2 o'clock) and 48% at the bottom of the pipe (4 o'clock-8 o'clock).

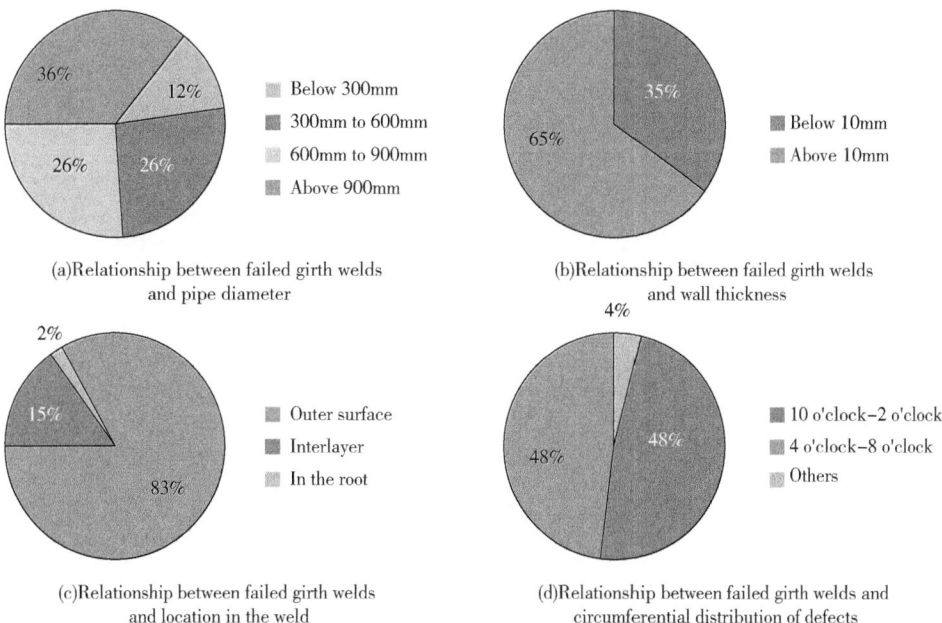

(a) Relationship between failed girth welds and pipe diameter

(b) Relationship between failed girth welds and wall thickness

(c) Relationship between failed girth welds and location in the weld

(d) Relationship between failed girth welds and circumferential distribution of defects

Fig.4 The effects of some main factors on the failure of the pipeline girth welds

Fig.5 shows the correlation coefficients of influencing factors to the failure risk of the girth weld. It can be seen that X1, X2, X3, X4, X5 and X15 show a negative correlation with the failure risk and others show a positive correlation. X18 (pressure), X13 (defect position along the depth direction), X14 (defect position along the circumference) and X1 6 (defect height) have the relative correlation coefficients with the weld failure risk.

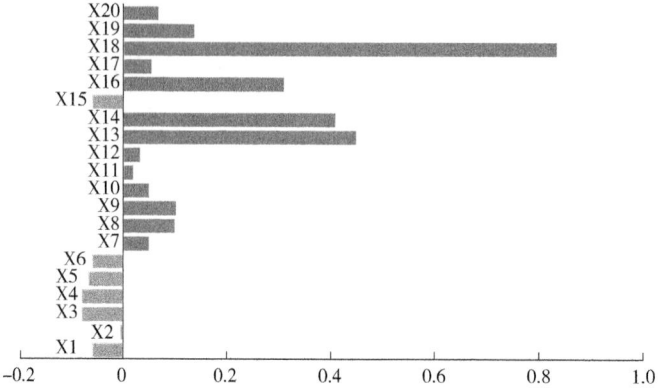

Fig.5 Correlation coefficients of influencing factors to the failure risk of the girth weld

Fig.6 shows the correlation between the influencing factors in different sample categories. The variability of the correlation coefficient matrices is not significant. X1 and X5, X1 and X6, X5 and X6, X12 and X13, X2 and X18, and X16 and X17 show strong correlations in the total samples. Differently, there are a strong correlation between X1, X2, X3, X5, X6 and X7 (all of these belong to pipe material and performance) in high-risk samples.

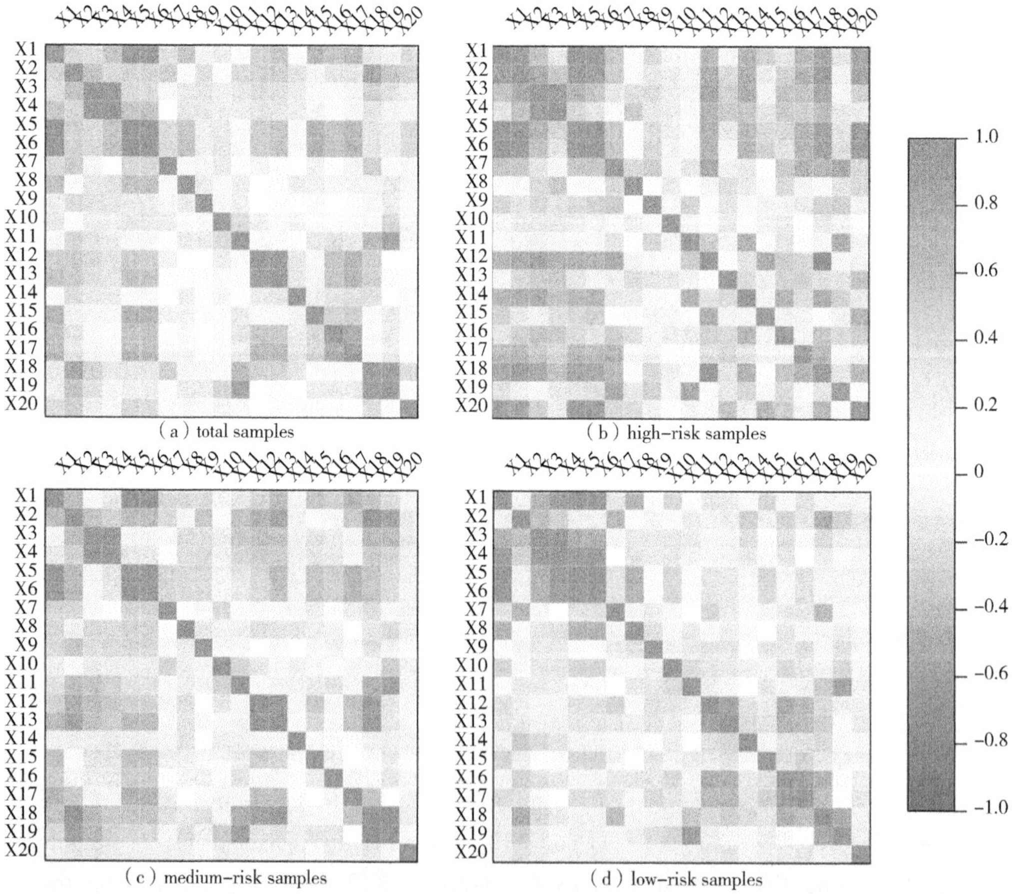

(a) total samples

(b) high-risk samples

(c) medium-risk samples

(d) low-risk samples

Fig.6 Correlation between the influencing factors in different sample categories

5.2 Model prediction performance

The BP neural network used in this study contains 3 layers: an input layer of 20 neurons, a hidden layer of 30 neurons and an output layer of 3 neurons. In the training process, the initial number of training samples K was set to 10 and the learning rate was set to 0.001. Table 5 and Table 6 shows the evaluation of the prediction results of the model. It can be seen that in the presence of extreme imbalance in the three risk levels of samples, a relatively accurate failure risk prediction can be achieved using the training sample selection method and neural network architecture established in this study. The prediction accuracy for all categories of girth welds is above 83%. In the case of high-risk welds, the accuracy reached 73.7% with only 25 training samples.

Table 5 Prediction results (including training samples) of the neural network for the failure risk prediction of the girth weld

Risk level	Correct identification	Incorrect identification	Success rate	Number of training samples
High risk	39	5	88.6%	25
Medium risk	1563	260	85.7%	247
Low risk	2533	417	85.9%	330
Overall accuracy			85.8%	602

Table 6 Prediction results (excluding training samples) of the neural network for the failure risk prediction of the girth weld

Risk level	Correct identification	Incorrect identification	Success rate	Number of training samples
High risk	14	5	73.7%	25
Medium risk	1316	260	83.5%	247
Low risk	2203	417	84.1%	330
Overall accuracy			83.8%	602

5.3 Parameter sensitivity

Fig.7 shows the sensitivity of the input parameters of the prediction model to the output. The sensitivity of different parameters varies considerably. X15 has the greatest sensitivity, followed by X2, X5 and X4, which means defect length, pipe diameter, yield strength, wall thickness and defect height have the greatest effect on the change in pipeline failure risk. In contrast, weld radiographic grade, geological area, repaired or not, steel grade and toughness have smaller effect on the change in risk. Therefore, the elimination of weld defects during the welding process and the inspection and monitoring of defects during the service life of the pipeline can play an important role in reducing the risks of failure of the girth welds.

6 Conclusions

This study focused on the failure risk of circumferential welds in oil and gas pipelines. Based on the collection of pipeline failure cases and field inspection data, a total of 20 main factors affecting the failure risk of the girth welds of pipelines and their normalization method were identified. The

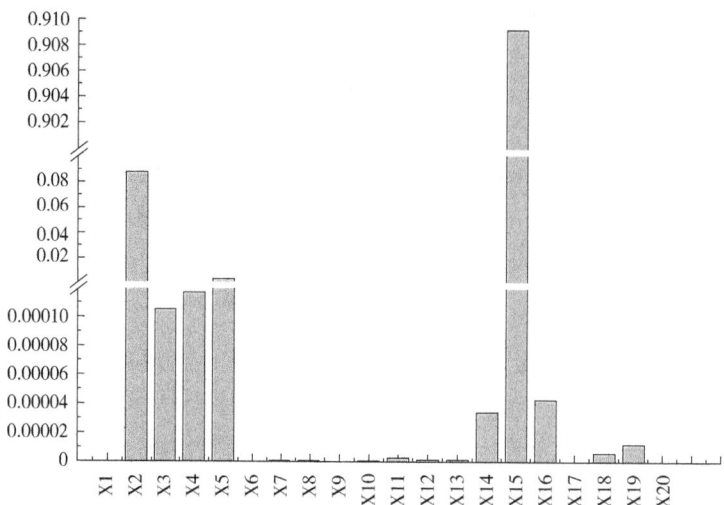

Fig.7 Sensitivity analysis of the input parameters on the output

samples were selected by the spatially orthogonal optimal method to effectively avoid the overfitting problem caused by the imbalance of classification samples. Using the BP neural network, a prediction model for the failure risk of girth welds in high-strength steel pipelines was established, and the prediction accuracy reached more than 83%, which improves the technical support for high-risk girth weld repair and maintenance decisions and avoids the failure of pipeline girth welds.

Acknowledgements

This work was supported by the Key R&D Plan of Shaanxi Province, China (Grant No. 2022ZDLSF07-08) and the Youth Science and Technology New Star Project of Shaanxi Province (Grant No. 2021KJXX65).

References

[1] Biezma, M. V., Andrés, M. A., Agudo, D., & Briz, E. (2020). Most fatal oil & gas pipeline accidents through history: A lessons learned approach. Engineering Failure Analysis, 110, 104446. https://doi.org/10.1016/j.engfailanal.2020.104446.

[2] Cao, J., Wang, K., Ma, W., Ren, J., Nie, H., Dang, W., Liang, X., Yao, T., & Zhao, X. (2022). Indentation creep deformation behavior of local zones for X70 girth weld.International Journal of Pressure Vessels and Piping, 199, 104776.

[3] Chang, Q., Cao, Y., Zhen, Y., Wu, G., & Li, F. (2023). Study on the effect of loading conditions on the fracture behavior of pipeline with girth weld.International Journal of Pressure Vessels and Piping, 203, 104940.

[4] Chen, P., Li, R., Fu, K., & Zhao, X. (2021). Research and Method for In-line Inspection Technology of Girth Weld in Long-Distance Oil and Gas Pipeline.Journal of Physics: Conference Series, 1986(1), 012052. https://doi.org/10.1088/1742-6596/1986/1/012052.

[5] Dai, L., Wang, D., Wang, T., Feng, Q., & Yang, X. (2017). Analysis and comparison of long-distance pipeline failures.Journal of Petroleum Engineering, 2017.

[6] Davis, P. M., Dubois, J., Olcese, A., Uhlig, F., Larivé, J. F., & Martin, D. E. (2006). Performance of European cross-country oil pipelines.Statistical Summary of Reported Spillages, 54.

[7] El-Abbasy, M. S., Senouci, A., Zayed, T., Mirahadi, F., & Parvizsedghy, L. (2014). Artificial neural network models for predicting condition of offshore oil and gas pipelines. Automation in Construction, 45, 50–65. https://doi.org/10.1016/j.autcon.2014.05.003.

[8] Feng, Q., Sha, S., & Dai, L. (2019). Bayesian survival analysis model for girth weld failure prediction. Applied Sciences, 9(6), 1150.

[9] Feng, Y., Ji, L., Chen, H., Jiang, J., Wang, X., Ren, Y., Zhang, D., Niu, H., Bai, M., & Li, S. (2021). Research progress and prospect of key technologies for high-strain line pipe steel and pipes. Natural Gas Industry B, 8(2), 146–153. https://doi.org/10.1016/j.ngib.2020.09.015.

[10] Guo, Y., Meng, X., Wang, D., Meng, T., Liu, S., & He, R. (2016). Comprehensive risk evaluation of long-distance oil and gas transportation pipelines using a fuzzy Petri net model. Journal of Natural Gas Science and Engineering, 33, 18–29. https://doi.org/10.1016/j.jngse.2016.04.052.

[11] He, T., Liao, K., He, G., Zhao, J., Deng, S., & Leng, J. (2022). Quantitative study on magnetic-based stress detection and risk evaluation for girth welds with unequal wall thickness of high-grade steel pipelines. Journal of Natural Gas Science and Engineering, 108, 104825.

[12] Kumari, P., Halim, S. Z., Kwon, J. S.-I., & Quddus, N. (2022). An integrated risk prediction model for corrosion-induced pipeline incidents using artificial neural network and Bayesian analysis. Process Safety and Environmental Protection, 167, 34–44. https://doi.org/10.1016/j.psep.2022.07.053.

[13] Li, H., Ren, X., & Yang, Z. (2023). Data-driven Bayesian network for risk analysis of global maritime accidents. Reliability Engineering & System Safety, 230, 108938. https://doi.org/10.1016/j.ress.2022.108938.

[14] Liao, K., Yao, Q., Wu, X., & Jia, W. (2012). A numerical corrosion rate prediction method for direct assessment of wet gas gathering pipelines internal corrosion. Energies, 5(10), 3892–3907.

[15] Markowski, A. S., & Mannan, M. S. (2009). Fuzzy logic for piping risk assessment (pfLOPA). Journal of Loss Prevention in the Process Industries, 22(6), 921–927. https://doi.org/10.1016/j.jlp.2009.06.011.

[16] Oliveira, N., Bisaggio, H., & Netto, T. (2016). Probabilistic analysis of the collapse pressure of corroded pipelines. International Conference on Offshore Mechanics and Arctic Engineering, 49965, V005T04A033.

[17] Parvizsedghy, L., & Zayed, T. (2015). Developing failure age prediction model of hazardous liquid pipelines.

[18] Parvizsedghy, L., & Zayed, T. (2016). Consequence of failure: Neurofuzzy-based prediction model for gas pipelines. Journal of Performance of Constructed Facilities, 30(4), 04015073.

[19] Ren, C., Qiao, W., & Tian, X. (2012). Natural gas pipeline corrosion rate prediction model based on BP neural network. Fuzzy Engineering and Operations Research, 449–455.

[20] Shahriar, A., Sadiq, R., & Tesfamariam, S. (2012). Risk analysis for oil & gas pipelines: A sustainability assessment approach using fuzzy based bow-tie analysis. Journal of Loss Prevention in the Process Industries, 25(3), 505–523.

[21] Shaik, N. B., Pedapati, S. R., Othman, A. R., Bingi, K., & Dzubir, F. A. A. (2021). An intelligent model to predict the life condition of crude oil pipelines using artificial neural networks. Neural Computing and Applications, 33(21), 14771–14792. https://doi.org/10.1007/s00521-021-06116-1.

[22] Shaik, N. B., Pedapati, S. R., Taqvi, S. A. A., Othman, A. R., & Dzubir, F. A. A. (2020). A Feed-Forward Back Propagation Neural Network Approach to Predict the Life Condition of Crude Oil Pipeline. Processes, 8(6), Article 6. https://doi.org/10.3390/pr8060661.

[23] Su, Y., Li, J., Yu, B., Zhao, Y., & Yao, J. (2021). Fast and accurate prediction of failure pressure of oil and gas defective pipelines using the deep learning model. Reliability Engineering & System Safety, 216, 108016. https://doi.org/10.1016/j.ress.2021.108016.

[24] Sun, J., Li, C., Wu, X.-J., Palade, V., & Fang, W. (2019). An Effective Method of Weld Defect

Detection and Classification Based on Machine Vision. IEEE Transactions on Industrial Informatics, 15(12), 6322-6333. https://doi.org/10.1109/TII.2019.2896357.

[25] Wu, K., Zhang, D., Feng, Q., Yang, Y., Dai, L., Wang, D., Zhang, H., Guo, G., & Liu, X. (2023). Improvement of fracture assessment method for pipe girth weld based on failure assessment diagram. International Journal of Pressure Vessels and Piping, 204, 104950.

[26] Xie, M., & Tian, Z. (2018). A review on pipeline integrity management utilizing in-line inspection data. Engineering Failure Analysis, 92, 222-239.

[27] Xu, W.Z., Li, C. B., Choung, J., & Lee, J.-M. (2017). Corroded pipeline failure analysis using artificial neural network scheme. Advances in Engineering Software, 112, 255-266. https://doi.org/10.1016/j.advengsoft.2017.05.006.

[28] Xu, Y., Wu, M., Nie, X., Feng, Z., Li, L., & Yang, F. (2022). Performance inspection and defect cause analysis of girth weld of high steel grade pipeline. Journal of Physics: Conference Series, 2262(1), 012006. https://doi.org/10.1088/1742-6596/2262/1/012006.

[29] Yang, D., Cui, Y., Yu, Z., & Yuan, H. (2021). Deep Learning Based Steel Pipe Weld Defect Detection. Applied Artificial Intelligence, 35(15), 1237-1249. https://doi.org/10.1080/08839514.2021.1975391.

[30] Zapata, J., Vilar, R., & Ruiz, R. (2011). Performance evaluation of an automatic inspection system of weld defects in radiographic images based on neuro-classifiers. Expert Systems with Applications, 38(7), 8812-8824. https://doi.org/10.1016/j.eswa.2011.01.092.

本论文原发表于《Processes》2023 年第 11 期。

Internal Localized Corrosion of X65-grade Crude Oil Pipeline Caused by the Synergy of Deposits and Microorganisms

Yuan Juntao[1]　Tian Lu[2]　Zhu Wenxu[3]　Tan Shuli[2]　Xin Tong[1,4]
Li Danping[1,5]　Feng Wenhao[1,6]　Zhang Huihui[2]　Li Xuanpeng[1]
Huang Jufeng[1]　Fu Anqing[1]　Feng Yaorong[1]

(1. State Key Laboratory of Performance and Structural Safety of Petroleum Tubular Goods and Equipment Materials, CNPC Tubular Goods Research Institute;
2. Engineering and Technology Research Institute, PetroChina Southwest Oil and Gas Field Company;
3. Sichuan Shale Gas Exploration and Development Co., Ltd, PetroChina Southwest Oil and Gas Field Company;
4. School of Materials Science and Engineering, Xi'an University of Science and Technology;
5. Xi'an Dexincheng Technology Co., Ltd, Xi'an 710075, China;
6. Shaanxi Jiuzhou Petroleum Engineering Technical Service Co., Ltd.)

Abstract: A case study of an X65-grade crude oil pipeline suffering from severe internal localized corrosion after short-time service was presented. The causes were analyzed via thickness measurement, microbial analysis, SEM, EDS, and XRD analysis. The results showed that a large number of corrosion pits were distributed in the area of 4~8 o'clock at the bottom of the pipeline, and severe localized wall thickness reduction was observed. Around the corrosion pits, extremely thick (approximately 3 mm) and complex products/deposits can be seen at the edge of the pit, while relatively thinner ones can be seen at the bottom of the pit. In addition, a large amount of SRB was detected in the pigging solids and products/deposits scarped from the inner pipe. Based on the results, the under-deposit microbial corrosion was proposed as the leading cause for the significant localized corrosion of the pipeline in a short-term service.

Keywords: Oil and gas engineering; Submarine pipeline; Steel; Microscopic characterization and microanalysis; Corrosion; Management deficiency; Pigging treatment and appropriate bactericide

1　Introduction

Subsea oil and gas pipelines are essential for offshore oil and gas production. Their failure

---

Corresponding author: Yuan Juntao, yuanjuntaolly@163.com; Fu Anqing, fuanqing@cnpc.com.cn.

consequences are severe, causing the loss of oil and gas resources, polluting the marine environment, and causing ecological damage[1-3]. Not only submarine pipelines but also onshore oil and gas pipelines often suffer from corrosion failure[4-7]. The Canada Energy Regulator (CER) statistical data indicates that corrosion is one of the three major causes of pipeline incidents[8]. Further investigation based on the statistical data by Alberta Energy Regulator (AER) shows that internal corrosion accounts for nearly 50%, of which almost 20% was related to microbiologically influenced corrosion (MIC)[9]. MIC involves the activities of microorganisms (e.g., microalgae, bacteria, and fungi), which tends to accelerate corrosion reactions and increase the susceptibility of materials to corrosion processes such as pitting, embrittlement, and under-deposit corrosion (UDC)[10-12]. MIC makes the corrosion process more complex[13].

Several pipeline failures have been attributed to SRB-induced localized corrosion, which is often coupled with other factors[14-16]. Corrosive factors like CO_2, H_2S, brine, and others are often present in pipeline fluids. Scale can be formed in high mineralized brines as well. Furthermore, the fluid may also carry solid particles, such as gravel, which can be deposited at the bottom of the pipeline by gravity. A key factor in SRB corrosion lies in the microenvironment created by biofilms, where oxygen content and pH may change and promote corrosion[12, 17-18]. Many corrosion factors exist in the internal environment of oil and gas pipelines, and their synergy is complex, so that the mechanism of pipeline corrosion failure remains unclear.

A case study of an X65-grade crude oil pipeline suffering from severe internal localized corrosion after short-time service was presented here. This 30km long pipeline was constructed for the transfer of producing crude oil between two platforms. The pipeline adopts the structure of "outer pipe–polyurethane insulation layer – inner pipe" to meet the bearing requirements of the submarine pipeline. During the operation of the pipeline, the inlet temperature was 80~90℃, the outlet temperature was 70~80℃, and the operating pressure was 3~4MPa. Inside the pipeline, crude oil containing 10%~50% (volume fraction) water (as shown in Table 1) and 1%~17% (mole fraction) CO_2 in the gas phase was transported. One year after the operation of the pipeline, an abnormal temperature drop occurred, and the external inspection found multiple inner and outer pipe ring hollow filling with water, indicating multiple breakage and leakage of the inner pipe. After several pipe cleaning treatments, a pipe section near the outlet was salvaged and used to analyze the cause of the inner pipe leakage.

Table 1 Chemical compositions of the formation water (mg/L)

Ions	Ca^{2+}	K^+	Na^+	Mg^{2+}	Cl^-	SO_4^{2-}	HCO_3^-
Content	541	392	12450	102	19400	878	439

Fig.1 shows the macro morphology of the inner surface of the inner pipe, indicating remarkable corrosion pits and significant amounts of corrosion products and deposits. More specifically, corrosion pits are concentrated at the bottom of the pipeline at 4~8 o'clock. In this paper, the internal localized corrosion of the inner pipe was carefully studied from the aspects of material characterization, identification of internal deposit, and microbial analysis of pigging solids from the inner pipe to clarify the localized corrosion mechanism and provide support for pipeline corrosion prevention.

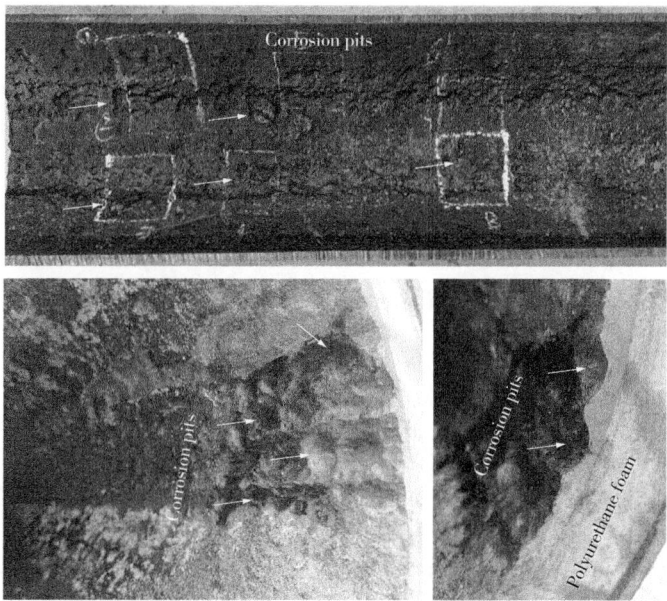

Fig.1 Macro morphology of the inner surface of the inner pipe

2 Experimental methods

2.1 Workflow for the case study

A total of thirteen pipe cleaning operations were conducted before the pipe was recovered from the seafloor, and each time one bag of pigging solids was collected for microbiological analysis. Afterwards, the pipes recovered were analyzed as follows. First, the outer pipe and polyurethane insulation were peeled off the outer surface of the inner pipe, and after the outer surface was cleaned, the wall thickness of the inner pipe was measured using ultrasonic technique. Second, for microbiological analysis and phase identification, the products were scraped from different locations on the inner surface of the inner tube. Third, specimens were cut from the bottom of the inner tube for microscopic and chemical analysis.

2.2 Measurement of wall thickness

The multi-channel portable ultrasonic A/B/C/TOFD scanning detection imaging system Table-UT (PAC, USA) was used to continuously detect the wall thickness of the 3~9 o'clock section at the bottom of the pipeline. The automatic movement range of the probe in the axial direction was 250mm. The moving steps of the probe in X and Y directions were 4mm.

2.3 Materials characterization

The chemical composition of the specimen was analyzed by ARL4460 spectrometer (Thermo Scientific, USA) according to ASTM A751-20 standard[19], and the nitrogen content in the specimen was analyzed by TC600 oxygen and nitrogen analyzer (LECO, USA).

According to ASTM E112[20] and ASTM E3[21], 2% nitric acid alcohol solution was used for etching. OLS4100 laser confocal microscope (OLYMPUS, Japan) was used to analyze the microstructure, grain size, non-metallic inclusions, and other information of both sides of the expanded bent inner pipe specimens.

2.4 Identification of the corrosion products/deposits

Specimens cut from the inner pipe contained the localized corrosion pits and surrounding area. They were embedded in the resin, ground, and polished to observe the cross-section morphology of the corrosion products and deposits via Scanning Electron Microscopy (SEM). The elemental composition was analyzed by the accompanying Energy Dispersive Spectrometer (EDS). In addition, an X-ray diffractometer (XRD) was used to analyze the phase composition of the corrosion products and deposits on the inner surface of the pipeline.

2.5 Microbial analysis

In order to clarify whether the localized corrosion of the inner pipe was related to microbial corrosion, microorganisms were identified and characterized in the pigging solids and scraped products on the inner pipe[22].

The biochemical culture method was used to cultivate corrosive bacteria such as SRB in thepigging solids as the source. Thirteen samples of pigging solids collected at different pigging stages were subjected to sulfate-reducing bacteria confirmation. The pigging solids for the crude oil pipeline were slimy, light in texture, and floated on the surface in an aqueous solution. Bacteria could not be extracted by centrifugation, and bacterial culture and quantitative testing could not be performed. Therefore, a small amount of the pigging solids was directly weighed, immersed in the SRB test bottle, and incubated at a constant temperature of 37℃. The positive reaction of the test bottle was continuously observed.

The real-time fluorescence quantitative PCR technique (qPCR) was employed to analyze SRB quantitatively in the internal attachments of the metallic pipe. This technique is based on the linear relationship between the C_t value of the index period template amplified by PCR and the initial copy number of the template. During the test, 0.5g of the attachment was placed in 5mL of STE buffer, vortexed and mixed, and left to stand for 1h. The upper suspension was taken, centrifuged at 7000r/min for 10min, and then collected bacteriophage cells. The genomic DNA was extracted according to the bacterial genomic DNA extraction kit. 12μL of reaction buffer 2xSYBR, 1.4μL each of primers APSA-F and APSA-R, 2μL of genomic DNA of each sample as a template, and made up to 20μL with sterilized water. Standard plasmids were extracted with the kit, and the concentration of standard plasmids was quantified to 10ng/μL (3×10^9 copies/mL). The serially diluted standard plasmids were obtained according to the 10-fold ratio gradient dilution method and used as templates for real-time quantitative PCR, and the standard curves were plotted. The SRB content in each sample was calculated from the standard curve.

The DNA was extracted by CTAB/SDS method, and the extracted DNA was used as a template for PCR amplification after quality control, and specific primers were selected. Professional software was used to analyze the sequencing data and clarify the microbial flora. During the experiment, the diluted genomic DNA was used as the template, and PCR was performed using specific primers with barcode, Phusion®(Thermo Scientific, USA) High-Fidelity PCR Master Mix with GC Buffer from New England Biolabs, and high - efficiency high fidelity enzymes selected according to the amplification region to ensure amplification efficiency and accuracy. PCR products were detected by agarose gel electrophoresis at 2% concentration, and equal amounts were mixed according to PCR product concentrations and then detected again by agarose gel electrophoresis at 2% concentration after thorough mixing. The target bands were recovered using the gel recovery kit provided by Qiagen. The library was constructed using the NEBNext® Ultra™ IIDNA Library Prep Kit, and it was quantified by Qubit and Q - PCR; after the library was qualified, it was sequenced using NovaSeq6000 (Illumina, USA).

3 Results and discussion

3.1 Wall thickness of the inner pipe

Fig.2 shows the pipe wall thickness's hot map and statistical data. The hot map in Fig.2a shows that the local thinning of wall thickness is concentrated in the area from 4 o'clock to 8 o'clock at the bottom of the pipe. The statistics in Fig.2b show that the minimum wall thickness is about 3mm, accounting for about 5% of the total. Wall thickness reduction of 30% or more accounts for 18%. This is consistent with the macroscopic morphology (Fig.1) and indicates that the inner pipe suffers from serious localized corrosion.

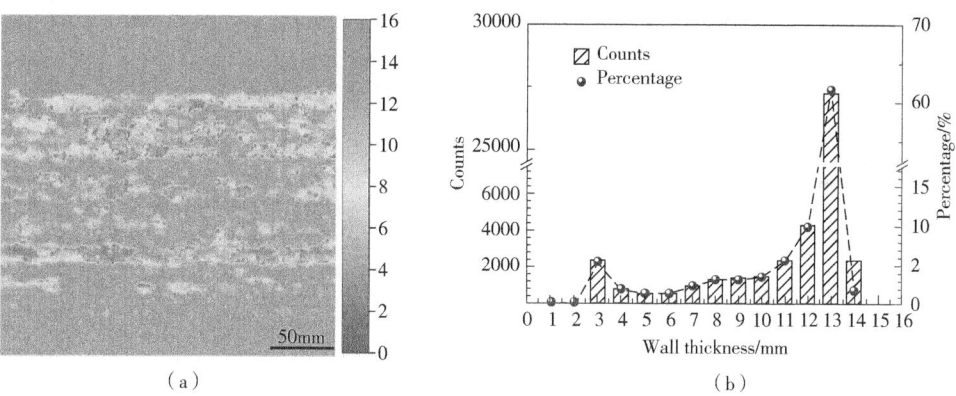

Fig.2 (a) Hot map and (b) statistical data of wall thickness of the corroded pipeline

3.2 Properties of the pipeline steel

Table 2 shows the chemical composition of the pipes. The elemental composition of the pipe material fully meets the technical requirements for X65 in the API Spec 5L standard[23]. Fig.3 shows the metallographic microstructure of the pipe, indicating the typical ferrite and pearlite microstructure.

3.3 Analysis of the corrosion products/deposits

Fig.4 shows the cross-section morphology under low magnification, where the regions indicated as R-I, R-II, and R-III were investigated carefully. R-I represents the pit edge region covered by a thick layer of corrosion products/deposits. R-II and R-III represent the bottom region of two independent pits covered with corrosion products/deposits.

Table 2 Chemical compositions of the pipe (%, mass fraction)

Element	Content	API Spec 5L requirement	Element	Content	API Spec 5L requirement
C	0.051	≤0.12	Ti	0.015	≤0.06
Si	0.17	≤0.45	Cu	0.014	≤0.50
Mn	1.46	≤1.65	B	0.0002	≤0.0005
P	0.0082	≤0.020	Al	0.019	≤0.06
S	0.0020	≤0.010	N	0.0050	≤0.010
Cr	0.017	≤0.50	CE	0.32	≤0.39
Mo	0.093	≤0.50	PCM	0.14	≤0.22
Ni	0.0077	≤0.50	(Nb+V+Ti)	0.06	≤0.12
Nb	0.023	≤0.06	(Al : N)	3.80	≥2 : 1
V	0.026	≤0.10			

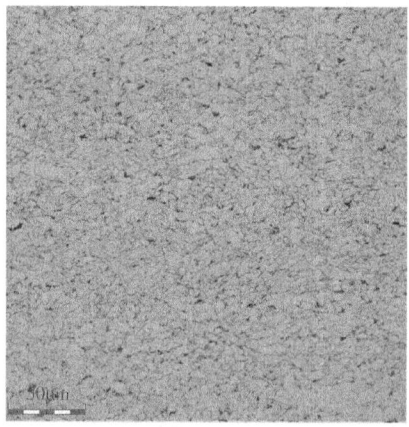

Fig.3 Metallographic microstructure of the inner pipe

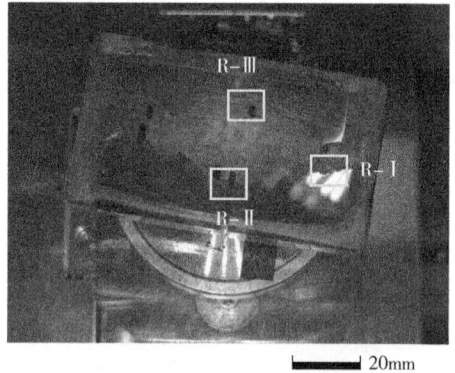

Fig.4 Cross-section SEM images of corrosion pits at low magnification

Fig.5 shows the cross-section BED morphology of the R-I as marked in Fig.4. The corrosion products/deposits here are not only very thick (about 3.67mm) but also highly complex (the distribution of multiple contrasts can be seen). In the backscatter morphology, the average atomic mass of corrosion products/deposits in the area with brighter contrast is generally believed to be more significant. On the contrary, the average atomic mass in the area with darker contrast is smaller. According to the EDS line analysis results shown in Fig.6, it can be seen that the light elements (e. g., C) content is low in the areas with brighter contrast, while the alloy elements (e.g., Fe and Ca) content is high. Also, some large holes can be observed in the product/deposition layer, and in the later inlaying process, the resin is filled in these holes. Chlorine can be seen in the EDS line scan results, which may originate from the chloride salt in the aqueous phase. Chlorine ions are generally considered aggressive ions and can induce localized damage to the passivation film, resulting in pitting corrosion.

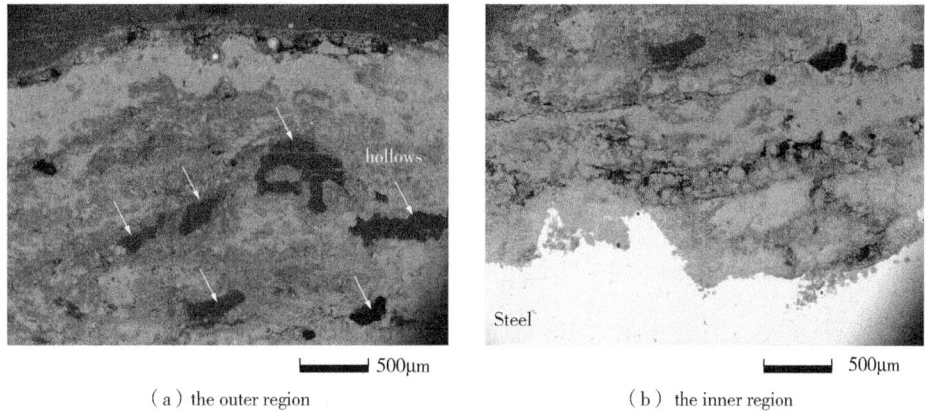

Fig.5 Cross-section BED morphology of corrosion scale at the location of R-I

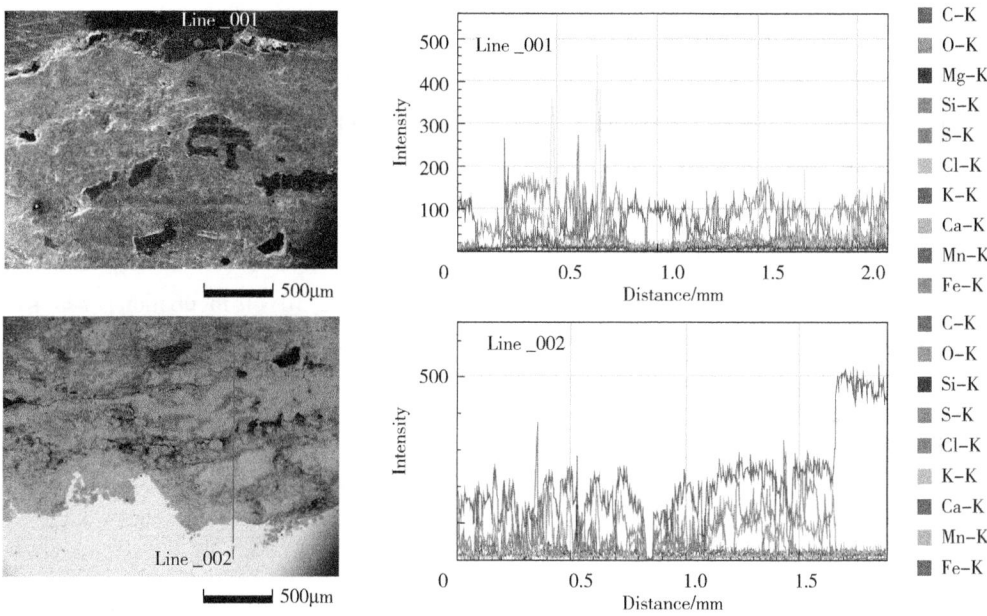

Fig.6 EDS line analysis of corrosion scale at the location of R-I

Fig.7 depicts the EDS line analysis of corrosion scales at the location of R-II and R-III. Compared with the morphology and composition at the pit edge (i.e., R-I), the corrosion products/deposits at the bottom of the pit are much thinner (100~1000μm). However, it still presents a complex composition. Some areas are rich in Si and O, some have an accumulation of Ca, O, and C, and some are composed of Fe and S. In addition, some areas comprise Fe, O, and C. Therefore, although the corrosion products/deposits composition is highly complex, it can be speculated that they may contain silicon oxide, calcium carbonate, iron sulfide, and ferrous carbonate.

Fig.8 presents the XRD pattern of the corrosion products/deposits scraped from the inner wall of the inner pipe. It is consistent with the EDS line analysis results and confirms the existence of $FeCO_3$, FeS, $FeO(OH)$, and SiO_2.

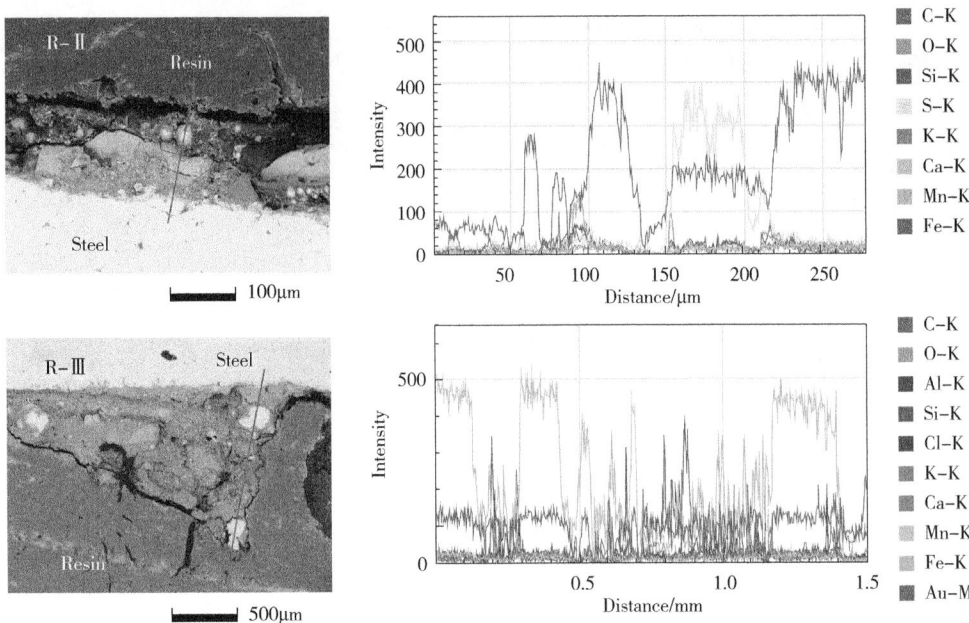

Fig.7 EDS line analysis of corrosion scale at the location of R-II and R-III

3.4 Identification of corrosive bacteria

3.4.1 Active SRB in pigging solids

Fig.9 illustrates the time of the positive reaction of the pigging solids in the SRB test bottle. It can be seen that eight samples showed positive reactions after 3d of incubation, two samples displayed positive reactions after 4d of incubation, and the remaining three samples remained non-positive after 14d of incubation. More than 75% of the samples showed positive reactions, indicating active SRB in these mucilaginous pigging solids.

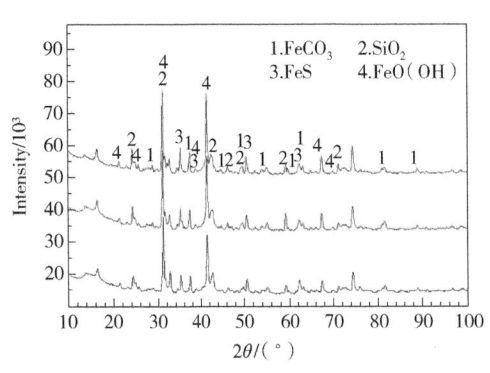

Fig.8 XRD patterns of the corrosion products/deposits scraped from the inner wall of the inner pipe

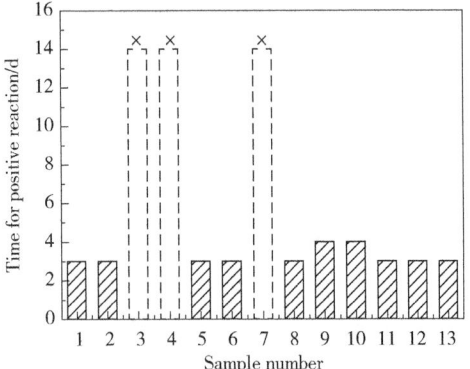

Fig.9 The time when the pigging solids begins to appear positive reaction in the SRB test bottles (× indicates no positive reaction within 14d of culture)

3.4.2 Quantity of SRB detected by qPCR

Fig.10 shows the quantity of SRB detected by qPCR methods. It is evident that a large number of SRB were detected in the corrosion products/deposits, although the number of SRB varied widely (10^5 copies/mL to 10^{14} copies/mL).

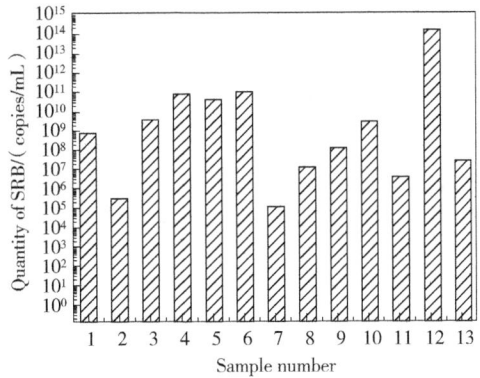

Fig.10 Quantity of SRB cells detected by qPCR

3.4.3 Microbial sequencing of the corrosion products/deposits

NGS provides a list of microbial genera and their relative abundance in the analyzed samples. This is achieved by sequencing the ribosomal 16S region amplified from the sample DNA. The sequences obtained are then compared to known microorganism sequences in the established database. Sample sequences that match database entries can be classified as belonging to a specific microbial family or genus. Unlike qPCR analysis, in which quantitative results are obtained for cells containing specific functional genes, NGS analysis provides only the relative abundance of genes in the sample. In addition, the tool functions only in the same way as the database used to analyze the data, as sample sequences that cannot be matched with sufficient confidence to sequences in the database will remain unclassified. However, despite these limitations, NGS technology offers an advantage that many other techniques lack. While tools such as bacterial test vials and qPCR provide quantitative information about specific types of cells, they provide the user with relatively limited information about the rest of the microbial community. In contrast, NGS allows the analyst to obtain detailed information about the microbial characteristics of a sample as a whole without looking for any specific microorganism or metabolic class.

Fig.11 shows the results of 16S rRNA sequencing, including a phylum-level ternary phase map and a genus-level species abundance map. The three vertices in the ternary phase diagram represent three samples (groups), and the circles represent species. The size of the circles is proportional to relative abundance. The closer the circle is to a vertex, the more abundant this species is in this sample (group). In the ternary phase diagram at the phylum level shown in Fig.11a, the circles are primarily in the center, indicating that the species and abundance of the three groups of samples are relatively similar. According to the circle size, Proteobacteria, Desulfobacterota, and Firmicutes are the three most dominant phyla. Regarding relative abundance at the genus level (see Fig.11b), the top ten species included Desulfotignum, a genus of sulfate-reducing bacteria, and Desulfovibrio, another genus of sulfate-reducing bacteria[24]. This also shows that the pipeline is very rich in microbial species, with a large number of other genera in addition to SRB. A comparison of bacterial test bottles, quantitative PCR, and 16S rRNA tests also showed agreement.

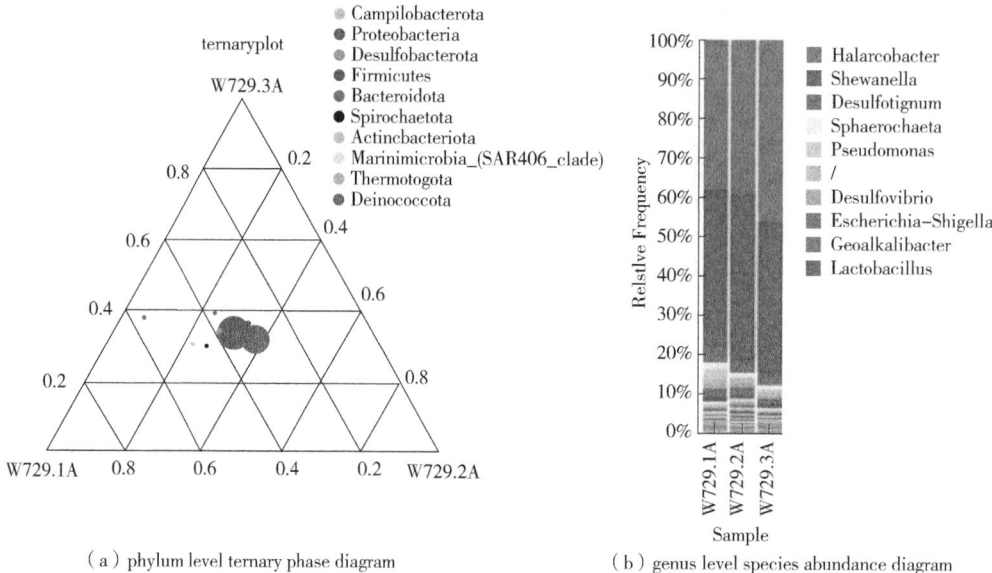

(a) phylum level ternary phase diagram (b) genus level species abundance diagram

Fig.11 Phylum level ternary phase diagram and genus level species abundance diagram

4 Mechanism for the localized corrosion

Corrosion is the process of degrading metal performance after interacting with the environment. Corrosion includes two factors: material and environment. Therefore, to analyze the causes of oil pipeline corrosion, we need to find the causes of both the pipe and the transport fluid.

4.1 Materials factors

No significant abnormalities were found in the chemical composition and microstructure of the X65-grade pipeline. However, carbon steels with poor corrosion resistance are prone to severe corrosion in oil and gas transmission fluids. According to the failure analysis of domestic oil and gas field gathering pipelines, the dominant failure mode in carbon steel pipelines is usually localized corrosion and perforation. Currently, there are no clear rules on the roles of chemical composition, microstructure, grain size, non-metallic inclusions, and other effects on the corrosion performance of carbon steel, so it is not yet possible to improve the corrosion resistance of carbon steel by limiting these aspects of the factors[25].

4.2 Transmission fluids

The CO_2 content of the fluid medium in the operation phase of the pipeline varied greatly, from 1% to 17%; the H_2S content increased during the operation phase, up to 160mg/L; the total iron content in the water samples showed an increasing trend. Therefore, the pipeline may suffer from CO_2 and H_2S corrosion during service[26-28]. This can also be verified from the internal attachment composition described earlier.

In addition to CO_2 and a small amount of H_2S, a large number of corrosive bacteria are present inside the pipeline. Although SRB was not detected in the pipeline water samples, a large amount of SRB was detected in the slimy pipeline cleanout and inorganic attachment, which indicated that corrosive bacteria would be involved in the pipe corrosion degradation. The low level of hydrogen sulfide in the gas phase and the fact that the addition of biocide would result in a significant reduction of hydrogen sulfide at the outlet end of the pipeline suggest that hydrogen sulfide may be derived from SRB metabolism in the pipeline[29].

Moreover, the deposit of calcium salts and sand also creates favorable conditions for UDC. Recently, some studies[12,30-31] have found the synergistic effect of microorganisms and sediments, which promote the localized corrosion of steel.

4.3 Flow factors

Based on parameters such as elevation, flow rate, water content, outlet pressure, and inlet temperature of the pipeline, it is calculated that the flow state in the riser pipe from the platform to the sea level is circular flow, while the flow state in the horizontal section of the pipeline laid on the sea bed surface is stratified flow. Therefore, during the pipeline operation, the water phase flows in the lower part of the pipeline due to its higher density. In contrast, the oil phase flows in the upper part of the pipeline. This is consistent with the pipeline wall thickness measurement and the pipeline observation after dissection. Furthermore, no flow-induced corrosion features such as groove shape were found inside the pipeline, so flow-induced corrosion was not considered a major cause in this case.

4.4 Corrosion degradation

4.4.1 CO$_2$/SRB corrosion

Carbon dioxide is a stable, inert, non-corrosive gas, but it dissolves in water and undergoes hydrolysis reactions as shown by Eq. (1) to Eq. (4)[32], forming acidic and corrosive solutions. For steel corrosion, iron dissolves according to Eq. (5) to form iron ions, which in turn react with CO_2 dissociation products (CO_3^{2-}) to form ferrous carbonate. $FeCO_3$ formation can be expressed as Eq. (6)[33]. The formation of a ferrous carbonate film and its density has a significant effect on corrosion. However, the nature of the ferrous carbonate film can be influenced by many factors, such as temperature, partial pressure of CO_2, etc. The thickness and porosity of the ferrous carbonate product film affect the uniform corrosion rate. The thickness has a more significant effect on the dense corrosion product film[33]. Only the formation of a thick and dense inner film can effectively protect the substrate from corrosion. The cross-section of the corrosion product film (as shown in Fig.5) showed that there were lots of pores in the product film, which made the product film significantly less resistant. Moreover, calcium ions in the aqueous phase will also affect CO_2 corrosion products. Calcium ions doped in $FeCO_3$ will reduce the crystallinity of the product film, which in turn reduces its protective properties[34]. As can be seen from Table 1, the aqueous phase in this case contains 541mg/L Ca^{2+}, which may reduce the protective properties of the product film.

$$CO_{2(g)} \rightleftharpoons CO_{2(aq)} \tag{1}$$

$$CO_{2(aq)} + H_2O_{(l)} \rightleftharpoons H_2CO_{3(aq)} \tag{2}$$

$$H_2CO_{3(aq)} \rightleftharpoons HCO_{3(aq)}^- + H_{(aq)}^+ \tag{3}$$

$$HCO_{3(aq)}^- \rightleftharpoons CO_{3(aq)}^{2-} + H_{(aq)}^+ \tag{4}$$

$$Fe_{(s)} \rightleftharpoons Fe_{(aq)}^{2+} + 2e^- \tag{5}$$

$$Fe^{2+}_{(aq)} + CO_3^{2-}_{(aq)} \rightleftharpoons FeCO_{3(s)} \tag{6}$$

For SRB corrosion, the cathodic reactions can be described as Eq. (7) and Eq. (8). As shown in Eq. (7), if hydrogen can be used as the electron source of SRB for sulfur reduction, then a large amount of SRB may promote the reduction of H_2CO_3 and HCO_3^- by consuming hydrogen, which in turn accelerates corrosion[35]. Under the condition of CO_2 and SRB coexistence, the production of $FeCO_3$ will be inhibited due to the production of FeS by SRB biomineralization. In this case, steel is more prone to localized corrosion[36].

$$SO_4^{2-} + 4H_2 \rightleftharpoons H_2S + 2H_2O + 2OH^- \tag{7}$$

$$SO_4^{2-} + 9H^+ + 8e^- \rightleftharpoons HS^- + 4H_2O \tag{8}$$

4.4.2 Acceleration of localized corrosion

Localized corrosion is a common cause of pipeline failure. Traditional localized corrosion theories are for passivation systems, which believe that aggressive anions (such as chloride ions) lead to the localized destruction of the passivation film, which induces pitting corrosion initiation. Then the autocatalytic effect of the occlusion cell occurs, leading to pitting corrosion. For corrosion failure of carbon steel pipelines, especially perforation or puncture leakage, localized corrosion is attributed to corrosion under deposits such as corrosion products or scales in most cases. It is commonly believed that in the presence of scales or deposits, localized corrosion is usually induced by the formation of a "large cathode" - "small anode" activation region due to the localized incompleteness of the scales/deposits[37-40].

The pitting mechanism of easily passivated metals is well understood, while the initiation and growth of pitting in the activation system are less understood. A key challenge is reproducibly and reliably generating surface pits without deformation and/or residual stresses on a realistic time scale so that pit growth and healing studies can be performed. A constant potential polarization method was used to induce corrosion pits (without deformation or residual stress) on X65 carbon steel with activated corrosion[41]. It indicated that applying anodic potential accelerated general corrosion and pitting, but its contribution to the total metal loss on the steel surface varies (depending on the applied potential). The average pit depth increased linearly with increasing applied potential, and the reproducible and deepest pits (i.e., 70μm) were generated under 150mV[41]. In addition, such pits can grow continuously. This corrosion characteristic is more similar to that in the present work.

In the present work, $FeCO_3$ in the internal attachment is the product of CO_2 corrosion, FeS is the product of SRB involvement in corrosion, and SiO_2 is the sand deposit. Therefore, CO_2, SRB, and deposits are the main factors leading to localized corrosion. In a CO_2-containing environment, the corrosion rate of carbon steel can be as high as approximately 2mm/a[42], which is insufficient to make a 9.7mm thickness reduction within 1~2 year. The presence of microorganisms can promote the occurrence of localized corrosion and also accelerate the rate of localized corrosion. Also, corrosion products, calcium and magnesium salts (e.g., $CaSO_4$, $MgSO_4$), sediment, and biological mucous film mixed to form a composite deposit layer will further significantly increase the localized corrosion rate under the deposit. The synergistic effect of deposits and microorganisms can intensify steel localized corrosion rates, and the magnitude of the increase would be several times more[12]. Thus, the synergistic effect of CO_2, deposits, and microorganisms is the primary cause of severe internal corrosion in the X65 pipeline.

5 Conclusions and recommendation

In this paper, the causes of significant localized corrosion of an X65-grade crude oil pipeline in a short period were analyzed carefully. The following conclusions and recommendations can be summarized.

(1) A large number of corrosion pits are distributed in the area of 4~8 o'clock at the bottom of the pipeline, and the wall thickness reduction of 30% or more accounts for 18%.

(2) There are very thick products/deposits with complex components at the edge of the pit and relatively thinner products/deposits with complex components at the bottom of the pit. These products/deposits mainly comprise ferrous carbonate, iron sulfide, silicon dioxide, and ferric hydroxide.

(3) A large amount of SRB was detected in the pigging solids and products/deposits scarped from the inner pipe, indicating that active sulfate-reducing bacteria participate in the corrosion of the pipeline.

(4) The inner pipe suffered from CO_2 corrosion in the transmission fluid. Simultaneously, the attachment of a large amount of SRB and the formation of deposits (such as silicon dioxide, calcium carbonate, ferrous carbonate, and oil sludge) leads to under-deposit microbial corrosion, resulting in significant localized corrosion and then remarkable wall thickness reduction in a short time.

(5) Although corrosion inhibitors and biocides were used in pipeline operation, they did not work effectively. Considering the synergistic effect of CO_2, deposits, and microorganisms, it is recommended that the pipeline be periodically synchronized with pipe cleaning and biocide batch filling to effectively remove the deposits in the pipeline and the corrosive bacteria remaining under the deposit, so as to prevent rapid corrosion of the deposits and microorganisms.

Acknowledgements

The authors would like to acknowledge the finical supports from the CNPC Scientific and Technical Project (No. 2021DJ2703), CNPC Basic Research and Strategic Reserve Technology Research Fund Project [No. 2020D-5008 (2020Z-08)], Innovation Capability Support Program of Shaanxi (No. 2019TD-038), and Ningbo Science and Technology Innovation 2025 Major Special Project (No. 2020Z108).

References

[1] G. Li, G. Li, J. Wang, T. Zhang, F. Zhang, X. Jiang, H. Liu, Microbiologically influenced corrosion mechanism and protection of offshore pipelines, J. Chin. Soc. Corros. Prot. 41 (2021) 429-438. https://doi.org/10.11902/1005.4537.2020.133.

[2] Y. Yang, F. Khan, P.Thodi, R. Abbassi, Corrosion induced failure analysis of subsea pipelines, Reliab. Eng. Syst. Safe. 159 (2017) 214-222.https://doi.org/10.1016/j.ress.2016.11.014.

[3] Y. Chen, S. Dong, Z. Zhang, M.Gao, H. Zhang, C. Ao, H. Liu, S. Ma, H. Liu, 2021. Collapse failure and capacity of subsea pipelines with complex corrosion defects. Eng. Fail. Anal. 123, 105266. https://doi.org/10.1016/j.engfailanal.2021.105266.

[4] J. Li, D. Wang, F. Xie, 2022. Failure analysis of CO_2 corrosion of natural gas pipeline under flowing conditions. Eng. Fail. Anal. 137, 106265. https://doi.org/10.1016/j.engfailanal.2022.106265.

[5] G. Wu, W. Zhao, Y. Wang, Y. Tang, M.Xie, 2021. Analysis on corrosion-induced failure of shale gas gathering pipelines in the southern Sichuan Basin of China. Eng. Fail. Anal. 130, 105796. https://doi.org/10.1016/j.engfailanal.2021.105796.

[6] L. Chen, B. Dong, W. Liu, F.Wu, H. Li, T. Zhang, 2022. Failure analysis of corrosion products formed during CO_2 pre-corrosion of X70 and 3Cr steels: Effect of oxygen contamination. Eng. Fail. Anal. 140, 106529. https://doi.org/10.1016/j.engfailanal.2022.106529.

[7] K. Liao, M. Qin, G.He, N. Yang, S. Zhang, 2021. Study on corrosion mechanism and the risk of the shale gas gathering pipelines. Eng. Fail. Anal. 138, 105622. https://doi.org/10.1016/j.engfailanal.2021.105622.

[8] B. Wei, J.Xu, C. Sun, Y. F. Cheng, 2022. Internal microbiologically influenced corrosion of natural gas pipelines: A critical review. J. Nat. Gas Sci. Eng. 102, 104581. https://doi.org/10.1016/j.jngse.2022.104581.

[9] R.B. Eckert, T.L.Skovhus, Failure Analysis of Microbiologically Influenced Corrosion, CRC Press, FL, 2022.

[10] Materials Performance, A Closer Look at Microbiologically Influenced Corrosion. https://www.materialsperformance.com/articles/chemical-treatment/2015/08/a-closer-look-at-microbiologically-influenced-corrosion.

[11] A. H. Alamri, 2020. Localized corrosion and mitigation approach of steel materials used in oil and gas pipelines - An overview. Eng. Fail. Anal. 116, 104735. https://doi.org/10.1016/j.engfailanal.2020.104735.

[12] W. Liao, J. Yuan, X. Wang, P.Dai, W. Feng, Q. Zhang, A. Fu, X. Li, 2023. Under-deposit microbial corrosion of X65 pipeline steel in the simulated shale gas production environment. Int. J. Electrochem. Sci. 18, 100069. https://doi.org/10.1016/j.ijoes.2023.100069.

[13] L.Procopio, 2022. Microbially induced corrosion impacts on the oil industry. Arch. Microbiol. 204, 138. https://doi.org/10.1007/s00203-022-02755-7.

[14] G. Wu, W. Zhao, Y. Wang, Y. Tang, M.Xie, 2021. Analysis on corrosion-induced failure of shale gas gathering pipelines in the southern Sichuan Basin of China. Eng. Fail. Anal. 130, 105796. https://doi.org/10.1016/j.engfailanal.2021.105796.

[15] R. Zhao, B. Wang, D. Li, Y. Chen, Q. Zhang, 2022. Effect of sulfate-reducing bacteria from salt scale of water flooding pipeline on corrosion behavior of X80 steel. Eng. Fail. Anal. 142, 106788. https://doi.org/10.1016/j.engfailanal.2022.106788.

[16] A.Bahrami, M. K. Khouzani, B. B. Harchegani, 2021. Establishing the root cause of a failure in a firewater pipeline. Eng. Fail. Anal. 127, 105474. https://doi.org/10.1016/j.engfailanal.2021.105474.

[17] J. Yang, Z.B. Wang, Y.X. Qiao, Y.G. Zheng, 2022. Synergistic effects of deposits and sulfate reducing bacteria on the corrosion of carbon steel. Corros. Sci. 199, 110210. https://doi.org/10.1016/j.corsci.2022.110210.

[18] M. Garcia, L.Procopio, Distinct profiles in microbial diversity on carbon steel and different welds in simulated marine microcosm, Curr. Microbiol. 77 (2020) 967-978. https://doi.org/10.1007/s00284-020-01898-4.

[19] ASTMA 751-20, Standard Test Methods and Practices for Chemical Analysis of Steel Products.

[20] ASTME 112 -13, Standard Test Methods for Determining Average Grain Size.

[21] ASTME 3-11, Standard Guide for Preparation of Metallographic Specimens.

[22] R. B. Eckert, T. L. Skovhus, Failure analysis of microbiologically influenced corrosion, CRC Press, New York, 2021.

[23] API Spec 5L, Specification for line pipe.

[24] S.J.Sagar-Chaparro, A. Darwin, A.H. Kaksonen, L.L. Machuca, 2020. Carbon Steel Corrosion by Bacteria from Failed Seal Rings at an Offshore Facility. Sci. Rep. 10, 12287. https://doi.org/10.1038/s41598-020-69292-5.

[25] A. Fu, J. Yuan, X. Li, X. Chen, W. Li, N.Lyu, L. Fan, L. Li, H. Li, W. Ma, F. Cao, C. Yin, Y. Feng, Gathering Pipeline Corrosion of Oil and Gas Field and Its Anti-corrosion Technologies, Petro. Tubul. Goods Instruments 7 (2021) 14-25.https://doi.org/10.19459/j.cnki.61-1500/te.2021.06.002.

[26] T. Tran, B. Brown, S.Nesic, Corrosion of Mild Steel in an Aqueous CO_2 Environment - Basic Electrochemical Mechanisms Revisited, CORROSION/2015, paper no. 5671 (Houston, TX: NACE, 2015).

[27] F. Pessu, R. Barker, A. Neville, Pitting and Uniform Corrosion of X65 Carbon Steel in Sour Corrosion Environments: The Influence of CO_2, H_2S, and Temperature, Corrosion 73 (2017) 1168-1183. https://doi.org/10.5006/2454.

[28] K. Addis, M. Singer, S.Nesic, A Corrosion Model for Oil and Gas Mild Steel Production Tubing, Corrosion 70 (2014) 1175-1176. https://doi.org/10.5006/1423.

[29] G. Ma, Y.Gu, J. Zhao, Research progress on sulfate-reducing bacteria induced corrosion of steels, J. Chin. Soc. Corros. Prot. 41 (2021) 289-297. https://doi.org/10.11902/1005.4537.2020.097.

[30] J. Yang, Z. B. Wang, Y. X.Qiao, Y. G. Zheng, 2022. Synergistic effects of deposits and sulfate reducing bacteria on the corrosion of carbon steel. Corros. Sci. 199, 110210. https://doi.org/10.1016/j.corsci.2022.110210.

[31] H.Liu, Z. Jin, Z. Wang, H. Liu, G. Meng, H. Liu, 2023. Corrosion inhibition of deposit-covered X80 pipeline steel in seawater containing Pseudomonas stutzeri. Bioelectrochemistry 149, 108279. https://doi.org/10.1016/j.bioelechem.2022.108279.

[32] A.M. EI-Sherik, Trends in oil and gas corrosion research and technologies: Production and Transmission, Woodhead Publishing, Duxford, 2017.

[33] C. Wang, X.Xu, C. Liu, X. Luo, Q. Hu, R. Zhang, H. Guo, X. Luo, Y. Hua, Y. Li, 2023. Improvement on the CO_2 corrosion prediction via considering the corrosion product performance. Corros. Sci. 217, 111127. https://doi.org/10.1016/j.corsci.2023.111127.

[34] X.Ren, Y. Lu, Q. Wei, L. Yu, K. Zhai, J. Tang, H. Wang, J. Xie, 2023. The influence of Ca^{2+} on the growth mechanism of corrosion product film on N80 steel in CO_2 corrosion environments. Corros. Sci. 218, 111168. https://doi.org/10.1016/j.corsci.2023.111168.

[35] H. Liu, C. Chen, X. Yuan, Y. Tan, G.Meng, H. Liu, Y.F. Cheng, 2022. Corrosion inhibition behavior of X80 pipeline steel by imidazoline derivative in the CO_2-saturated seawater containing sulfate-reducing bacteria with organic carbon starvation. Corros. Sci. 203, 110345. https://doi.org/10.1016/j.corsci.2022.110345

[36] M.M. Fan, H.F. Liu, Z.H. Dong, Microbiologically influenced corrosion of X60 carbon steel in CO_2-saturated oilfield flooding water, Mater. Corros. 64 (2013) 242-246. https://doi.org/10.1002/maco.201106154.

[37] P. Li, Y. Zhao, B. Liu, G.Zeng, T. Zhang, D. Xu, H. Gu, T. Gu, F. Wang, Experimental testing and numerical simulation to analyze the corrosion failures of single well pipelines in Tahe oilfield, Eng. Fail. Anal. 80 (2017) 112-122.https://doi.org/10.1016/j.engfailanal.2017.06.014.

[38] H.Mansoori, R. Mirzaee, F. Esmaeilzadeh, A. Vojood, A.S. Dowrani, Pitting corrosion failure analysis of a wet gas pipeline, Eng. Fail. Anal. 82 (2017) 16-25.https://doi.org/10.1016/j.engfailanal.2017.08.012.

[39] X. Li, X. Wang, X. Chen, W. Yuan, Z. Chen, P. Cui, J. Yuan, 2022. Under deposit corrosion of low alloy steel in saturated CO_2 formation water. Int. J. Electrochem. Sci. 17, 220760. https://doi.org/10.20964/2022.07.66.

[40] X. Li, Q. Chen, C. Li, X. Zhang, G. Tong, P. Cui, L. Lu, Q. Ma, J. Yuan, A. Fu, C. Yin, Y.Feng, 2022. Insights into the perforation of the L360M pipeline in the liquefied natural gas transmission process. Eng. Fail.

Anal. 140, 106566.https://doi.org/10.1016/j.engfailanal.2022.106566.

[41] S.Mohammed, Y. Hua, R. Barker, A. Neville, Investigating pitting in X65 carbon steel using potentiostatic polarisation, Appl. Surf. Sci. 423 (2017) 25-32.https://doi.org/10.1016/j.apsusc.2017.06.015.

[42] L. Shi, C. Wang, C.Zou, Corrosion failure analysis of L485 natural gas pipeline in CO_2 environment, Eng. Fail. Anal. 36 (2014) 372-378.https://doi.org/10.1016/j.engfailanal.2013.11.009

本论文原发表于《Engineering Failure Analysis》2023 年第 149 卷。

Failure Analysis of the Leakage in Girth Welds of Bimetal Composite Pipe

Zhang Shuxin[1,2]　Xie Faqin[1]　Li Xianming[3]　Luo Jinheng[2]
Su Gege[2]　Zhu Lixia[2]　Chen Qingguo[3]

(1. School of Civil Aviation, Northwestern Polytechnical University;
2. Tubular Goods Research Institute, China National Petroleum Corporation & State Key Laboratory for Performance and Structure Safety of Petroleum Tubular Goods and Equipment Materials;
3. Tarim Oilfield Company, PetroChina Company Limited)

Abstract: Abimetal mechanically - composite pipe leaked at the entrance of processing station. The failure location was the 6 o'clock direction of the girth weld connecting the gathering pipe and pipe in the station. In order to figure out the root cause of the failure, a series of experiments were carried out. The macroscopic analysis and metallographic test results show that the leakage point is located at the 6 o'clock direction of the girth weld, the root welding and the base pipe of the bimetallic composite pipe are corroded, resulting in perforation. The end of the bimetallic composite pipe is not seal-welded, the girth weld has misalignment and at 6 o'clock, the root weld was concave, and the wall thickness of the root weld and the lining is less than 1.5mm. The $FeCl_3$ pitting corrosion test results of the girth weld show that the pitting corrosion resistance of the root weld is lower than that of the liner and pure stainless-steel pipe. The galvanic corrosion test shows that the galvanic corrosion grades of weld-pure material, weld-liner, carbon steel-weld, carbon steel-liner are all severe corrosion, and the galvanic corrosion of carbon steel-weld is the most serious. The root cause of the failure of girth weld puncture leakage: the end of the bimetallic composite pipe is not seal-welded. At 6 o'clock, the girth weld has misalignment, and the root weld is concave, resulting in that the thickness connecting the root weld and the liner being small. In the medium environment containing Cl^- and CO_2, local corrosion occurs, and then the medium enters the base pipe and the liner interlayer, and the galvanic corrosion aggravates the corrosion of the base pipe. To avoid the reoccurrence of such failure, the mitigation measures were proposed.

Keywords: Bimetal composite pipe; Seal-weld; Galvanic corrosion; Leakage

---

Corresponding author: Zhang Shuxin, wolfzsx@163.com.

1 Introduction

The bimetallic compositepipe[1-2] is a new type of steel pipe with carbon steel pipe as the base pipe and corrosion-resistant steel pipe as the lining pipe. It combines the high strength of carbon steel with the good corrosion resistance of stainless steel. Compared with traditional carbon steel pipe or corrosion-resistant alloy steel pipe, it is more economical and with good corrosion resistance property. It is widely used as gathering pipe and offshore pipe in petroleum industry.

There are two types of bimetal pipes, mechanically (named as lined pipe) and metallurgically (named as clad pipe) composite pipes. The manufacturing cost of mechanically-composite pipe is low, so it is more widely used. The manufacturing process of the mechanically composite pipe include underwater explosion compositing method and mechanically spinning compositing method, both of which are made by external force to make the liner fit to the base pipe. In this case, after removing the external force, the base pipe and liner has no bonding force. Due to this special structure, a series of problems have arisen in the application of bimetal composite pipes, and many scholars have done related researches.

The biggest problem faced is liner collapse. Due to the thin liner, this thin-walled structure is prone to buckling when subjected to external compression and bending forces. When the composite pipe is laid in the subsea pipeline, the composite pipe is prone to collapse under bending load. Yuan et al.[3-6] carried out finite element studies to verify that when the bimetallic composite pipe is subjected to axial compression and bending load, it will first produce wrinkles, and finally form collapses. The failure cases reported mainly include corrosion, liner collapse, and liner cracking. The corrosion problem is mainly due to the incompatibility of the liner material and the medium, resulting in corrosion of the liner and weld. Zhang et al.[7] reported the low cycle fatigue cracking of collapsed liner under pressure fluctuations. Fu et al.[8] studied a girth weld cracking of a bimetal composite pipe, found that there is martensite structure in the weld of the base pipe, under the external force the girth weld cracked. Li et al.[9] researched post internal-welding process of girth weld of the mechanical lined pipe, and obtained a joint with mechanical property and poor corrosion resistance property.

At present, there is a scarcity on the girth weld failure of bimetal composite pipe and stainless-steel pipe.

In this study, a bimetal mechanically-composite pipe leaked at the entrance of processing station. After excavation, it was confirmed that the failure location was the 6 o'clock direction of the girth weld connecting the gathering pipe and pipe in the station. The macroscopic appearance of the failed pipe is shown in Fig 1. The gathering pipe is a bimetal mechanically composite pipe, with a length of 1872m and a specification of ϕ114mm× (base pipe thickness 12mm+liner thickness 1.5mm). The base pipe is made of 20 steels, according to the standard GB 6479—2000. The design temperature and operating temperature are 70℃ and 46℃ respectively, the design pressure and operating pressure are 20MPa and 12.4MPa respectively, and the conveying medium is oil and gas containing H_2S. The pipeline has been in operation for 14 years before failure. The pipe in station is made of S31603 stainless steel. The girth welding process was a three-pass weld, including root, transition, cap welds, welding consumable used for the three-pass was ER309LMo.

The failed sample submitted for inspection is shown in Fig.1. From Fig.1(a), it can be seen that the sample submitted for inspection is about 55cm long, witha slight yellow rust on the surface of the sample, and a girth weld at one end. Observing the inner side of the girth weld, the girth welds at 6'o clock were severely corroded, form grooves and pits, and the girth welds at 12'o clock were intact, without corrosion marks, and had obvious weld excess.

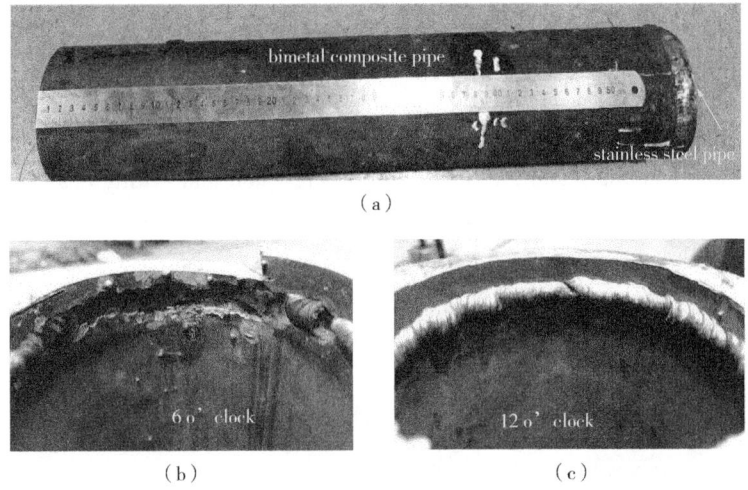

Fig.1 Macroscopic appearance of the failed pipe section

2 Experiment

In order to find out theroot cause of leakage, a series of experiments were carried out on the girth weld of the bimetallic composite pipe. X-ray nondestructive testing was performed on the failed girth weld according to the standard ASTM E94-2017, and the girth weld was divided into 6 equal parts. The chemical composition was analyzed by ARL 4460 direct reading spectrometer according to ASTM A751-14a standard. The base pipe and liner pipe of the composite pipe, stainless steel pipe and girth weld were sampled and tested. Metallographic microstructure of the base pipe, liner, girth weld, and corrosion failure zone were tested to verify whether there is abnormal structure to affect their corrosion resistance. TESCAN VEGA II scanning electron microscope and OXFORD INCA350 energy spectrum analyzer was adopted to characterize the composition of different regions of welds, the morphology of the corrosion pit. Before SEM test, the sample were cleaned in acetone and removed the surface corrosion product by cellulose acetate. $FeCl_3$ pitting corrosion test were carried out on the girth weld samples according to method A in ASTM G48-2015 standard. The size of the sample was 35mm×10mm×1.5mm, in which the sample with the excess height of the weld removed was marked as 1#, and the sample without the excess height of the weld was removed, marked as 2#. The test temperature is 50℃ and the test time is 48h. Galvanic corrosion test was carried out according to T/CSTM 00046.12-2018/T/CSCP 0035.12-2017 on electrochemical workstation. The dimension of the sample was 5mm×5mm×1.5mm. Four groups of samples were tested, 1# weld - stainless steel pipe, 2# weld -liner, 3# carbon steel-weld, and 4# carbon steel-liner. The test medium used was the water medium produced by the well (Table 1), and the test temperature was 50℃.

Table 1 water medium produced by well (mmol/L)

Composition	HCO_3^-	Cl^-	SO_4^{2-}	Ca^{2+}	Mg^{2+}	Ba^{2+}	Sr^{2+}	K^+	Na^+	pH
Content	1.38	4980	5.911	116.9	18.85	5.79×10^{-2}	4.006	69.57	4698	6.03

3 Experiment results

3.1 Visual inspection

The macroscopic morphology of the bimetallic composite pipe is shown in Fig. 2. The inner surface of liner shows metallic luster, without uniform corrosion and pitting characteristic (Fig.2a). The outer wall is covered with slight oil stains, and no corrosion (Fig.2b). The inner wall of the base pipe is smooth and there is no trace of corrosion (Fig.2c).

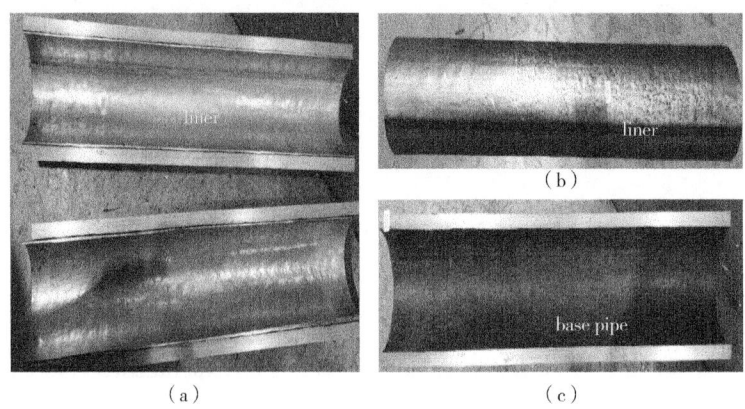

Fig.2 Macroscopic appearance of failed pipe

The macroscopic morphology of the failed girth weld is shown in Fig.3. There is no corrosion trace on the liner of the bimetallic composite pipe and the stainless-steel pipe. Two corrosion zones can be observed at the 6 o'clock of the girth weld, and root weld material is corroded to form pits. The other is near the liner melting zone, where the base tube is corroded and formed a leakage tunnel.

3.2 Non-destructive test

The girth weld was divided to 6 parts to preform X-ray non-destructive test, the results are shown in Fig.4. There are obvious black areas in 1~4 zone of the girth weld, indicating that there is volume damage there, and zone 3 is located at the leak, so the black area indicates that the area is corroded. The non-destructive testing results show that there is only one leakage tunnel in the girth weld.

3.3 Chemical composition

The chemical composition analysis results are shown in Table 2. The content of S element in the base pipe of the bimetallic composite pipe is slightly higher than the upper limit (0.015%) of GB 6479-2013 requirement for 20 steels, and the content of Ni element in the liner pipe is slightly lower than the lower limit (10.0%) of API 5LC-1998 for S31603. The chemical composition of stainless-steel pipe material meets the requirements of API 5LC-1998 for S31603. The C element content at the weld is higher than upper limit of the required content (0.03%) of AWS A5.9-93 for ER309LMo, the content of Cr, Mo and Ni are slightly lower than requirement of AWS A5.9-93.

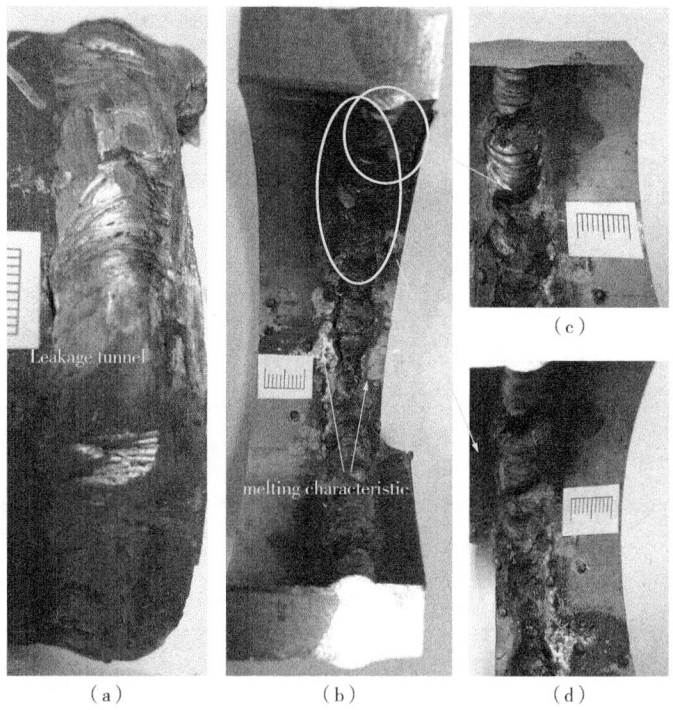

Fig.3 Macroscopic appearance of failed girth weld at 6 o'clock

Fig.4 Radiographic inspection results of girth weld

Table 2 Chemical composition of buffer tank base metal (%, mass fraction)

Element	Base metal	GB 6479-2013 requirement for 20 steels	liner	Stainless steel pipe	API 5LC-1998 requirement for S31603	weld	AWS A5.9-93 requirement for ER309LMo
C	0.196	0.17~0.23	0.019	0.027	<0.030	**0.085**	<0.03
Si	0.21	0.17~0.37	0.74	0.42	<0.75	0.54	0.3~0.65
Mn	0.45	0.35~0.65	1.17	1.13	<2.00	1.27	1.0~2.5
P	0.017	<0.025	0.023	0.037	<0.040	0.024	<0.030
S	**0.016**	<0.015	0.0025	0.0024	<0.030	0.0076	<0.030
Cr	0.056	<0.25	16.27	16.98	16.0~18.0	**21.15**	23.0~25.0

continued

Element	Base metal	GB 6479-2013 requirement for 20 steels	liner	Stainless steel pipe	API 5LC-1998 requirement for S31603	weld	AWS A5.9-93 requirement for ER309LMo
Mo	0.003	—	1.90	2.09	2.0~3.0	0.32	2.0~3.0
Ni	0.042	<0.25	9.83	12.3	10.0~15.0	11.44	12.0~14.0
Nb	<0.001	—	0.015	0.012	—	0.0056	—
V	0.002	—	0.077	0.041	—	0.048	—
Ti	0.001	—	0.043	0.0027	—	0.017	—
Cu	0.135	<0.20	0.24	0.64	—	0.24	<0.75
B	<0.001	—	0.011	0.001	—	0.0005	—
Al	<0.001	—	0.012	0.008	—	0.0095	—
N	0.008	≤0.008	0.037	—	<0.16	—	—

3.4 Tensile test

The tensile test results are shown in Table 3. The tensile strength, yield strength and elongation after fracture of the base pipe and liner meet the standard requirements.

Table 3 The tensile test results

Sample	Width×Gauge/ mm×mm	Tensile strength R_m/MPa	Yield strength $R_{t0.5}$/MPa	Elongation A/%
Base pipe	25×50	495	317	35
		489	313	32
		483	292	34
Requirement of GB 6479-2000		410~550	≥245	≥24
Liner	25×50	671	470	41
		655	410	42
		650	465	37
Requirement of ASTM A312/312A-17		≥485	≥170	—

3.5 Microstructure

Microstructure test results are shown in Fig.5. The metallographic structure of the base pipe is "ferrite+pearlite", and the grain size is 8.5 grade. The metallographic structure of liner is austenite, and the grain size is 8.5 grade. The stainless pipe is austenite steel with a grain size of 4.5 grade, the grain size is much larger than that of the liner.

The macroscopic view of the girth weld at 12 o'clock zone is shown in Fig.6. The girth weld is a three-pass weld, including root weld, transition weld, and cap. The internal and external weld heights are about 3mm. From the cross-section point of view, the pipe end of the bimetallic composite pipe is not seal-welded. The metallographic structure of the weld is dendrite austenite.

The microstructure of the heat-affected zone on the stainless-steel pipe side is austenite, and the structure does not change under the welding heat input. The microstructure of the heat-affected zone on the base pipe side is F(ferrite)+P(pearlite)+B(bainite)+a small amount of WF(Widmanite). Compared with the microstructure of the base metal of the base pipe, a small amount of WF (Widmanite) and B (bainite) appear(Fig.7).

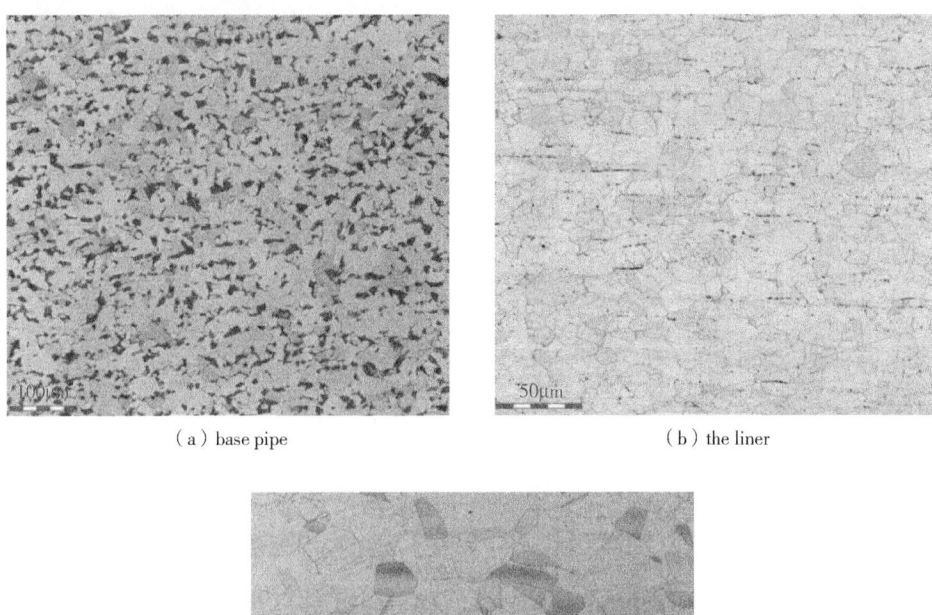

(a) base pipe

(b) the liner

(c) the stainless-steel pipe

Fig.5　Microstructure of the bimetal composite pipe

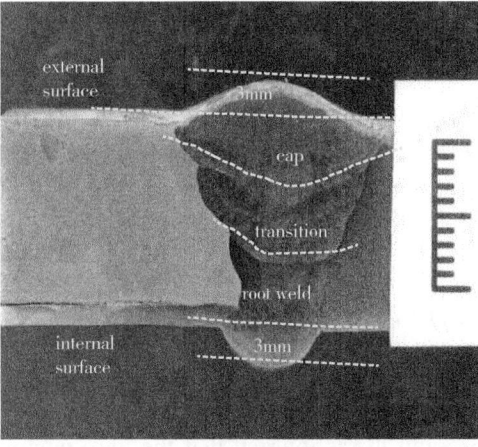

Fig.6　Macroscopic view of the girth weld at 12 o'clock zone

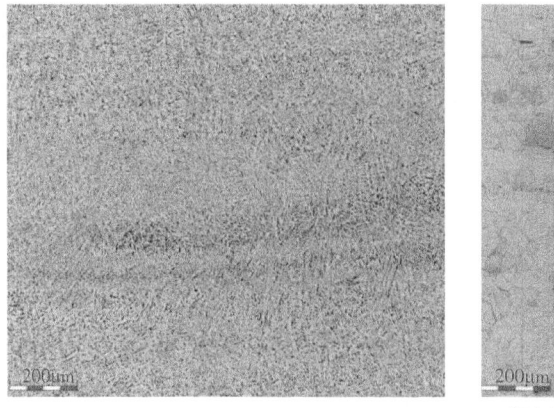
(a) weld

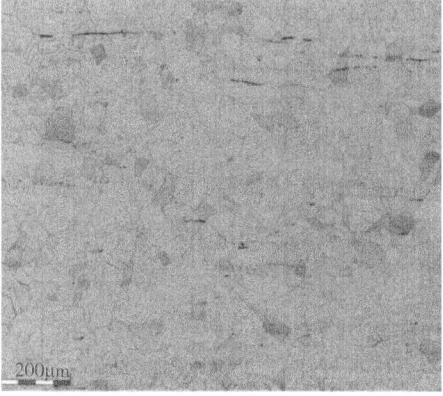
(b) heat-affected zone of stainless-steel pipe side

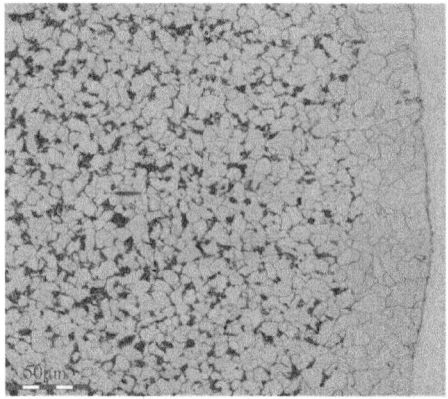
(c) heat-affected zone on the base pipe side

Fig.7 Microstructure of the girth weld

The macroscopic image of the girth weld failed zone is shown in Fig.8. There is an obvious misalignment (about 1.8mm, 2mm on each side of the failed girth weld sample) in the weld. The root weld material was corroded, forming pits, and the bimetallic composite pipe liner and stainless-steel pipe on both sides were basically intact. It can be deduced that the root weld material has poor corrosion resistance and be corroded. The base pipe near the girth weld has been completely corroded, and a yellowish-brown corrosion product is formed, and leakage holes appear near the outer wall.

3.6 Scanning electron microscope analysis of girth welds

The morphology of the corrosion pit of the girth weld is shown in Fig.9. There are obvious corrosion pits on the inner wall and root weld. The parts where the base pipe and the weld are connected are corroded through the entire wall thickness, and holes appear on the side near the outer wall. Bright white substances can be seen around the corrosion zone. It might be iron oxides, which have poor conductivity and appear bright white under the electron microscope. Under high magnification, a large number of "ulcer-like" pits were distributed on the corrosion matrix, with a diameter of about $10\sim50\mu m$, which was a typical CO_2 corrosion morphology[10]. Energy dispersive spectroscopy result is shown in Fig.10 and Table 4. Corrosion products contain C, O, Cl, and Fe elements in different positions, and it was inferred that the base pipe was corroded by CO_2.

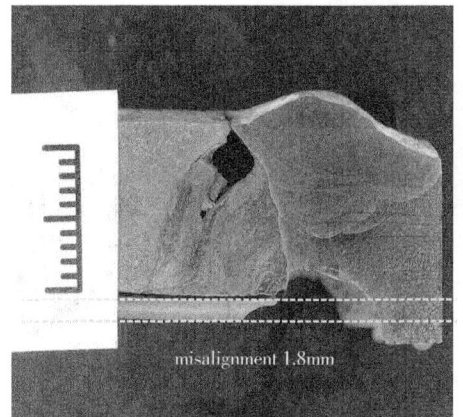
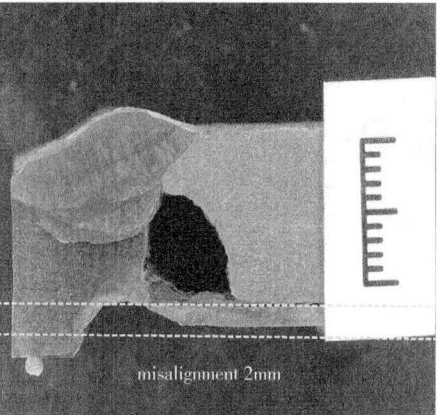

Fig.8　Macro view of girth weld defect

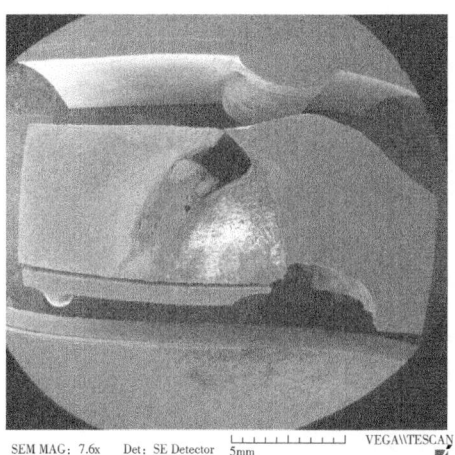
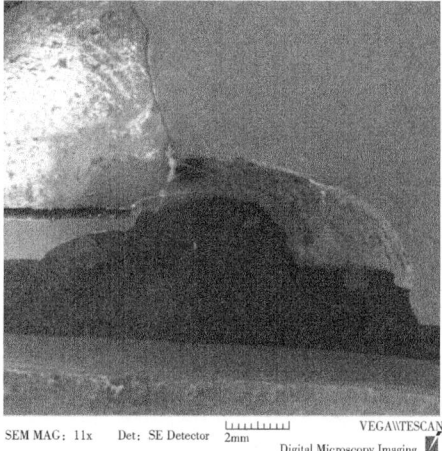

（a）overall view　　　　　　　　　　（b）inner wall root weld corrosion pit

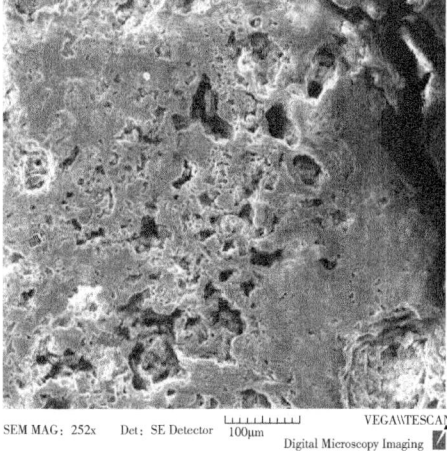

（c）high magnification SEM image

Fig.9　Scanning electron microscope image of girth weld defect 1

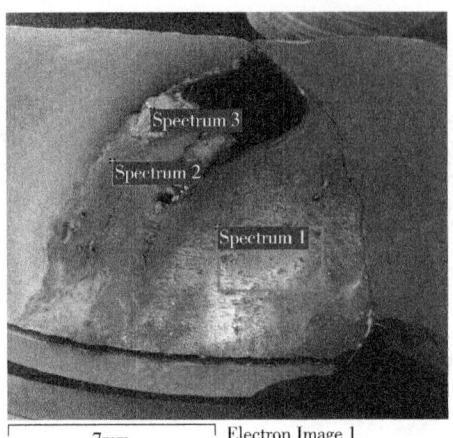

Fig.10 Schematic diagram of energy spectrum in different regions of corrosion of substrate

Table 4 The EDS results of the weld (%, mass fraction)

Spectrum	C	O	Na	Si	Cl	Cr	Fe
Spectrum 1	18.92	23.04	9.67	1.01	15.52	—	31.84
Spectrum 2	17.08	34.71	—	—	8.03	—	40.17
Spectrum 3	11.86	14.45	—	—	38.85	1.36	33.49

At the 6 o'clock direction of the girthweld (Fig.11), the welding wave is obviously concave and show scale-like characteristic. At the root welding position, there is local corrosion. There are melting characteristic on both sides of liner and stainless-steel material.

The product energy spectrum characterization at the liner corrosion site (Fig.12, Table 5) shows that the corrosion products mainly contain C, O, Cl, and Fe elements, and it is inferred that the site is corroded by CO_2/Cl^-.

Table 5 Energy spectrum analysis results of liner corrosion products

Element	Weight/%	Atomic/%	Element	Weight/%	Atomic/%
C	10.39	22.08	S	1.81	1.44
O	29.39	46.90	Cl	5.42	3.90
Na	1.52	1.68	K	0.65	0.43
Si	0.53	0.48			

3.7 FeCl₃ pitting test

The pitting test results are shown in Fig.13 and Table 6. There is no obvious corrosion on the liner and stainless-steel pipe, the front and back surfaces of the root weld are severely pitted for both samples, and the corrosion products are loose and easy to peel off. Compared with the 1# sample, the surface of the 2# sample was relatively intact. While according to the weight-loss results before and after corrosion, the weight loss of the 2# sample is greater than that of the 1# sample, there should be serious corrosion inside the root weld of the 2# sample, forming a corrosion cavity.

From the FeCl$_3$ pitting corrosion test, the corrosion resistance of root welding is much lower than that of liner and stainless-steel pipe.

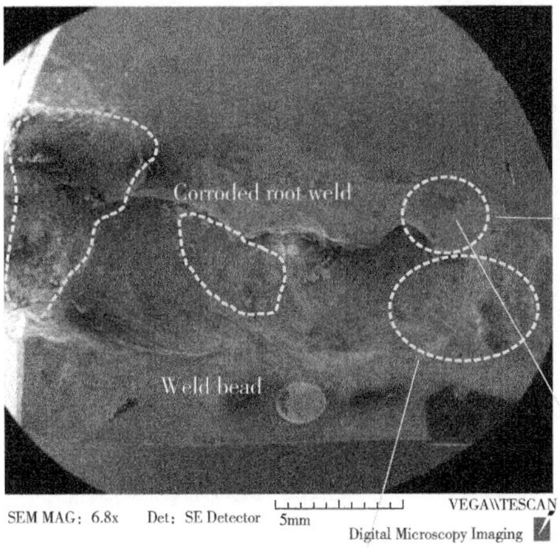

(a) overall view

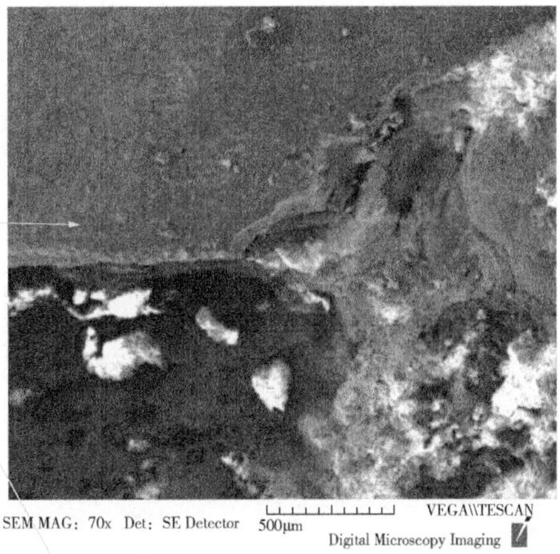

(b) corrosion pit close to outer wall

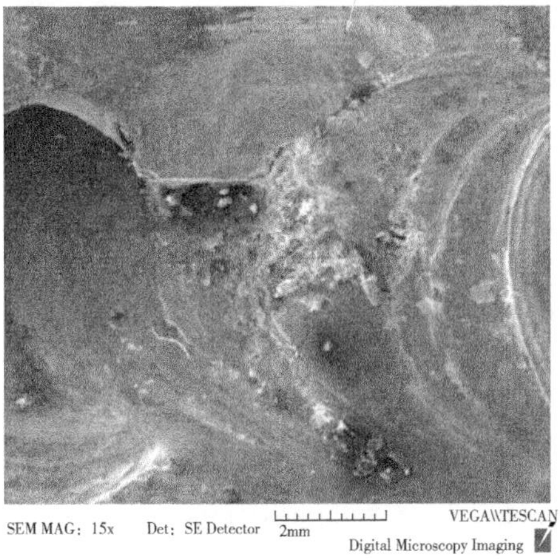

(c) corrosion pit of root weld on inner wall

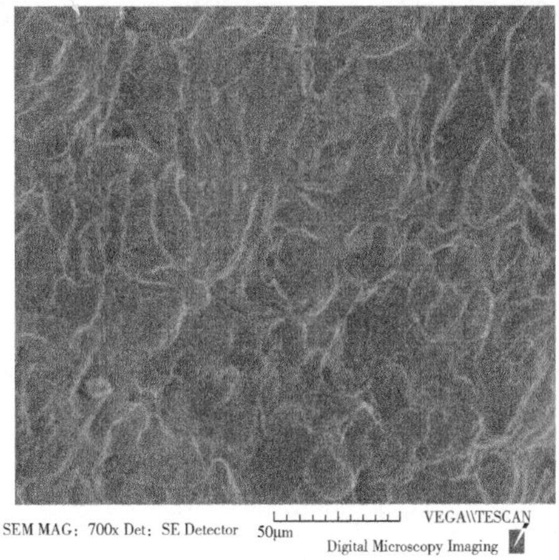

(d) corrosion morphology of liner

Fig.11 Scanning electron microscope image of corroded specimen at 6 o'clock of girth weld

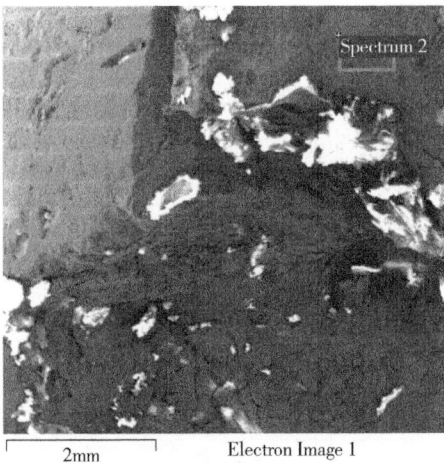

Fig.12 Schematic diagram of energy spectrum position of liner corrosion products

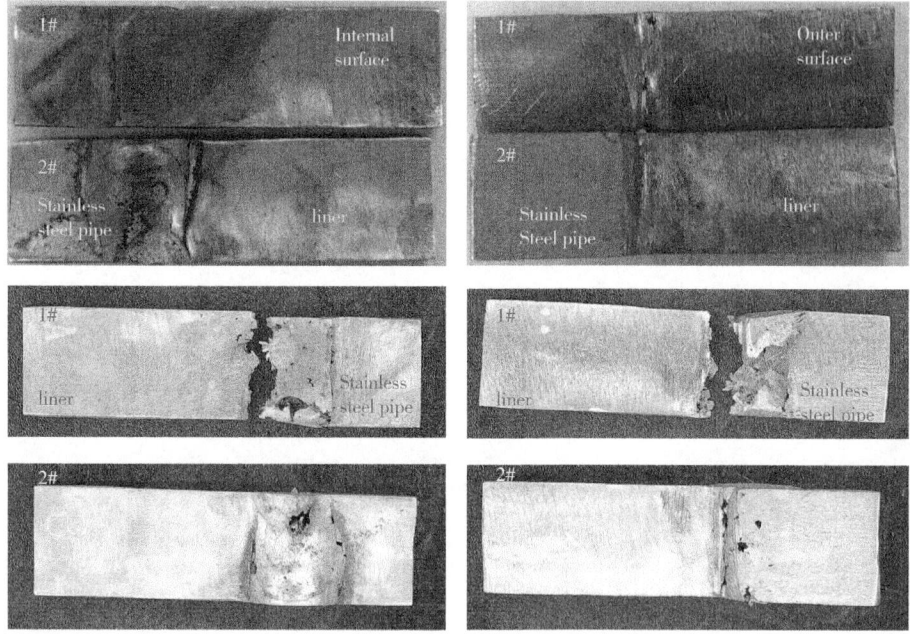

Fig.13 Macroscopic morphology of samples before and after pitting corrosion test

Table 6 Weight loss in pitting test

Sample	Before test/g	After test/g	Weight loss/g	Corrosion rate/(mil/a)
1#- remove excess weld height	4.5080	3.3635	1.1445	1243
2#- remain excess weld height	5.8391	4.5760	1.2631	1372

3.8 Galvanic corrosion test

The open circuit potential of the base metal, weld, liner, stainless pipewas tested, and the results are shown in Fig.14 (a). The standard required that when the galvanic potential difference does not exceed 50mV, galvanic corrosion does not occur. Therefore, there is a galvanic corrosion tendency among the S31603 stainless steel pipe, the S31603 liner and the base carbon steel pipe.

Further, the galvanic corrosion grade was evaluated by using the galvanic corrosion current density. In Fig.14(b), the galvanic corrosion current changes with time, the relatively stable value of 1000s was used for evaluation, 1#~4# are 2.91μA, 3.22μA, 12.33μA, 4.76μA, respectively. The galvanic corrosion tendency was evaluated, and the galvanic corrosion current density of 1#~4# was 11.64μA/cm^2, 12.88μA/cm^2, 49.32μA/cm^2, 19.0460μA/cm^2. According to the standard requirements (Table 7), 1# weld-stainless steel pipe pair, 2# weld-liner pair, 3# carbon steel-weld pair, 4# carbon steel-liner pair galvanic corrosion grades are all severely corroded and cannot be contact, and the carbon steel-weld corrosion current density is the largest.

Table 7 Evaluation of galvanic corrosion grades

No.	Current density/(μA/cm^2)	Grade	Corrosion grade	Advice for usage
1	$I \leqslant 0.3$	A	Negligible corrosion	Can contact
2	$0.3 < I \leqslant 1.0$	B	Less slight	Contactunder certain conditions

continued

No.	Current density/($\mu A/cm^2$)	Grade	Corrosion grade	Advice for usage
3	$1.0<I\leqslant3.0$	C	Slight	Do not contact, use after protection
4	$3.0<I\leqslant10.0$	D	Lesssevere	Do not contact, use after protection
5	$I\geqslant10.0$	E	Severe	Do not contact, use after protection

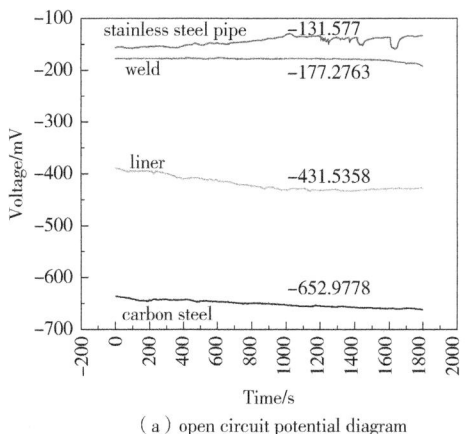
(a) open circuit potential diagram

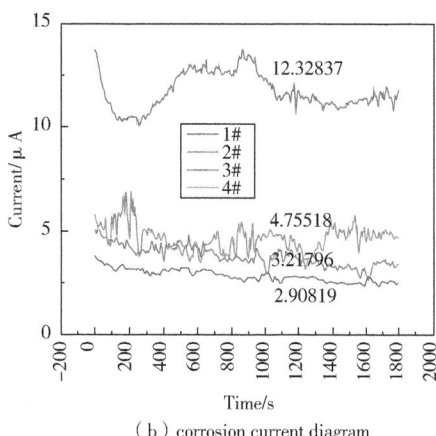
(b) corrosion current diagram

Fig.14 Galvanic corrosion test result

4 Comprehensive analysis

Macroscopic analysis and scanning electron microscope analysis show that the girth weld has a weld excess height of about 2mm at 12 o'clock, and the weld is concave at 6 o'clock, with undercuts and local pits. From melting characteristic, it is inferred that the welding heat input is large, and with the effect of gravity, the weld protruding at 12 o'clock and concave at 6 o'clock. The inner wall of the girth weld is oxidized and blackened, mainly because the girth weld is the connection between the main line pipe and the pipeline in the station, and cannot be protected by argon gas. The root weld at the perforated part and the base pipe connected to it are corroded, forming a leakage tunnel.

The experiment results showed that the chemical composition of the base pipe 20 steel and the liner pipe S31603 of the bimetallic composite pipe did not meet the standard requirements, in which the S element of the base pipe slightly exceeded the standard, and the Ni element of the liner pipe was slightly lower than lower limit of API 5LC-1998 requirement. The chemical composition of S31603 stainless steel pure material meets the requirements of API 5LC-1998 standard. The tensile properties of liner and base pipe of the bimetal composite pipe meet the standard requirements. The metallographic structure of the bimetal composite pipe and stainless steel pipe shows no abnormalities.

In order to find the root cause of the leakage failure of the girth welds, it can be comprehensively explained from the aspects of welding process and material corrosion resistance.

4.1 Welding process

According to the survey, the pipeline was put into operation in 2007, which was the initial stage of the application of bimetallic composite pipes in the oil field. The girth weld adopts three-pass welding, root, transition, cap welding. Usually, stainless steel welding wire is used for sealing welding, root welding and transition welding, and carbon steel welding wire is used for filling cap. The failure part is the girth weld of the bimetallic composite pipe and the pure stainless-steel pipe in the station, not the conventional girth weld. According to the chemical composition analysis and the reference welding process provided by the manufacturer, the root welding, transition and cover surface are all made of ER309LMo welding wire.

From the macroscopic appearance of the girth weld, there is a weld excess height of about 2mm at 12 o'clock, and the weld is concave at 6 o'clock, and the bimetallic composite pipe has melting marks on the liner, which can be inferred that the welding heat input is large, and under the action of gravity, the weld seam protrudes at 12 o'clock, and the weld seam is concave at 6 o'clock. Typically, the 5G location is a relatively difficult-to-weld location. The inner wall of the girth weld is oxidized and blackened, mainly because the girth weld is the connection between the main line pipe and the pipeline in the station, and cannot be protected by argon gas.

4.2 Material corrosion resistance

The corrosion test results show that the liner of the bimetallic composite pipe and the pure stainless steel have good corrosion resistance and no pitting traces, but the root welding has poor corrosion resistance, resulting in a large number of loose corrosion products;

The PREN (pitting resistance equivalent number PREN = Cr% + 3.3Mo% + 16N%) was calculated for the seam, and the PREN = 23.877 for the pure stainless-steel pipe, the PREN = 23.132 for the liner, and the PREN = 22.206 for the weld. Therefore, the pitting corrosion resistance is pure stainless-steel pipe > liner > welding seam, and the welding seam pitting corrosion resistance is the worst, which is consistent with the results obtained by the pitting corrosion test. There is a tendency of galvanic corrosion between the weld and the liner and carbon steel, and the carbon steel-weld has the highest corrosion current density.

Scanning electron microscope analysis and energy spectrum analysis results show that the joints between the base pipe and the weld are corroded by CO_2.

In summary, when the bimetallic composite pipe is girth welded with the stainless-steel pipe in the station, the content of C and Cr elements exceeds the standard, and the Mo content is less than AWS A5.9-93 requirements for ERS31603 welding wire, resulting in a decrease in the pitting corrosion resistance of the weld. In the 6 o'clock direction, the weld is concave, resulting in a wall thickness of less than 1.5mm connecting the root weld and the lining. Local corrosion occurs in with high Cl^- and CO_2 environments. Further, the medium enters the base pipe, the liner, and the root weld interlayer, and galvanic corrosion occurs. The base pipe forms holes near the root weld, and eventually leaks. The schematic diagram of leak formation is shown in the Fig.15.

5 Conclusions and recommendations

In this study, a bimetal mechanically-composite pipe leaked at the entrance of processing station. The failure location was the 6 o'clock direction of the girth weld connecting the gathering pipe and pipe in the station. In order to figure out the root cause of the failure, a series of

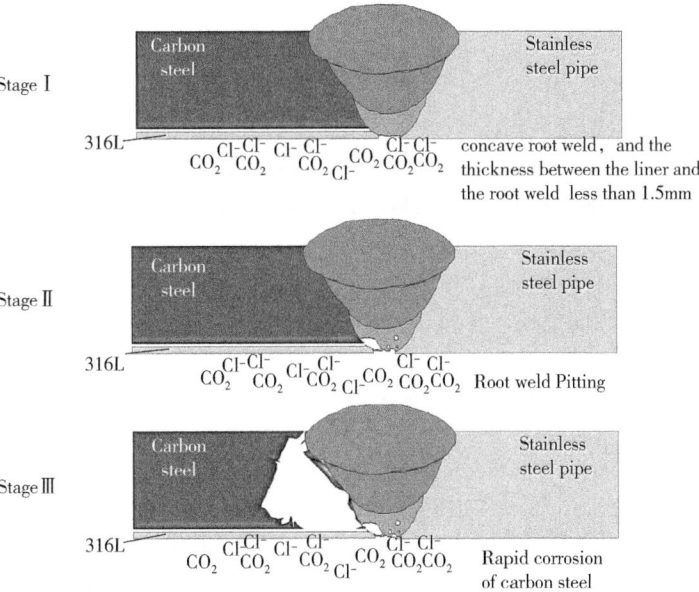

Fig.15 Schematic diagram of failure causes

experiments were carried out, and the following conclusion can be drawn.

(1) The macroscopic analysis and metallographic test results show that the leakage point is located at the 6 o'clock direction of the girth weld, the root welding and the base pipe of the bimetallic composite pipe are corroded, resulting in perforation. The end of the bimetallic composite pipe is not seal-welded, the girth weld has misalignment and at 6 o'clock, the root weld was concave, and the wall thickness of the root weld and the lining is less than 1.5mm.

(2) The $FeCl_3$ pitting corrosion test results of the girth weld show that the pitting corrosion resistance of the root weld is lower than that of the liner and pure stainless-steel pipe. The galvanic corrosion test shows that the galvanic corrosion grades of weld-pure material, weld-liner, carbon steel-weld, carbon steel-liner are all severe corrosion, and the galvanic corrosion of carbon steel-weld is the most serious.

(3) The root cause of the failure of girth weld puncture leakage: the end of the bimetallic composite pipe is not seal-welded. At 6 o'clock, the girth weld has misalignment, and the root weld is concave, resulting in that the thickness connecting the root weld and the liner is small. In the medium environment containing Cl^- and CO_2, local corrosion occurs, and then the medium enters the base pipe and the liner interlayer, and the galvanic corrosion aggravates the corrosion of the base pipe.

To avoid reoccurrence of such failure, it is recommended that strengthen the quality control of on-site welding, use an endoscope to conduct visual inspection of the welding seam, and perform sealing welding on the pipe end of the bimetallic composite pipe. Girth welding consumable should have a higher pit corrosion resistance than base pipe.

Acknowledgement

The authors are grateful to the fund support of CNPC Fundamental Research Project [Grant No.

2021DQ03(2022Z-21)]. Sincere thanks to my wife Ms. Yan Xi and my son Zhang Jiahe for their tremendous support to me.

References

[1] A. Kalaki, M. Eskandarzade, S. Barghani, M. Mohammadpour, Experimental and numerical evaluation of influencing parameters on the manufacturing of lined pipes, International Journal of Pressure Vessels and Piping. 169 (2019) 71-76.https://doi.org/10.1016/j.ijpvp.2018.11.014.

[2] L. Liying, X. Jun, H. Bin, W. Xiaolei, Microstructure and mechanical properties of welded joints of L415/316L bimetal composite pipe using post internal-welding process, International Journal of Pressure Vessels and Piping. (2020) 10.

[3] L. Yuan, S. Kyriakides, Liner wrinkling and collapse of bi-material pipe under bending, International Journal of Solids and Structures. 51 (2014) 599-611.https://doi.org/10.1016/j.ijsolstr.2013.10.026.

[4] L. Yuan, S. Kyriakides, Liner wrinkling and collapse of bi-material pipe under axial compression, International Journal of Solids and Structures. 60-61 (2015) 48-59.https://doi.org/10.1016/j.ijsolstr.2015.01.029.

[5] D. Vasilikis, S.A. Karamanos, Buckling Design of Confined Steel Cylinders Under External Pressure, Journal of Pressure Vessel Technology. 133 (2011) 011205. https://doi.org/10.1115/1.4002540.

[6] L. Yuan, S. Kyriakides, Liner buckling during reeling of lined pipe, International Journal of Solids and Structures. 185-186 (2020) 1-13.https://doi.org/10.1016/j.ijsolstr.2019.08.017.

[7] S. Zhang, Q. Ma, C. Xu, L. Li, M. Wang, Z. Zhang, S. Wang, L. Li, Root cause analysis of liner collapse and crack of bi-metal composite pipe used for gas transmission, Engineering Failure Analysis. 132 (2022) 105942.https://doi.org/10.1016/j.engfailanal.2021.105942.

[8] A.Q. Fu, X.R. Kuang, Y. Han, C.H. Lu, Z.Q. Bai, C.X. Yin, J. Miao, Y.R. Feng, Y.G. Wei, Q. Tang, Y. Yang, Failure analysis of girth weld cracking of mechanically lined pipe used in gasfield gathering system, Engineering Failure Analysis. (2016) 12.

[9] L.Y. Li, J. Xiao, B. Han, X.L. Wang, Microstructure and mechanical properties of welded joints of L415/316L bimetal composite pipe using post internal-welding process, International Journal of Pressure Vessels and Piping. 179 (2020) . https://doi.org/10.1016/j.ijpvp.2019.104026.

[10] G. Cui, Z. Yang, J. Liu, Z. Li, A comprehensive review of metal corrosion in a supercritical CO_2 environment, International Journal of Greenhouse Gas Control. (2019) 17.

本论文原发表于《Engineering Failure Analysis》2023 年第 143 卷。

Corrosion Behavior of Tubing in High-salinity Formation Water Environment Containing H₂S/CO₂ in Yingzhong Block

Zhao Xuehui[1]   Liu Junlin[2]   Yao Baisheng[3]   Li Cheng[2]
Xia Xue[2]   Fu Anqing[1]

(1. State Key Laboratory for Performance and Structure Safety of Petroleum Tubular Goods and Equipment Materials, CNPC Tubular Goods Research Institute.;
2. Research Institute of Drilling and Production Technology, PetroChina Qinghai Oilfield Company;
3. The No.1 Oil Production Plant, Changqing Oilfield Company of Petrochina Co., Ltd.)

Abstract: To clarify the corrosion behavior of P110SS material under the synergistic action of multiple factors such as a CO_2/H_2S coexistence environment, a high temperature, and high-salinity formation water, a series of simulation tests and analyses were carried out in this paper. High-temperature high-pressure autoclaves, scanning electron microscopy, and a three-dimensional microscope were used to analyze and evaluate the changing trend of the corrosion performance of P110SS tubing material under different temperatures and a H_2S/CO_2 partial pressure ratio in a high-salinity formation water environment, and the corrosion cracking sensitivity and pitting sensitivity of the material with stress were compared and analyzed. The results indicate that the average corrosion rate of P110SS material without stress increases with the rising test temperature, and the corrosion damage worsens with an increase in the $H_2S : CO_2$ partial pressure ratio. The highest corrosion rate for P110SS material is 0.99 mm/a. When the test temperature varies from 80℃ to 180℃ and $p_{H_2S} : p_{CO_2} = 0.53 : 0.17$, the P110SS material with a loading stress of 85% YS_{min} is not susceptible to stress corrosion cracking (SCC). Although surface pitting nucleation is evident at a high temperature of 180℃, no expansion-induced cracking or fracture phenomena occur.

Keywords: CO_2/H_2S corrosion; High salinity; Stress corrosion sensitivity; Pitting; Synergistic effect

1 Introduction

As the unique channel for oil and gas well production, the oil casing string is essential for oil

---

Corresponding author: Zhao Xuehui, zhaoxuehui@cnpc.com.cn.

and gas exploration, development, and safe production. At present, with China's continuous expansion in the field of oil and gas exploration and development, the service environment for oil casing strings has become increasingly complex and harsh[1-2]. Especially in deep and complex exploration and development conditions, the pipe string is not only subjected to tension, pressure, bending, torsion, and other combined stress loads, but also suffers from the synergistic effect of extreme corrosion factors such as high temperature, high pressure, high salinity, and CO_2/H_2S, which poses severe challenges to the long-term safe and reliable service of the pipe string as well as the efficient development of oil and gas resources[3]. The operating environment of high-temperature and high-pressure (HTHP) oil and gas wells in the Yingzhong Block of the Qinghai Oilfield is relatively harsh, mainly manifested in the following "four high characteristics": deep reservoir burial (≥ 6000 m), high formation temperature (≥ 190℃), high formation pressure ($\geq$ 110.9MPa), and high fluid salinity (≥ 290g/L). The reservoir contains both H_2S and CO_2 corrosive gases, and the corrosive environment is extremely complex. In particular, the concentration of Cl^- in formation water is as high as 186g/L, almost reaching the brine saturation point. This rare and highly saline and corrosive medium poses a severe challenge for material selection and string application optimization both domestically and internationally.

At present, a lot of research is being conducted on the corrosion law and mechanism of oilfield-produced water and the mineralization degree at home and abroad[4-6], mainly reflected in an environment with a low temperature, a Cl^- content not exceeding 60g/L, and a salinity not exceeding 200g/L. Wang Shutao et al.[7] investigated the sulfide stress corrosion cracking susceptibility of P110SS casing material under conditions of a high H_2S/CO_2 content in formation water with a salinity of 67g/L in the Puguang gas field, and found that the material has a high stress corrosion cracking sensitivity at a low wellhead temperature of 50℃. Deng Hu et al.[8] studied the environmental cracking behavior of C110 casing at a low temperature of 60℃ in oil field formation water containing H_2S/CO_2, and compared and analyzed the stress sensitivity of the material under different conditions, and the results show that the stress sensitivity of the C110 material was relatively high, and the surface W-Ni coating and heat treatment could effectively reduce the environmental cracking sensitivity. Huang Shilin et al.[9] studied the factors affecting the corrosion of P110SS steel in the simulated environment of annular fluid in a sulfur-bearing gas well. The solution medium was prepared using NaCl and Na_2S, and the results show that the main factors affecting the corrosion rate of P110SS steel were temperature, followed by PH value and chloride ion content. Liu Junlin et al.[10] studied the corrosion behavior of P110/P110SS oil casing string in a high-salinity annular protective fluid environment, indicating that the corrosion rate of materials in an environment containing a small amount of dissolved oxygen was relatively increased, and the material was not sensitive to stress corrosion cracking at a high temperature of 180℃. Li Mingxing et al.[11] studied the corrosion law of tubing in the 80S of formation water with a Cl^- concentration of 36.6g/L and containing H_2S/CO_2, and determined that when the H_2S concentration was below the range of 0.1MPa, the material corrosion rate decreased with the increase in the H_2S concentration, and the corrosion rate showed an upward trend with the increase in temperature. Wang Yunfan et al.[12] studied the corrosion law of P110SS steel with high H_2S and CO_2 contents in a formation water environment with a salinity of 67g/L, and showed that with the increase in the H_2S, CO_2 partial pressure, and temperature, the corrosion rate of the P110SS steel first decreased and then increased. When the bottom hole temperature is 130℃, the SSCC sensitivity of the material decreases, and the temperature plays a major role. Therefore, the produced formation water environment in the Yingzhong Block is relatively harsh. In a relatively high-salinity environment, the oil casing string has a serious corrosion failure risk under the synergistic action of multiple factors such as high temperature corrosion, tensile stress, CO_2/H_2S and scale. In particular, the stress corrosion cracking sensitivity of the string under an H_2S environment and a high-temperature environment is of great importance[13-14].

This paper is based on the safety requirements of pipe string service; it is more important to actively study the corrosion performance of oil casing string in a field environment containing H_2S/CO_2 and the coupling effect of multiple factors in the oilfield. With the increase in the well depth and the increase in the pipe string's service temperature, it is especially necessary to clarify the influence of high-salinity formation water combined with H_2S, high temperature, and other factors on the corrosion behavior of pipe string. This paper provides a reliable theoretical basis and technical support for the optimization and selection of test production wells in Yingzhong Block.

2 Experimental Procedure

2.1 Material and Solution

The material used for the testwas a commercial P110SS tubing extracted from the oil field, of which the chemical composition (by mass fraction) was 0.20% C, 0.21% Si, 0.51% Mn, 0.0013% S, 0.0092% P, 0.52% Cr, 0.03% Ni, 0.73% Mo, and 0.046% Cu, and the remaining part was composed of Fe.

There were two types of samples, no-stress state and applied stress state. Specifically, the size of specimens was 40mm×10mm×3mm, with a ϕ6mm hole on one end. The stressed samples were 115mm×15mm×3mm in size, in which the stress, equal to 85% of the nominal yield strength (YS_{min}) of the material, was applied using the four-point bending method, and stress loading was carried out according to the loading mode and stress calculation method of method E in standard GB/T 4157—2017. Before the test, all the samples were carried out in accordance with the treatment procedures specified in the standard GB/T 15970.2—2000 (ISO 7539-2) so as to avoid and reduce the existence of residual stress. After polishing, the samples were cleaned with distilled water and ethanol, dried under cool air, and stored in a dry N_2 atmosphere[15]. The high-salinity formation water extracted from the oilfield site was adopted as the test solution, and the ion concentration of the water sample was analyzed, as shown in Table 1. Before the test, nitrogen was fed into the solution for deoxygenation treatment, and nitrogen was continuously connected to the autoclave after the sample was installed to remove the oxygen. Then, the solution was heated to the required temperature, and the CO_2 and H_2S gases were added to the required partial pressure values. The test parameters are shown in Table 2, where H_2S partial pressure (p_{H_2S}) is divided into two types, $p_{H_2S}=$ 0.35MPa and $p_{H_2S}=$ 0.28MPa, which were to simulate the oil field's environment with different H_2S and CO_2 partial pressure ratios. Based on the actual situation of the oil field, the produced fluid comprises three phases of oil, gas, and water. As the produced fluid flows through the tubing, it may come into contact with either the gas phase or formation water (no consideration was given to the contact with crude oil to slow down corrosion here). Therefore, it is necessary to consider installing the sample in a gas and liquid phase environment for evaluation during the simulation testing and to control the amount of solution added according to the volume of the autoclave during the test. The schematic diagram of sample installation is shown in Figure 1.

Table 1 Analysis of ion composition of field formation water (g/L)

Cl^-	Mg^{2+}	Ca^{2+}	K^+	Na^+	SO_4^{2-}	HCO_3^-	pH	Salinity
186	0.035	0.11	7.05	104	24.3	1.07	6.28	292

Table 2 Simulation test parameters

Items	Test Temperature/℃	H_2S Content/MPa	CO_2 Content/MPa	Total Pressure/MPa	Test Period/h
Parameters	80	0.53	0.17	10	168
	120				
	180	0.28			

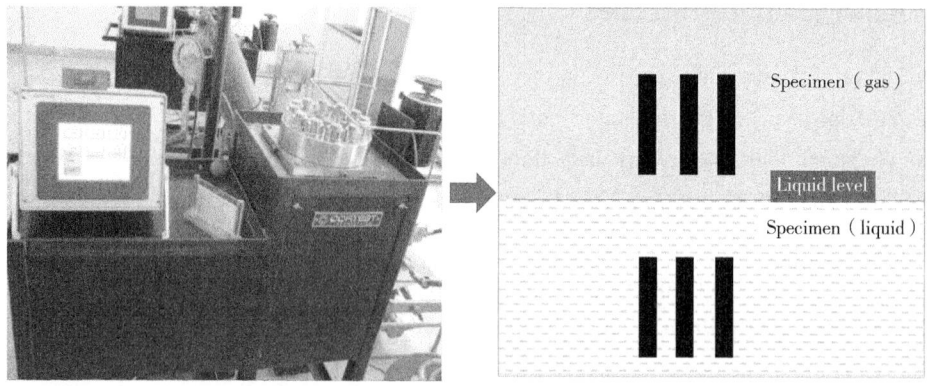

Fig. 1 Installation diagram of samples in autoclave

2.2 Weight Loss Tests

Weight loss tests were conducted in an autoclave to investigate corrosion rate in oilfield formation water. Prior to the weight loss tests, the samples were cleaned with distilled water and acetone, dried, and then weighed using a balance with a precision of 0.1mg. The weight values were recorded as the original weight (W_{0i}, $i = 1, 2, \cdots$). After the tests, all samples were taken out and immediately cleaned using distilled water and acetone. Then, the corrosion product scale on the sample's surface was removed with the film removal solution at room temperature and then rinsed and dried. After that, the samples were weighed again to obtain the final weight. The corrosion rate (v_{corr}) was calculated in mm/a (average corrosion thinning depth in years) from the weight loss by using Equation (1)[16] as follows:

$$v_{corr} = 8.76 \times 10^6 \times (W_{0i} - W_{1i})/(t \times \rho \times S), i = 1, 2, \cdots \quad (1)$$

Where W_{0i} and W_{1i} are, respectively, the original and final weights of the samples in g, S is the exposed surface area of the samples in mm^2, t represents the immersion time in h, and ρ is the steel density equaling $7.8 \times 10^{-3} g/mm^3$. An average corrosion rate of the three different samples for each test condition was reported as an overall corrosion rate for each set of conditions.

2.3 SCC Testing

Stress corrosion cracking testing was carried out using the immersion method under high temperature and high pressure conditions in an autoclave. In these tests, all sample surfaces required precision machining and a surface roughness of $R_a \leq 0.2\mu m$. The stress ($85\% YS_{min}$) was applied using the four-point bending method according to the GB/T 4157—2017, and stress loading can be achieved by Equation (2) as follows:

$$\sigma = 12E \times t \times y / (3H^2 - 4A^2) \tag{2}$$

Where H represents the distance between the two outermost fulcrums, E is the elastic modulus, A represents the distance between the inner and outer fulcrums, t is the sample thickness, and σ represents the stress value of the loading. Using these known parameters to calculate the maximum deflection y between the outer fulcrums, the stress is loaded by detecting the deflection change of the specimen. The schematic diagram of the four-point bending loading device is shown in Figure 2[17].

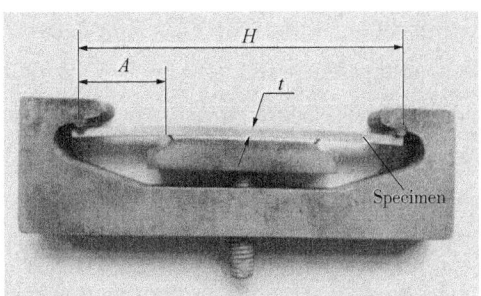

Fig. 2　Schematic diagram of stress application using four-point bending method

2.4 Characterization

The corrosion products were studied using scanning electron microscopy (SEM) and optical microscopy (OM). The elemental composition of the corrosion products was quantitatively analyzed using EDS spectrometer. For the samples under stress, it is necessary to observe any crack morphology and corrosion microstructure.

3　Results and Analysis

3.1　Average Corrosion Rate

After the test, the average corrosion rate of the material under two test conditions at different temperatures was calculated using the weight loss method. Figure 3 shows the trend of the average corrosion rates (v_{corr}) of the material. It can be seen from Figure 3 that the corrosion rate of the material under $p_{H_2S} = 0.53MPa$ is significantly higher than that under $p_{H_2S} = 0.28MPa$ when the test temperature is exceeding 120℃, indicating that increasing the H_2S partial pressure under the same condition can intensify the material corrosion. As the test temperature increases,

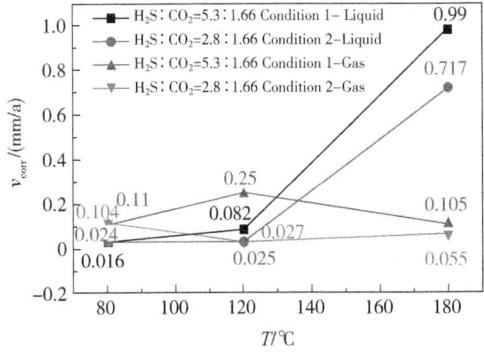

Fig. 3　Variation trend of corrosion rate of P110SS tubing material under different temperature environments

the average corrosion rate of the material in the liquid phase environment gradually increases. The corrosion level of the material was evaluated according to NACE SP0775-2023. Thus, it is indicated that the material exhibits low corrosion at the test temperature of 80℃, moderate corrosion at 120℃, and high corrosion at 180℃. Therefore, when the service temperature of the P110SS pipe is exceeding 120℃, it is recommended to take anti-corrosion measures, such as adding a matching corrosion inhibitor or coating, to prevent or slow down the corrosion damage of the pipe.

By comparison, it can be learned that there is no significant change in the corrosion of materials ina gas phase environment with the temperature. This relates to the extent of random adsorption and condensation of moisture on a material's surface in a gas phase environment. When the temperature is not exceeding 120℃, the materials in a gas phase environment are more susceptible to corrosion than those in a liquid phase environment. However, the corrosion of materials in a liquid phase environment is significantly higher than that in a gas phase environment at a temperature of 180℃, indicating that the effect of a high temperature increases the surface activity of the material and accelerates the electron transfer between the material and the solution medium, thereby rapidly strengthening the electrochemical interaction of the material with CO_2 and H_2S dissolved in water, further speeding up the corrosion[18-20].

3.2 Corrosion Morphology Characteristics

3.2.1 Macroscopic Corrosion Morphology under $p_{H_2S} = 0.53$ MPa

Figure 4 shows the macroscopic corrosion morphology of P110SS material under a simulated high-pressure gas phase environment of oilfield formation water at various temperature conditions. It can be observed that the sample's surface loses its metallic luster and is relatively rough, presenting an uneven spot corrosion. Due to the varying degrees of condensation of moisture containing corrosive gases on the surface of the sample in a gas phase environment, the corrosion varies in different areas. In particular, the local corrosion caused by moisture condensation is more obvious at the high temperature of 180℃, resulting in relatively large differences in the spotted color on the surface. By comparing the macroscopic corrosion morphology of P110SS materials in a liquid phase environment under the same condition (Figure 5), it can be seen that the surface of the sample is greyish black, flat, and uniform in general, but it is uniform and significantly rough at the high temperature of 180℃, without obvious pitting corrosion. This indicates that the surface of the material in the liquid phase environment has undergone uniform corrosion due to the sufficient contact with the corrosive media.

(a) 80℃　　　　　　　(b) 120℃　　　　　　　(c) 180℃

Fig. 4　Macroscopic corrosion morphology of materials in
gas environments under different temperatures

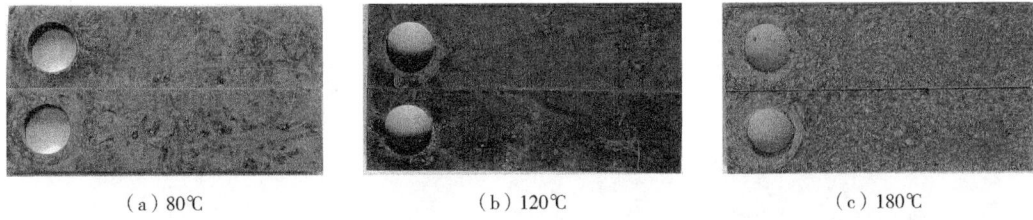

(a) 80℃　　　　　　　　(b) 120℃　　　　　　　　(c) 180℃

Fig. 5　Macroscopic corrosion morphology of materials in liquid environments under different temperatures

3.2.2　Micro-Corrosion Morphology when $p_{H_2S} = 0.53$ MPa

A scanning electron microscope was used to observe the microscopic corrosion morphology of P110SS material in gas and liquid phase environments. Figure 6 displays the corrosion morphology of the material under gas phase conditions at different temperatures, where uneven and rough corrosion product films are clearly visible, and there is variation in the accumulation of products across different regions. This indicates that the moisture is adsorbed and condenses differently on the surface of the sample, with droplets condensed locally, causing obvious local corrosion, and with thin infiltration films in some areas, resulting in varying corrosion. Figure 7 shows the microscopic corrosion morphology of P110SS material in a liquid phase environment, from which it can be seen that the surface of the specimen is relatively flat at the temperature not exceeding 120℃ compared to that in the gas phase environment. The material is uniformly exposed to a corrosive medium in an environment of solution immersion, with an equal probability of corrosion occurring. The corrosion product film peels off locally on the surface. At the high temperature of 180℃, the corrosion product film is especially rough and relatively thick, with the surface film partially peeled off, so that the bottom layer of the corrosion product film can be observed, without obvious pitting corrosion.

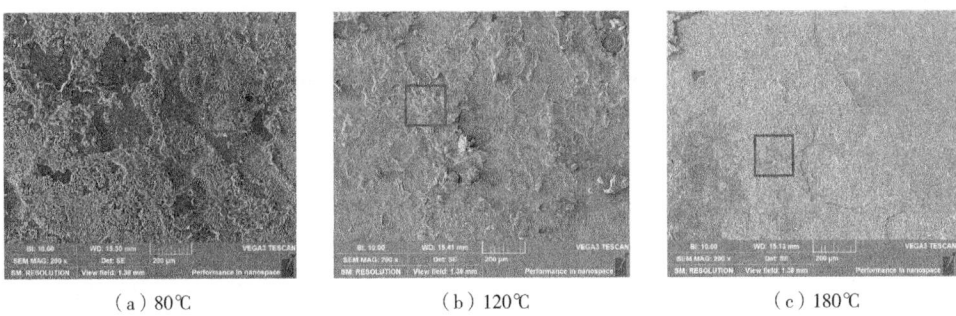

(a) 80℃　　　　　　　　(b) 120℃　　　　　　　　(c) 180℃

Fig. 6　Microscopic corrosion morphology of materials in gas environments under different temperatures

EDS was used to analyze the elemental composition of the corrosion product film on the surface of the material (see the red box area in Figure 6 and Figure 7). The results show that the corrosion product film on the surface of the gas phase and liquid phase are mainly composed of O, S, and Fe elements (Table 3), and with the increasing test temperature, the content of S in the corrosion product film increases gradually. However, no calcium and magnesium ions that are prone to scaling were found, so it was concluded that the corrosion product mainly consisted of a mixture of FeS and $FeCO_3$.

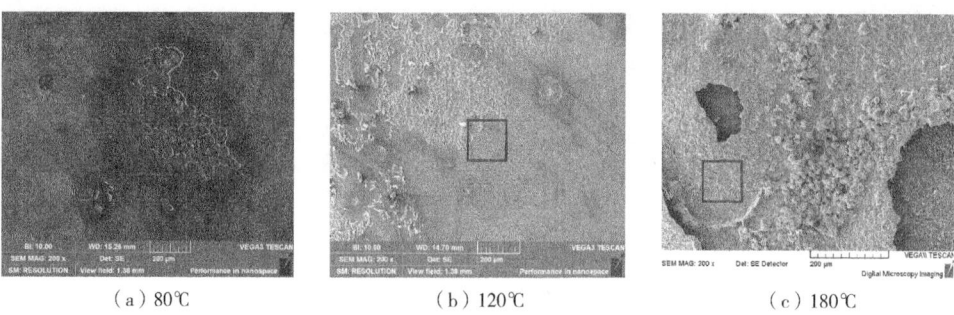

(a) 80℃　　　　　　　　(b) 120℃　　　　　　　　(c) 180℃

Fig. 7　Microscopic corrosion morphology of materials in liquid environments under different temperatures

Table 3　EDS analysis results of product films on the surfaces of materials under different test conditions　　　(%, mass fraction)

Ambient Temperature Elements	80℃		120℃		180℃	
	Gas	Liquid	Gas	Liquid	Gas	Liquid
O	7.12	8.82	8.69	9.86	9.29	8.48
S	29.09	10.21	32.49	18.64	29.41	32.47
Fe	63.79	80.09	58.82	73.49	59.77	59.06

By comparing the element contents of the surface corrosion products under the liquid immersion environment, it can be seen that the content of S shows an increasing trend with the gradual increase in the temperature, indicating that the high-temperature environment promotes the intensification of H_2S and substrate corrosion, and that the FeS content in the corrosion film increases, leading to more severe corrosion[21-22].

Figure 8 shows the cross-sectional morphology of the product film at the high temperature of 180℃, where the product film is in greyish black. EDS analysis was performed on the corrosion products in the red box. The result shows that the film is mainly composed of O, S, and Fe elements. As shown on the cross section, the substrate under the film is uneven due to corrosion, indicating that corrosion will further grow to the deeper layers when the corrosion product film on the surface is unable to resist the penetration of the corrosive media. The corrosion morphology indicates that the material constantly changes in the dynamic balance of " corrosion product film formation-film peeled off-film formation". The formation and densification of the film can slow down the further growth of the corrosion. On the contrary, when the film is loosened, a similar form of under-scale corrosion will occur[23-27].

3.2.3　Microscopic Corrosion Morphology under $p_{H_2S}=0.28\mathrm{MPa}$

Based on the change in the partial pressure of the corrosive gases in the produced fluid at the oil field, the corrosion performances of the materials at the decreasing H_2S partial pressure were analyzed comparatively. Figure 9 shows the surface corrosion morphology under different gas phase conditions at different temperatures. By comparison, it can be seen that the different areas on the surface are in different states due to the adsorption and condensation of moisture on the specimen

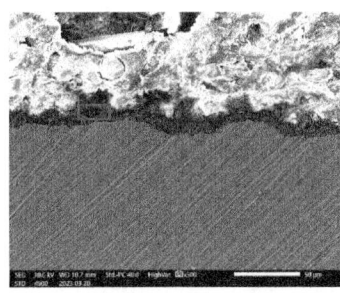
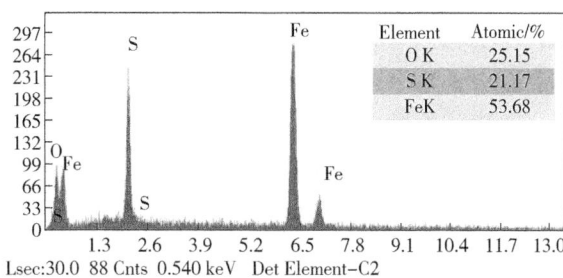

Fig. 8 Morphology and energy spectrum analysis on cross section of product film under ultra-high temperature of 180℃

surface. Generally, the spotted morphology is presented due to the accumulation of surface product, with local roughness as a result of corrosion. Figure 10 shows the micro-corrosion morphology of the sample in a liquid phase environment with p_{H_2S} = 0.28MPa. Compared with p_{H_2S} = 0.53MPa, the surface of the sample is relatively flat, especially when the simulated temperature is not exceeding 120℃, when the surface of the material is flat and uniform, and when there is no obvious local corrosion phenomenon. It can be seen that the corrosion degree is relatively reduced from the morphology when the H_2S content is relatively reduced, which is consistent with the change trend of the corrosion rate above. It shows that the presence and concentration of H_2S have a great influence on the erosivity of solution system, which is of great significance for the adaptability evaluation of anti-sulfur materials (such as P110SS) and the definition of critical indicators in an application environment. Similarly, the relatively thick corrosion product film is visible at a high temperature of 180℃, and a surface film and a bottom product film form after the surface film peels off. Hence, the corrosion is still severe in high-temperature environments, and the film's roughness is relatively reduced compared to that in the high-concentration H_2S environment with p_{H_2S} = 0.53MPa.

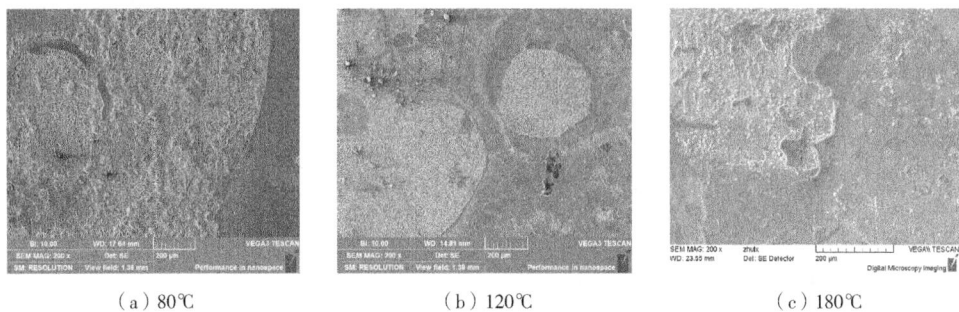

(a) 80℃　　　　　　　　(b) 120℃　　　　　　　　(c) 180℃

Fig. 9 Microscopic corrosion morphology of materials under gas phase environments at different temperatures

EDS was used to analyze the corrosion product films on the surfaces of the materials under liquid phase conditions at different temperatures (Figure 11), and it was found that they were mainly composed of O, S, and Fe elements. At the same time, with the increase in the simulated test temperature, the product films were relatively thicker at 180℃, and a second product film was visible at the shed area, indicating that the bare substrate continued to be corroded, and the content of S in the product films gradually increased. These results indicate that in the co-existence environment of CO_2/H_2S, corrosive gases cooperate with the high temperature to accelerate the electrochemical action of the solution system and promote the intensification of corrosion. Although the increasing temperature decreases the solubility of corrosion gas, the activity of the solution system increases with the increasing temperature, accelerating electron transfer. The Cl^- ion in the product film is the residue of the solution medium.

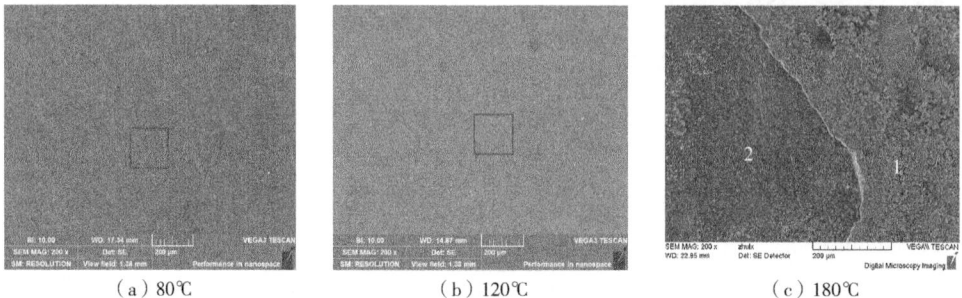

(a) 80℃　　　　　　　(b) 120℃　　　　　　　(c) 180℃

Fig. 10　Microscopic corrosion morphology of materials under the liquid phase environments at different temperatures

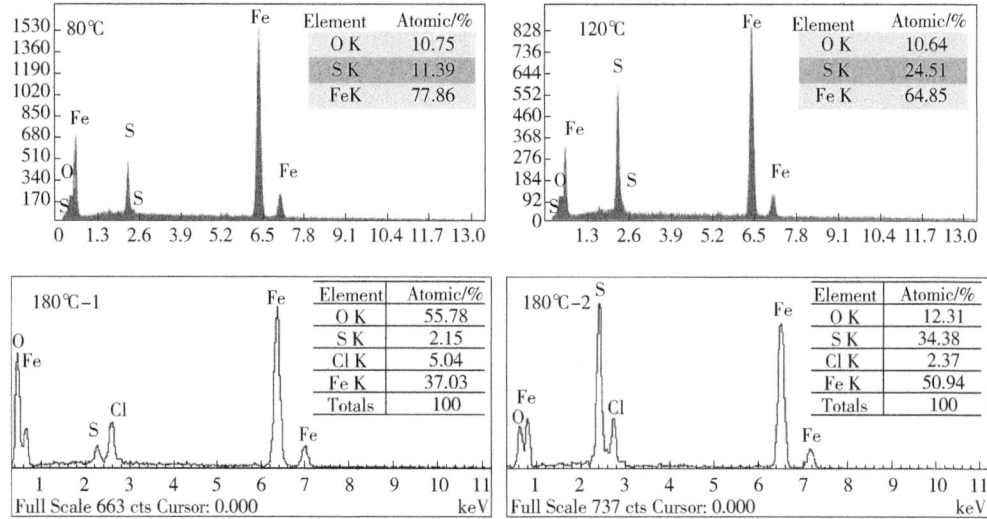

Fig. 11　EDS analysis results of product films on the surfaces of materials under the liquid phase environments

3.3　SCC in H_2S-Containing Environment

In order to study the stress corrosion cracking sensitivity of P110SS material in the high-salinity and high-temperature environments containing H_2S, the test was conducted by simulating the formation water environment using the high-temperature and high-pressure device. The test parameters are shown in Table 4.

Table 4　Specific test parameters (total pressure = 10MPa)

Items	Temperature/℃	H_2S Content/MPa	CO_2 Content/MPa	TestCycle/h	State	Applied Stress
Parameters	80, 120, 180	0.53	0.17	720	Liquid	$85\%YS_{min}$

After 720h of testing, no cracking or fractures were found in the samples. Figure 12 shows the surface corrosion morphology of the samples at different temperatures, with obvious grey corrosion products on the surface of the samples under low temperature conditions (80℃). However, no obvious local corrosion or cracking was found in the stress concentration area of the samples. Under the high temperature condition of 120℃, the stress concentration area on the surface of the sample remains flat, but visible uniform micro-pitted morphology can be observed. This indicates that the synergistic effect of the corrosive media and stress causes the pitting nucleation on the material's surface as the temperature increases. When the temperature rises to 180℃, the product scale in the surface stress concentration area is rough, loose, and partially detached. Additionally, it is observed by a 30× stereo microscope in the red box area of Figure 12(c) that the morphology of corrosion pits is present in the local detachment area (Figure 13). As observed by the stereo microscope, the depth of the pits is immeasurable, indicating the relatively shallow depth of the pits. This shows that the pitting nucleation appears on the material's surface with the increasing temperature under the condition of high-salinity formation water containing H_2S/CO_2 corrosive gas. In addition, the corrosion pits are relatively obvious at the high temperature of 180℃, but the pitting does not expand to cause SCC in the specimens.

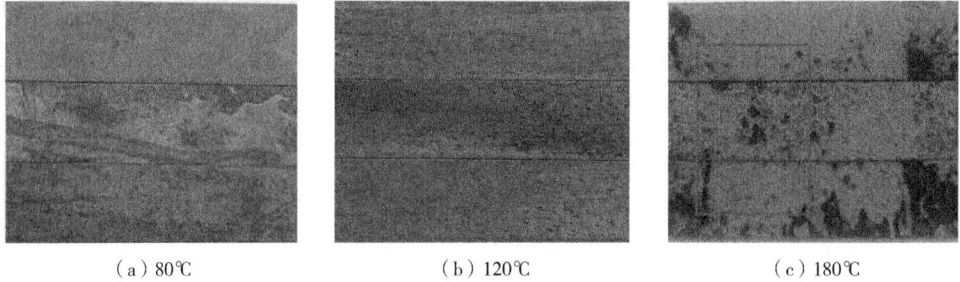

(a) 80℃   (b) 120℃   (c) 180℃

Fig. 12 Stress-sensitive corrosion morphology of formation water environment at different temperatures (×30 times)

Fig. 13 Macroscopic morphology of surface corrosion pits on the stressed specimen in 180℃ formation water environment

3.4 Discussion

3.4.1 Corrosion Mechanism

In the simulated environments of high temperature and high pressure, the gas phase differs significantly from the liquid phase. When the material is in the wet environment containing CO_2/H_2S, some areas of the surface become soaked by the moisture and form a liquid film, resulting in relatively uniform corrosion. However, in other areas, the local liquid film gradually accumulates into droplets with high concentrations of dissolved CO_2/H_2S corrosive gas. This leads to increased corrosion in those droplet areas and creates obvious local rough porphyritic corrosion morphology[28]. On the other hand, the H_2S/CO_2 corrosion gas dissolves and ionizes in the solution, and the ionized H^+ gives full play to the effect of cathode depolarization, promoting the rapid electrochemical action between the matrix material and the solution medium[29]. At the same time, the salinity in the solution medium is relatively high, which is conducive to the transfer, gain, and loss of electrons. In recent years, corrosion failures occurred frequently in high–well–depth and high–temperature environments, so the influence of temperature has been in the main position[9]. With the increasing temperature, the activity of the solution system is increased, and the surface activity of the material is strengthened, so the synergistic effects of multiple factors are promoted to deeper severe corrosion. The corrosion degree of the P110SS material increases significantly with the increasing temperature. At the high temperature of 180℃, the corrosion product film is relatively thick, and a second film can be seen at the shedding place, indicating that the exposed matrix has a further electrochemical interaction with the solution medium rapidly after the shedding of the film layer, resulting in an increase in the corrosion rate. Therefore, in high–temperature solution systems, a dynamic equilibrium of "corrosion–film layer shedding–film layer formation" occurs on the material surface. When the corrosion product film is dense, it can alleviate further corrosion expansion; however, if the film is loose, under–scale corrosion may occur.

The EDS analysis of the corrosion product film showed no calcium–magnesium ion scaling phenomenon, indicating that although the solution medium has a high salinity, the scaling trend is weak in the test system containing CO_2/H_2S. In the simulated test system, the H_2S and CO_2 corrosion gases dissolve and ionize a large amount of H^+, which makes the acidity of the solution increase relatively, which affects the formation of the calcium and magnesium scale layer in the high–salinity test system. So, the corrosion failure is mainly the form of under–scale corrosion caused by the accumulation of corrosion products.

3.4.2 Stress Sensitivity Analysis

According to the test results on the stress cracking sensitivity of the material, the synergistic effects of high temperature and stress induce the gradual pitting nucleation of the material. As the temperature increases from 80℃ to 180℃, the P110SS material's surface exhibits a trend of "no nucleation–small nucleation–pitting corrosion growth". Generally, the synergistic effects of high temperature and stress not only promote the intensification of corrosion, but also accelerate the Cl^- ions penetrating the film layer, resulting in local corrosion under the film layer. However, due to local pitting nucleation and limited deep expansion under the test environment, no cracking or fracture of the material was caused. Therefore, P110SS material is insensitive to sulfide stress corrosion cracking when p_{H_2S} is 0.53MPa at the high temperature of 180℃.

4 Conclusions

(1) In the simulated formation water environment with a high salinity of 292g/L and H_2S/CO_2

corrosion gas, the average corrosion rate of the P110SS material gradually increases as the temperature rises from 80℃ to 180℃, and when $p_{H_2S}=0.53$MPa and $p_{CO_2}=0.17$MPa, the corrosion rate of P110SS can reach up to 0.99 mm/a.

(2) In both liquid and gas phase environments, the morphology of the corrosion product film on P110SS varies due to the different states of the material that are in contact with the corrosive medium. In a gas phase environment, a locally rough and spotty-shaped corrosion product film is formed as a result of varying degrees of aggregation in the liquid film.

(3) When the simulated test temperature gradually increases from 80℃ to 180℃ with a $p_{H_2S}:p_{CO_2}$ ratio of 0.53:0.17, the P110SS material shows no sensitivity to SCC when loaded with stress at 85% YS_{min}. However, as the temperature increases, the material becomes more susceptible to pitting corrosion. At high temperatures (180℃), pitting occurs, but does not lead to cracking or fracturing of the material.

(4) When the simulated test temperature is exceeding 120℃, the P110SS material experiences extremely severe corrosion, Therefore, it is recommended to take appropriate anti-corrosion measures when using P110SS in an environment with a temperature of exceeding 120℃.

Author Contributions

X.Z. conducted the research and result analysis, J.L. and X.X. contributed to the design of the experiments and morphology analysis during the research activities, and B.Y., A.F., and C.L. contributed to image processing and data calculation. All authors have read and agreed to the published version of the manuscript.

Acknowledgement

This research was funded by the National Natural Science Foundation (52071338), the China National Petroleum Corporation for Science Research, and the technology development project (2021ZZ01-04).

References

[1] Zhang, Z.; Zheng, Y.S.; Li, J.; Liu, W.Y.; Liu, M.Q.; Gao, W.X.; Shi, T.H. Stress corrosion crack evaluation of super 13Cr tubing in high-temperature and high-pressure gas wells. Eng. Fail. Anal. 2019, 95, 263.

[2] Zhao, M.F.; Fu, A.Q.; Qin, H.-D.; Xie, J.-F.; Xie, G.; Long, Y.; Li, Y.; Wang, H. Overview and Future Research Prospect of Tubing String Corrosion of High Pressure and High Temperature Gas Well. Surf. Technol. 2018, 47, 44-50.

[3] Fu, A.Q.; Feng, Y.R.; Cai, R.; Yuan, J.T.; Yin, C.X.; Yang, D.M.; Long, Y.; Bai, Z.Q. Downhole corrosion behavior of NiW coated carbon steel in spent acid & formation water and its application in full-scale tubing. Eng. Fail. Anal. 2016, 66, 566-576.

[4] Chudyk, I.; Poberezhny, L.; Hrysanchuk, A.; Poberezhna, L. Corrosion of drill pipes in high mineralized produced waters. Procedia Struct. Integr. 2019, 16, 260-264.

[5] Gao, S.; Dong, C.; Fu, A.; Xiao, K.; Li, X. Corrosion behavior of the expandable tubular in formation water. Int. J. Miner. Metall. Mater. 2015, 22, 149-156.

[6] Zhang, P. Study on Formation Water and Wellbore Corrosion and Scaling after CO_2 Flooding in Changqing. Ph. D. Thesis, China University of Petroleum (Beijing), Beijing, China, 2022.

[7] Wang, S.; Zheng, X.; Li, M.; Huang, X.; Guan, J.; Zheng, S.; Chen, C.; Chen, Y. Stress corrosion cracking sensitivity of sulfide-resistant casing steel P110SS in hyperbaric H_2S/CO_2 enviroments. Corros. Prot. 2013, 34, 189-192.

[8] Deng, H.; Li, Y.; Zhang, Z.; Lu, Q.; Hou, D. Environment cracking and surface protection of C110 casing in sour and deep well. J. Southwest Pet. Univ. 2021, 43, 118-128.

[9] Huang, S.; Ke, Y.; Liu, J.; Fan, Z. Analysis of Corrosion Influencing Factors of P110SS Steel in Simulated Annular Fluid Environment of Gas Well Containing Sulfur. Mater. Prot. 2022, 55, 46-51.

[10] Liu, J.; Wang, Y.; Li, G.; Wang, Q.; Wang, F.; Luo, J.; Zhao, X. Study on the Corrosion Behavior of Oil Casing String in Annular Protection Fluid Environment of Kunteyi Block. Mater. Prot. 2022, 55, 46-53.

[11] Li, M.; Zhang, Z.; Li, Y.; Dong, X.; Li, Q. Study on the Influence Factors of Corrosion Behavior of Gas Well Tubing in Environments Containing $CO_2-H_2S-Cl^-$. Mater. Prot. 2022, 55, 41-45.

[12] Wang, Y. Corrosion rule of P110SS under high H_2S and CO_2 conditions. Fault-Block Oil Gas Field 2017, 24, 863-865.

[13] Zhang, N.Y.; Zhang, Z.; Zhao, W.T.; Liu, L.; Shi, T.H. Corrosion Evaluation of Tubing Steels and Material Selection in the CO_2/H_2S Coexistent Environment. In CORROSION 2018; NACE International: Phoenix, AZ, USA, 2018.

[14] Bueno, A.H.S.; Moreira, E.D.; Gomes, J.A.C.P. Evaluation of stress corrosion cracking and hydrogen embrittlement in an API grade steel. Eng. Fail. Anal. 2014, 36, 423.

[15] Xue, H.; Feng, Y.; Tang, S.; Zhang, J. Electrochemical Corrosion Behavior of 15Cr-6Ni-2Mo Stainless Steel with/without Stress under the coexistence of CO_2 and H_2S. Int. J. Electrochem. Sci. 2018, 13, 6296-6309.

[16] JB/T (1999). GB/T Standard 7901; Metal Materials-Uniform Corrosion-Methods of Laboratory Immersion Testing. Institute of Integrated Technology and Economics of Mechanical Industrial Instruments and Meters: Beijing, China, 2001.

[17] Zhao, X.; Huang, W.; Li, G.; Feng, Y.; Zhang, J. Effect of CO_2/H_2S and Applied Stress on Corrosion Behavior of 15Cr Tubing in Oil Field Environment. Metals 2020, 10, 409.

[18] Liu, Q.Y.; Mao, L.J.; Zhou, S.W. Effects of chloride content on CO_2 corrosion of carbon steel in simulated oil and gas well environments. Corros. Sci. 2014, 84, 165-171.

[19] Zhang, G.A.; Zeng, Y.; Guo, X.P.; Jiang, F.; Shi, D.Y.; Chen, Z.Y. Electrochemical corrosion behavior of carbon steel under dynamic high pressure H_2S/CO_2 environment. Corros. Sci. 2012, 65, 37.

[20] Kittel, J.; Ropital, F.; Grosjean, F.; Sutter, E.M.M.; Tribollet, B. Corrosion mechanisms in aqueous solutions containing dissolved H_2S. Part 1: Characterisation of H_2S reduction on a 316L rotating disc electrode. Corros. Sci. 2013, 66, 324-329.

[21] Zhao, X.H.; Han, Y.; Bai, Z.Q.; Wei, B. The experiment research of corrosion behavior about Ni-based alloys in simulant solution containing H_2S/CO_2. Electrochim. Acta 2011, 56, 7725.

[22] Jiang, W.J. Piping Material Selection for Wet H_2S Environment of Hydrotreating Unit. Shandong Chem. Ind. 2019, 48, 118.

[23] Lin, P. Study on Mechanism and Inhibition Strategy of Under-Deposit Corrosion of Carbon Steel Pipeline in Oil and Gas Fields. Ph.D. Thesis, University of Science and Technology of China, Jiangsu, China, 2021.

[24] Zhu, Y.; Qiu, Y.; Guo, X. Underscale corrosion behavior of carbon steel in a NaCl solution using a new occluded cavity cell for simulation. J. Appl. Electrochem. 2009, 39, 1017-1023.

[25] He, S.; Luo, S.; Zhao, H.; Xue, P. Review of Carbon Steel Under Deposit Corrosion in CO_2 Environment.

Surf. Technol. 2023, 52, 148-157.

[26] Liu, Y.; Chang, Z.; Zhao, G.; Xue, Y.; Niu, K. Corrosion Behavior of Super 13% Cr Martensitic Stainless Steel Under Ultra-deep, Ultra-high Pressure and High Temperature Oil and Gas Well Environment. Mater. Heat Treat. 2012, 41, 71-75.

[27] Zhao, G.; Chen, C.; Lu, M.; Li, H. The formation of product scale and mass transfer channels during CO_2 corrosion. J. Chin. Soc. Corros. Prot. 2002, 22, 363-366.

[28] Li, M.; Hu, Z.; Yang, G.; Wu, C. CO_2 Corrosion Behaviour of Three kinds of Pipeline Steels with Different Cr Contents. Corros. Prot. 2013, 34, 586-589.

[29] Zakroczymski, T.; Glowacka, A.; Swiatnicki, W. Effect of hydrogen concentration on the embrittlement of a duplex stainless steel. Corros. Sci. 2005, 47, 1403-1414.

本论文原发表于《Coatings》2023 年第 13 期。

Failure Analysis of S13Cr-110 Telescopic Tube Used in an Ultra-deep Gas Well

Zhu Lixia[1] Luo Jinheng[1] Long Yan[1] Li Lei[1] Xie Junfeng[2]

(1. State Key Laboratory of Performance and Structural Safety for Petroleum Tubular Goods and Equipment Material, CNPC Tubular Goods Research Institute;
2. Tarim Oilfield Company, PetroChina Company Limited)

Abstract: The failure reason and stress corrosion cracking mechanism of S13Cr-110 telescopic tube used in an ultra-deep gas well are discussed based on physical and chemical properties, operating conditions and construction processes. The results show that: stress corrosion cracking occurred in S13Cr-110 telescopic tube during acid fracturing, the crack initiated from the bottom of the corrosion pit in the inner wall of the telescopic tube and the crack propagation mode is inter-granular cracking. The reason for cracking is that S13Cr-110, the telescopic tube material, is exposed to high-chlorine-containing acid environment for a long time in the acid fracturing process, and the corrosion inhibitor used at the temperature of ultra-deep well has poor corrosion inhibition effect, resulting in inter-granular crack originated from the bottom of the corrosion pit on the inner wall of the telescopic tube. During the subsequent acidizing operations, the telescopic tube sustained a high circumferential tensile stress, stress corrosion occurs, and the stress corrosion cracks expand rapidly. It is suggested that the S13Cr-110 pipe string used in acidic environments with high chloride concentrations should be selected with appropriate corrosion inhibitor based on well depth and formation temperature.

Keywords: S13Cr-110; Stress corrosion cracking; Acid fracturing; Failure; Ultra-deep gas well

1 Introduction

With the continuous increasing global demand for oil and gas, and the continuous progress of drilling and completion technology, oil and gas exploitation are gradually developing towards deep and ultra-deep wells. These environments usually contain CO_2, H_2S and high salinity formation water, which puts forward higher requirements for the performance of oil well tube materials. Ordinary carbon steels such as J55, N80 and P110 can no longer meet the requirements of harsh working conditions. High strength stainless steels are increasingly used in oil and gas field industry.

Corresponding author: Zhu Lixia, zhulx@ cnpc. com. cn.

JFE in Japan and Baosteel in China have developed super 13Cr martensitic stainless steel (S13Cr-110) by reducing carbon content to inhibit the precipitation of chromium carbide phase and increasing nickel content to obtain single-phase martensitic structure[1]. S13Cr-110 has good comprehensive mechanical properties, good corrosion resistance and low cost, which is more and more widely used in high CO_2 well, deep well and ultra-deep wells.

Currently, majority of high-temperature and high-pressure (HPHT for short) gas wells adopt acid fracturing technology for stimulation and reconstruction. However, several failure cases of S13Cr-110 have been reported in recent years, all of which are related to S13Cr-110 contacting acidizing fluid during application. Fu [2] et al. introduced the stress corrosion cracking failure of S13Cr-110 in high-temperature and high-pressure gas wells, which was caused by the synergistic effect of KH_2PO_4 phosphate and CO_2; Lei [3] et al. showed that S13Cr-110 was sensitive to SCC during acidification, so the acidification process must be precisely controlled; Qi [4] et al. studied the effect of acidizing process on the stress corrosion cracking of HP-13Cr stainless steel in the ultra-depth well environment, and believed that the acidizing process significantly increased the stress corrosion cracking susceptibility of HP-13Cr stainless steel and induced the fracture mode to the brittle characteristic in the high temperature and CO_2 pressure environment; Lei[5] et al. showed that when the temperature of high pressure gas well is 90℃, the pitting corrosion of S13Cr-110 stainless steel mainly occurs in the returned acidizing fluids stage; Meng [6] et al. believed that the main reason for stress corrosion cracking was that the existing acid corrosion inhibitor was ineffective in eliminating the stress corrosion of S13Cr-110 material tube caused by acidic and high-concentration brine under high temperature environment; Yao [7] et al. studied the effects of pH value on the stress corrosion cracking of S13Cr tubing steel and showed that the tendency of the stress corrosion cracking increased with decreasing pH value in acidic medium; Zhao [8] et al. show that the corrosion behavior of S13Cr stainless steel in acidizing solution is different from that of ordinary 13Cr stainless steel, which requires special acidizing corrosion inhibitor. All the above studies indicate that continuous attention should be paid to acid corrosion of S13Cr-110 pipe string in the acidizing fracturing stage of HPHT gas wells.

This paper systematically analyzes the cracking reason of S13Cr-110 telescopic tube used in a HPHT gas well in western China. The well depth of the HPHT gas well is 8882m, and the packer is located at 8582.65m. The well was fractured between 8737m and 8750m, and a total of 56m³ base fluid of fracturing fluid was squeezed into the wellbore. When 35.5m³ of gelled acid was pumped (corresponding well depth is about 5243m), it took about 14 hours to rectify the wellhead due to tube blocking, then the acidizing operation was continued. During acidizing operation, it is found that the oil pressure decreases and the casing pressure increase. After pressure relief, the string was pulled out for inspection, and it was found that the first telescopic tube at the upper part of the packer cracked and failed (as shown in Fig. 1). The failed telescopic tube was located at depth of 8561.92m, with an outer diameter of 127.76mm and a wall thickness of 7.34mm, and is produced according to API 5CRA-2010《Specification for corrosion resistant alloy seamless tubes for use as casing, tubing and coupling stock》[9] and the oilfield's supplementary technical specifications (STS for short). The main component of fracturing fluid is 20%KCl + assistant + clean water, which has a neutral pH. The main component of gelled acid is 20%HCl + assistant, the pH value is less than 1. The formation temperature at the well depth of 5243m is about 110℃, and 170℃ at 8561.92m.

2 Failure analysis

2.1 Material Analysis

Chemical composition analysis, tensile test, Charpy impact test and hardness test were carried out on the telescopic tube according to ASTM A751, ASTM A370, ASTM E23 and ASTM E18 respectively using Optical spectrum analyzer, tensile testing machine, impact testing machine and Rockwell hardness tester. Microstructures were analyzed using OLS4100 laser scanning confocal microscopy (LSCM). The chemical composition analysis results are shown in Table 1, and the mechanical properties are shown in Table 2. The metallographic structure is tempered martensite, as shown in Fig. 2. The material meets the manufacturing requirements of S13Cr-110.

Fig. 1　Failure appearance of telescopic tube　　　Fig. 2　Microstructure of telescopic tube

Table 1　Results of chemical composition analysis(%, mass fraction)

Element	C	Si	Mn	P	S	Cr	Mo	Ni	V
Tube	0.020	0.24	0.34	0.0093	0.0009	12.8	2.08	6.0	0.13
API 5CR3-2010 & STS	≤0.03	≤0.5	≤0.5	≤0.015	≤0.005	11.5~13.5	1.5~3	4.5~6.5	≤0.5

Table 2　Mechanical properties

Test project	Tensile property			Impact toughness(-10℃)	Hardness/HRC		
	Tensile Strength/MPa	Yield Strength/MPa	Elongation after fracture/%	Absorbed energy/J	Outer surface	Wall thickness Center	Inner surface
Tube	893	808	30	232	29	28	28
	890	798	29	203			
	885	800	29	193			
API 5CR3-2010 & STS	≥827	758~965	≥16	≥140	≤32		

2.2 Macroscopic analysis of crack

The macroscopic morphology of the outer wall of the failed telescopic tube is shown in Fig. 3

and Fig. 4. The outer wall of the telescopic tube cracks along the axial direction, with a crack length of 111.34mm and a maximum width of 2.52mm. There are many small cracks visible at the ends of the crack, which are arranged along the axial direction of the tube and have the characteristics of step like propagation. The telescopic tube is cut along the axial direction, and the macroscopic morphology of the inner wall is shown in Fig. 5 and Fig. 6. The crack is 164.02mm long and 2.76mm wide, and dense axial shallow cracks can be observed on the inner wall. According to the length and width of the crack, it can be judged that the crack originated from the inner wall.

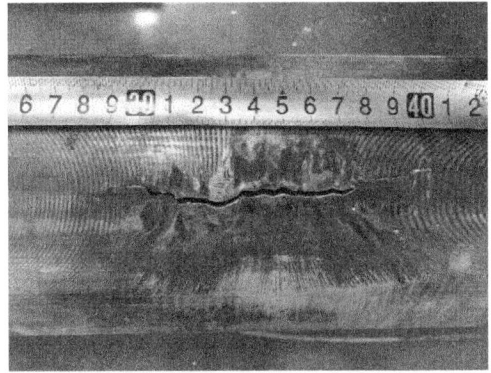

Fig. 3 Macroscopic appearance of the outer tube

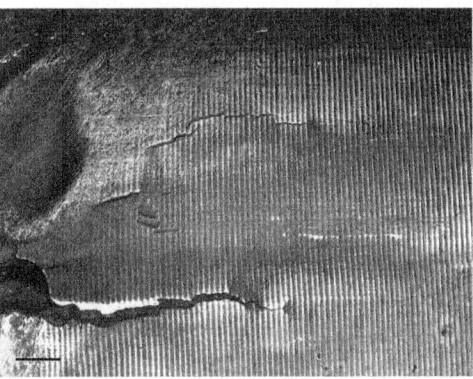

Fig. 4 Morphology of small cracks

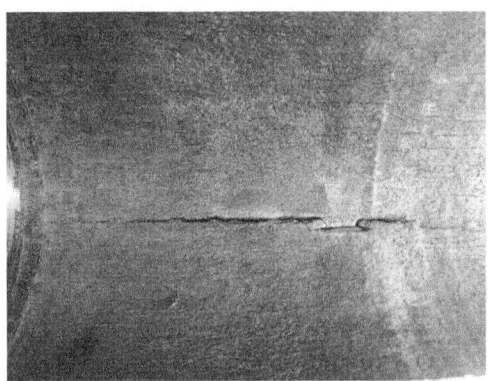

Fig. 5 Morphology of inner wall

Fig. 6 Shallow crack morphology of inner wall

2.3 Microanalysis of cracks

Samples were taken from the cracking part of the telescopic tube as shown in Fig. 7, the cracks were observed by TESCAN-VEGAII scanning electron microscope and OLS4100 LSCM respectively. The fracture surface was covered by a large number of corrosion products, and the morphology of inter-granular could be observed locally, as shown in Fig. 8 and Fig. 9. Energy dispersive spectroscopy analysis at the crack tip shows that the main elements were Fe, O, Ca, Cr, Mg and Cl, as shown in Fig. 10. Metallographic analysis (Fig. 11) shows that there are a large number of cracks spreading from the inner wall to the outer wall, the root of the cracks is wider, most of the cracks originated from the corrosion pit bottom of inner wall. The microstructure around the crack was tempered martensite. No abnormal structure or deformation was observed. As shown in Fig. 11

(b) and Fig. 11 (c), the main crack has a large number of cracks branches distribute interlacedly, and the crack branches in many areas along the main crack are inter-granular. Combined with the inter-granular fracture characteristics under the corrosion products in Fig. 9, it is believed that the cracks preferentially initiate along the grain boundary, gradually expand into large cracks and form a network under the corrosion environment. The big intergranular crack becomes wide enough to allow the corrosive environment that created the pits into the crack. The fracture with this feature is essentially inter-granular fracture, and the fracture is caused by this crack expansion. This is a typical feature of stress corrosion cracking. Due to "post-fracture corrosion", the inter-granular corrosion crack surface was relatively obscured.

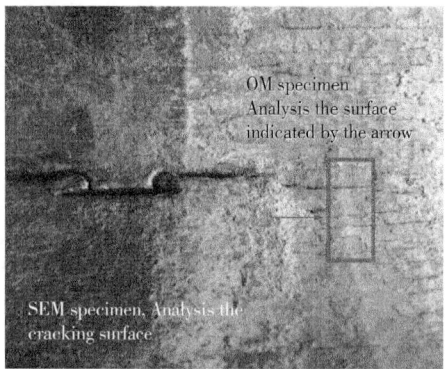

Fig. 7 Sampling location

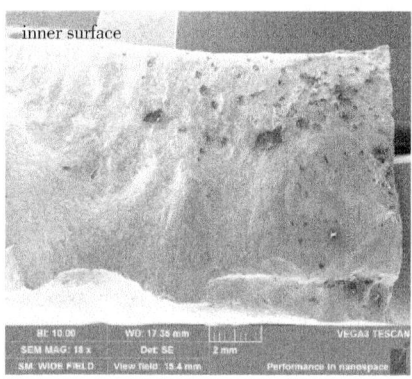

Fig. 8 Macroscopic morphology of fracture

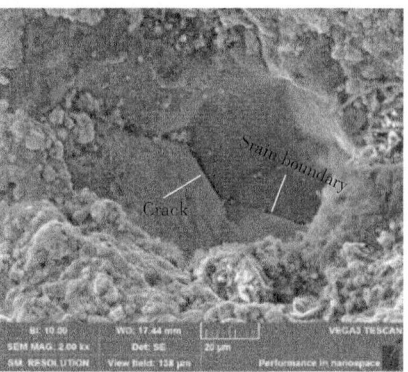

Fig. 9 Microscopic morphology of inter-granular fracture

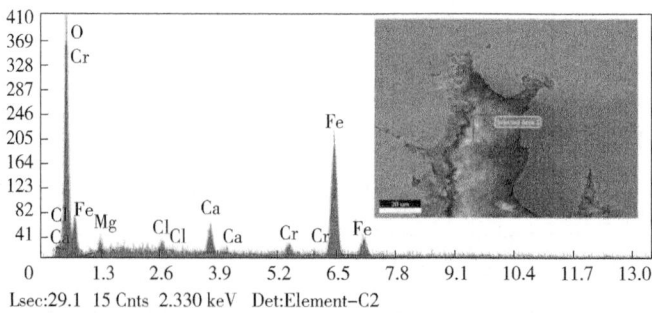

Fig. 10 Position and spectrum of energy spectrum analysis of crack tip

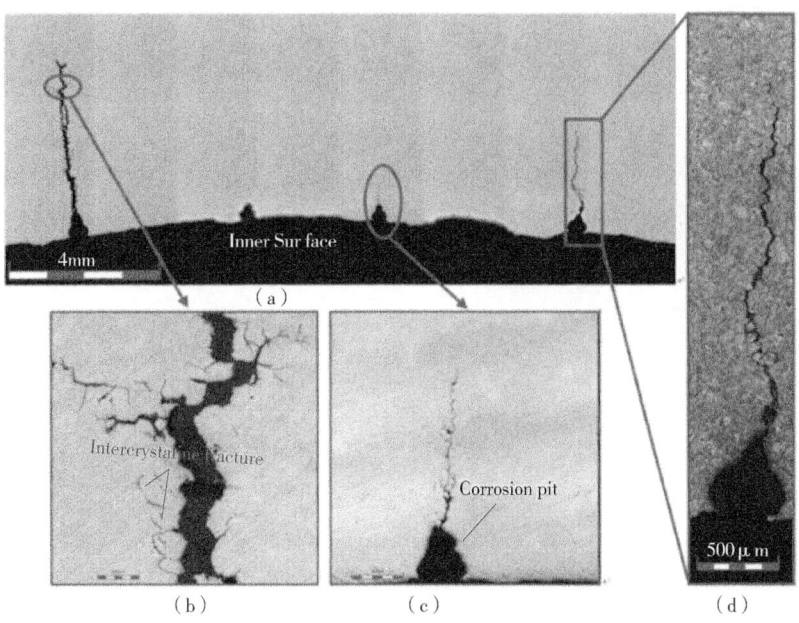

Fig. 11 Appearance of inner wall crack propagation

2.4 Acid corrosion simulation test

The failure of the telescopic tube occurred in the process of acidizing. Therefore, a simulated corrosion test was carried out with reference to the corrosive environment used in acidizing (20%HCl +2% corrosion inhibitor, total pressure is 10MPa, test temperature 170℃). The relevant testing standards are shown in Reference[10]. The corrosion inhibitor used in the test is consistent with the acidizing corrosion inhibitor used in the field operation, Sichuan Beide brand, BD1-20. The test was carried out in high-pressure autoclave with corrosion coupon weight-loss method, and the corrosion time was 4h. The test results are shown in Table 3, and the morphology of sample after acid corrosion is shown in Fig. 12 to Fig. 14. The results show that the sample is thinned and the surface is covered with corrosion products. The average corrosion rate is 133.46g/(m² · h), and the corrosion inhibitor has poor inhibition effect. The cross-section analysis of the sample after the test shows that there are many corrosion pits on the sample surface, and there are cracks at the bottom of most of the corrosion pits. The cracks are tortuous and branched, and the inter-granular morphology can be observed at the crack tip. It showed that after acid corrosion, the uncracked part of the telescopic tube has a crack morphology similar to that of the cracked part.

Table 3 Corrosion test results of HTHP acid solution

Weight before test/ g	Weight after test/ g	Corrosion temperature/ ℃	Corrosion rate/ [g/(m² · h)]	Average corrosion rate/ [g/(m² · h)]
8.8652	8.2961		130.4682	
9.0559	8.4403	170	138.5469	133.46
9.1128	8.5239		131.3728	

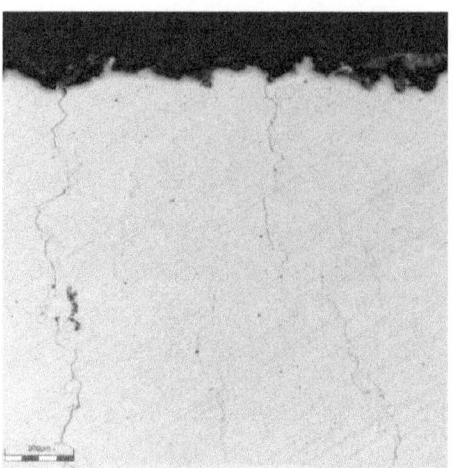

Fig. 12 Morphology after corrosion   Fig. 13 Surface crack

Fig. 14 Intergranular morphology of crack tip

2.5 Residual stress test

A 300mm pipe ring was taken from the uncracked part of the telescopic tube, and the residual stress was tested by ring cutting method[11-12]. After the pipe ring was cut and placed for 24 hours, the opening gap was measured to be 1.34mm. The circumferential residual stress of the telescopic tube is calculated to be 36.7MPa, which is tensile stress.

3 Results and discussion

The physical and chemical properties of the telescopic tube showed that its microstructure was tempered martensite. The chemical composition, tensile, Charpy impact and hardness test results all met the requirements of the standard for S13Cr-110.

The crack of the telescopic tube is axial, and there are many small cracks distributed axially and propagating along a step shape at each end of the crack, as is shown in Fig. 7. The crack width of the inner wall is larger than that of the outer wall, which can be judged that the crack originated from the inner wall, as is shown in Fig. 11(c). Besides, the cracks originated at the bottom of the inner wall corrosion pit, as is shown in Fig. 11(a), the overall crack morphology is arborization, and the fracture is inter-granular, which presented typical characteristics of stress corrosion cracking, indicating that the stress corrosion cracking occurred in the telescopic tube.

The material used for the telescopic tube is S13Cr-110, which is a typical alloy of Fe-Cr-Ni-C type. The research shows that this material is sensitive to stress corrosion and is prone to stress corrosion crack. Chloride is the sensitive medium of stress corrosion cracking of Fe-Cr-Ni-C alloy [13]. In the environment containing Cl^-, when concentration of Cl^- is less than 15%, the degree of stress corrosion of S13Cr-110 steel is relatively light [14]. However, with the increase of Cl^- concentration in the environment, the mechanical properties, SCC resistance decrease and SCC tendency increase of S13Cr-110 steel [15], and the acidification process obviously promoted the SCC susceptibility [4].

According to the failure background, acidizing fracturing was carried out in the well where the failure telescopic tube used. When 56 cubic meters of fracturing fluid (20%KCl + assistant + clean water) was squeezed into the wellbore and 35.5 cubic meters of gelling acid (20%HCl + assistant) was pumped (the gelled acid is located at the corresponding to the well depth of about 5243m, the formation temperature is about 110℃, and the basic fracturing fluid is below the string), the gelling acid and the fracturing fluid remained in the string for about 14 hours due to the wellhead rectification due to construction obstruction. However, the failed telescopic tube was located at a depth of 8561.92m (the formation temperature is about 170℃), and the gel acid trapped in the string did not reach the depth of the telescopic tube. With the increase of the retention time, the liquid temperature in the wellbore will gradually increase and approach the formation temperature, and there is a temperature difference between the liquids with large well depth difference. Under the action of temperature and concentration difference, there will be convection between gelling acid and fracturing fluid and gradually mix: (1) The mixed fluid will be acidic, so that the telescopic tube which located at the well depth of 8561.92m will be immersed in the acidic mixed fluid containing HCl; (2) With the increase of the retention time, the temperature of the acid mixture gradually increases to 170℃, and the telescopic tube is immersed in the high-temperature acid mixture. Therefore, S13Cr-110 steel is in the acidic environment with High Cl^- which has the medium conditions for stress corrosion. Although 2% corrosion inhibitor was added into the gelled acid, the acid corrosion simulation test results showed that the corrosion inhibitor could not effectively inhibit the occurrence of corrosion of S13Cr-110 in this environment, resulting in the initiation of inter-granular cracks in the inner wall of telescopic tube. According to the real-time working condition results, it can be concluded that the inter-granular corrosion cracks have occurred in the telescopic tube before acid fracturing operation, but the cracks did not penetrate the wall thickness.

The circumferential stress of telescopic tube is mainly the superposition of internal pressure and circumferential residual stress. According to the pumping pressure and casing pressure during fracturing and acidification before the telescopic tube fails (Table 4), it can be calculated that the internal pressure sustained by the telescopic tube during fracturing (after squeezing gelled acid) is 30.2 ~ 66.4MPa (corresponding to the circumferential tensile stress 261 ~ 574MPa). The calculation formula of internal pressure is shown in Equation (1). In the process of acidizing, the internal pressure is 41.5~70MPa (corresponding to the circumferential tensile stress of 241~606MPa). The residual stress test shows that the circumferential residual stress sustained by the outer cylinder of the telescopic tube is 36.7MPa. Therefore, when the two are combined, the actual circumferential tensile stress sustained by the telescopic tube is 297.7~610.71MPa in fracturing construction stage (after squeezing in gelled acid) and 277.7~642.7MPa in acidizing construction stage. The calculation formula of actual circumferential tensile stress is shown in Equation (2). Due to the pressure test report of this batch of S13Cr-110 tube shows that its ultimate circumferential internal pressure is 106.7MPa (corresponding to the circumferential tensile stress of 923MPa), it can be seen that the circumferential tensile stress sustained by the telescopic tube during acidification construction is in the higher stress range of circumferential bearing capacity.

Table 4 Pump pressure and casing pressure during fracturing and acidizing operation

Process	Pumping procedure	Pumping pressure/MPa	Casing pressure/MPa	Internal pressure/MPa
Fracturing	Fracturing fluid	64.0~110.1	51.0~59.2	50.8~59.1
	Gelling acid	108.5~118.0	51.6~58.2	59.8~66.4
	Fracturing fluid	55.0~91.3	59.2~61.1	30.2~32.1
	Gelling acid	91.1~97.0	58.5~60.3	36.7~38.5
Acidification	Gelling acid	12.5~117.3	47.9~57.8	59.5~69.4
	Fracturing fluid	99.6~122.9	52.9~58.1	41.5~70.0
	Cross-linked acid	99.6~118.6	53.2~58.0	41.6~65.4
	Fracturing fluid	114.3~118.5	52.9~58.1	56.2~65.6
	Gelling acid	108.1~116.1	52.8~54.5	53.6~63.3

$$\text{Internal pressure} = |\text{Pumping pressure} - \text{Casing pressure}| \tag{1}$$
$$\text{Actual stress} = \text{Tensile stress} + \text{Residual stress} \tag{2}$$

In conclusion, combined with physical and chemical properties, operating conditions and construction process, the failure mode of S13Cr-110 telescopic tube is stress corrosion cracking. The crack initiated from the inner wall of the telescopic tube, and the crack propagation mode is inter-granular cracking. The reason for failure is that: when the acid fracturing construction is blocked, the telescopic tube was immersed in the acidic mixed solution containing HCl for a long time, and the temperature of the mixed solution gradually rises to the formation temperature, the corrosion inhibition effect of the inhibitor decreases [16]. The inter-granular crack originated from the bottom of the corrosion pit on the inner wall of the telescopic tube. In the subsequent acidizing construction, the telescopic tube sustained a high circumferential tensile stress, the stress corrosion cracks expand rapidly along the axial direction, and the cracks merge in the process of expansion to form a main crack, which finally penetrates the wall thickness to cause cracking. Crack expansion rate of SCC increases with pH value (pH<7), which is due to the gradual consumption of Cl⁻ and the slowing down of acidification. Therefore, the crack growth rate of SCC is negatively correlated with pH value. The initial crack is caused by corrosion. Then, when stress is introduced, the double effect of corrosion and stress shortens the crack initiation period, which makes the material easier to fracture, that is, it shows larger primary cracks and denser secondary cracks.

4 Conclusion

(1) Chemical composition and mechanical properties (including tensile, hardness and impact properties) of S13Cr-110 telescopic tube meets the standard requirements. The microstructure of S13Cr-110 tube is tempered martensite.

(2) The fracture failure of S13Cr-110 telescopic tube is caused by cracks initiated from the bottom of the corrosion pits on the inner surface. The crack spreads from the inner surface to the outer surface in a "arborization" shape and the crack tip has inter-granular characteristics, which demonstrate the SCC characteristics.

(3) The failure mode of the telescopic tube is stress corrosion cracking. The failure reason is that S13Cr-110 has been in contact with acid for a long-time during acid fracturing. With theincrease of acid temperature, the inhibition effect of the inhibitor decreases. The inner wall of the telescopic tube initiated inter-granular cracks due to corrosion. In the subsequent acidification process, due to high circumferential tensile stress, stress corrosion occurs, and the stress corrosion cracks expand rapidly, resulting cracking.

(4) It is recommended that the S13Cr-110 pipe string used in the acid environment with high chloride concentration should select the appropriate corrosion inhibitor according to the well depth and formation temperature.

References

[1] K. H. Lo, C. H. Shek, J. K. L. Lai. Recent developments in stainless steels[J]. Materials Science and Engineering R, 65 (2009) 39-104. https://doi.org/10.1016/j.mser.2009.03.001

[2] Fu Anqing, Long Yan, Liu Hongtao, et al. Stress corrosion cracking behavior of super 13Cr tubing in phosphate packer fluid of high pressure high temperature gas well[J]. Eng. Failure Anal., 2022, 139. https://doi.org/10.1016/j.engfailanal.2022.106478

[3] X. W. Lei, Y. R. Feng, A. Q. Fu, et al., Investigation of stress corrosion cracking behavior of super 13Cr tubing by full-scale tubular goods corrosion test system[J], Eng. Failure Anal., 50 (2015) 62-70. https://doi.org/10.1016/j.engfailanal.2015.02.001

[4] Qi Wenlong, Zhao Yang, Zhang Tao, et al., Effect of Acidizing Process on the Stress Corrosion Cracking of HP-13Cr Stainless Steel in the Ultra-depth Well Environment[J], Frontiers in Materials, 8 (2021). https://doi.org/10.3389/FMATS.2021.732931

[5] Lei Bing, Ma Yuantai, Li Ying, et al., Pitting behavior of HP2-13Cr in simulated high pressure and high temperature gas well environment [J]. Corrosion Science and Protection Technology, 2013, 25 (2): 100-104.

[6] Meng Xuangang, Wu Wen, Peng Fen, et al., Analysis and countermeasures of corrosion cracking of an oil pipe [J]. Drilling Fluid & Completion Fluid, 2021, 38(3): 380-384.

[7] X. F. Yao, F. Q. Xie, X. Q. Wu, et al., Effects of pH Value on the Stress Corrosion Cracking of Super 13Cr Tubing Steel[J]. Advanced Materials Research, Volume 1916, Issue. 2012, 1916 (557-559): 127-130. https://doi.org/10.4028/www.scientific.net/AMR.557-559.127

[8] Y. Zhao, W. L. Qi, J. F. Xie, et al., Investigation of the failure mechanism of the TG-201 inhibitor: Promoting the synergistic effect of HP-13Cr stainless steel during the well completion [J]. Corrosion Science,

2020, 166: 108448-108448. https://doi.org/ 10.1016/j.corsci.2020.108448
[9] American Petroleum Institute. ANSI/ API SPEC 5CRA: 2010, Specification for Corrosion Resistant Alloy Seamless Tubes for Use as Casing, Tubing and Coupling Stock [S].
[10] NACE International. NACE RP0775-2005, Standard Recommended Practice Preparation, Installation, Analysis, and Interpretation of Corrosion Coupons in Oilfield Operations [S].
[11] International Organization for Standardization. ISO/TR 10400: 2007 (E), Petroleum and natural gas industries-Equations and calculations for the properties of casing, tubing, drill pipe and line pipe used as casing or tubing [S].
[12] Xiong Maoxian, Yang Shasha, Xie Junfeng, etc. Discussion about correlation between blind-hole technology and the ring method in testing residual-stress of oil casing[J]. Phys. Exam. Test., 37 (2019) 33-35.
[13] Toshio Shibata, Taro Takeyama. Dissolution and Film Formation of Fe-Cr-Ni Alloys in Boiling Magnesium Chloride Solution in Relation to Stress Corrosion Cracking [J]. Corrosion Engineering. 1974, 73(8): 379-383. https://doi.org/
[14] Jin-Jin Zhao, Xian-Bin Liu, Shuai Hu, et al. Effect of Cl^- Concentration on the SCC Behavior of 13Cr Stainless Steel in High-Pressure CO_2 Environment[J]. Acta Metall., 32 (2019) 1459-1469.
[15] Xiaofei Yao, Faqin Xie, Xiangqing Wu, et al., Effect of Cl- concentration on stress corrosion cracking behaviors of super 13Cr tubing steels [J]. Mater. Rep., 26 (2012) 38-41+45. https://doi.org/10.1007/s11783-011-0280-z
[16] S. Sakamoto, K. Maruyama. Corrosion Property of API and modified 13Cr steels in oil and gas environment [C]. CORROSION96, Houston, TX: NACE 1996.

本论文原发表于《Engineering Failure Analysis》2023 年第 146 卷。

四、其他

A Novel Inner Wall Coating-insulated Oil Pipeline for Scale and Wax Prevention

Cao Jing [1] Ma Wenhai [2] Huang Weiming [2]
Su Zhanfei [2] Zhu Yunbo [3] Wang Jianjun [1]

(1. State Key Laboratory for Performance and Structure Safety of Petroleum Tubular Goods and Equipment Materials, CNPC Tubular Goods Research Institute;
2. Daqing Oilfield Production Technology Institute;
3. CNPC Daqing Drilling & Exploration Engineering Company)

Abstract: During the production of deep oil and gas, scaling, waxing, hydrate ice plugging and other problems easily occur. To solve these problems, reducing the temperature loss of oil pipelines is a feasible method. In order to protect the outer wall coating from being damaged and losing its thermal insulation performance, this paper proposes a developed technology for a novel inner wall coating-insulated oil pipeline. A new temperature - resistant and heat - insulating material aerogel was optimized and developed, and it has an extremely low thermal conductivity of less than $0.055 W/(m \cdot K)$. The influence of different coating processes on the thermal insulation coefficient was analyzed, and a novel inner wall coating-insulated oil pipeline was developed. A testing and evaluation platform for its thermal insulation effect was built, and a finite element model was established to analyze the temperature field distribution. When the thickness of the inner coating was 0.5mm, the thermal insulation effect of the new oil pipeline improved by about 29%. This technology could be widely used in the production of deep oil and gas, high salinity oil and water reservoirs, thin oil reservoirs, etc., to alleviate scaling and waxing phenomena.

Keywords: Coating material; Oil pipeline; Thermal insulation performance; Numerical model

1 Introduction

During the exploitation of high-salinity oil and water reservoirs, heavy oil reservoirs and thin oil reservoirs, etc., the inner walls of oil pipelines are prone to scaling and waxing, resulting in a sharp decline in production. In order to solve the problems of scaling, waxing and hydrate ice plugging, the best way is to increase the temperature of the upper part of oil pipeline. The most convenient way

Corresponding authors: Cao Jing, caojing412@126.com; Wang Jianjun, wangjianjun005@cnpc.com.cn.

of increasing the temperature of pipeline is to effectively utilize the bottom temperature of wells and reduce temperature loss during fluid production[1-3]. Therefore, it is urgent to put forward a new technology of wellbore heat insulation, which not only can effectively reduce heat loss in the wellbore but also has simple surface technology and strong applicability.

Zhang et al. prepared an aerogel insulation blanket by using glass fiber felt and aerogel under normal pressure drying[4]. The insulation blanket can be used as an outer covering part. When the thickness of the insulation layer is 10mm, its apparent thermal conductivity reaches 0.024W/(m·℃). The disadvantage is that the external pressure resistance of the outer covering of the insulation pipe is low, which easily causes external extrusion deformation, resulting in external fluid infiltration into the insulation medium and reducing the insulation effect[5]. The sacrifice of annulus space would lead to jamming when lifting tubing during the well repair process. In addition, in the process of running tubing into the well, it is also necessary to cover the coupling on site, extending working hours.

Xing et al. analyzed and summarized the thermal insulation system and material of pipes, and they found that vacuum thermal insulation pipe was a widely used wellbore thermal insulation technology[6]. The product quality was reliable, and its performance was stable. Compared with the apparent thermal conductivity of traditional insulation pipes, the apparent thermal conductivity was reduced by an order of magnitude. However, after hydrogen permeation failure, the heat flow density of the vacuum-insulated pipe was very large, i.e., more than 10 times of that before the failure, and the heat leakage at the tubing coupling was large, which needed further improvement. Wang et al. established a mathematical model of heat transfer in the insulating layer by testing the apparent thermal conductivity of the insulating oil pipe[7]. The influence of vacuum degree on the performance of vacuum-insulated tubing was analyzed. The main conclusions were as follows: when the internal pressure was 0~20Pa or greater than 40Pa, the thermal conductivity increased with the increase in pressure; when the internal pressure was between 20Pa and 40Pa, the apparent thermal conductivity decreased with the increase in pressure; when the internal pressure was between 30Pa and 40Pa, the apparent thermal conductivity of the insulating oil pipe was small. From a technical and economic point of view, vacuum-insulated pipe is not suitable for deep well operation due to its high price, complex structure and heavy weight.

In addition to the commonly used insulation blanket and vacuum-insulated pipe, many scholars have also carried out studies on thermal insulation coating. Komkov et al. studied the ecological problems of the superheated steam injection method in the process of heavy oil exploitation[8], and they proposed that the most effective thermal insulation material was a highly porous material based on basalt fiber. The cylindrical insulation pipe coating was prepared by filtering and depositing short basalt fiber and alumina. The thickness of the oil pipe was determined according to the thermophysical properties of basalt fiber and the technical characteristics of manufacturing highly porous coating insulation material. During a long period of air extraction at a working temperature of 400℃, it was found that the surface temperature of the thermal insulation coating did not exceed 60℃. This technology could greatly reduce the negative impact of heavy oil exploitation on the biosphere.

Lin et al. analyzed three methods of heat transfer in the annulus of heavy oil wells[9], and they proposed a method of combining thermal insulation coating with vacuum technology: the outer wall of thermal insulation tubing is treated with thermal insulation coating; the high-temperature and high-pressure resistant packer is placed at the designated position outside the oil pipeline; the vacuum pump is used to vacuum the oil jacket annulus; and the heat injection operation could be started later. This method can minimize the heat loss of the wellbore and has better heat insulation. The thermal insulation technology combining thermal insulation coating and vacuum technology has the advantages of simple processes, convenient equipment and good thermal insulation, which has good application prospects in offshore heavy oil thermal recovery.

In order to reduce the heat transfer coefficient, diminish heat loss and maintain the temperature of the injected steam, Afra adopted two types of thermal insulators: nanosilicon-based and nanoceramic-based insulators[10]. The results show that when the injection temperature was raised from 119℃ to 145℃, the heat transfer coefficient increased by about two times, resulting in a rapid decrease in temperature during the injection process. The heat loss of steam injected at 145℃ could be reduced by 45% and 33% using 5mm thick nanosilicon-based and nanoceramic-based materials, respectively.

The above coatings are all sprayed on the outer walls of oil pipelines. In practical engineering applications, the outer wall coating at oil pipe coupling is easy to damage, losing thermal insulation performance and affecting thermal insulation continuity[5]. The main function of the existing internal coating of pipeline is to prevent corrosion. Therefore, it is urgent to develop a novel inner wall coating-insulated oil pipeline, providing an important guarantee for the extraction of oil and gas in high salinity oil, heavy oil and thin oil reservoirs, etc.

2 Materials and Methods

Due to the limitation of oil well annulus size and wellbore inclination, the outer diameters of heat-insulating oil pipelines is 88.9mm, and the wall thickness is 6.45mm. Based on the SY/T 6717-2016 standard, the thickness of inner-wall coating cannot exceed 0.5mm, so the selected heat-insulating material should have a high thermal insulation performance to reduce temperature drop.

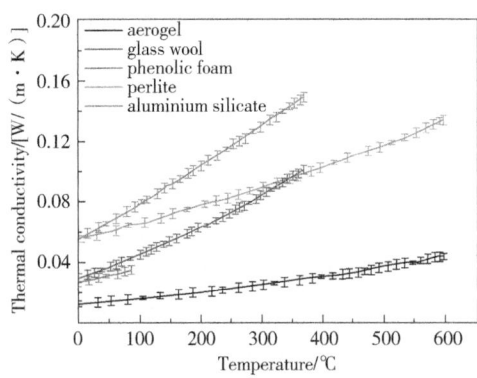

Fig. 1 Thermal conductivity curves of different materials with temperature

By comparing thermal conductivity of the common materials, such as glass wool, phenolic foam, perlite, aluminum silicate, aerogel, etc., as shown in Figure 1, it is found that aerogel heat-insulating material is not only applicable to a wide range of temperatures but also has the lowest thermal conductivity at different temperatures. Aerogel has good compressive strength, tensile strength, super hydrophobic properties and high fire resistance. Moreover, aerogel material has a super hydrophobicity with the contact angle 160° after special processing and has corrosion resistance properties. Therefore, nanoporous aerogels are the promising choice for the heat-insulating layer for oil pipelines.

Through nearly 10 years of research and optimization, aerogel heat-insulating material has been successfully applied in petrochemical, marine vehicles, construction and other engineering fields, and has achieved outstanding results. Among them, silica aerogel is considered a promising heat-insulating material because of the nano porous structure (1~100nm), low density (200kg/m^3), low thermal conductivity [0.013~0.055W/(m·K)], high porosity (80%~99.8%) and high specific surface area (200~1000m^2/g)[11-12].

Figure 2 shows the heat-insulating coating model, from which the heat-insulating mechanism of aerogel can be analyzed. The main components of aerogel are resin base, aerogel powder, hollow particles, coating filler, etc. The pore size of aerogel powder is lower than the average free path of air molecules under normal pressure. The air molecules in pores of aerogel are nearly static. The high-strength hollow particles added to the coating components can reduce coating density, improve compressive strength of the coating and reduce the coating damage under the external force. Therefore, the aerogel material could achieve the insulation effect unmatched by other materials[13-14].

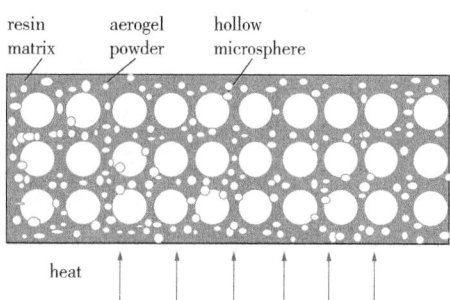

Fig. 2 Heat conduction model of thermal insulation coating

In this paper, comparative research and analysis are carried out according to the common material classification: inorganic and organic aerogel. The photos of aerogel powder and slurry are shown in Figure 3. The density of the aerogel is 0.5~0.7g/cm^3, and the thermal conductivity is 0.013~0.055W/(m·K). By adding hollow particles, the thermal conductivity of coating is reduced and the compressive strength is improved.

(a) powder

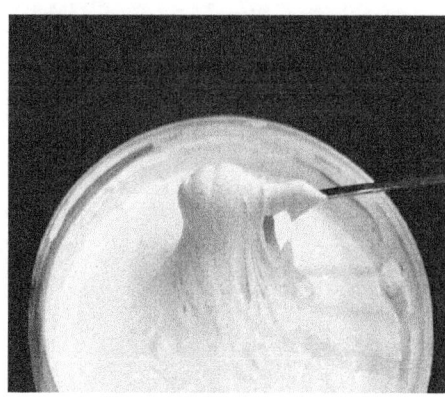
(b) slurry

Fig. 3 Schematic diagram of aerogel

3 Results and Discussion

3.1 Microstructure and Thermal Conductivity of the Aerogel Material

The microstructures of different materials are studied by SEM. The microscopic images are shown in Figure 4. The organic aerogel is composed of resin, aerogel powder and hollow particles. The hollow particles are spherical with a diameter of about 40 μm. Inorganic aerogel is made of SiO_2 aerogel powder and inorganic binder system. A large number of fibrous structures can be seen in the structure, which can increase the compressive and tensile strength of aerogel coating.

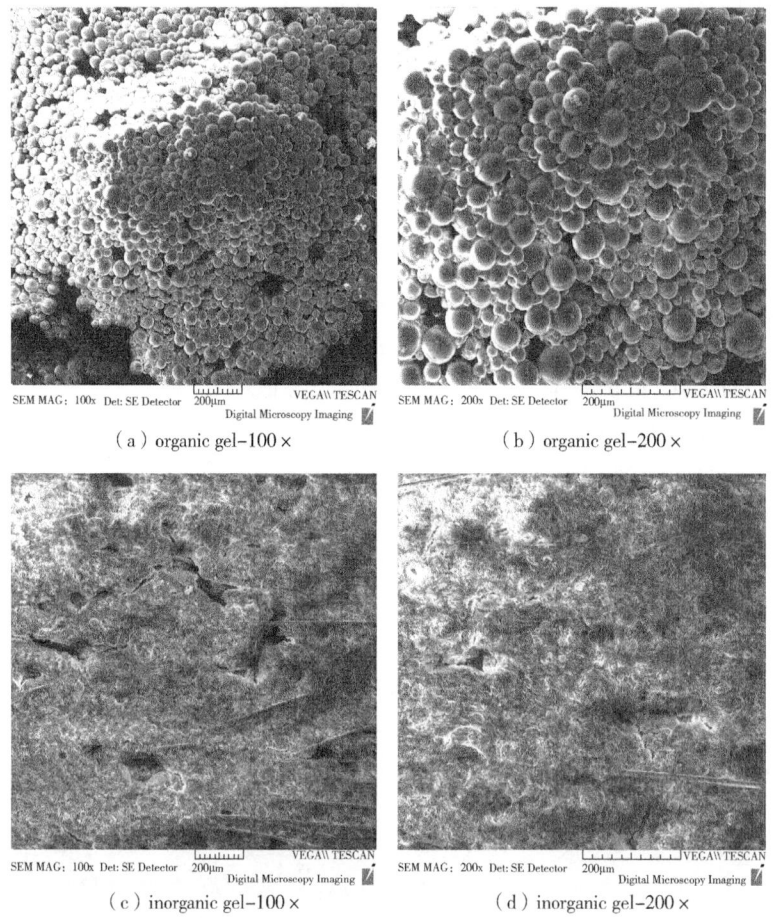

(a) organic gel-100× (b) organic gel-200×
(c) inorganic gel-100× (d) inorganic gel-200×

Fig. 4 Microstructure of aerogels with different compositions

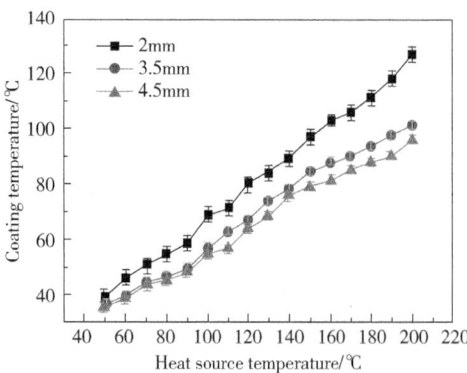

Fig. 5 Temperature variation curve of coating with different thicknesses

Thermal conductivity is measured by using the transient flat plate heat source method based on standard ISO22007-2. The probe of thermal conductivity meter is a power output source and also a temperature detector. The probe outputs a certain amount of power to heat the sample, and the temperature rise of sample is recorded by the probe. According to Fourier diffusion thermal conductivity equation, the power is equal to the thermal conductivity multiplied by temperature rise and divided by distance. Therefore, the thermal conductivity can be calculated. Through the thermal conductivity test of aerogel coating, it is found that the thickness has a greater impact on thermal insulation performance, as shown in Figure 5. With the increase in thickness, the coating temperature decreases, the temperature difference increases and the thermal insulation effect is better. When coating thickness increases from 2mm to 4.5mm, the coating surface temperature reduces from 84℃ to 68.6℃, the temperature difference increases from 46℃ to 61.4℃, so the thermal insulation performance can be improved by 33.5%.

The heat-insulating aerogel is coated on the inner walls of oil pipelines in this paper. The heat-insulating effects of coating thicknesses are studied by spraying or brushing process according to standard SY/T 6717-2016, the achievements can provide a technical support for oil field production.

3.2 Construction of Thermal Conductivity Test Platform

This study proposes a heat-insulating performance test device for inner wall-coated oil pipelines, including heating device, centrifugal pump, heat transfer tubing, circulation pipeline, temperature recorder, etc., as shown in Figure 6. The heating device comprises a temperature controller, an oil bath box, thermal circulating oil, a heating rod, and a thermocouple. The preset temperature of heating rod is set by temperature controller, and heating rod is connected by a thermocouple to monitor the actual oil temperature of oil bath box in real time. The centrifugal pump is connected to oil bath tank through a steel pipe, and the hot oil in oil bath tank is pumped out to circulating pipeline.

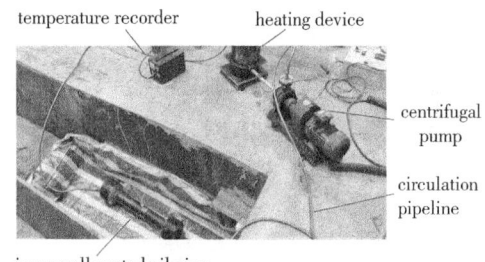

Fig. 6 Thermal insulation performance test platform

The heat insulation oil pipeline is a kind of inner wall-coated one. A new effective heat insulation material, i.e., aerogel is sprayed on inner walls of oil pipes. During spraying construction, the thickness of a single layer coating is 50μm. When the thickness is sprayed to 200μm, the next layer of coating is sprayed after the surface drying. The drying method is to dry at 60℃ for 10h, or dry naturally at room temperature for 48h. The two ends of circulation pipeline are connected to centrifugal pump and heat insulation oil pipe, respectively. Hot oil circulation inside the oil pipe is formed and a stable temperature value 130℃ is reached. The temperature recorder is connected to insulation oil pipe through thermocouples, and multiple thermocouples can simultaneously monitor the temperature values of the inner and outer walls of oil pipe, forming a temperature-time curve to calculate the thermal insulation of coatings.

3.3 Thermal Conductivity Evaluation of the Insulated Oil Pipelines

The thermal conductivity test platform in Section 3.2 can be used to evaluate the thermal insulation performance of inner wall-coated oil pipelines. The main steps include the following:

Step 1: Use the temperature controller to set the preset temperature of thermal circulating oil at 130℃ (the temperature value is determined according to service condition of oil pipe), heat the thermal circulating oil in oil bath through heating rod and monitor the actual temperature of thermal circulating oil in real time through the thermocouple. When the actual temperature reaches 130℃, the temperature lasts for 10min until it is stable.

Step 2: Turn on the centrifugal pump motor, pump the hot oil in the oil bath tank to circulation pipeline and inject the oil into insulating oil pipe to form the hot oil circulation. The hot oil flow rate is 1.2m³/h, and oil pipe length is 550mm. When the insulating oil pipe is full of hot oil and the temperature is stable at 130℃, turn off the heating device and centrifugal pump.

Step 3: The temperature recorder monitors the inner wall temperature value T_i and outer wall temperature value T_o until the circulating oil in oil pipes is heated to the target temperature value.

Step 4: Wait for the circulating oil temperature in insulated oil pipe to room temperature, and the temperature recorder records series of temperature values that change with time.

Step 5: By comparing the temperature-time curve 1 of internally coated pipe test and the temperature-time curve 2 of pure pipe test, the temperature difference after cooling for a period of time is calculated to determine the thermal insulation performance of inner wall-coated oil pipelines.

The inner wall-coated oil pipelines studied in this paper are divided into organic-aerogel-coated and inorganic-aerogel-coated ones, as shown in Figure 7. Among them, a 0.2mm thick organic oil pipe coating is formed by spraying and 0.5 mm thick inorganic coating is formed by brushing due to the existence of fiber. The coating is dried naturally after spraying or brushing, and then hot oil circulation is carried out by welding plugs at both ends of the coated oil pipelines.

(a) coating oil pipe　　(b) after welding plug

Fig. 7　Picture of thermal insulation oil pipe

Figure 8 shows the temperature change curve of coated pipes during the physical temperature rise tests. The temperature-time curves of organic aerogel-coated pipe, inorganic-aerogel-coated pipe and pure pipe are presented in temperature rise processes. The ambient temperature is 20℃. When the inner walls of pure pipe are heated to 130℃, the outer wall temperatures of pure pipe, inorganic-aerogel-coated pipe and organic-aerogel-coated pipe are 117.8℃, 115.5℃ and 113.2℃, respectively. It is found that the thermal insulation effect of organic-aerogel-coated pipe is the best one. For the coating of greater thickness, 0.5mm, a three step spraying method can be used, i.e., firstly, a 0.2mm thick coating is applied; after the surface drying, another 0.2mm thick coating is applied; and finally, the 0.1mm thick coating is applied.

When the thickness of organic aerogel coating is 0.2mm, the temperature difference between inner and outer walls of pipe is 16.8℃, which is 38% higher than that of pure pipe. When the thickness of inorganic aerogel coating is 0.5mm, the temperature difference between inner and outer walls of pipe is 14.5℃, which is 19% higher than that of pure pipe.

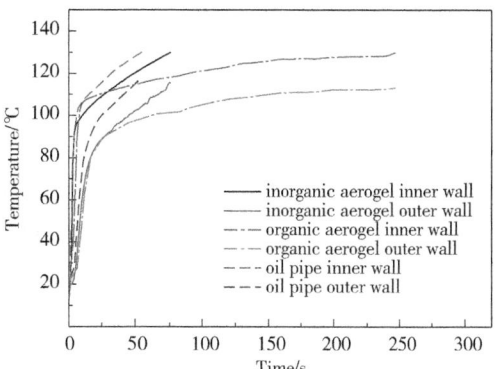

Fig. 8　Temperature curves of oil pipe heating process

For relatively long oil pipes, there are certain difficulties inthe inner wall spraying process. Therefore, in order to systematically study the insulation performance of inner coating insulation oil pipes with different coating thicknesses, the finite element simulation method is adopted, which is validated using existing experimental data and can also intuitively represent the changes of temperature fields. Based on the ABAQUS platform, a finite element model of thermal insulation pipe is established to analyze the temperature change of inner wall-coated pipes. The FEA governing equations include equilibrium equations, geometric equations, and constitutive equations. The equilibrium equation is as follows:

$$\frac{\partial \sigma_{ij}}{\partial x_j}+f_i=0, \quad i=1, 2, 3 \tag{1}$$

Where σ_{ij} is the stress component and f_i is the external force.

The geometric equationis as follows:

$$\varepsilon_{ij}=\frac{1}{2}\left(\frac{\partial u_i}{\partial x_j}+\frac{\partial u_j}{\partial x_i}\right) \tag{2}$$

Where u_i is the displacement component and ε_{ij} is the strain component.

The constitutive equations are as follows:

$$\dot{\varepsilon}_{ij}=\dot{\lambda}\sigma_{ij}' \tag{3}$$

$$\dot{\lambda}=\frac{3}{2}\frac{\dot{\bar{\varepsilon}}}{\bar{\sigma}'} \tag{4}$$

Where σ_{ij}' is deviatoric stress tensor; $\bar{\sigma}$ is the equivalent stress, $\bar{\sigma}=\sqrt{\frac{3}{2}(\sigma_{ij}'\sigma_{ij}'')}$; and $\dot{\bar{\varepsilon}}$ is equivalent strain rate, $\dot{\bar{\varepsilon}}=\sqrt{\frac{2}{3}(\dot{\varepsilon}_{ij}\dot{\varepsilon}_{ij})}$.

The FEA governing equations for heat transfer problems include Fourier's heat transfer law and energy conservation law. The transient temperature field of an object is $T(x, y, z, t)$, governing equation is as follows:

$$\frac{\partial}{\partial x}\left(\kappa_x\frac{\partial T}{\partial x}\right)+\frac{\partial}{\partial y}\left(\kappa_y\frac{\partial T}{\partial y}\right)+\frac{\partial}{\partial z}\left(\kappa_z\frac{\partial T}{\partial z}\right)+\rho Q(x, y, z, t)=\rho c_T\frac{\partial T}{\partial t'} \tag{5}$$

Where ρ is material density, and the unit is kg/m^3; c_T is specific heat of material, and the unit is J/

(kg · K); κ_x, κ_y, κ_z are thermal conductivity coefficients along the x, y, and z directions, respectively, and the unit is W/(m · K); $Q(x, y, z, t)$ is heat source intensity inside the object, and the unit is W/kg; T is temperature; and t is time.

The diameter of oil pipe is 88.9mm, wall thickness is 6.45mm and pipe length is 550mm; hexahedral grids are used, and the number of both grids in the diameter direction of the oil pipe and coating is 5. The minimum grid size of the oil pipe is 1.2mm, the minimum grid size of the inner-wall coating is 0.04mm and the number of coating grids is 20100. The thermal expansion coefficient of the oil pipeline is $1.3 \times 10^{-5} ℃^{-1}$, and that of the coating is $4.5 \times 10^{-5} ℃^{-1}$. The environment temperature outside the oil pipe is 20℃ and the inner wall temperature of coating is 130℃. The film coefficient between the oil pipe and the coating is 15 W/(m² · ℃). The bottoms of the oil pipes and coating are fixed in the U2 direction. The Von Mises yielding criterion and isotropic hardening law are adopted. Considering the heat transfer between oil pipe and coating, heat transfer (transient) solver is adopted.

As shown in Figure 9, during the process of temperature elevation, the temperature field distribution of the 0.2mm thick coating of organic aerogel is presented. When the inner wall of pipe reaches 130℃, there is an obvious temperature gradient from the inner to the outer wall of the pipe. The main parameters of the finite element model can be determined by combining the finite element model with the indoor full-scale test. By adjusting thermal conductivity of the coating material in finite element model, different outer wall temperature values of the pipe can be simulated. When the outer wall temperature value is equal to test temperature 113.2℃, the thermal conductivity of coating material is 0.052W/(m · K). Other parameters of the finite element model, including thermal expansion coefficient, elastic modulus and density, are shown in Table 1.

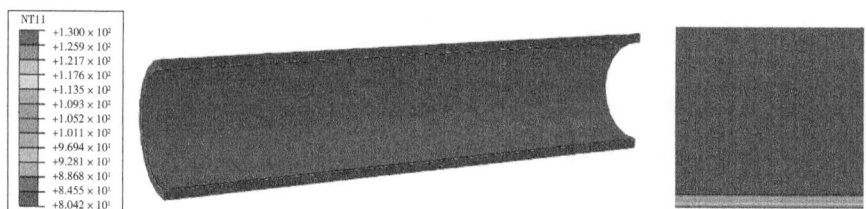

Fig. 9 Temperature distribution graph with the inner wall 130℃

Table 1 Finite element simulation parameters

Parameter	Oil Pipeline	Coating
Thermal conductivity/[W/(m · K)]	20	0.052
Thermal expansion coefficient/$10^{-5}℃^{-1}$	1.3	4.5
Elastic modulus/GPa	210	0.002
Density/(g/cm³)	7.85	0.7

Figure 10 shows the temperature change curves of inner-coated pipes during the physical cooling tests. When the temperature in pipes increases to 130℃, there is a certain temperature difference between the inner and outer walls due to coating, as shown in Figure 10a. The temperature recorder is used to record the series of temperature values that change with time. It is found that the temperature of the inner wall of the oil pipe decreases exponentially with the change of time. When the time is 180min, the thermal insulation performance of 0.2mm organic aerogel coating increases by about 10%.

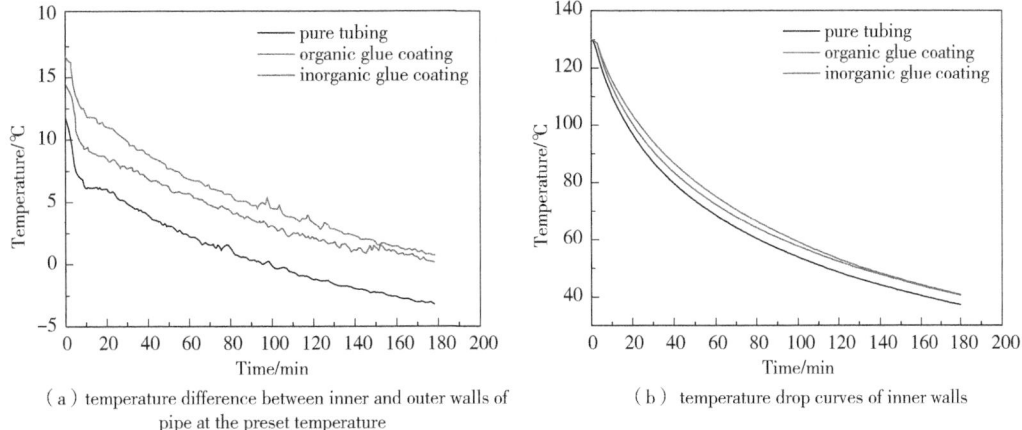

(a) temperature difference between inner and outer walls of pipe at the preset temperature

(b) temperature drop curves of inner walls

Fig. 10　Temperature change curve of coated tubing

By establishing the finite element model of cooling process, the temperature field change law is presented, as shown in Figure 11. When the inner wall of tubing is heated to 130℃, the pump is turned off and the coated pipe is cooled naturally that we might study the temperature change law. When the inner wall of pipe is sprayed with 0.2mm organic aerogel coating, the temperature of inner wall is 102℃ and that of the outer wall is 40℃ after 40min.

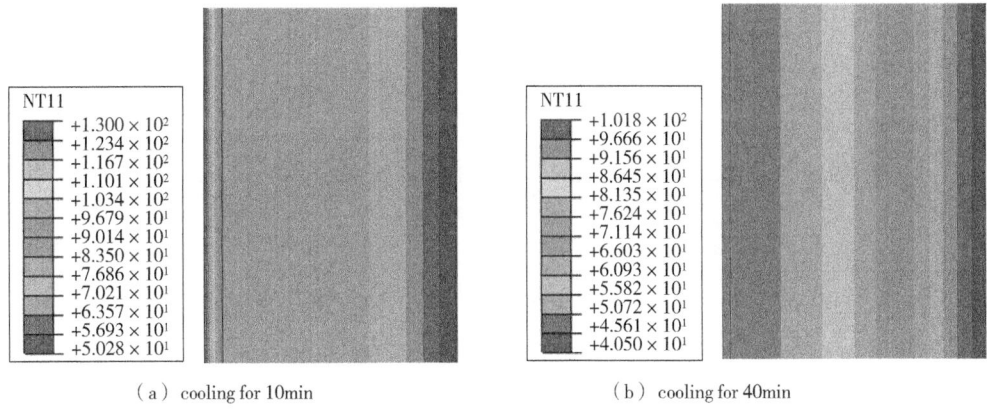

(a) cooling for 10min

(b) cooling for 40min

Fig. 11　Temperature change process of pipe with coating thickness 0.2mm

Based on the finite element simulation platform, the temperature variation of organic coating with thickness 0.5mm is studied. When the coating thickness is 0.5mm and the other finite element model parameters are the same as 0.2mm, as shown in Figure 12, it is found that when the inner wall temperature of oil pipe is 130℃ and the heating time is 246s, the outer wall temperature of 0.2mm coated oil pipe is 113.2℃, while the outer wall temperature of the 0.5mm coated oil pipe is only 81.7℃, indicating that thickness has a great impact on the thermal insulation effect of coating. Under the condition that the process is satisfied, the thermal insulation effect of coating could be improved by increasing coating wall thickness as much as possible. This technology may acquire good industrial application.

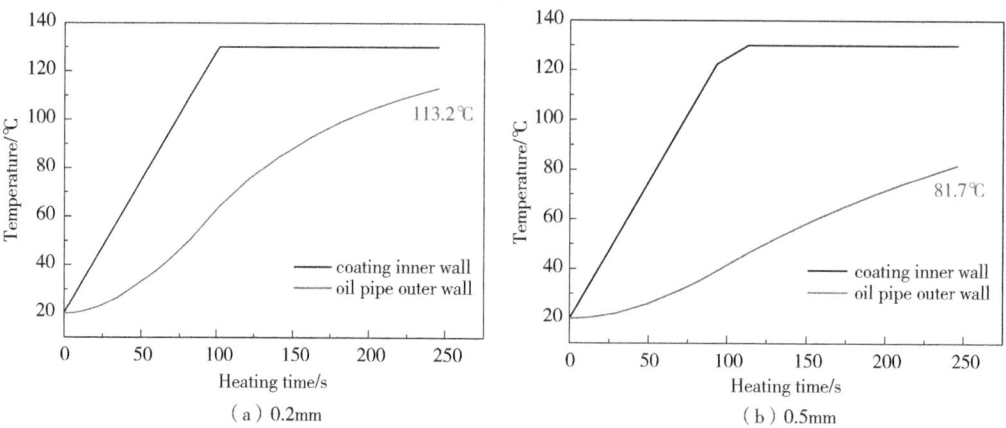

Fig. 12 Temperature rise process of tubing under different coating thickness

Comparing the indoor test and numerical simulation data of cooling process of the 0.2mm thick coating, as shown in Figure 13a, the temperature drop rate gradually decreases with the increase in time. When the time is 180min, the difference between test and simulation value is 2.4℃; the error rate is about 6%, which proves the reliability of the finite element model.

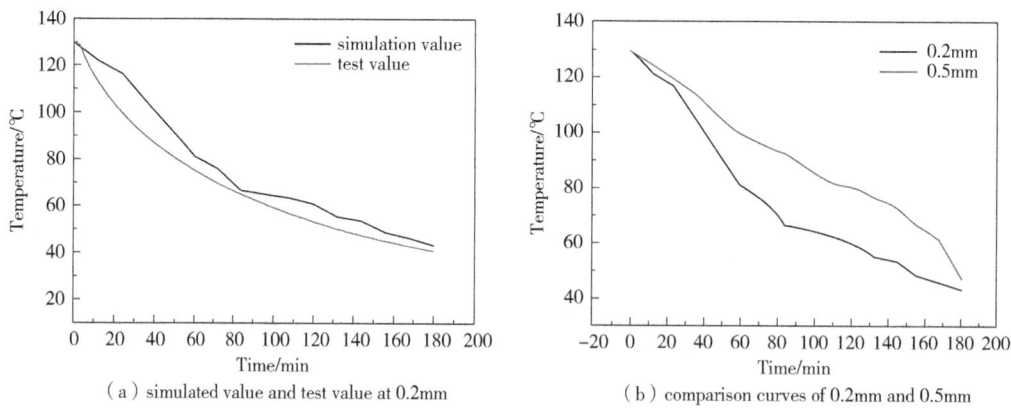

Fig. 13 Cooling process of tubing with different coating thicknesses

Further, afinite element model is established to analyze the cooling law of the inner wall of the coated pipe when the coating thickness is 0.5mm, as shown in Figure 13b. Compared with 0.2mm, the temperature drop rate is small. After 180min, the temperature of pipe inner wall is 48℃; compared with the 43.2℃ of the 0.2mm coated pipe, the temperature is increased by about 5℃; compared with the 37.3℃ of the uncoated pure pipe, the thermal insulation performance is improved by 29%, which can reduce the temperature loss near the wellhead position and effectively make use of the downhole temperature. The heating and cooling temperature fields of the 0.5mm coated oil pipe is shown in Figure 14. The temperature difference of the pipe wall is large between the inside and outside parts, indicating that the thermal insulation performance of aerogel thermal insulation coating is good. Since the inner wall temperature of 0.5mm inner-coated tubing is 48℃, which is greater than the scaling and waxing point of most oil fields, the inner coating can effectively prevent the scaling and waxing of oil pipelines.

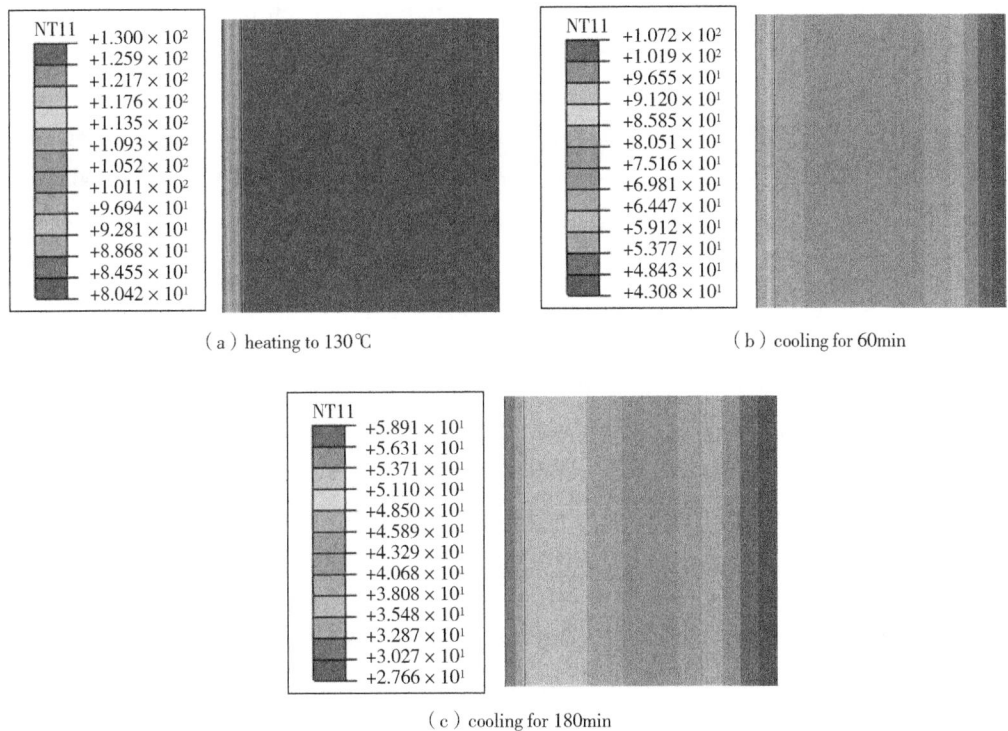

Fig. 14 Temperature change process of tubing with coating thickness 0.5mm

To sum up, through the development of new heat-insulating coating materials, new thermal insulation pipe products and the establishment of a test and evaluation platform for the thermal insulation effect, it is determined that the thermal insulation effect of the new thermal insulation pipe has increased by more than 20%, which can thus be widely used in the development of deep oil and gas, high-salinity oil and water wells, thin oil wells, etc., to alleviate the ice blocking, scaling, waxing and other phenomena caused by natural gas hydrate. In our subsequent research, a layer of 50μm thick corrosion-resistant high-temperature primer was first sprayed on the inner wall of oil pipe; then, the 500μm thick aerogel thermal insulation coating was sprayed upside the primer layer; finally, a layer of wear-resistant and anti-corrosion topcoat was sprayed. After the topcoat drying, the friction coefficient is small, so the entire insulated tubing has a smooth surface, reducing the surface roughness of oil pipelines. Therefore, the inner-wall coating of the oil pipe could not affect the fluid flow rate, and meanwhile, it has strong corrosion resistance. The downhole temperature could effectively be used to increase the storage and production of group company, as well as to also provide technical support for cost reduction and efficiency increases.

4 Conclusions

To solve the problems of scaling and waxing caused by high temperature loss near the wellhead part during the mining process of high-salinity oil and water, heavy oil and thin oil reservoirs, etc., a novel inner wall coating-insulated oil pipeline for scale and wax prevention is proposed in this paper. A high thermal insulation coating material – organic aerogel for the inner wall of the oil pipeline has been optimized and developed, which has an extremely low thermal conductivity of less than 0.055W/(m · K). The thermal insulation effect increases with increasing coating thickness. Based on the coating densification, the appropriate coating process of the inner – wall coating is determined to be spraying. For the coating thickness of 0.5mm, a three-step spraying method could be used. Then, an indoor test platform for the thermal insulation performance of inner wall coating-insulated oil pipeline was built, including a heating device, centrifugal pump, heat transfer tubing, circulation pipeline, temperature recorder, etc. A test and evaluation method for the thermal insulation coating material performance is established, and a finite element model is established to analyze the temperature field distribution. In comparing the outer temperature of the inner wall – coated oil pipeline with that of the uncoated oil pipeline, it is found that when the coating thickness is 0.5mm, the thermal insulation performance is improved by 29%. Since the inner wall temperature of 0.5mm coated pipe is 48℃, which is greater than the scaling and waxing points of most oil fields, the inner-coated oil pipelines can effectively prevent scaling and waxing phenomena.

Author Contributions

Conceptualization, J. C. and J. W.; methodology, J. C.; software, J. C.; validation, W. M., W. H. and Z. S.; formal analysis, Y. Z.; investigation, J. C.; resources, J. W.; data curation, J. C.; writing-original draft preparation, J. C.; writing-review and editing, J. C.; visualization, W. M.; supervision, J. W.; project administration, J. C.; funding acquisition, J. C. All authors have read and agreed to the published version of the manuscript.

Acknowledgements

This research was funded by the China National Petroleum Corporation [2021DJ2705, 2021DQ03(2022Z-10), 2020B-4020] and the Study on Key Technologies of Production Increase and Transformation of Gulong Shale Oil (2021ZZ10-04). The APC was funded by the China National Petroleum Corporation [2021DQ03(2022Z-10)].

References

[1] Liang, W. T.; Zhu, L. Q.; Li, W. P.; Yang, X.; Xu, C.; Liu, H. C. Bioinspired composite coating with extreme underwater superoleophobicity and good stability for wax prevention in the petroleum industry. Langmuir 2015, 31, 11058-11066.

[2] Boissonnet, G.; Chalk, C.; Nicholls, J. R.; Bonnet, G.; Pedraza, F. Phase stability and thermal insulation of YSZ and erbia-yttria co–doped zirconia EB–PVD thermal barrier coating systems. Surf. Coat. Technol. 2020, 389, 125566.

[3] Yao, H. H.; Xu, F. F.; Wang, X. Z.; Zeng, Y.; Tan, Z.; He, D. Y.; Yang, Y. G.; Liu, Y. B.; Zhou, Z. Thermal transport property correlated with microstructure transformation and structure evolution of Fe-based amorphous coating. Surf. Coat. Technol. 2023, 457, 129298.

[4] Zhang, J. Z.; Liu, Y. W.; Long, C.; Liu, M.; Su, X. J.; He, J. H. Preparation and Properties of SiO_2 Aerogel Insulation Felt. Contemporary Chemical Industry. 2021, 50, 1337-1341.

[5] Yatsenko, E. A.; Goltsman, B. M. Study of synthesis processes of heat-insulating silicate materials for external

protection of steel oil pipelines. CIS Iron Steel Rev. 2020, 20, 33-36.

[6] Xing, Y.; Liu, L.; Cao, C.; Fan, W. D.; Yan, K. J.; Cheng, Z. F. A Review of Research Status and Prospect of Vacuum Insulated Tubing Insulation System. In E3S Web of Conferences; EDP Sciences: Lisbon, Portugal, 2020; Volume 155, p. 01007.

[7] Wang, Z. H.; Zhao, H. Q. Research of vacuum degree influence on performance of vacuum heat insulation oil pipe. Adv. Mater. Res. 2013, 732-733, 172-175.

[8] Komkov, M. A.; Moiseev, V. A.; Tarasov, V. A.; Timofeev, M. P. Minimization of the negative influence on the biosphere in heavy oil extraction and ecologically clean technology for the injection of the steam with supercritical parameters in oil strata on the basis of new ecologically clean tubing pipes with heat resistant coatings. Atmos. Ocean. Phys. 2015, 51, 819-825.

[9] Lin, T.; Zhang, W.; Zou, J.; Sun, Y. T.; Song, H. Z.; Liu, H. T. Heat insulation coating and vacuum technology and its application in heavy oil thermal production. Pet. Geol. Eng. 2016, 30, 127-129.

[10] Afra, M.; Peyghambarzadeh, S. M.; Shahbazi, K.; Tahmassebi, N. Experimental study of implementing nano thermal insulation coating on the steam injection tubes in enhanced oil recovery operation for reducing heat loss. J. Pet. Sci. Eng. 2020, 189, 107012.

[11] Xu, C.; Xia, T.; Li, X.; Zhang, A.; Chen, Y.; Wang, C.; Lin, R.; Li, Z.; Dai, P.; Zhou, Y.; et al. Covalent binding of holey Si-SiC layer on graphene aerogel with enhanced lithium storage kinetics and capability. Surf. Coat. Technol. 2021, 420, 127336.

[12] Shan, Y. N.; Wu, L.; Jiang, Y. J. Preparation of SiO_2/cellulose composite aerogel and exploration of their property. New Chem. Mater. 2018, 46, 250-255.

[13] Koebel, M.; Rigacci, A. Aerogel-based thermal superinsulation: An overview. J. Sol-Gel Sci. Technol. 2012, 63, 315-339.

[14] Du, A.; Zhou, B.; Zhang, Z. H.; Shen, J. A Special Material or a New State of Matter: A Review and Reconsideration of the Aerogel. Materials 2013, 6, 941-968.

本论文原发表于《Processes》2023 年第 11 期。

Effect of Carbon-foam Composite-coated Electrode on the Power Storage Performance of Soluble Lead Flow Batteries

Ji Dongdong[1]　Liu Zheng[2]　Li Liwei[1]　Jiang Long[1]
Li Le[1]　Liu Xiaoxu[1]　Yang Yuanbo[1]　Li Tiantian[1]

(1. State Key Laboratory for Performance and Structure Safety of Petroleum Tubular Goods and Equipment Materials, CNPC Tubular Goods Research Institute;
2. School of Materials Science and Engineering, Xi'an University of Technology, No. 5 South Jinhua Road)

Abstract: In this study, we meticulously fabricated a carbon-foam composite-coated electrode through a plating electroless copper technique, which was subsequently followed by electrodeposition of lead on a carbon-foam matrix. The investigation focused on exploring the physicochemical properties of the electrodes and the cycling performances of soluble lead flow batteries. In comparison to a conventional graphite electrode, the carbon-foam composite-coated electrode exhibited superior characteristics, including a higher specific surface area characterized ($282m^2/g$) and a higher conductivity ($915.37S/cm$). The utilization of cyclic voltammetry and electrochemical impedance spectroscopy demonstrated a notable enhancement in the reversible activity of the Pb/Pb^{2+} couple on the novel electrode. The porous structure of the novel electrode facilitated efficient charge dispersion between the poles and effectively suppressed the dendrite growth of the negative lead. As indicated by the result, the battery incorporating the novel electrode exhibited an energy efficiency exceeding 90.0%, an extended charge duration of 6h, and an elevated charge/discharge rate of $60mA/cm^2$. The cell employing the CFCC-E had been successfully cycled for over 650 cycles with an average coulombic efficiency of 94.5% when cycled at $60mA/cm^2$ for 1h.

Keywords: Carbon-foam composite coated electrode; Graphite electrode; Soluble lead flow battery; Cycle performance; Dendrite growth

1　Introduction

The aggressive development of wind and photovoltaic renewable energies represents a fundamental strategy employed by China to establish new clean power systems and attain its carbon peaking and

---

Corresponding author: Ji Dongdong, E-mail address: jidongdong@cnpc.com.cn; Tel: +86(0)2981887927.

carbon neutrality objectives. Nevertheless, the inherent instability associated with renewable energies imposes limitations on their large-scale integration into the grid. Consequently, the deployment of high-capacity long-duration energy storage technologies has emerged as a crucial means to facilitate local consumption of renewable energy, mitigate output fluctuations, enable peak load management, and optimize the utilization of renewable resources[1-2]. As depicted in Fig. 1, soluble lead flow batteries (SLFBs) exhibit significant potential for commercial applications compared to existing electricity energy storage technologies. SLFBs possess several advantages, including the utilization of a single electrolyte, high capacity, flexible module design, and a safe and reliable operational profile[3-4]. Notably, the power and capacity of SLFBs can be expanded through parallel or series arrangements of battery stacks and by adjusting the volume of the electrolyte[5]. Upon charging, PbO_2 and Pb are simultaneously deposited on the positive and negative electrodes, respectively. During the discharge process, the Pb and PbO_2 deposits undergo dissolution, returning to the soluble Pb^{2+} through oxidation and reduction, respectively[6]. It is important to note that no other heterogeneous reactions occur within the electrolyte medium. The electrode reactions and standard potentials are described as follows[7-8]:

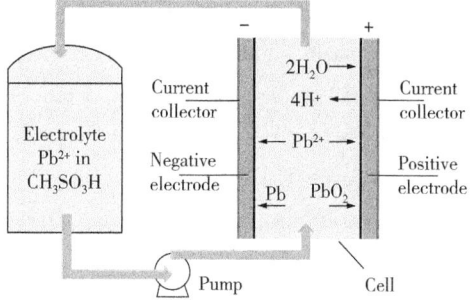

Fig. 1 Principle of soluble lead flow batteries (SLFBs) (the cell is shown for charging)

Negative electrode:

$$Pb^{2+} + 2e^- \xrightleftharpoons[\text{Discharge}]{\text{Charge}} Pb \qquad (1)$$

$$E^o_{(-ve)} = -0.130V \text{ vs. SHE}$$

Positive electrode:

$$Pb^{2+} + 2H_2O \xrightleftharpoons[\text{Discharge}]{\text{Charge}} PbO_2 + 4H^+ + 2e^- \qquad (2)$$

$$\alpha\text{-}PbO_2\, E^o_{(+ve)} = +1.468V \text{ vs. SHE}$$

$$\beta\text{-}PbO_2\, E^o_{(+ve)} = +1.460V \text{ vs. SHE}$$

Overall reaction:

$$2Pb^{2+} + 2H_2O \xrightleftharpoons[\text{Discharge}]{\text{Charge}} Pb + PbO_2 + 4H^+ \qquad (3)$$

$$E^o_{cell} = +1.581V$$

The conductivity, surface area, corrosion resistance, and electrochemical stability of the electrode are significant factorsfor the cell performance[9]. In contrast to conventional electrochemical batteries, the electrode in a SLFB serves as a platform for facilitating the charging and discharging redox processes without actively participating in the electrochemical reactions. Furthermore, the redox reaction of the Pb^{2+}/Pb couple in SLFBs is characterized by its rapid kinetics, exhibiting minimal overpotentials. Moreover, the dissolution and deposition ratio of lead during the redox process is notably high[10]. However, the graphite electrodes (G-Es) commonly used in batteries exhibit several drawbacks, such as large impedance values, small electrode surface areas, and poor structural uniformity. Moreover, prolonged cycling in acidic electrolytes leads to corrosion of the G-E surface. After a long period of charge and discharge cycle, combined with the presence of micro-cracks, the acidic electrolyte is allowed to penetrate the electrode, compromising the binder and causing the electrode to expand and bulge[8, 11]. The negative Pb is susceptible to dendrite growth on the edge of a G-E, and the adhesion between the deposit and the G-E is poor, resulting in weak adhesion between the Pb deposits and the electrode during cycling. The above disadvantages of the G-E affect the distribution of ions, the rates of the electrochemical reactions, and the deposit's electrochemical activity, finally altering the cycling performance of the battery. To address these issues, researchers, such as Pletcher et al. and Wills et al., have proposed various modifications to G-Es. They have heat fused nickel foam, reticulated vitreous carbon, and titanium meshes onto G-E surfaces, resulting in electrodes with larger specific surface areas and improved current distribution uniformity[12-13]. Oury et al. designed a three-dimensional honeycomb-shaped positive PbO_2 electrode that was sandwiched between two planar negative electrodes, which significantly reduced the charge/discharge overpotentials[14]. A tin(II) methanesulfonate additive could increase the number of crystal nuclei, change the nucleation type of lead from three-dimensional instantaneous nucleation to three-dimensional progressive nucleation, thereby effectively inhibiting the formation of lead dendrites[15]. However, despite these existing research efforts, the comprehensive improvement of the various disadvantages associated with G-Es has not been achieved. Therefore, it is imperative to explore composite electrodes with novel structures and coatings to fulfill multiple functions, such as enhancing electrode reactivity, inhibiting Pb dendritic growth, and improving battery capacity and efficiency.

In this study, we aimed to develop an alternative to conventional G-Es for the negative electrode of SLFB. We prepared a carbon-foam composite-coated electrode (CFCC-E) and systematically investigated its physiochemical and electrochemical characteristics, as well as the morphologies of the deposits formed on the electrode and the charge/discharge performance of the SLFBs. Additionally, we analyzed the mechanisms by which the CFCC-E influenced the characteristics of the deposits and the cycling performance of the SLFBs. The main objective of this study was to fabricate composite electrodes with novel structures and coatings that exhibit outstanding specific surface areas, redox reversibility, conductivities, and hydrogen evolution characteristics, while effectively inhibiting dendrite growth. By achieving these goals, we aimed to significantly enhance the charge/discharge power, depth, efficiency, and cycling stability of SLFBs when utilizing the CFCC-E as the negative electrode.

2 Experimental details

We employed a carbon foam with a porosity of 50 ppi (ppi denotes the number of apertures per inch) as the matrix for the deoiling, coarsening, sensitizing, and activating pretreatment processes[16]. The degreasing step involved subjecting the carbon foam to a lye solution composed of 30~40g/L NaOH, 25~30g/L Na_2CO_3, and 20~30g/L Na_3PO_4 at 60~70℃ for 20min. Subsequently, coarsening was achieved by immersing the carbon foam in 65% nitric acid with in an ultrasonic cleaner for 15min at 25℃. The carbon foam was sensitized for a duration of 10min in a continuously stirred sensitizing solution at 25℃. This step was crucial to ensure the uniform adsorption of reducing catalytic water film. The sensitizing solution consisted of 50mL/L concentrated HCl, 40g/L $SnCl_2$, as well as a trace quantity of tin particles. Palladium ($PdCl_2$, 0.6g/L and 50mL/L concentrated HCl) was deposited onto the carbon foam surface, which can form a catalytic activation center in the activation solution. The pretreated carbon foam was subjected to electroless copper plating in a copper sulfate solution. Finally, the carbon foam copper-plated sample underwent electrodeposition of lead in a lead fluoborate system to produce the CFCC-E[17].

The cellconfiguration, as depicted in Fig. 2[18], consisted of CFCC-E and G-E (40mm×30mm ×5mm) as negative electrodes. These electrodes were paired with G-Es to construct SLFBs. The electrodes were bonded with a copper plate as a current collector by employing Ag conductive adhesive by curing at 60℃ for 5h. To serve as current collectors, the electrodes were affixed to a copper plate using Ag conductive adhesive and cured at 60℃ for 5 hours. Each electrode was then mounted onto a polypropylene plate and sealed with silica gel after ensuring its alignment with the polypropylene surface. The cell assembly was secured between two aluminum plates, establishing an inter-electrode gap of 0.8cm to facilitate electrolyte flow. The electrolyte (500mL) was circulated through the cell at a linear flow rate of 2.5cm/s by a peristaltic pump. The electrolyte was prepared with deionized water ($\leqslant 1\mu S/cm$), methanesulfonic acid (Chengdu KeLong Chemical Co. Ltd., 98%, mass fraction), and lead oxide (Tianjin Fuchen Chemicals Reagent Factory, 99%, mass fraction) through the following reaction[19]:

$$PbO+2CH_3SO_3H =\!=\!= Pb(CH_3SO_3)_2 + H_2O \tag{4}$$

The electrolyte referred to throughoutthis study comprised 1.5M $Pb(CH_3SO_3)_2$ + 0.9M CH_3SO_3H. The charge/discharge cycle tests of the flow cells were carried out at different current densities by employing a battery testing system (CT2001A, Wuhan LANHE).

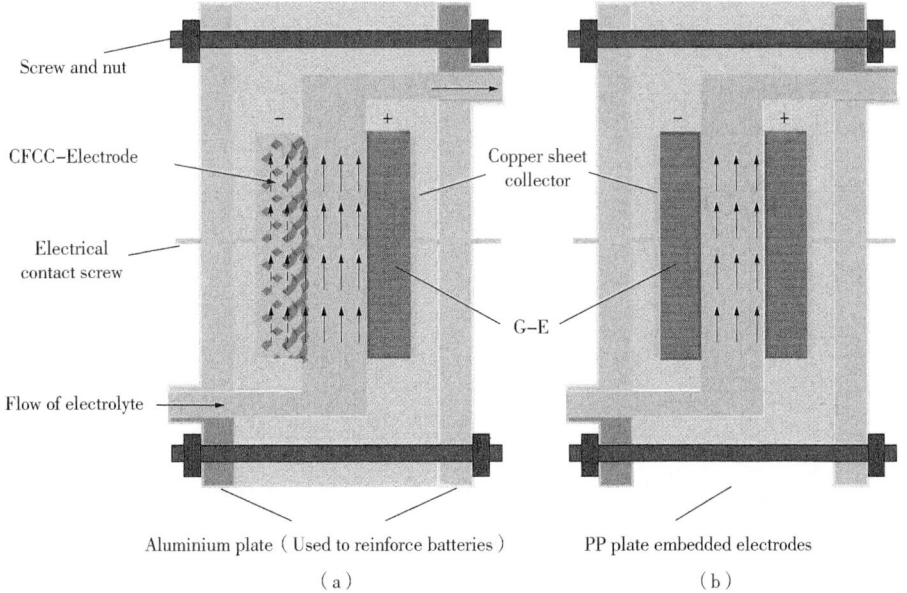

Fig. 2 (a) Sideview of the flow cell with different negative electrodes.
(a) carbon-foam composite-coated electrode (CFCC-E) with flow-through structure.
(b) graphite electrode (G-E) with flow-by structure

For the electrochemical characteristic measurements, a standard three-electrode system was employed, employing an electrochemical workstation (Donghua DH7000, Jiangsu). CFCC-E (10mm× 10mm), platinum (10mm × 10mm), and saturated Hg/Hg_2SO_4 served as working, counter, and reference electrodes, respectively. Varying cyclic voltammetry scan rates were used (5mV/s, 10mV/s, 20mV/s, 50mV/s, and 100mV/s) from −1.3V to 0V for the Pb^{2+}/Pb couple. Electrochemical impedance spectroscopy (EIS) spectra were acquired over a frequency range of 10^{-1} Hz to 10^5 Hz using an alternating current (AC) amplitude of 5mV in a 120mL solution. The impedance analysis of the Pb^{2+}/Pb couple was performed utilizing the ZSimDemo software. Linear sweep voltammetry (LSV) experiments were conducted at a scan rate of 5mV/s, spanning from −0.6V to −1.5V vs. Hg/Hg_2SO_4, employing a 1.5M $Pb(CH_3SO_3)_2$ + 0.9M CH_3SO_3H solution. Potentiodynamic polarization curves were recorded at a scan rate of 0.1mV/s, encompassing the range of open circuit potentials (E_{op}) of ±300mV. The corrosion current density of the electrode was determined through Tafel curve extrapolation after Tafel fitting[20].

3 Results and discussion

3.1 Physical properties of the CFCC-E

As depicted in Fig. 3(a) ~ (c), the carbon foam provided both a porous structure and matrix skeleton. The intermediate copper plating increased the electrode conductivity, and the bond strength with electrodeposited lead was strengthened by metal bonds and solid solution strengthening[21] facilitated by copper and lead having the same face-centered cubic crystal structure and similar atomic radius[22]. Lead is a superpotential metal with a large hydrogen overpotential[23]. Remarkable bonding can be achieved between the identical metals, the electrodeposited lead coating, and the negative lead after charging, which is beneficial for reducing the discharge capacity loss caused by the shedding of a negative lead. As depicted in Fig. 3(d), there are a large number of pores with different sizes on a single skeleton of CFCC-E, which will greatly increase the specific surface area of the electrode.

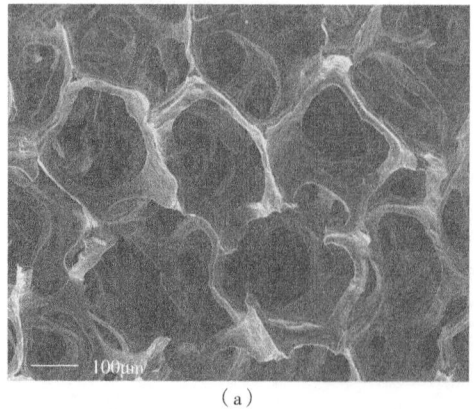 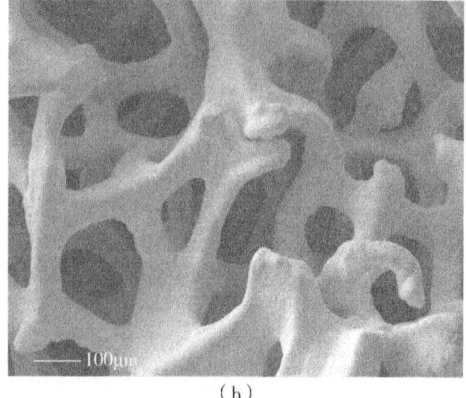

(a) (b)

Fig. 3 Scanning electron microscopy (SEM) images.
(a) carbon foam, (b) CFCC-E, (c) cross section, and (d) the single skeleton of CFCC-E

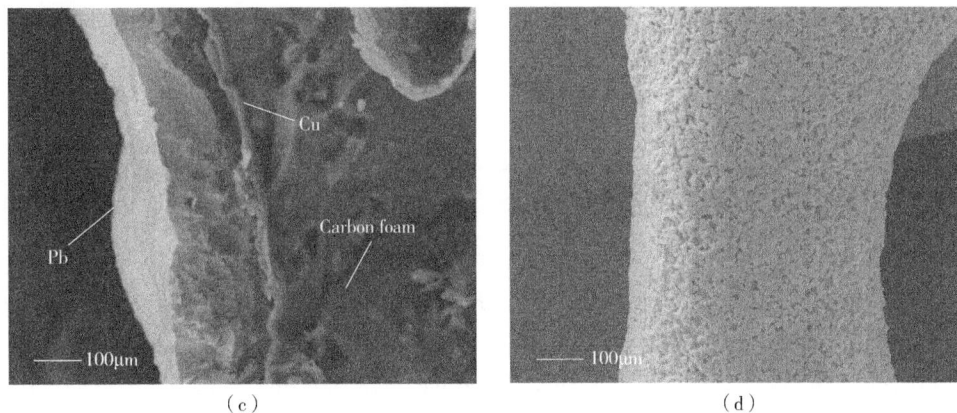

Fig. 3 Scanning electron microscopy (SEM) images.
(a) carbon foam, (b) CFCC-E, (c) cross section, and (d) the single skeleton of CFCC-E(continued)

The physicochemical properties of the electrode affected both the electrochemical characteristics and the cycling performance of the cell. Fig. 4(a) and Fig. 4(b) display the representative N_2 adsorption/desorption isotherm and corresponding pore size distribution of the CFCC-E and G-E. As shown in the Fig. 4(a), the S_{BET} of the CFCC-E is significantly enhanced and exhibit typical IV-type isotherms accompanied by the emergence of hysteresis loops (H3 - type), indicating their macropore structures[24]. The S_{BET} of CFCC-E is $282 m^2/g$, which is much larger than that of G-E ($27 m^2/g$) (in Table 1). In addition, the pore size distribution maps Fig. 4(b) reveal that the two materials are mainly composed of macropores. A specific pore size of 1.2μm is obtained for CFCC-E, which is much smaller than the 3.7μm of G-E. The high S_{BET} and abundant macropore of CFCC-E is in favor of boosting the its activity of the electrode reaction by increasing the contact area and the active sites. Meanwhile, the porous-structured CFCC-E offers good contact with electrolytes, and these pores strongly favor immediate electron and ion transmission. Due to the conventional G-E contained polyethylene adhesive materials, its overall conductivity had been measured to be only 16.89S/cm. The conductivities of carbon foam matrix, the pure copper and the pure lead were 3.41×10^{-2} S/cm, 5.8×10^5 S/cm and 9.6×10^4 S/cm, respectively[25]. Therefore, the conductivity of the CFCC-E, measured by a four-probe tester, was up to 915.37S/cm, which was fifty times that of the G-E (in Table 1).

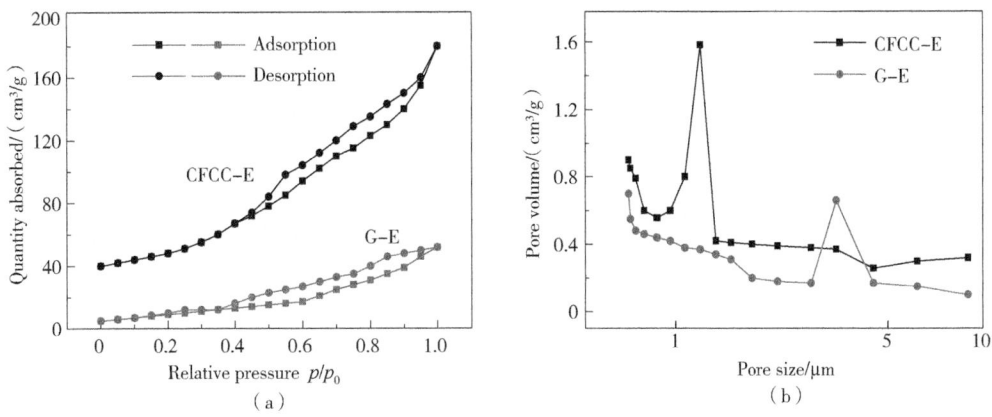

Fig. 4 (a) Nitrogen adsorption-desorption isotherms and
(b) pore size distribution curves for CFCC-E and G-E

Table 1 Conductivities, BET specific surface areas and pore size of the two different electrodes

Electrode	Conductivity/(S/cm)	BET specific surface area/(m^2/g)	Pore size/μm
CFCC-E	915.37	282	1.2
G-E	16.89	27	3.7

3.2 Deposit morphology analysis

Cells with different negative electrodes were charged at 40mA/cm^2 for 1h in a 1.5M Pb(CH_3SO_3)$_2$ + 0.9M CH_3SO_3H solution. As depicted in Fig. 5, the electrode material significantly affected the morphologies of both the positive and negative deposits after charging. The overall lead deposits on the CFCC-E exhibited a smooth and tightly coated electrode skeleton, as visually demonstrated in Fig. 5(a). Notably, no dendrites were observed in the cross-sectional view of the CFCC-E. In contrast, the lead deposits on the G-E appeared loose and lacked density, as depicted in Fig. 5(b). The growth of dendrites was clearly evident in the cross-sectional analysis of the G-E, with the particle size gradually increasing along the vertical surface of the electrode, particularly in the magnified region depicted in Fig. 5(b). The above-mentioned phenomenon can be attributed to the significantly higher conductivity of the CFCC-E (measured at 915.37S/cm) resulting from the integration of copper and lead coatings, surpassing that of the G-E. The elevated electrode resistance in the G-E induced substantial Ohmic polarization, thereby promoting excessive growth of the initial lead particles and ultimately leading to the formation of coarse deposited particles[7]. The growth of dendrites in the negative lead deposit on the conventional G-E proved to be a significant factor influencing the performance of the SFLB. The detachment of dendritic crystals caused irreversible losses in cell capacity and lead(II) concentration. Furthermore, the substantial growth of dendritic crystals could easily lead to contact short circuits between the positive and negative electrodes[26]. In contrast, the morphology of the lead deposit on the CFCC-E exhibited limited dendrite growth, thanks to the composite coating mitigating the point effect of the electrode.

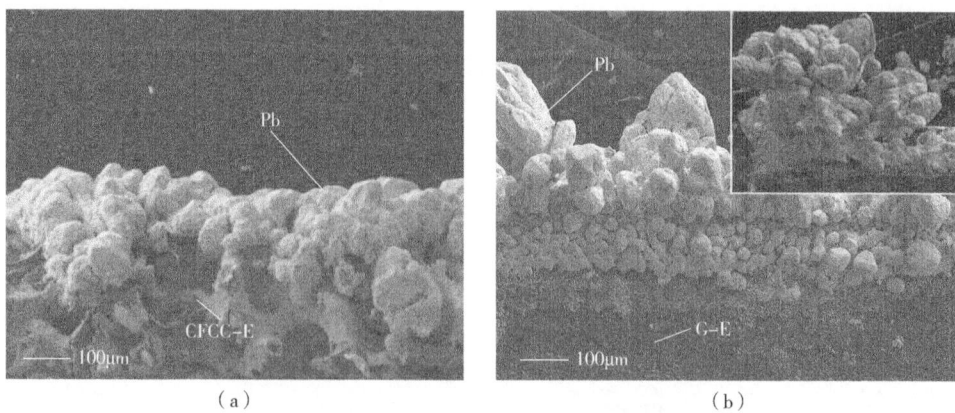

Fig. 5 SEM cross-sectional images of Pb deposits on
(a) CFCC-E and (b) G-E

The preferred orientation of the crystal growth and the nonuniform distribution of the electric field are two factors determining the formation of dendrites[27]. The spatial distribution of electric field lines on the surface of the G-E is depicted in Fig. 6(a). It is evident that the flux of ions within the spherical electric field was notably higher compared to that within the planar electric field. Consequently, a larger number of metal ions tended to aggregate at the edges and corners of the G-E, resulting in a localized point effect that can promote dendrite growth in the lead deposits. In contrast, the interconnected porous structures of the CFCC-Es exhibited a more uniform distribution of ions and possessed skeletons with greater curvature radii. As depicted in Fig. 6(b), within the magnified circular region, the incorporation of copper and lead layers successfully eliminated the burrs present in the matrix material. Consequently, the transmission of charge across the porous framework of the CFCC-E could be promptly and effectively dispersed. This dispersion ensured that the intensity of the electric field was evenly distributed across the electrode surface, preventing excessive concentration. Consequently, the CFCC-E surface exhibited dense and smooth lead deposits, as illustrated in Fig. 5(a). On the other hand, distinct dendritic crystals were observed to deposit on the corners and edges of the G-E, as depicted in Fig. 5(b).

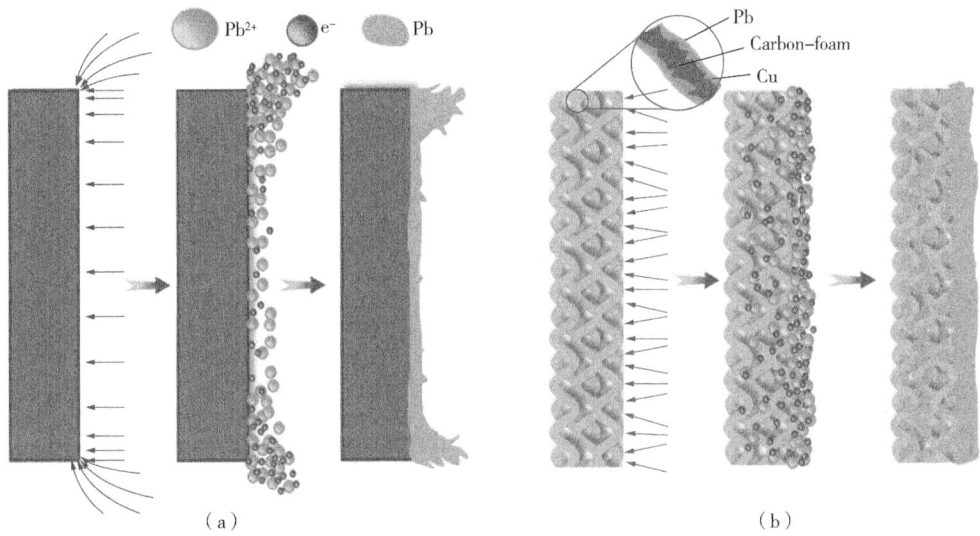

Fig. 6 Schematic diagrams of cross section of electric field lines on surfaces of
(a) G-E and (b) CFCC-E

When the CFCC-E was employed as the negative electrode, the crystal particles of PbO_2 deposited on the positive G-E exhibited a dense arrangement without notable gaps, as depicted in Fig. 7(a). Conversely, when the G-E was utilized as the negative electrode, the crystal particles of PbO_2 formed distinct clusters resembling cotton balls, exhibiting uneven sizes. Notably, the larger interstitial spaces between these clustered particles revealed deposits that were prominently protruding and sunken, as illustrated in Fig. 7(b). This phenomenon can be attributed to the non-uniform distribution and intensity discrepancy of the electric field between the positive and negative electrodes during the charging process[28]. The formation of PbO_2 involves adsorption and oxidation processes, with hydroxyl radicals (OH_{ads}) playing a pivotal role[29]. The uniformity and intensity of the electric field distribution directly determine the uniformity and quantity of OH_{ads} generated through the ionization of H_2O. This, in turn, affects the nucleation state on the positive electrode and consequently influences the final morphology of the deposited PbO_2. The high resistivity of G-E leads to a significant ohmic voltage drop, resulting in an uneven distribution and weakened intensity of the electric field between the electrodes. In contrast, the utilization of CFCC-E effectively mitigates excessive concentration of the electric field, leading to higher intensities of a uniform electric field.

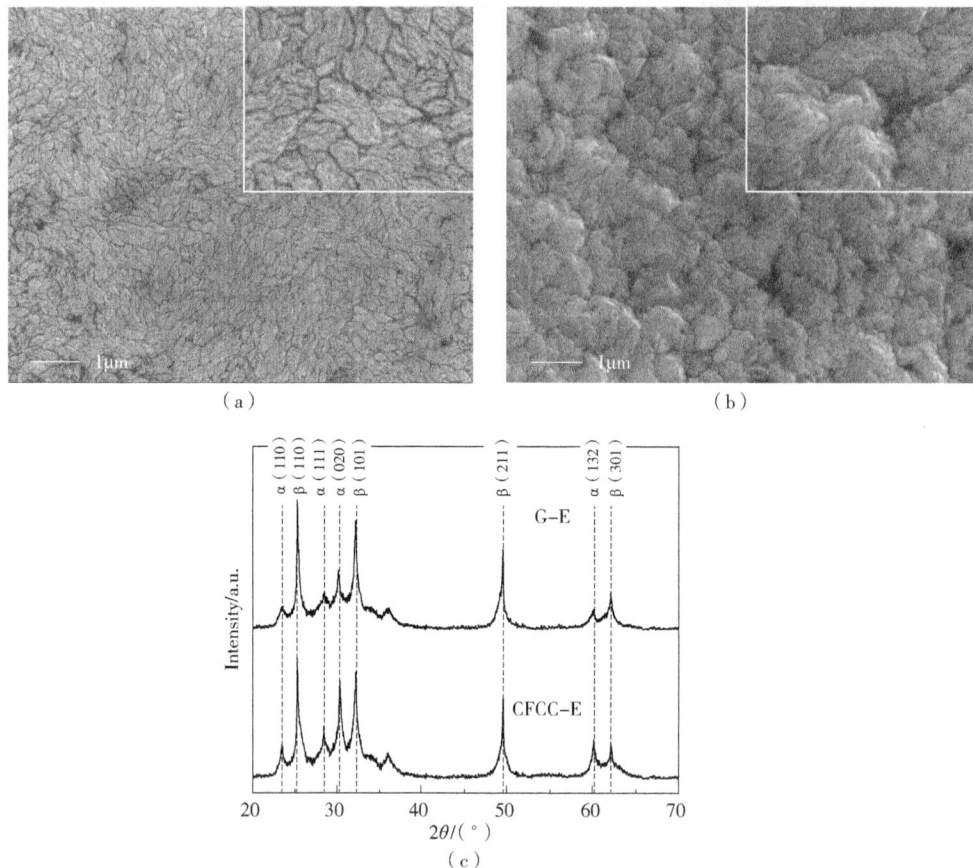

Fig. 7　SEM images of PbO_2 deposits on positive G-E with
(a) CFCC-E and (b) G-E as negative electrode; (c) XRD spectra of PbO_2 deposits on positive G-E with different negative electrode

We investigated the phase composition and the average grain sizes by measuring the diffraction patterns of PbO_2 deposits (via XRD) formed on positive G-E with the CFCC-E and the G-E as negative electrodes. Fig. 7(c) shows that the mixture of α-PbO_2 and β-PbO_2 existed in the PbO_2 deposits. The typical peaks of β-PbO_2(110), (101), (211) and (301) crystal planes appear at 25.4°, 31.9°, 49.0° and 62.4°. The peaks of α-PbO_2(110), (111), (020) and (132) crystal planes appear at 23.4°, 28.5°, 30.1° and 60.0°. The peak intensity of α-PbO_2(020) on positive G-E with CFCC-E as negative electrode is higher than that with G-E as negative electrode, which exhibits a preferred orientation in the α-PbO_2(020). There is no significant difference between the other crystal planes of α-PbO_2 and β-PbO_2. The average grain sizes of α-PbO_2 and β-PbO_2 are calculated by Scherrer's equation, and the relative content ratio (%) of two phases is quantitatively calculated by RIR value (Reference intensity) (Table 2). The average grain sizes of α-PbO_2 and β-PbO_2 deposited on the positive G-E, when utilizing CFCC-E as the negative electrode, were larger compared to those when G-E was used as the negative electrode. The relative content ratios (%) of α-PbO$_2$: β-PbO$_2$ deposited on the positive G-E, with CFCC-E and G-E as the negative electrode, were 11 : 89 and 19 : 81, respectively. This difference can be attributed to the high conductivity of CFCC-E, which can reduce the overpotential caused by ohmic polarization within the system. The reduced overpotential subsequently lowered the nucleation rate and enhances nucleus growth, resulting in the formation of PbO$_2$ with larger grain sizes. Furthermore, the low electrode potential observed on G-E when CFCC-E was used as the negative electrode promotes the formation of β-PbO$_2$. This is due to the fact that the electrode potential of Pb^{2+}/β-PbO$_2$ is lower than that of Pb^{2+}/α-PbO$_2$. Generally, β-PbO$_2$ exhibits superior electrochemical performance, while α-PbO$_2$ contributes to maintaining the mechanical strength of the PbO$_2$ deposits.

Table 2 Average grain sizes and relative content ratios of crystal phases formed on positive G-E with the CFCC-E and the G-E as negative electrodes

Electrode	Average grain sizes / nm		α-PbO$_2$: β-PbO$_2$ (Relative content ratio) / %
	α-PbO$_2$	β-PbO$_2$	
CFCC-E	28	21	11 : 89
G-E	34	25	19 : 81

3.3 Electrochemical tests

Voltammograms depicting the behavior of the Pb^{2+}/Pb couple were recorded for the different working electrodes at different scan rates [Fig. 8(a) and Fig. 8(b)]. As depicted in Table 3, the anodic peak current density (i_{pa}) of the Pb^{2+}/Pb couple containing the CFCC-E increased from 0.068A/cm^2 to 0.203A/cm^2, whereas the cathodic peak current density (i_{pc}) increased from 0.049A/cm^2 to 0.169A/cm^2 as the scan rate increased from 5mV/s to 100mV/s. The difference between the oxidation peak potential and the reduction peak potential (ΔV) increased with the rise of the scan rate from 5mV/s to 100mV/s. The variation tendencies of the peak current density and peak potential of the Pb^{2+}/Pb couple with G-E aligned with those with the CFCC-E. Specifically, the peak current densities and ΔV increased with the scan rate. Fig. 8(c) depicts the ratios of $-i_{pc}/i_{pa}$ of the Pb^{2+}/Pb couple for the different electrodes. The value of $-i_{pc}/i_{pa}$ of the Pb^{2+}/Pb couple with G-E decreased with the scan rate. In contrast, the value of $-i_{pc}/i_{pa}$ for the CFCC electrode increased with the rise of the scan rate. The above results indicated that the Pb^{2+}/Pb couple on the CFCC-E demonstrated superior oxidation and reduction reversibility than the G-E compared to high-current-density operating conditions ($\geqslant$20mA/cm^2), and the opposite trend was observed at low current densities.

Table 3 Peak potentials and peak current densities of the Pb^{2+}/Pb couple over CFCC-E and G-E of varying scan rates

Electrode	Scan rate/(mV/s)	E_{pa}/V	E_{pc}/V	ΔV_p/V	i_a/(A/cm^2)	i_c/(A/cm^2)
CFCC-E	5	-0.434	-1.021	0.587	0.068	-0.049
	10	-0.405	-1.052	0.647	0.101	-0.078
	20	-0.381	-1.083	0.702	0.148	-0.101
	50	-0.369	-1.102	0.733	0.178	-0.136
	100	-0.358	-1.164	0.806	0.203	-0.169

Continued

Electrode	Scan rate/(mV/s)	E_{pa}/V	E_{pc}/V	ΔV_p/V	i_a/(A/cm^2)	i_c/(A/cm^2)
G-E	5	-0.531	-1.002	0.471	0.034	-0.028
	10	-0.497	-1.024	0.527	0.064	-0.038
	20	-0.453	-1.038	0.585	0.091	-0.059
	50	-0.432	-1.085	0.653	0.119	-0.077
	100	-0.392	-1.106	0.714	0.145	-0.098

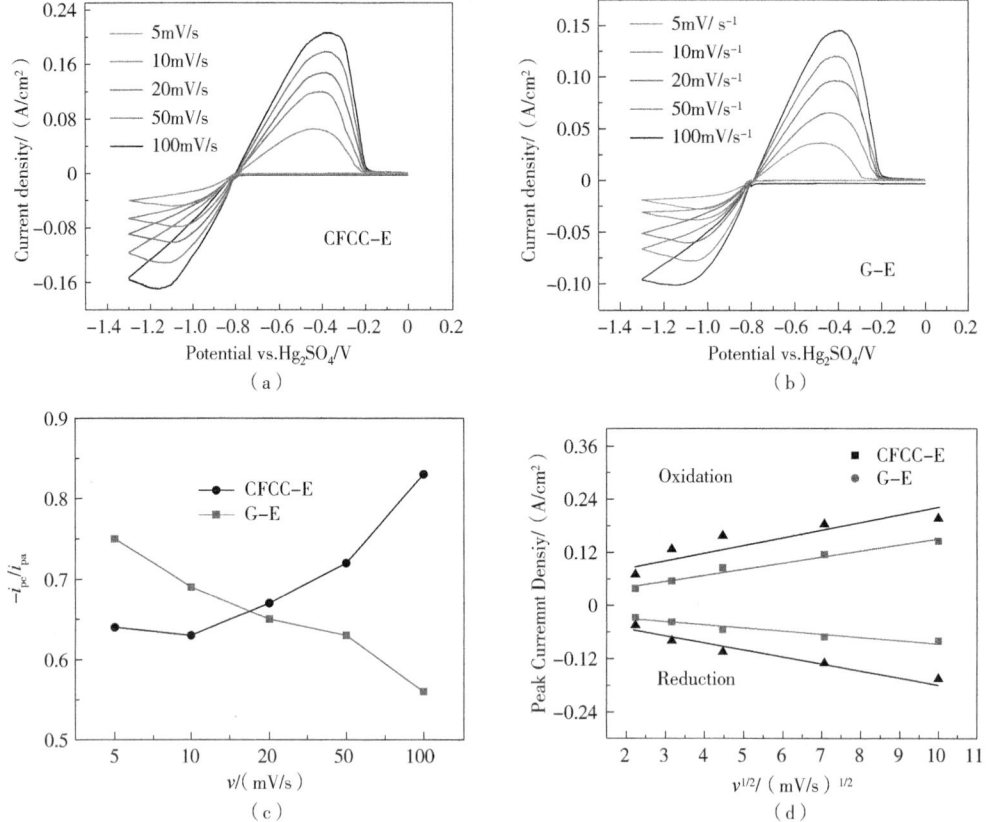

Fig. 8 Cyclic voltammograms of Pb^{2+}/Pb couple on:
(a) CFCC-E and (b) G-E recorded at scan rates 5mV/s, 10mV/s, 20mV/s, 50mV/s, and 100mV/s; (c) $-i_{pc}/i_{pa}$ values of Pb^{2+}/Pb couple at various scan rates; (d) Fitting plots of Pb^{2+}/Pb couple peak current density vs. square root of scan rate

Fig. 8(d) presents the correlation between the peak value of the reduction (or oxidation) current density and the square root of the scan rate. Notably, a strong linear relationship was observed between i_p of the reduction (or oxidation) process and $v^{1/2}$ of the Pb^{2+}/Pb couple on the G-E across different scan rates. However, the goodness of fit between i_p and $v^{1/2}$ on the CFCC-E was low (Table 4). The $|\Delta V_p|$ values of the Pb^{2+}/Pb couple for the two electrodes were greater than 57n^{-1}mV (with n being the number of reaction electrons, $n = 2$), and $|\Delta V_p|$ increased with the scanning rate. Given that $|\Delta V_p| > 57n^{-1}$ and considering the linear correlation between i_p and $v^{1/2}$ for the Pb^{2+}/Pb couple on the G-E, it can be deduced that redox processes are irreversible and diffusion-controlled[30]. The correlation between the peak current and the scan rate can be described as follows[31]:

$$i_p = (2.99 \times 10^5) \alpha^{1/2} A C_0 D^{1/2} v^{1/2} \tag{5}$$

Table 4 Goodness of fit between i_p and $v^{1/2}$ of the Pb^{2+}/Pb couple for the two electrodes under study

Electrode	Pb^{2+}/Pb couple	
	$R\text{-}Square_{oxidation}$	$R\text{-}Square_{reduction}$
CFCC-E	0.92953	0.93938
G-E	0.99365	0.99654

Where i_p denotes the peak current, A; α represents the transfer coefficient; A is the working electrode surface area, cm^2; C_0 is the molar concentration of the solute, mol/cm^3; D is the diffusion coefficient, and v expresses the scanning rate, V/s. Since $|\Delta V_p| > 57n^{-1}$ and based on the nonlinear correlation between i_p and $v^{1/2}$ for the Pb^{2+}/Pb couple on the CFCC-E, were concluded that the redox processes were quasi-reversible processes[32] and not controlled by diffusion. The correlation between the peak currents and scan rates can be described as follows[33]:

$$i_p = i_{p(rev)} K(\Lambda, \alpha) \tag{6}$$

Where i_p denotes the peak current, A; $i_{p(rev)}$ is the peak current of the reversible process, A; $i_p = 2.69 \times 10^5 n^{3/2} A C_0 D^{1/2} v^{1/2}$, and $K(\Lambda, \alpha)$ expresses a function describing the reversibility of the electrode.

To investigate the influence of different electrodes on impedance and the mechanism of electrode reactions, EIS was conducted. As illustrated in Fig. 9(a), the EIS spectrum of the G-E exhibited a semicircle at high frequencies along with a straight line at low frequencies in the Nyquist plot. In contrast, the CFCC-E's EIS spectrum solely displayed a semicircle at high frequencies. This observation suggests that the redox process of the Pb^{2+}/Pb couple on the G-E involves simultaneous electrochemical polarization and concentration polarization, indicating a combined influence of charge transfer and diffusion processes[34]. Conversely, the reactions of the Pb^{2+}/Pb couple on the CFCC-E are solely governed by the charge transfer process. This conclusion aligns with the analysis of reversibility obtained from the cyclic voltammograms of the Pb^{2+}/Pb couple. The Nyquist plots were fit to the different basic equivalent circuit models shown in Fig. 9. The difference between the equivalent circuit models of the two electrodes was the diffusion resistance, which can be interpreted by using the Warburg diffusion element (W). According to the data fitting performed by the ZsimpWin software with the equivalent circuit, the charge transfer resistance R_{ct} of the Pb^{2+}/Pb couple on the CFCC-E was 1.82Ω, significantly lower than that on the G-E (5.56Ω). The charge transfer process of the Pb^{2+}/Pb couple was notably influenced by the electrode employed. The deposition and dissolution of negative Pb on the CFCC-E were considerably facilitated, exhibiting enhanced ease and speed compared to the G-E. This can be attributed to the exceptional conductivity of the CFCC-E and the reduced activation energy necessary for the nucleation of lead (II) during the charging process.

To investigate the hydrogen evolution reaction (HER) on various electrodes, linear sweep voltammetry (LSV) was conducted within the potential range of -1.5V to -0.6V vs. Hg/Hg_2SO_4,

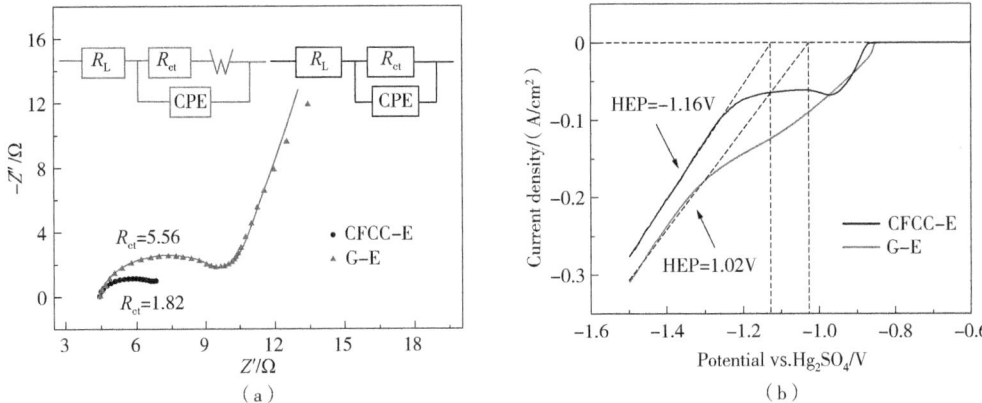

Fig. 9 (a) Nyquist plots from EIS measurements for different negative electrodes at −0.5V vs. Hg$_2$SO$_4$; (b) LSV results of cells with different electrodes recorded at scan rate of 5mV/s from −1.5V to −0.6V

employing a scan rate of 5mV/s. As illustrated in Fig. 9(b), the HER current densities of the different electrodes exhibited a significant increase as the negative scan shifted from −0.6V to −1.5V. Notably, the CFCC-E demonstrated lower HER current density compared to the G-E within the potential range of −1.0V to −1.5V. This observation confirms that the HER rate on the CFCC-E was considerably inferior to that on the G-E at the same potential. The HER potential was the potential at the inflection point after taking the derivative of the LSV curve[8] [Fig. 9(b)]. The HER potential on the CFCC-E was −1.16V vs. Hg/Hg$_2$SO$_4$, whereas it was −1.02V vs. Hg/Hg$_2$SO$_4$ on the G-E. According to the potential and current density results for the HER, we can conclude that this reaction was less likely on the CFCC-E compared with the G-E, avoiding the loss of H$^+$ in the electrolyte and the decrease in the cell capacity. The composition of the CFCC-E surface containing superpotential lead metal with a large hydrogen overpotential accounted for this result.

The corrosion current density of the electrode is another important factor to estimate the cycling performance of a flow cell. Fig. 10 presents the potentiodynamic polarization curves in the range of open circuit potentials of ±300mV, the corrosion current densities, and the surface corrosion morphologies of the different electrodes in 1.5M Pb(CH$_3$SO$_3$)$_2$ + 0.9M CH$_3$SO$_3$H solutions at 25℃. The corrosion current densities of the different electrodes were determined based on the Tafel linear epitaxy method[35]. The corrosion current density on the CFCC-E was 0.71×10^{-6} A/cm^2, and it was less than that on the G-E, at 0.35×10^{-5} A/cm^2. The above-described results revealed the corrosion was faster on G-E compared with the CFCC-E. There were some micro-sized corroded holes on the surface of the G-E after 200 cycles, although no evident corrosion phenomenon was reported on the skeleton of the CFCC-E. Lead on the CFCC-E surface formed a protective film of lead salts in acidic solutions, which can potentially explain the above-mentioned results[36]. Graphite is easily oxidized to form electrochemical corrosion in an acidic solution[37], and the infiltration of the electrolyte in the G-E would result in the easy expansion and failure of the electrode. The cycling stability of the flow cell was improved by using the CFCC-E due to the low corrosion current density.

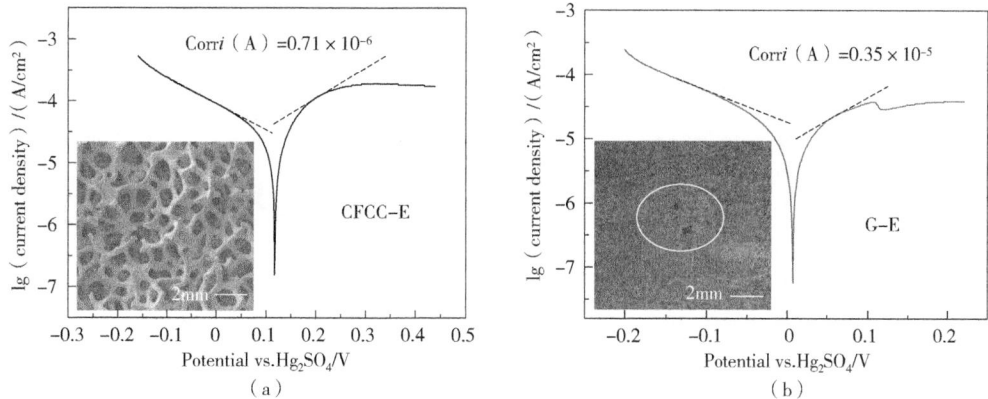

Fig. 10 Potentiodynamic polarization curve, corrosion current density, and corrosion morphology of: (a) CFCC-E and (b) G-E of flow cell

3.4 Cycling performance tests

Different thicknesses of lead coating and copper coating significantly affect the overall efficiency of the cells. Due to the limited activation points of the foam carbon matrix after activating, the maximum thickness of copper coating approached 5μm. As depicted in Fig.11(a), when the thickness of the electrodeposited lead coating was fixed at 8μm, the conductivity of the electrode significantly increased with the increase of copper coating thickness (as depicted in the inserted SEM after copper plating). However, the CEs of the cells increased pronouncedly and then stabilized, and the efficiency reached the maximum value when the copper coating was 3~4μm. The main reason may be that the continuous thickening of the copper coating is no longer the dominant factor affecting the efficiency of cell. As depicted in Fig.11(b), for a fixed copper coating thickness of 3μm, the cells exhibited maximum efficiency when the lead coating thickness ranged between 10μm and 12μm (as depicted in the inserted SEM). As the lead layer thickness increased beyond this range, the cell efficiency decreased. When the lead coating on the carbon foam electrode was thin, the bonding strength between the electrode and the negative lead after charging was inadequate. This can result in the negative lead becoming prone to detachment after charging, consequently leading to a loss of battery capacity and efficiency. On the other hand, when the electrodeposited lead coating on the foam carbon electrode is excessively thick, it will tend to generate coarse lead particles. Furthermore, besides affecting the bonding strength of the negative lead after charging, the excessive lead coating thickness can reduce the electrode's porosity and specific surface area, ultimately diminishing the cell's capacity and efficiency.

CFCC-E and G-E were selected to evaluate the effects of the different negative electrodes on the cell charge/discharge performances. The cells were charged for 2h and subsequently discharged until the voltage dropped to 1.0V at 10mA/cm^2 in a solution of 1.5M Pb(CH$_3$SO$_3$)$_2$ + 0.9M CH$_3$SO$_3$H. The cell voltage responses over time for charge/discharge cycles by employing different negative electrodes were recorded and presented in Fig.12(a). The charge voltage of the cell incorporating the CFCC-E was found to be lower compared to that of the cell with the G-E, whereas the discharge voltage exhibited the opposite trend. This observation suggests that the polarization level of the CFCC-

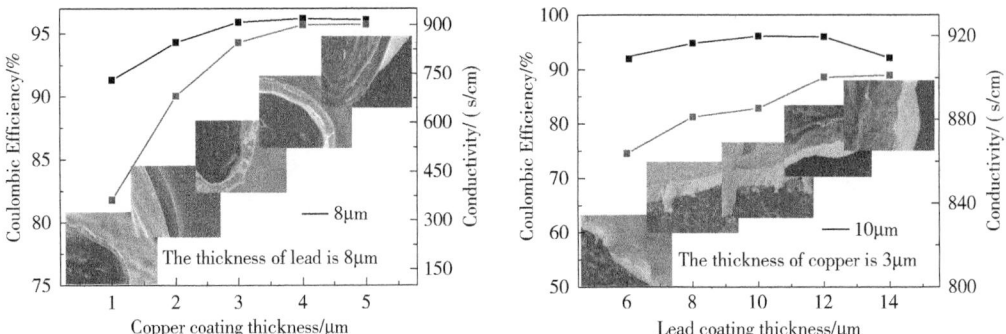

Fig. 11 (a) Coulombic efficiency (CE) of Cell and the conductivity of CFCC-E as functions of the thickness of copper coating, and inserted SEM images of electrodes; (b) Coulombic efficiency (CE) of Cell and the conductivity of CFCC-E as functions of the thickness of lead coating, and inserted SEM images of electrodes

E during the charging and discharging processes was lower than that of the G-E. The enhanced conductivity of the CFCC-E contributed to significantly reduced ohmic polarization overpotentials. Furthermore, the larger specific surface area of the CFCC-E facilitated the dispersion of concentration polarization and accelerated the reaction rate of the electrode. Consequently, the CFCC-E exhibited a smaller polarization overpotential ($\Delta\phi_-$) which effectively lowered the voltage during charging while increasing it during discharging.

Fig. 12(b) shows the coulombic efficiency (CE), voltage efficiency (VE), and energy efficiency (EE) over 500 cycles for a flow cell containing different negative electrodes. The CE of the CFCC-E was stable at 97.5%±1.0%, while that of the G-E was 92.5%±2.0%. The CE of the flow cell was determined by the charging and discharging capacities, which were further determined by both the state of the deposits and the reaction process of the electrode[38]. Based on the evaluation of deposit morphology and cyclic voltammograms presented in Section 3.2 and Section 3.3, it was observed that the negative lead species formed on the CFCC-E surface after charging exhibited smaller particle sizes and higher redox activity compared to those on the G-E. Furthermore, the outer composite lead layer of the CFCC-E exhibited a higher HER overpotential in comparison to graphite. This characteristic played a crucial role in reducing the occurrence of HER side reactions. Conversely, the performance of the flow cell by employing the G-E was significantly compromised due to acid loss resulting from HER and the elevated internal resistance of the battery. The above-described results can partially explain the lower capacities and efficiency of the cell containing the G-E compared with the battery containing the CFCC-E. In addition, the lead deposits after charging combined with the outer composite lead of the CFCC-E via metal bonds. Therefore, the enhanced interfacial bonding strength between the electrode and the deposits and the lack of evident dendrite growth on the CFCC-E could significantly reduce the irreversible loss of the cell capacity during charging and discharging.

The VE of the flow cell with the CFCC-E remained stable at 92%±0.5%, which was higher than that with the G-E, at 87.5%±1.0%. The VE is the ratio of the charging voltage to the discharging voltage. As depicted in the voltage curve presented in Fig. 12(a), the CFCC-E exhibited lower polarization overpotential and faster electrode reaction rates, leading to higher VE

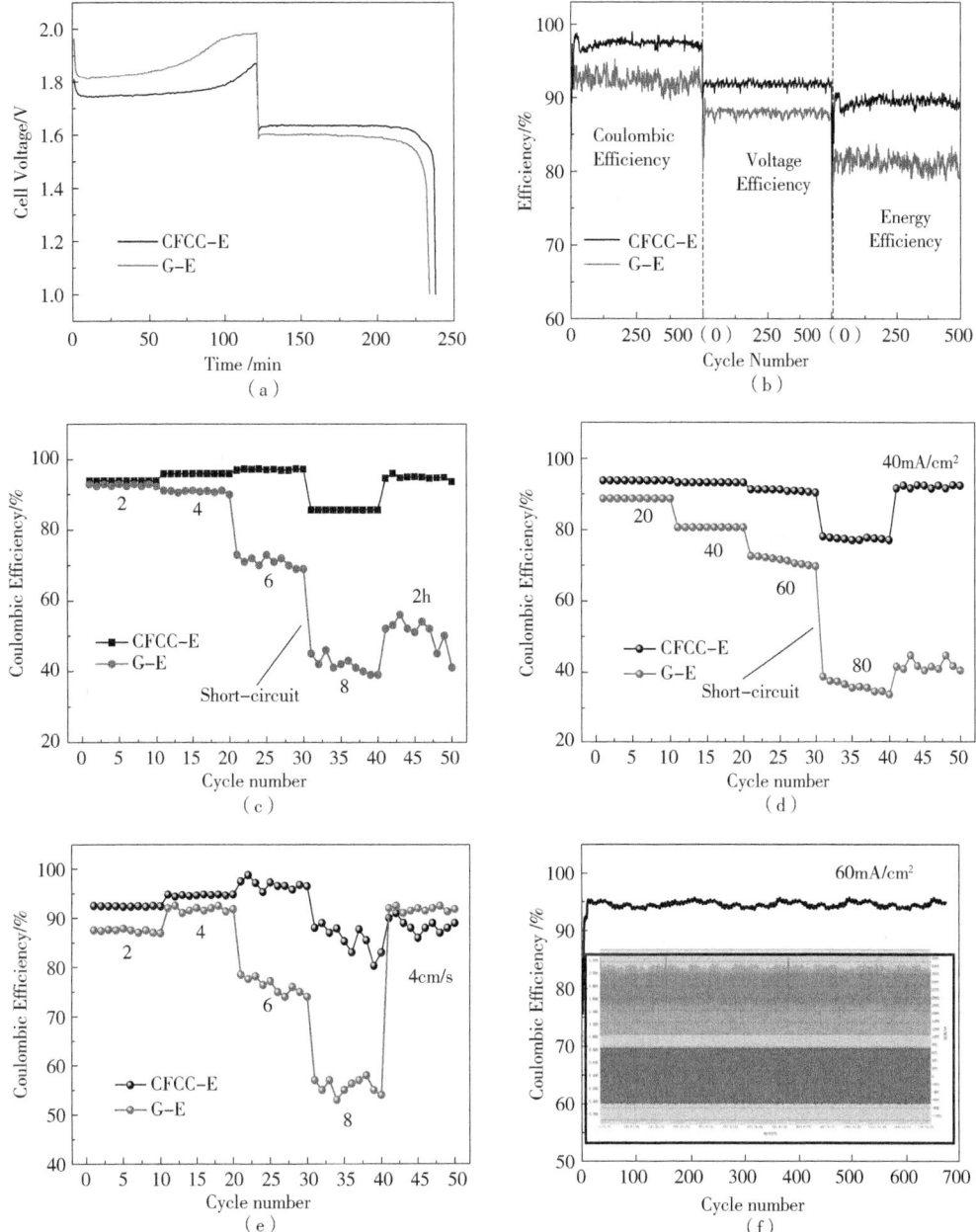

Fig. 12 (a) Cell voltage vs. time responses for charge-discharge cycles; (b) Coulombic efficiency (CE), voltage efficiency (VE), and energy efficiency (EE) as functions of the cycle number; CEs of the SLFBs by employing the CFCC-E and the G-E at (c) different charge times, (d) different charge/discharge current densities, (e) different flow rates, and (f) CEs of the cell long cycle by employing the CFCC-E at 60mA/cm^2

values. Since EE is the product of CE and VE, the remarkable performance of the flow cell utilizing CFCC-E as the negative electrode, characterized by superior CE and VE, contributed to higher EE values (90%±1.0%). In contrast, the flow cell employing G-E as the negative electrode displayed a stabilized EE value of 82%±2.0%. Moreover, the fluctuation ranges of the three efficiencies of the flow cell with the CFCC-E as the negative electrode were significantly lower than those of the G-

E. In contrast Liu et al. and Jaiswal et al. reported that the CEs, VEs, and EEs of the SLFBs by employing G-Es and graphite felt were roughly 92% ~ 95%, 85% ~ 88%, and 74% ~ 76%, respectively, at ambient temperature[8, 39].

With the increase of the charging time, the dissolution activity of the deposits and the cycling efficiency of the flow cell was reduced. The cells were charged from 2h to 8h at a constant current density of 10mA/cm^2 [Fig. 12(c)]. The maximum CE of the cell with the CFCC-E as the negative electrode was stable at 98.1%±0.2% after being charged for a duration of 6h. The CE of the cell, with the G-E serving as the negative electrode, exhibited a continuous decrease from 94.4%±0.5% (charge time of 2h) to 90.3%±0.6% (charge time of 4h). Subsequently, it experienced a rapid decline to 71.5%±1.2% at 6h, ultimately resulting in a short-circuiting scenario. The porous structure of the CFCC-E facilitated the placement of additional lead deposits. The remarkable performance of CFCC-E in terms of the electrode reaction rate, redox activity, HER, and suppression of dendrite formation, significantly augmented the deep charging ability of the SLFB. The cells were subjected to testing under current densities ranging from 20mA/cm^2 to 80mA/cm^2. The CE of the cell with the CFCC-E as the negative electrode remained consistently high at 96.5%±0.8% across a current density range of 20mA/cm^2 to 60 mA/cm^2 [Fig. 12(d)]. In contrast, in the study conducted by Ho et al., each cycle of the cell was 15min of charging and discharging at 15mA/cm^2. The average CE was 88%, the average VE was 70%, and the EE was only 62%[40]. The porous structure possessed by the CFCC-E effectively facilitated charge dispersion, particularly at high current densities, thus mitigating electrochemical polarization and minimizing capacity loss. The CE of the cell with the G-E as the negative electrode decreased continuously from 91.8%±0.2% at 20mA/cm^2 to 72.6%±1.0% at 60mA/cm^2, ultimately leading to a short-circuit scenario at 80mA/cm^2. The growing dendrites of lead making contact with the positive electrode at high current densities accounted for the observed results. As depicted in Fig. 12(c) and Fig. 12(d), the CEs of the cells by employing CFCC-E can return to the initial state after alternating changes in charging time (from 8 h to 2 h) and current density (from 80mA/cm^2 to 40mA/cm^2), whereas the cell by employing G-E did not.

With the increase of the electrolyte flow rate from 2cm/s to 8cm/s in Fig. 12(e), the cell with the CFCC-E as the negative electrode reached the maximum CE of 96.3%±0.6% at the flow rate 6cm/s, while the cell with G-E as the negative electrode reached the maximum CE of 92.6%±0.6% at the flow rate 4cm/s. The porous structure of CFCC-E was conducive to the uniform diffusion of electrolyte, so the electrode reaction rate and the cycle efficiency could be improved by accelerating the flow rate. Nevertheless, upon decreasing the flow rate from 8cm/s to 4cm/s, the CE of the cell utilizing the G-E as the negative electrode nearly returned to its initial state, whereas the CE of the CFCC-E did not exhibit the same recovery. This discrepancy can be attributed to the shedding of lead deposits from the CFCC-E surface during charging and discharging under high flow rates. Consequently, the lead coating on the CFCC-E surface was compromised, thereby impacting the subsequent cycling performance of the cell. To validate the enhanced current density and prolonged cycling performance of the CFCC-E, as depicted in Fig. 12(f), extended cycling tests were conducted on cells employing CFCC-E as the negative electrode, operating at a current density of 60mA/cm². Remarkably, the cell utilizing CFCC-E exhibited over 600 charging and discharging cycles, surpassing a total duration of 1300 hours, while maintaining a stable CE of approximately 94.5%. The utilization of CFCC-E resulted in excellent stability and a substantial increase in the power density of the SLFB system.

4 Conclusions

By improving the electrode materials and structure, the CFCC - E showed an excellent conductivity, electrochemical specific surface area, and porous structure was prepared in this study. Compared with conventional G-E, even and fine lead particles were formed after charging on the CFCC-E, with no significant growth of dendrites being observed in the cross section of the electrode thanks to its porous structure dispersing charges effectively. The Pb^{2+}/Pb reaction on the CFCC-E was solely governed by the charge transfer process, exhibiting superior oxidation and reduction reversibility compared to the G-E, particularly at high current densities. HER and corrosion current density were limited on CFCC-E compared with G-E, avoiding the loss of H^+ in the electrolyte and the corrosion of the electrode. The SLFB by employing the CFCC-E had excellent charge/discharge cycling performances owing to the smaller polarization overpotential and ohmic resistance, and the higher bonding strength hindered the capacity loss. The charging voltage exhibited a low value of 1.7436V, which increased gradually during the process. Conversely, the discharge platform was higher at 1.6359V and demonstrated good stability. The CE, VE, and EE reached values of 97.5%± 1.0%, 92%±0.5%, and 90%±1.0%, respectively. The utilization of CFCC-E effectively enhanced the battery's capacity, enabling it to be charged to 6 hours, and improved the charge/discharge current density to 60mA/cm² with the stable CE of 97.5%±0.5%. The cell employing the CFCC-E had been successfully cycled for over 650 cycles and a cumulative test duration of approximately 1300h with an average coulombic efficiency of 94.5% when cycled at 60mA/cm² for 1h. These notable enhancements in the battery's cycling performance can be attributed to the synergistic effect resulting from the numerous and excellent physicochemical characteristics of CFCC-E.

Acknowledgements

Financial support from the projects "Development of low-cost and long-time vanadium flow battery energy storage technology" and "Key technology research and equipment development of intrinsic safety flow battery" is gratefully acknowledged. The above - mentioned projects have benefited from CNPC and CNPC Tubular Goods Research Institute, respectively. The authors would like to thank the technical support of The Laboratory of Energy Storage Technology from Xi'an University of Technology.

References

[1] K. Tan, T. Babu, V. Ramachandaramurthy, et al. Empowering smart grid: A comprehensive review of energy storage technology and application with renewable energy integration[J]. Journal of Energy Storage, 2021, 39: 102591.

[2] J. Sousa, J. Lagarto, E. Carvalho, et al. SWHORD simulator: A platform to evaluate energy transition targets in future energy systems with increasing renewable generation, electric vehicles, storage technologies, and

hydrogen systems[J]. Energy, 2023, 271: 126977.

[3] A. Oury, A. Kirchev, Y. Bultel, et al. PbO_2/Pb^{2+} cycling in methanesulfonic acid and mechanisms associated for soluble lead-acid flow battery applications[J]. Electrochimica Acta, 2012, 71(1): 140-149.

[4] D. Pletcher, R. Wills. A novel flow battery-A lead acid battery based on an electrolyte with soluble lead(II) III. The influence of conditions on battery performance[J]. Journal of Power Sources, 2005, 149: 96-102.

[5] D. Pletcher, H. Zhou, G. Kear, et al. A novel flow battery—A lead-acid battery based on an electrolyte with soluble lead(II) V. Studies of the lead negative electrode[J]. Journal of Power Sources, 2008, 180: 621-629.

[6] X. Li, D. Pletcher, F. Walsh. A novel flow battery: A lead acid battery based on an electrolyte with soluble lead(II) Part VII. Further studies of the lead dioxide positive electrode[J]. Electrochimica Acta, 2009, 54: 4688-4695.

[7] M. Krishna, E. Fraser, R. Wills, et al. Developments in soluble lead flow batteries and remaining challenges: An illustrated review[J]. Journal of Energy Storage, 2018, 15: 69-90.

[8] Z. Liu, D. Ji, J. Shi, et al. Weatherability of soluble lead flow batteries using the electrolytes of $Pb(CH_3SO_3)_2$ and CH_3SO_3H/HBF_4[J]. Journal of Energy Storage, 2022, 53: 105221.

[9] A. Hazza, D. Pletcher, R. Wills. A novel flow battery: A lead acid battery based on an electrolyte with soluable lead(II) Part I. Preliminary studies[J]. Journal of Chemical Physics, 2004, 6: 1773-1778.

[10] A. Hazza, D. Pletcher, R. Wills. A novel flow battery: A lead acid battery based on an electrolyte with soluble lead(II) IV. The influence of additives[J]. Journal of Power Sources, 2005, 149: 103-111.

[11] Z, Liu, B Sun, X Zhang, et al. Improvement of commercial carbon/high density polyethylene conductive composite plate with an aqueous silane treatment for soluble lead flow batteries[J]. Journal of Energy Storage, 2021, 41: 102967.

[12] D. Pletcher, R. Wills. A novel flow battery: A lead acid battery based on an electrolyte with soluble lead (II). Part II. Flow cell studies[J]. Physical Chemistry Chemical Physics, 2004, 8: 1779-1785.

[13] E. Fraser, R. Wills, A. Cruden. The use of gold impregnated carbon- polymer electrodes with the soluble lead flow battery[J]. Energy Reports, 2020, 6(5): 19-24.

[14] A. Oury, A. Kirchev, Y. Bultel. A numerical model for a soluble lead-acid flow battery comprising a three-dimensional honeycomb-shaped positive electrode[J]. Journal of Power Sources, 2014, 246: 703-718.

[15] Z. Liu, J. Shi, D. Ji, et al. Deposition Behavior of Lead in Lead Methanesulfonate Flow Batteries with the Addition of Tin(II) Methanesulfonate[J]. Journal of The Electrochemical Society, 2021, 168(7): 1-7.

[16] Q. Zhou, G. Li, Y. Qu, et al. Preparation and antioxidation of Cu-Sn composite coating on the surface of Cu-coated carbon fibers[J]. Surfaces and Interfaces, 2022, 31: 102016.

[17] A. Abdel-Aziz, A. El-Zomrawy, M. El-Sabbah, et al. Electrodeposition of lead and lead-tin alloy on copper using an eco-friendly methanesulfonate plating bath[J]. Journal of Materials Research and Technology, 2022, 18: 2166-2174.

[18] D. Ji, Z. Liu, B. Jiang, et al. Temperature adaptability of the lead methanesulfonate flow battery[J]. Journal of Energy Storage, 2020, 28: 101218.

[19] D. Ji, Z. Liu, B. Jiang, et al. Temperature adaptability of the lead methanesulfonate flow battery: Optimization of electrolytic composition based on solubility, conductivity, viscosity and cycle performance of battery[J]. Journal of Energy Storage, 2021, 34: 101989.

[20] Z. Liu, X. Luo, D. Ji. Effect of phase composition of PbO_2 on cycle stability of soluble lead flow batteries [J], Journal of Energy Storage, 2021, 38: 102524.

[21] H. Zhang, X. Deng, G. Zhang. Preparation and properties of multiphase solid-solution strengthened high-performance W-Cu alloys through alloying with Mo, Fe and Ni[J], Materials Science and Engineering: A, 2023, 871: 144909.

[22] A. Abdel-Aziz, A. El-Zomrawy, M. El-Sabbah, et al. Electrodeposition of lead and lead-tin alloy on

copper using an eco-friendly methanesulfonate plating bath[J], Journal of Materials Research and Technology, 2022, 18: 2166-2174.

[23] A. Singh, K. Ansari, I. Ali, et al. Inhibition of hydrogen evolution and corrosion protection of negative electrode of lead-acid battery by natural polysaccharide composite: Experimental and surface analysis[J], Journal of Energy Storage, 2023, 57: 106272.

[24] C. Tang, K. Zhao, Y. Tang, et al. Forest-like carbon foam templated rGO/CNTs/MnO_2 electrode for high-performance supercapacitor[J], Electrochimica Acta, 2021, 375: 137960.

[25] ASM, ASM Metals Handbook[Z], ASM International, 2001.

[26] R. Wills, J. Collins, D. Stratton-Campbell, et al. Developments in the soluble lead-acid flow battery[J]. Journal Of Applied Electrochemistry, 2010, 40: 955-965.

[27] K. Wang, Y. Xiao, P. Pei, et al. A Phase-Field Model of Dendrite Growth of Electrodeposited Zinc[J]. Journal of The Electrochemical Society, 2019, 166(10): D389-D394.

[28] F. Beck. Cyclic behaviour of lead dioxide electrodes in tetrafluorborate solutions[J]. Journal of Electroanalytical Chemistry, 1975, 65(1): 231-243.

[29] A. Velichenko, D. Girenko, F. Danilov. Mechanism of lead dioxide electrodeposition[J]. Journal of Electroanalytical Chemistry, 1996, 405(1): 127-132.

[30] Y. Noritomi, T. Kuboki, H. Noritomi. Promotion of the redox reaction at horseradish peroxidase modified electrode combined with ionic liquids under irreversible electrochemical conditions[J]. Results in Chemistry, 2022, 4: 100666.

[31] J. Dong, X. Wu, Y. Chen, et al. A study on Pb^{2+}/Pb electrodes for soluble lead redox flow cells prepared with methanesulfonic acid and recycled lead[J]. Journal of Applied Electrochemistry, 2016, 46: 861-868.

[32] V. Mirceski, L. Stojanov, B. Ogorevc. Step potential as a diagnostic tool in square-wave voltammetry of quasi-reversible electrochemical processes[J]. Electrochimica Acta, 2019, 327: 134997.

[33] F. Garay, M. Lovriĉ. Square-wave voltammetry of quasi-reversible electrode processes with coupled homogeneous chemical reactions[J]. Journal of Electroanalytical Chemistry, 2002, 518(2): 91-102.

[34] M. Nandanwar, S. Kumar. Charge coup de fouet, phenomenon in soluble lead redox flow battery[J]. Chemical Engineering Science, 2016, 154: 61-71.

[35] A. Gupta, C. Srivastava. Electrodeposition current density induced texture and grain boundary engineering in Sn coatings for enhanced corrosion resistance[J]. Corrosion Science, 2022, 194: 109945.

[36] O. Nevgasimov, V. Bohomaz, S. Petrushenko, et al. Morphology of island structures formed by self-organization processes during melting of lead films[J]. Materials Today: Proceedings, 2022, 62(9): 5787-5795.

[37] H. Liu, Q. Xu, C. Yan. On-line mass spectrometry study of electrochemical corrosion of the graphite electrode for vanadium redox flow battery[J]. Electrochemistry Communications, 2013, 28: 58-62.

[38] C. Zhang, S. Sharkh, X. Li, et al. The performance of a soluble lead-acid flow battery and its comparison to a static lead-acid battery[J]. Energy Conversion & Management, 2011, 52(12): 3391-3398.

[39] N. Jaiswal, H. Khan, K. Ramanujam. The combined impact of trimethyloctadecylammonium chloride and sodium fluoride on cycle life and energy efficiency of soluble lead-acid flow battery[J]. Journal of Energy Storage, 2022, 54: 105243.

[40] Y. Ho, M. Kim, G. Kim, et al. Study on electrochemical properties of Pb(BF4)2 electrolyte for improvement of cycle lifetime and efficiency in soluble lead flow batteries[J]. Journal of Saudi Chemical Society, 2022, 26(3): 101472.

本论文原发表于《Journal of Energy Storage》2024年第78卷。

Cracking Failure Analysis of a Steel Wire Reinforced Thermoplastic Composite Pipe Used in an Oily Sewage Conveying System

Kong Lushi   Li Houbu   Wei Bin   Zhang Zhao

(State Key Laboratory for Performance and Structure Safety of Petroleum Tubular Goods and Equipment Materials, Tubular Goods Research Institute, China National Petroleum Corporation)

Abstract: A steel wire reinforced thermoplastic composite pipe used in the oily sewage conveying system cracked after an 8 year service. Macroscopic and microscopic observations, FTIR, density and hardness measurements, Differential Scanning Calorimeter (DSC), Vicat Softening Temperature (VST), and the stability of no splitting for pressed composite pipes were conducted to analyze the causes and prevent such cases from happening again. The results show that the size of the pipe and the wire winding meet the requirements of the standard of CJ/T 189−2007, but the cracks along the axial direction of the pipe appeared on the outside surface of the pipe during the pressing test, which is similar to the cracks in the failed pipe, indicating that the pressure bearing performance of the pipe decreased. The direct cause of the crack was found that due to the performance degradation of the hot glue wrapped around reinforced steel wire, cracks induced by stress first appeared in the hot glue matrix, and developed in both the inner and outer layers of the PE matrix. The outer layer was thinner than the inner layer and was penetrated first, showing that the number of cracks on the outer surface was significantly more than that of the inner surface. The crack of the outer matrix induced the corrosion and even fracture of the reinforced steel wire, resulting in a decrease in the pressure-bearing capacity of the pipe. The pipe expanded due to wire corrosion or even fracture under pressure conditions, which resulted in the further development of cracks in the inner layer matrix, and eventually, penetrated the inner layer matrix. Finally, leakage occurred.

Keywords: Steel wire reinforced; Polyethylene; Hot glue; Performance degradation

1 Introduction

The corrosion of metal pipelines in oil and gas gathering and transportation systems has received

---

Corresponding author: Kong Lushi, kongls@cnpc.com.cn.

ongoing attention. Oil and gas environments that are dominated by CO_2, H_2S gas, and microorganisms are the primary contributors to the deterioration of pipelines[1-3]. Much effort has gone into finding effective solutions to the corrosion problem, including the addition of anti-corrosion chemical agents[4-5], the choice of anti-corrosion alloy[6-7], or the use of non-metallic and composite pipes[8], etc.

Non-metallic and composite pipes are an effective alternative to address corrosion issues in metal pipes [9-10]. Many types of non-metallic and composite pipes are used in the oil field, including fiberglass-reinforced pipes, reinforced thermoplastic pipes (RTP), steel-skeleton-reinforced thermoplastic pipes, and others. Among them, steel-skeleton-reinforced thermoplastic pipes rank among the top non-metallic pipes used in China's oil fields. The steel wire reinforced thermoplastic (SRTP) composite pipe is a typical kind of steel-skeleton-reinforced thermoplastic pipe. Its structure typically contains three layers: an inner layer of high density polyethylene (HDPE), layers of spiral-wound steel reinforcement, and an external coating of polyethylene. The inner and outer layers are composed of HDPE, which contains fluid and offers external protection, respectively. Whereas the intermediate layers include steel wires as the reinforced body and hot glue as the matrix[11-12]. The above structure and materials endow SRTP with high specific strength and good corrosion resistance. SRTP has been used in many fields, especially under extreme conditions including convective slurry in salt lakes for minerals, subsea pipelines for freshwater transport, and sewage transportation in the oil field. SRTP in such an application is typically subjected to combined loads including internal pressure, bending, etc, where the ambient temperature or operation temperature fluctuates widely simultaneously. As for the SRTP in such severe conditions or improper operation during installation including over-welding at the joint and mechanical damage, structural failures occasionally occur, and they are almost inevitable although SRTP shows excellent mechanical performance[11-15].

In this work, we report a new failure mode that occurred in an SRTP service in the oily sewage conveying system cracked after an 8 year service. Macroscopic and microscopic observations, FTIR, density and hardness measurements, Differential Scanning Calorimeter (DSC), Vicat Softening Temperature (VST), and the stability of no splitting for pressed composite pipes were conducted to analyze the causes and prevent such cases from happening again.

2 Background of the incident

The pipeline was put into operation on October 10, 2013, with a total length of 13.2km and service conditions: the transporting medium is oily sewage, with a volume of $60m^3/h$, a pressure of 0.36~1.07MPa and temperature of 20~30℃. In November 2021, the cracking failure occurred. Figure 1 shows the failed pipe and its cross-section morphology. The mark on the pipe is the following "SRTP-L-PE Dn 200 PN 1.6MPa CJ/T 189-2007". It indicates that the pipe specification was DN 200mm and PN 1.6MPa. The standard implemented in this pipe was a China standard CJ/T 189-2007 Steel wire reinforced thermoplastic (PE) composite pipe and fitting.

Fig. 1 Photographs of the failed SRTP

3 Failure description

To conduct a complete visual examination, internal and external surface cleaning was performed on the failed pipe. Figure 2a~d shows cracks distributed at the outer surface along the axial direction. It can be seen from Figures 2e~g that 3 cracks penetrated the pipe wall with lengths of 14.96mm, 1059mm, and 8.01mm, respectively. Figure 2h indicates that at the corresponding zone at the internal surface, the number of cracks is obviously less than that of the external surface. It was inferred that the crack originated near the outer surface and developed from outside to inside, eventually penetrating the tube wall.

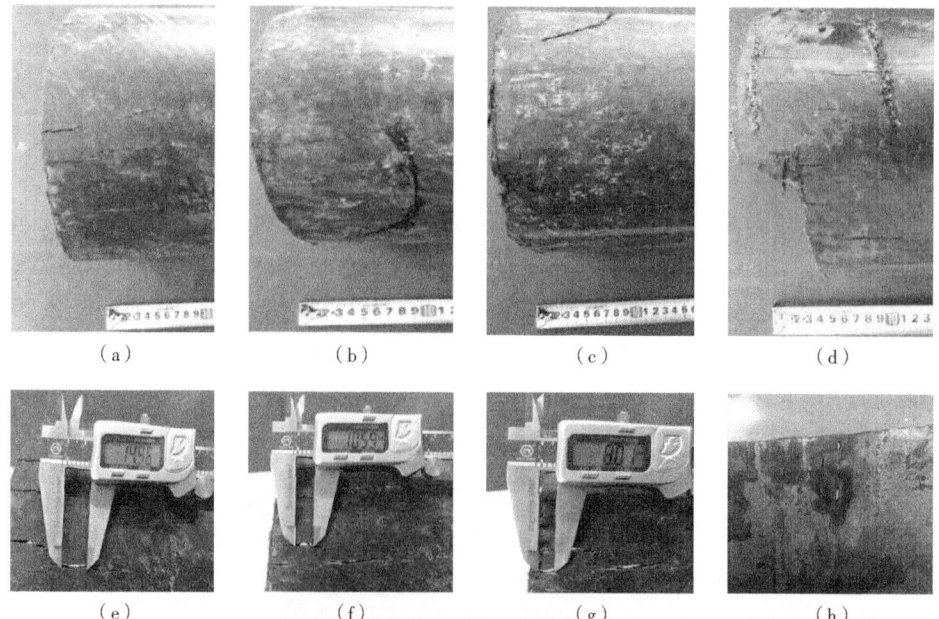

Fig. 2 Photographs of the failed zone in the failed SRTP

4 Methodology

In this work, the failure analysis contains three parts. First, based on the background information and failure description, the causes that might lead to the failures of the liner were proposed. Then, probable causes of the failure were systematically analyzed by kinds of measurements, including microscopic observation, Fourier Transform infrared spectroscopy (FTIR), density and hardness measurements, tensile test, and gel permeation chromatography (GPC). Finally, a comprehensive analysis based on the above results was conducted and countermeasures were offered.

To analyze the causes of the failure, the samples used for chemical, physical, and mechanical tests were cut from the area close to the failed zone. Before testing, all samples were first washed with kerosene to remove the oil pollution on the surface of the samples. Then, the samples were washed with ethanol and deionized water under ultrasonic treatment. Finally, the samples were dried in an oven at 60℃ for 4h.

Zeiss Smart Zoom 5 digital light microscope was used to observe the microscopic morphology. FTIR spectra of the samples via the Attenuated total reflection model were collected by using a Nicolet is 50 IR spectrometer. The density of the samples was measured by an ET-12SL electronic densitometer. The hardness of the samples was obtained by a TIME5410 Shore A durometer. An RV-300FW VST Tester was used to assess VST of specimens. A sample with the size of 10mm×10mm was selected from the pipe. For the VST test, the inner surface was used. The stability of no splitting for pressured composite pipe tests and tensile test of the steel wire were evaluated by an AGS-X10kN testing machine. The molecular weight and distribution of liner pipe material were tested by high-temperature gel permeation chromatography (GPC) with 1, 2, 4-trichlorobenzene (TCB) as a solvent, Mixed A and B columns as filtration columns, a differential detector, and polystyrene as a calibration standard. The sample was first dissolved by shaking at 160℃ for 4h, the test temperature was 160℃ and the flow rate was 1mL/min. An AQ200 differential scanning calorimeter was used to perform the DSC test to evaluate the pipe's degree of crystallinity and the oxidation induction temperature (OIT). Samples with dimensions of 10mm×10mm were collected from the pipe. The temperature program was set to warm at a rate of 10K/min from room temperature to 400℃. At a rate of 50mL/min, nitrogen flushing was done before 200℃, and at a rate of 50mL/min, oxygen flushing was done following 200℃. The melting enthalpy of a hypothetical 100% crystalline polyethylene [$\Delta H_m (100\%)$] was 125.4J/g, and the degree of crystallinity X_c was derived according to Eq. (1)[16]. The actual ΔH_m of the specimen was obtained after the integration of the melting curve in the DSC with a tangential reference plane. For crystallinity, each liner sample underwent a minimum of three separate DSC measurements.

$$X_c = \frac{\Delta H_m}{\Delta H_m(100\%)} \times 100\% \tag{1}$$

5 Results and discussions

5.1 Probable cause of the crack

The surface morphology of the penetrating crack, as shown in Figure 3, shows that there is no obvious damage at the surface of the pipe near the crack, and no reinforced wires are seen in the penetrating crack, it can be assumed that the wire at the crack zone has been fractured. In Figure

3a, the cross-section morphology of the crack shows that the fracture surface is clean and smooth near the outer layer and reinforcement layer, exhibiting a brittle fracture morphology. The reinforced steel wire has a clear corrosion fracture morphology. The part near the inner surface shows a ductile fracture pattern similar to water ripple, as shown by the dashed box mark in Figure 3a. According to the morphology, it can be judged that the fracture direction is from outer to inner. These results are consistent with the phenomenon that the number of cracks on the outer surface of the pipe is significantly greater than that on the inner surface, which confirms that cracks are being developed from the exterior to the interior. As shown in Figure 3b~e, the cross-sectional morphology of the crack that did not penetrate the wall of the pipe shows that the crack starts in the hot glue matrix of the reinforcement layer. There is direct morphological evidence in Figure 3c that the crack occurred at the interface between the hot melt and the wire. The speculation of the cause of failure is further verified.

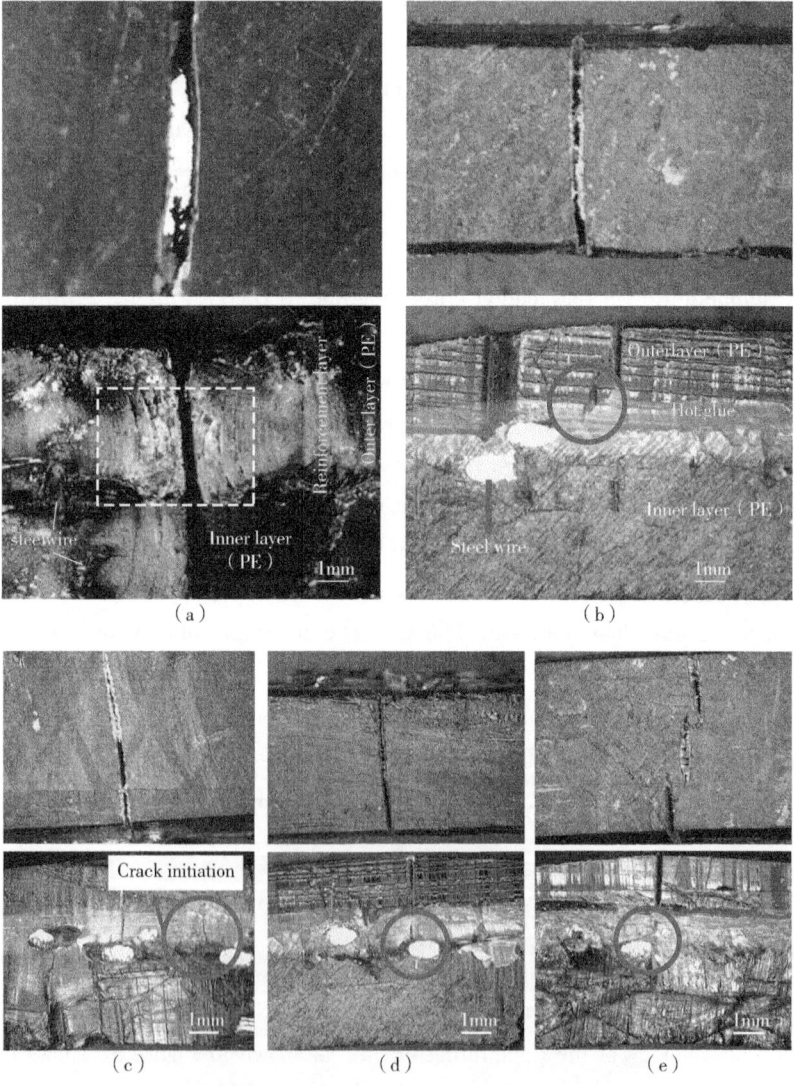

Fig. 3 Surface and cross-section morphology of cracks

5.2 Stability of no splitting for pressed pipe

Based on CJ/T 189-2007 requirements, a 100mm±10mm length pipe specimen was taken for the stability of no splitting for the pressed pipe test, and the specimen was positioned between the pressure plate of the test machine for a slow downward press pressure of 10~15s to 50% of the nominal outer diameter of the pipe, see Figure 4a and Figure 4b. The standard requirements state that during the test, the pipe body should be free of crevices and cracks. Figure 4c shows that cracks appear on the outer wall of the pipe, and the cracks lie along the axial direction of the pipe with different lengths, similar to the failure morphologies. The corresponding inner surface shows no cracks, indicating that the cracks had not penetrated the wall of the pipe. From the above results, it is proved that the failure of the pipe is generated by the part near the outer surface and develops from the outer to the inner along the radial direction. The stability of no splitting for pressed SRTP does not meet the requirements CJ/T 189-2007, indicating that the SRTP is not suitable for continued use.

Fig. 4 (a) and (b) Stability of no splitting for pressed pipe and (c) external and (d) internal surface morphology of pressed SRTP

5.3 FTIR, Hardness, and density

In Figure 5, the FTIR spectra show that both the matrix and the hot glue material are polyethylene, with the characteristic peaks at $2915cm^{-1}$, $2547cm^{-1}$, $1462cm^{-1}$, and $717cm^{-1}$[17], indicating the absence of any significant changes in chemical structure. The FTIR spectra of the inner and outer surface show that after 8 years of service, no other new characteristic peaks are present, indicating that the chemical structure of polyethylene on the inner and outer surfaces has not changed in a significant way. The FTIR spectrum of hot glue shows the absorption peaks at $1088cm^{-1}$, $1021cm^{-1}$ and $872cm^{-1}$, presumably for two reasons: one is that the main component of

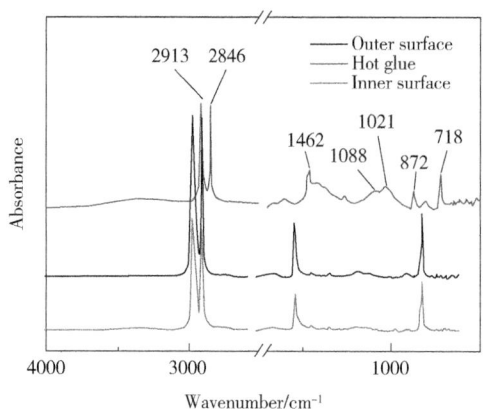

Fig. 5 FTIR spectra of the external surface, internal surface, and hot glue in the failed SRTP

the hot glue is polyethylene, but small amounts of other components are also present or PE was modified; the other is that the hot glue has undergone structural changes during the longterm service and new absorption peaks have emerged. Research speculations on hot glue products suggest that hot glue may be the Maleic anhydride graft-modified linear LDPE.

The density and hardness of the failed pipe matrix were measured respectively. The density of the pipe matrix is 0.962g/cm^3, as shown in Table 1, meeting the density requirements of HDPE in CJ/T 189-2007. The hardness of the sample from the external surface and internal surface of the pipe matrix are 54.4 and 63.1, respectively (see Table 2), matching the hardness of HDPE. It can be inferred that the transporting medium penetrated into the pipe matrix may act as a plasticizer and soften the liner, causing hardness to decrease.

Table 1 Density of SRTP matrix

No.	1	2	3	Mean
Density/(g/cm^3)	0.961	0.965	0.960	0.962

Table 2 Hardness of external surface and internal surface

Position	1	2	3	4	5	Mean
Outer surface	64.9	64.1	63.5	61.9	61.3	63.1
Inner surface	53.7	52.3	55.9	54.7	55.3	54.4

5.4 Vicat softening temperature (VST)

As shown in Table 3, the VST of the interior and exterior surfaces of the SRTP matrix material is 122.94℃ and 124.51℃, respectively, which is higher than the standard requirement of no lower than 120℃ in CJ/T 189-2007. It shows that the inner surface is slightly lower than the outer surface, likely due to the inner surface being in contact with the transporting medium for a long time, there is a slight reduction in thermal dimensional stability, so VST became lower [18].

Table 3 VST of SRTP matrix

Position	No.	Initial temperature	VST/℃	Mean/℃
Inner surface	1	R.T.	122.72	122.92
	2	R.T.	122.97	
	3	R.T.	123.06	
Outer surface	1	R.T.	123.44	124.15
	2	R.T.	124.94	
	3	R.T.	123.56	
Requirement in CJ/T 189-2007	—	—	—	≥120.00

5.5 Oxidation induction temperature and crystallinity

Oxidation-induced temperature (OIT) was used to characterize the thermal stability of the polymeric material. Figure 6 shows that the OIT of the inner and outer surfaces of the matrix and the hot glue were 239.01℃, 237.47℃, and 226.05℃, respectively, and there was little difference in the thermal aging performance of the inner and outer surfaces. The OIT of hot glue is 11℃ lower than that of the matrix, which indicates that the thermal aging performance of hot gule is worse than that of the matrix. The peak melting temperature of the outer surface of the failed matrix material was found to be 130.32℃ with a melting enthalpy of 155.5 J/g; the melting peak of the inner surface was found to be 130.23℃ with an enthalpy of melting of 150.5J/g; the peak melting temperature of the hot glue was 135.57℃ with an enthalpy of fusion of 149.4J/g.

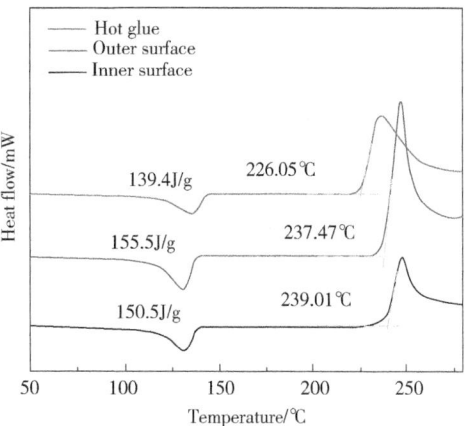

Fig. 6   DSC curves of SRTP matrix and hot glue

The crystallinity can be calculated according to Eq. (1). The crystallinity of the interior and exterior surfaces of the matrix material and hot glue are calculated to be 63.0%, 65%, and 62.5% respectively. The inner surface has a lower crystallinity than the outer surface, which could be due to the interaction of the inner surface with the transporting medium during the long-term service, leading to a small change in condensed structure and decreased crystallinity; and the lower crystallinity of the hot glue which can result in lower mechanical properties of the hot glue compared to the matrix material.

5.6 Molecular weight and distribution

Gel permeation chromatography (GPC) was used to investigate the molecular weight and molecular weight distribution of the pipe matrix resin as well as the hot glue, and the results are shown in Table 4. It can be seen from the test results that the molecular weight of the hot glue is lower than that of the matrix resin and the distribution of molecular weights is broader. Polyethylene materials with a low molecular weight and a broad molecular weight distribution have been shown to have worse mechanical and thermal properties than polyethylene materials with a high molecular weight and a narrow molecular weight distribution. The results indicates that the hot glue is the performance shortcoming in the pipe, which is more prone to yield or fracture failure than the matrix resin when subjected to the deformation of the material caused by internal pressure or external forces.

Table 4   Molecular weight and polymer dispersity index (PDI) of SRTP matrix and hot glue

Sample	Mp	Mn	Mw	Mz	PDI
Matrix	136778	84974	271881	702365	3.19958
Hot glue	105508	42153	201462	646586	4.7793

5.7 Winding mode of steel wire

Figure 7 shows that the winding mode of the steel wire is determined by the morphology after stripping the outer layer and the cross-sectional morphology of the pipe. The parameters related to the winding mode of the steel wire are shown in Table 5, and the steel wire winding mode satisfies CJ/T 189-2007 requirements. Figure 7 shows that the lap of steel wire is not welded, and when the pipe is subjected to internal pressure or other external forces, because of the force, the steel wire will slip relatively, causing the interface between the steel wire and the hot glue to become a point of stress concentration.

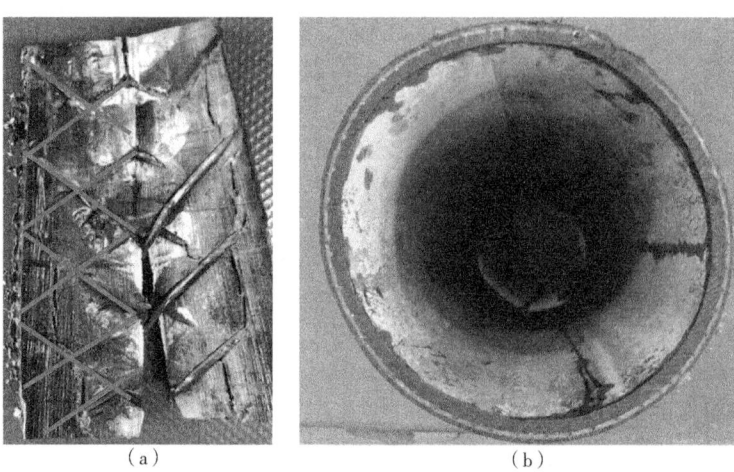

(a)　　　　　　　　(b)

Fig. 7　Winding mode of the steel wire in the failed SRTP

Table 5　Information of winding mode of steel wire in the failed SRTP

Item	Measured	Requirements in CJ/T 189-2007
Min diameter of steel wire/mm	0.98	0.5
Winding angle/(°)	55	54.7~60
Winding direction	Left-handed + right-handed	Left-handed + right-handed
Number of steel wires in cross section	94	>86

5.8 Surface morphology and tensile properties of steel wire

An area without cracks on the surface was chosen for stripping of the outer layer matrix, and the wire surface morphology was shown in Figure 8a. The surface of the wire was smooth with a metallic sheen and there were no signs of corrosion. Simultaneously, the morphology of the cross-section shows that a clear interfacial separation between the hot glue and the wire is occurring, and the hot glue loses its bonding effect causing the wire to move with respect to the hot glue when the wire is under tension. An area with cracks on the surface was chosen for stripping of the outer layer matrix, and the wire surface morphology was shown in Figure 8b, it can be seen that the hot glue has obvious cracks, and corrosion of the reinforced steel wire was obvious. Corrosion of the reinforced steel wire will inevitably lead to a decrease in mechanical properties, which ultimately results in a decrease in the pressure-carrying capacity of the SRTP. Based on CJ/T 189-2007 requirements for steel wire, the elongation and tensile strength of the steel wire is expected to meet the GB/T 14450 requirements. The tensile properties of the steel wire in the reinforcement layer with no cracked area on the surface of the failed SRTP were tested according to the GB/T 14450, and the results are shown in Table 6. Tensile strength and elongation at breakage of steel wire meet standard requirements.

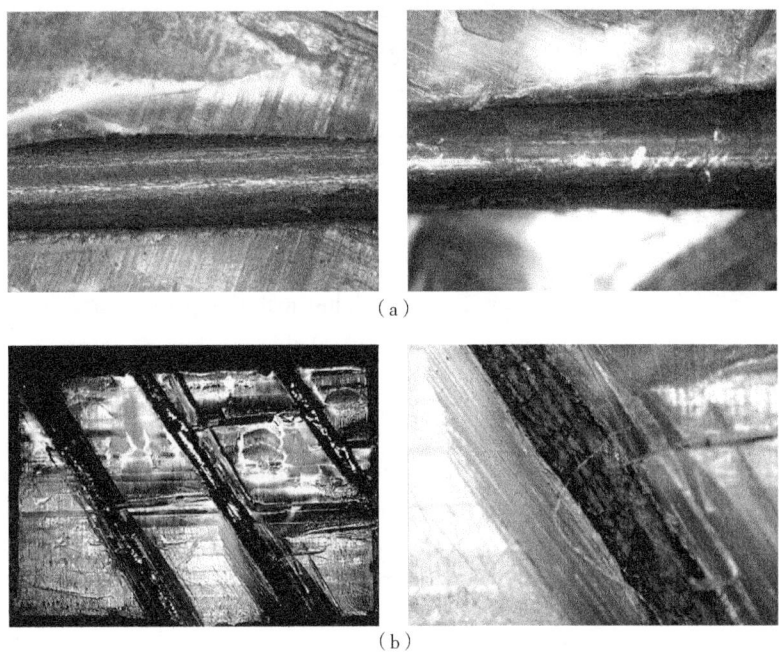

Fig. 8 (a) Surface morphology of steel wire in the area without cracks and (b) Surface morphology of steel wire in the area with cracks in the failed SRTP

Table 6 Tensile properties of steel wire in the area without cracks in the failed SRTP

Properties	1	2	3	Mean	GB/T 14550(NT)
Tensile strength/MPa	1950	1996	2002	1983	1850
Elongation at break/%	7.2	9.6	4.5	7.1	≥5.0

6 Conclusions and recommendations

6.1 Conclusions

The hot glue and the steel wire have a significant interface separation, that is, the hot glue loses the role of bonding the steel wire. When the internal pressure or other external loads are applied to the pipe, because the lap of steel wire is not welded and fixed, the steel wire in the load bearing process would slip. The sliding of the steel wire induced the cracks at the interface between the hot glue and steel wires, and cracks developed to the inner and outer layer of matrix at the same time. The outer layer is thinner and penetrates first, exhibiting a significantly greater number of cracks on the outer surface compared to the inner one. Cracking of the outer matrix induces corrosion and even fracture of the reinforcing steel wire, leading to a decrease in the pressure-carrying capacity of the SRTP. Under pressure conditions, the SRTP would inflate further due to corrosion or even fracture of the steel wire, leading to further development of cracks to the inner matrix, and finally, cracks penetrate the pipe wall and leakage failure occurred.

6.2 Recommendations for failure prevention

(1) The rest of the SRTP should be inspected visually, focusing on the parts where the topography is prone to change or the fluid flow conditions is prone to change, and if similar cracks are found, pipeline operation must be halted. The results of the failure analysis show that the portion of the SRTP had cracks on the surface, indicating that the pressure-carrying capacity of the SRTP would be significantly reduced. It is no longer able to meet pressure – bearing performance requirements under existing working conditions, the risk of failure is very high, it is recommended to replace.

(2) For the portion of the pipeline that does not exhibit surface cracks, it is recommended that the SRTP should be carried out the performance evaluation, including the pressure – carrying capacity of the entire pipe, matrix material and hot glue material performance, and based on the evaluation results, to comprehensively assess the safety of this pipeline.

Declaration of Competing Interest

The authors declare that they have no known competing financial interests or personal relationships that could have appeared to influence the work reported in this paper.

Acknowledgment

This research was funded by the National Natural Science Foundation of China (NSFC), grant number 52104072, the Shaanxi Provincial Natural Science Basic Research Program, grant number 2022JQ-559, and China National Petroleum Corporation Basic Research and Strategic Reserve Technology Research Fund Project, grant number 2021DQ03 (2022Z-04).

References

[1] Askari, M.; Aliofkhazraei, M.; Ghaffari, S.; Hajizadeh, A., Film Former Corrosion Inhibitors for Oil and Gas Pipelines – a Technical Review. Journal of Natural Gas Science and Engineering 2018, 58, 92-114.

[2] Wang, Z.; Zhang, L.; Yu, T.; Xu, M., Study of Corrosion Behavior of Oil Gathering Facilities with Co2 Flooding in Low-Permeability Oilfields. Chemistry and Technology of Fuels and Oils 2018, 53, 933-942.

[3] Yuli Panca, A., The Roles of H2s Gas in Behavior of Carbon Steel Corrosion in Oil and Gas Environment: A Review. Jurnal Teknik Mesin Mercu Buana 2018, 7, 37-43.

[4] Al Jahdaly, B. A.; Maghraby, Y. R.; Ibrahim, A. H.; Shouier, K. R.; Alturki, A. M.; El-Shabasy, R. M., Role of Green Chemistry in Sustainable Corrosion Inhibition: A Review on Recent Developments. Materials Today Sustainability 2022, 20, 100242.

[5] El-Katori, E. E.; El-Saeed, R. A.; Abdou, M. M., Anti-Corrosion and Anti-Microbial Evaluation of Novel Water-Soluble Bis Azo Pyrazole Derivative for Carbon Steel Pipelines in Petroleum Industries by Experimental and Theoretical Studies. Arabian Journal of Chemistry 2022, 15, 104373.

[6] Escrivà-Cerdán, C.; Ooi, S. W.; Joshi, G. R.; Morana, R.; Bhadeshia, H. K. D. H.; Akid, R., Effect of Tempering Heat Treatment on the Co2 Corrosion Resistance of Quench-Hardened Cr-Mo Low-Alloy Steels for Oil and Gas Applications. Corrosion Science 2019, 154, 36-48.

[7] Liu, H.; Xie, J.; Zhao, M.; Feng, C.; Yu, Y.; Li, G.; Song, S.-y.; Yin, C.-x., Influence of Ru on

Structure and Corrosion Behavior of Passive Film on Ti – 6al – 4v Alloy in Oil & Gas Exploration Conditions. 2022.

[8] de Leon, A. C. C.; da Silva, Í. G. M.; Pangilinan, K. D.; Chen, Q.; Caldona, E. B.; Advincula, R. C., High Performance Polymers for Oil and Gas Applications. Reactive and Functional Polymers 2021, 162.

[9] Alabtah, F. G.; Mahdi, E.; Eliyan, F. F., The Use of Fiber Reinforced Polymeric Composites in Pipelines: A Review. Composite Structures 2021, 276, 114595.

[10] Bukhari, A.; Bashar, M.; Aladawy, A.; Goh, S.; Sarmah, P. In Review of Non-Metallic Pipelines in Oil & Gas Applications - Challenges & Way Forward, International Petroleum Technology Conference, OnePetro: 2022.

[11] Bai, Y.; Liu, S.; Han, P.; Ruan, W.; Tang, G.; Cao, Y., Behaviour of Steel Wire-Reinforced Thermoplastic Pipe under Combined Bending and Internal Pressure. Ships and Offshore Structures 2018, 13, 696-704.

[12] Xiong, H.; Bai, Y.; Fang, Q.; Tang, G., Analysis on the Ultimate Bearing Capacity of Plastic Pipe Reinforced by Cross-Helically Winding Steel Wires under Internal Pressure. Ships and Offshore Structures 2018, 13, 264-272.

[13] Shi, J.; Zhong, S.; Nie, X.; Shi, J.; Zheng, J., Study on Steel Wire Reinforced Thermoplastic Pipes under Combined Internal Pressure and Bending Moment at Various Temperatures. Thin-Walled Structures 2021, 169.

[14] Zhang, X.; Hou, L.; Li, H.; Qi, G.; Qi, D., Leakage Analysis of Steel Wire Reinforced Polyethylene Composite Pipe Used for Waste Water Transportation. Engineering Failure Analysis 2021, 130, 105750.

[15] Shi, J.; Miao, Y.; Li, X.; Li, G.; Wan, Y.; Lu, P. In Investigation of Failure Behavior of Polyethylene Pipe Reinforced by Winding Steel Wires Subject to Inner Pressure and Bending, ASME 2020 Pressure Vessels & Piping Conference, 2020.

[16] Wunderlich, B.; Cormier, C., Heat of Fusion of Polyethylene. Journal of Polymer Science B Polymer Physics 1967, 5, 987-988.

[17] Zhang, Y.; Lin, Y.; Gou, H.; Feng, X.; Zhang, X.; Yang, L., Screening of Polyethylene-Degrading Bacteria from Rhyzopertha Dominica and Evaluation of Its Key Enzymes Degrading Polyethylene. Polymers 2022, 14, 5127.

[18] Kong, L.; Qi, D.; Li, H.; Ding, N.; Ge, P.; Xu, Y.; Zhang, C.; Pan, C.; Fan, X., Aging of Polyethylene of Raised Temperature Resistance Pipe Liner after a Four-Year Service in a Crude Oil Gathering System. Journal of Failure Analysis and Prevention 2021, 21, 1323-1330.

本论文原发表于《Journal of Failure Analysis and Prevention》2023 年第 23 卷第 1 期。

Oxidation Behavior of Boron-containing (Zr, Ti) $C_x B_y$ Solid Solution Ceramic at 1600℃ in Air

Lun Huilin[1,2] **Zeng Yi**[2] **Xiong Xiang**[2] **Li Houbu**[1]

(1. State Key Laboratory of Performance and Structural Safety for Petroleum Tubular Goods and Equipment Materials, CNPC Tubular Goods Research Institute;
2. State Key Laboratory of Powder Metallurgy, Central South University)

Abstract: Multicomponent boron-containing carbide coating [i.e., (Zr, Ti) $C_x B_y$] on a C/C composite shows good ablation resistance. However, the high temperature oxidation behavior of this new type of boron-containing (Zr, Ti) $C_x B_y$ solid solution ceramic has not been clarified yet. The present work fabricated (Zr, Ti) $C_x B_y$ solid solution block ceramic by spark plasma sintering and its oxidation behavior at 1600℃ in air (80% N_2 + 20% O_2) was investigated for the first time. The effects of boron on the oxidation resistance of (Zr, Ti) $C_x B_y$ ceramic were examined. The results indicate that (Zr, Ti) $C_x B_y$ ceramic displays a good oxidation resistance with a parabolic rate law described oxidation process. After trace solution of boron (0.5%, mass fraction) into (Zr, Ti) C_x, the oxidation resistance of carbide ceramic is significantly enhanced, leading to a decrease of 30% of oxidation rate constant. The formed oxide scale in (Zr, Ti) $C_x B_y$ ceramic is dense and the interlayer shows a stronger ability of inhibiting the inward diffusion of oxygen. In addition, the introduction of boron leads to a more negative binding energy of (Zr, Ti) $C_x B_y$ and improves the oxidation resistance of carbide.

Keywords: Ultra-high temperature ceramics; (Zr, Ti) $C_x B_y$ solid solution; Oxidation behavior; Oxidation resistance

1 Introduction

Ultra-high temperature ceramics (UHTCs) usually refer to the ceramics that can work above 2000℃ whose melting point is above 3000℃ [1], such as diborides and carbides consisting of group IV-V metals. UHTCs are the potential candidates for application in hypersonic aircraft nose cones, wing leading edges, and the combustion chamber used for thermal protection systems [2-4]. Among these UHTCs, diborides (e.g., ZrB_2, TiB_2 and HfB_2) display a better oxidation resistance than those of the corresponding carbides with the same metal component (e.g., ZrC, TiC and HfC) due to the presence of boron atoms [5-6]. Generally, carbides have higher melting points (T_m) than the

---

Corresponding authors: Lun Huilin, lunhl@cnpc.com.cn; Zeng Yi, zengyi001@csu.edu.cn.

diborides (e.g., $T_{m(ZrC)}$ = 3540℃, $T_{m(ZrB_2)}$ = 3000℃), indicating that carbides are more thermal stable and would be better suited to use in extreme high temperature environments. Unfortunately, the poor oxidation resistance of carbides may lead to failure and hence restricts their range of applications[7-9]. For example, Rama Rao and Venugopal[10] found that ZrC starts to oxidize at 277℃ in low oxygen partial pressure (5kPa). The formed ZrO_2 scales are generally porous below 1200℃, which leads to a weak ability to slow oxygen diffusion[11-12].

To overcome this problem, efforts have been made to improve the oxidation resistance of carbides. Some secondary phases, such as SiC[13-14] and borides[15], are added into carbides to form a composite to improve its oxidation resistance. Although the addition of secondary phase improves the oxidation resistance of carbides to a certain extent, the multiphase ceramics are generally at the micron or submicron level, and a homogeneity problem may remain and result in an increased risk of material failure. In addition, introducing the metal elements into the binary carbide to form a multicomponent solid solution to avoid the homogeneity problem of multiphase ceramics is another promising way to improve the oxidation resistance. An introduction of TaC into ZrC to form a (Ta, Zr)C increases the melting point (approximately 3930℃, 8TaC–1ZrC, mol fraction) and improves the oxidation and ablation resistance of (Ta, Zr)C[16-18]. The oxidation resistance of non-stoichiometric (Zr, Hf, Ti)C is better than those of their constituent binary carbides[19]. Carbide stoichiometry also has an important effect on the oxidation resistance of carbides. Wuchina et al.[20] found that $HfC_{0.67}$ has a thinner oxide scale than $HfC_{0.98}$ after oxidation below 1600℃ in air, suggesting a slower oxidation rate. Meanwhile, among $(Zr_{0.8}Ti_{0.2})C_x(x=C/M, x=0.7\sim1.0)$[21], $(Zr_{0.8}Ti_{0.2})C_{0.8}$ with a nominal carbon concentration ($x=0.8$) displays a good oxidation resistance below 900℃ in air. This is attributed to the formation of dense $t-(Zr, Ti)O_2$ oxide solid solution with fewer cracks as an oxygen barrier during the oxidation.

Importantly, a novel solid solution carbide [i.e., $(Zr_{0.8}Ti_{0.2})C_{0.74}B_{0.26}$] coating on C/C composite displays superior ablation resistance in the range of 2000~3000℃[22], being a potential ultra-high temperature thermal protection material. A material loss rate of this composite [C/C–$(Zr_{0.8}Ti_{0.2})C_{0.74}B_{0.26}$] is over 12 times better than that of pure ZrC ceramic at 2500℃[22]. This $(Zr_{0.8}Ti_{0.2})C_{0.74}B_{0.26}$ contains two transition metal elements and boron and has a single-phase face-centered cubic (FCC) structure similar to ZrC[22]. Unfortunately, a clear understanding of the oxidation behavior of $(Zr, Ti)C_xB_y$ (x and y are independent of each other) solid solution block ceramic has not been achieved, which limits its potential applications in a wide temperature range. It is noted that materials used in aerospace thermal protection systems potentially need to withstand oxidation over a wide temperature range (e.g., 1500~3000℃)[23]. Moreover, the factors and mechanism of ablation performance are different from those of oxidation performance due to the physical-chemical coupled environment in the ablation process, such as high temperature reactions, volatilization of oxide phases and gas flow scouring[24]. The ablation behavior of C/C–$(Zr_{0.8}Ti_{0.2})C_{0.74}B_{0.26}$ had been systematically studied in the range of 2000~3000℃, whereas the oxidation behavior of $(Zr, Ti)C_xB_y$ at a relatively low temperature (<1800℃) has not been understood. In our recent work[25], the single-phase $(Zr, Ti)C_xB_y$ solid solution powders are successfully fabricated using solid-state diffusion of boron atoms. In order to fully understand the high temperature oxidation of this new type of boron-containing carbide ceramic, this work is the first to investigate the oxidation behavior of (Zr, Ti)C_xB_y solid solution block ceramic at high temperature.

In this study, the (Zr, Ti)C_xB_y solid solution block ceramic was fabricated by a spark plasma sintering (SPS) apparatus. The isothermal oxidation behavior of (Zr, Ti)C_xB_y ceramic was investigated in 80% N_2 + 20% O_2 at 1600℃. Combining the experimental and first-principles calculation results, the effects of boron on the oxidation resistance of (Zr, Ti)C_xB_y ceramic were assessed.

2 Experimental procedure

2.1 Materials preparation

The as-synthesized (Zr, Ti)C_x and (Zr, Ti)C_xB_y powders were utilized as raw materials for sintering (Zr, Ti)C_x and (Zr, Ti)C_xB_y ceramics, both of which were single-phase solid solutions with FCC structure [21,25]. The preparation of (Zr, Ti)C_x solid solution powder was used elemental powders as starting materials, including Zr [a purity > 99.5% (mass fraction), an impurity Hf < 0.5% (mass fraction), particle size ≤ 50μm], Ti [>99.5% (mass fraction), O < 0.2% (mass fraction), H < 0.2% (mass fraction), ≤ 50μm] and carbon [graphite, 99.95% (mass fraction), 3 μm, Macklin Biochemical Co., Ltd, China]. It was fabricated by a modified SPS apparatus and the details were reported in the prior work[21]. The (Zr, Ti)C_xB_y solid solution powder was synthesized using (Zr, Ti)C_x powder and B_2O_3 (Aladdin, amorphous) as boron source by solid-state diffusion of boron atoms, and the details were shown in the work[25].

The above two powders were at a composition of ($Zr_{0.75}Ti_{0.25}$)$C_{0.82}$ and ($Zr_{0.75}Ti_{0.25}$)$C_{0.82}B_{0.06}$, respectively. The atomic ratio of Zr/Ti was designed to 3/1 due to its good oxidation resistance, which was also related to the good ablation resistance of C/C-($Zr_{0.8}Ti_{0.2}$)$C_{0.74}B_{0.26}$ composite[22], where ($Zr_{0.8}Ti_{0.2}$)$C_{0.74}B_{0.26}$ owned a high atomic ratio of Zr/Ti. Further, the (Zr, Ti)C_x displayed a good oxidation resistance when the x is approximately 0.8[21]. Therefore, the compositions of samples were designed as nonstoichiometric ($Zr_{0.75}Ti_{0.25}$)$C_{0.82}$ and ($Zr_{0.75}Ti_{0.25}$)$C_{0.82}B_{0.06}$.

The ($Zr_{0.75}Ti_{0.25}$)$C_{0.82}$ and ($Zr_{0.75}Ti_{0.25}$)$C_{0.82}B_{0.06}$ powders were loaded into a graphite die for sintering by a SPS apparatus (HP D 25-3, FCT Systeme GmbH, Germany) at 1950℃ for 7min under vacuum (<5Pa) with a heating rate of 100℃/min. The pellets diameter was 20mm. The applied uniaxial pressure was 64MPa. After sintering, the ceramics fabricated by ($Zr_{0.75}Ti_{0.25}$)$C_{0.82}$ and ($Zr_{0.75}Ti_{0.25}$)$C_{0.82}B_{0.06}$ powders were denoted as ZTC and ZTCB ceramics, respectively, and their relative density (R.D.) was 99.2%±0.2% and 97.0%±0.5%, respectively. Table 1 shows the compositions of the ZTC and ZTCB ceramics. The oxygen contents of ZTC and ZTCB ceramics were low, at 1.07%±0.06% (mass fraction) and 0.82%±0.05% (mass fraction) respectively, which were close to those of ZrC fabricated by self-propagting high-temperature synthesis (SHS) with low oxygen content (0.94%, mass fraction)[26], indicating that a slight oxidation happened in the ceramics after sintered.

Table 1 Compositions of the carbide solid solution ceramics

Samples	Ceramic compositions	Phases according to XRD	Elemental analysis/% (mass fraction)					Relative density (R.D)/%
			Zr	Ti	C	B	O	
ZTC	$(Zr_{0.75}Ti_{0.25})C_{0.82}$	$(Zr, Ti)C_x$	74.8 ± 0.4	13.1± 0.2	10.8± 0.3	—	1.07 ± 0.06	99.2 ± 0.2
ZTCB	$(Zr_{0.75}Ti_{0.25})C_{0.82}B_{0.06}$	$(Zr, Ti)C_xB_y$	74.7± 0.3	13.3± 0.1	10.7± 0.3	0.5± 0.05	0.82 ± 0.05	97.0 ± 0.5

2.2 Oxidation test

Isothermal oxidation tests of the ceramics (6mm×3mm×2mm) were performed at 1600℃ in flowing dry air [a mixture of N_2/O_2(80:20, volume fraction)] within a muffle furnace. The mass of samples used for oxidation experiment was approximately 0.22g. Before the test, the surfaces of all samples were polished using a 0.5μm diamond paste. The rate of air flow was kept constant at 200mL/min during the whole oxidation process. After the furnace was heated up to 1600℃, the samples were placed in Al_2O_3 crucible with curved bottom to provide line contacts between the samples and crucible, making all surfaces of the sample were in contact with air. The samples together with crucible were put into the furnace for 5min, 20min, 40min and 60min, respectively. Afterwards, they were taken out and cooled at room temperature in air to measure their weights using an electronic balance (a sensitivity of 0.1mg). The weight changes (mean ± standard deviation) were measured from four samples after oxidation under the same conditions. The oxidation behavior of ceramics was determined using the calculated weight gain per unit area as a function of time. In that case, whether the oxidation behavior of ceramics followed a linear rate law or a parabolic rate law was determined by the following two equations:

$$(\Delta W/A) = k_1 \cdot t \tag{1}$$

$$(\Delta W/A)^2 = k_p \cdot t \tag{2}$$

Where ΔW is the weight change of ceramics, mg; A is the area of ceramics, cm^2; t is the oxidation time; k_1 and k_p are the linear and parabolic oxidation rate constants, respectively.

2.3 Characterization

The X-ray diffraction (XRD) patterns were obtained by an Advance D8 X-ray diffraction meter (Bruker, Germany) using Cu-Kα radiation at a scanning rate of 0.5°/min from 5° to 90° of 2θ. The Rietveld refinement was carried out using the general structure analysis system (GSAS) software[27]. The R.D. of ceramics was calculated according to the ratio of the tested density to the theoretical density. The tested density of ceramics was measured by the Archimedes method from 5 samples sintered under the same conditions to obtain an average value using deionized water as an immersing medium, whereas the theoretical density was calculated by the Rietveld refinement. The morphology of ceramics before and after oxidation was observed by a scanning electron microscope (SEM, FEI NOVA NanoSEM230, USA) with an X-ray energy-dispersive spectrometer (EDS) analyzer. The oxygen content was analyzed by Leco TCH-600 N/H/O analyzer and Leco CS-600 analyzer was used to test the carbon content. The contents of Zr, Ti and B in ceramics were measured by inductively coupled plasma-optical emission spectroscopy (ICP-OES). Element mapping analysis was carried out by electron probe microanalysis (EPMA) with a JEOL JXA-8530F system.

2.4 First-principle calculations

In the traditional carbides (e.g. ZrC and TiC, denoted as MC), metal atoms occupy the FCC lattice and carbon atoms fill the octahedral sites[28]. Among the possible forms of vacancy replacement in MC, besides the MC with 1/1 ratio, M_6C_5 also has high structural stabilities, which is resulted from the vacancy ordering along <112> direction in the close-packed (111) plane of carbon layer[29]. The existence of vacancies can change the atomic distributions of the nearest neighbors shells, which will differ from the average structure of carbides and result in various chemical short range orders (CSROs) that often characterize the local atomic distributions[29-31]. Based on CSROs, a cluster-plus-glue-atom model is proposed to describe such local distributions of atoms, i.e., any phase structure is regarded as being composed of a nearest-neighbor cluster part and a glue atom part located outside of the clusters, which can be expressed in terms of the cluster formula [cluster] (glue atoms)[32-33]. This cluster structural unit can provide information on composition, atomic configuration, and electronic structure, just like molecular formulas. For instance, TiC can be expressed with a cluster formula $[C-Ti_6]C_5$, one $[C-Ti_6]$ cluster matching with 5 extra carbon atoms (Fig. 1a and Fig. 1b), in which the octet rule is fully satisfied since there are eight electrons for each carbon atom. When one vacancy substitutes for one carbon atom in the glue sites (Fig. 1c), the cluster structural unit of $[C-M_6](C_{4\square1})$ ($= M_6C_5$) is constructed, in which the symbol of $\square$ represents the vacancy. When one boron atom fills one vacancy can result in a cluster structural unit of $[C-M_6](C_5B_1)$ (Fig. 1d), corresponding to $M_6C_5B_1$. These cluster units were then taken as the model input for the first principle calculations.

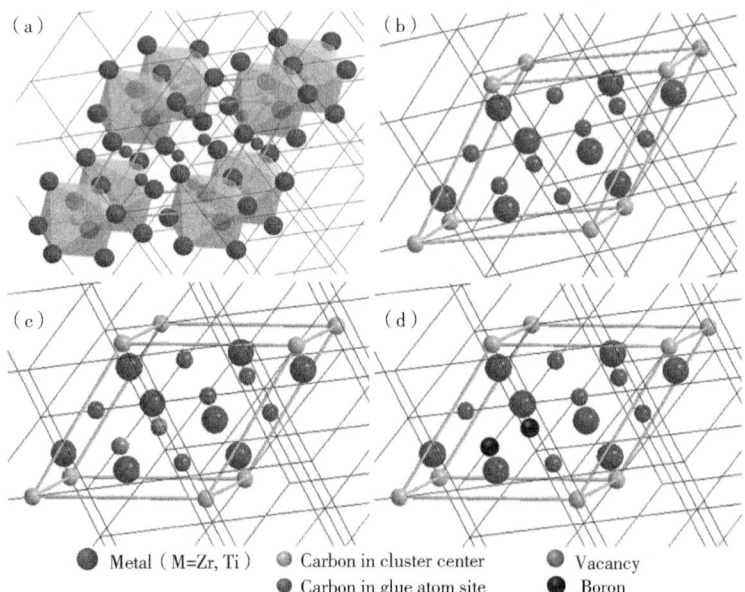

● Metal (M=Zr, Ti) ○ Carbon in cluster center ● Vacancy
● Carbon in glue atom site ● Boron

Fig. 1 Cluster model and the input structural models for the first principle calculations
(a) cluster model, (b) MC, (c) M_6C_5 and (d) $M_6C_5B_1$

The first principle calculations were performed within the framework of density functional theory (DFT) formulated within the generalized gradient approximation (GGA) by Perdew-Burke-

Ernzerhof (PBE)[34] for the exchange-correlation functional. The Kohn-Sham equations were solved using the Quantum-ESPRESSO first-principles code[35]. The projector augmented wave (PAW)[36] pseudo potentials were employed to treat the valence electrons for Ti ($3s^2 3p^6 4s^2 3d^2$), Zr($4s^2 4p^6 5s^2 4d^2$), C ($2s^2 2p^2$) and B ($2s^2 2p^1$). Brillouin-zone integrations were performed using Monkhorst and Pack k-point meshes[37]. In order to obtain the most equilibrium crystal structure and to ensure the accuracy of subsequent calculation, the convergence tests for the total energy were strictly implemented. For MC (M=Zr, Ti) crystal system, the k-point meshes fixed at 11×11×11, and cut-off energy was 800eV.

3 Results and discussion

3.1 Densification and characterization of (Zr, Ti)C_x and (Zr, Ti)$C_x B_y$ ceramics

In order to prepare the compact ceramics, the powders are sintered at 64MPa from room temperature to 1950℃. Fig. 2 shows the sintering parameter curves of ZTCB and ZTC ceramics illustrated by the force, displacement and displacement rate associated with sintering temperature as a function of the sintering time. Four distinct stages are identified by the characteristic peaks in the sintering parameters of ZTCB ceramic as marked in Fig. 2a. At the first stage, the displacement increases negatively, which is attributed to thermal expansion of ZTCB powder together with the expansion of graphite die. When the temperature increases to 1000℃, the force begins to be applied and then the displacement increases gradually from negative to positive. This is attributed to the rearrangement of ZTCB powder caused by the force. This stage takes place up to 1720℃, denoted as the second stage. At the third stage, a fast

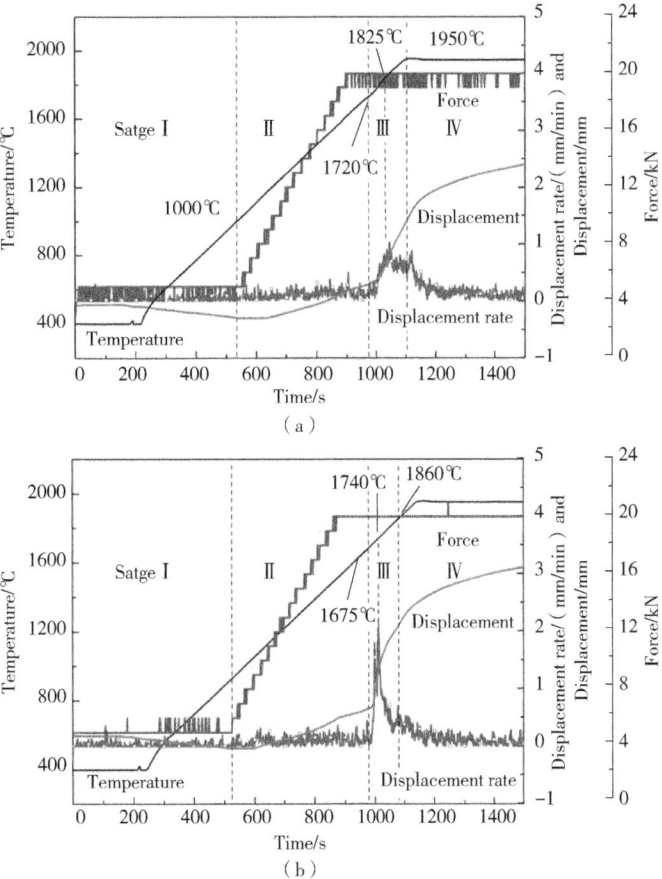

Fig. 2 Temperature, force, displacement and displacement rate curves of (a) ZTCB and (b) ZTC ceramics during SPS

densification process is in a temperature range of 1720~1950℃. The peak of displacement rate is observed at 1825℃, contributing a significant shrinkage. With temperature increase to 1950℃, the displacement increases steadily at the fourth stage. The displacement curve shows a remarkable slope down at this isothermal heating stage. The thermal expansion of dense ceramic and graphite die overwhelms the sintering shrinkage. After sintering at 1950℃ and 64MPa with a holding time of 7min, the ZTCB ceramic has a R. D. of 97.0%±0.5%.

The four distinct stages identified by the characteristic peaks in the sintering parameters of ZTC ceramic are similar to those in ZTCB ceramic. As shown in Fig. 2b, the second stage attributed to the rearrangement of ZTC powder is up to 1675℃, which is lower than that of ZTCB (1720℃). At the third stage, a fast densification process is in a temperature range of 1675~1860℃. The peak of displacement rate is at 1740℃, lower than that 1825℃ of ZTCB ceramic. Therefore, the densification temperatures of ZTC ceramic are lower than those of ZTCB ceramic, indicating that the carbide needs higher sintering temperatures after the solution of boron.

Fig. 3a shows the XRD patterns of ZTC and ZTCB ceramics. Both of them contain only one set of diffraction peaks, indicating that Zr, Ti, C and/or B basically react with each other as a carbide solid solution. Compared with ZTC ceramic, the peak location of ZTCB ceramic moves toward a lower angle, as observed for the (311) plane reflection found at 67.3° (Fig. 3b), lower than that 67.5° of ZTC ceramic, due to its larger lattice parameters (0.4617nm, Table 2). This increase of the lattice parameters is due to the introduction of B atoms into carbides with a larger atomic radius (0.98Å)[38] than C atoms (0.76Å)[39]. The B atoms may distribute through the defect channels or fill in the carbon vacancies in $(Zr_{0.75}Ti_{0.25})C_x$, similar to the $(Zr_{0.8}Ti_{0.2})C_{0.74}B_{0.26}$ phase in the C/C-$(Zr_{0.8}Ti_{0.2})C_{0.74}B_{0.26}$ composite[22].

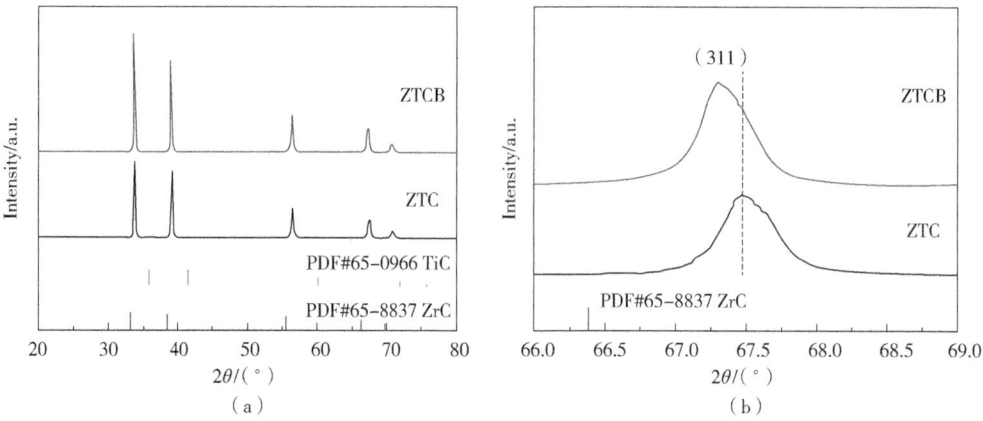

Fig. 3 XRD patterns of (a) ZTC and ZTCB ceramics and (b) enlargement of XRD patterns from 66° to 69°

Table 2 shows the Rietveld crystal structure parameters of ZTC and ZTCB ceramics. They are generated in a single-phase of FCC structure with $Fm\bar{3}m$ space group. The weighted profile residual factor (R_{wp}) and the profile residual factor (R_p) are the values to evaluate the degree of convergence during Rietveld refinement by GSAS software. The smaller these values are, the more convergent for the Rietveld refinement and the more reliable for the calculated results are. The values of R_{wp} and R_p of ZTCB are converged to a low value, such as R_{wp} of 11.10% and R_p of 7.68%, indicating a close fit to the experimental data.

Fig. 4 shows the morphology of ZTC and ZTCB ceramics. ZTC ceramic is almost completely dense (Fig. 4a), whereas some pores exist inside grains and at grain boundaries of ZTCB ceramic

(Fig. 4b). The element mapping results of Zr, Ti, C and B (Fig. 4c) show a uniform distribution of each element, indicating that ZTCB ceramic is a solid solution.

Table 2 Crystal structure parameters of ceramics from Rietveld refinement

Samples	Space group	a/nm	Volume/ nm^3	Density/ (g/cm^3)	Selected carbon-metal bond length/pm	R_{wp}/%	R_p/%
ZTC	$Fm\overline{3}m$	0.4613	0.0982	6.25	C1-Zr1 = 230.7(1) C1-Ti3 = 230.7(1)	11.83	8.85
ZTCB		0.4617	0.0985	6.13	C1-Zr1 = 230.9(1) C1-Ti3 = 230.9(1) B4-Zr1 = 230.9(1) B4-Ti3 = 230.9(1)	11.10	7.68

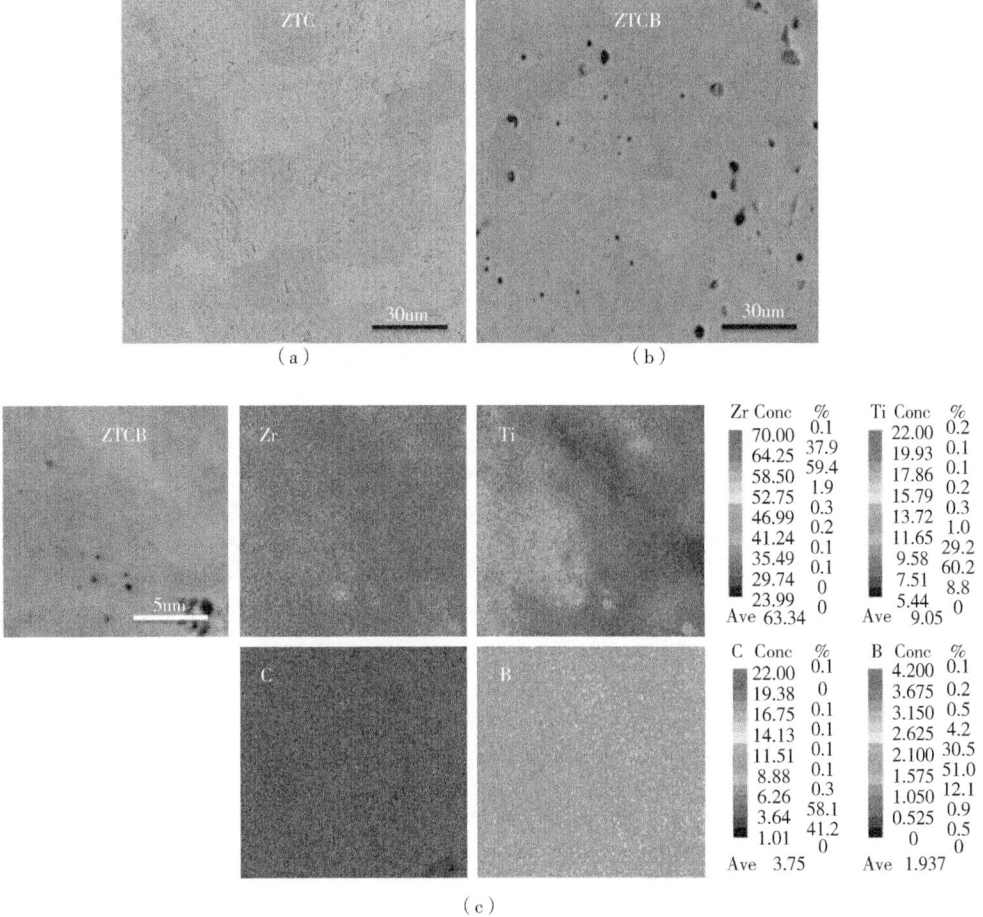

Fig. 4 SEM images of (a) ZTC and (b) ZTCB ceramics and (c) EPMA elements mapping analysis of ZTCB ceramic

3.2 Oxidation behavior of $(Zr, Ti)C_xB_y$ ceramics

The specific weight change of ZTC and ZTCB ceramics as a function of oxidation time during isothermal oxidation at 1600 ℃ is shown in Fig. 5. It can be seen that the isothermal oxidation of

ZTC and ZTCB ceramics at 1600℃ is a weight gain process. The weight of ZTC ceramic does not increase obviously after oxidation for 40min at 1600℃ (Fig. 5a), indicating that the ZTC ceramic has been completely oxidized at this time, whereas the ZTCB ceramic is completely oxidized after 60min.

Fig. 5b displays the plots of the square of the specific weight change of ZTC and ZTCB ceramics as a function of oxidation time at 1600℃, from which a good linear relationship between the square of the specific weight change and oxidation time is clearly seen. It is suggested that the oxidation behavior of ZTC and ZTCB ceramics at 1600℃ follows a parabolic rate law. That is to say, the oxidation of ZTC and ZTCB ceramics is controlled by a diffusion process that the oxygen diffuses into the oxide scale[40-41]. The k_p is calculated and listed in Table 3. It clearly demonstrates that, with addition of B atoms into ZTC ceramic, the k_p of ZTCB ceramic is (46.9 ± 2.7) mg^2/(cm^4 · min), being smaller than that (67.0 ± 5.9) mg^2/(cm^4 · min) of ZTC ceramic. The k_p of ZTC ceramic decreases nearly 30% after the solution of boron, which is a significant enhancement of oxidation resistance because only trace boron (0.5%, mass fraction) dissolves into ZTCB ceramic. In general, ZrB$_2$ displays an excellent oxidation resistance at high temperatures (<1800℃) since its surface is coated with a film of liquid B$_2$O$_3$, which fills the pores formed by the oxides and slows the inward diffusion of oxygen[6]. However, the boron content of ZrB$_2$ is as high as 19.1% (mass fraction). Therefore, the introduction of trace B atoms greatly improves the oxidation resistance of carbide ceramics.

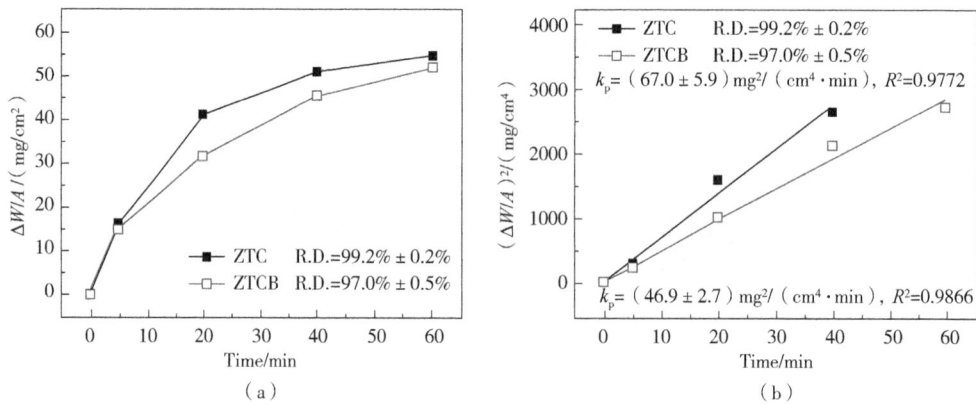

Fig. 5 (a) Weight gain per unit surface area as a function of oxidation time and (b) square of weight gain per unit surface area as a function of oxidation time in air at 1600℃ of ZTC and ZTCB ceramics

Table 3 Weight gain per unit surface area, oxide scale and interlayer thickness of ceramics after oxidation in air at 1600℃ for different time

Samples	Oxidation time/ min	Weight gain per unit surface area/(mg/cm^2)	Oxidation rate constant k_p/[mg^2/(cm^4 · min)]	Oxide scale thickness/μm	Interlayer thickness/μm
ZTC	5	16.5± 0.7	67.0± 5.9	80± 4	150± 7
	20	41.0± 1.6		450± 22	280± 14
	40	51.1± 2.0		—	—
	60	54.8± 2.2		—	—

Continued

Samples	Oxidation time/ min	Weight gain per unit surface area/(mg/cm^2)	Oxidation rate constant k_p/[mg^2/(cm^4·min)]	Oxide scale thickness/μm	Interlayer thickness/μm
ZTCB	5	14.6± 0.6	46.9± 2.7	50± 2	100± 5
	20	31.8± 1.3		310± 15	220± 11
	40	45.7± 1.8		—	—
	60	52.2± 2.1		—	—

In addition, Table 3 compares the oxide scale thickness of ceramics after isothermal oxidation in air at 1600℃ for different time. The thickness of the oxide scale of ZTC ceramic is (80±4) μm and (450±22) μm after oxidation for 5min and 20min, respectively, whereas those of ZTCB ceramic are thinner, as (50±2) μm and (310±15) μm. It is worth noting that an interlayer is observed between the oxide scale and boron-containing carbides in ZTCB ceramic after oxidation. The interlayer may be an oxycarbide since the oxycarbide is also observed during the oxidation of (Zr, Ti)C$_x$[21], (Zr, Hf, Ti)C$_x$[19] and (Hf, Ta, Zr, Nb)C[42]. For ZTCB ceramic, the thickness of interlayer after oxidation at 5min and 20min is (100±5) μm and (220±11) μm, respectively, whereas those of ZTC ceramic are thicker, thus indicating that the introduction of B atoms will reduce the thickness of oxide scale and interlayer of carbides.

Fig. 6 shows XRD patterns of ZTC and ZTCB ceramics after isothermal oxidation tests for 60min at 1600℃. After oxidation, the ZTC monocarbide phase is not detectable in the XRD. The oxidation products of ZTC ceramic are composed of $m-ZrO_2$ (JCPDS 83-0944) and $t-(Zr, Ti)O_2$ solid solution. For the oxidation products of ZTCB ceramic, it contains $m-ZrO_2$ and $t-(Zr, Ti)O_2$. Moreover, two week peaks are observed approximately at 23.1° and 43.6°, which are ascribed to B_2O_3 (JCPDS 76-0781). The low intensity of these peaks of B_2O_3 is due to the low content of B in ZTCB ceramic (0.5%, mass fraction) and/or low crystallinity caused by rapid cooling after oxidation. Hence, the oxidation products of ZTCB ceramic are composed of $m-ZrO_2$, $t-(Zr, Ti)O_2$ and B_2O_3.

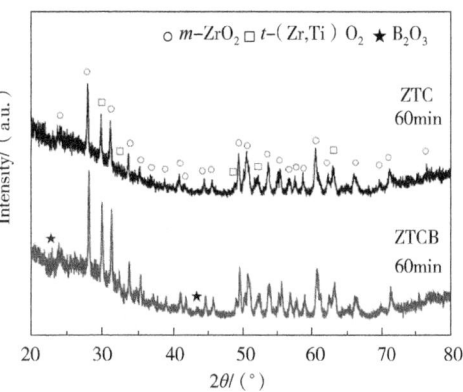

Fig. 6 XRD patterns of ZTC and ZTCB ceramics after oxidation in air at 1600℃ for 60min

The surface morphology of ZTC and ZTCB ceramics after isothermal oxidation in air at 1600℃ for different time is shown in Fig. 7. When the oxidation time is 5min, the surface of ZTC ceramic is dense (Fig. 7a). When the time is extended to 20min, the surface pores are observed obviously (Fig. 7c). During oxidation, CO/CO_2 is generated from carbon oxidation in the carbide, and when the gas is discharged outward, pores are formed on the surface of ceramic. O_2 can directly participate in the oxidation reaction through the pores. With the extension of oxidation time, the gas increases rapidly in the oxide layer. When the gas pressure increases to a certain extent, a large amount of gas discharges out, causing the oxide protrusion near the pores. Many cracks are also observed on the surface of ZTC ceramic after oxidation for more than 20min (Fig. 7c, Fig. 7e and Fig. 7g). These cracks are probably caused by volume difference between oxide and carbide or growth stress due to the formation of oxide[43]. The appearance of cracks greatly reduces the oxidation resistance of ceramics.

A few pores and cracks are also observed on the surface of ZTCB ceramic after isothermal oxidation at 1600℃ in air for 5min (Fig. 7b). However, the overall morphology of the oxidized ZTCB ceramic is compact. As the oxidation time increased to 20min, the cracks on the surface increased slightly (Fig. 7d). Compared with the morphology of ZTC ceramic after oxidation, the surface of ZTCB ceramic is generally denser, which may be due to the formation of B_2O_3 in the process of oxidation. The melting point and boiling point of B_2O_3 is approximately 450℃ and 1860℃[44], respectively. At 1600℃, where the temperature is above the melting point and below the boiling point of B_2O_3, the B_2O_3 will be liquid. The liquid B_2O_3 fills the pores left by carbon oxidation into CO/CO_2, leading to the surface of ZTCB ceramic with fewer pores and cracks than that of ZTC ceramic, as shown in Fig. 7. Therefore, the formed liquid B_2O_3 makes the oxide scale denser.

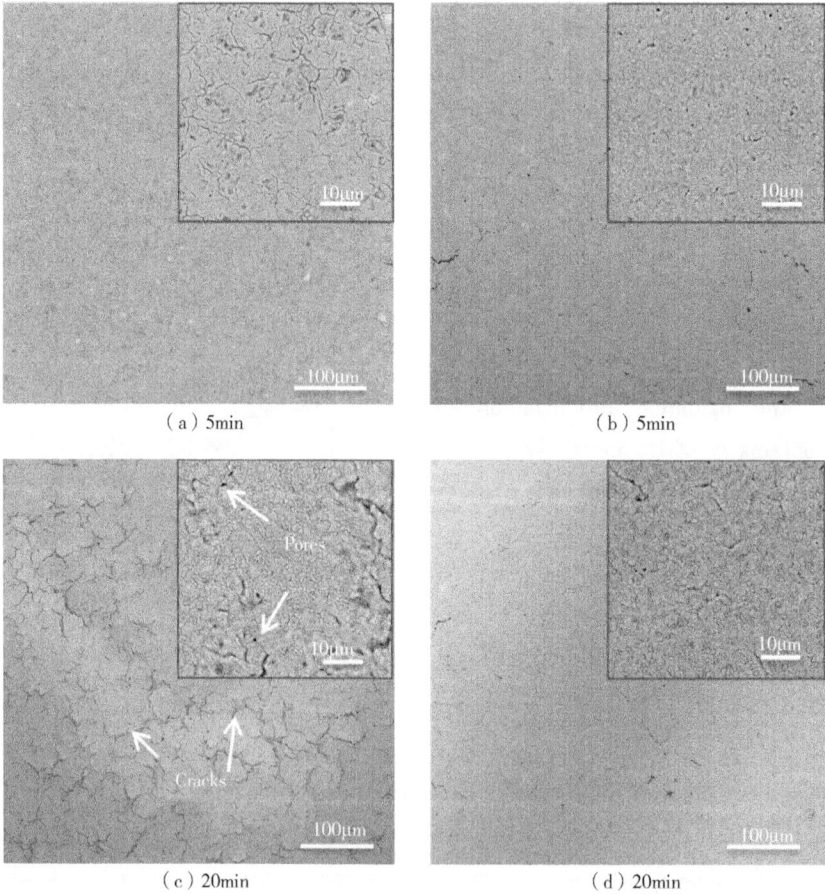

Fig. 7 Surface morphology of (a, c, e and g) ZTC and (b, d, f and h) ZTCB ceramics after oxidation in air at 1600℃ for different time

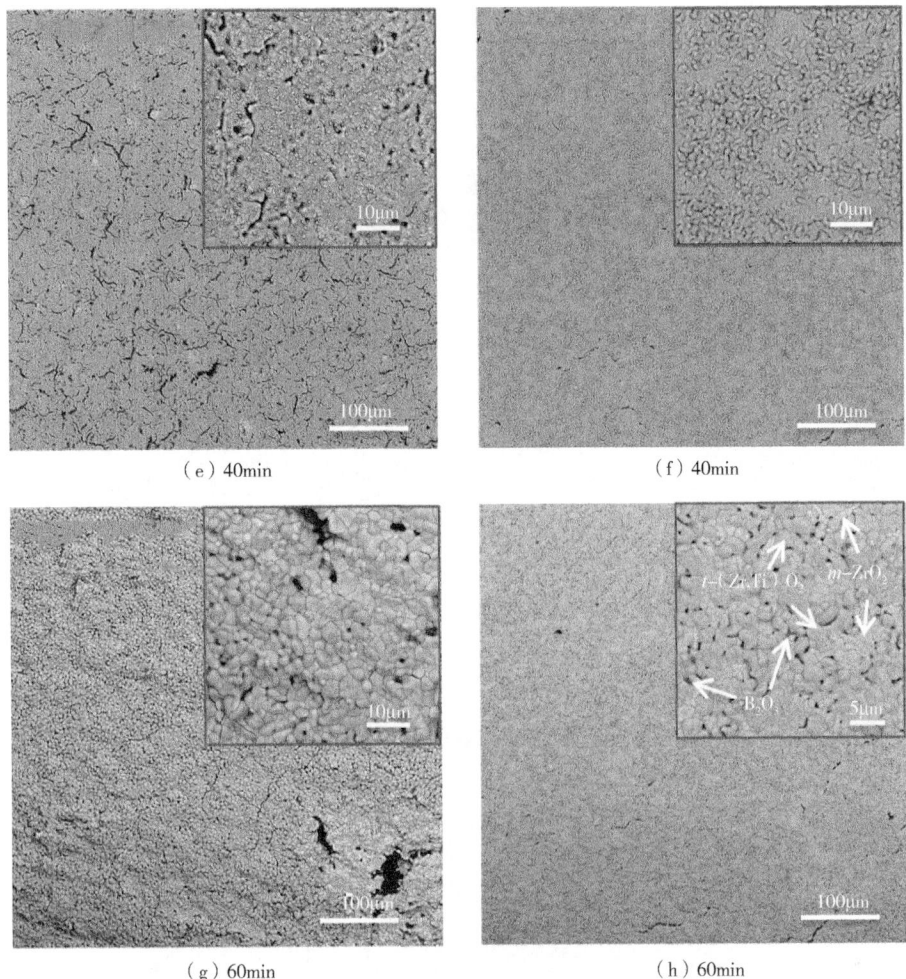

Fig. 7 Surface morphology of (a, c, e and g) ZTC and (b, d, f and h) ZTCB ceramics after oxidation in air at 1600℃ for different time(continued)

Fig. 8 shows the cross section morphology of ZTC ceramic after isothermal oxidation at 1600℃ in air for 5min. The ceramic exhibits three layers after oxidation, which is oxide scale, interlayer layer and carbide matrix, respectively. As shown in Fig. 8a, the thickness of oxide scale is approximately 78μm. A few pores are observed inside the oxide (Fig. 8b). The result of element line scanning analysis (Fig. 8a) shows that the O content decreases slightly from the ceramic surface to the interior. The O content is 27.4% (atom fraction) at the interface between the oxide scale and the interlayer and is 26.0% (atom fraction) at the interface between the interlayer and the carbide matrix. This result indicates that the interlayer layer formed in ZTC ceramic has a weak ability to prevent the oxygen diffusion into the carbide matrix.

In addition, according to the result of element mapping analysis (Fig. 8f), the oxide particles are enriched in Ti in the interlayer layer, and the oxide is presumed to be $t-(Zr, Ti)O_2$. Backman et al.[45-46] reported a preferential oxidation law of metal elements in the oxidation process of carbide at 1700℃. According to this law, Zr oxidizes prior to Ti in Zr-Ti alloys or carbides, which leads to the enrichment of residual Ti in the matrix material. It is inferred that when the ZTC ceramic is

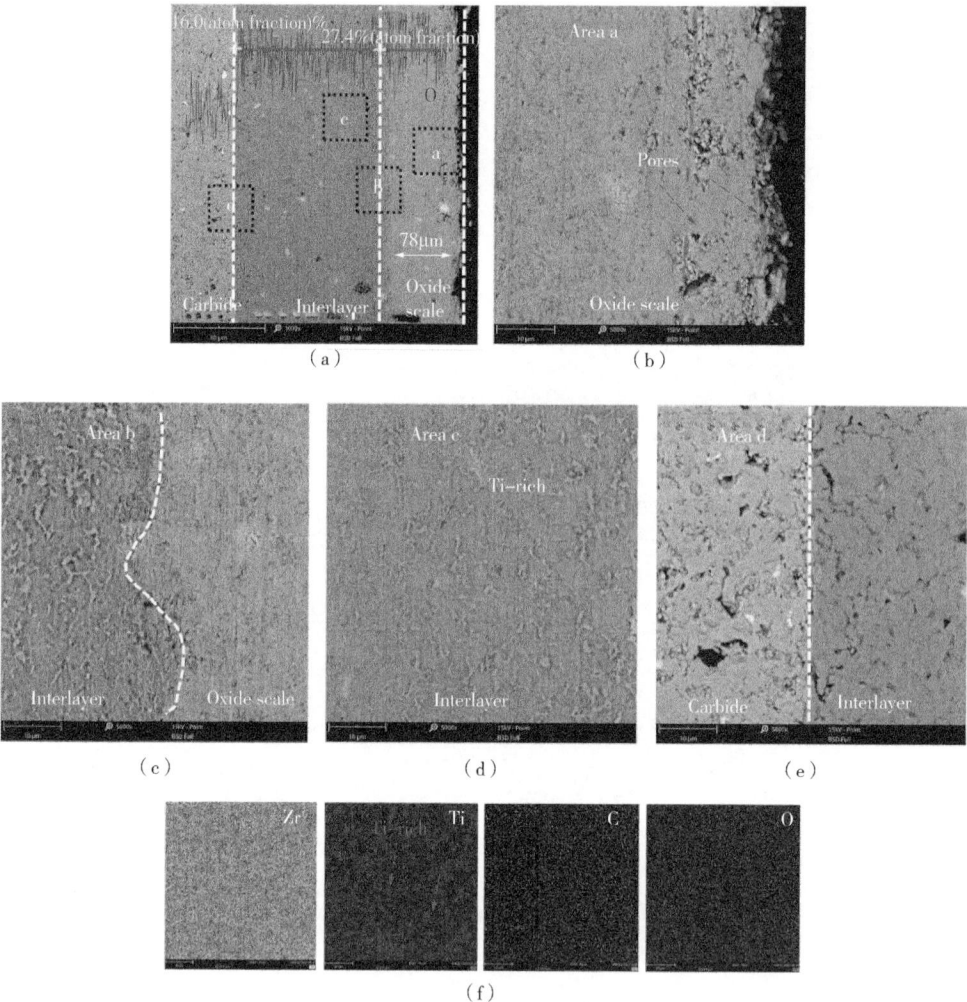

Fig. 8 Cross section morphology of ZTC ceramic after oxidation in air at 1600℃ for 5min. (a) SEM image. The insert line in (a) is O line analysis. (b, c, d and e) high magnification images of areas a, b, c and d in (a), respectively. (f) EDS elements mapping analysis of (d)

oxidized at 1600℃ in this work, Zr will be oxidized to ZrO_2 or ZrC_xO_y prior to Ti. The Ti content will be increased accordingly in the carbides, resulting in the formation of Ti-rich oxides between the Zr oxide particles during oxidation.

Fig. 9 shows the cross section morphology of ZTCB ceramic after oxidation at 1600℃ in air for 5min. In Fig. 9a, the ceramic exhibits a three-layer structure after oxidation. This is similar to ZTC ceramic after oxidation, which is oxide scale, interlayer and boron-containing carbide matrix, respectively. The O content gradually decreases from the surface to the interior of the ceramic, as shown in the inset in Fig. 9a. The O content at the interface of the oxide scale and the interlayer is 16.5%(atom fraction) and it is 11.5%(atom fraction) at the interface between the interlayer and the boron-containing carbide matrix, indicating that the interlayer layer has a strong inhibition effect on oxygen diffusion through the interlayer. The reduction of O content is larger than that in the interlayer of ZTC ceramic (Fig. 8a), which means a stronger inhibition ability of oxygen diffusion during oxidation. After oxidation at 1600℃ for 5min, B_2O_3 exists in the oxide scale. Some small black phases are observed in the oxide scale, as shown in Fig. 9b. According to elements analysis of Spot A (black phase, Fig. 9c), the content of B, C and O of black phase is 9.9% (atom fraction), 68.8% (atom fraction) and 17.9% (atom fraction), respectively, suggesting that the black phase is composed of B_2O_3 and carbon. Moreover, the content of B of Spot B (white phase) is tiny, whereas the content of O, Zr and Ti is high. The white phase is inferred as $(Zr, Ti)O_2$. In addition, Ti enrichment in oxide particles is also observed in the interlayer (Fig. 9e and Fig. 9g), which is related to the oxidation product ZrO_2 or ZrC_xO_y formed by the oxidation of Zr prior to Ti in the boron-containing carbides.

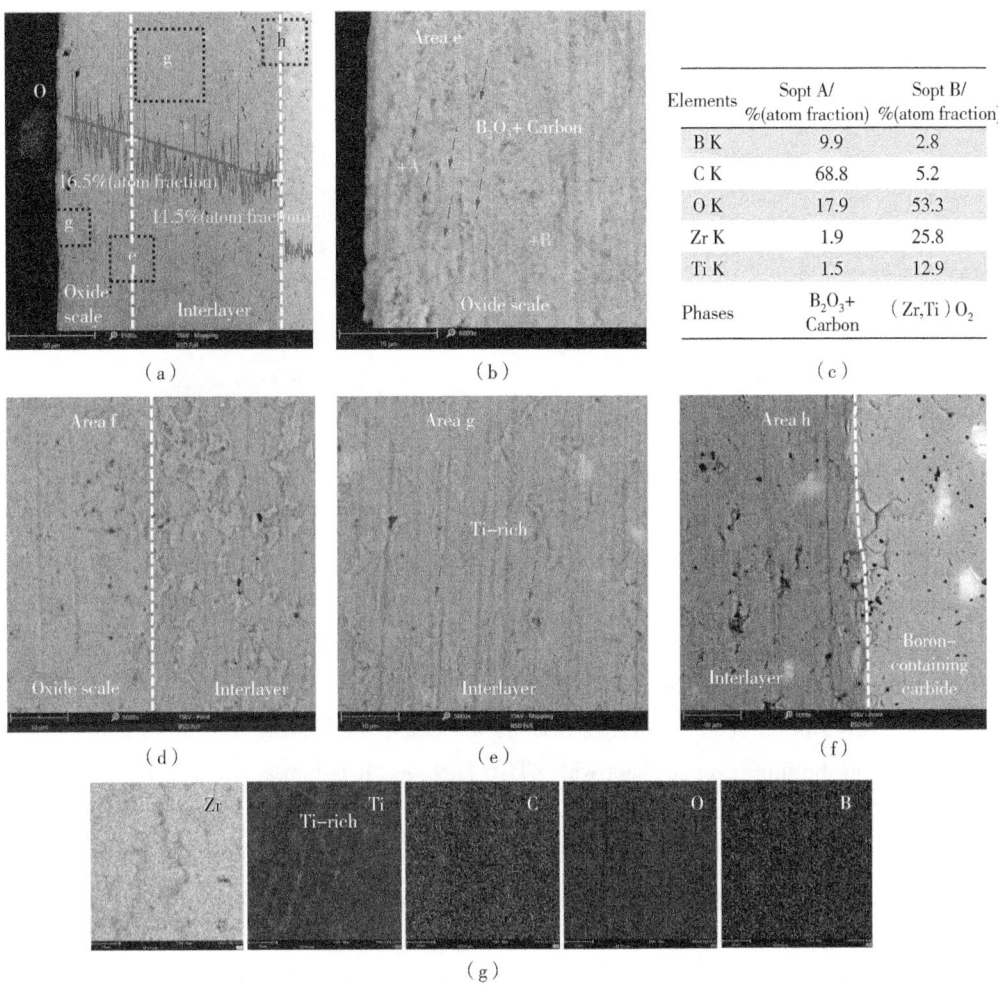

Fig. 9 Cross section morphology of ZTCB ceramic after oxidation in air at 1600℃ for 5min. (a) SEM image. The insert line in (a) is O line analysis. (b, d, e and f) high magnification images of areas e, f, g and h in (a), (c) EDS elements analysis of Spots A and B in (b), (g) EDS elements mapping analysis of (e)

Fig. 10a shows the XRD patterns of ZTCB ceramic at the corresponding different depths after oxidation at 1600℃ in air for 5min. The oxides on the surface of the oxidized ceramic are $m-ZrO_2$, $t-(Zr, Ti)O_2$ and B_2O_3 (position a in Fig. 10b). In addition, carbon is found at the interface between oxide scale and outer interlayer (position b in Fig. 10b). At the interface between the inner interlayer and boron-containing carbides (position c in Fig. 10b), the diffraction peak of t-(Zr, Ti)O_2 becomes weak and there is no detectable carbon phase in the XRD, whereas the peak of ZTCB appears. At the position d in Fig. 10b, only diffraction peaks of ZTCB and m-ZrO_2 are observed. Therefore, during the oxidation, the ZTCB ceramic first forms m-ZrO_2 and then t-(Zr, Ti)O_2, carbon and B_2O_3. Finally the carbon is oxidized to be gas at 1600℃ in air. The oxidation process of ZTCB ceramic at 1600℃ in air is depicted by following reactions:

$$(Zr_{0.75}Ti_{0.25})C_{0.82}B_{0.06} + 0.795O_2(g) = 0.5\ m\text{-}ZrO_2 + 0.25\ t\text{-}(Zr, Ti)O_2 + 0.82C + 0.03B_2O_3(l) \quad (3)$$

$$2C + O_2(g) = 2CO(g) \quad (4)$$

$$C + O_2(g) = CO_2(g) \quad (5)$$

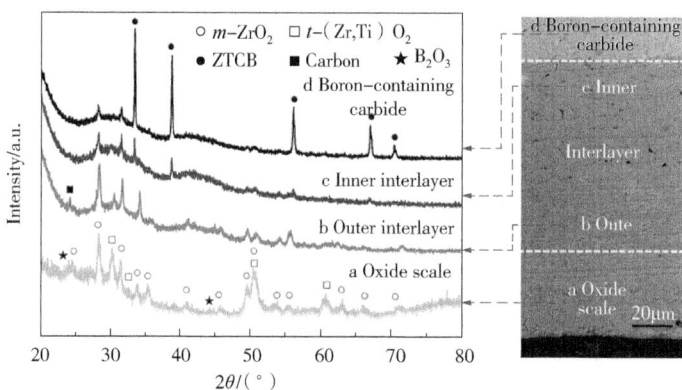

Fig. 10 XRD patterns of ZTCB ceramic at the corresponding different depths after oxidation in air at 1600℃ for 5min

3.3 Effect of boron on the oxidation resistance of (Zr, Ti)C_xB_y ceramic

The results of isothermal oxidation tests show that the oxidation resistance of ZTCB ceramic is stronger than that of ZTC ceramic, which is due to the following three aspects.

First, the oxide scale generated in ZTCB ceramic is denser and has fewer cracks, as shown in Fig. 7 and Fig. 9b. The O_2 needs to diffuse through the dense oxide scale until the interface between the oxide scale and the interlayer to react with ZTCB carbide. It is harder for O_2 diffused through the oxide scale in ZTCB ceramic than in ZTC ceramic due to less cracks in the oxide scale of ZTCB ceramic. This is confirmed by the lower O content (16.5%, atom fraction) at the interface between the oxide scale and the interlayer in ZTCB ceramic than that 27.4% (atom fraction) in the oxide scale of ZTC ceramic. Compared to ZTC ceramic, the oxide scale of ZTCB ceramic is composed of B_2O_3, m-ZrO_2 and t-(Zr, Ti)O_2. At 1600℃, the B_2O_3 will be liquid since the temperature is below its boiling point (1860℃). The formed liquid B_2O_3 at 1600℃ could fill the pores and reduce the cracks in the oxide scale, and hence enhanced the oxidation resistance of carbide. When the oxidation temperature is over the boiling point of B_2O_3 (1860℃), the volatilization of B_2O_3 will reduce the surface oxidation temperature by absorbing heat, which helps to improve oxidation resistance of ZTCB ceramic.

Second, the ability of interlayer layer in ZTCB ceramic to prevent oxygen diffusion into the carbide is stronger. The O content decreases from 16.5% (atom fraction) to 11.5% (atom fraction)

after diffusing through the interlayer in ZTCB ceramic and from 27.4% (atom fraction) to 26.0% (atom fraction) in ZTC ceramic, as shown in Fig. 8a and Fig. 9a. The interlayer in ZTCB ceramic is composed of B_2O_3, carbon, $m-ZrO_2$ and $t-(Zr, Ti)O_2$, whereas the phase assembly in ZTC ceramic is without the B_2O_3. The good oxidation resistance is due to the low oxygen permeability (OP) of the interlayer. The OP of B_2O_3 is 8.6×10^{-11} g/(cm·s) at 1000℃ [47]. When the temperature increases to 1200℃, the OP of B_2O_3 is extrapolated to approximately 1.1×10^{-10} g/(cm·s) [47], which is lower than the OP of $t-ZrO_2$ (approximately 1.7×10^{-9}, at 1200℃ [47]), as shown in Table 4. Note that the OP of $t-(Zr, Ti)O_2$ cannot be referred. Although the oxygen anion diffusion coefficient of B_2O_3 (2.16×10^{-11}, at 1200℃ [48]) is higher than those of $t-ZrO_2$ (1.03×10^{-14}, at 1300℃ [49]) and $t-TiO_2$ (6.3×10^{-15}, at 1200℃ [50]), it is believed that the lower OP of formed interlayer containing B_2O_3 in ZTCB ceramic versus in ZTC ceramic leads to a larger decrease of O content, and improves the oxidation resistance of carbide.

Table 4 Oxygen permeability and oxygen anion diffusion coefficient in the oxides[47-51]

Oxides	Oxygen permeability/ [g/(cm·s)]	Diffusion coefficient/(cm²/s)
$t-ZrO_2$	approximately 1.7×10^{-9}, at 1200℃ [47]	1.03×10^{-14}, at 1300℃ [49]
		1.78×10^{-13}, at 1600℃ [49]
$t-TiO_2$	approximately 2.6×10^{-12}, at 1200℃ [47,51]	6.3×10^{-15}, at 1200℃ [50]
B_2O_3	8.6×10^{-11}, at 1000℃ [47] approximately 1.1×10^{-10}, at 1200℃ [47]	2.16×10^{-11}, at 1200℃ [48]

Third, ZTCB carbide has a higher structural stability than ZTC carbide, which is more difficult to be oxidized to form oxides. Comparing the thickness of interlayer in ZTC and ZTCB ceramics in Table 3, it is found that the thickness of the interlayer of ZTCB ceramic is thinner than that of ZTC ceramic after oxidation for 5min and 20min. The interlayer is the oxidation product in the onset stage of oxidation of carbide ceramics. Combined with the microstructure and phase analysis results of the oxides, the oxidation reaction of ZTCB carbide to form the interlayer is described in Eq. (3). During the oxidation process, O atoms need to replace the lattice positions of C and B atoms in ZTCB carbide, and then forms oxides. Therefore, the thickness of interlayer is depended on the difficulty in breaking the M–C and M–B bonds in boron-containing carbides. Binding energy (E_b) is a parameter that measures the bonding strength. In general, the lower the E_b is, the higher the energy is required to break the bonds. The E_b of M_xC_y and $M_xC_yB_z$ (M = Zr or Ti) is expressed with Eq. 6 and Eq. 7:

$$E_{b(M_xC_y)} = \frac{E_{M_xC_y} - xE'_M - yE'_C}{x+y} \quad (6)$$

$$E_{b(M_xC_yB_z)} = \frac{E_{M_xC_yB_z} - xE'_M - yE'_C - zE'_B}{x+y+z} \quad (7)$$

Where $E_{M_xC_y}$ is the total energy of M_xC_y, $E_{M_xC_yB_z}$ is the total energy of $M_xC_yB_z$, E'_M is the energy of a single Zr or Ti atom, E'_C is the energy of a single C atom, E'_B is the energy of single B atom.

Table 5 shows the calculated E_b of boron-containing carbides. The E_b of $(Zr_5Ti_1)_6C_5B_1$ is −245.6kJ/mol, which is lower than that of $(Zr_5Ti_1)_6C_5$ as −208.9kJ/mol, indicating that the M−C and M−B bonds in $(Zr_5Ti_1)_6C_5B_1$ require a higher energy to break to form M−O bonds. Meanwhile, this trend also exist in the E_b of $(Zr_4Ti_2)_6C_5B_1$. The E_b of $(Zr_4Ti_2)_6C_5B_1$ is −65.5kJ/mol, which is also lower than that of $(Zr_4Ti_2)_6C_5$. Therefore, the introduction of B atoms into carbide makes the E_b of $(Zr, Ti)C_xB_y$ to be more negative. The M−C and M−B bonds in $(Zr, Ti)C_xB_y$ are more difficult to break and thus enhance the structural stability of $(Zr, Ti)C_xB_y$, decreasing the thickness of interlayer, and finally improving the oxidation resistance.

Table 5 Binding energies calculated from different models

Models	E_b/(kJ/mol)
$(Zr_5Ti_1)C_5$	−208.9
$(Zr_5Ti_1)C_5B_1$	−245.6
$(Zr_4Ti_2)C_5$	−17.3
$(Zr_4Ti_2)C_5B_1$	−65.5

4 Conclusion

In this work, the isothermal oxidation behavior of $(Zr, Ti)C_xB_y$ ceramics are investigated at 1600℃ in air. The oxidation resistance of ZTC ceramic is significantly enhanced by the introduction of trace B atom (0.5%, mass fraction). The k_p of ZTCB ceramic decreases 30% to be $(46.9±2.7)$ mg^2/(cm^4 · min), smaller than that of ZTC as $(67.0±5.9)$ mg^2/(cm^4 · min). A denser oxide scale with less pores and cracks is formed in the ZTCB ceramic after oxidation. Moreover, the interlayer in ZTCB ceramic greatly inhibits the inward diffusion of oxygen and the O content decreases from 16.5% (atom fraction) to 11.5 % (atom fraction). The introduction of B atom reduces the E_b of carbide and finally improves the oxidation resistance. The results obtained herein provide fundamental information for design advanced oxidation-resistant boron-containing carbides.

Acknowledgements

This work was supported by the National Natural Science Foundation of China (No. 5207021797), and the Scientific Research and Technology Development Project of China National Petroleum Corporation Limited [No. 2020E-2804(JT)]. The authors would like to thank Qing Wang (Dalian University of Technology) for the assistance in first-principle calculations.

Declaration of competing interest

The authors have no competing interests to declare that are relevant to the content of this article.

References

[1] Guan J, Li D, Yang Z, et al. Ta(B, C, N) and (Ta, Mi)(B, C, N) (Mi = Nb, W) ceramics by high-energy ball milling: processing and solution mechanisms. J Am Ceram Soc 2023, 106: 699−708.

[2] Ni D, Cheng Y, Zhang J, et al. Advances in ultra-high temperature ceramics, composites, and coatings. J Adv Ceram 2022, 11: 1-56.

[3] Xiang H, Xing Y, Dai F-z, et al. High-entropy ceramics: Present status, challenges, and a look forward. J Adv Ceram 2021, 10: 385-441.

[4] Fahrenholtz W G, Hilmas G E. Ultra-high temperature ceramics: Materials for extreme environments. Scripta Mater 2017, 129: 94-99.

[5] Nisar A, Ariharan S, Venkateswaran T, et al. Effect of carbon nanotube on processing, microstructural, mechanical and ablation behavior of ZrB_2-20SiC based ultra-high temperature ceramic composites. Carbon 2017, 111: 269-282.

[6] Parthasarathy T A, Rapp R A, Opeka M, et al. A model for the oxidation of ZrB_2, HfB_2 and TiB_2. Acta Mater 2007, 55: 5999-6010.

[7] Shimada S. Oxidation and mechanism of single crystal carbides with formation of carbon. J Ceram Soc Jpn 2001, 109: S33-S42.

[8] Shimada S. A thermoanalytical study on the oxidation of ZrC and HfC powders with formation of carbon. Solid State Ionics 2002, 149: 319-326.

[9] Ye Z, Zeng Y, Xiong X, et al. Elucidating the role of preferential oxidation during ablation: Insights on the design and optimization of multicomponent ultra-high temperature ceramics. J Adv Ceram 2022, 11: 1956-1975.

[10] Rama Rao G A, Venugopal V. Kinetics and mechanism of the oxidation of ZrC. J Alloys Compd 1994, 206: 237-242.

[11] Voitovich R F, Pugach É A. High-temperature oxidation of ZrC and HfC. Powder Metall Met Ceram 1973, 12: 916-921.

[12] Zhao L, Jia D, Duan X, et al. Oxidation of ZrC-30vol% SiC composite in air from low to ultrahigh temperature. J Eur Ceram Soc 2012, 32: 947-954.

[13] Wang H, Cao Y, Liu W, et al. Oxidation behavior of $(Hf_{0.2}Ta_{0.2}Zr_{0.2}Ti_{0.2}Nb_{0.2})$C-xSiC ceramics at high temperature. Ceram Int 2020, 46: 11160-11168.

[14] Tan Y, Liao W, Xia Y, et al. Understanding the oxidation kinetics of $(Ti_{0.8}Nb_{0.2})$C and $(Ti_{0.8}Nb_{0.2})$C-SiC composite in high-temperature water vapor. Corros Sci 2022, 200: 110248.

[15] Chen S, Zhang C, Zhang Y, et al. Preparation and properties of carbon fiber reinforced $ZrC-ZrB_2$ based composites via reactive melt infiltration. Compos Part B 2014, 60: 222-226.

[16] Vorotilo S, Sidnov K, Mosyagin I Y, et al. Ab-initio modeling and experimental investigation of properties of ultra-high temperature solid solutions $Ta_xZr_{1-x}C$. J Alloys Compd 2019, 778: 480-486.

[17] Kurbatkina V V, Patsera E I, Levashov E A, et al. SHS processing and consolidation of Ta-Ti-C, Ta-Zr-C, and Ta-Hf-C carbides for ultra-high-temperatures application. Adv Eng Mater 2018, 20: 1701075.

[18] Andrievskii R A, Strel'nikova N S, Poltoratskii N I, et al. Melting point in systems ZrC-HfC, TaC-ZrC, TaC-HfC. Soviet Powder Metall Met Ceram 1967, 6: 65-67.

[19] Lun H, Zeng Y, Xiong X, et al. Oxidation behavior of non-stoichiometric (Zr, Hf, Ti)C_x carbide solid solution powders in air. J Adv Ceram 2021, 10: 741-757.

[20] Wuchina E, Opila E, Opeka M, et al. UHTCs: Ultra-high temperature ceramic materials for extreme environment applications. Electrochem Soc Interface 2007, 16: 30-36.

[21] Lun H, Yuan J, Zeng Y, et al. Mechanisms responsible for enhancing low-temperature oxidation resistance of nonstoichiometric (Zr, Ti)C. J Am Ceram Soc 2022, 105: 5309-5324.

[22] Zeng Y, Wang D, Xiong X, et al. Ablation-resistant carbide $Zr_{0.8}Ti_{0.2}C_{0.74}B_{0.26}$ for oxidizing environments up to 3,000℃. Nat Commun 2017, 8: 15836.

[23] Huang Q. Fabrication, Structure and Application of High-performance Carbon/Carbon Composites. Changsha (China): Central South University Press, 2010.

[24] Hu D, Zhang Y, Dong Z, et al. Relationship analyses on environmental factors-ablation performance based on ZrC-TaC system: Oxygen partial pressure and gas flow scouring. J Eur Ceram Soc 2023, 43: 2331-2344.

[25] Zeng Y, Lun H, Xiong X, et al. A new method for solid-state diffusion of boron atoms into powders of a multicomponent carbide. J Am Ceram Soc 2020, 103: 23-27.

[26] Li J, Fu Z, Wang W, et al. Preparation of ZrC by self-propagting high-temperature synthesis. J Chi Ceram Soc 2010, 38: 979-985.

[27] Kotnana G, Jammalamadaka S N. General structure analysis system (GSAS). J Appl Phys 2015, 117: 562.

[28] Smith C J, Yu X X, Guo Q, et al. Phase, hardness, and deformation slip behavior in mixed $Hf_xTa_{1-x}C$. Acta Mater 2018, 145: 142-153.

[29] Korzhavyi P A, Pourovskii L V, Hugosson H W, et al. Ab initio study of phase equilibria in TiC_x. Phys Rev Lett 2002, 88: 015505.

[30] Xiang J Y, Hu W T, Liu S C, et al. Spark plasma sintering of the nonstoichiometric ultrafine-grained titanium carbides with nano superstructural domains of the ordered carbon vacancies. Mater Chem Phys 2011, 130: 352-360.

[31] Hu W, Xiang J, Yang Z, et al. Superstructural nanodomains of ordered carbon vacancies in nonstoichiometric $ZrC_{0.61}$. J Mater Res 2012, 27: 1230-1236.

[32] Dong C, Wang Q, Qiang J B, et al. From clusters to phase diagrams: Composition rules of quasicrystals and bulk metallic glasses. J Phys D Appl Phys 2007, 40: R273.

[33] Han G, Qiang J, Li F, et al. The e/a values of ideal metallic glasses in relation to cluster formulae. Acta Mater 2011, 59: 5917-5923.

[34] Perdew J P, Burke K, Ernzerhof M. Generalized gradient approximation made simple. Phys Rev Lett 1996, 77: 3865-3868.

[35] Giannozzi P, Baroni S, Bonini N, et al. QUANTUM ESPRESSO: a modular and open-source software project for quantum simulations of materials. J Phys-Condens Mat 2009, 21: 395502.

[36] Kresse G, Joubert D. From ultrasoft pseudopotentials to the projector augmented-wave method. Phys Rev B 1999, 59: 1758-1775.

[37] Monkhorst H J, Pack J D. Special points for Brillouin-zone integrations. Phys Rev B 1976, 13: 5188-5192.

[38] Li G, Liu M, Wang H, et al. Effect of the rare earth element yttrium on the structure and properties of boron-containing high-entropy alloy. JOM 2020, 72: 2332-2339.

[39] Gendre M, Maître A, Trolliard G. Synthesis of zirconium oxycarbide (ZrC_xO_y) powders: Influence of stoichiometry on densification kinetics during spark plasma sintering and on mechanical properties. J Eur Ceram Soc 2011, 31: 2377-2385.

[40] Ye B, Wen T, Chu Y. High-temperature oxidation behavior of $(Hf_{0.2}Zr_{0.2}Ta_{0.2}Nb_{0.2}Ti_{0.2})C$ high-entropy ceramics in air. J Am Ceram Soc 2019, 103: 500-507.

[41] Ye B, Wen T, Chu Y. Low-temperature oxidation behavior of $(Zr_{1/3}Nb_{1/3}Ti_{1/3})C$ solid-solution ceramics in air. Mater China 2020, 39: 918-923.

[42] Wang Y, Zhang R Z, Zhang B, et al. The role of multi-elements and interlayer on the oxidation behaviour of (Hf-Ta-Zr-Nb)C high entropy ceramics. Corros Sci 2020, 176: 109019.

[43] Shimada S, Ishii T. Oxidation kinetics of zirconium carbide at relatively low temperatures. J Am Ceram Soc 1990, 73: 2804-2808.

[44] Fahrenholtz W G. The ZrB_2 volatility diagram. J Am Ceram Soc 2010, 88: 3509-3512.

[45] Backman L, Gild J, Luo J, et al. Part I: Theoretical predictions of preferential oxidation in refractory high

entropy materials. Acta Mater 2020, 197: 20−27.

[46] Backman L, Gild J, Luo J, et al. Part II: Experimental verification of computationally predicted preferential oxidation of refractory high entropy ultra-high temperature ceramics. Acta Mater 2020, 197: 81−90.

[47] Sheehan J E, Buesking K, Sullivan B. Carbon-carbon composites. Annual Review of Materials Science 1994, 24: 19−44.

[48] Tokuda T, Ito T, Yamaguchi T. Seif diffusion in a glassformer melt oxygen transport in boron trioxide. Z Naturforsch A 1971, 26: 2058−2060.

[49] Park K, Olander D. Oxygen diffusion in single–crystal tetragonal zirconia. J Electrochem Soc 1991, 138: 1154−1159.

[50] Gruenwald T B, Gordon G. Oxygen diffusion in single crystals of titanium dioxide. J Inorg Nucl Chem 1971, 33: 1151−1155.

[51] Buckley J D, Edie D D. Carbon − Carbon Materials and Composites. New Jersey (USA): Noyes Publications, 1993.

本论文原发表于《Journal of Advanced Ceramics》2023 年第 12 卷第 10 期。

Research on Property and Burning Behavior of Flammable Casing for Underground Coal Gasification

Ren Xiangyi[1]   Wu Jianjun [2]   Wang Cankun[3]   Xie Junfeng[4]
Wang Jianjun[1]   Liu Mingtao[5]   Han Lihong[1]

(1. State Key Laboratory of Performance and Structural Safety for Petroleum Tubular Goods and Equipment Materials, CNPC Tubular Goods Research Institute;
2. Department of engineering technology, Xinjiang Oilfield Company;
3. Research Institute of Engineering and Technology, PetroChina Coalbed Methane Company Limited;   4. Oil and Gas Engineering Research Institute of Tarim Oilfield Company;
5. Drilling & Production Technology Research Institute of Liaohe Oilfield Company)

Abstract: In this work, the comprehensive properties of flammable casing for underground coal gasification is systematically investigated, including flammable casing material physical, chemical and mechanical properties and full-size flammable casing mechanical properties and burning behavior. The flammable casing material consists of magnesium alloy matrix and rare earth particles, thermal conductivity and expansion property of which are weak. Results of high-temperature tensile test reveal that flammable casing material has good high temperature strength which declines by 30% at 300℃. Corrosion rate of flammable casing material is relatively high without extra protection. The full-size flammable casing possesses considerable mechanical property, thread property and high temperature collapse resistance. Burning of flammable casing is safe and stable. Burning rate of flammable casing material can be effectively controlled by water flow. Combustion product of flammable casing presents powder condition, which has no risk of blocking the gasification channel. To sum up, flammable casing is necessary to the realization of underground coal gasifying process, which plays the significant role of the development and application of underground coal gasification technology.

Keywords: Flammable casing; Underground coal gasification; Horizontal well; Burning behavior

1 Introduction

1.1 Technology of underground coal gasification

The underground coal gasification (UCG) is a kind of technology which can transform

Corresponding authors: Ren Xiangyi, mmerenxiangyi@126.com; Han Lihong, hanlihong@cnpc.com.cn.

conventional coal resource into H_2, CO, CH_4 or other gas resource through underground controllable chemical reaction. Coal resource with large depth underground cannot be mined directly. With the application of UCG technology, large-depth coal resource can be effectively developed in the form of flammable gas mixture. UCG integrates the technologies of oil drilling, coal mining and gasifying, which realizes the environmentally friendly use of coal resource. China is rich in coal resource. The proven exploitable quantity of coal is 3.77×10^{12} t approximately in the depth of 1000~3000m. Large scale development of coal resource can effectively solve the problem of fossil energy shortage. However, traditional mining technology can only obtain less than one third of this coal resource. UCG technology can utilize coal resource with large depth that traditional coal mining technology cannot reach. If 40% of the coal resource aforesaid is transformed into flammable gas, quantity of the gas is two times larger than that of conventional natural gas[1-2].

1.2 Research status of UCG technology

China is one of the countries that takes many researches on UCG technology[3]. For example, China University of Mining and Technology established the Researching Center of Underground Coal Gasification engineering for the further investigation of U-type UCG technology, which initiated the new technology of "long-channel, big-section, double-stage"[4-5]. Mazhuang coalmine in Xuzhou, China, finished the field test of UCG channel in which the length of horizontal well is 32m. This channel kept producing raw gas for two months. Total volume and average calorific value of the raw gas is $1.6 \times 10^5 m^3$, $4196.6 kJ/m^3$, respectively[6-8]. Exploration of UCG technology started early in 1960s[9]. Australia is rich in various mine resource included coal. Linc Energy Company carried on more than ten UCG projects in the Basin of Dongara, Gippsland, Wallo way, St. Vincent and eastern Sydney area. All of the projects successfully finished the channel constructing process, some of which produced raw gas for months[9]. Canada Swan Hill Synthetic Fuel Company successfully finished the deepest UCG field test in the world through utilizing controlled retraction injection point (CRIP) technology during 2009—2011. The target coal seam is 7~8m thick in the depth of 1400m. However, this project was stopped because of the fire accident on the ground resulted by explosion inside the injection well[10]. With the development of prospecting, mining, chemical and mechanical industry, material and information control, and the previous research, most of the technological bases of UCG technology has been established[11-13]. However, because of the environmental problems and limitations of key technologies for industrial application, UCG technology is still in the experimental stage, no entire industrial chain is built[14-19]. Fig. 1 schematically presents the U-type well string structure based on CRIP UCG technology. Injection well and production well can be realized through conventional drilling and completion[20-22]. Horizontal well is the most important part of UCG well string, which remarkably affects the safety, stability and efficiency of coal gasifying process[23-24]. Fig. 2 shows the gasifying process of underground coal resource. Gasifying equipment is moved from injection wellhead to the bottom of production well. During gasification, equipment should be moved synergistically with coal burning area (CRIP technology)[25]. If steel casing is used during horizontal well completion, casing wall will make the flame of gasifying equipment unavailable to touch the coal seam, so the UCG process cannot be started. Some researchers suggest that casing should have enough through-holes on the surface so that the flame of gasifying equipment can touch the coal seam[26]. However, existence of casing limits the burning efficiency of coal seam because flame cannot touch the coal seam completely[27]. Additionally, holes have the risk of being blocked by coal ash, soil and broken rock. If some kind of casing possess both high strength and flammability, the CRIP UCG process will be completely realized[28-29].

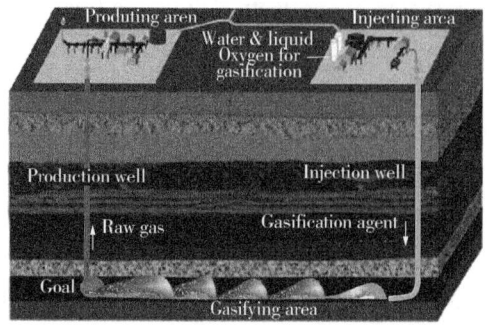 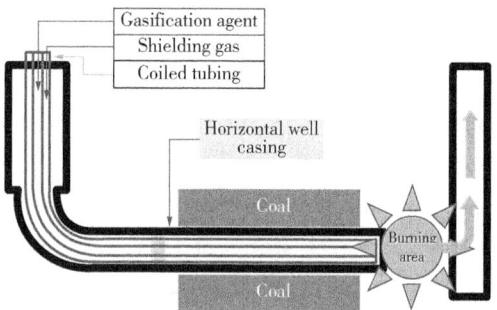

Fig. 1 Schematic of CRIP UCG channel structure

Fig. 2 Schematic of coal gasifying process based on CRIP UCG technology

1.3 Proposal of flammable casing

Generally, Materials with excellent high temperature strength, such assuperalloy, cannot be burned out under high temperature. However, flammable materials possess weak strength especially high temperature strength[30]. For solving this problem, this research proposed the concept of "flammable casing". The flammable casing is made of special metal material with high strength and flammability. Simultaneously, this research aims at development and property evaluation of flammable casing for UCG horizontal well. The mechanical properties, thermal properties, corrosion resistance, microstructure and burning behavior of casing material and the strength, thread property and burning behavior of full-size casing are systematically investigated. This research will provide theoretical consideration and technical guidance to the industrial application of UCG technology in the future.

2 Experimental procedure

2.1 Materials and specimens

The 100kg induction furnace was used to produce the flammable casing material whose raw materials are Gd - Y RE ferrosilicon, ferroneodymium, ferrozirconium, zinc, aluminum and magnesium. The furnace was located in the 600m^2 workshop, smelting form of which is induction heating in the vacuum or under Ar atmosphere. This research chose the form of Ar atmosphere smelting to avoid volatilization of raw materials during smelting process. All of the raw materials were smelted in the furnace after being crushed into particles with the shape of 15mm, and the melting temperature is 800℃[30-31]. After 6min preservation, the liquid alloy was poured to pro-heated columnar sand mold. Before heat treatment, the ingot need accurate lathing to remove the surface defect. Heat treatment of the ingot is 5h preservation under 530℃, air cooling. Ingot after heat treatment was made into flammable casing by plastic processing. The rolling load, rolling temperature, rolling ratio and rolling rate were 12000kN, 580℃, 8 and 1mm/s respectively. The aging treatment was executed to improve the property of the flammable casing, parameters of which are 60h preservation under 200℃. The produced flammable casing was used for casing experiments, which is presented in Fig. 3. Specimens for material experiments were cut from the flammable casing. Some of the casing was machined thread on each end. Types of the thread are shown in Fig. 4.

Fig. 3　Full-size samples of flammable casing

(a) round thread　　　　　　　　　　　　　　(b) trapezoidal thread

Fig. 4　Thread types of flammable casing

2.2　Characterizations and experiments

X-ray flourescence analysis (XRF) was used to determine the chemical composition of flammable casing material. The optical microscope (OM) and scanning electron microscope (SEM) were used to characterize the microstructure and fracture morphology of specimens. All of the metallurgical specimens were mechanical polished and then etched by nital. The thermal conductivity and thermal expansion experiment were carried out to test the thermal property of flammable casing material. Mechanical property under room temperature and high temperature was tested by high-temperature tensile experiment. The autoclave experiment was used for testing the corrosion resistance under the atmosphere of UCG production gas (H_2, CO_2 and H_2S included) and formation water environment of flammable casing material, parameters of which is shown in Table 1[20, 25, 27-28]. The burning behavior of flammable casing material was investigated by material burning test. Specimen of burning test is 10mm×10mm×

10mm cube. The full-size test of flammable casing consists of thread test, mechanical test and burning behavior test. Thread test includes torque experiment and over-torque experiment, which was finished by hydraulic tong. Mechanical test includes tensile test and high-temperature collapse test, which was finished by 1500t compound loading equipment.

The underground environment simulating system (shown in Fig. 5) can create high-temperature, high-pressure and oxygen-enriched atmosphere which is similar to UCG working condition. Utilization of this system can realize the simulated burning behavior of flammable casing during UCG process. Specimen for casing burning test consists of 2 short knots and a steel coupling, length of which is 2.8m. Schematic of the specimen is presented in Fig. 6. Connecting torque of short knots and coupling is 2000N · m. During the test, specimen was ignited from one end. For obtaining the temperature distribution of specimen, 7 temperature measuring holes were made along the axial direction, distance between each two adjacent holes was 300mm. Temperature curves during the test were recorded by Pt-Rh-Pt thermocouple which was stuck in each hole. Location of temperature measuring holes is shown in Fig. 7. Parameters of the test are shown in Table 2. After the test, the axial burning rate r_a and mass burning rate r_m were calculated by formula 1 and formula 2 in which L refers to length of the specimen (Unit: m), t refers to burning time (Unit: s), m_1 and m_2 refer to mass of the specimen before and after ignition respectively.

Fig. 5   Underground environment simulating system

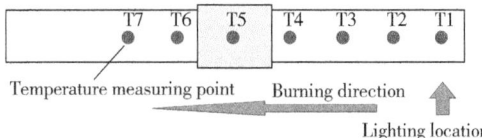

Fig. 6   Specimen for casing burning test

Fig. 7   Location of temperature measuring points

$$r_a = L/t \tag{1}$$
$$r_m = (m_1 - m_2)/t \tag{2}$$

Table 1   Experimental parameters of autoclave test

Parameter types	Value	Specification
Temperature/℃	80, 160, 240	—
Pressure/MPa	10	0.002MPa H_2S, 1.29MPa CO_2, 0.48MPa H_2, 8.228MPa N_2
Time/h	168	—

Continued

Parameter types	Value	Specification
Simulated formation water composition/(mg/L)	10000	Cl^-
	1390	HCO_3^-
	85	SO_4^{2-}
	3102	K^+ and Na^+
	111	Ca^{2+}
	55	Mg^{2+}

Table 2 Main experimental parameters of casing burning behavior test

Parameter types	Specification
Atmosphere	0~30% O_2 (volume fraction, adjustable), N_2 the rest
O_2 pouring place	Inside the casing
Pressure	2MPa
Ignition temperature	700~800℃
total mass	21400g

3 Results and discussion

3.1 Material properties of flammable casing

The XRF testing results of flammable casing material are shown in Table 3, which indicates that composition of the casing material meets the designing expectation. The microstructure morphology of flammable casing material is presented in Fig. 8. It is seen that the matrix of magnesium alloy consists of grains with different shape and RE phase particles. Casing heat treatment temperature is higher than the recrystallization temperature of casing material, thus matrix grains possess the isotropic distribution through recrystallization. Besides, distribution of RE particles are along the rolling direction. This type of microstructure is beneficial to the improvement of mechanical property of casing[30]. Fig. 9 shows the testing results of thermal conductivity of flammable casing material. With the increasing of testing temperature, thermal conductivity of the material ascends obviously. Table 4 shows the thermal conductivities of several kinds of metals and alloys, by comparing with the thermal conductivities of steel, aluminum alloy, titanium alloy and flammable casing material at different temperatures, it is found that thermal conductivity of flammable casing material is the worst[31]. With this characteristic, flammable casing can possess high temperature gradient, tools and equipment inside the casing can be protected from burning down during coal gasification.

Table 3 Chemical composition of the flammable casing material (%, mass fraction)

Gd	Nd	Zn	Y	Zr	Si	Pd	Cu	Fe	Mg
4.00	1.53	0.83	0.78	0.40	0.08	0.05	0.04	0.02	92.27

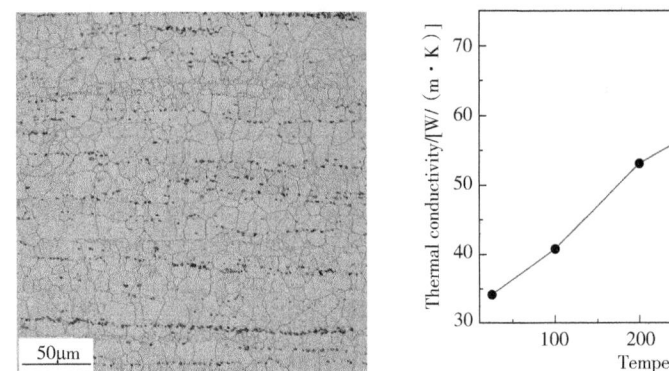

Fig. 8　Morphology of flammable casing material　　　Fig. 9　Results of thermal conductivity test

Table 4　Thermal conductivities of metals and alloys

Alloys	Thermal conductivity/[W/(m·K)]
Silver	411
Al-Cu alloy	177
Copper	398
Aluminum	237
6061 aluminum alloy	155
Low-carbon steel	54
Magnesium	172
Nickel	93
Stainless steel	16
Iron	73

The thermal expansion testing result of flammable casing material is shown in Fig. 10. Thermal expansion coefficient of flammable casing material needs to be low enough in case horizontal well string deforms severely because of high temperature[24, 27]. According to the result, it is found that the expansion behavior of flammable casing material is similar to that of normal metal materials such as steel and other non-ferrous alloy[32-33].

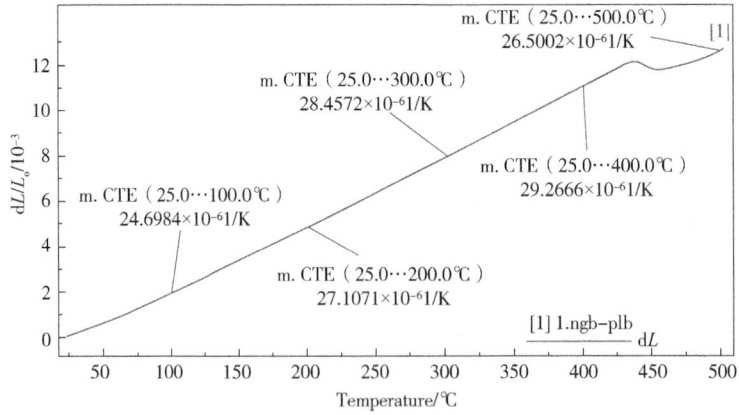

Fig. 10　Result of thermal expansion test

High temperature mechanical property is one of the most significant properties of flammable casing material. The high temperature tensile test of flammable casing material was carried out, results of which are shown in Fig. 11. The results of yield strength, tensile strength and elongation can be seen in this graph. It is indicated that strength under 100℃ of flammable casing material reaches the steel grade of 40ksi (275MPa). With the increasing of temperature, yield strength and tensile strength of casing material decrease smoothly when the testing temperature is lower than 300℃. When the temperature exceeds 300℃, strength of casing material decreases severely. Besides, elongation of casing material descends first and then rises obviously. It is because atomic binding force and grain boundary strength descend at temperature higher than 100℃. Descending of atomic binding force and grain boundary strength increases the trend of atom sliding, thus plasticity of casing material is improved. When the temperature is 100℃ approximately, grain boundary is found RE segregation which decreases the plasticity of casing material[34-35], Through observing the morphology of fracture section in Fig. 12, together with the variation of elongation in Fig. 11, it is proved that fracture mechanism of flammable casing material changes from brittle fracture to plastic fracture when the temperature exceeds 100℃. In Fig. 12a and Fig. 12b, fracture sections in extended area and instantaneous broken area present neither plastic nor brittle characteristic. It is indicated that flammable casing material at room temperature possesses slight brittleness. In Fig. 12c, dimple morphology can be obviously observed, which proves the good plasticity of flammable casing material at high temperature. Only in the instantaneous broken area (Fig. 12d), edge of dimples becomes irregular.

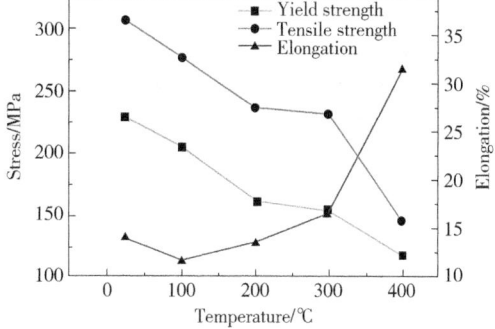

Fig. 11　Result of high temperature tensile test

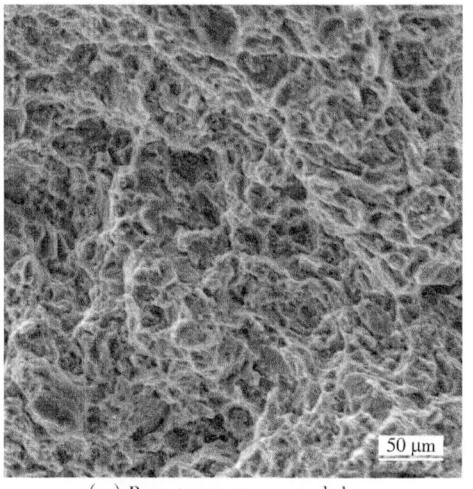 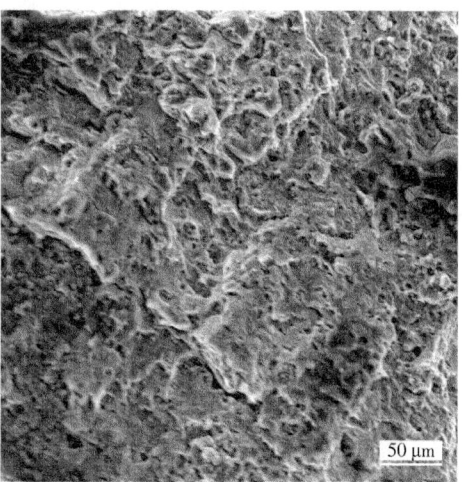

　　　(a) Room temperature, extended area　　　　　　(b) room temperature, instantaneous broken area

Fig. 12　SEM morphology of fracture sections of flammable casing material

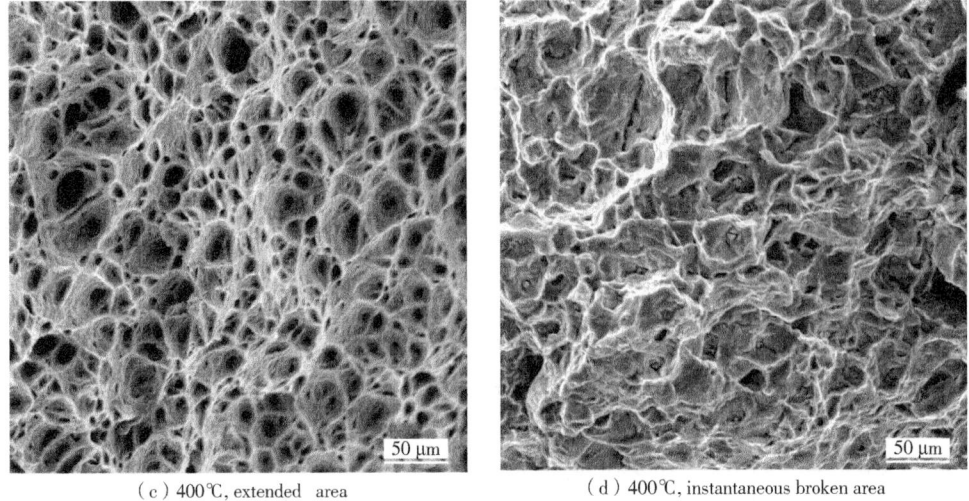

(c) 400℃, extended area　　　　　　　　(d) 400℃, instantaneous broken area

Fig. 12　SEM morphology of fracture sections of flammable casing material (continued)

Generally, magnesium alloy is a kind of metal material with bad corrosion resistance[34]. For the application of UCG working condition, flammable casing material should possess enough corrosion resistance in order to avoid leak or collapse caused by corrosion before being burned. Results of corrosion rate of flammable casing material are shown in Table 5. It is seen that under simulated underground working condition, flammable casing material possesses relatively low corrosion rate. By comparing with normal steel casing, wear resistance of the material is much worse than that of J55 and N80 casing material[35-36]. Morphologies of corrosion specimens before and after experiment are shown in Fig. 13. By observing Fig. 13a, Fig. 13b and Fig. 13c, it is found that specimens after experiment become incomplete, surfaces of which are covered with thick corrosion production layer. Existence of HCO_3^- made the surface of specimen passivated during experiment. Within the range of 80~160℃, the higher temperature specimens are at, the higher passivating rate specimens possess[36-37]. Consequently, in Fig. 13b and Fig. 13c, corrosion production layer of specimen at 80℃ is thicker than that of specimen at 160℃. Besides, at 240℃, protection from passivation becomes ineffective, specimens are completely corroded, which is unable to be used for corrosion rate calculation.

Table 5　Results of autoclave experiment

Testing temperature/℃	general corrosion rate/(mm/a)	average corrosion rate/(mm/a)
80	24.5566	23.88
	22.7445	
	24.3372	
160	28.3019	27.23
	25.2844	
	28.1040	
240	—	

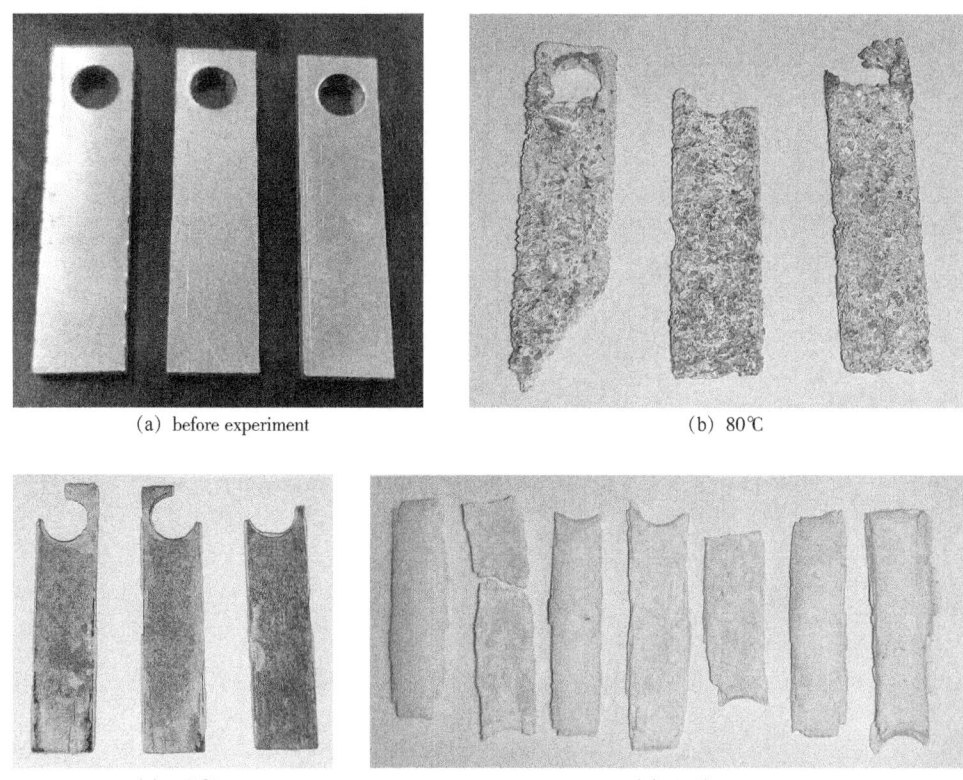

(a) before experiment (b) 80℃
(c) 160℃ (d) 240℃

Fig. 13 Morphologies of corrosion specimens

Combustion characteristic is the most significant for flammable casing, which includes ignition point, flame temperature, burning stability and condition of combustion products. This research studied some of the characteristics by material burning experiment, results of which are shown in Fig. 14. During coal gasification process, water spray is one of the gasifiers[11], consequently, burning process of flammable casing should not be affected by water spray. Furthermore, when the burning needs to be slowed or stopped by water flow, the water flow should be effective. In Fig. 14a, it is observed that flame color of flammable casing material is similar to that of magnesium. Besides, burning of casing material can be sustained under water spray condition. In Fig. 14b, material during burning is stopped by water flow. By observing Fig. 14c in which combustion product of casing material is shown, it is seen that product possesses the powder form which can be completely dispersed by water flow. Combustion product with powder form can effectively avoid blocking the gasification channel, which contributes to the safety and stability of gasification channel.

3.2 Mechanical properties of full-size flammable casing

The safety and stability of coal gasification process depend on the comprehensive property of flammable casing, including thread property, mechanical property and burning behavior. The surface condition of thread after one-time make-up is shown in Fig. 15. No adhesion is observed on the surface, which indicates that thread of flammable casing and steel coupling connect smoothly. The over-torque experiment of flammable casing and coupling is carried out, result of which is shown in Fig. 16. It is seen that thread of the casing passes through 60% length of the coupling

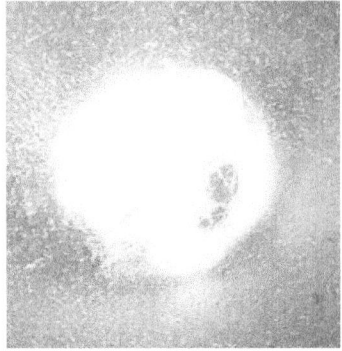
(a) burning under water spray

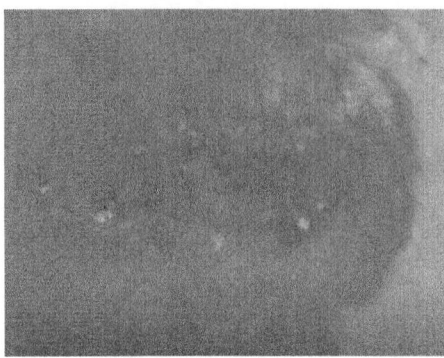
(b) drenching during burning

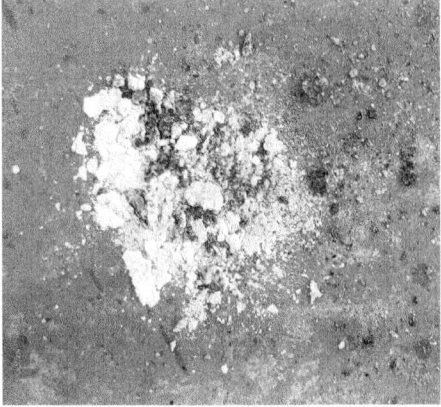
(c) burning production

Fig. 14 Burning condition of flammable casing material

without cracking. This result indicates that casing thread possesses good toughness. Besides, the highest torque measured is 4300N · m(Fig. 17).

Fig. 15 Morphology of thread after make-up   Fig. 16 Conditions of thread and coupling after over-torque experiment

The tensile property of full-size flammable casing is tested, before which the diameter and thickness of flammable casing is measured. Results of size measuring and tensile test is shown in Table 6.

According to the tensile test results of full-size casing and casing material in Fig. 11, tensile strength of flammable casing is slightly lower than that of casing material, which indicates that strength of trapezoidal thread is practically equivalent to that of flammable casing material. By observing the fracture which appears in the middle of the thread, no obvious plastic deformation is observed. This result fits the SEM tensile fracture morphology observing result of flammable casing material.

Table 6 Results of size measuring and tensile test of flammable casing

Diameter/mm	Thickness/mm	Fracturing load/kN	Section area/mm^2	Tensile strength/MPa
114.3	9.65	920	3171	290.2

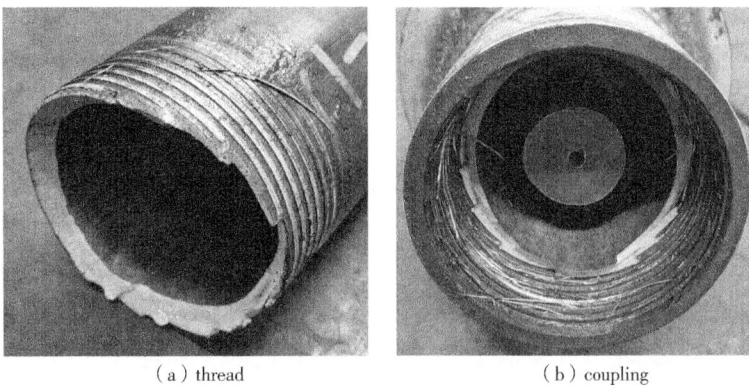

(a) thread (b) coupling

Fig. 17 Morphology of casing after tensile test

Fig. 18 presents the result curve of high-temperature collapse test of full-size flammable casing. According to the result, flammable casing can bear approximately 3100psi (22.1MPa) at 200℃. Reference [12] put forward that stratigraphic load of coal resource in the depth of 600~800m is 15~20MPa. That means the highest available depth of flammable casing is over 800m. The available depth of flammable casing exceeds most of the depths of coal resource which cannot be developed by general methods. Besides, with the cooling effect of water flow, temperature of horizontal well is less than 200℃[9]. Consequently, collapse resistance of flammable casing is enough for the application.

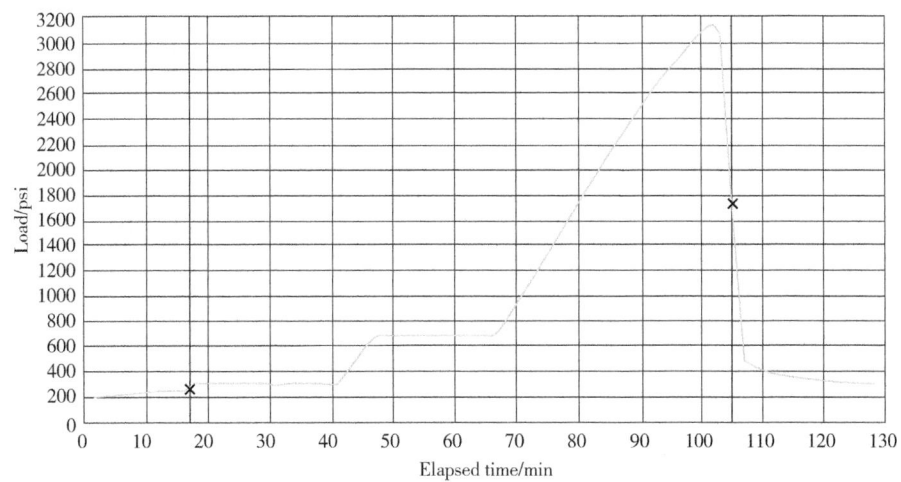

Fig. 18 Result of high-temperature collapse test

3.3 Burning behavior of flammable casing under UCG working condition

The gasifying efficiency of underground coal resource mainly depends on the burning process of flammable casing. Furthermore, burning behavior of flammable casing has significant effect on safety and stability of gasifying process. This part of research aims at gathering the combustion characteristics of full-size flammable casing such as ignition point, temperature distribution, flame temperature and burning rate. Research on combustion characteristics of flammable casing can establish significant technical theory for the application of UCG technology.

Table7 presents the experimental results of full-size flammable casing burning test under simulated UCG working condition. The casing kept burning for 2640s since being ignited, then burned out with coupling left. Fig. 19a shows the casing which has been burned to powder which is similar to the result in Fig 14. In Fig. 19b, it is observed that coupling has no deformation after the casing is burned out, surface of which is covered with MgO (white) and Mg_3N_2 (yellow) production. Besides, coupling cannot stop the burning process of flammable casing, indicating that distribution and material of couplings of UCG horizontal well string can be ignored. Different coal resource needs various moving speed of gasifying equipment, thus the calculated mass burning rate r_m and axial burning rate r_a provides the effective reference and basis for the designing of horizontal well string axial consuming rate.

Table 7 Experimental results of flammable casing burning test under simulated UCG working condition

Result types	Value
Specimen length L	2800mm
Mass before test m_1 (coupling not included)	21400g
Mass after test m_2 (coupling not included)	0g
Ignition point	584℃
Burning time	2640s
Axial burning rate r_a	1.06mm/s
Mass burning rate r_m	8.11g/s
Coupling deformation	No
Burning production condition	powder

(a) casing

(b) coupling

Fig. 19 Specimen condition after burning process under simulated UCG working condition

Fig. 20 shows the temperature distribution during the whole burning process of the full-size flammable casing. The highest measuring limit of thermal couple is 1450℃, when the internal temperature is over 1450℃, the thermal couple will be destroyed. Measuring point T1 and ignition point are located in the same axial place. Consequently, when the burning started, the temperature of flame can be measured by thermal couple at point T1 before the thermal couple is destroyed. It is seen that flame temperature which is measured by thermal couple at point T1 is 1450℃ at least. This temperature is slightly higher than gasifying temperature of coal[15], which indicates that heat of flammable casing flame is beneficial to the saving of

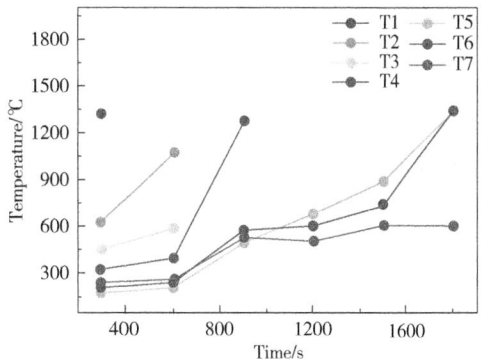

Fig. 20 Temperature distribution curves of flammable casing during burning process under simulated UCG working condition

gasifier. Furthermore, measured temperature of point T2 to T6 ascends gradually and obviously. With the increasing of burning time, thermal couples at point T2, T3 and T4 detect the flame temperature and stopped working within 1000s. However, thermal couples at point T5, T6 and T7 keep working for over 1800s. Besides, ascending trends of point T5, T6 and T7 are quite slow, when the casing keeps burning for 1800s, temperature near point T7 is still less than 600℃. It is the bad thermal conductivity of flammable casing material that attributes the huge temperature gradient. According to the time when each point detects the flame temperature, the axial burning rate r_a of flammable casing can be determined, which is given in Table 7.

4　Conclusions

The flammable casing material consists of magnesium alloy matrix and rare earth particles. It possesses bad thermal conductivity and expansion property. The results of high-temperature tensile test reveal that flammable casing material has good high temperature strength which declines by 30% at 300℃. Corrosion rate of flammable casing material is over 20mm/a, which needs extra protection in future application. Burning of flammable casing material can be maintained in water spray and stopped by water flow. Combustion product of flammable casing presents powder condition. The full-size flammable casing possesses excellent comprehensive properties. Highest torque of the thread reaches 4300N·m. Thread after one-time make-up keeps no deformation, tensile strength of which reaches the same level of flammable casing material. Collapse strength of flammable casing at 200℃ is over 22MPa, which meets the requirement of underground working condition.

The burning behavior of full-size flammable casing string under simulated UCG working condition reveals that the burning process is continuous and orderly. Burning rate of the casing string is controllable. Combustion product with powder condition has no risk of blocking the gasification channel. Besides, temperature gradient of the casing string is high enough, which can protect the internal tools and equipment from destroyed by high temperature.

Acknowledgement

This work was supported by CNPC project of scientific research and technological development (2019E-2502, 2021DJ2705, 2022ZG06) and CNPC project of basic research and strategic reserve (2021DQ03).

References

[1] C. Ma, L. Yu, J. Liang. Development of underground coal gasification technology of China. Chinese Energy, 2003, 2: 11-15.

[2] L. Xin, Z. Wang, G. Wang, et al. Technological aspects for underground coal gasification in steeply inclined thin coal seams at Zhongliangshan coal mine in China. Fuel, 2017, 191: 486-494.

[3] X. Hu, D. Zhao, Y. Guo, et al. Present development situation an prospect of underground coal gasification (UCG) technology in China. Unconventional oil and gas, 2017, 4(1): 108-115.

[4] J. Xi. Study on the evolution law of characteristic fields in the context of underground coal gasification. Beijing: China university of mining and technology, 2016.

[5] J. Liu. Study on the combustion cavity growth and stability of the roof during underground coal gasification. Beijing: China university of mining and technology, 2014.

[6] J. Wang, Z. Wang, H. Wang. Design and application of roadway filling technology in underground coal gasification gasifier. Coal engineering, 2017, 50(10): 74-79.

[7] J. Liang, S. Liu, L. Yu, et al. Study on stability controlling methods of underground coal gasification process. Journal of China university of mining and technology, 2002, 31(5): 358-361.

[8] L. Yang. Study on moving velocity of burning front in underground coal gasification. Journal of China coal society, 2000, 25(5): 496-500.

[9] S. Chen, L. Li, J. Cui, et al. Advances of underground coal gasification (UCG) and industrial development. Resources and industries, 2014, 16(5): 129-135.

[10] R. Mandapati, P. Ghodke. Modeling of gasification process of Indian coal in perspective of underground coal gaisifcation (UCG). Environment, development and sustainability, 2020, 22: 6171-6186.

[11] C. Zou, Y. Chen, L. Kong, et al. Underground coal gasification and its strategic significance to the development of natural gas industry in China. Petroleum exploration and development. 2019, 46(2): 195-204.

[12] Z. Wang, J. Liang, K. Liang, et al. Correlation characteristics of reaction zone distribution and technical parameters during UCG of Ezhuang bituminous coal. Journal of China coal society. 2015, 40(7): 1677-1683.

[13] J. Liang, Y. Cui, Z. Wang, et al. Gasifier type and technique of underground coal gasification. Coal science and technology, 2013, 41(5): 10-15.

[14] S. Ge. Chemical mining technology for deep coal resources. Journal of China university of mining and technology. 2017, 46(4): 679-691.

[15] J. Liang, L. Yu. Study of two-state underground coal gasification in counter directions. Journal of China coal society. 1996, 21(1): 68-72.

[16] P. Thakur. CO_2 sequestration and underground coal gasification with horizontal wells. Coal bed methane, 2020: 319-333.

[17] Y. Makabe, A. Hamanaka, K. Itakura, et al. Experimental study on Co-axial system with a horizontal well of underground coal gasification to evaluate gasification efficiency //International symposium on earth scence and technology, Fukuoka, Japan, 2020.

[18] J. Grabowski, K. Korczak, A. Tokarz. Aquatic risk assessment based on the results of research on mine waters as a part of a pilot underground coal gasification process. Process safety, 2021, 148: 548-558.

[19] G. Pivnyak, V. Falshtynskyi, R. Dychkovskyi, et al. Conditions of suitability of coal seams for underground coal gasification. Key engineering materials, 2020, 844: 38-48.

[20] M. Wiatowski, K. Kpusta. Evolution of tar compounds in raw gas from a pilot-scale underground coal gasification (UCG) trial at Wieczorek mine in Poland. Fuel, 2020, 276: 118070.

[21] S. Kashyap, P. Vairakannu. Movable injection point-based syngas production in the context of underground coal gasification. International journal of energy research, 2020, 44(5): 2574-3586.

[22] A. Strugala-Wilczek, K. Stanczyk, K. Bebek. Comparison of metal adsorption from aqueous solution on coal and char remaining after in-situ underground coal gasification (UCG). Mine water and the environment, 2020, 39(2): 369-379.

[23] G. Perkins. A. O-dimensional cavity growth submodel for use in reactor models of underground coal gasification. International journal of coal science and technology, 2019, 3: 334-353.

[24] L. Kong, X. Zhu, E. Zhan, et al. Suggestions on China's natural gas self-sufficiency by deep coal underground gasification technology. International petroleum economics, 2018, 26(6): 85-94.

[25] W. Huang, Z. Wang, T. Duan, et al. Effect of oxygen and steam on gasification and power generation in industrial tests of underground coal gasification. Fuel, 2020, 289: 119855.

[26] X. Min, X. Lin, W. Liu, et al. Study on the physical properties of coal pyrolysis in underground coal gasification channel. Powder technology, 2020, 376: 573-592.

[27] J. Xie, L. Xin, X. Hu, et al. Technical application of safety and cleaner production technology by underground coal gasification in China. Journal of cleaner production, 2020, 250: 1-14.

[28] G. Walowski. The method to assess the gas flow of a porous bed product derived from underground coal gasification technology. Energy, 2020, 199: Issue C.

[29] Q. Irum, S. Khan, A. Uppal, et al. Galerkin finite element based modeling of one dimensional packed bed reactor for underground coal gasification (UCG) process. IEEE Access, 2020, 8: 130-139.

[30] X. Ren. Research on borocarbide regulation and wear resistance of high boron multicomponent alloy. Xi'an: Xi'an Jiaotong University, 2019.

[31] Y. Bai. Investigation on Fabrication, Dry Sliding and Erosive Wear Behavior of Al_2O_3/Fe_3Al Composites. Xi'an: Xi'an Jiaotong University, 2013.

[32] T. Zhao, E. Guo, Y. Feng, et al. Microstructure and mechanical properties of as cast and heat-treated Al-Si-Cu-Ni-Ce-Cr. Journal of materials research. 2019, 33(8): 588-596.

[33] Q. Zhao. The influence of Co and Ni element on microstructure and high temperature tensile mechanical properties of Al-Mg-Si alloy. Harbin: Harbin institute of technology, 2017.

[34] J. Li. Study on microstructures and mechanical behaviors of high strength and heat resistant Mg-Gd-Y-Zr alloys. Hefei: University of science and technology of China, 2019.

[35] C. Wang, A. Ma, H. Liu, et al. Research progress on heat resistance of magnesium-rare earth alloys reinforced by long period stacking ordered phase. Materials reports. 2019 33(8): 588-596.

[36] S. Wang, X. Guo, Z. Wang, et al. Effect of heat treatment on microstructure and properties of VW63Z magnesium alloy. Aerospace Manufacturing Technology. 2019, 4: 1-5.

[37] N. Zhang, R. Cheng, H. Dong, et al. Application and research progress of strontium in heat-treatment magnesium alloy. Materials reports. 2019, 33(8): 2565-2571.

本论文原发表于《Heliyon》2023年第9期。

Study on Spray Cooling of Ultra-high Temperature Production Wellbore in Underground Coal Gasification

Wang Jianjun[1] Zhao Huanzhen[1,2] Zhang Chao[3]
Zhou Sheng[1,2] Hao Xuelei[3] Ren Xiangyi[1]

(1. CNPC Tubular Goods Research Institute, State Key Laboratory of Performance and Structural Safety for Petroleum Tubular Goods and Equipment Materials;
2. Mechanical Engineering College, Xi'an Shiyou University;
3. The First Gas Plant of PetroChina Changqing Oilfield Company)

Abstract: In order to analyze the influencing factors of the cooling effect of water injection in underground coal gasification production wells, the finite element analysis method was used to simulate the temperature field of the produced gas under different mass flow, spray cone angle, atomized particle diameter and spray cycle. The results show that after the produced gas is cooled by spray, the gas temperature gradually increases along the axial direction at the outlet and decreases along the depth at the wall. On the premise of the same other conditions, increasing the mass flow of spray liquid can significantly reduce the temperature of high-temperature gas and increase the cooling rate; When the atomization effect is improved and the diameter of atomized particles is reduced, the cooling effect will be significantly improved; In a certain period of time, when the total amount of spraying is certain, the periodic spraying can significantly improve the cooling effect, but when the cycle time increases, the temperature at the wall will rise to a certain extent. Smaller spray cone angle has less influence on the cooling effect. When the spray cone angle increases to more than 40 °, the casing wall temperature will decrease to a certain extent with the increase of spray cone angle; The research results provide a technical basis for the selection of spray parameters of underground coal gasification.

Keywords: Underground coal gasification; High temperature; Numerical simulation; Spray cooling

1 Introduction

In recent years, the development of underground coal gasification (UCG) has provided a new method for the utilization of coal resources in my country, and has become one of the important

Corresponding author: Wang Jianjun, wangjianjun005@cncp.com.cn.

research directions in the field of energy in my country.

When the high-temperature gas formed by the redox reaction of the coal seam enters the production well, its maximum temperature can be close to 900℃. At this temperature, the current API casing strength is difficult to meet the working conditions, which poses a serious threat to the safety and sealing of the wellbore. To this end, it is necessary to use the spray water injection process to cool the high-temperature gas, thereby reducing the overall temperature of the wellbore, accurately predicting the fluid temperature distribution, and providing a basis for the subsequent wellbore design of coal gasification production wells.

With the development of CFD (Computational Fluid Dynamics) technology, there are more and more studies on numerical simulation of fluid heat transfer problems using numerical calculation methods. Yu Qingyuan[1] et al. Characteristic analysis; Deng Xiaoye[2] et al. simulated the temperature distribution of the fluid in the corrugated tube, and analyzed the influence of the wave crest diameter and arc length on the heat transfer characteristics of the fluid; Zhang Peng[3] et al. The heat transfer characteristics of liquid nitrogen phase change in the container were simulated. Wang Wensong[4] used numerical simulation method to analyze the fluid temperature and flow state in the shell heat exchanger. Wang Jun[5] et al. used the numerical simulation calculation method to analyze the change of seawater temperature field and the pressure change caused by the pressure test in the submarine pipeline. Yan Mingyu[6] and others conducted numerical simulations on the heat transfer efficiency of a two-phase closed siphon under different wall temperatures and liquid filling rates in the evaporation section. Wang Changbin[7] and others used numerical calculation method to calculate the temperature drop along the buried hot oil pipeline, and analyzed the influence of pipeline buried depth, pipeline radius, pipeline flow, thermal insulation layer and unsteady state on pipeline heat transfer. Wang Yongxing[8] simulated the flow of two fluids and the flow of two fluids when they were mixed. You Jiang[9] used the finite element method to approximate the film flow in the packing area of the countercurrent closed cooling tower with the droplet flow.

In this paper, the finite element analysis method is used to first establish a fluid model to simulate the temperature field of the fluid in the coal gasification production well, and then change the mass flow rate of the spray liquid, the spray cone angle, the diameter of the atomized particles and the spray cycle to discuss the parameters. Influence on the cooling effect of high temperature gas.

2 Establishment of Mathematical Models

This paper makes the following assumptions when establishing the model[10-13]:

(1) Realizable k-ε turbulence model is used as the turbulence model;

(2) The heat transfer of the liquid in the spray pipe and the influence of the fluid temperature on the physical properties of the fluid are ignored;

(3) The atomization model adopts the droplet breakup model;

(4) The nozzle model adopts the pressure swirl nozzle;

(5) All the walls are considered as adiabatic and non-slip walls.

2.1 Fluid governing equations

For continuous phase fluid, considering its temperature change and composition change, the governing equation is as follows[14]:

$$\frac{\partial \rho}{\partial t} + \text{div}(\rho \boldsymbol{u}) = 0 \tag{1}$$

$$\frac{\partial(\rho u)}{\partial t} + \text{div}(\rho u \boldsymbol{u}) = \text{div}(\mu \text{ grad } u) - \frac{\partial p}{\partial x} + S_u \tag{2}$$

$$\frac{\partial(\rho v)}{\partial t} + \text{div}(\rho v \boldsymbol{u}) = \text{div}(\mu \text{ grad } v) - \frac{\partial p}{\partial y} + S_v \tag{3}$$

$$\frac{\partial(\rho w)}{\partial t} + \text{div}(\rho w \boldsymbol{u}) = \text{div}(\mu \text{ grad } w) - \frac{\partial p}{\partial z} + S_w \tag{4}$$

$$\frac{\partial(\rho T)}{\partial t} + \text{div}(\rho \boldsymbol{u} T) = \text{div}\left(\frac{k}{c} \text{grad } T\right) + S_T \tag{5}$$

$$\frac{\partial(\rho c_s)}{\partial t} + \text{div}(\rho \boldsymbol{u} c_s) = \text{div}[D_s \text{ grad }(\rho c_s)] + S_s \tag{6}$$

$$\frac{\partial(\rho k)}{\partial t} + \frac{\partial(\rho k u_i)}{\partial x_i} = \frac{\partial}{\partial x_j}\left[\left(\mu + \frac{\mu_l}{\sigma_k}\right)\frac{\partial k}{\partial x_j}\right] + G_k - \rho \varepsilon \tag{7}$$

$$\frac{\partial(\rho \varepsilon)}{\partial t} + \frac{\partial(\rho \varepsilon u_i)}{\partial x_i} = \frac{\partial}{\partial x_j}\left[\left(\mu + \frac{\mu_l}{\sigma_\varepsilon}\right)\frac{\partial k}{\partial x_j}\right] + \rho C_1 E \varepsilon - \rho C_2 \frac{\varepsilon^2}{k + \sqrt{V \varepsilon}} \tag{8}$$

$$\frac{d\boldsymbol{u}_p}{dt} = f_D(\boldsymbol{u} - \boldsymbol{u}_p) + \frac{g(\rho_p - \rho)}{\rho_p} + f_i \tag{9}$$

2.2 Fog model

In this paper, the model of droplet breakup is selected, which assumes that in the process of droplet breakup, the disturbance and the surface wave interact and influence each other. The mechanism is applicable to the spray model at high relative velocities, and the mechanism is applicable to describe the rapid growth of large surface waves on the leeward side of the droplet due to the rapid deceleration of the droplet, causing the parent droplet to break up and form child droplets. Its fastest growing surface wave frequency is:

$$\Omega_{\max} = \sqrt{\frac{2}{3}\frac{g_t}{\sqrt{3\sigma}}\frac{|\rho_p - \rho|^{1.5}}{\rho_p + \rho}} \tag{10}$$

The wavenumber used for the fastest increasing surface wave frequency is:

$$n = \sqrt{\frac{g_t |\rho_p - \rho|}{3\sigma}} \tag{11}$$

The breaking time is:

$$t_a = c_1 \frac{1}{\Omega_{\max}} \tag{12}$$

The wavelengths are:

$$\Lambda = c_2 \frac{\pi}{n} \tag{13}$$

2.3 Nozzle model

In this model, the pressure swirl nozzle based on the LISA model is selected as the atomizing nozzle model. In this model, when the liquid flows through the swirl groove in the nozzle, it accelerates into the swirl chamber, and the fluid flows on the wall of the swirl chamber under the action of centrifugal force. After the hollow liquid cone is formed, it leaves the nozzle outlet as a continuous thin liquid film, and then the liquid film is broken to form flocs and drops. The nozzle mass flow is:

$$q_m = c_d F_0 \sqrt{2\rho_p \Delta p} \tag{14}$$

The spray angle is:

$$\tan \frac{\alpha}{2} = \frac{v_{\text{penzui}}}{u_{\text{penzui}}} \tag{15}$$

3 Numerical calculations

3.1 Physical model and meshing

Considering the actual wellbore structure of the coal gasification production well, the production casing is a P110 casing string with an outer diameter of 7 inches and a wall thickness of 12.65mm. Figure 1 shows the geometric model of the area where the produced gas flows through the spray pipe, and the structure size of the fluid area is shownin Table 1. Simplify the model into a two-dimensional symmetrical model and perform structured mesh division. The surface mesh is divided by quadrilateral mesh, and the boundary mesh division method is defined by quantity. The growth ratio is 1. Figure 2 is a schematic diagram of the local mesh model, which is divided There are 155000 grids, 157081 nodes, and the average grid quality is 0.99812, which meets the calculation requirements.

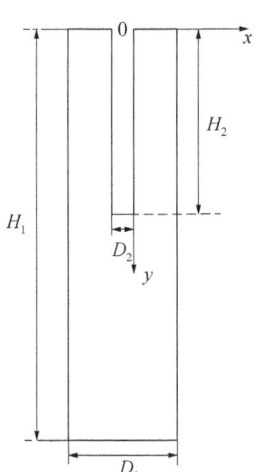
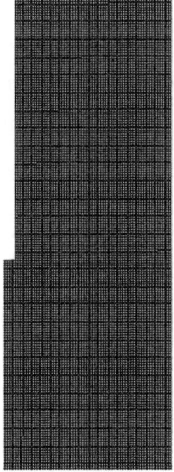

Fig. 1 geometric model   Fig. 2 local grid model

Table 1 Structural parameters of geometric model

Inner diameter of casing D_1/mm	Casing length H_1/mm	Sprinkler pipe diameter H_2/mm	Sprinkler pipe length D_2/mm
152.5	2000	10	1000

3.2 Solution settings

In this paper, the pressure-based solver, transient and turbulent models are used to select the Realizabe k-epsilon model, and the standard wall function is used to deal with the near-wall region. In the solution control, the energy equation, the component transport model and the discrete phase model are simultaneously enabled; The speed coupling method adopts the SIMPLE algorithm, and the convection phase adopts the Second Order Upwind mode[15].

3.3 Boundary condition settings

(1) Inlet: The inlet type adopts Velocity-inlet, the velocity is set to 75m/s, the temperature is 900 ℃, and the discrete phase BC type is reflect.

(2) Outlet: The outlet type adopts Pressure-outlet, the reflux temperature is 25℃, and the discrete phase BC type is escape.

(3) Wall: the wall type is adiabatic non-slip wall, and the discrete phase BC type is reflect.

4 Results and Discussion

4.1 Analysis of Fluid Temperature Field

Figure 3 shows the distribution of the temperature field of the produced gas at different spray times when the mass flow rate is 0.1kg/s, the spray cone angle is 45°, and the particle radius is 1mm. It can be seen from the figure that the temperature of the output gas decreases rapidly within 0.2s. Since the spray liquid continues to absorb heat during the heat exchange process, the temperature difference between the high temperature gas and the liquid gradually decreases, and the heat exchange effect gradually decreases, resulting in the high temperature gas at the nozzle. The cooling effect is the most obvious, and the temperature can be reduced to below 200℃; while the cooling effect at the wall is the worst, and the gas temperature is still around 600℃ after sufficient heat exchange. Due to the influence of factors such as collisions between particles, mutual interference between gas and liquid phases, and changes in gas components, the irregular movement of the fluid in the flow field causes the temperature of the high-temperature gas to fluctuate up and down within a certain range, and it is impossible to achieve stability.

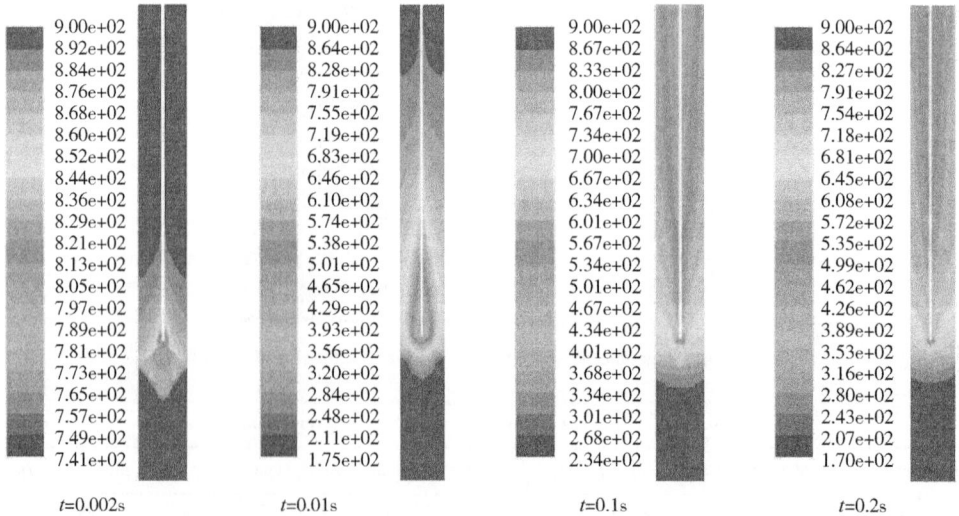

Fig. 3 Fluid temperature field

4.2 Influence of mass flow on cooling effect

In order to analyze the influence of mass flow on the cooling effect, under the same conditions (particle diameter 1mm, spray cone angle 45°, continuous spray), the mass flow rates were 0.05kg/s, 0.1kg/s and 0.15kg/s respectively The fluid temperature field is simulated at kg/s. Figure 4 shows the distribution of the gas temperature at the inner wall of the casing and the gas temperature at the outlet of the model at each mass flow rate at 0.2s. It can be seen from the figure that when the mass flow rate is 0.05kg/s, the gas temperature at the wall is the lowest at 785℃, and the gas temperature at the outlet is The minimum gas temperature is 653℃; when the mass flow rate increases to 0.15kg/s, the minimum gas temperature at the wall can be reduced to 390℃, and the minimum gas temperature at the outlet is 358℃. Increasing the mass flow not only improves the cooling effect, but also increases the cooling rate of the high-temperature gas; at the same time, under the larger mass flow, the temperature distribution interval at the outlet is significantly reduced.

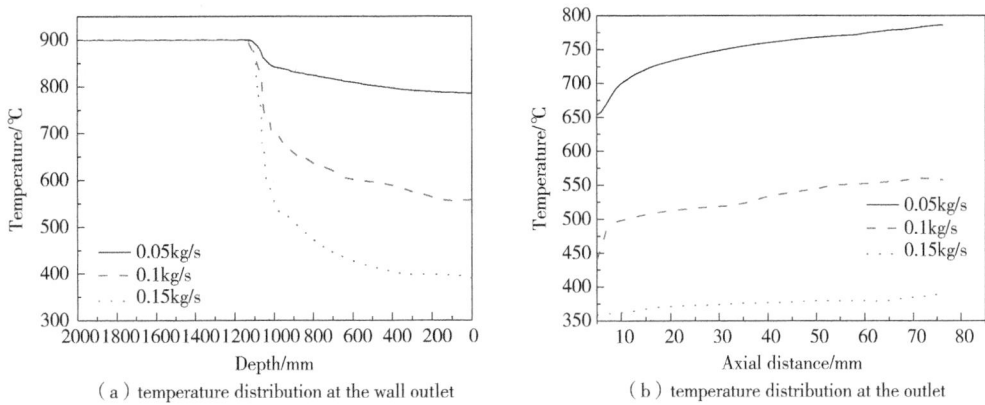

Fig. 4 Effect of spray mass flow on cooling effect

4.3 Influence of spray cone angle on cooling effect

Figure 5 shows the distribution of the gas temperature at the wall and the gas temperature at the outlet under the same conditions (mass flow rate 0.1kg/s, particle diameter 1mm, continuous spraying) at different spray cone angles for 0.2s. When the spray cone angle is small, increasing the spray cone angle has almost no effect on the cooling effect. When the spray cone angle reaches more than 40°, increasing the spray cone angle will have a small impact on the cooling effect: in Under the spray cone angle of 40°, the minimum wall temperature is 589℃; when the spray cone angle increases to 60°, the wall temperature drops to a minimum of 542℃; at the same time, when the spray cone angle increases from 30° to 60°, the cooling depth increased from 1158mm to 1088mm, and the cooling depth decreased by about 6%.

4.4 Influence of atomized particle diameter on cooling effect

Figure 6 shows the gas temperature distribution at the wall and the gas temperature distribution at the outlet under different atomization degrees. It can be seen from the figure that under certain other conditions (mass flow rate 0.1kg/s, spray cone angle 45°, continuous spraying), when the degree of atomization is increased to reduce the particle diameter, the cooling of the gas at the wall and the outlet The effect will be significantly increased. When the spray liquid is atomized into

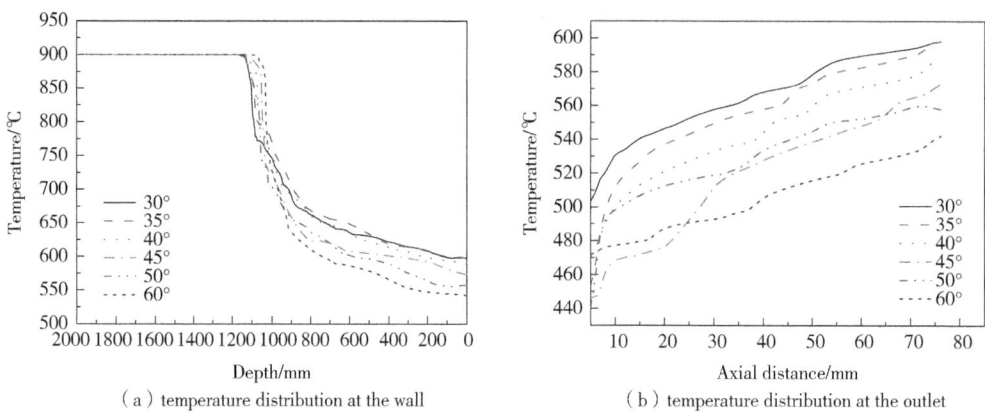

Fig. 5 Effect of spray cone angle on cooling effect

particles with a diameter of 3mm, the minimum wall temperature is 821℃; when the particle diameter is atomized to 1 mm, the gas temperature at the wall can be reduced to a minimum of 557℃; for the outlet gas temperature distribution, When the particle diameter decreased from 3mm to 2mm, the minimum gas temperature at the outlet increased from 607℃ to 617℃, but when the degree of atomization continued to increase, the minimum gas temperature at the outlet decreased accordingly. It can be seen that a larger particle diameter will have a stronger cooling effect in the initial stage of spraying, but a smaller particle diameter will improve the overall cooling effect.

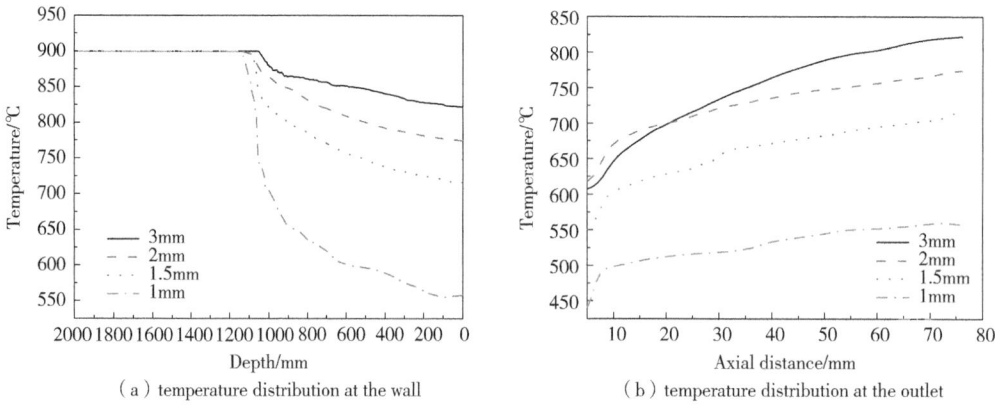

Fig. 6 Effect of particle diameter on cooling effect

4.5 Influence of spray cycle on cooling effect

In order to discuss the influence of the spray cycle on the cooling effect, Figure 7(a) shows the different spray cycles (0.002s interval, 0.004s interval) when the mass flow rate is 0.1kg/s, the particle diameter is 1mm, and the spray cone angle is 45°. And 0.005s interval and 0.05kg/s, particle diameter of 1mm, and the spray cone angle of 45°. The gas temperature distribution at the wall surface. It can be seen from the figure that, under certain other conditions, continuous spray cooling can only reduce the wall temperature to a minimum of 785℃; while the wall temperature can be as low as 716℃ when periodic spray cooling is used. Figure 7(b) shows the gas temperature distribution on the lower wall under different spray cycles (0.004s interval, 0.005s interval and 0.01s interval) when the mass flow rate is 0.2kg/s; the particle diameter is 1mm, and the spray cone angle is 45°; the same mass flow rate of 0.1kg/s, particle diameter of 1mm, and continuous spraying at a spray cone angle of 45° are correct. The results show that when the spraying period is 0.01s, a better cooling effect can be achieved. In addition, a large spray cycle will lead to insufficient heat exchange of some high-temperature gases, so that the temperature of the wall surface will rise to a certain extent.

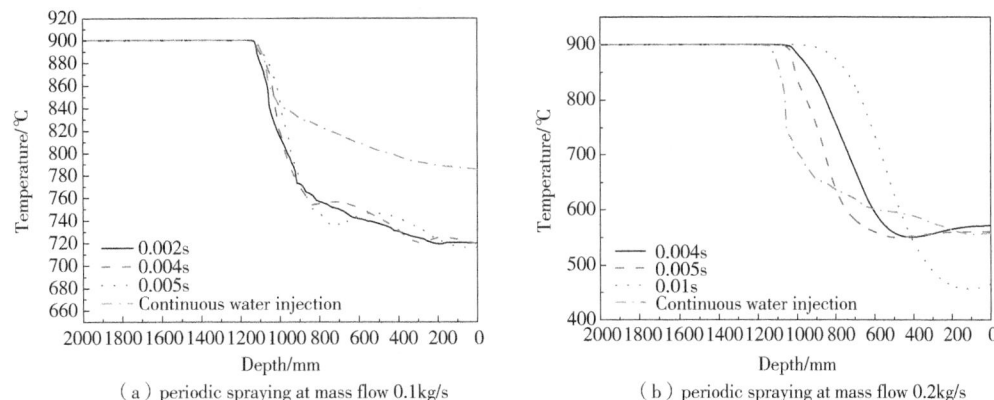

Fig. 7 Gas temperature distribution at the wall under the influence of spraying cycle

5 Conclusion

This paper analyzes the effect of spray water injection on the cooling effect of the gas produced in the UCG production well under different parameters. The results show that the mass flow rate of the spray liquid has the greatest influence on the cooling effect, followed by the diameter and periodicity of the atomized particles. The influence of spraying method, and the influence of spray cone angle on the cooling effect is not obvious. It is recommended to select parameters such as 0.2kg/s mass flow, spray cone angle above 45°, atomized particle radius of about 1mm, and 0.01s periodic spray to cool the production wellbore. The research results provide a technical basis for the selection of spray parameters for UCG.

By comparing the temperature distribution of the wall surface under the periodic spraying method and the continuous spraying method, it can be determined that under the condition of a certain total water injection amount, appropriate periodic spraying can significantly improve the cooling effect of the produced gas. However, with the increase of the spray cycle interval, some high-temperature gases cannot be cooled sufficiently, resulting in an increase in temperature to a certain extent. The selection of the spray cycle under different working conditions needs to be further studied.

Considering the complex downhole conditions of coal gasification production wells, physical experiments are needed for the construction of atomizing nozzles and the overall sprinkler system to prove the reliability and feasibility of the conclusions in this paper.

Nomenclature

ρ continuous phase density, kg/m³

ρ_p	particle density, kg/m^3
u、v、w	component of the velocity vector in the direction
μ	dynamic viscosity, Pa·s
S_u、S_v、S_w	eneralized source term in the direction
c_p	specific heat capacity at constant pressure, J/(kg·K)
k	heat transfer coefficient
S_T	viscous dissipation term
c_s	volume concentration of the component, %
D_s	diffusion coefficient
f_i	additional acceleration term
$\boldsymbol{u}$	continuous phase velocity, m/s
$\boldsymbol{u}_p$	particle velocity, m/s
f_D	drag function
g_t	droplet motion deceleration, m/s^2
c_1、c_2	model parameters
σ	droplet surface tension
Δp	differential pressure of the nozzle
c_d	flow coefficient
F_0	cross-sectional area of the spout

Acknowledgments

The project is supported by CNPC Major Project(2019E-24).

References

[1] Yu Qingyuan, Zhao Pengcheng, Ma Yugao. Research on high temperature heat pipe characteristics based on CFD method [J/OL]. Nuclear Power Engineering: 1-7 [2022-03-15].

[2] Deng Xiaoye, Peng Peiying, Zhu Hairong, Lu Hongliang. Research on fluid flow and heat transfer characteristics in corrugated tube based on Fluent [J]. Hebei Industrial Science and Technology, 2020, 37(3): 144-150.

[3] Zhang Peng, Wu Zhilin. Numerical simulation of liquid nitrogen phase transition heat transfer based on FLUENT [J]. Cryogenics and Superconductivity, 2014, 42(8): 26-29.

[4] Wang Wensong, Liu Shuang. Numerical simulation analysis of shell and tube heat exchanger based on FLUENT [J]. Piping Technology and Equipment, 2019(6): 30-31, 55.

[5] Wang Jun, Li Youxing, Zheng Qiuming, Yuan Chaogang, Zheng Songxian, Luo Zhao. Analysis of heat transfer characteristics of pressure test submarine pipeline based on FLUENT [J]. Petroleum Engineering Construction, 2017, 43(1): 31-33, 37.

[6] Yan Mingyu, Sun Tie, Yang Xuefeng, Meng Qingjuan. Numerical analysis of optimal heat transfer conditions for two-phase closed thermosiphon [J]. Journal of Petrochemical Universities, 2015, 28(5): 91-94.

[7] Wang Changbin, Zhao Yue, Jia Xuesong, Liu Zhaodong, Shen Yanxia. Numerical Simulation of Thermal Oil Pipeline Temperature Field [J]. Pipeline Technology and Equipment, 2011(3): 9-12.

[8] Wang Yongxing. Simulation calculation of flue gas mixing flow field [J]. Shandong Chemical Industry, 2018,

47(15): 177-179.

[9] You Jiang, ZhouYasu, Zhao Jingde, Zhang Hengqin. Numerical simulation of thermal characteristics of countercurrent closed cooling tower [J]. Building Thermal Energy Ventilation and Air Conditioning, 2010, 29(3): 13-16.

[10] Versteeg H K H K, Malalasekera W W. An introduction to computational fluid dynamics: the finite volume method / H. K. Versteeg and W. Malalasekera [J]. epfl.

[11] Tao Wenquan. Numerical Heat Transfer [M]. Xi'an: Xi'an Jiaotong University Press, 2001: 353.

[12] Liu Richao, Le Jialing, Yang Shunhua, Zheng Zhonghua, Song Wenyan, Huang Yuan. Application of KH-RT model in the process of jet atomization under the action of lateral incoming flow [J]. Propulsion Technology, 2017, 38(7): 1595-1602.

[13] Xu Feng. Design and Research of Water Ramjet Nozzle [D]. Shanghai Jiaotong University, 2007.

[14] Fang Daxian, Wang Yanhua, Wang Wei, Jin Xian, Xiao Jianqiang, Zhu Wenjin. Fluid Mechanics [M]. Nanjing Southeast University Press, 201801. 258.

[15] He Chuan. CFD Foundation and Application [M]. Chongqing University Press: Excellent Teaching Materials for General Higher Education, 201511. 186.

本论文原发表于《Proceedings of the International Field Exploration and Development Conference 2022》2022 年。

Permeability Model of Liquid Microcapsule Based on Multiple Linear Regression Method

Xu Xiuqing   Li Fagen   Zhao Xuehui   Yang Fang

(CNPC Tubular Goods Research Institute)

Abstract: The release rate of liquid core material from microcapsules is crucial for the surface properties of self-protective metal/liquid microcapsule composite plating coating. However, there is not a method to accurately predict release rate of microcapsule. In this paper, the permeability experiments of different shell membranes and core materials were carried out by weight loss method and the permeability model of liquid microcapsule was studied based on multiple linear regression method. The results show that three-variable mathematical model C including membrane porosity, viscosity of core material and membrane thickness is suitable to describe permeability and the membrane thickness is the most significant influence factor. Additionally, the accuracy of the model C was experimentally verified and the error of permeation rate is about 2.06 % between predictive and experimental values.

Keywords: Liquid microcapsule; Multiple linear regression; Permeability experiments; Permeability model; Predictive accuracy

1 Introduction

Nowadays, composite materials containing microcapsules have attracted more and more attention due to their magical self-repairing properties[1-6]. Self-healing microcapsule as a new material has the characteristics of easy dispersion in the matrix and intelligent self-healing. The repair principle is that repaired material is stored inside microcapsules and enclosed with the capsule shell membrane. The internal core material is released gradually or quickly to modify the properties of materials in service[7]. In our previous research, the functional Cu-based and Ni-based composite coatings containing liquid microcapsules were prepared by electrolytic co-deposition[8-11]. Actually the process of core material arriving coating surface is divided into two steps, as seen in Figure 1: (1) the core material of microcapsules diffuse into the matrix through the shell material; (2) the released core material from microcapsules diffuses to the coating surface. The diffusion rates of core material in shell and matrix are crucial to realize controllable release. If the release rule is mastered, the control release of microcapsule will become possible. Unfortunately, there is not a method to accurately

---

Corresponding author: Xu Xiuqing, Tel: 86-29-81887905; E-mail address: xuxiuqing@cnpc.com.cn

predict release properties of microcapsule. Herein, the first step was mainly discussed in this research.

As is known, the release of microcapsules are closely related to many factors such as the porosity of shells, viscosity of core material and shell thickness. Yuan[12] studied the chloride ion transport model in a chloride-activated self-healing concrete system. Zhu[13] synthesized the self-healing microcapsule by in-situ polymerization with dicyclopentadiene (DCPD) as capsule core and urea-formaldehyde resin as shell material. And he built the quantitative structure-property relationship (QSPR) model by SPSS statistical analysis tool software. In our previous research, hydrophobic agent, organosilicon resin and lubricant were usually used as the liquid core materials and polyvinyl alcohol (PVA), gelatin or methylcellulose (MC) as shell materials of microcapsule[14].

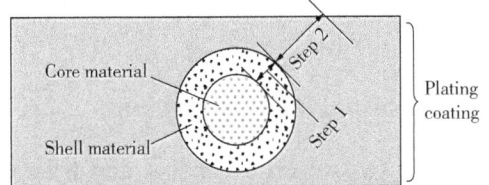

Fig. 1　The diffusion process of microcapsule in plating coating

In this paper, the shell and core materials above are used as research objects and a series of permeability experiments were carried out under ambient temperatures and moistures. We attempt to seek a statistics rule about permeability without considering the change of environmental factors. Multiple linear regression (MLR) method is an effective tool for quick predictions[15-17]. Here, it is firstly introduced to constitute a suitable permeability model based on the permeability data obtained in experiments. This paper provides a new method to analyze experimental data and is favorable to the establishment of mathematical models.

2　Materials and Experiments

2.1　Preparation of shell membranes of microcapsule

2%(mass fraction) methyl cellulose (MC), polyvinyl alcohol (PVA) and gelatin aqueous solutions were prepared respectively, which are purchased from YiLi fine chemicals Co. Ltd of Beijing. After deformed, the calculated solutions were poured on the plastic mold with a certain area and dried at room temperature for 24 hours in vacuum drying oven. Hence, the films with 5μm, 15μm and 30μm thickness were obtained.

2.2　Characterization of membranes

The cross-section photographs of shell membranes were characterized by scanning electron microscopy (SEM, HITACHI S-530, 20kV). And the cross-section fractal dimensions of membranes were used to characterize their porosity.

2.3　Permeability of shell membranes

The core materials permeability of films was measured by weighing. BH-102 hydrophobic agent and L-MH46 lubricating oil were adopted as the core materials of microcapsules and their measured viscosities were $15m^2/s$ and $32m^2/s$ respectively. The orifices (ϕ12mm) of the cylindrical tube were closed by different shell membranes, in which were filled with 3mL core materials. These tubes were placed downward in air at 25℃ and weighed every 2 hours. The residual core material around orifice was cleaned using hexane solution before weighing. Then the weight loss of the total tube was calculated and each test was repeated at least three times to ensure reproducibility. The permeability data were analyzed by multiple linear regression method with the software of Statistical Package for the Social Science (SPSS).

3 Results and discussion

3.1 Cross-section images and fractal dimensions D_f of membranes

The SEM graphs of the cross section of the shell membranes are shown in Figure 2. As observed, the cross sectional samples have different porosity which are characterized by the cross-section fractal dimensions D_f. In this study, the area dimension D_f can be determined by the box-counting method[18-19] which is based on the membrane image analysis of a sufficiently large section. SEM images of shell membranes are JEPG format with true color and then should be transferred to gray-scale formation with Adobe Photo software. Before analyzing an image, a threshold has to be determined in order to distinguish pores from the background obtaining a binary image. And then the cross-section fractal dimensions D_f are obtained by Matlab program according to the box-counting method. Simultaneously, the pore areas of films are also calculated. Table 1 displays the fractal dimensions D_f and pore areas of these three membranes.

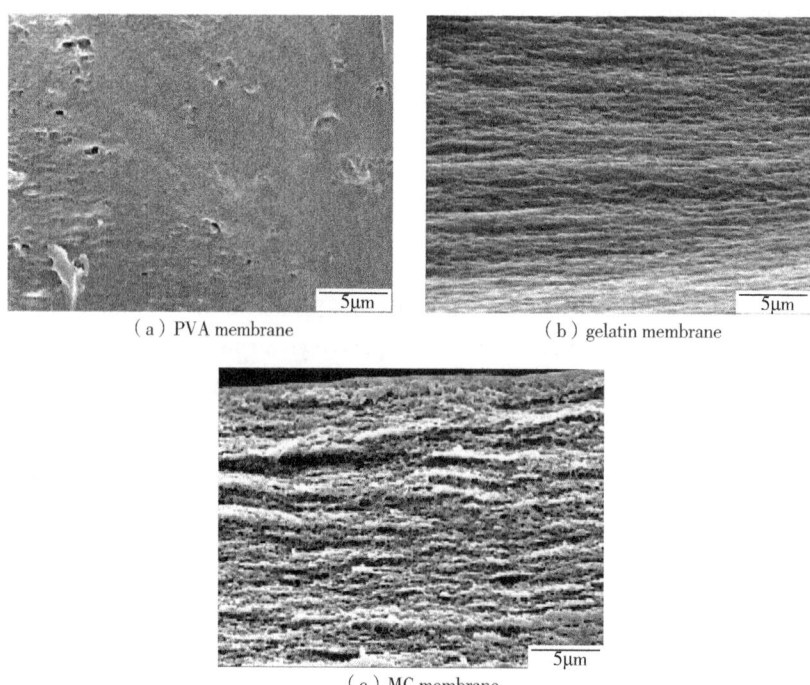

(a) PVA membrane (b) gelatin membrane

(c) MC membrane

Fig. 2 SEM images

Table 1 Calculated cross-section fractal dimension D_f and pore area of membranes

Shell materials	Pore area/%	Cross-section fractal dimension D_f
PVA film	0.0019	2.00
Gelatin film	5.3120	2.29
MC film	16.7062	2.51

3.2 Permeability analysis of shell membranes

Figure 3 depicts the permeability properties of three shell membranes with different thickness membranes under different systems. As seen, the mass loss all increase linearly with the time. The slope of fitting line represents the permeation rate of film, which is almost a fixed value. These values of permeation rates are listed in Table 2.

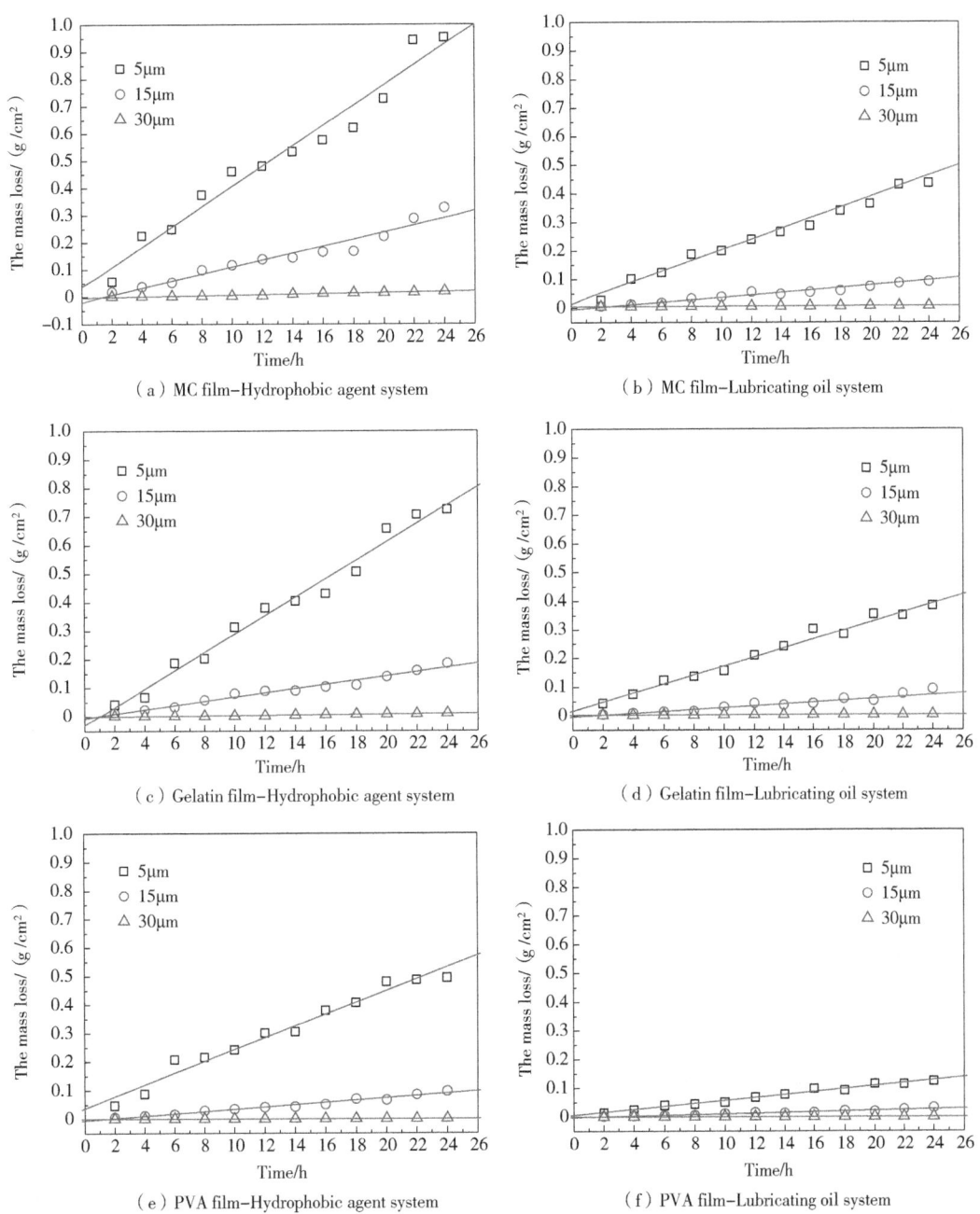

Fig. 3 Permeability curves of different thickness membranes under different systems

Table 2 Permeation rates of different thickness membranes under different conditions

Fractal dimension (D_f)	Viscosity of core/ (m²/s)	Membrane thickness/ μm	Permeation rate/ [mg/(cm²·h)]
MC (D_f = 2.51)	15	5	40
	15	15	8
	15	30	0.75
	32	5	23
	32	15	3.2
	32	30	0.14
Gelatin (D_f = 2.29)	15	5	32
	15	15	6.4
	15	30	0.36
	32	5	16
	32	15	2.4
	32	30	0.06
PVA (D_f = 2.00)	15	5	26
	15	15	2.53
	15	30	0.0053
	32	5	5
	32	15	0.2
	32	30	0.0014

3.3 Multiple linear regression theory

Multiple linear regression analysis is an effective tool to confirm the relationship of dependent variable and several independent variables. The attained expression is called regression equation or mathematical model, which can predict the value of dependent variable. MLR method requires a certain linear relationship between dependent variable and explanatory variables.

Usually, the MLR model is typically expressed as[20]

$$Y_i = \beta_0 + \sum_{i=1}^{p} \beta_i x_i \qquad (1)$$

Where Y_i represents the observed value of dependent variable; x_i and β_i denote the explanatory variables and the regression coefficients for the constant and the variables, respectively; p is the number of explanatory variables.

3.4 Model selection and assessment

We try to present the permeability mathematical model according to the 18 sets of experimental data in Table 2. The three measured explanatory variables are used in this work including membrane porosity (fractal dimension D_f), viscosity of core material (*VCM*) and membrane thickness (*MT*). The variables should be properly transformed to satisfy the requirement of the MLR method when it is necessary. After transformations, three variables are finally identified to constitute the models: D_f, 1/*VCM*, 1/[sqrt(*MT*)].

Three models are obtained after MRL and the statistic results of the models for the experimental data are listed in Table 3. The correlation coefficient R shows the serial correlations between dependent variable and explanatory variables. As observed, the significant serial correlation is obtained for the models based on the whole data set ($N = 18$). Comparing these three models, the correlation coefficient $R_3 = 0.920$ in model C shows the optimal serial correlation. Moreover, Durbin-Watson value (DW) is 2.180 which is a statistic to test the serial correlation with a first-order autoregression. It illustrates the linear independence among the explanatory variables. Consequently, D_f, $1/VCM$ and $1/[\text{sqrt}(MT)]$ in this study are linear independent.

Table 3 shows the anova analysis results of models after MLR. It can be seen that the F values are 38.184 [Sig. $(P) = 0 < 0.05$] for model A, 28.951 ($P = 0 < 0.05$) for model B and 25.879 ($P = 0 < 0.05$) for model C. All the three models show the significant linear relationship. Based on the discussion above, it can be considered that the model C is more suitable to describe the relationship of dependent variable and independent variables. It is also concluded from standardized coefficients that the influence of $1/[\text{sqrt}(MT)]$ on the dependent variable is most significant. Hence, the optimal permeability mathematical model is obtained as follows

$$Y = -40.4 + 13.5 D_f + \frac{207.2}{VCM} + \frac{584.5}{MT^2} \tag{2}$$

Table 3 The statistic results of the models for the complete data

Model	Variables	R	Unstandardized Coefficients B	Standardized Coefficients	F	Sig.	Durbin-Watson
A	Constant	0.839	0.360		38.184	0	
	$1/[\text{sqrt}(MT)]$		584.476	0.839			
B	Constant	0.891	-9.784		28.951	0	
	$1/VCM$		207.197	0.299			
	$1/[\text{sqrt}(MT)]$		584.476	0.839			
C	Constant	0.920	-40.414		25.879	0	2.180
	D_f		13.513	0.230			
	$1/VCM$		207.197	0.299			
	$1/[\text{sqrt}(MT)]$		584.476	0.839			

To validate the rationality of the model we built, three models are investigated. The actually measured values Y and the prediction values calculated by model A, B and C are shown in Figure 4. The results exist a certain differences among the prediction values of three models. Especially, the predictions made by model C is more consistent with the actually measured values. This implies that the model C predicts the permeation rate Y more accurately than model A and B.

A further model evaluation based on indices is implemented, but only part of the results are presented in Table 4 for brevity. The second block of Table 4 represents the goodness of fit of the three models which are theaverage of the observed and predicted values respectively. S_p denotes the standard errors of the predicted values. MAE (mean absolute error) and $RMSE$ (root mean squared error) are respectively defined as[21]

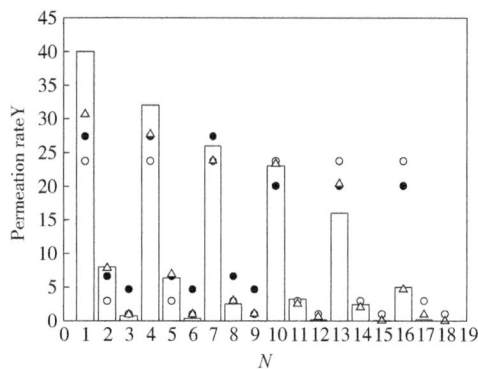

Fig. 4 Comparison of the predictive capacity of model A, model B and model C. Bar: measured Y on the whole $N = 18$; open circle: predictions made by model A; filled circle: predictions made by model B; open triangle: predictions made by model C

$$MAE = \sum_{i=1}^{N} |P_i - O_i| / N \qquad (3)$$

$$RMSE = \left[\sum_{i=1}^{N} (P_i - O_i)^2 / N \right]^{1/2} \qquad (4)$$

Where P_i and $O_i (i=1, 2\cdots, N)$ are predictions and observations, respectively. The results indicate that model C fits better than model A and B with a higher R^2 and lower standard error S_P, which might be attributed to the slight change of the actual permeation rate data. On the other hand, model C has a more satisfactory predictive capacity with the lower levels of MAE and $RMSE$. These results are in accordance with that of in Figure 4. Therefore, it is reasonable to deduce that model C [Eq. (2)] is more suitable to predict the permeation rate of shell membranes of microcapsules under ambient temperature and moisture.

Table 4 Comparison of the predictive capacity of model A, model B and model C ($N=18$)

Model	R^2	$\overline{O}$	$\overline{P}$	S_P	MAE	RMSE
A	0.705	9.22	9.25	7.0674	3.96	6.66
B	0.794	9.22	9.24	6.0929	4.19	5.56
C	0.847	9.22	9.21	5.4344	1.45	2.72

In the course of linear regression, we hypothesize the residual follows normal distribution. Figure 5 depicts the normal P-P plot of regression standardized residual. As seen, all the scattered points show linear relationship which illustrates dependent variable indeed coming from normal population.

To validate the good predictive capacity of the model C, a new verification test was carried out at ambient temperatures and moistures. The MC membrane with 21 μm thickness and organosilicon resin is adopted as the experimental membrane and permeability media, respectively. The measured viscosity of organosilicon resin under room temperature is about $36m^2/s$. The true permeability curve and fitted line of MC membrane-organosilicon resin system are shown in Figure 6. As observed, the permeation quantity increases linearly with the time. The fitted line depicts the high correlation coefficient of 0.997 and its slope about 0.559 is calculated.

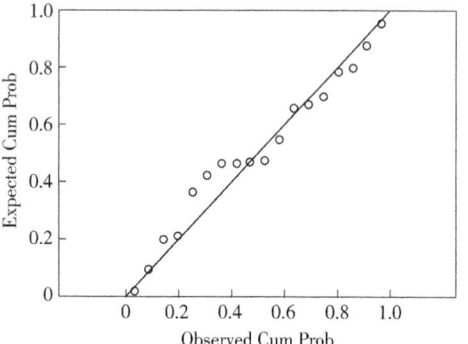

Fig. 5 The normal P-P plot of regression standardized residual

Meanwhile, the predictive permeation rate of 0.566 is also calculated by Eq. (2). Comparing the true and predictive values, it can be found that the error is about 2.06%. This result illuminates the well predictive capability and the reliability of model C.

Further study will focus on the permeability process of released core material through the metal coatings. It would be also interesting to build another complex mathematical model based on the coating thickness and structure data. It will be beneficial to the implementation of controlled release.

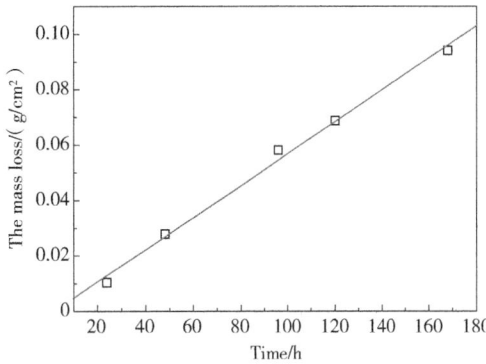

Fig. 6 Permeability curve of MC membranes-organosilicon resin system ($MT=21\mu m$)

4 Conclusion

For studying the permeability rule of liquid microcapsules within composite plating coatings, the permeability experimental data of different core materials and shell membranes were obtained by weight loss method. Permeability model of liquid microcapsule are established and verified. The results could be summarized as follows:

(1) Cross-section fractal dimension D_f of PVA, Gelatin and MC membrane are 2.00, 2.29 and 2.51 respectively according to the box-counting method.

(2) Multiple linear regression is firstly used to constitute permeability model of liquid microcapsules used for electrolytic co-deposition.

(3) Three-variable mathematical model C including membrane porosity, viscosity of core material and membrane thickness is suitable to describe permeability and the membrane thickness is the most significant influence factor.

(4) The accuracy of the model C was experimentally verified and the error of permeation rate is about 2.06% between predictive and experimental values.

References

[1] Wang, Y. L.; Zhang, Z. Z.; Chu, F. J.; Liu, M.; He, Y. H.; Li, P. L.; Yuan, J. Y.; Yang, M. M. Designing polydopamine-capped [BMIm]PF_6/NaL microcapsule optimize the wear-resistance of polymer composite liner. Tribology International 2023, 179, 108104.

[2] Nisar, M.; Bernard, F. L.; Duarte, E.; Chaban, V. V.; Einloft, S.. New polysulfone microcapsules containing metal oxides and ([BMIM][NTf_2]) ionic liquid for CO_2 capture. Journal of Environmental Chemical Engineering 2021, 9.

[3] Choi, Y. W.; Oh, S. R.; Kim, C. G.; Lim, H. S. The experimental study on preparation characteristics of self-healing microcapsules for mixing cement composites utilizing liquid inorganic materials. J. Korean Recycled Construction Resources Institute 2018, 6, 236.

[4] Shi, T.; S. Livi, S.; Duchet, J.; Gerard, J. F. Ionic liquids-containing silica microcapsules: a potential tunable platform for shaping-up epoxy-based composite materials. Nanomaterials 2020, 10, 881.

[5] Wu, Y.; Han, T. L.; Huang, X. F.; Lin, X. R.; Hu, Y. F.; Chen, Z. H.; Liu, J. Y. A Ga-Sn liquid alloy-encapsulated self-healing microcapsule as high-performance Li-ion battery anode. J. Electroanalytical Chem. 2022, 922.

[6] Liu, Z.; Liu, D. Application of microcapsule technology in urea-formaldehyde resin adhesive for wood-based panels manufacturing. China Wood-Based Panels 2019, 3, 30.

[7] Fang, G. H.; Chen, J. M.; Wang, Y. S.; Zhang, Y. Y.; Xing, F.; Dong, B. Q. Research progress of microencapsulated self-healing concrete. J. Chinese Ceramic Soc. 2023, 51, 1.

[8] Xu, X. Q.; Zhu, L. Q.; Li, W. P.; Liu, H. C. A variable hydrophobic surface improves corrosion resistance of electroplating copper coating. Appl. Surf. Sci. 2011, 257, 5524.

[9] Xu, X. Q.; Guo, Y. H.; Li, W. P.; Zhu, L. Q. Electrochemical behavior of different shelled microcapsule composite copper coatings. Inter. J. Minerals, Metallurgy and Materials 2011, 18, 377.

[10] Xu, X. Q.; Liu, H. C.; Li, W. P.; Zhu, L. Q. A novel corrosion self-protective copper/liquid microcapsule composite coating. Mater. Lett. 2011, 65, 698.

[11] Xu, X. Q.; Zhu, L. Q.; Li, W. P.; Liu, H. C. Microstructure and deposition mechanism of electrodeposited Cu/liquid microcapsule composite. Transactions of nonferrous metals society of China 2011, 21, 2210.

[12] Yuan, Y. B. Chloride ion transport properties in self-healing concrete system. Shenzhen University 2019.

[13] Zhu, M. H. Preparation and performance study on composite materials for microcapsule self-healing. Harbin Institute of Technology 2012.

[14] Zhu, L. Q. Functional coatings by polymer microencapsulation. WILEY-VCH, Germany, 2006, Ch. 9.

[15] Jiang, Z. Q.; Du, Y. P.; Cheng, F. P.; Zhang, F. Y.; Yang, W. Y.; Xiong, Y. R. A simple multiple linear regression model in near infrared spectroscopy for soluble solids content of pomegranate arils based on stability competitive adaptive re-weighted sampling. J. Near Infrared Spectroscopy 2021, 3.

[16] Piramanayagam; Sethu, R. K.; Mayandi, N.; RajiniAbdul, K. S.; Abdul, M.; Rajesh, K.; Sikiru, O. S.; Suchart, M.; FaruqAl-Lohedan; Hamad, A. Experimental investigation and statistical analysis of additively manufactured onyx-carbon fiber reinforced composites. J. Appl. Polymer Sci. 2021, 18.

[17] Xu, Z. J. Properties of phase change/heat reflection microcapsules and temperature control coating on concrete surface. Harbin Institute of Technology 2020.

[18] Xiong, W. T.; Fan, J. J.; Wang, Y. Y.; Xiao, S. H. Progress of fractal dimension measurement method of material fracture. Failure Analysis and Prevention 2019, 14, 66.

[19] Cao, J. H.; Hou, Z. B.; Guo, D. W.; Guo, Z. G.; Tang, P. Morphology characteristics of solidification structure in high-carbon steel billet based on fractal theory. J. Mater. Sci. 2019, 54, 12851.

[20] Draper, N. R.; Smith, H. Applied regression analysis, second ed. Wiley, New York, 1980.

[21] Rice, J. A. Mathematical statistics and data analysis (Third edition). China Machine Press: Beijing, 2011.

本论文原发表于《Coatings》2023 年第 13 期。

Experimental Study on Erosion Behavior of Fracturing Pipelines Involving Fluctuating Stress

Yang Siqi[1,2]　Fan Jianchun[2]　Zhao Sheng[2]　Dai Siwei[2]
Han Lihong[1]　Wang Jianjun[1]　Yang Shangyu[1]　Zhang Laibin[2]　Li Jiao[3]

(1. State Key Laboratory for Performance and Structure Safety of Petroleum Tubular Goods and Equipment Materials, CNPC Tubular Goods Research Institute;
2. China University of Petroleum-Beijing, Key Laboratory of Oil and Gas Safety and Emergency Technology, Ministry of Emergency Management China;
3. Beijing Construction Engineering Environmental Remediation Co, Ltd.)

Abstract: Erosive wear due to particle impact serves as the main cause of pipeline failure. Particularly, during hydraulic fracturing, where fracturing pipelines inevitably suffer from erosion damage under high fluctuating stress. In this study, a novel erosion test rig is developed in order to reproduce erosion scenes under complicated fracturing conditions. Then the erosion experiments involving fluctuating stress are carried out for the first time, and the erosion characteristics under varying conditions are obtained. Furthermore, the erosion mechanism with consideration of applied stress is subsequently revealed according to micro-structural characterization as well as the rheological spring-slider model. The complex erosion behavior of fracturing pipelines behavior under severe conditions is well unraveled at the laboratory scale.

Keywords: Erosive wear; Fluctuating stress; Micro - structure; Stress - erosion mechanism

1 Introduction

Erosive wear is a tribological phenomenon that depicts the material removal caused by particle impact on solid surface, which has garnered the interest of researchers for many years[1-3]. In the process of fluid transportation, the pipe walls are always impacted by particles entrained in flowing fluid. As a result, the corresponding erosive wear may be detrimental to pipe wall structural integrity[4]. In the process of shale gas development, hydraulic fracturing is known to be a widely used treatment for well stimulation. The basic principle of hydraulic fracturing is to inject high-pressure fluid containing proppant particles from the ground into the wellbore, thus forming artificial micro-fractures in deep formation[5]. During the operation, the pulsating high-pressure of the

---

Corresponding authors: Fan Jianchun, fjc19090@126.com; Han Lihong, hanlihong@copc.com.cn.

fracturing fluid output by the upstream multi-cylinder plunger pump makes the fracturing pipeline bear high pulsating stress. In light of the interaction of erosive wear caused by high-speed impacting particles as well as extremely severe fluctuating stress, vulnerable components within the pipeline system (e.g., elbows, tees and valves) may be inevitably damaged by the corresponding erosion. As a result, associated cracks may thus be induced, eventually leading to the fracture of pipe fittings under continued dynamic action[6-9]. The morphology of pipe failure caused by erosive wear can be shown in Fig. 1. In such event, the flowing fluid in the pipe fittings can leak and seriously threaten the safety of personnel and the integrity of equipment. In addition, leakage of fracturing fluid can also give rise to environmental pollution. Unfortunately, a dearth of research exists pertaining to erosion behavior of materials under actual fracturing conditions; hence, it is currently impossible to gain a deeper understanding of the erosion wear mechanism of fracturing pipelines and carry out early damage prevention. Therefore, fracturing pipelines are one of the weakest links in the entire operation process, and the catastrophic failure of fracturing pipelines serves as a major challenge for high-pressure hydraulic fracturing.

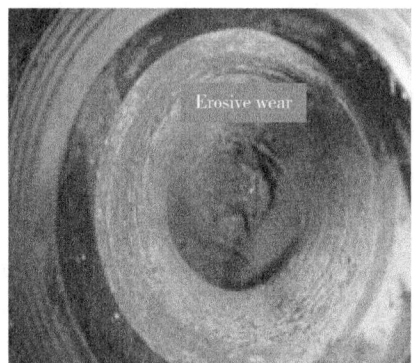 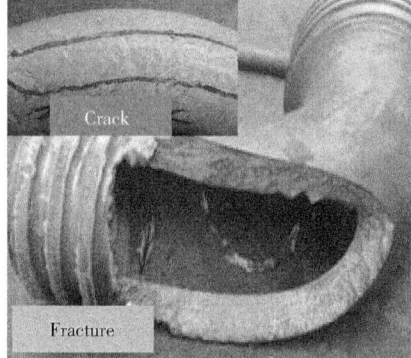

Fig. 1 Morphology of pipe failure caused by erosive wear

Severity of erosive wear is influenced by numerous factors, of which the relationships between different factors are considered to be quite complicated. As yet it is not fully understood all the influencing factors governing erosive wear. It has been widely accepted that particle impact velocity, impact angle, particle characteristics and target material characteristics are the main parameters that affect erosion severity. For the ductile metals, the erosion rates increase exponentially with the particle impact velocity. The velocity index usually ranges from 1.6 to 2.8, due to the different experimental conditions[2, 10-13]. With the increase of particle impact angle, the erosive wear of ductile metal increases first and then decreases. The most typical impact angle causing the maximum erosion wear of plastic metal is generally 20°~40°[14-16]. Islam et al.[17] conducted a series of gas-solid erosion tests under different flow velocities and impact angles and concluded that plastic deformation, delamination, micro-ploughing and micro-cutting are the mechanisms of erosive wear. In terms of particle characteristics, the larger size, higher concentration and sharper angularity of particles may result in more severe erosive wear[18-22]. In regard to target material characteristics, the erosion resistance of steels is largely dependent on the mechanical properties and chemical composition of materials. Furthermore, some studies have found that the pearlite and bainite structures are more effective in resisting cutting, ploughing and deformation compared to that of ferrite[23-24].

Although numerous studies have been committed to investigating the influential mechanism of various parameters on erosive wear, almost all erosion tests in existent literature were carried out at approximately atmospheric pressure. The erosion characteristics obtained in this manner fail to consider the effect of applied stress. However, in real-world engineering situations, there are very few equipment is not subject to applied stress during the service period, especially in the oil and gas field, where almost all pipelines must be subjected to mechanical stress caused by internal pressure or external load[25]. Surprisingly, only a few related tests involving high stress have been conducted. Wang et al.[26] performed wear experiments of various alloy materials under simulated hydrostatic pressure in deep-sea environment, in which the wear rates of 316 steel, Inconel 625 and Hastelloy C-276 were demonstrated to increase significantly with increasing hydrostatic pressure. Additionally, some researchers have performed the experimental studies of solid particle erosion in consideration of static tensile stress, which found that tensile stress may play a crucial role in aggravating erosive wear rate[27-30]. While erosive wear experiments involving static tensile stress have been carried out, investigations that consider pulsating stress have yet to be performed. Note that in actual high-pressure operations, such as hydraulic fracturing, the failure of pipe fittings due to erosion damage is usually characterized by sudden bursts rather than piercing. This may be due to the fact that pipe fittings are simultaneously subjected to the coupling effect of erosive wear and tensile fatigue under high pulsating stress, which is entirely different under static stress.

Therefore, it is a new challenge to study the erosion behavior of target materials while considering applied fluctuating stress. In this study, a new type of erosive weartest apparatus was firstly designed, which can exert the controllable fluctuating load onto the testing specimens during the process of erosion to simulate the actual fluctuating stress of fracturing pipelines. Erosive wear experiments of fracturing pipelines under high fluctuating stress were carried out. In addition, the erosion rates of specimens at different impact angles and under different impact velocities were investigated, and the effect of pulsating frequency on erosion performance was studied. By using scanning electron microscope (SEM), the micro-structures of the eroded surfaces and its erosion characteristics were then examined. Furthermore, the nature of the complicated mechanisms in relation to the stress-erosion coupling effect were revealed according to micro-structural characterization as well as the rheological mechanical model.

2 Experimental methodology

2.1 Erosive wear test apparatus

In order to truly simulate the erosive wear phenomenon of fracturing pipelines under operation conditions, a novel slurry jet erosion test apparatus including the function of applying the pulsating load was proposed that was developed from the previous apparatus[7, 28-29], as shown in Fig. 2. The proposed apparatus was an automated, circulated and closed experimental system. The testing specimens placed within the erosion chamber were impacted by slurry flow during the erosion process. The two sides of the specimens were then connected with a servo cylinder and tension sensor, respectively [see Fig. 2(c)]. The hydraulic servo system was able to continuously supply pulsating hydraulic oil to the servo cylinder. The specimens were then subjected to adjustable sine-wave tensile stress, which was monitored by the tension sensor and fed back to the control & data acquisition system so as to form a closed loop.

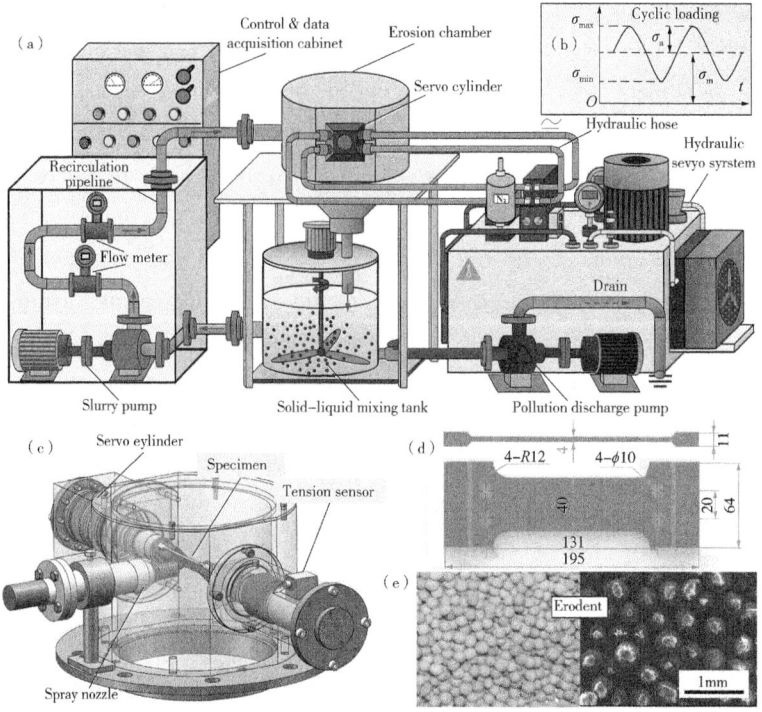

Fig. 2 Experimental set-up: (a) schematic diagram of erosion test apparatus; (b) curve of applied sinusoidal stress; (c) detail view of erosion chamber; (d) shape and dimensions (mm) of testing specimen; (e) morphology of ceramsite sand

2.2 Material and specimen preparation

The designed flat tensile specimens were used for erosion tests [Fig. 2(d)]. The erosion target was 35CrMo steel after quenching and tempering, which is known to be a typical material commonly used in oil and gas industry, such as in fracturing pipelines, drill pipes and risers, in light of its high performance. The mechanical properties and nominal chemical composition of this steel are shown in Fig. 3.

(a) Mechanical properties of 35CrMo steel

Yield strength/ MPa	Tensile strength/ MPa	Elastic modulus/ GPa	Brinell hardness	Reduction of area/ %
835	985	207	229	45

(b) Nominal chemical composition of 35CrMo steel

C	Si	Mn	Cr	Mo	P	S	Ni	Cu
0.32~0.34	0.17~0.37	0.40~0.70	0.80~1.10	0.15~0.25	≤0.035	≤0.035	≤0.030	≤0.030

Fig. 3 Mechanical properties and nominal chemical composition of 35CrMo steel

The ceramsite sand is a kind of commonly used fracturing proppant, which was served as erodent in this study [Fig. 2(e)]. The mechanical properties and nominal chemical composition of ceramsite sand used are shown in Fig. 4.

(a) Mechanical properties of ceramsite sand

Density/(kg/m^3)	Sphericity	Roundness	Mohs hardness	Breakage rate/%
1800	0.9	0.9	8	≤8(86MPa)

(b) Nominal chemical composition of ceramsite sand

Al_2O_3	SiO_2	Fe_2O_3	TiO_2	MnO	K_2O	CaO	Others
65.82%	14.94%	9.43%	3.88%	2.94%	0.85%	0.77%	1.37%

Fig. 4 Mechanical properties and nominal chemical composition of ceramsite sand

2.3 Test setup

Prior to conducting the test, the ceramsite sand was sieved into a nominal size of (400±100) μm, then the sand and water were mixed at a particle concentrations of 10% (mass fraction). The velocities were measured from the flow meters (average value), ranging from 7.5 ~ 20m/s. The slurry flow impacted the testing specimens through a nozzle, the impact angles can be adjusted from 0° to 90°. The inner diameter of the nozzle was 8 mm and the distance from the nozzle outlet to specimens was 20mm. The average value of the applied fluctuating tensile stress varied from 0MPa to 500MPa, and the loading frequency used for the tests were 0Hz, 5Hz and 10Hz. The experiment was repeated three times under the same conditions and the experimental period was 60min. The testing specimens were treated by ultrasonic cleaning. The weight loss of specimens was obtained by a microelectronic balance with an accuracy of 0.0001g (MS104TS/02, Mettler Toledo Inc.). The erosion rates were characterized by the ratio of the mass loss of specimens to the impacting ceramsite sand mass. For post-examination, the surface morphology of wear specimens was obtained using a laser displacement sensor (ZX-LD40, Omron Inc.), and micro-graphs of the eroded surface at different conditions were analyzed using SEM (FEI Quanta 200F, FEI Inc.).

3 Results and discussion

3.1 Effect of impact angle during stress-erosion

Fig.5(a) describes the erosion rates of specimens under different impact angles and stress states. The flow velocity was set to 17.5m/s. The loading frequency of the applied fluctuating stress was set to 5Hz. It can be indicated that the erosion rate increases with impact angle from 15° to 30°, reaching a maximum value at impact angle of 30°. The erosion rate was then noted to decrease gradually until the impact angle was 90°. It is noteworthy that the erosion rate increased with the average stress. Although the aggravation of erosion caused by the rise in applied stress was evident at different impact angles, the trend in variation of erosive wear with impact angles remained unchanged under the same stress state. In addition, the surface morphology of erosion scars at different impact angles under an average stress of 500MPa is depicted in Fig.5(b). With an increase in impact angle, the route on the material surface in which the particles slid became narrower, causing the longitudinal length of the wear area to be shorter.

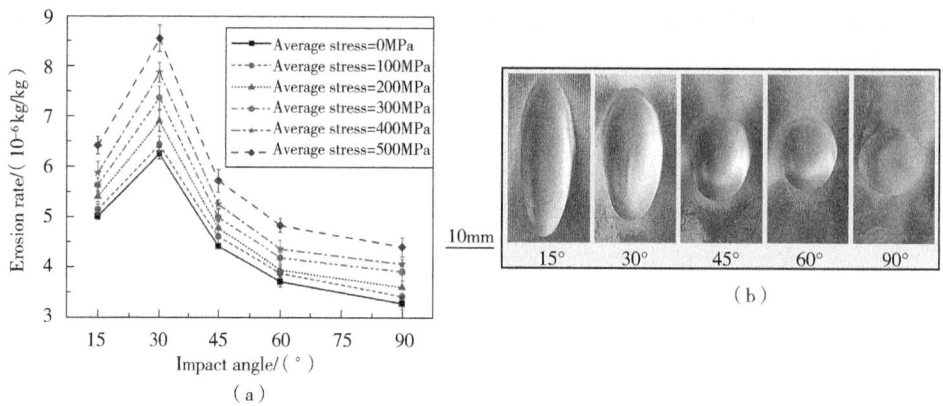

Fig. 5 Erosion rate and surface morphology at different impact angles
(a) erosion rate, (b) surface morphology of erosion scars

3.2 Effect of flow velocity during stress-erosion

Fig. 6 and Fig. 7 illustrate the variation of erosion rate and surface morphology with different flow velocities and applied stress, respectively. The impact angle was set to be 30°. The loading frequency of the applied fluctuating stress was set to be 5Hz. The erosion severity increased rapidly with a rise in flow velocity, which was due to the fact that higher velocity particles impacting the material with a higher kinetic energy, resulting in more severe erosive wear. Furthermore, the plots indicated that increasing applied stress can effectively aggravate the erosive wear of the testing specimens at the same flow velocity. Therefore, the coupling effect of stress state and flow velocity evidently served as a crucial factor in regard to the erosion failure prevention of fracturing pipelines.

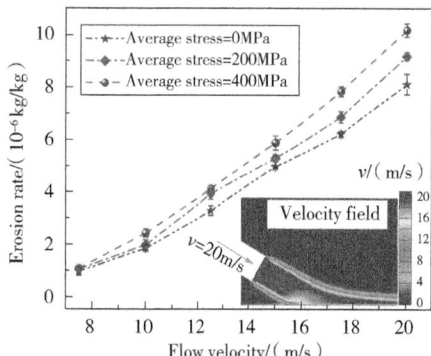

Fig. 6 Erosion rate at different flow velocities

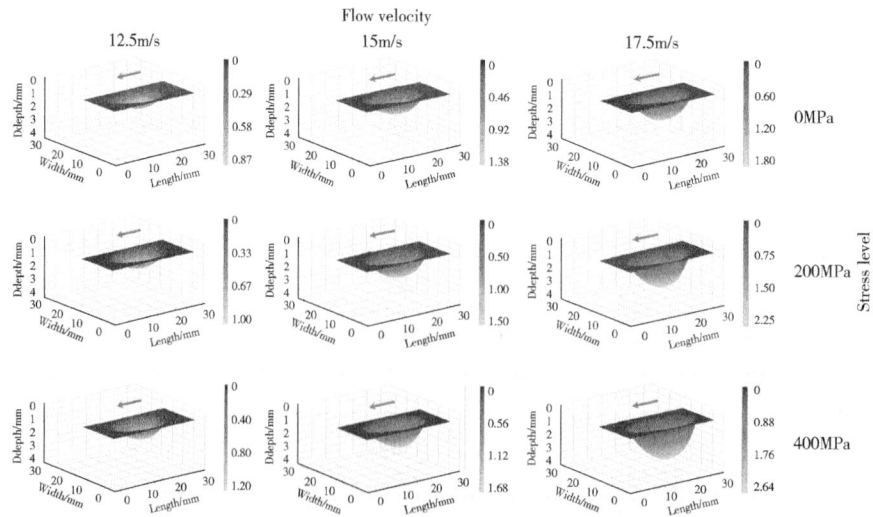

Fig. 7 Surface morphology of erosion scars; brown arrow indicates the flow direction

3.3 Effect of loading frequency during stress-erosion

In order to investigate the effect of loading frequency on erosive wear, erosion experiments under different loading frequencies and average stress were carried out. The changes in erosion rate and surface morphology of the target materials are illustrated in Fig. 8. The impact angle was set to 30°. The flow velocity was set to 17.5m/s. According to Fig. 8(a), with a rise in applied stress, the erosion rate increased and resulted in a more material removal. Moreover, when the average value of fluctuating stress reached a relatively high degree (over 400MPa), erosion slightly increased with a rise in the loading frequency.

Interestingly, Fig. 8(b) shows that the fatigue cracks appeared on the centerline of the erosion scars under high pulsating stress, which was not noted under static tensile stress (loading frequency = 0Hz). Notably, it is the first time that the fatigue cracks were observed in the process of erosive wear. The findings of the phenomenon of erosive wear – tension fatigue interaction may be very helpful. The fracturing pipelines suffer from the coupling action of erosion and fatigue under the impact of abrasive particles and high pulsating stress[5, 31-32]. Beneath these conditions, the pipe fittings may be prone to unexpected burst, which is one of the most dangerous accident in the field of high-pressure operation.

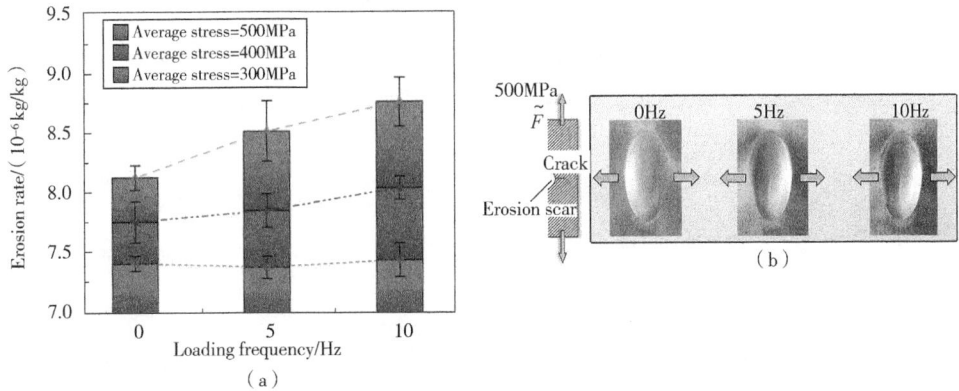

Fig. 8 Erosion rate and surface morphology at different loading frequencies

(a) erosion rate, (b) surface morphology of erosion scars

4 Stress-erosion mechanism

4.1 Erosive wear mechanism under stress state

A detailed study was performed on the eroded specimens to understand the erosive wear mechanism under stress state. The surface micro-morphology of eroded scars (center area) of the erosion specimens at different impact angles and under different stress states was obtained using SEM, as shown in Fig. 9 and Fig. 10. At low impact angles (15°~30°), the tangential component of the kinetic energy of the impacting particles was consumed by micro-cutting on the ploughed surface, and its value was higher than the normal component which was consumed by penetrating the surface[33-34]. The narrow furrows were formed by the plastic flow of the metal of the eroded surface along the flow direction. The lips can be observed on both sides and front end of the furrows, which were essentially the accumulation of metal caused by particle extrusion. These fragile lips can be prone to fracture under subsequent particles impact. It can be indicated that from these plots that the furrows generated at 30° impact angle [Fig. 9(d) ~ (f)] were deeper and shorter than those generated at 15°[Fig. 9(a) ~ (c)], which may be related to the differences in the distribution of the components of the particle's initial kinetic energy under different impact angles. In addition, it can be observed that the erosion characteristics of the eroded area were different under the condition of applying stress and not applying stress, while the width of the furrows evidently increased with a rise in stress levels. To further quantitatively analyze the micro-structure of the eroded region affected by applied stress, the distribution and statistical results of the width of furrows in the eroded region were obtained, as shown in Fig. 11. The results indicated that the size of furrows become much wider along the tension direction under applied stress.

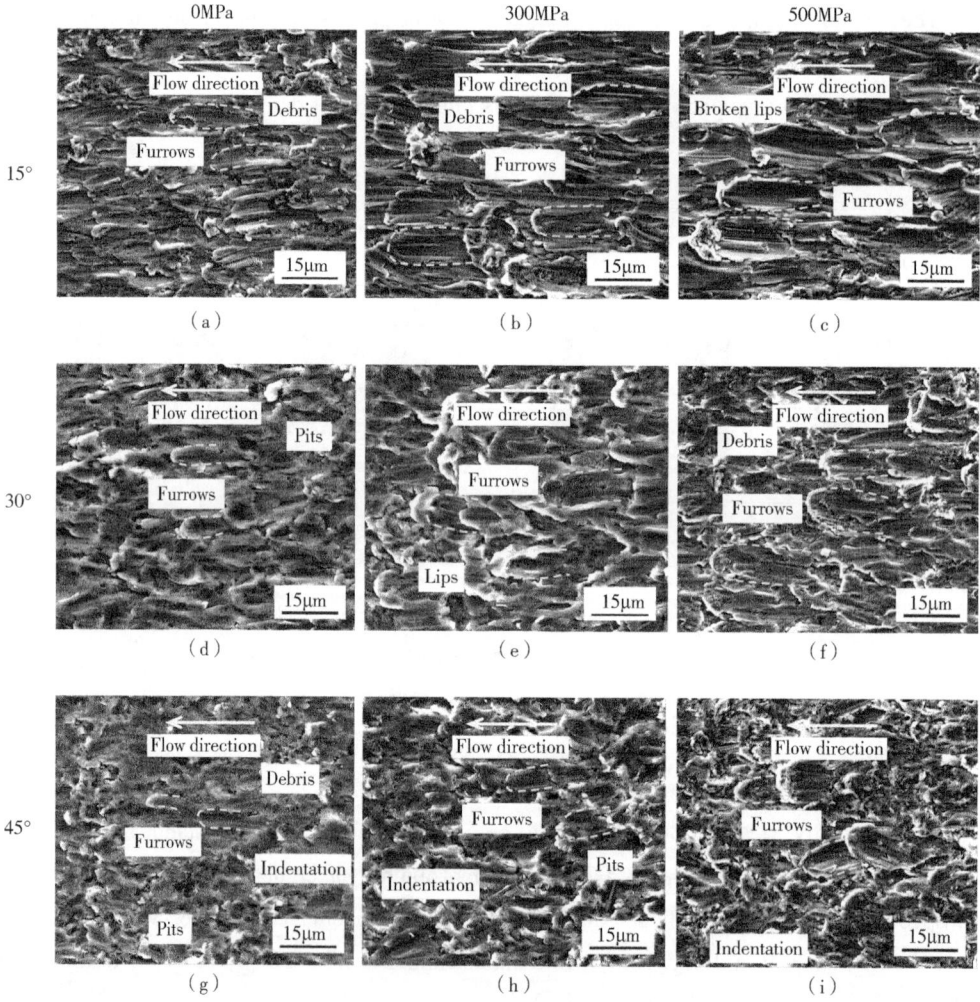

Fig. 9 SEM micro-graphs of the erosion surface of specimens at low and intermediate angles under different applied stress

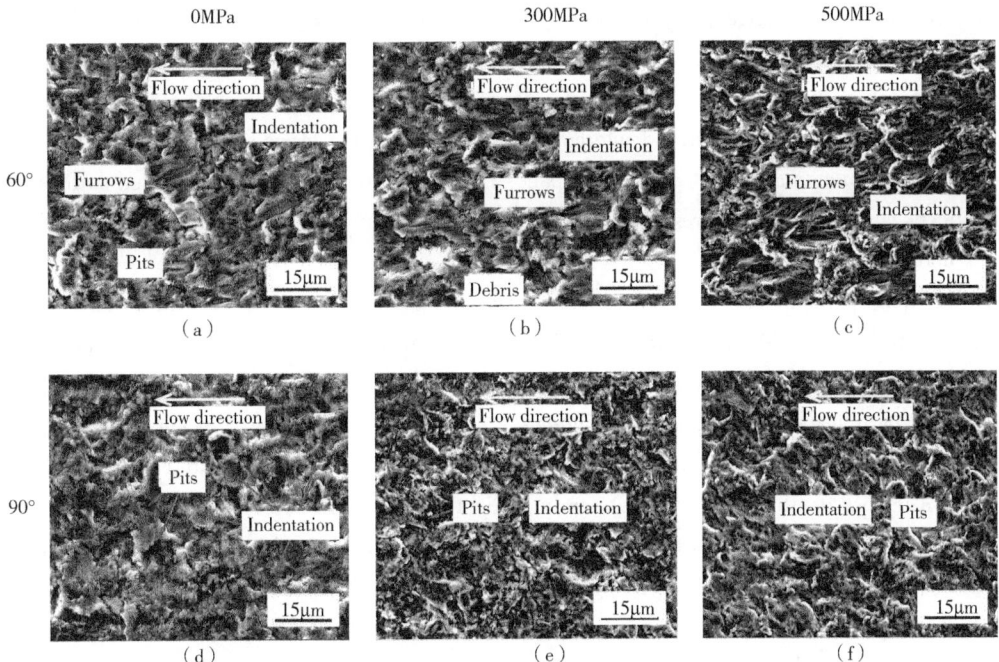

Fig. 10　SEM micro-graphs of the erosion surface of specimens at high angle under different applied stress

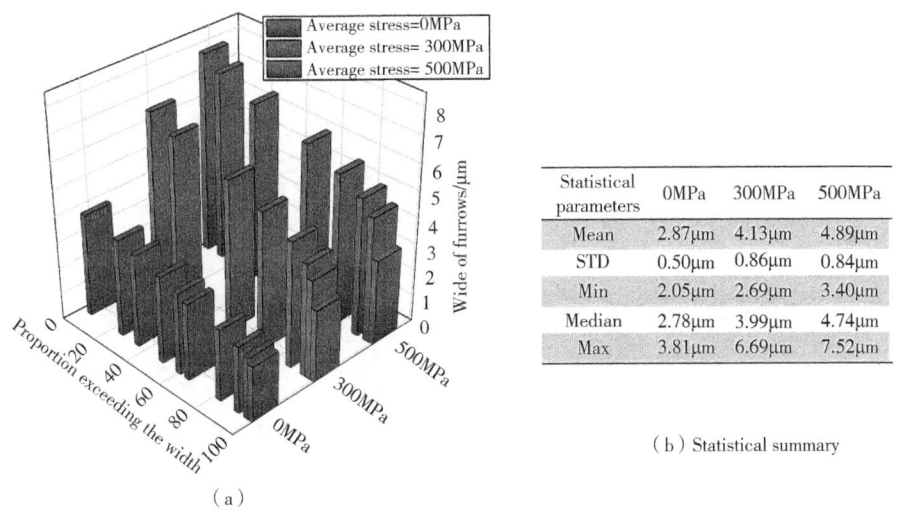

Statistical parameters	0MPa	300MPa	500MPa
Mean	2.87μm	4.13μm	4.89μm
STD	0.50μm	0.86μm	0.84μm
Min	2.05μm	2.69μm	3.40μm
Median	2.78μm	3.99μm	4.74μm
Max	3.81μm	6.69μm	7.52μm

(b) Statistical summary

(a)

Fig. 11　Distribution and statistical results of the width of furrows in the eroded region (impact angle = 30°) under different applied stress

At intermediate impact angle (45°), the furrows produced by the oblique impact of abrasive particles became shorter than that at lower impact angle. In addition, the number of furrows in the eroded region was shown to slightly decrease, accompanied by a few indentation craters [Fig. 9(g) ~ (i)]. Moreover, the size of furrows was observed to be affected by applied stress, which was consistent with that under low impact angle. At higher impact angle (60° ~ 90°), the sliding distance of the particles on the surface of specimens was shown to decrease gradually, and the number of furrows in the eroded region obviously decreased [Fig. 10(a) ~ (c)]. Stronger indentations was noted at a higher impact angles, especially in regard to the indention craters produced by frontal impact of particles

· 561 ·

covering the full erosion range [Fig. 10(d) ~ (f)]. In such a mode of erosion, due to the repeated extrusion of impacting particles, ridge-shaped metal accumulation can be formed around the indention craters. Under the continuous particles impact, micro-cracks tend to appear on the accumulated metal, which may be gradually developed and interconnect, resulting in the formation of flake-like debris, and may then fall off under subsequent particles impact. The SEM micro-graphs of the eroded surface showed no significant changes between stressed and unstressed at head-on impact, which was different from that at other impact angles.

In addition to impact angle, flow velocity also undoubtedly serves as a crucial factor affecting erosion severity[10, 35-36]. Fig. 12 shows the SEM micro-graphs of eroded scars (center area) of the erosion specimens at different impact velocities and stress (at impact angle of 30°). At higher flow velocity, the particles before impact have higher kinetic energy. Therefore, when the particles with higher flow velocity ploughed over the specimen surface, the abrasive particles slid a longer distance and penetrated a deeper depth. Thus, it was clearly evident that particles with higher flow velocity caused longer and deeper furrows. Furthermore, the width of furrows of the eroded surface significantly increased with applied stress. Therefore, by investigating the erosion mechanism of eroded surface morphology under the effects of flow velocity and applied stress, it was extrapolated that pipeline is more prone to serious erosive wear under high flow velocity and high-pressure operating conditions.

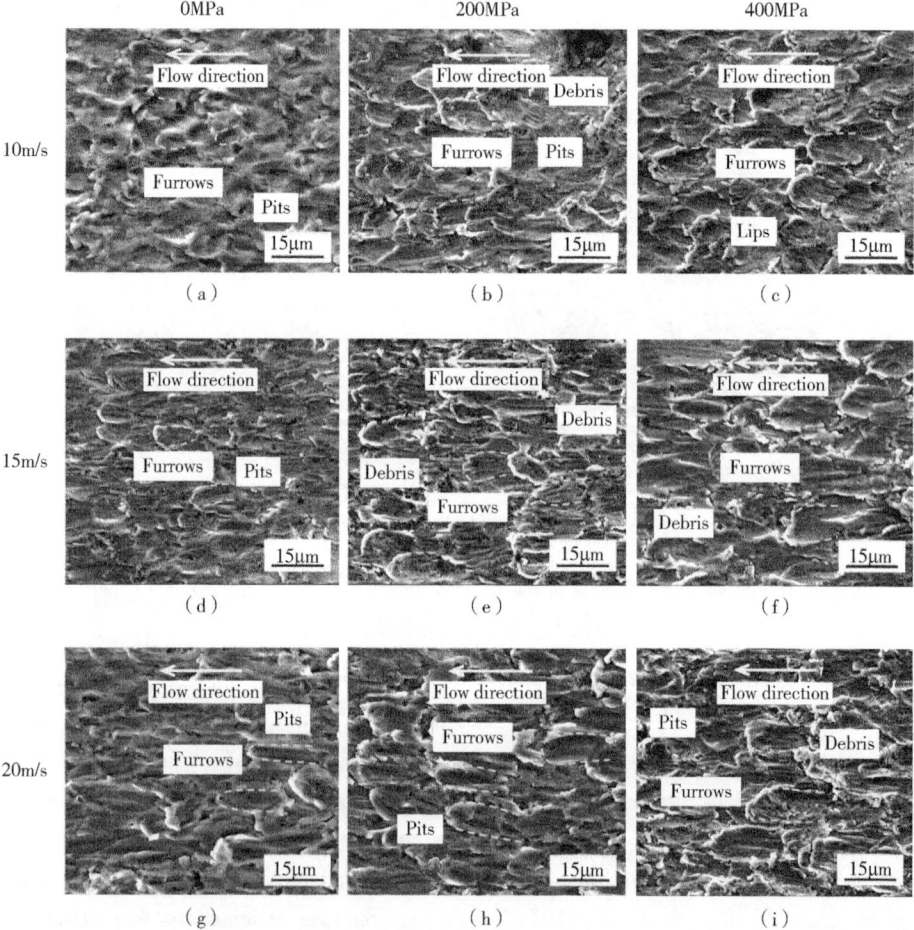

Fig. 12 SEM micro-graphs of the erosion surface of specimens under different impact velocities and applied stress

4.2 Fatigue crack characteristics

The metal material removal caused by particle impact leads to local stress concentration of the specimens. Therefore, fatigue cracks caused by externally applying fluctuating tensile stress may form at the center of the erosion scar in which local stress concentration occurs, as discussed in section 3.3. Fig. 13 demonstrates the surface morphology of the eroded region where the fatigue crack occurs under the coupling effect of erosive wear and tension fatigue. The fatigue crack initiated from the deepest part of the erosion pit and then propagated in the direction opposite to the flow direction [Fig. 13(a)]. A large area of cluster-like structures caused by plastic deformation was noted near the initiation of the crack [Fig. 13(b) ~ (e)], in which its micro-appearance characteristics differed from the furrows [Fig. 13(f)]. These fragile micro-structures may be prone to breaking and detaching from the surface of specimens under repeated particle impact. Meanwhile, SEM micro-morphology demonstrated the presence of fatigue striations on both sides of the fatigue crack [Fig. 13(g) ~ (i)], which may be due to the combined action of the dislocation motion from tension fatigue and plastic flow caused by erosive wear.

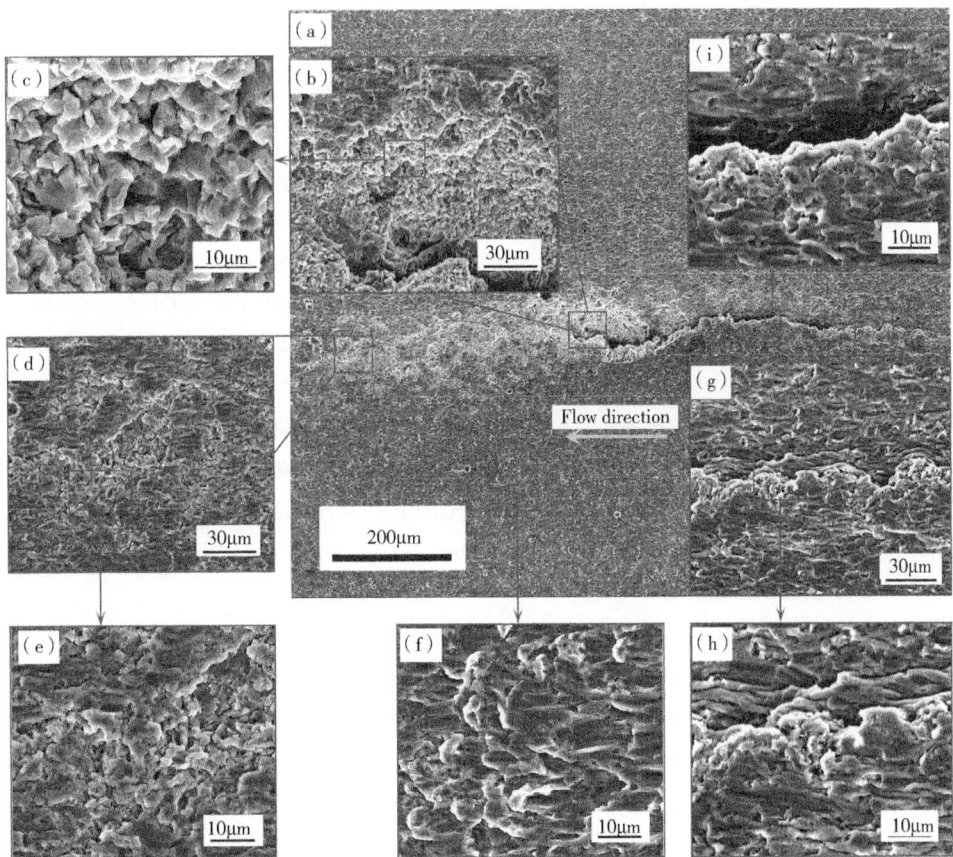

Fig. 13 SEM micro-graphs of the fatigue crack in the eroded region
(impact angle = 30°, flow velocity = 17.5m/s, average stress = 500MPa, loading frequency = 5Hz)

4.3 Analysis of fracture morphology

In contrast to the erosion tests that were performed at atmospheric pressure, the ultimate failure mode of the specimen subjected to stress-erosion was evident by the specimen fracture along the centerline of the erosion pit, as illustrated in Fig. 14. Likewise, the failure of fracturing pipelines under the coupling effect of erosive wear and high fluctuating stress is generally presented as an accidental pipe burst. Note that the erosion tests involving the external fluctuating stress were shown to reproduce the actual failure behavior of field high-pressure fracturing pipelines for the first time at the laboratory scale.

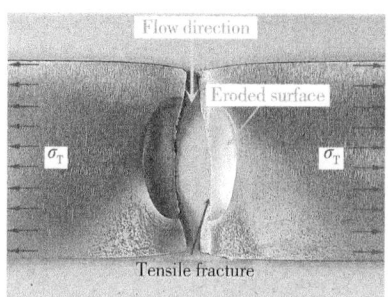 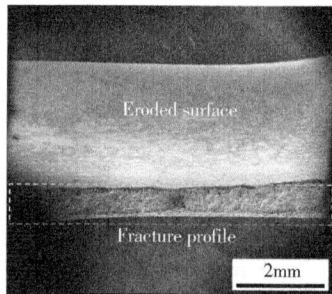

Fig. 14　Macro-image of tensile fracture of the wear specimen

Fig. 15 depicts the SEM results of the fracture morphology of the erosion specimen. As illustrated in Fig. 15(b), the furrows covered the eroded surface above the fracture section, for which the entire fracture section can be divided into four parts. Fatigue striations were formed beneath the eroded surface, and micro-cracks began to form fatigue cracks, which initially grew downwards at stage Ⅰ [Fig. 15(c)]. According to Fig. 15(d), it can be demonstrated that there were numerous micro-dimples, which were elongated into a parabola shape and may be attributed to uneven stress[37-38]. At stage Ⅱ, when the cracks were propagating steadily, the size of the dimples became larger due to the increased growth time, as shown in Fig. 15(e). The SEM micro-graph demonstrated that intact oval-shape dimples existed, and the tear ridges were clearly visible. In Fig. 15(f), the size of the dimples was found to gradually decrease at stage Ⅲ, which was the final stage of fatigue crack propagation. When the fluctuating stress reached the fracture strength, the final fracture occurred in stage Ⅳ. As shown in Fig. 15(g), the SEM morphology exhibited a step-like tear stripe and appeared in this region, while the number and size of dimples dropped sharply.

4.4　Stress-erosion model

A stress-erosion model was proposed to illustrate the complicated mechanisms of stress-erosion. Fig. 16 represents a schematic diagram that depicts the erosive wear failure mechanism with the consideration of applied stress. When the impact particles caused erosion damage to the surface of the specimen, the contact stress between the particles and specimen can be divided into three components, namely, σ_{P_x}, σ_{P_y}, σ_{P_z}. Particles can slide and penetrate the surface of the specimen due to the action of σ_{P_x} and σ_{P_z}, respectively[17, 39-40]. Meanwhile, σ_{P_y} was consumed in the extrusion between particles and both sides of the furrow, resulting in the formation of lips[41]. The proportion of the three components of contact stress changed with the impact angles. Thus, the morphology characteristics of the eroded surface due to different impact angles were relatively diverse, which was also confirmed by Fig. 9 and Fig. 10. Under the combined action of particle impact and tensile stress, an influence area under multi-stress was formed where the particles and

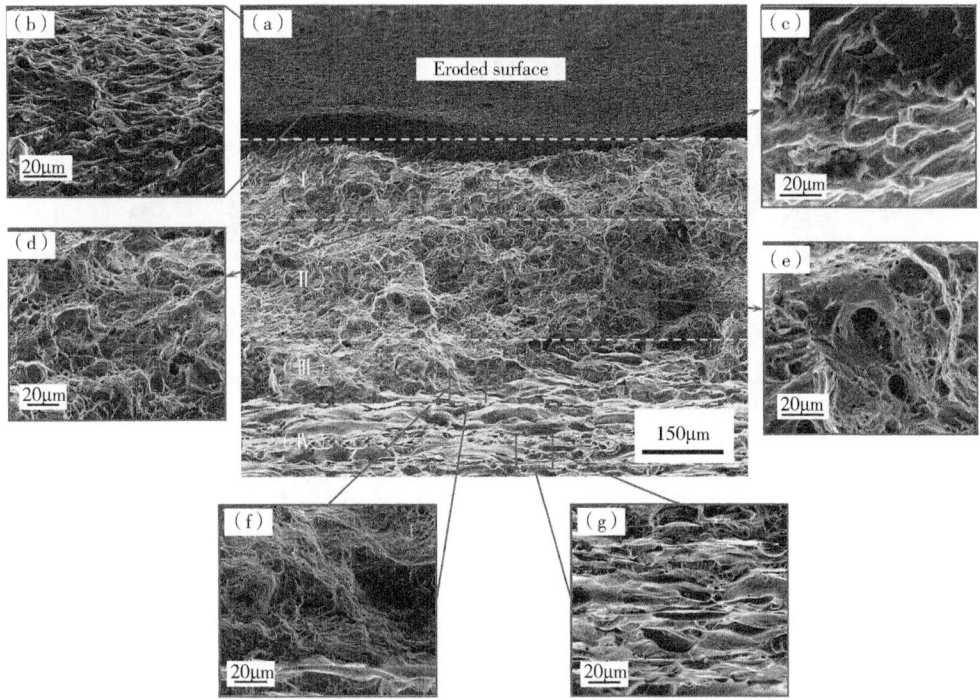

Fig. 15 SEM micro-structure of the fatigue fracture surface:
(a) overall image, (b) eroded furrows, (c)~(d) stage Ⅰ, (e) stage Ⅱ, (f) stage Ⅲ, (g) stage Ⅳ

specimen met. In this area, the stress state of the elements can be analyzed using the rheological spring-slider model[42-43]. When particles impact, the metal produced a plastic strain due to the slip transmission of grain boundaries[44]. Notably, the plastic strain of elements in this area intensified with the action of external tensile stress, which may explain the difference in surface morphology of the erosion area with or without applying stress. Apparently, at higher stress, the intermolecular force of the material was more easily overcome by the kinetic energy of impacting particles, resulting in more severe erosive wear of pipeline.

Furthermore, under the external fluctuating stress, the fatigue crack nucleation may appear at the position in which erosive wear was most severe at a certain time during the erosion process. The fatigue crack began to grow from a micro scale[45], after which the micro-crack gradually became a more visible size, eventually leading to tensile fracture of the eroded specimen during the final cycle of fatigue life. Therefore, it can be inferred that under the action of high fluctuating stress, fracturing pipelines carrying high-pressure fluid may be peculiarly prone to burst failure due to the interaction of erosive wear and tension fatigue.

5 Conclusions

In this work, erosive wear of fracturing pipelines under high fluctuating stress was investigated. A novel, advanced erosion test apparatus was initially developed, which was able to apply controllable fluctuating stress to specimens during the erosion process. Then, a series of erosive wear experiments involving fluctuating stress were conducted to analyze the coupling effect of average

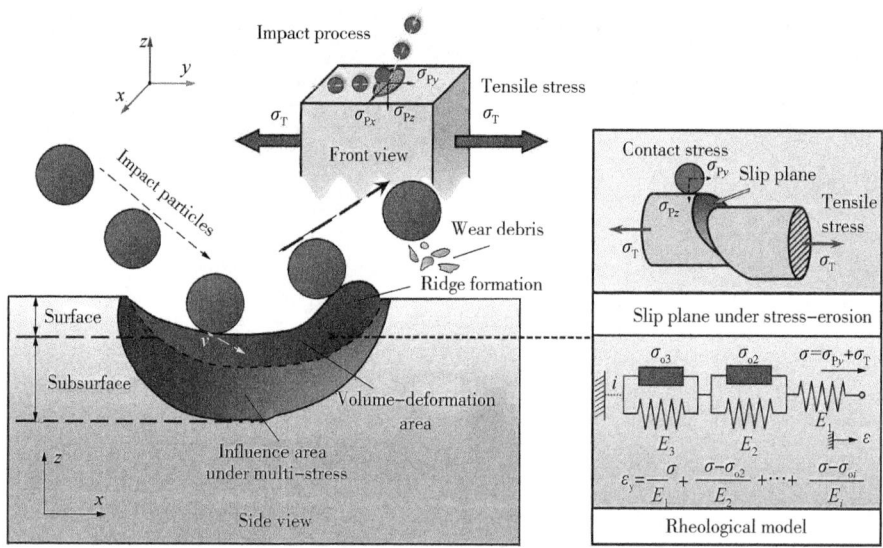

Fig. 16 Schematic diagram of the mechanism of stress-erosion. σ_T, tensile stress; ε, total strain; σ, total stress; E_1, E_2, $\cdots$, E_i, elastic modulus of each spring; σ_{o2}, σ_{o3}, $\cdots$, σ_{oi}, yield strength of each slider; σ_{Px}, σ_{Py} and σ_{Pz}, component of contact stress in x, y, and z direction, respectively

stress, impact angles, flow velocity and loading frequency on erosive wear during stress-erosion. Furthermore, stress - erosion mechanisms were unraveled according to micro - structural characterization as well as the rheological mechanical model. The following important conclusions can be drawn:

(1) According to the erosive wear experiments involving fluctuating stress, the erosive wear initially increases, after which it decreases rapidly with the increase of impact angle, with a maximum value appearing at 30°. Flow velocity is a crucial factor influencing erosive wear, for which the erosion rate significantly increases with a rise in flow velocity. When the specimen is subjected to a relatively high stress, erosion slightly increases with a rise in loading frequency. Furthermore, when other conditions remain unchanged, the erosion rate increases with a rise in average stress, which proves that in order to prevent the failure of pipe fittings under high-pressure operating conditions, the coupling effect of stress and other erosion parameters needs to be paid attention to.

(2) Under high fluctuating stress, fatigue cracks can be observed in the center of the eroded area during the process of erosive wear, which is a novel finding. After cracks appear, the specimen can be rapidly prone to fracture due to the combined action of the dislocation motion caused by tension fatigue and plastic flow caused by erosive wear. The finding of the phenomenon of erosive wear-tension fatigue interaction may be very helpful in understanding the failure mode of fracturing pipelines in actual high-pressure operating conditions.

(3) It can be found from the surface micro-morphology that the failure mode of erosive wear varies from micro-cutting to plastic deformation with a rise in impact angle. Moreover, high-velocity particles can cause longer and deeper furrows in the eroded region. Furthermore, under the action of stress-erosion, the width of the furrows evidently increases with a rise in stress levels, which

illustrated that erosive wear damage becomes increasingly more severe with applied stress. Combined with micro-structural characterization and rheological spring-slider model, the stress-erosion mechanism can be well explained.

Acknowledgements

This research is supported by National Natural Science Foundation Project of China (No. 52175208 and No. 52105237) and Scientific Research and Technology Development Project of CNPC (No. 2021DJ2705).

References

[1] A Correspondent. Materials: Abrasive Wear. Nature. 1969; 223: 351-352.

[2] Finnie I. Erosion of surfaces by solid particles. Wear 1960; 3: 87-103.

[3] Parsi M, Najmi K, Najafifard F, Hassani S, McLaury BS. A comprehensive review of solid particle erosion modeling for oil and gas wells and pipelines applications. J Nat Gas Sci Eng 2014; 21: 850-873.

[4] Wang QL, Jia BB, Yu MQ, He M, Li XC, Sridhar K. Numerical simulation of the flow and erosion behavior of exhaust gas and particles in polysilicon reduction furnace. Sci Rep 2020; 10: 1909.

[5] Josifovic A, Roberts JJ, Corney J, Davies B, Shipton ZK. Reducing the environmental impact of hydraulic fracturing through design optimisation of positive displacement pump. Energy 2016; 115: 1216-1233.

[6] Zhang X, Zhang LB, Hu JQ. Real-time risk assessment of a fracturing manifold system used for shale-gas well hydraulic fracturing activity based on a hybrid Bayesian network. J Nat Gas Sci Eng 2019; 62: 79-91.

[7] Zhang JX, Kang J, Fan JC, Gao JC. Research on erosion wear of high-pressure pipes during hydraulic fracturing slurry flow. J Loss Prev Process Indust 2016; 43: 438-48.

[8] Yang SQ, Zhang LB, Fan JC, Sun BC. Experimental study on erosion behavior of fracturing pipeline involving tensile stress and erosion prediction using random forest regression. J Nat Gas Sci Eng 2021; 87: 103760.

[9] Panda JN, Orquera EY, Mohanty AA, Egberts P. Tribo-corrosion inhibition of AISI 4715 steel pipe carrying hydraulic fracturing fluid. Tribol Int 2021; 161: 107066.

[10] Oka YI, Okamura K, Yoshida T. Practical estimation of erosion damage caused by solid particle impact. Part 1: effects of impact parameters on a predictive equation. Wear 2005; 259: 95-101.

[11] Huang C, Chiovelli S, Minev P, Luo J, Nandakumar K. A comprehensive phenomenological model for erosion of materials in jet flow. Powder Technol 2008; 187: 273-9.

[12] Gupta R, Singh SN, Seshadri V. Prediction of uneven wear in a slurry pipeline on the basis of measurements in a pot tester. Wear 1995; 184: 169-178.

[13] Sapate SG, RamaRao AV. Effect of erodent particle hardness on velocity exponent in erosion of steels and cast irons. Mater Manuf Process 2003; 18 (5): 783-802.

[14] Desale GR, Gandhi BK, Jain SC. Effect of erodent properties on erosion wear of ductile type materials. Wear 2006; 261(7): 914-21.

[15] Nguyen QB, Nguyen VB, Lim CYH, Trinh QT, Sankaranarayanan S, Zhang YW, Gupta M. Effect of impact angle and testing time on erosion of stainless steel at higher velocities. Wear 2014; 321: 87-93.

[16] Abedini M, Ghasemi HM. Synergistic erosion-corrosion behaviour of Al-brass alloy at various impingement angles. Wear 2014; 319: 49-55.

[17] Islam MA, Alam T, Farhat Z. Construction of erosion mechanism maps for pipeline steels. Tribol Int 2016; 102: 161-173.

[18] Desale GR, Gandhi BK, Jain SC. Particle size effects on the slurry erosion of aluminium alloy (AA 6063).

Wear 2009; 266: 1066-1071.

[19] Abouel-Kasem A. Particle size effects on slurry erosion of 5117 steels. J Tribol 2011; 133: 014502.

[20] Raask E. Tube erosion by ash impaction. Wear 1969; 13: 301.

[21] Levy AV, Chik P. The effects of erodent composition and shape on the erosion of steel. Wear 1983; 89: 151-162.

[22] Tsai W, Humphrey JAC, Cornet I, Levy AV. Experimental measurement of accelerated erosion in a slurry pot tester. Wear 1981; 68: 289-303.

[23] Islam MA, Alam T, Farhat ZN, Mohamed A, Alfantazi A. Effect of microstructure on the erosion behavior of carbon steel. Wear 2015; 332-333: 1080-1089.

[24] Liu WM, Xia YQ, Han N, Xue QJ. A comparative study of the friction and wear behaviour of untreated mild steel and its boron-permeated counterpart under lubrication of liquid paraffn containing zinc dialkyldithiophosphate. Tribol Int 2002; 35: 497-502.

[25] Wang HK, Yu Y, Yu JX, Wang ZY, Li HD. Development of erosion equation and numerical simulation methods with the consideration of applied stress. Tribol Int 2019; 137: 387-404.

[26] Wang JZ, Chen J, Chen BB, Yan FY, Xue QJ. Wear behaviors and wear mechanisms of several alloys under simulated deep-sea environment covering seawater hydrostatic pressure. Tribol Int 2012; 56: 38-46.

[27] Imrek H, Bagci M, Khalfan OM. Solid Particle Erosion as Influenced by Tensile Axial Loads. Tribol Trans 2011; 54: 779-783.

[28] Sun BC, Fan JC, Wen D, Chen YY. An experimental study of slurry erosion involving tensile stress for pressure pipe manifold. Tribol Int 2015; 82: 280-286.

[29] Zhang JX, Kang J, Fan JC, Gao JC. Study on erosion wear of fracturing pipeline under the action of multiphase flow in oil & gas industry. J Nat Gas Sci Eng 2016; 32: 334-46.

[30] Weroński A, Hejwowski T. Effect of stress on abrasive and erosive wear of steels and sprayed coatings. Vacuum 2008; 83: 229-233.

[31] Frosell T, Fripp M, Gutmark E. Investigation of slurry concentration effects on solid particle erosion rate for an impinging jet. Wear 2015; 342-343: 33-43.

[32] Zhang JX, Bai YQ, Kang J, Wu X. Failure analysis and erosion prediction of tee junction in fracturing operation. J Loss Prev Process Indust 2017; 46: 94-107.

[33] Hutchings IM. Prediction of the resistance of metals to erosion by solid particles. Wear 1975; 35(2): 371-4.

[34] Al-Bukhaiti MA, Ahmed SM, Badran FMF, Emara KM. Effect of impingement angle on slurry erosion behaviour and mechanisms of 1017 steel and highchromium white cast iron. Wear 2007; 262(9-10): 1187-98.

[35] Yoganandh J, Natarajan S, Kumaresh Babu SP. Erosive wear behavior of nickel-based high alloy white cast iron under mining conditions using orthogonal array. J Mater Eng Perform 2013; 22: 2534-41.

[36] Rajesh JJ, Bijwe J, Venkataraman B, Tewari US. Effect of impinging velocity on the erosive wear behaviour of polyamides. Tribol Int 2004; 37: 219-26.

[37] Blach J, Falat L, Ševc P. Fracture characteristics of thermally exposed 9Cr-1Mo steel after tensile and impact testing at room temperature. Eng Fail Anal 2009; 16: 1397-403.

[38] An T, Li SJ, Qu JL, Shi J, Zhang S, Chen LQ, Zheng SQ, Yang F. Effects of shot peening on tensile properties and fatigue behavior of X80 pipeline steel in hydrogen environment. Int J Fatigue 2019; 129: 105235.

[39] Okonkwo PC, Shakoor RA, Ahmed E, Mohamed AMA. Erosive wear performance of API X42 pipeline steel. Eng Fail Anal 2016; 60: 86-95.

[40] Javaheri V, Porter D, Kuokkala VT. Slurry erosion of steel-Review of tests, mechanisms and materials. Wear 2018; 408-409: 248-73.

[41] Hutchings IM. A model for the erosion of metals by spherical particles at normal incidence. Wear 1981; 70: 269-81.

[42] Dowling DE. Mechanical Behavior of Materials: Engineering Methods for Deformation, Fracture, and Fatigue (4nd Edition). Pearson Education Inc, New Jersey, 2012.

[43] Iwan WD. On a Class of Models for the Yielding Behavior of Continuous and Composite Systems. J Appl Mech-T ASME 1967; 34: 612-17.

[44] Weaver JS, Li N, Mara NA, Jones DR, Cho H, Bronkhorst CA, Fensin SJ, Gray GT. Slip transmission of high angle grain boundaries in body-centered cubic metals: Micropillar compression of pure Ta single and bi-crystals. Acta Mater 2018; 156: 356-68.

[45] Pegues JW, Roach MD, Shamsaei N. Influence of microstructure on fatigue crack nucleation and micro-structurally short crack growth of an austenitic stainless steel. Mat Sci Eng A 2017; 707: 657-67.

本论文原发表于《Wear》2023 年第 518~519 卷。

Quantum Dots Bridge Enabling Highly Efficient Carbon-based HTM-free Perovskite Solar Cells

Yang Yuanbo[1,2] Wang Shuo[1] Li Simiao[1] Li Tiantian[1,2]
Chen Peng[1] Zhao Qian[1] Li Guoran[1]

(1. School of Materials Science and Engineering, Nankai University;
2. State Key Laboratory for Performance and Structure Safety of Petroleum Tubular Goods and Equipment Materials, CNPC Tubular Goods Research Institute)

Abstract: Realization of improved charge transport with suppressed recombination at the interface of perovskite and carbon electrode is the main key for remarkably increasing the power conversion efficiency of carbon-based hole transport material (HTM)-free perovskite solar cells (PSCs). Here, we demonstrate a strategy that builds a perovskite quantum dot (PQD) interlayer, for the first time, to bridge the perovskite absorber and carbon electrode for solving the interface issue in HTM-free PSCs. It is found that the introduced PQD interlayer concurrently functions as morphology changer, defect passivator and photogenerated hole extractor. Compared with the pristine perovskite film, the PQD-modified perovskite absorber shows increased contact area and high compatibility with carbon electrode, prolonged carrier lifetime, deduced defect density as well as suppressed recombination. These positive effects, combined with a heterostructure created by perovskite bulks and PQDs facilitating hole transport at the interface, enable an improvement in device efficiency from 16.71% to 17.93%.

Keywords: Quantum dot; Carbon electrode; Perovskite solar cell; Interlayer; Charge transport

1 Introduction

Organic-inorganic lead halide perovskite solar cells (PSCs) have attained significant attractiveness and remarkable commercial advances havebeen made all over the world to enhance the power generation and lower the cost of production[1]. The main elements of a high-performance PSC typically include transparent conducting substrate, electron transport material (ETM), perovskite absorber, hole transport material (HTM), and metal back electrode, where the last two components cover more than 50% of device cost.[2] Therefore, a carbon-back electrode HTL-free PSC is developed for low-cost, high-stability, and printable perovskite photovoltaics. However, along with

Corresponding authors: Zhao Qian, qian.zhao@nankai.edu.cn; Li Guoran, guoranli@nankai.edu.cn.

all provided benefits, such a device structure brings new vulnerabilities like poor contact at the perovskite/carbon interface and inefficient hole collection/extraction owing to the misalignment of energy levels at the perovskite/carbon interface.[3]

Since the first carbon-based HTM-free PSC with an efficiency of 6.64% was reported by Han et al. in 2013, considerable efforts have been made for the improvement of charge transport at the interface between perovskite layer and carbon back electrode[4]. One efficient way to enhance the power conversion efficiency (PCE) of carbon-based HTM-free PSCs through interface optimization is to control the nucleation and growth of perovskite film for achieving high-quality absorber layer in PSCs. For example, Ye et al. introduced perfluorotetradecanoic acid with a carbonyl unit and carbon fluorine bonds to suppress the ion migration and reduce the crystal defects in perovskites, delivering a high efficiency of 18.9%[5]. Another way to improve the contact between perovskite and carbon layers is the modification to the carbon paste. Liu et al. employed oxidized cellulose additives inside carbon electrode to yield a better contact at the perovskite/carbon interface by coordinating carbonyl group in cellulose to the Pb in perovskite, which demonstrates a PCE of 15.5% for biomass-based carbon-electrode PSCs[6]. Nevertheless, the effectiveness of additives in carbon paste for planar carbon-based devices is much lower than that for mesoporous carbon-based PSCs. Therefore, directly introducing additives at the perovskite/carbon interface can be considered a more efficient and simpler way to solve the interface problems in planar carbon-based HTM-fee PSCs, which also avoids the high complexity of controlling the process of perovskite crystallization and minimize the need for reoptimizing the device processing conditions.

Perovskite quantum dot (PQD), a type of nano-sized perovskite-structured materials, possesses superior optoelectronic properties of perovskites such as strong light absorption, long intrinsic recombination lifetime, and small exciton binding energy, as well as favorable characteristics of QDs such as widely tunable bandgap, multiple exciton generation, high photoluminescence quantum yield (PLQY), and narrow emission linewidth[7]. PQDs giving functional elements with surface ligands not only have been used as light harvester in photovoltaics, but also can be employed as interfacial passivator, modifier and charge extractor for improving the performance of PSCs. The PQDs introduced on the surface of perovskite bulk (PBK) films can modify the surface morphology, reduce non-radiative recombination losses and generate built-in potentials determined by the quasi-Fermi level spitting, which benefits charge carrier kinetics in PSCs[8]. The organic ligands binding on the surface of PQDs render the perovskite film's surface hydrophobic inhibiting moisture penetration and blocking the escape of organic cation from perovskite crystalline, and thus yield superior device long-term stability[9]. In addition, by incorporating one PQD layer with appropriate valence band maximum (VBM) offset engineering at the interface between PBK and HTM films, the hole transport and collection are well-enhanced, leading to significant improvements in PCE of PSCs[10].

In this work, we unitize these advantageous features of PQDs to build an interlayer bridging perovskite absorber and carbon electrode in carbon-based HTL-free PSC, which is found to have noticeable positive effect on device performance. Then, we systematically characterize the fabricated devices to understand the origin of the performance improvement, revealing the role of the PQD interlayer as morphology changer, defect passivator and photogenerated hole extractor in PSCs. Particularly, the increase in device efficiency is not only resulted from the increased contact area of PQD-modified perovskite film and the high compatibility of PQDs with carbon electrode, but also ascribed to the prolonged TRPL lifetime, deduced defect density and suppressed recombination as well as improved hole transport in the PBK/PQD heterostructure. These findings highlight the multiple functions of PQDs in solving the interfacial issues at the perovskite/carbon interface for effectively improving performance of carbon-based HTL-free PSCs.

2 Results and Discussions

Colloidal $FAPbI_3$ PQDs are synthesized via a hot injection method and purified following the previous reports.[11] The details on synthetic methodology of the $FAPbI_3$ PQDs are described in the Experimental Section. Oleic acid (OA) and oleylamine (OAm) are used as surface ligands bonding to PQD surfaces, rendering them hydrophobic.

The morphologies, structural and optical properties of as-synthesized PQDs are first investigated as shown in Figure 1. The TEM images show the cubic shape of PQDs with a lattice spacing of 0.32nm that can be ascribed to (002) plane of perovskite crystallites with cubic phase (Figure 1a and Figure 1b)[12]. The size analysis of PQDs from TEM images shows a mean size of

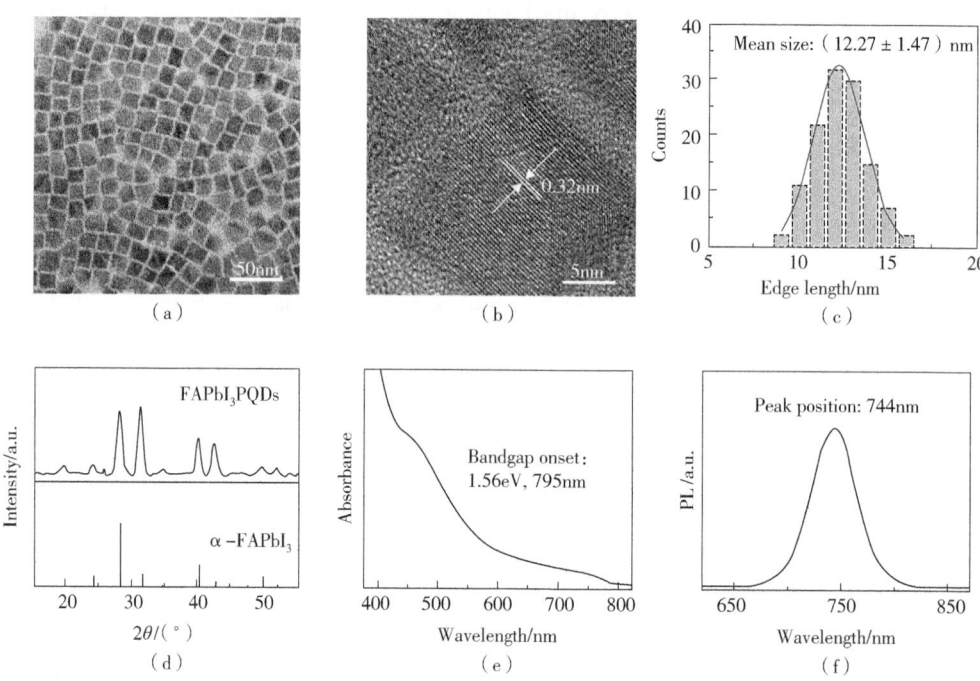

Fig.1 (a) TEM images, (b) high-resolution TEM images, (c) size distribution, (d) XRD patterns, (e) UV-Vis absorption spectra, and (f) photoluminescence emission spectra of as-synthesized $FAPbI_3$ PQDs

12.27nm with narrow size distribution (Figure 1c and Figure 2). The crystalline phase of $FAPbI_3$ PQDs are again identified to cubic alpha phase by XRD patterns as presented in Figure 1d. For the absorbance and emission spectra of PQDs, a bandgap onset at 795nm (1.56eV) and a PL emission peak at 744nm are observed in Figure 1e and Figure 1f, respectively. These results are consistent with previous studies showing high reproducibility and high quality of the obtained FAPbI$_3$ PQDs in this work[11,13]. To deliver a uniform and controlled modification at PBK/carbon interface, the as-synthesized PQDs are dispersed in hexane to prepare an additive solution with various PQD concentrations (1mg/mL, 2mg/mL, 3mg/mL, and 4mg/mL). The photographs of the prepared PQD additive solutions are displayed in Figure 3, demonstrating their high colloidal stability and high luminescence.

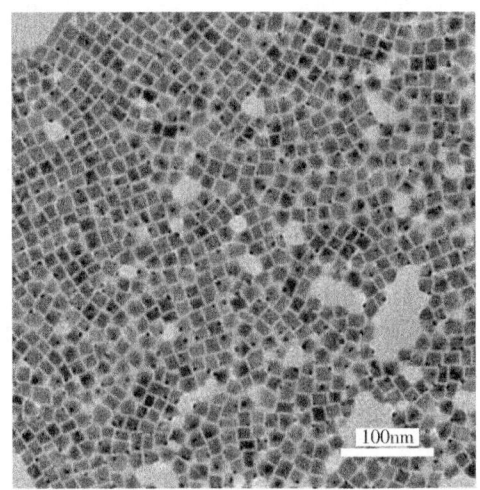

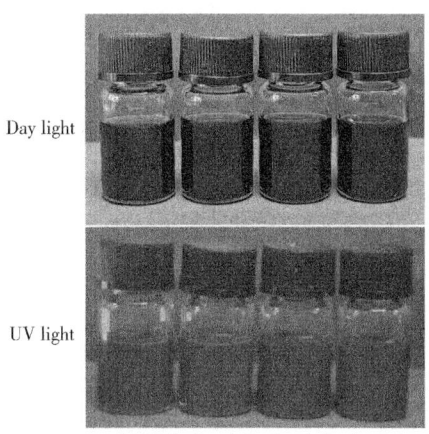

Fig. 2　TEM images of as-synthesized FAPbI$_3$ PQDs

Fig. 3　Photographs of as-synthesized FAPbI$_3$ PQD additive solutions with different concentrations under UV and day light

Through a comparison using such a series of PQD additive solutions, the effectiveness of PQD interlayer on performance of carbon-based HTL-free PSCs are evaluated. Figure 4a presents the used device structure of FTO glass/TiO$_2$/PBK film/PQD film/carbon. A pristine device without PQD interlayer (0mg/mL) is fabricated as a reference. For PQD-modified devices, the prepared PQD solution is directly spin coated on top of PBK layer forming a solid PQD film after solvent evaporates. It can be seen that the concentrations used here are quite low so that the thickness of the deposited PQD layer is difficult to measure. To roughly quantify its thickness, we spin coat a thick PQD film using a highly-concentrated PQD solution (100mg/mL). As shown in Figure 4b (right panel), a PQD interlayer with a thickness of approximately 200 nm is achieved. When we use 2mg/mL PQD additive solution, the thickness of the produced PQD film should be less than several nanometers, by that analogy and following the equation for calculating the film thickness of Newtonian fluids after spin coating[14]. Therefore, we estimate that only one discontinuous monolayer of PQDs is deposited on the PBK film when the PQD concentration is set in the range of 1~5mg/mL because of the size of PQDs around 12nm. The influence of the PQD interlayer on photovoltaic performance of carbon-based HTL-free PSCs is demonstrated in Figure 4c~e. 2mg/mL PQD additive solution exhibits the largest improvement in PCE of devices, contributed from the enhancements of J_{sc} and V_{oc}. There is no significant change in fill factor (FF) as presented in Figure 5. From the analysis of $J-V$ characteristics (in reverse scans), the PCE increases from

16.71% to 17.93%, the J_{sc} increases from 22.01mA/cm to 22.52mA/cm, and the V_{oc} increases from 1.01V to 1.06V (Figure 4f and Table 1), after introducing a PQD interlayer with appropriate thickness (using 2mg/mL PQD solution). The $J-V$ curves in the forward scan direction (from short circuit to open circuit) for the pristine and PQD-modified (2mg/mL) devices are shown in Figure 6. The integrated J_{sc} values calculated from EQE spectra are consistent with the results obtained from $J-V$ curves (Figure 4g). The enhancement in the long wavelength range of EQE can be attributed to the enhanced hole collection and the passivated defects at the PBK/carbon interface. The champion device exhibits a stabilized power output (SPO) of 16.9% at 0.88 V for 500s (Figure 4h).

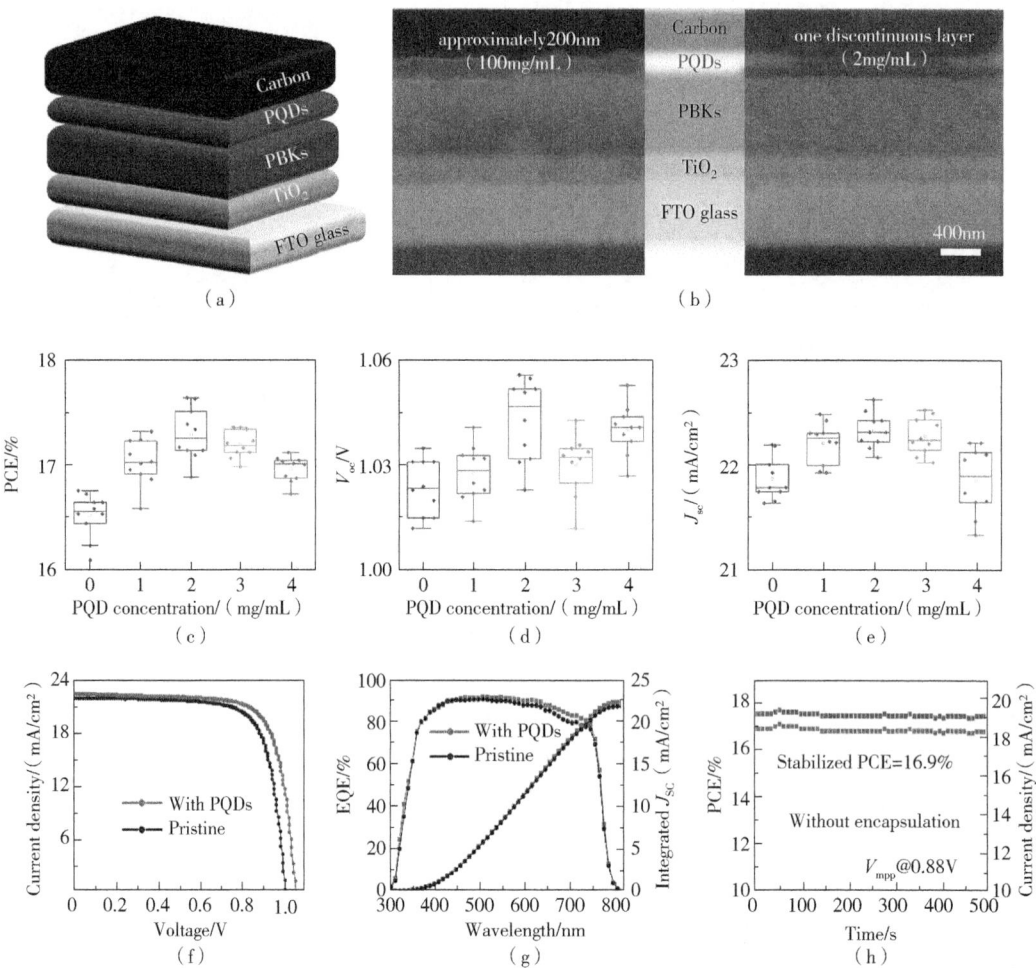

Figure 4 (a) Device structure of fabricated carbon-based HTL-free PSCs. (b) Cross-sectional SEM of devices with a thick (left panel) and an ultra-thin (right panel) PQD interfacial layer. (c~e) Parameter distributions of (c) PCE, (d) V_{oc} and (e) J_{sc} for the devices fabricated with PQD concentrations of 0mg/mL, 1mg/mL, 2mg/mL, 3mg/mL, and 4mg/mL. (f) $J-V$ curves and (g) EQE spectra of pristine and PQD-modified (2mg/mL) devices. (h) Stabilized PCE and J_{sc} of the champion carbon-based HTL-free PSCs fabricated using 2mg/mL PQD additive solution

Table 1 Parameters of *J-V* curves for the pristine and PQD-modified carbon-based HTM-free PSCs

Condition	Scan mode	J_{sc}/ (mA/cm^2)	V_{oc}/ V	FF/ %	PCE/ %	R_s/ Ω	R_{sh}/ Ω
Pristine	Reverse scan	22.01	1.01	75.20	16.71	45.46	10301.01
	Forward scan	22.09	1.02	65.99	14.91	54.47	12404.51
With PQDs	Reverse scan	22.52	1.06	75.13	17.93	45.13	13845.43
	Forward scan	22.66	1.05	66.56	15.83	54.75	17341.75

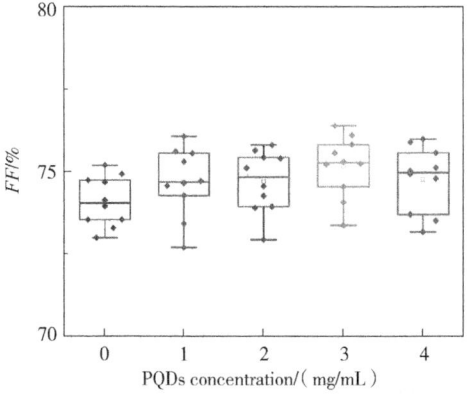

Fig. 5 Parameter distributions of FF for the devices fabricated with PQD concentrations of 0mg/mL, 1mg/mL, 2mg/mL, 3mg/mL, and 4mg/mL

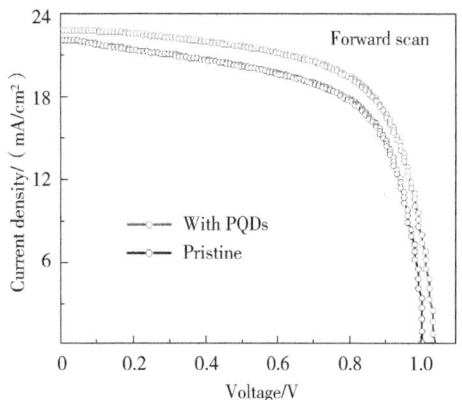

Fig. 6 *J-V* curves in the forward scan direction (from short circuit to open circuit) for the pristine and PQD-modified (2mg/mL) devices

Given the results on the performance of devices and the unique properties of PQDs, the PQD interlayer may concurrently play a few different roles: (1) it changes the surface morphologies of absorber layer and thereby changes the contact area between absorber and carbon counter electrode, (2) it electronically bridges the PBK and carbon layers by forming heterostructures, which improves hole transport and thus increases the built-in potential in PSCs, (3) it prevents backflow of electrons from conduction band of PBK to carbon direction and suppresses the undesired interfacial nonradiative recombination by passivating the PBK surface, (4) the hydrophobic organic long-chain ligands capping on the PQDs shields the perovskite structure from atmospheric moisture improving device stability, but may be resulted from the decrease in PCE because of the low carrier mobility of the PQD interlayer especially when highly concentrated PQD solution is used.

The characterizations of film morphologies as well as electronic, optical, optoelectronic and electrochemical properties are carried out to explain the improvements of device performance and to prove the effects of the PQD interlayer proposed above. In Figure 7a, the SEM images of the PBK film surface with covering PQDs show that the formed PQD interlayer is a discontinuous monolayer of PQDs as expected when using 2mg/mL PQD solution. The AFM images demonstrate that the PQDs are deposited into the holes located at the grain boundaries in PBK films, but the roughness increases after inserting the PQD layer since the PQDs cannot entirely fill the holes (Figure

7b). The increased roughness leads to an increase in contact area between perovskite absorber and carbon layer, which is beneficial to charge transport of PSCs. The UV-Vis spectra and XRD patterns of the PBK film with and without covering PQDs are shown in Figure 8 and Figure 9, respectively. No significant differences in absorbance and crystal structure of perovskite absorber are observed. In another aspect, to investigate the compatibility of PQDs with carbon electrode, we deposited PQD solution onto the carbon electrode, and performed SEM/EDS mapping on pristine and PQD-deposited carbon films. Figure 7c demonstrates that the PQDs can be successfully attached onto the carbon electrode, with a uniform distribution throughout the carbon surface. As demonstrated in Figure 7d, a discontinuous PQD region is produced on the carbon electrode as in the case of the deposition on the PBK films. The PQD-deposited carbon film under UV light irradiation exhibits a red-light emission, which indicates the PQDs keep the perovskite structure and emission features, showing their high stability when attaching onto carbon electrode surface.

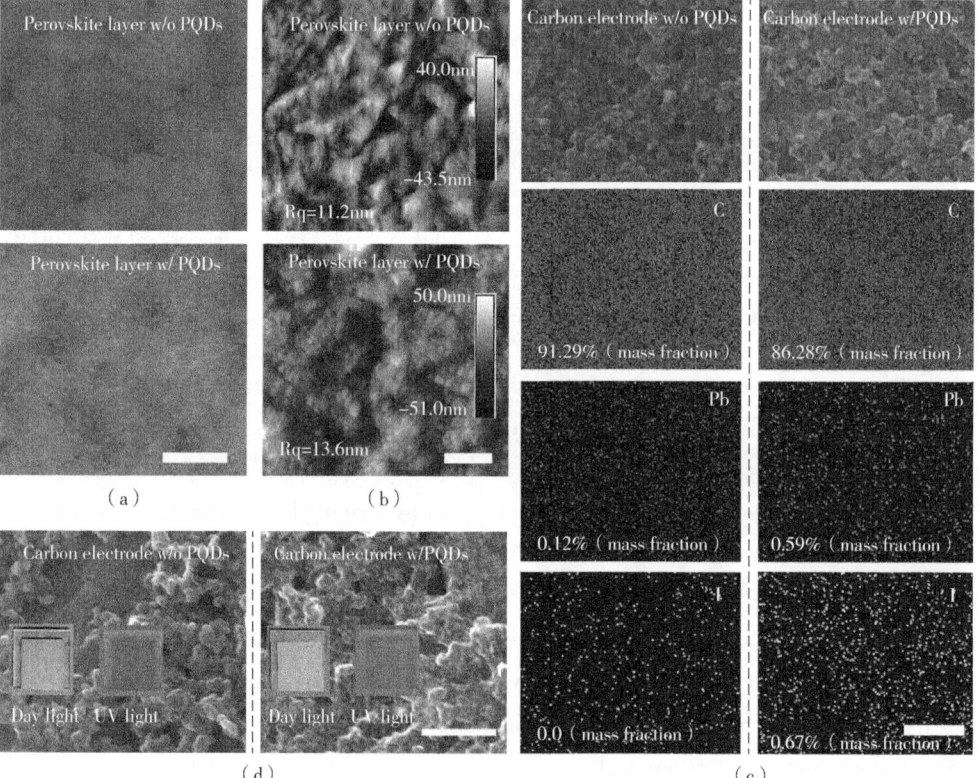

Fig. 7 (a) SEM and (b) AFM images of the PBK film surface w/o (upper panel) and w/ (down panel) PQD modification. Scale bar is 400nm. (c) Low-resolution SEM images and the corresponding EDS mappings of the carbon electrode w/o (left panel) and w/ (right panel) PQDs. Scale bar is 1μm. (d) High-resolution SEM images and photographs of carbon film w/o (left panel) and w/ (right panel) PQDs under day (left in inset) and UV (right in inset). Scale bar is 400nm. The concentration of the used PQD solution is 2mg/mL

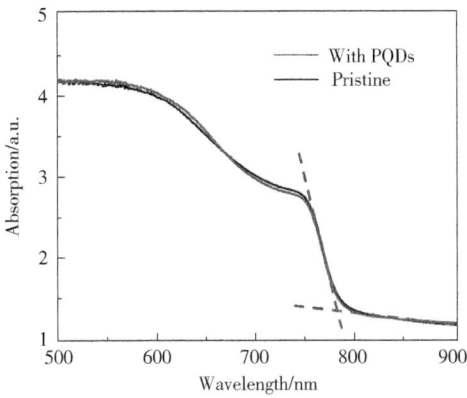
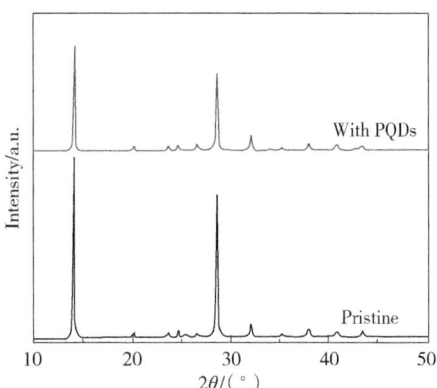

Fig. 8 UV-Vis spectra of pure PBK and PQD-modified PBK films

Fig. 9 XRD patterns of pure PBK (pristine) and PQD-modified PBK films

The work function (W_F) and VBM of MAPbI$_3$ PBKs and FAPbI$_3$ PQDs are determined from UPS spectra (Figure 10a and Figure 10b). Combined with the bandgaps extracted from Figure 1e (1.56eV for PQDs) and Figure 8 (1.58 eV for PBKs), a band alignment diagram of the carbon-based HTM-free PSCs with PQD interlayers can be built as demonstrated in Figure 10c. The obtained band position values are consistent with the previously published data[15]. As a result of such a band alignment, the photogenerated hole in PBKs are extracted into the PQD interlayer, while the photogenerated electron transfers to the TiO$_2$ layer, which favors the charge separation and collection thus enhancing the device efficiency[16]. The introduced PQD interlayer also could act as an electron blocking layer, suppressing the undesired recombination pathways of photogenerated carriers from the PBK layers. To further demonstrate the charge transfer that occurs at the PBK/PQDs interface, we measured the PL and TRPL decay of the samples with the structure of glass/perovskite and glass/perovskite/carbon, where the perovskite absorber was modified with or without PQDs (Figure 10d and Figure 10e). The parameters of TRPL decay are calculated and are shown in Table 2. The PQD-modified glass/perovskite sample exhibits higher PL intensity and longer TRPL lifetime than the pristine counterpart, which could be resulted from the suppressed nonradiative recombination. As for the glass/perovskite/carbon structure, PQD-modified sample show lower PL intensity with shorter TRPL lifetime when compared to the pristine sample, indicating that the holes are captured from the PBK/PQDs interface leading to fast quenching of the PL spectra.

Table 2 Parameters of TRPL of the samples using the structure of glass/perovskite absorber and glass/perovskite absorber/carbon with and without PQD modifications

Sample structure	Modification	A_1	τ_1/ns	A_2	τ_2/ns	τ_{avg}/ns
glass/perovskite	Pristine	162.58	4.84	0.66	47.51	6.48
	With PQDs	36.15	6.36	0.51	86.32	19.21
glass/perovskite/carbon	Pristine	273.84	4.60	0.63	34.14	5.10
	With PQDs	13730.83	2.79	1.28	18.33	2.80

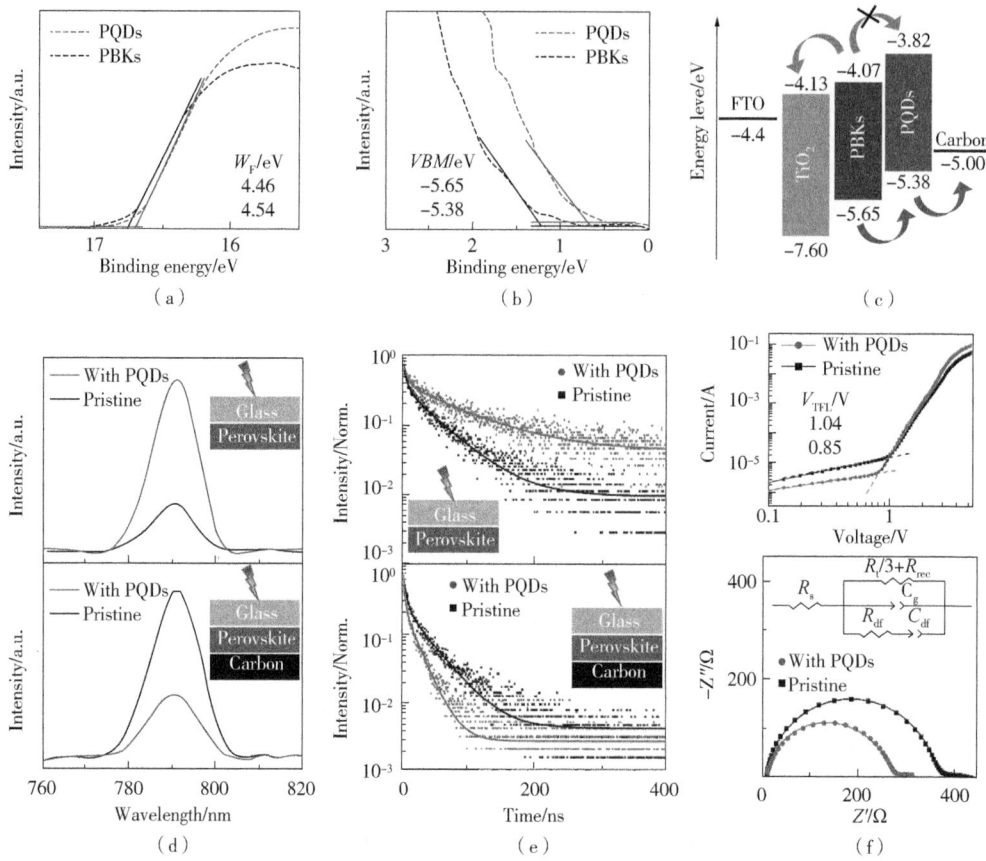

Fig. 10 (a~b) UPS spectra of MAPbI$_3$ PBK and FAPbI$_3$ PQD films. (c) Band alignment of the fabricated carbon-based HTM-free PSCs with PQD interlayers. (d) PL and (e) TRPL spectra of the samples with the structure of glass/perovskite and glass/perovskite/carbon. (f) SCLC and (g) EIS data for the pristine and PQD-modified devices. The inset of Figure 10g is the equivalent circuit diagram used in the analysis of the EIS spectra

In order to verify the effect of the PQD interlayer on passivating the PBK surface defect states, the space-charge limited current (SCLC) and electrical impedance spectroscopy (EIS) experiments are performed, as shown in Figure 10f and Figure 10g, respectively. Using a device structure of glass/FTO/NiO/perovskite/carbon, the defect density, N_{trap}, is calculated following the equation as $N_{trap} = \frac{2V_{TFL}\varepsilon_r\varepsilon_0}{eL^2}$, where V_{TFL} is the trap-filled limit voltage, e is the elementary charge, L is the film thickness, ε_0 is the vacuum dielectric constant, and ε_r is the relative permittivity[17]. The V_{TFL} values for the pristine and PQD-modified devices are determined to be 1.04V and 0.85V, respectively. After introducing the PQD interlayer, the calculated N_{trap} decreases from 5.745×10^{15} cm^{-3} (for the pristine one) to 4.696×10^{15} cm^{-3}. Furthermore, the field-dependent SCLC can be expressed using the equation as $\mu = \frac{8JL^3}{9\varepsilon_r\varepsilon_0V^2}$, where J is the measured current density, and μ is the charge mobility. The J vs V^2 for the pristine and PQD-modified samples are plotted and linearly fitted in Figure 11. The zero-field charge mobility of the PQD-modified device calculated from the fitted line shows an increase from 5.124×10^{-3} cm^2/(V·s) (for the pristine one) to 9.011×10^{-3} cm^2/(V·s). These results allow us to conclude that the introduction of PQD interlayer could passivate the surface defect and suppress the nonradiative recombination for the perovskite absorber. In the XPS spectra of the perovskite films with and without the PQD modification in Figure 12, there is no observable peak shift in I 3d and Pb 4f peaks. It suggests that the passivation works primarily by capping the PBK surface with organic ligands of PQDs, rather than directly interacting with the PQD itself. Further evidence for the suppressed recombination of surface defects come from the EIS analyses using an equivalent circuit presented in inset of Figure 10g, where R_s is the series resistance, R_t is the charge transport resistance, R_{rec} represents the recombination resistance, C_g is a geometrical capacitance accounting for the device dielectric behavior, C_{df} and R_{df} represent a capacitor and a resistor accounting for the low frequency processes such as ion migration or carrier accumulation at the interfaces, respectively[18]. The calculated parameter from EIS are presented in Table 3. The device with the introduced PQD interlayer shows a remarkable decrease in R_{rec} compared to the pristine device, further strengthening the fact that the PQD interlayer is in favor of the suppression of trap-assisted recombination for the perovskite absorber in PSCs.

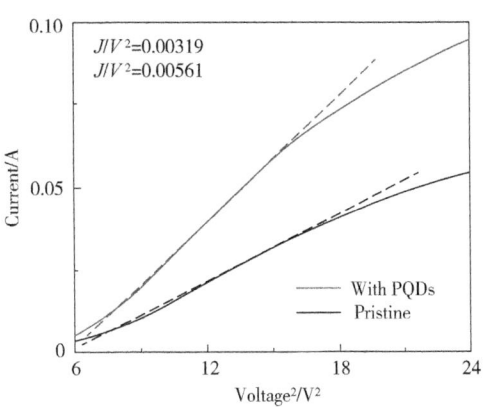

Fig. 11 The plotted J vs V^2 for the pristine and PQD-modified samples in SCLC analysis

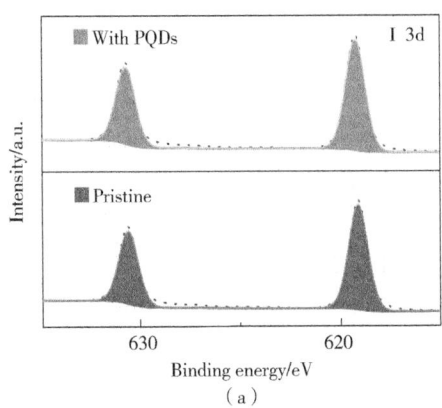
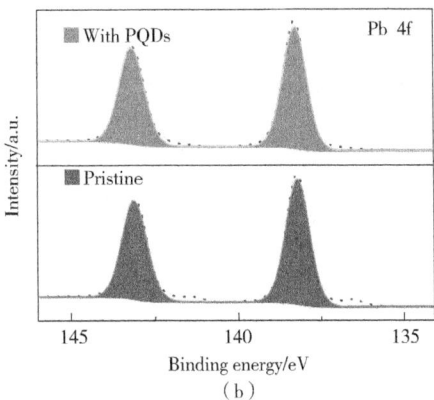

Fig. 12 (a) I 3d and (b) Pb 4f XPS spectra for pure PBK (pristine) and PQD-modified PBK films

Table 3 Parameters of EIS spectra for the pristine and PQD-modified carbon-based HTM-free PSCs

Condition	R_s/Ω	$R_t/3+R_{rec}/\Omega$	C_g-T/F	C_g-P	R_{df}/Ω	C_{df}-T/F	C_{df}-P
Pristine	9.82	429.0	8.21×10^{-8}	0.94	1646	1.38×10^{-4}	0.30
With PQDs	9.91	295.5	4.74×10^{-8}	0.91	1631	3.98×10^{-5}	0.34

3 Conclusion

In summary, astrategy through the introduction of a PQD interlayer bridging the perovskite absorber and carbon electrode in carbon-based HTM-free PSCs has been demonstrated for the first time. SEM and AFM characterizations show an increased roughness after the deposition of the PQDs on PBK film, which favors the charge transport by increasing the contact area between perovskite absorber and carbon layer. The PQD deposited on carbon film exhibits stable perovskite structure and good emission characteristics, demonstrating its high compatibility with carbon electrode. Importantly, the inserted PQD layer facilitates the hole transport by creating a heterostructure with the PBK layer. On the other hand, the PQD-modified perovskite absorber demonstrates prolonged TRPL lifetime, deduced defect density as well as suppressed recombination, resulting in an increase in device PCE from 16.71% to 17.93%. This work explores the functionality of PQDs and provides valuable insights into carbon-electrode PSCs as well as other optoelectronic applications toward further performance improvement.

4 Experimental Section

Materials: Lead iodide (PbI_2) and methylammonium iodide (MAI) were purchased from Advanced Election Technology Co., Ltd. and Xi'an Polymer Light Technology Corp., respectively. Titanium tetrachloride ($TiCl_4$, >98%) was bought from Adamas-beta. Dimethyl sulfoxide (DMSO), dimethylformamide (DMF), and methyl acetate (MeOAc, anhydrous, 99.5%) were purchased from J&K Scientific. 1-octadecene (1-ODE, 90%) and Oleylamine (OAm, 80%~90%) were purchased from Acros Organics. Oleic acid (OA, 80%~90%) was purchased from Alfa Aesar. Formamidinium acetate (FAAc) and hexane were purchased from TCI and Macklin, respectively. Low temperature carbon electrode paste was obtained from Shanghai MaterWin New Materials Co., Ltd. All of them were used directly without further purification.

$FAPbI_3$ PQD synthesis: $FAPbI_3$ QDs were synthesized via a hot-injection method. In details, FA-oleate precursor was prepared by mixing 0.521g FAAc with 10mL OA and then degassing under vacuum at 90℃ for 30min. The temperature was subsequently increased to 110℃ to get a clear solution. After that, the mixture was cooled to 90℃ and maintained in N_2 for injection. For PbI_2 precursor, a mixture of 0.344g PbI_2 and 20mL 1-ODE was degassed under vacuum at 120℃ for 30min. Then, a mixture of 6mL OA and 2mL OAm, preheated at 130℃, was injected into the PbI_2 precursor. After PbI_2 was fully dissolved, the temperature of the PbI_2 mixture was reduced to 80℃ under N_2. 5mL FA-oleate precursor solution was swiftly injected into the PbI_2 precursor solution. After around 15s, the reaction was quenched using an ice-water bath. When the reaction was cooled to room temperature, 9mL MeOAc was added to the mixtures and centrifuged at 8000r/min for 30min. The residual precipitate was dispersed in 9mL hexane, re-precipitated with 10mL MeOAc, and centrifuged at 8000r/min for 10min. The final precipitate was dispersed in 3 mL hexane and stored in a refrigerator at 5℃. Before use, the $FAPbI_3$ PQD solution was centrifuged at 8000r/min for 5min and the supernatant was filtered through a 0.45μm PTFE filter.

Solar cell fabrication: At first, fluorine-doped tin oxide (FTO, 15Ω/cm) substrates were sequentially sonicated with water, ethanol and isopropanol for 30min each, and blew dry under N_2-flowing. The compact TiO_2 layers were deposited on the surface via a three-steps hydrolysis process and the details were shown as below. $TiCl_4$ solutions with the concentrations of 40mmol/L, 50mmol/L, and 240mmol/L were obtained from liquid $TiCl_4$ solutions (98%) diluted with ice-water mixtures. After that, the cleaned FTO substrates were treated with low-temperature oxygen plasma at 300W for 300s and sequentially immersed into 40mmol/L $TiCl_4$ solutions for 30min, 240mmol/L $TiCl_4$ solutions for 20min, and 50mmol/L $TiCl_4$ solutions for 30min at 80℃. After that, the substrates with compact TiO_2 layers were washed with deionized water and then annealed at 150℃ for 10min. Then, mesoporous TiO_2 layers were spin-coated using the commercial 30NR-D paste at 4000r/min for 20s and then annealed at 150℃ and 500℃ for 30min, sequentially. Perovskite precursor solutions were prepared by dissolving 159mg of MAI and 477mg of PbI_2 into the mixed solvent of DMF and DMSO (500μL DMF with 145μL DMSO). Perovskite solutions were spin-coated in N_2-glovebox via two steps: 1000r/min for 10s and then 4000r/min for 24s, in which 125μL anisole were quickly dropped at 20 s in the whole spin-coating process. A solid $MAPbI_3$ bulk film was formed after annealing at 150℃ for 7min in ambient air. Lastly, an approximately 50μm carbon electrode was deposited by using a commercial carbon paste and annealed at 150℃ for 7min. As for the devices with the $FAPbI_3$ PQD interlayer, a $FAPbI_3$ PQD solution with the concentration of 1mg/mL, 2mg/mL, 3mg/mL or 4mg/mL was spin-coated on $MAPbI_3$ bulk film at 4200r/min for 30s.

Characterizations: X-ray diffraction (XRD) patterns were obtained by Rigaku miniFlex Ⅱ with Cu Kα radiation ($\lambda = 1.5418$Å) and UV-visible absorption spectra (UV-Vis) were measured using Varian Cary 100 spectrometer in absorbance mode. Steady-state photoluminescence (PL) emissions and time-resolved photoluminescence (TRPL) were carried out by Edinburgh FS5 and all samples were excited at 475nm. Scan electron microscope (SEM) and transmission electron microscope (TEM) images were recorded using JEM-7800F and JSM-2800, respectively. X-ray photoelectron spectroscopy (XPS) and ultraviolet photoelectron spectroscopy (UPS) were conducted on Thermo Scientific ESCALAB 250Xi spectrometer. Electrochemical impedance spectroscopy (EIS) was recorded on IM6Ex. The photocurrent density-voltage ($J-V$) curves of all devices were obtained from Keithley 2612 Source Meter under AM1.5 G illumination (100mW/cm) with a Newport/Oriel Sol 3A solar simulator, in which the light intensity was calibrated by a Newport/Oriel instruments PV reference cell system (model 91150). The area of the mask was controlled as 0.1cm^2. External quantum efficiency (EQE) measurements were conducted using a QE system (EnliTech) with a monochromatic light of 20Hz. Both $J-V$ curves and EQE measurements were conducted under ambient conditions.

Acknowledgements

Y. Y., and S. W. contributed equally to this work. This work is supported by the National Natural Science Foundation of China (52102266) and the China Postdoctoral Science Foundation (2020M680861).

References

[1] a) J.-P. Correa-Baena, M. Saliba, T. Buonassisi, M. Grätzel, A. Abate, W. Tress, A. Hagfeldt, Science 2017, 358, 739; b) H. S. Jung, N.-G. Park, Small 2015, 11, 10; c) J. Y. Kim, J.-W. Lee, H. S. Jung, H. Shin, N.-G. Park, Chem. Rev. 2020, 120, 7867.

[2] J. Gong, S. B. Darling, F. You, Energy Environ. Sci. 2015, 8, 1953.

[3] T. Binyamin, L. Etgar, Sol. RRL 2022, 6, 2200295.

[4] a) Z. Ku, Y. Rong, M. Xu, T. Liu, H. Han, Sci. Rep. 2013, 3, 3132; b) Y. Wang, L. Li, Z. Wu, R. Zhang, J. Hong, J. Zhang, H. Rao, Z. Pan, X. Zhong, Angew. Chem., Int. Ed. Engl. 2023, 62, 202302342.

[5] T. Ye, Y. Hou, A. Nozariasbmarz, D. Yang, J. Yoon, L. Zheng, K. Wang, K. Wang, S. Ramakrishna, S. Priya, ACS Energy Lett. 2021, 6, 3044.

[6] C. Liu, C. Gao, W. Wang, X. Wang, Y. Wang, W. Hu, Y. Rong, Y. Hu, L. Guo, A. Mei, H. Han, Sol. RRL 2021, 5, 2100333.

[7] a) Q. Zhao, R. Han, A. R. Marshall, S. Wang, B. M. Wieliczka, J. Ni, J. Zhang, J. Yuan, J. M. Luther, A. Hazarika, G.-R. Li, Adv. Mater. 2022, 34, 2107888; b) Q. Zhao, S. Wang, Y.-H. Kim, S. Mondal, Q. Miao, S. Li, D. Liu, M. Wang, Y. Zhai, J. Gao, A. Hazarika, G.-R. Li, Green Energy Environ. 2023, 131, 6050; c) F. Liu, Y. Zhang, C. Ding, S. Kobayashi, T. Izuishi, N. Nakazawa, T. Toyoda, S. Ohta, S. Hayase, T. Minemoto, K. Yoshino, S. Dai, Q. Shen, ACS Nano 2017, 11, 10373; d) H.-C. Wang, Z. Bao, H.-Y. Tsai, A.-C. Tang, R.-S. Liu, Small 2018, 14, 1702433.

[8] W. Yang, R. Su, D. Luo, Q. Hu, F. Zhang, Z. Xu, Z. Wang, J. Tang, Z. Lv, X. Yang, Y. Tu, W. Zhang, H. Zhong, Q. Gong, T. P. Russell, R. Zhu, Nano Energy 2020, 67, 104189.

[9] a) M. Wang, L. Gao, P. Yu, Q. Wang, C. Yu, X. Zhang, Y. Wang, W. Zheng, J. Zhang, J. Mater. Chem. C 2022, 10, 5134; b) X. Zheng, J. Troughton, N. Gasparini, Y. Lin, M. Wei, Y. Hou, J. Liu, K. Song, Z. Chen, C. Yang, B. Turedi, A. Y. Alsalloum, J. Pan, J. Chen, A. A. Zhumekenov, T. D. Anthopoulos, Y. Han, D. Baran, O. F. Mohammed, E. H. Sargent, O. M. Bakr, Joule 2019, 3, 1963; c) S. Wang, Q. Zhao, A. Hazarika, S. Li, Y. Wu, Y. Zhai, X. Chen, J. M. Luther, G. Li, Nat. Commun. 2023, 14, 2216.

[10] M. Cha, P. Da, J. Wang, W. Wang, Z. Chen, F. Xiu, G. Zheng, Z.-S. Wang, J. Am. Chem. Soc. 2016, 138, 8581.

[11] a) A. Hazarika, Q. Zhao, E. A. Gaulding, J. A. Christians, B. Dou, A. R. Marshall, T. Moot, J. J. Berry, J. C. Johnson, J. M. Luther, ACS Nano 2018, 12, 10327; b) X. Zhang, H. Huang, L. Jin, C. Wen, Q. Zhao, C. Zhao, J. Guo, C. Cheng, H. Wang, L. Zhang, Y. Li, Y. Maung Maung, J. Yuan, W. Ma, Angew. Chem., Int. Ed. Engl. 2023, 62, 202214241.

[12] L. Protesescu, S. Yakunin, S. Kumar, J. Bär, F. Bertolotti, N. Masciocchi, A. Guagliardi, M. Grotevent, I. Shorubalko, M. I. Bodnarchuk, C.-J. Shih, M. V. Kovalenko, ACS Nano 2017, 11, 3119.

[13] J. Xue, J.-W. Lee, Z. Dai, R. Wang, S. Nuryyeva, M. E. Liao, S.-Y. Chang, L. Meng, D. Meng, P. Sun, O. Lin, M. S. Goorsky, Y. Yang, Joule 2018, 2, 1866.

[14] U. G. Lee, W.-B. Kim, D. H. Han, H. S. Chung, Symmetry 2019, 11, 1183.

[15] a) J. Xue, R. Wang, L. Chen, S. Nuryyeva, T.-H. Han, T. Huang, S. Tan, J. Zhu, M. Wang, Z.-K. Wang, C. Zhang, J.-W. Lee, Y. Yang, Adv. Mater. 2019, 31, 1900111; b) A. K. Jena, A. Ishii, Z. Guo, M. A. Kamarudin, S. Hayase, T. Miyasaka, ACS Appl. Mater. Inter. 2020, 12, 33631; c) C. Hanmandlu, S. Swamy, A. Singh, H.-A. Chen, C.-C. Liu, C.-S. Lai, A. Mohapatra, C.-W. Pao, P. Chen, C.-W. Chu, J. Mater. Chem. A 2020, 8, 5263; d) H. Wang, Y. Song, S. Dang, N. Jiang, J.

Feng, W. Tian, Q. Dong, Sol. RRL 2020, 4, 1900468.

[16] a) F. Li, S. Zhou, J. Yuan, C. Qin, Y. Yang, J. Shi, X. Ling, Y. Li, W. Ma, ACS Energy Lett. 2019, 4, 2571; b) Q. Zhao, A. Hazarika, X. Chen, S. P. Harvey, B. W. Larson, G. R. Teeter, J. Liu, T. Song, C. Xiao, L. Shaw, M. Zhang, G. Li, M. C. Beard, J. M. Luther, Nat. Commun. 2019, 10, 2842.

[17] a) P. N. Murgatroyd, J. Phys. D: Appl. Phys. 1970, 3, 151; b) R. Han, Q. Zhao, A. Hazarika, J. Li, H. Cai, J. Ni, J. Zhang, ACS Appl. Mater. Inter. 2022, 14, 4061; c) V. M. Le Corre, E. A. Duijnstee, O. El Tambouli, J. M. Ball, H. J. Snaith, J. Lim, L. J. A. Koster, ACS Energy Lett. 2021, 6, 1087.

[18] A. J. Riquelme, K. Valadez-Villalobos, P. P. Boix, G. Oskam, I. Mora-Seró, J. A. Anta, Phys. Chem. Chem. Phys. 2022, 24, 15657.

本论文原发表于《Solar RRL》2023年。

Understanding Hydrogen Diffusion Mechanisms in Doped α-Fe through DFT Calculation

Zhu Lixia[1]　Luo Jinheng[1]　Zheng Shunli[2]
Yang Shuaijun[3]　Hu Jun[3]　Chen Zhong[4]

(1. Tubular Goods Research Institute of China National Petroleum Corporation;
2. College & Hospital of Stomatology, Anhui Medical University, Key
Lab. of Oral Diseases Research of Anhui Province; 3. School of Chemical
Engineering, Northwest University; 4. School of Materials Science
and Engineering, Nanyang Technological University)

Abstract: There has been sufficient experimental evidence that hydrogen embrittlement is detrimental to structural metals during applications. Herein, we explore the hydrogen diffusion mechanisms in doped α-Fe using first-principles calculations. We demonstrate that the trap hydrogen on the α-Fe (110) surface is a thermodynamically spontaneous process, and doping will decrease the hydrogen adsorption energy due to the change of the adsorption sites. Furthermore, hydrogen diffusion from surface to subsurface will determine the overall diffusion rate, while hydrogen diffusion in the bulk is more time-consuming. Mo, Mn and C are beneficial, which can be able to increase the energy barrier of hydrogen diffusion from the surface to subsurface as well as in the bulk. The current work provides a promising path towards enhancing the hydrogen diffusion barrier in α-Fe.

Keywords: Hydrogen embrittlement; Hydrogen diffusion; First - principles calculations; Doping

1 Introduction

Hydrogen embrittlement often leads to premature and catastrophic failure of structural metals during the industrial applications. Understanding the embrittlement effects of hydrogen is critical to the development of more reliable metals exposed to hydrogen environment, and this topic has attracted enormous research attentions recently [1-2]. Various mechanisms have been proposed to explain hydrogen embrittlement, including hydrogen - induced phase changes to form metal hydrides[3], hydrogen-enhanced localized plasticity[4], and cohesion reduction[5-6], H-induced

---

Corresponding authors: Luo Jinheng, luojh@cnpc.com.cn; Zheng Shunli, zhengshunli1986@126.com; Chen Zhong, ASZChen@ntu.edu.sg.

dislocation emission[7] and initiation of fissure from hydrogen blister[8] and so on. In general, these mechanisms are not mutually exclusive, and it is always difficult to validate these molecular mechanisms. Up to date, the hydrogen embrittlement mechanisms remain unclear and more work is needed to shed light into this outstanding issue.

Density functional theory (DFT) has been extensively used for the hydrogen embrittlement research to prove the corresponding mechanism step by step with its continuous development and advancement. For example, Yin et al. found the embrittlement of silver nanowires is governed by the hydrogen-induced suppression of dislocation nucleation at the free surface[9]. Matsumoto et al. found the hydrogen atoms are trapped at dislocation cores and along a slip plane in the vicinity of a dislocation core using molecular dynamics simulation to study the mode I crack propagation in α-Fe single crystal, which can facilitate the subsequent hydrogen enhanced decohesion[10].

Despite the uncertainty in the hydrogen embrittlement mechanisms, it certainly involves the diffusion of hydrogen atoms. H_2 molecule can firstly separate into two hydrogen atoms on the surface, and the adsorbed hydrogen atoms on the surface can move into the bulk. Understanding the whole process of hydrogen diffusion is of great importance for and the future development of hydrogen embrittlement resistant materials. Mohammadi et al. adopted a strategy using gradient microstructure on a high-entropy alloy to enable both resistance to hydrogen embrittlement on the surface and high mechanical strength in the bulk [11]. X70 and X80 are the most potential materials in hydrogen transportation in China. As we all know, it is very important for the hydrogen embrittlement mechanism of X70 and X80 materials in the large-scale hydrogen storage and transportation. So far, only a few studies has been published on the effects of hydrogen embrittlement with doped elements of X70 and X80.

Herein, we demonstrate a comprehensive theoretical analysis on the hydrogen diffusion behavior ofX70 and X80. The results are able to unveil the hydrogen diffusion from surface to bulk, as well as the change of the activation energy and Gibbs free energy in the whole process, which can help to understand hydrogen embrittlement mechanism from atomic scale.

2 Computational and experimental details

The CASTEP module from Accelrys Inc. was employed for theDFT calculations. As is known, X70 and X80 is consisted of α-Fe with different ion-doping. At first, α-Fe (space group: 186, $a = b = c = 2.8608$Å, $\alpha = \beta = \gamma = 90°$) was employed to construct the model and implement comprehensive evaluation. Considering the symmetry, the (001), (110), and (111) surfaces were established from the optimized bulk structures, and a vacuum region was set as 15Å thickness[12].

After building those models, self-consistent periodic DFT was adopted to explore the hydrogen diffusion process on the facets. Ionic cores were represented by an ultrasoft pseudopotential. The Perdew-Burke-Ernzerhof (PBE) approximation with the Generalized Gradient Approximation (GGA) method was used to calculate the exchange-correlation energy. The Broyden-Fletcher-Goldfarb-Shanno (BFGS) scheme was applied as the minimization algorithm. The k-space sampling points were set as 9×9×4 for 1×1×2 cell of bulk α-Fe, and 4×4×1, 5×5×1, 4×4×1 for (001), (110) and (111) surfaces respectively. Furthermore, bottom four layers were fixed in order to simulate the bulk property. The energy cutoff is 330 eV and the SCF tolerance is 5.0×10^{-7}

eV/atom. The optimization ends when the energy, maximum force, maximum stress and maximum displacement are smaller than 5.0×10^{-6} eV/atom, 0.01eV/Å, 0.02GPa and 5.0×10^{-4} Å, respectively. LST (Linear Synchronous Transit)/QST (Quadratic Synchronous Transit) transition state search algorithm was used to search for the transition state based on nudged elastic band method[13-14]. The calculation setting had been verified by our previous calculation[15], and the results are largely consistent with previous experimental works[16-17].

The free energy of the adsorption atomic hydrogen (in bulk called solution energy) is obtained by Eq. (1), where ΔE_{ZPE} is denoted as the zero-point energy of the system and simplified with the value of 0.05eV. The term $-T\Delta S_H$ is the contribution from entropy in the temperature form of K and is 0.20eV at 298K [18]. ΔE_H, described as the energy, is needed to increase the coverage by one hydrogen atom, and can be calculated with Eq. (2), where $E[\text{surface}+H]$ is the total energy of the system, including the adsorbed molecules and the surface; $E[\text{surface}]$ is the energy of surface; $E[H_2]$ represents the total energy of one H_2 molecule in the gas phase.

$$\Delta G_H = \Delta E_H + \Delta E_{ZPE} - T\Delta S_H \tag{1}$$

$$\Delta E_H = E[\text{surface}+H] - E[\text{surface}] - 1/2E[H_2] \tag{2}$$

The surface energy (γ) can be calculated by:

$$\gamma = \frac{1}{2A}(E_{[\text{surface}]} - nE[\text{bulk}]) \tag{3}$$

Where $E[\text{bulk}]$ is the total energy per unit cell of the bulk, n is the number of unit cells that the slab model contains, and A is the surface area of the slab model[19-20].

3 Results and discussion

3.1 Geometric Structure of α-Fe and Low-index Surfaces

The geometric structure of α-Fe and low-index surfaces are shown in Fig. 1.

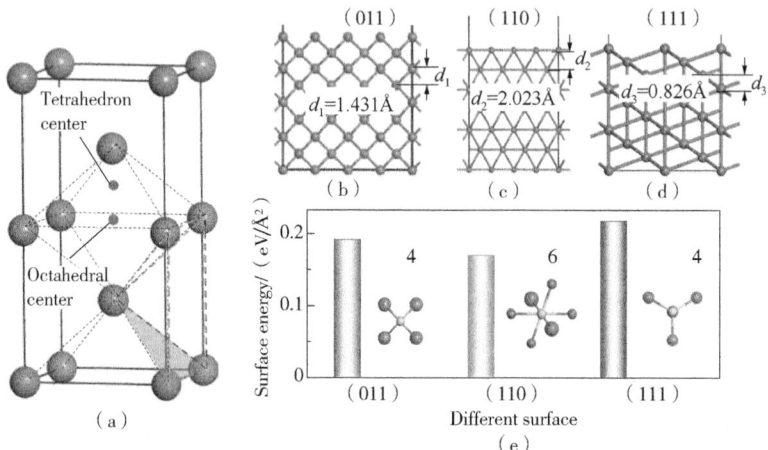

Fig. 1 (a) The geometric structure of 1×1×2 α-Fe and low index surfaces, Fe atoms are shown in dark blue spheres; (b) Side view the geometric structure of α-Fe (001); (c) Side view the geometric structure of α-Fe (110) and (d) Side view the geometric structure of α-Fe (111); (e) Calculated surface energies of α-Fe (001), α-Fe (110) and α-Fe (111), the inset geometric structures are the top view to illustrate the coordinate number of surface Fe atom (yellow)

From Fig. 1, the lattice type of α-Fe is body centered cubic, which is composed of coplanar octahedrons. The tetrahedron and octahedral center sites are indicated by the red dots. Based on the symmetry of α-Fe, the following surfaces are equivalent: (001), (00$\bar{1}$), (010), (0$\bar{1}$0), (100), and ($\bar{1}$00); (110), (101), (1$\bar{1}$0), (10$\bar{1}$), (011), (0$\bar{1}$1), (01$\bar{1}$), (0$\bar{1}\bar{1}$) ($\bar{1}$10), ($\bar{1}$0$\bar{1}$), ($\bar{1}$10), and ($\bar{1}\bar{1}$0); (111), (11$\bar{1}$), (1$\bar{1}$1), (1$\bar{1}\bar{1}$), ($\bar{1}$11), ($\bar{1}$1$\bar{1}$), ($\bar{1}\bar{1}$1) and ($\bar{1}\bar{1}\bar{1}$). Therefore, the (001), (110), (111) surfaces were selected during calculation.

Among thesurfaces above, the surface energies for (100), (110) and (111) are 0.189eV/Å^2, 0.167eV/Å^2 and 0.215eV/Å^2, respectively. It can be found that the energy of the (110) surface is the lowest. The surface energy is mainly determined by the termination position. In general, the bulk Fe has eight coordination positions. Fe has four coordination positions on the (001) and (111) surfaces and six coordination positions on the (110) surface (Fig. 1e). This means that four bonds (8 bonds −4 bonds) need to be cut off on the (001) and (111) surfaces while only two bonds (8 bonds −6 bonds) need to be cut off on (110) surface to form the free surface. Terminating fewer bonds will consume little energies, which will generate the more stable surface. Furthermore, the (110) surface has the largest plane spacing (2.023Å), and the plane spacing has negative correlation relationship with the surface energy. This indicated that larger surface spacing can make bonding weaker and more conducive to forming stable surface. Stable surface is easy to highly exposed during synthesis. Therefore, the (110) surface was selected in the subsequent calculations.

3.2 Hydrogen Adsorptionon α-Fe (110) Surfaces

Chemical adsorption of hydrogen on Fe surface is the first step, which plays an important role in the hydrogen embrittlement evolution[21]. Based on the chemical compositions of pipeline steel of X70 and X80, the atoms of Mo, Ni, Co, Mn, Cr, V, Si and Nb were considered as the substitutional dopants, while C atom stays in the interstitial sites of the Fe structure[22]. Fig. 2 displays the adsorption energies of hydrogen on α-Fe (110) surfaces at different sites while C atom stays in the interstitial sites of the Fe structure.

Forα-Fe (110), the surface is flat and the metal sites on the first layer form a triangular structure. The center of triangular structure (hollow sites) is the most stable adsorption position for hydrogen (−1.31eV) because the top and bridge sites will change into hollow sites after optimization. Furthermore, the surface doping leads to the decrease of adsorption of hydrogen. The lowest absorption energies are −0.75eV, −0.86eV, −0.83eV, −0.78eV, −0.81eV, −0.88eV, −0.91eV and 0.72eV for C, Mo, Ni, Mn, Cr, V, Si and Nb doped structures, respectively. The main reason for the above absorption energy change is that the center of triangular structure is the stable site for hydrogen adsorption and doping changes the electron density at this site. For clean surface, three Fe atoms on the triangular structure have the same electron density because of their same elements and coordination environment. Therefore, the hydrogen can adsorb in the center of triangular structure. After doping, the electron density around three atoms on the triangular structure changed due to the different elements or different coordination environments. This can make the hydrogen adsorption sites slightly change and the adsorption energies decrease.

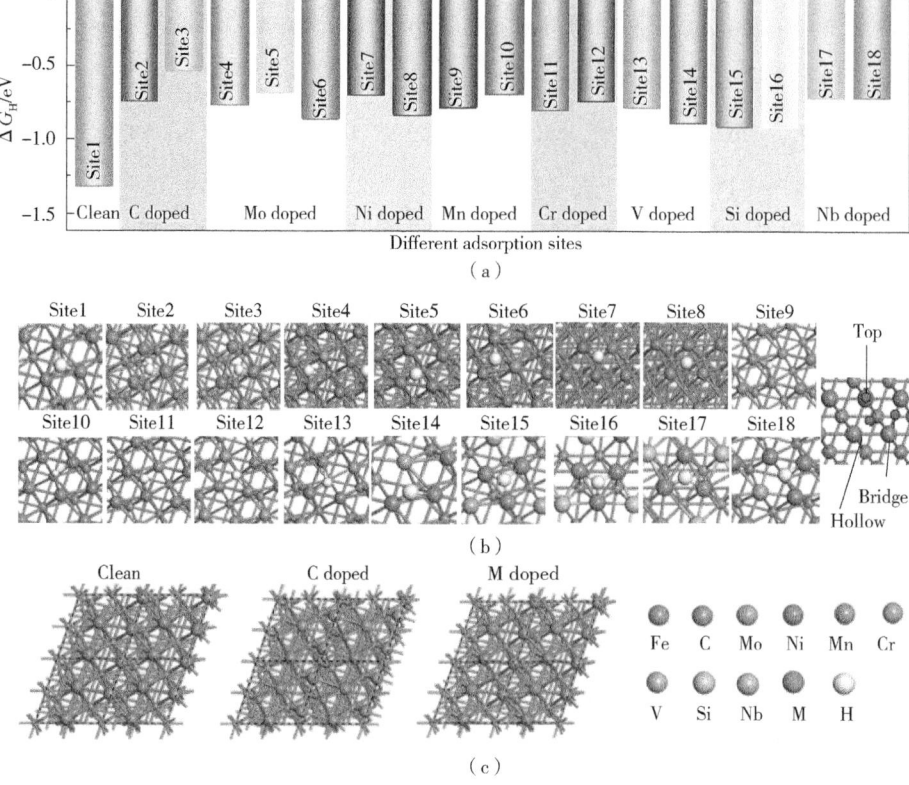

Fig. 2 (a) The adsorption energy of α-Fe (110) with different dopants and (b) the corresponding geometric structure of adsorption where top, bridge and hollow sites are considered; (c) the top view of the surfaces; M stands for doped elements of Mo, Ni, Mn, Cr V, Si and Nb

The negative adsorption energy on the surface means trapping hydrogen is a spontaneous process from the perspective of thermodynamics. Under the hydrogen environment, metals can spontaneously absorb hydrogen atoms on the surface during manufacture and/or service. To reduce the hydrogen adsorption on the surface, a thin coating or surface modification is needed.

3.3 Diffusion of Hydrogen from the Surface to Bulk

After adsorption, the next step is the diffusion of H atom from the surface to bulk. Fig. 3 illustrates the detailed information about the thermodynamics and dynamics of the process.

From thermodynamics aspect, the diffusion of hydrogen atom from surface to bulk is non-spontaneous process because the energies for finial states are positive. As indicated, Cr, V, Nb doping will decrease the reaction energies when compared with the clean α-Fe (110) (about 0.88eV), while C, Mo, Ni and Si doping will increase the reaction energies. It is well known that thermodynamics energy can only judge the tendency of the reaction while kinetic process determines the diffusion rate. The relatively small thermodynamic energy barrier needs further discussion on the dynamic energy barrier in order to judge the diffusion rate. Therefore, transition state energies are also calculated to understand the dynamics. It was found that the activation energy for clean α-Fe (110) is about 1.17eV. C, Mo, Mn and Si doping will increase the dynamics energies while Ni, Cr and V doping will decrease the dynamics energies when comparing with clean α-Fe (110). The increased activation energies of C, Mo, Mn and Si doping are about 0.02eV, 0.08eV, 0.40eV and 0.75eV, respectively. It is likely observed that Mn and Si will greatly increase the activation energies, which will make the H difficult enter into the bulk. So it is beneficial in reducing the hydrogen embrittlement. It is interesting that Si doping has the highest increase in the activation energy among the elements investigated. This finding is in agreement with previous experimental results, where Fe-30Mn-0.6C steel was more sensitive to hydrogen embrittlement than Fe-30Mn-3Al3Si steel[23]. Besides alloying, ion implantation containing silicon can also be effective to inhibit the diffusion of hydrogen from the surface to bulk. The main reason is that Si doping makes the stable adsorption site unstable in sub-layer (inset Fig. 3h). So it needs to consume more energy to cross the sub-layer.

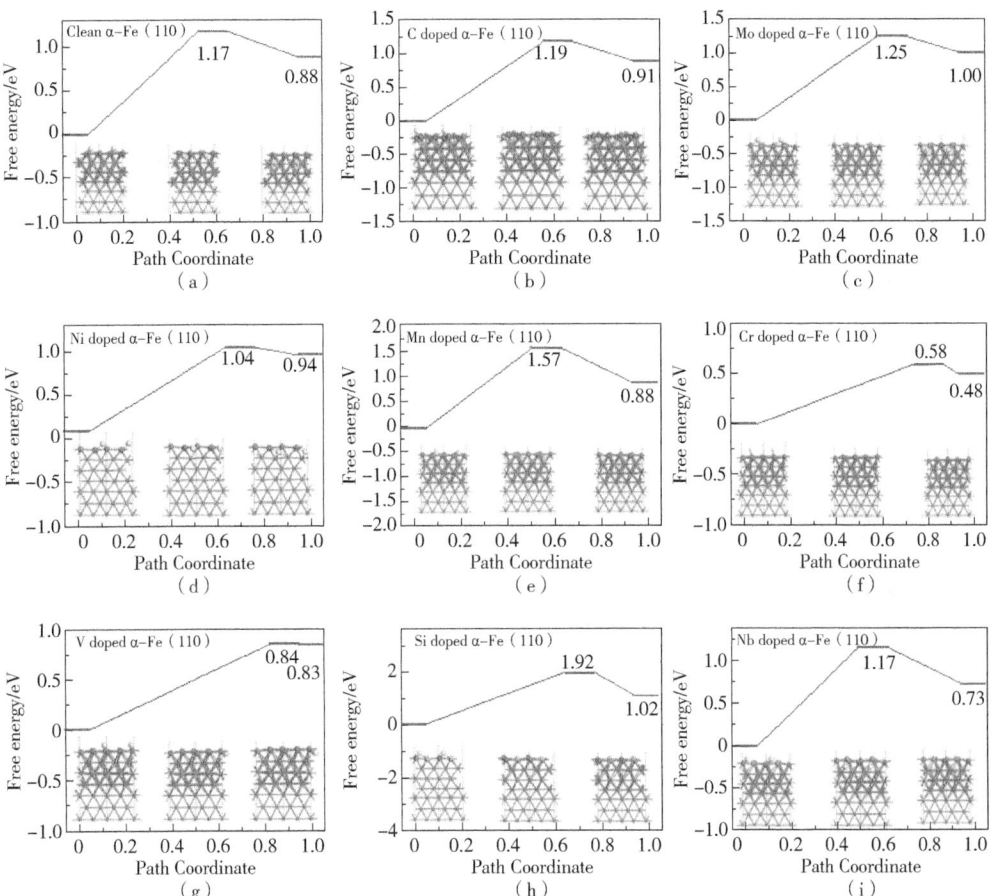

Fig. 3 The energy barrier of H penetrated into α-Fe (110) in different doping and corresponding geometric structure. (a) clean; (b) C doping; (c) Mo doping; (d) Mn doping; (e) Ni doping; (f) Cr doping; (d) V doping; (e) Si doping; (f) Nb doping

3.4 HydrogenAtom Soluble Properties of Bulk

After thehydrogen diffusion from the surface to bulk, the hydrogen will diffuse in bulk from shallow to deep. So the hydrogen soluble properties of the bulk are equally important towards understanding the hydrogen embrittlement. Fig. 4 shows the soluble energies and structures of hydrogen in bulk α-Fe.

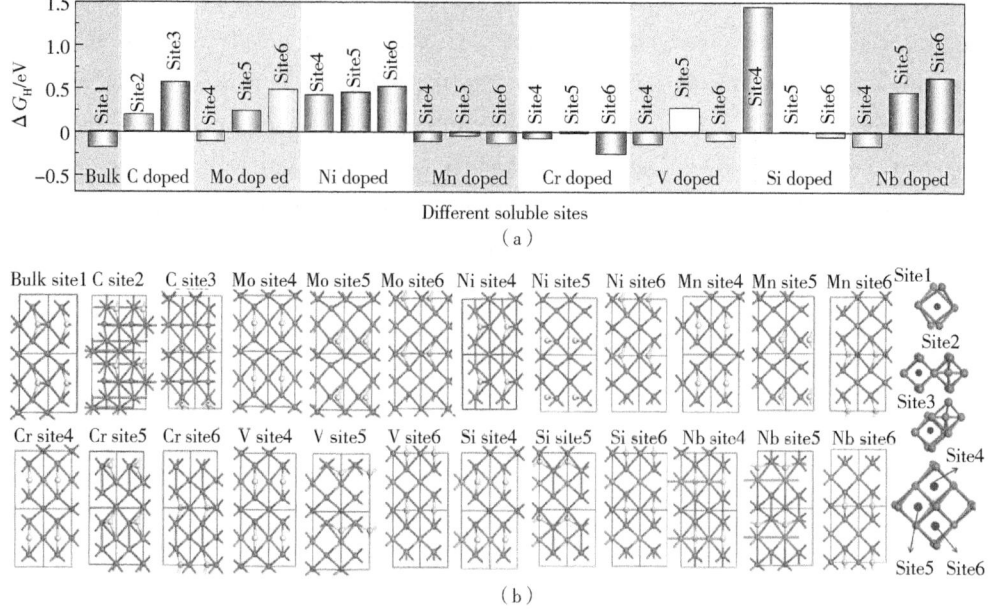

Fig. 4 (a) The soluble energy of bulk α-Fe with different doping atoms and (b) the corresponding geometric structures

At first, we considered octahedral and tetrahedron centers for hydrogen dissolution. After optimization, it was found that H diffuses from octahedral center to tetrahedron center and tetrahedron center is the site for stable hydrogen adsorption, named site 1 in Fig. 4. Single point energy calculation indicates the soluble energies for octahedral center is 0.69eV while for tetrahedron centers is -0.19eV. Therefore, tetrahedron centers are the preferential sites for hydrogen dissolution, which are consistent with the previous calculation[24].

After doping, otherdissolution sites need to be considered, because the atomic properties of octahedral vertices are different. Site 2 and site 3 are the sites where the neighbor position of C is different. Site 2 shares the same vertex with carbon doping sites and site 3 is coplanar with carbon doping sites. Site 4, site 5 and site 6 are the sites where the atom on octahedral vertices is different. Site 4 is the octahedral cell with four dopants atoms, site 5 is the octahedral cell with one dopant atom, and site 6 is the octahedral cell without dopants. These sites can be further divided into octahedral center and tetrahedron center based on the stable H adsorption.

As manifested, most adsorption sites are still tetrahedron centers after optimization. While octahedral center with four dopants atoms of Mo, Mn, Cr, V, Vb can stabilize adsorption H. This indicates doping can change the stable hydrogen adsorption sites. The main reason is that doping metal atoms have repulsive interaction with H atoms, which can make hydrogen run away and change the adsorption sites to octahedral center.

Compared with bulk Fe without dopant, the soluble energies increase except for Cr doping in site 6. Furthermore, some soluble energies become positive with atoms (such as of C, Mo, Ni, Si and Nb) doping in some sites. The increased energies indicate the doping tends to reduce the hydrogen fixation capacity in the bulk.

3.5 The Hydrogen Diffusion in Bulk

The hydrogen diffusion in bulk also plays an important role in hydrogen embrittlement, which can be calculated by the transition state search algorithm, as shown in Fig. 5.

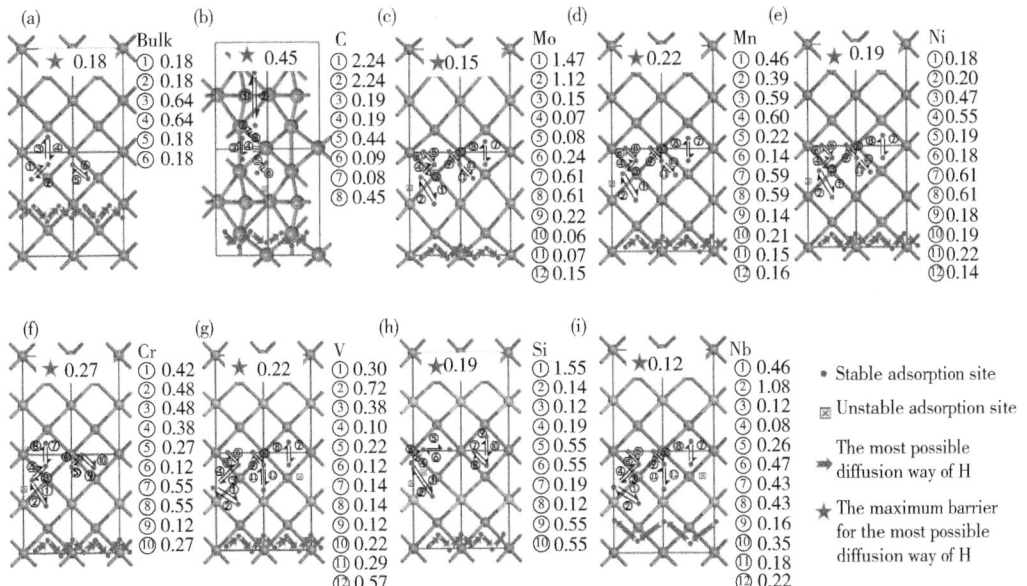

Fig. 5 The diffusion path for path 1 to path 12 with the corresponding geometric structure. The transition state energies are manifested below the geometric structure with unit of eV, and the major role in hydrogen diffusion based on the transition states energies indicated by red arrows (a) bulk; (b) C doping; (c) Mo doping; (d) Mn doping; (e) Ni doping; (f) Cr doping; (g) V doping; (h) Si doping; (i) Nb doping

During the calculation, six hydrogen diffusion paths were considered. Path 1 and path 2 are the hydrogen bypass coplanar structures while path 3 to path 6 are the hydrogen diffusion from one site to another in the same octahedron cell. As manifested, the transition state energies for path 3 and path 4 are 0.64eV while that for path 1, path 2, path 5, path 6 are 0.18eV for the bulk. Based on the results, it is the most possible way for the zig-zag diffusion along path 1, path 2, path 5, path 6 with the maximum barrier of 0.18eV.

After doping, it can be found that some sites can stably adsorb H atoms while some sites can't due to the atom change. Therefore, the diffusion ways for stable sites to neighbourly stable sites were considered. It is interesting to noted that all doping will increase the transition state energies near the doped atoms (path 1 and path 2). This indicates doped octahedron cell can increase the diffusion energy. The diffusion energies for C, Mo, and Si are 2.24eV, 1.47eV, 1.55eV, respectively, which are significantly increased in diffusion energy. In this model, the doping concentration is only 25%. If other properties are not considered, it is a good choice to reduce hydrogen diffusion rate by increasing dopants of C, Mo, and Si.

However, doping will affect the undoped octahedron cell due to its slightly changed structure after doping. Therefore, the hydrogen diffusion along the undoped octahedron cell should be considered. As indicated, the transition state energies for undoped octahedron cell are greatly smaller than that of doped one. It can be judged from the energy that the major diffusion path is undoped octahedron cell with zigzag. During diffusion, the biggest transition state energies with the corresponding major diffusion path will determine the diffusion energy. The biggest values for the major diffusion path were 0.18eV, 0.45eV, 0.15eV, 0.22eV, 0.19eV, 0.27eV, 0.22eV, 0.19eV, and 0.12eV for bulk, C, Mo, Mn, Ni, Cr, V, Si and Nb doping, respectively. It can be also found that C, Mn, Ni, Cr, V and Si doping will increase diffusion resistance for the hydrogen diffusion, especially for the C doping.

During the entire hydrogen diffusion process, hydrogen can be chemically adsorbed on the surface at first, and then diffused into the sub-surface and the bulk. The whole process for the hydrogen diffusion can be obtained based on the energies at different steps. We find that the hydrogen diffusion from surface to subsurface will consume more energy than other steps. Therefore, it is the rate-determining step in the whole process, which can play a critical role in the hydrogen embrittlement evolution. At this step, C, Mo, Mn and Si doping can increase the activation energies. Furthermore, hydrogen diffusion in the bulk also plays an important role because it is a time-consuming process. After hydrogen diffusing into the bulk, it will weaken the metal bonds, or accumulate on internal defects or crack to form hydrogen ballooning. All of them will deteriorate the mechanical properties of metals. C, Mn, Ni, Cr, V and Si doping will greatly increase the activation energies. Based on the findings above, it can be concluded that C, Mn and Si are beneficial elements in the whole process, which will increase the energy barrier against hydrogen diffusion both from surface to bulk and in bulk.

3.6 The Eeffect of Adsorbed Hydrogen on Hydrogen Embrittlement

To understand the impact of surface and bulk adsorbed with hydrogen on the embrittlement of carbon steel, the partial density of states (PDOS) is shown in Fig. 6.

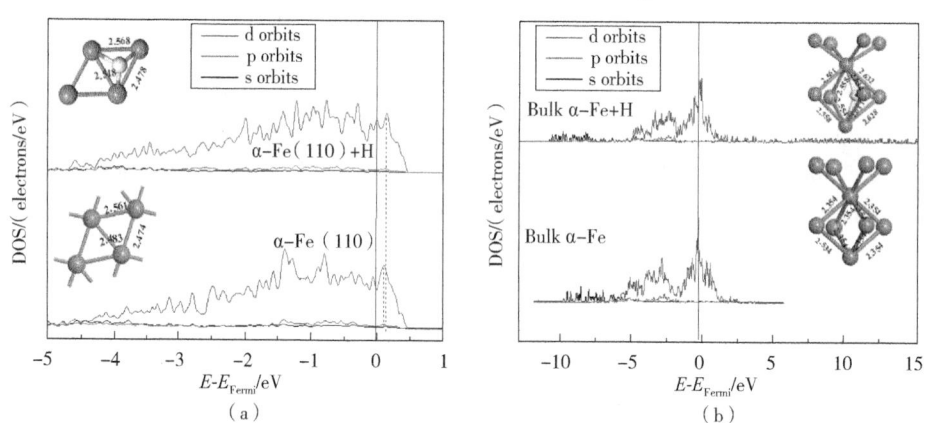

Fig. 6 (a) The calculated total PDOS of Fe on α-Fe (110), hydrogen adsorbed α-Fe (110) surfaces;
(b) The calculated total DOS of Fe in bulk α-Fe, hydrogen adsorbed bulk α-Fe. The insets are
the theoretical bond lengths of Fe-Fe with unit of Å

As indicated, p orbital electrons in the PDOS peaks of hydrogen adsorption surface will move to a shallower energy level above the Fermi level, which indicates there is a certain repulsion because orbits above the Fermi level represent antibonding orbitals. Furthermore, hydrogen adsorption on the surface will lengthen the Fe-Fe bond. The average bond length of Fe-Fe bond on α-Fe (110) plane changes from 2.506Å to 2.521Å, while that in bulk changes from 2.354Å to

2.595Å. The value change indicated the surface hydrogen adsorption can weaken the Fe-Fe bond. Moreover, it is different for the degree of the effect of the hydrogen adsorption on surface and bulk. The change of the average bond length is 6.08% and 9.27% for surface and bulk, respectively, which implied that the bulk hydrogen has a higher weakening degree to the structure than the surface hydrogen, probably enhancing the fracture of Fe-Fe bonds and subsequent formation of new surface.

4 Conclusions

In this paper, hydrogen diffusion behavior onundoped, as well as C, Mo, Ni, Co, Mn, Cr, V, Si and Nb doped α-Fe were analyzed by DFT calculation. The whole diffusion process of hydrogen from surface to bulk were considered step by step. The results show that the hydrogen adsorption on α-Fe(110) surface is a spontaneous process from the perspective of thermodynamics, and doping will decrease the hydrogen adsorption energy due to the change in the adsorption sites. C, Mo, Mn and Si doping will increase the dynamics energies while Cr, Ni and V doping will decrease the dynamics energies when comparing with the energy for pristine α-Fe(110). Furthermore, doping will change the diffusion path in the bulk, along with an increase in the energy barrier. When considering the whole diffusion process, it is found that hydrogen diffusion from the surface to subsurface is the rate-determining step. Mo, Mn, and C are beneficial elements to slow down the entire hydrogen diffusion process. The current results provide new insights into the hydrogen diffusion mechanisms in α-Fe and suggest effective alloying elements to prevent hydrogen embrittlement.

Credit authorship contribution statement

L. X. Zhu and J. Hu conceived the idea, designed the framework, L. X. Zhu and S. L. Zheng drafted the manuscript. S. J. Yang performed the calculation. J. H. Luo and Z. Chen assisted in drafting the manuscript.

Acknowledgements

This research is supported by the Science & Technology Project of CNPC (2021DJ5402), National Natural Science Foundation of China (No. 21676216), Special project of Shaanxi Provincial Education Department (No. 20JC034), Grants for Scientific Research of BSKY (No. XJ201918) from Anhui Medical University, 2022 Disciplinary Construction Project in School of Dentistry, Anhui Medical University (No. 2022xkfyts09) and Hefei Advanced Computing Center.

References

[1] X. C. Ren, W. Y. Chu, Y. J. Su, et. al. , The effects of atomic hydrogen and flake on mechanical properties of a tyre steel, Corros. Sci. , 491, 1-2 (2008), 164-171.

[2] J. Li, Y. P. Xie, Y. X. Chen, et. al. First-principles study of the hydrogen adsorption and diffusion on ordered NiFe(111) surface and in the bulk3, Corros. Sci. , 44 (2014), 64-72.

[3] D. Westlake, Generalized model for hydrogen embrittlement, ASM, Trans. Quart, 1969.

[4] P. Sofronis, I. M. Robertson, Transmission electron microscopy observations and micromechanical/continuum models for the effect of hydrogen on the mechanical behaviour of metals, Philosophical Magazine A, 82: 17-18

(2002), 3405-3413.

[5] R. A. Oriani, Hydrogen embrittlement of steels, Ann. Rev. Mater. Sci, 8 (1978): 327.

[6] R. A. Oriani, A mechanistic theory of hydrogen embrittlement of steels, Ber. Bunsenges. Phys. Chem., 76 (1972), 848857.

[7] T. P. Chapman, V. A. Vorontsov, A. Sankaran, D. Rugg, T. C. Lindley, D. Dye, The dislocation mechanism of stress corrosion embrittlement in Ti-6Al-2Sn-4Zr-6Mo, Metall. Mater. Trans. A 47 (1) (2016), 282-292.

[8] W. Y. Chu, K. W. Gao. Initiation of fissure from hydrogen blister in rail ateel. Corros.: J. Sci. Engineering, 2000, 56(10): 1046-1049.

[9] S. Yin, G. M. Cheng, T. Chang, et. al. Hydrogen embrittlement in metallic nanowires, Nat. Commun., 10 (2019), 2004.

[10] R. Matsumotoab, S. Taketomiab, S. Matsumotoa, N. Miyazakiab, Atomistic simulations of hydrogen embrittlement, Int. J. Hydrogen energy, 34 (2009), 9576-9584.

[11] A. Mohammadi, MarcNovelli, M. Arita, J. W. Bae, et. al. ThierryGrosdidier, K. Edalati, Gradient-structured high-entropy alloy with improved combination of strength and hydrogen embrittlement resistance, Corros. Sci., 200 (2022), 110253.

[12] R. Crespo-Otero, A. Walsh, Variation in Surface Ionization Potentials of Pristine and Hydrated $BiVO_4$, J. Phys. Chem. Lett, 6 (2015), 2379-2383.

[13] G. Henkelman, H. Jónsson, Improved tangent estimate in the nudged elastic band method for finding minimum energy paths and saddle points, J. Chem. Phys., 113 (2000), 9978-9985.

[14] Y. Tan, Y. C. Zhu, X. F. Cao, et. al. Discovery of hydrogen spillover based binary electrocatalysts for hydrogen evolution: from theory to experiment. ACS Catal., 12 (2022), 11821-11829.

[15] J. Hu, S. L. Zheng, X. Zhao, et. al. A theoretical study on the surface and interfacial properties of Ni_3P for hydrogen evolution reaction, J. Mater. Chem. A, 6 (2018), 7827-7834.

[16] J. K. Nørskov, T. Bligaard, A. Logadottir, et. al. Stimming, Trends in the exchange current for hydrogen evolution, J. Electrochem. Soc., 152 (2005), J23-J26.

[17] T. L. Tan, L. L. Wang, D. D. Johnson, et. al. Hydrogen deposition on Pt (111) during electrochemical hydrogen evolution from a first-principles multiadsorption-site study, J. Phys. Chem. C, 117 (2013), 22696-22704.

[18] Q. Tang, D. Jiang, Mechanism of hydrogen evolution reaction on $1T-MoS_2$ from first orinciples, ACS Catal., 6 (2016), 4953-4961.

[19] J. Hu, W. Chen, X. Zhao, et. al. Anisotropic electronic characteristics, adsorption, and stability of low-index $BiVO_4$ surfaces for photoelectrochemical applications, ACS Appl. Mater. Interfaces, 10 (2018), 5475-5484.

[20] Z. Łodziana, N.-Y. Topsøe, J. K. Nørskov, A negative surface energy for alumina, Nat. Mater., 3 (2004), 289-293.

[21] S. S. Sohn, S. Y. Han, S. Y. Shin, et. al. Analysis and estimation of the yield strength of API X70 and X80 linepipe steels by double-cycle simulation tests, Metals and Mater, Int., 19 (2013), 377-388.

[22] L. Chen, S. Antonov, K. Song, et. al. Effect of solute atoms (C, Al and Si) on hydrogen embrittlement resistance of high-Mn TWIP steels, Corros. Sci., 203 (2022) 110376.

[23] Y. Ma, Y. Shi, H Wang, et al. A first-principles study on the hydrogen trap characteristics of coherent nano-precipitates in α-Fe, Int. J. Hydrogen Energy, 45, 2020, 27941-27949.

本论文原发表于《International Journal of Hydrogen Energy》2023 年第 48 卷。

第二篇　成果篇

一、2023 年获得省部级(含社会力量)科技奖励二等奖以上成果

2023 年获得省部级(含社会力量)科技奖励见表1。

表1 2023 年获得省部级(含社会力量)科技奖励概览

序号	成果名称	颁奖机构	获奖等级	获奖类型
1	第三代超大输量低温高压管线用钢关键技术开发及产业化	中国钢铁工业协会、中国金属学会	特等奖	冶金科学技术奖
2	大口径高压力非金属复合管及其制备方法	中国国家知识产权局	优秀奖	中国专利奖
3	能源开发用钛合金石油管材料、配套技术研发及推广应用	辽宁省人民政府	一等奖	科技进步奖
4	油气田集输管网服役安全关键技术与工业化应用	中国石油和化学工业联合会	一等奖	科技进步奖
5	复杂苛刻深井用油套管完整性技术研发及规模化应用	陕西省人民政府	二等奖	科技进步奖
6	复杂油气井管材及工具关键技术研究与应用	陕西省人民政府	二等奖	科技进步奖
7	CO_2 驱注采环境管材腐蚀防控技术研究与应用	陕西省人民政府	二等奖	科技进步奖
8	石墨烯技术在石油管材表面处理中的应用基础研究	中国石油天然气集团有限公司	二等奖	基础研究奖
9	460 机组高质量钢管轧制技术和新孔型开发	中国宝武钢铁集团有限公司	二等奖	技术创新重大成果奖
10	陆上复杂油气开采用非 API 标准规格特殊螺纹油套管开发与应用	中国宝武钢铁集团有限公司	二等奖	技术创新重大成果奖

1. 第三代超大输量低温高压管线用钢关键技术开发及产业化(中国钢铁工业协会、中国金属学会冶金科学技术特等奖)

提高天然气等清洁燃料在一次能源中的比例是我国实现"环境友好、经济可持续发展"的重要战略举措。近年来,我国大力发展"西气东输、北气南送、海上登陆"能源通道建设,东北、西北高寒地区年输量高达 $380×10^{12} m^3$ 以上超大输量管道急需的大口径 X80 管材对钢铁行业提出了严峻挑战。为保障国家能源安全,钢铁行业齐心协力,在中国钢铁协会指导下,调动全行业并联合中国石油组成产—学—研—用优势力量,通过多年来的攻关、发展和完善,形成了完整的管线钢/管理论、技术和产品体系,有力支撑了西气东输系列、中俄管道等重大工程建设。主要创新成果如下。

(1)建立了高钢级管线钢低温断裂韧性的冶金学原理及控制理论。

在国际上率先发现了针状铁素体管线钢中相邻晶粒{100}解理面夹角不小于35°晶界所包围的面积是低温解理断裂的有效晶粒尺寸;发现了"超细晶铁素体+针状铁素体"双相组织

具有更细化的有效晶粒尺寸,为特厚管线钢低温断裂韧性控制奠定了理论基础。

(2) 集成了特宽幅 X80 钢板和超大口径埋弧焊管全流程核心制造技术。

创新开发了特宽幅 X80 钢板,基于高渗透轧制的温控形变、低负荷快速叠加形变、连续变速控冷等全流程关键制造技术,解决了特宽幅厚规格 X80 钢板总压缩比小、心部变形渗透少、截面组织性能不均匀、超宽钢板轧制负荷高等系列难题。

(3) 突破了超厚规格 X80 热轧钢带和螺旋焊管全流程关键制造技术。

攻克了"高过热度+二冷强冷+大压下量"连铸坯中心偏析控制、全流程组织超细化和均匀化控轧、超厚高强钢带头部冷却控制、超厚/宽钢带卷型控制等关键制造技术,突破了大口径螺旋缝埋弧焊管低应力成形和专用焊接工艺技术。

(4) 首创了低温 X80 管材质量评价和标准体系。

国内首次建立了低温环境管线钢本构模型,确立了温度及壁厚耦合效应下的 X80 钢管断裂韧性确定方法;形成了低温服役条件管线工程用 X80 管材、焊材产品标准,制定了完善的低温 X80 钢管质量评价方案。

成功开发出最大直径 1422mm,最大壁厚 35.2mm 的 X80 钢板/带、直缝/螺旋埋弧焊管系列产品,实现了钢级、管径、壁厚规格全覆盖;累计供货 170 余万吨,钢板销售收入超 85 亿元,累计经济效益达 15.35 亿元。此外,本项目获授权专利 21 项,发表论文 50 余篇,实现全球首发 2 项、国内首发 3 项,有力支撑了国家重大工程建设,保障了国家能源安全。

2. 大口径高压力非金属复合管及其制备方法(中国专利奖优秀奖)

大口径金属集输干线管道的运行安全及环保问题成为油田地面工程亟待重点解决的生产难题。与金属管道相比,非金属管道制造过程节能节水,其优异的耐蚀性也为解决金属管道的腐蚀问题,进而保障油气管道的运行安全及环保提供了有力支撑。参评专利围绕大口径(>150mm)非金属管材承压性能极为有限(≤4MPa)的实际现状,开发出大口径高压力非金属复合管产品及连接方式和制备方法,填补了国内外空白,具有很强的新颖性和创造性。

本专利原创性突出,解决了本领域诸多关键共性技术难题,与同类技术相比不仅同时提高了管材口径和压力等级,还有效改善了产品性能并拓展了应用范围,全面指导了油田非金属石油管材的科学选用。与此同时,还通过申请 16 项专利、5 项标准、2 项软件著作权,建立多项规章制度等形式对参评专利实现了多维度、多系统的保护,全面引领了油气工业非金属管材行业发展,并取得了显著的经济效益和社会效益。

本专利产品在油气工业领域可代替传统钢制管道,从本质上保障油气输送管道的服役安全,大幅降低资源消耗及环境污染风险。该专利技术具备较高的推广应用价值和市场需求。

3. 能源开发用钛合金石油管材料、配套技术研发及推广应用(辽宁省科技进步一等奖)

随着我国能源安全重要性的不断提升,能源开采不断向超深地层、深水、地热及可燃冰等非常规领域拓展,对高性能石油管材的需求巨大。目前已有的铁镍基石油管材已经无法满足未来越来越苛刻的能源开发工况,对轻质高耐蚀、高性能的钛合金管材提出了巨大且迫切的需求。面对能源开发环境的巨大差异,已有钛合金材料用于能源开发还有诸多问题,如能源工况下钛合金选材不清,钛合金管耐蚀及强韧性不足,钛合金管加工制备不成熟、钛合金石油管设计、评价等配套技术缺乏,以及国际上钛合金石油管标准体系空白等,限制了钛合金管材在能源开发领域的应用。

本项目从能源开发用钛合金石油管技术需求入手，弥补现有技术和工程应用的差距，经过十余年"产学研用"多学科联合攻关，取得以下创新成果：

(1) 首次建立能源开发用钛合金材料服役损伤机理和选材体系。首次明确了钛合金材料在多种能源开发工况中的服役损伤机理，通过大量模拟工况试验评价和总结分析建立了基于服役工况的能源开发用钛合金材料选材指南，为能源行业选用钛合金材料提供了指导。

(2) 填补国际上钛合金石油管材料—加工—产品—应用的标准体系空白。针对国际上油气开发钛合金石油管材的标准空白，首次制定并发布了钛合金石油管材料—加工—产品—应用标准体系，形成了2个国家标准、4项行业标准及2项企业标准并均已发布实施，且成功立项国际标准《石油天然气工业用钛合金油套管》，为我国在国际上钛合金石油管领域技术和标准引领奠定了基础。

(3) 显著提升了国内钛合金石油管性能水平，开发出满足工况需求的系列化钛合金石油管产品并得到规模化应用。基于能源行业钻井、采输等不同用途对管材的性能需求，自主研发出高耐蚀钛合金材料、高强高韧钛合金管材，提升了国内能源开发用钛合金石油管的性能水平。开发出钛合金油管、套管、钻杆、集输管、防砂筛管、膨胀管等系列化钛合金石油管产品，在南海可燃冰、陆上超深层钻采、高含硫天然气采输、稠油地热等多领域实现规模化应用。

(4) 形成了高效成形一体化钛合金石油管工业化制造成套技术。突破了大规格钛合金石油管材熔炼—斜轧穿孔—轧制/挤压高效成形技术，攻克管材工业化制备工艺—组织—性能—质量一体化控制技术难题，形成了高效优质的钛合金石油管工业自动化制造成套技术，市场占有率超过70%。

(5) 首次建立完善的钛合金石油管设计、评价、应用配套技术体系。针对钛合金石油管应用方面的一系列"卡脖子"难题，在钛合金螺纹设计、评价、表面处理、抗黏扣技术、实物性能预测计算方法、钻采管柱动力学等多个方面进行技术突破，建立并形成了完善的钛合金石油管设计、评价和应用配套技术体系，为钛合金石油管的应用扫清了障碍。

本项目发布国家标准2项、行业标准4项、企业标准2项，出版2本专著、发表30余篇论文(其中SCI/EI收录22篇)、授权14项专利，其中发明专利9项。本项目开发出系列化钛合金石油管产品，成功应用于中国石油大庆油田、西南油气田、大港油田、海洋工程有限公司、大庆钻探、渤海钻探，中国石化西北局等可燃冰商业试采、陆上高含硫油气田、超深井钻井及老油田侧钻水平井开采等，打开了钛合金石油管系列产品在我国规模化应用局面，产生了显著的经济和社会效益，实施以来近三年新增经济收入近2.82亿元，新增利润7177.9万元，间接经济效益超12亿元。

4. 油气田集输管网服役安全关键技术与工业化应用(中国石油和化学工业联合会科技进步一等奖)

集输管网是油气田地面工程重要组成部分，承担着采集、输送、注入等功能，集输管网安全平稳运行是保障油气产量的重要前提。中国石油已建集输管网$37.3×10^4$km，介质腐蚀性强，管道失效率最高达400次/(10^3km·a)，相比国外先进水平存在巨大差距。要实现集输管网安全平稳运行，亟需解决三大技术难题：一是适用油气田管道特点的缺陷检测技术体系尚不健全；二是缺乏科学有效的风险评价方法；三是缺乏适用于复杂工况下的内外腐蚀防护技术。

针对以上技术问题，在国家科技重大专项、863计划、国家科技支撑计划及中国石油科

技项目支持下,历经15年研究,取得重大技术突破,主要创新成果如下:

创新点1:针对集输管网拓扑结构复杂,现有内检测器通过性差,缺乏焊缝复合缺陷检测技术与工具等难题,创新两大缺陷检测技术。研制高通过性超高清漏磁、变形检测、IMU位置检测三位一体智能内检测技术与装备,具有1.5D弯头和变径管道通过能力,实现同一时间轴对齐的几何变形、金属损失等复合缺陷数据采集与精准量化;创新应用缺陷端点超声波衍射技术和回波幅度对比技术(ACT),研制了焊缝复合缺陷窄脉冲宽频探头,建立了在役管道焊缝缺陷检测和复合缺陷评价方法;系列成果为集输管网监检测提供了可靠的技术方法。

创新点2:针对集输管网介质流态多样、腐蚀成因复杂、服役安全状态缺乏评价方法等问题,创新基于失效数据的风险评价技术。研发7大类24小类集输管道失效分类方法,构建三级识别模型,开发了行业最大规模集输管道失效数据库;首创基于失效数据的集输管网风险评价技术,实现了复杂介质环境下集输管网的风险精准识别和科学评价。创建了多因素耦合的原位pH值计算模型,研发了酸性气田、页岩气田,以及低渗透气田湿气管道内腐蚀直接评价系列技术。系列成果指导中国石油年均识别出高后果区和高风险级管道$5×10^4$km。

创新点3:针对集输管网服役工况日益严苛,现有腐蚀控制措施适应性差的难题,创新系列高效防腐技术。研发了CT2-19系列缓蚀剂和XY-618水溶油分散缓蚀杀菌一体化药剂,有效解决了酸性气田、含CO_2-SRB油田集输管道内腐蚀控制难题;研制在线内挤涂技术与质检装备,国际首次实现了内涂层全程电火花检测。相关技术在长庆油田、大庆油田、吉林油田、西南油气田等9家油气田应用3万余千米。

授权发明专利32件(国际专利1件)、实用新型专利16件、软件著作权10项、技术秘密4项,发布标准16项、专著9部,发表SCI/EI论文40篇。成果在中国石油16家油气田推广应用,年失效次数降低了68%,减少油气泄漏$5.6×10^4$t,取得了良好的安全、环保及社会效益。项目研发新产品实现合同收入21575万元,大庆油田、西南油气田、长庆油田和吉林油田4家应用单位累计获得经济效益42.99亿元。

二、授权发明专利目录

专利号	名称	授权日期
2020111680729	轴向表面型裂纹钢管爆破试验装置及其断裂阻力评价方法	2023-01-10
2021100105982	一种抗剪切气密封的螺纹接头	2023-01-10
2021104623853	聚乙烯管电熔焊接接头缺陷无损检测仿真试块的制作方法	2023-01-24
2021104610266	一种非金属管道本体缺陷无损检测仿真试块的制作方法	2023-01-24
2021104847341	一种非均匀加载应力腐蚀试验装置及其试验方法	2023-01-24
2021104759393	一种控制晶粒尺寸的低温增材制造用丝材及制备和应用	2023-02-21
2020109978941	天然气场站压缩机出口管道防腐层服役可靠性预测方法	2023-02-10
202011403159X	一种抽油杆表面防腐耐磨复合涂层及其制备方法	2023-02-28
2020103743459	一种抗高钙采出水腐蚀的油气田集输管线缓蚀剂	2023-02-10
2020102521850	一种油气井水泥环密封性能检测评价装置及评价方法	2023-04-25

续表

专利号	名称	授权日期
2020103230392	一种管道焊缝微区拉伸试样加工及测试方法	2023-04-25
2020104698802	一种穿透型裂纹钢管爆破试验方法	2023-04-25
2020101771505	一种玻璃钢管材服役寿命预测方法	2023-05-26
202010252187X	一种预测油气环境下热塑性塑料服役寿命试验方法	2023-05-26
2021105121040	一种抗腐蚀的氮化钛耐磨涂层及其制备方法和包含该涂层的制品	2023-05-02
2020107855453	一种焊缝特征区域变形损伤演化规律实验方法	2023-04-25
2020109979041	一种页岩气井水泥环在剪切载荷下破碎形态分级装置及方法	2023-04-07
201811051866X	一种评价钢管表面状态对开裂行为影响的试验装置及方法	2023-06-30
2019108774910	一种含表面划伤复合凹陷油气管道极限内压的预测方法	2023-06-30
2020114622513	一种评估爆炸危害引起管道变形程度的方法	2023-04-25
2019107754394	一种基于螺纹接头压缩效率的环空压力计算方法	2023-09-26
2021100776652	一种地下储气库注采管柱密封性能检测方法	2023-04-25
2019107754464	一种基于管柱接头压缩能力的油套环空压力控制方法	2023-10-31
2021101696848	一种不锈钢管材耐水线腐蚀性能测试装置和方法	2023-04-25
2020100539205	绘制各向异性材料断裂成形极限图的方法及其使用方法	2023-05-26
202110064977X	一种连续油管疲劳试验装置及方法	2023-01-10
2020100819808	一种气藏型储气库注采管柱冲蚀失效风险测定方法	2023-06-30
2020101577971	一种基于天然气热辐射的安全距离确定方法及系统	2023-06-30
2021100043577	一种油套管实物断面形貌的检测方法及检测装置	2023-06-30
2021104624038	一种聚乙烯管热熔焊接接头缺陷仿真试块制作方法	2023-06-02
2018108636263	一种高温实时监测溶解氧浓度的腐蚀评价装置及方法	2023-07-25
2020102096235	一种硫化氢腐蚀管材的模拟装置及模拟方法	2023-07-25
2019102943686	一种用于高钢级输气管道的钢套筒止裂器设计方法	2023-08-22
2019113127766	一种套管孔眼冲蚀试验系统及方法	2023-08-22
2020106634627	一种油套环空污染环境中管材应力腐蚀开裂敏感性评价装置及方法	2023-04-07
2021100762876	一种用于三通管件和环焊缝裂纹的在役修复套筒及方法	2023-06-30
2021104505236	修复三通支管连接环焊缝缺陷的套筒及加工、修复方法	2023-02-21
2021100776629	一种用于三通支管与对接钢管间环焊缝缺陷修复的B型套筒及其安装方法	2023-04-25

续表

专利号	名称	授权日期
2020106415649	一种含裂纹金相试样的浸蚀及图像处理方法和设备	2023-04-25
2020106711252	油气管道环焊缝缺陷修复用B型套筒承载能力检验方法	2023-04-07
2021102456748	环状流流动液膜厚度及界面波的三维实时测量装置及方法	2023-08-22
2020109218661	一体式远程控制水泥头	2023-01-24
2020110969661	一种动力猫道用纵向排放多根管柱送钻柱装置及输送方法	2023-01-24
2020110348524	一种用于深海采矿船的管船联接装置	2023-01-24
2020113578299	一种铁钻工自动定位钻杆接箍高度的方法及系统	2023-01-28
2020113763872	石油钻机K型井架卧装对角尺寸差井口中心对正的方法	2023-03-10
2020109827528	一种用于压裂混砂设备的混合罐	2023-03-10
2020111918934	一种固井车用多功能用计量水罐	2023-03-10
201910574682X	一种实体钻机数字化监测方法	2023-03-24
2020110775295	可调式通用石油钻机平移轨道	2023-03-14
2020110334911	一种高位夹持直线机械手	2023-03-10
20191920755.6	一种具有内衬层的钢管气密封结构	2023-03-14
202110671283.2	一种X100级耐低温耐腐蚀厚壁无缝管线管及其制造方法	2023-07-11
202110613116.2	一种低成本调质型连续油管用钢、热轧钢带、钢管及其制造方法	2023-08-11
202110509391.X	一种调质型抗酸管线用钢板及其制造方法	2023-08-11
202111072403.3	一种高精度螺旋埋弧焊管用高强韧板卷制造方法	2023-04-07
202211172545.1	一种经济型氢气输送管线钢及生产方法	2023-06-03